Statistical Signal Processing

Detection, Estimation, and Time Series Analysis

Louis L. Scharf
University of Colorado at Boulder

with Cédric Demeure collaborating on Chapters 10 and 11

ADDISON-WESLEY PUBLISHING COMPANY
Reading, Massachusetts • Menlo Park, California • New York
Don Mills, Ontario • Wokingham, England • Amsterdam
Bonn • Sydney • Singapore • Tokyo • Madrid • San Juan

This book is in the Addison-Wesley Series in Electrical and Computer Engineering: Digital Signal Processing

Consulting Editor: Richard Roberts

Library of Congress Cataloging-in-Publication Data

Scharf, Louis L.
    Statistical signal processing : detection, estimation, and time
series analysis / Louis L. Scharf.
        p.   cm.
    Includes bibliographical references and index.
    ISBN 0-201-19038-9
    1. Signal processing—Statistical methods.   I. Title.
TK5102.5.S3528   1990
621.382'2—dc20   90-42747
                                                              CIP

Reprinted with corrections July, 1991

2 3 4 5 6 7–MA–939291

To Carol, Greg, and Heidi

Preface

The field of statistical signal processing embraces the many mathematical procedures that engineers and statisticians use to draw inference from imperfect or incomplete measurements. The major domains of the field are detection, estimation, and time series analysis.

Abstractly, statistical signal processing is a theory for using experimental measurements to transform a prior model for a signal into a posterior model that may be used to make informed decisions. The quality of the decision is measured by a loss function that depends on ground truth and the decision taken. It is the intricate interplay between prior models, measurement schemes, loss functions, and decision rules that gives statistical signal processing its great variety.

ORGANIZATION AND USE

This book is my personal statement about the fundamental ideas in statistical signal processing. The book breaks down along four distinct topical lines: mathematical and statistical preliminaries; detection theory; estimation theory; and time series analysis. There is enough material to support a two-semester course in statistical signal processing, but the book may be used for separate one-semester courses in detection theory, estimation theory, or time series analysis. In a detection theory course, Chapters 1 through 5 may be covered in their entirety. In an estimation theory course, Chapters 1 through 3 and 6 through 8 may be covered. In a time series course, Chapters 1 through 3 and 9 through 11 are appropriate. Chapter 9 on least squares is a swing chapter that may be treated as a topic in estimation theory or time series analysis.

A GUIDED TOUR OF THE BOOK

Mathematical and Statistical Preliminaries

I begin in Chapter 2 with a fairly comprehensive review of linear algebra, matrix theory, and multivariate normal theory. Linear algebra, and the geometrical pictures that bring life to it, forms the basis for our prior structural information about a signal. Matrix theory provides an algebra for manipulating and composing linear transformations, and multivariate normal theory provides the statistical methodology for computing the distribution of linear and quadratic forms in normal random vectors. When teaching from this chapter, I pick and choose from the topics, making sure to develop the ideas of linear independence, subspaces and their spans, orthogonal subspaces, QR factorizations, projections, the singular value decomposition, the multivariate normal distribution, and the distribution of quadratic forms in projection operators. In Chapter 3, I develop the main results in the theory of sufficient statistics and show the fundamental role they play in the computation of minimum variance unbiased estimators.

Detection Theory

Chapters 4 and 5 are dedicated to detection theory. In Chapter 4 I treat the many faces of the Neyman-Pearson theory of hypothesis testing. I cover the rudiments of decision theory, discuss the roles of sufficiency and invariance in hypothesis testing, and develop the theory of uniformly most powerful tests. I then apply the principles of sufficiency and invariance to signal detection when the signal model, or the noise model, is incompletely known. This produces matched filters, CFAR matched filters, matched subspace filters, and CFAR matched subspace filters. The final two sections of Chapter 4 treat reduced-rank detectors and linear discriminant functions for detecting Gaussian signals in Gaussian noise. Chapter 4 is long, so some instructors may wish to omit the linear discriminant functions and, perhaps, the sections on matched subspace filtering, although the latter is very important in the modern study of detection theory. In Chapter 5, I treat the Bayesian theory of hypothesis testing, wherein a prior distribution is assigned to the hypotheses under test. Minimax tests are constructed as Bayes tests against least favorable priors. The study of M-orthogonal symbolling produces insights into channel capacity, and the study of composite matched filters produces insights into associative memories.

Estimation Theory

Chapters 6 through 9 are devoted to estimation theory. I begin in Chapter 6 with the maximum likelihood theory of parameter estimation, where I discuss the roles that sufficiency and invariance play in the maximum likelihood theory and discuss the Cramer-Rao bound for the variance of unbiased estimators. Nuisance parameters are explored in depth. In the last several sections of the chapter, I apply maximum likelihood theory to the identification of subspaces, to the identification of ARMA parameters, and to the identification of structured covariance matrices. Chapter 6 is long, so some instructors may want to give nuisance parameters a once-over-lightly, and select just a few of the applications. In Chapter 7 parameters are endowed with prior distributions, and the Bayes theory is developed for turning prior distributions into posterior distributions. The Bayes theory produces the Gauss-Markov theorem for multivariate normal parameters and measurements. I interpret the Gauss-Markov theorem by showing how it transforms a channel model for measurements into an inverse channel model, or estimator-plus-noise model. The Gauss-Markov theorem is applied to sequential Bayes estimators, the Kalman filter, and the Wiener filter. In Chapter 8 we explore in more detail minimum mean-squared error estimators and the role that the conditional mean estimator plays. I derive conditional mean estimators to solve a number of problems in signal processing: low-rank Wiener filters, linear predictors, Kalman filters, low-rank approximations of random vectors, scalar and block quantizers, and reduced-rank block quantizers. The last application produces rate-distortion formulas of the Shannon variety. In Chapter 9 we develop the theory of least squares for estimating parameters in the linear statistical model and stress the singular value decomposition for the insight that it brings to least squares problems. We then study the performance of least squares when errors are normally distributed. This study produces order selection rules for reducing the rank of least squares estimators. The middle sections of the chapter are devoted to special topics such as sequential, weighted, constrained, and total least squares. The last several sections are devoted to applications: inverse problems, mode identification, parameter estimation in ARMA systems, linear prediction, and the identification of structured covariance matrices. Chapter 9 is long, so some instructors may want to select just a few of the applications.

Time Series Analysis

In Chapters 10 and 11 we cover linear prediction and modal analysis. In our treatment of linear prediction we begin with the classical stationary theory of Wold and Kolmogorov and establish the connection between linear prediction and maximum entropy extension of correlation sequences. We then develop the nonstationary theory of fitting order-increasing whiteners to finite covariance matrices, paying special attention to the Levinson and Schur recursions for computing the reflection coefficients that keep the recursions going. We study the least squares theory of linear prediction and derive the lattice recursions for computing reflection coefficients. Linear prediction in ARMA time series produces the MSK algorithms for fast Kalman filtering. In the last two sections of the chapter we apply linear prediction to the computation of likelihood and the design of a differential PCM system. In Chapter 11 we draw a distinction between linear prediction and modal analysis. We study Prony's method, exact least squares, the total least squares of Golub and Van Loan, the principal components of Tufts and Kumaresan, and MUSIC of Bienvenu and Schmidt as the most prominent techniques for estimating modes. We outline pencil methods and then present Kumaresan's procedure for estimating modes from frequency domain data.

ACKNOWLEDGMENTS

Since the early 70s I have had the opportunity to talk with Henry Cox and Ben Friedlander about many problems in signal processing. Both have generously volunteered interesting problems to explore, shared their own elegant solutions, and offered insightful interpretations of others' solutions. In 1980 Claude Gueguen spent six months at Colorado State University, where he taught his course on Parametric Signal Processing. In this course, and in our collaborative research, I gained my first appreciation for the power of linear algebraic models as structural models for signals. I want to thank Claude for sharing with me his ideas and his deep understanding of system theory and signal processing. Since 1985 I have resided in the office adjacent to Tom Mullis. This has provided me with the opportunity to follow Tom's courses in linear algebra and spectrum analysis, to collaborate with him on several pieces of writing, and to attend his research seminars. I thank him for sharing his gift for clear thinking with me and for getting me out of a few tight spots in the final stages of writing this manuscript.

I first conceived this book in 1984, while teaching a graduate course in signal processing at the University of Rhode Island. During my stay at the University of Rhode Island I profited immensely from my association with Don Tufts. His insights profoundly influenced my own thinking, and I would like to acknowledge my intellectual debt to him. Ramdas Kumaresan improved my understanding of modal analysis, and Steve Kay directed me to a deeper understanding of signal detection in the linear statistical model. Many other people at the University of Rhode Island made my stay there a happy time of intellectual growth and professional development. I would like to acknowledge their collegiality.

Dick Roberts, late Professor of Electrical and Computer Engineering at the University of Colorado, reviewed my writing of several sections of the book. He encouraged me to complete it and recommended that I use his editor, Tom Robbins at Addison-Wesley. I followed Dick's advice, and as a consequence I have had the pleasure of working with Tom for the past 18 months, as I put the final touches on the manuscript. During this period, Lynn Kirlin used the manuscript at the University of Victoria and offered many improvements to the presentation.

For the past 20 years or so I have been able to maintain a single-minded interest in statistical signal processing because four program directors at the Office of Naval Research have supported our work. I thank Drs. Bruce McDonald, Doug Depriest, Ed Wegman, and Neil Gerr for their consistent management of ONR's programs in signal processing and mathematical statistics.

I thank my many students for their contributions to my understanding of signal processing. Cédric Demeure helped me write Chapters 10 and 11, so much so that they are as much his as mine. Finally, let me express my sincere thanks to Julie Fredlund, secretary to the group in digital signal processing at the University of Colorado, for her masterful and patient preparation of a demanding manuscript.

Louis Scharf
Boulder, CO

Contents

CHAPTER **3**

Sufficiency and MVUB Estimators   77

CHAPTER **4**

Neyman-Pearson Detectors   103

CHAPTER **5**

Bayes Detectors   179

CHAPTER **6**

Maximum Likelihood Estimators   **209**

CHAPTER **7**

Bayes Estimators   277

CHAPTER **8**

Minimum Mean-Squared Error Estimators   323

CHAPTER **9**

Least Squares   359

CHAPTER **10**

Linear Prediction   423

CHAPTER **11**

Modal Analysis   483

Introduction

Statistical signal processing is the branch of digital signal processing that deals with detection, estimation, and time series analysis. The theory of statistical signal processing draws upon ideas from probability theory, mathematical statistics, linear algebra, Fourier analysis, complex analysis, systems theory, and digital signal processing. In this opening chapter we list the domains of application for statistical signal processing, discuss three illustrative problems, and establish the notation that will be used throughout the book. In Chapter 2 we lay down the mathematical foundations for statistical signal processing, and in Chapter 3 we begin our subject in earnest.

1.1 STATISTICAL SIGNAL PROCESSING AND RELATED TOPICS

Detection theory is an engineering term for what the statistician calls *hypothesis testing* or *decision making*. The problem is one of taking measurements and then estimating in which of a finite number of states an underlying system resides. Often the state is indexed, or labeled, with a real or complex vector of finite dimension. Detection theory is used to

1. detect signals in sonar and radar,

2. decode the symbols that are used to communicate data or analog information in digital communication systems,

1

3. search codebooks in low data-rate systems that employ vector quantizers,

4. recognize speech,

5. classify boundaries and other features of pictures and images, and

6. detect resonances in physical systems.

Quite generally, detection theory is used to select the physical or mathematical model, from a finite class of models, that best describes measured phenomena.

Estimation theory is an engineering term for what the statistician calls *parameter estimation* or *point estimation*. The problem is one of taking measurements and estimating the numerical value of a real or complex vector that describes the system under study. When the vector assumes discrete values in a finite-dimensional space, then there is no distinction between detection theory and estimation theory. Estimation theory is used to

1. identify linear and nonlinear systems,

2. fit polynomial and power series models to data,

3. identify communication channels and characterize propagation media,

4. identify resonant vibration modes in physical systems,

5. estimate range and Doppler shift in radar and sonar systems,

6. identify pitch period and filter parameters in analysis and synthesis systems for low bit-rate speech transmission,

7. estimate features of images and pictures for classification and low bit-rate transmission, and

8. estimate source directions in multisensor arrays.

Generally, estimation theory is used to identify unknown, information-bearing parameters in a physical or mathematical model.

Time series analysis is the statistician's term for a collection of techniques that the engineer uses to *predict, filter, smooth*, and *spectrum analyze* discrete-time sequences. Prediction, filtering, and smoothing are used to extract information about the future, present, and past from a random waveform that is known to have some regularity or smoothness. This information is used to

1. control stochastic systems,

2. predict source outcomes in data compression systems, and

3. track trajectories of spacecraft, aircraft, surface vehicles, and underwater vehicles.

Spectrum analysis is used to determine which frequency bands are important in the spectral representation of a waveform. This information is used to build models for time series analysis in order to

1. fingerprint vibration data,

2. analyze the structure of acoustic and electromagnetic fields, and

3. find resonant modes in geophysical data.

Statistical inference		
Hypothesis testing		Parameter estimation
Detection	Estimation	Time series analysis
Signal detection and classification	Signal modeling and system identification	Filtering and spectrum analysis

Figure 1.1 Terminology.

A slightly different perspective on terminology is given in Figure 1.1, where it is suggested that the general category of *statistical inference* includes hypothesis testing and parameter estimation as subsets. Hypothesis testing and parameter estimation, in turn, include detection, estimation, and time series analysis as subsets. The latter terms are interchangeable with *signal detection and classification, signal modeling and system identification,* and *filtering and spectrum analysis.*

The paragraphs to follow offer rudimentary and nonstatistical treatments of three problems. Our purpose is to illustrate the types of problems that commonly arise in statistical signal processing. Some of the techniques employed in these problems will be foreign to you. The mystery will be cleared up in ensuing chapters.

1.2   THE STRUCTURE OF STATISTICAL REASONING

Most of the reasoning in statistical signal processing goes like this: There exists some random or deterministic function $S(\cdot|\theta)$ that we would like to estimate by estimating the parameter θ. The quality of the estimate $S(\cdot|\hat{\theta})$ is measured by a criterion such as mean-squared error, probability of error, or maximum deviation. The parameter θ also indexes a distribution function $F(\mathbf{x}|\theta)$ that governs the distribution of experimental measurements $\mathbf{x}$. Sometimes our knowledge of the indexing is incomplete, in which case θ may index only several moments of the measurements. From the measurement $\mathbf{x}$ and our knowledge of $F(\mathbf{x}|\theta)$, we estimate θ in order to find an estimate $S(\cdot|\hat{\theta})$ that optimizes the criterion of quality. This structure for statistical reasoning is illustrated in Figure 1.2. The essential problem in statistical signal processing is to fill in the box (labeled with a question mark) that determines how measurements are to be mapped into parameter estimates.

The following examples illustrate the application of statistical reasoning to detection, estimation, and time series analysis.

1. If $S(\cdot|\theta) = m$ whenever $\theta \in \Theta_m$, $m = 0, 1, \ldots, M - 1$, then the problem is to estimate in which of M classes θ lies. This kind of problem characterizes detection theory as it applies to data communication. The classes Θ_m represent symbols, and the criterion of optimality is often the minimum probability of misclassifying θ.

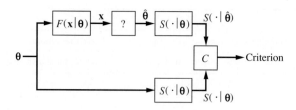

Figure 1.2 Structure of statistical reasoning.

2. If $S(\cdot|\mathbf{\theta}) = \mathbf{\theta}$, then the problem is to estimate the parameter itself. This is a parameter estimation problem that underlies much of maximum likelihood theory, Bayes theory, and conditional mean theory. The criterion of optimality is typically minimum variance unbiased, minimum mean-squared error, or some related criterion.

3. If $S(\cdot|\mathbf{\theta}) = F(\mathbf{x}|\mathbf{\theta})$, then the problem is one of estimating the distribution function of the data by estimating the parameter $\mathbf{\theta}$.

4. If $S(\cdot|\mathbf{\theta}) = \mathbf{H\theta}$, then the problem is to estimate the linear model $\mathbf{H\theta}$. The criterion of optimality is often minimum mean-squared error.

5. If $S(\cdot|\mathbf{\theta})$ is the power spectral density $S(e^{j\omega}|\mathbf{\theta})$, then the problem is to estimate a parametrically-described power spectral density. A typical criterion of optimality is minimum variance for a specified resolution.

1.3   A DETECTION PROBLEM

To illustrate a few of the key ideas in detection theory, we will begin with a simple example. Let $x_1, x_2, \ldots, x_N$ denote N scalar measurements. Assume that the measurements take the form

$$x_t = \theta s_t + n_t; \qquad t = 1, \ldots, N, \qquad\qquad 1.1$$

where s_t, $t = 1, \ldots, N$, is a known sequence of numbers and θ is a scalar parameter. If the system that generates the measurements is in its *zero* state, then we say $\theta \in \Theta_0$ and that hypothesis H_0 is in effect. We abbreviate this statement as follows:

$$H_0 : \theta \in \Theta_0.$$

If the system is in its *one* state, then we say $\theta \in \Theta_1$ and that alternative H_1 is in force:

$$H_1 : \theta \in \Theta_1.$$

The detection problem is one of observing x_t, $t = 1, \ldots, N$, and deciding between H_0 and H_1.

Assume that $\Theta_0 = \{-\mu\}$ and $\Theta_1 = \{\mu\}$. Then we have a model for a binary communication system in which the binary digit 0 is transmitted by the signal $-\mu s_t$, $t = 1, \ldots, N$, and the binary digit 1 is transmitted by the signal μs_t,

$t = 1, \ldots, N$. The signals $\pm \mu s_t$, $t = 1, \ldots, N$, are called *symbols*. They are assumed to be known by the sender and the receiver (also called the *modulator* and the *demodulator* in a *modem* set). The interval $t = 1, \ldots, N$ is called the *baud interval*.

If the fluctuations n_t, $t = 1, \ldots, N$, are noiselike, a term that will be made precise in subsequent chapters, then a useful statistic would be the following correlation statistic:

$$c_N = \sum_{t=1}^{N} s_t x_t. \qquad\qquad 1.2$$

The value of c_N will be composed of two terms:

$$c_N = \theta \sum_{t=1}^{N} s_t^2 + \sum_{t=1}^{N} s_t n_t. \qquad\qquad 1.3$$

If there is no correlation between s_t and n_t, we might expect the sum of the $s_t n_t$ terms to be small. Then c_N should take on values near θE_s when the symbol $\{s_t\}_1^N$ has energy E_s:

$$c_N \approx \theta E_s \qquad \text{when} \quad \sum_{t=1}^{N} s_t^2 = E_s. \qquad\qquad 1.4$$

When $\theta = -\mu$, indicating that a binary digit 0 has been transmitted, then c_N should be near $-\mu E_s$; when $\theta = \mu$, then c_N should be near μE_s. A reasonable strategy would be to select H_0 and H_1 according to the rule

$$\phi(c_N) = \begin{cases} 1 \sim H_1, & c_N > 0 \\ 0 \sim H_0, & c_N \le 0. \end{cases} \qquad\qquad 1.5$$

This equation is read as follows: "The test statistic ϕ assumes the value 1, indicating our belief that alternative H_1 is in force, whenever the correlation statistic c_N exceeds the threshold of zero; otherwise, it assumes the value 0, indicating our belief that hypothesis H_0 is in force." The test statistic is called a *detector*. A diagram of the detector is given in Figure 1.3(a).

In subsequent chapters, we will establish the following: If the fluctuations n_t are realizations of independent, identically distributed random variables, each with mean 0 and variance σ^2, then the term $\sum_{t=1}^{N} s_t n_t$ has mean 0 and variance $\sigma^2 E_s$. If the fluctuations n_t are realizations of normal random variables, then c_N is also normal. The values of c_N are then normally distributed around μE_s if a 1 is transmitted and around $-\mu E_s$ if a 0 is transmitted. As illustrated in Figure 1.3(b), the probability of detecting a 1 when, in fact, a 0 was transmitted is just the probability that $c_N > 0$ when H_0 is in force; this is represented by the shaded area under the curve in Figure 1.3(b). Can you identify the probability of detecting a 0 when a 1 is transmitted? Can you see that the performance improves—that is, the error probabilities decrease—as the "signal-to-noise" ratio $\mu^2 E_s / \sigma^2$ increases? All of this will be clarified in Chapters 4 and 5.

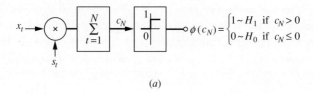

(a)

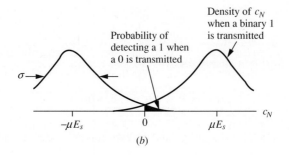

(b)

Figure 1.3 Correlation detector. (a) Implementation.
(b) Performance.

The correlation statistic c_N may be computed with a digital filter. To see this, consider the following convolution:

$$b_t = \sum_{n=1}^{t} h_{t-n} x_n; \qquad t = 1, \dots, N. \qquad 1.6$$

This is the response of a filter, whose impulse response is h_n, $n = 1, \dots, N$, when it is excited by the input x_n, $n = 1, \dots, N$. Choose the impulse response to be the time reversal of s_n, namely,

$$h_n = s_{N-n}. \qquad 1.7$$

$$x_t \rightarrow \boxed{H(z)} \xrightarrow{b_t} \underset{t=N}{\times} \xrightarrow{b_N} \boxed{\begin{array}{c} 1 \\ \hline 0 \end{array}} \!\!-\!\!\circ \phi(b_N) = \begin{cases} 1 \sim H_1 & \text{if } b_N > 0 \\ 0 \sim H_0 & \text{if } b_N \le 0 \end{cases}$$

$$H(z) = \sum_{n=0}^{N-1} h_n z^{-n}$$

$$h_n = s_{N-n}$$

Figure 1.4 Detector implemented as a digital filter.

Then the filter response looks like this:

$$b_t = \sum_{n=1}^{t} s_{N-t+n} x_n.$$

1.8

At time $t = N$, this equals the correlation statistic:

$$b_N = \sum_{n=1}^{N} s_n x_n = c_N.$$

1.9

The implementation of the correlation statistic using a digital filter is illustrated in Figure 1.4. This example is treated more fully in Problem 1.1 and in Chapter 4.

1.4   AN ESTIMATION PROBLEM

Let x_t, $t = 1, \ldots, N-1$, denote $N-1$ observations of a scalar variable that consists of the constant value θ, plus fluctuating errors:

$$x_t = \theta + n_t.$$

1.10

The problem is to estimate θ. A plausible estimator is the sample mean:

$$\hat{\theta}_{N-1} = \frac{1}{N-1} \sum_{t=1}^{N} x_t.$$

1.11

The subscript $N-1$ indicates that data up to time $N-1$ is used to estimate θ. When one more measurement is taken, we may construct the new estimate of θ as follows:

$$N\hat{\theta}_N = \sum_{t=1}^{N} x_t = (N-1)\hat{\theta}_{N-1} + x_N = N\hat{\theta}_{N-1} + x_N - \hat{\theta}_{N-1}.$$

1.12

This means that the estimate $\hat{\theta}_N$ may be written recursively:

$$\hat{\theta}_N = \hat{\theta}_{N-1} + \frac{1}{N}(x_N - \hat{\theta}_{N-1}).$$

1.13

The performance of the estimator $\hat{\theta}_N$ is measured by the error

$$\epsilon_N = \hat{\theta}_N - \theta = \frac{1}{N} \sum_{t=1}^{N} (x_t - \theta) = \frac{1}{N} \sum_{t=1}^{N} n_t.$$

1.14

If the errors n_t are independent, identically distributed random variables with mean zero and variance σ^2, then the mean of ϵ_N is zero and the mean of the estimator $\hat{\theta}_N$ is θ:

$$E(\epsilon_N) = 0$$
$$E(\hat{\theta}_N) = \theta.$$

1.15

The mean of the squared error is the variance of $\hat{\theta}_N$:

$$E\epsilon_N^2 = E(\hat{\theta}_N - \theta)^2$$
$$= \text{var } \hat{\theta}_N = \frac{1}{N}\sigma^2.$$

1.16

We call $\hat{\theta}_N$ an unbiased estimator because its mean is θ. We say that it is consistent because its variance, σ^2/N, converges to zero as N increases without bound. If the n_t, $t = 0, 1, \ldots, N - 1$, are independent, identically distributed normal random variables (denoted $N[0, \sigma^2]$), then the estimator $\hat{\theta}_N$ is distributed as $N[\theta, \sigma^2/N]$. These statistical concepts may be entirely foreign to you. If so, be patient, for we will return to this example in later chapters.

If the final number N is replaced by the variable t, the estimates $\hat{\theta}_t$ may be generated as illustrated in Figure 1.5(a). The errors ϵ_t may be plotted as in Figure 1.5(b). The envelopes $\pm\sigma/\sqrt{t}$ are standard deviations. As t increases, the standard deviation $\sigma/\sqrt{t}$ decreases, and the estimates $\hat{\theta}_t$ are more and more likely to lie within some fixed ϵ of θ. In fact, as Chebyshev's inequality shows us, the probability that $|\hat{\theta}_t - \theta| > \epsilon$ is less than or equal to the variance of $\hat{\theta}_t$ divided by ϵ^2:

$$P[|\hat{\theta}_t - \theta| > \epsilon] \le \frac{\text{var } \hat{\theta}_t}{\epsilon^2} = \frac{\sigma^2}{\epsilon^2 t}.$$

1.17

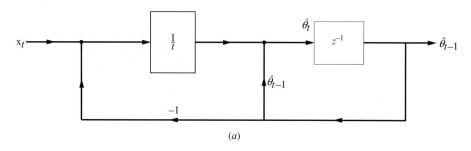

(a)

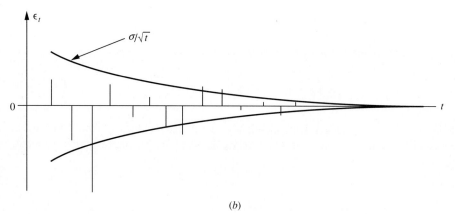

(b)

Figure 1.5 Recursive estimation of a constant. (a) Recursive filter. (b) Performance.

With high probability, $\hat{\theta}_t$ lies within $\pm\epsilon$ of θ as t increases. This example is treated more fully in Problem 1.2 and in Chapter 6.

1.5   A TIME SERIES PROBLEM

This example relates to time series analysis, but, like many such problems, it has detection and estimation interpretations as well. Let x_t, $t = 0, 1, \ldots, N-1$, denote a scalar measurement sequence that contains a discrete-time cosine of radian frequency θ, plus fluctuations:

$$x_t = s_t + n_t$$

$$s_t = \cos\theta t = \frac{1}{2}(e^{j\theta t} + e^{-j\theta t}).$$

1.18

The cosinusoidal sequence $\{s_t\}$ is illustrated in Figure 1.6.

The sequence x_t, $t = 0, 1, \ldots, N - 1$, may be zero-padded to produce the sequence x_t, $t = 0, 1, \ldots, N - 1, N, \ldots, M - 1$, where

$$x_t = \begin{cases} s_t + n_t, & t = 0, 1, \ldots, N - 1 \\ 0, & t = N, \ldots, M - 1. \end{cases}$$

1.19

The discrete Fourier transform (DFT) of this zero-padded sequence produces the following transform variables:

$$X_m = \sum_{t=0}^{N-1} x_t e^{-j2\pi mt/M}; \qquad m = 0, 1, \ldots, M - 1.$$

1.20

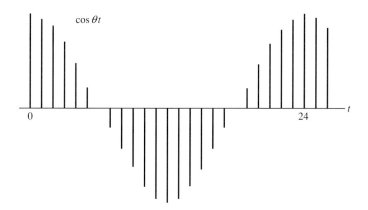

Figure 1.6 Cosinusoidal sequence.

The signal component of this variable is

$$S_m = \sum_{t=0}^{N-1} s_t e^{-j2\pi mt/M}$$

$$= \frac{1}{2} \sum_{t=0}^{N-1} e^{jt(\theta - 2\pi m/M)} + \frac{1}{2} \sum_{t=0}^{N-1} e^{-jt(\theta + 2\pi m/M)}$$

$$= \frac{1}{2} e^{j[(N-1)/2](\theta - 2\pi m/M)} \frac{\sin N(\theta - 2\pi m/M)}{\sin(\theta - 2\pi m/M)}$$

$$+ \frac{1}{2} e^{-j[(N-1)/2](\theta + 2\pi m/M)} \frac{\sin N(\theta + 2\pi m/M)}{\sin(\theta + 2\pi m/M)}.$$

1.21

Think of this as the function

$$S(\phi) = \frac{1}{2} e^{j[(N-1)/2](\theta - \phi)} \frac{\sin N(\theta - \phi)}{\sin(\theta - \phi)} + \frac{1}{2} e^{-j[(N-1)/2](\theta + \phi)} \frac{\sin N(\theta + \phi)}{\sin(\theta + \phi)} \qquad 1.22$$

evaluated at the discrete value $\phi = 2\pi (m/M)$. Then the two terms of S_m may be plotted as in Figure 1.7. The number of values within each mainlobe is $\lfloor M/N \rfloor$, where $\lfloor x \rfloor$ denotes the largest integer less than or equal to x.

By selecting a value of M arbitrarily large, we can find a value for m such that $2\pi m/M$ is arbitrarily close to θ:

$$\frac{2\pi m}{M} \cong \theta \qquad \text{for some } m. \qquad 1.23$$

The corresponding value of S_m is arbitrarily close to the value $N/2$. However, there are $\lfloor M/N \rfloor$ values of S_n in the mainlobe, within the neighborhood of m, that produce values near to $N/2$. In the presence of noise fluctuations, these values will fluctuate, and the determination of m will be imprecise. So, although the numerical resolution of the

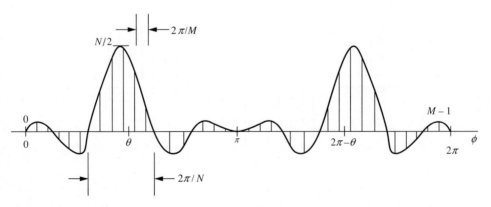

Figure 1.7 DFT variables.

function $\sin N(\theta - \phi) / \sin(\theta - \phi)$ at values of $\phi = 2\pi m/M$ may be made arbitrarily fine by zero-padding the data sequence, this fine-grained reading does nothing to reduce the ambiguity in the mainlobe peak. This ambiguity can be reduced only by increasing the size of the original observation interval N or by resorting to more sophisticated methods of analysis. The ambiguity of $2\pi/N$ is called the *Rayleigh limit to resolution*.

No statistical argumentation has entered into these examples. Consequently, the interpretation of the results is incomplete. On the other hand, each algorithm that we have discussed is fundamental in its respective domain of statistical signal processing. Each can be generalized, and each can be analyzed for its performance. Under carefully stated assumptions, each is an optimum solution of a statistical signal processing problem.

1.6   NOTATION AND TERMINOLOGY

In statistical signal processing we deal with sequences of multivariable measurements. We denote a sequence of M measurements by $\mathbf{x} = [\mathbf{x}_0, \mathbf{x}_1, \ldots, \mathbf{x}_{M-1}]$, where each $\mathbf{x}_m$ is an $N \times 1$ vector. We call $\mathbf{x}_m$ the mth measurement vector. Unless otherwise specified, $\mathbf{x}$ is real. When there is a single measurement vector, then $\mathbf{x} = \mathbf{x}_0$, and we often drop the subscript 0, making $\mathbf{x}$ an $N \times 1$ vector. We often denote this as $\mathbf{x} \in R^N$, meaning that $\mathbf{x}$ is a point in the N-dimensional space R^N. When $\mathbf{x}$ is a scalar we use the notation x.

Probability Distributions

When the measurement vector $\mathbf{x}_m$ is a realization of a random variable, we denote the random variable by $\mathbf{X}_m$. The corresponding sequence of M measurement random vectors is $\mathbf{X} = [\mathbf{X}_0, \mathbf{X}_1, \ldots, \mathbf{X}_{M-1}]$. The distribution of an $N \times 1$ random vector $\mathbf{X}$ is written as $F_\theta(\mathbf{x})$ or as $F(\mathbf{x}|\theta)$, with θ a $p \times 1$ vector that parameterizes the distribution. We often denote this as $\mathbf{X} : F_\theta(\mathbf{x})$ or $\mathbf{X} : F(\mathbf{x}|\theta)$. If the mean and covariance of $\mathbf{X}$ are $\mathbf{m}$ and $\mathbf{R}$, we write $\mathbf{X} : (\mathbf{m}, \mathbf{R})$ where $\mathbf{X}$ and $\mathbf{m}$ have dimension $N \times 1$ and $\mathbf{R}$ has dimension $(N \times N)$. If $\mathbf{X}$ is normally distributed, then we write $\mathbf{X} : N[\mathbf{m}, \mathbf{R}]$. When no confusion can result and the context is clear, we make no distinction between a random vector $\mathbf{X}$ and its realization $\mathbf{x}$. Thus, we sometimes denote a random vector by the lower-case symbol $\mathbf{x}$. Again, when $\mathbf{X}$ is a scalar random variable, we use the notation X.

When we write $\mathbf{X} : F(\mathbf{x}|\theta)$, or $\mathbf{X} : F_\theta(\mathbf{x})$, we mean only that the random vector $\mathbf{X}$ has a distribution function that is parameterized by the parameter θ. Nothing is said about the parameter θ being deterministic or random. However, when it *is* random and it is jointly distributed with $\mathbf{x}$ according to the joint distribution $F(\mathbf{x}, \theta)$, we take $F(\mathbf{x}|\theta)$ to be the conditional distribution whose conditional density or probability mass function is

$$f(\mathbf{x}|\theta) = \frac{f(\mathbf{x}, \theta)}{f(\theta)}. \qquad\qquad 1.24$$

The density $f(\boldsymbol{\theta})$ is the marginal density of $\boldsymbol{\theta}$:

$$f(\boldsymbol{\theta}) = \int f(\mathbf{x}, \boldsymbol{\theta}) \, d\mathbf{x}. \qquad\qquad 1.25$$

When we need to differentiate between the distribution for a random vector $\mathbf{X}$ and a random vector $\mathbf{Y}$, we write $\mathbf{X} : F_{\mathbf{X}}(\mathbf{x}|\boldsymbol{\theta})$ and $\mathbf{Y} : F_{\mathbf{Y}}(\mathbf{y}|\boldsymbol{\theta})$. Otherwise, we drop the subscript, so that $F(\mathbf{x}|\boldsymbol{\theta})$ is understood to be the distribution of the random vector $\mathbf{X}$, parameterized by $\boldsymbol{\theta}$ or conditioned on $\boldsymbol{\theta}$, depending upon whether $\boldsymbol{\theta}$ is random or not.

Linear Models

When $\mathbf{x}$ is a linear combination of vectors $\mathbf{h}_1, \mathbf{h}_2, \ldots, \mathbf{h}_p$, we write $\mathbf{x} = \mathbf{H}\boldsymbol{\theta}$, with $\mathbf{H}$ the $N \times p$ matrix $\mathbf{H} = [\mathbf{h}_1 \cdots \mathbf{h}_p]$ and $\boldsymbol{\theta}$ a $p \times 1$ parameter vector. We say $\mathbf{H} \in R^{N \times p}$ and $\boldsymbol{\theta} \in R^p$, meaning that the matrix $\mathbf{H}$ is a point in $R^{N \times p}$ and $\boldsymbol{\theta}$ is a point in R^p. In Chapter 2 we establish that $\mathbf{x} = \mathbf{H}\boldsymbol{\theta}$ lies in a p-dimensional subspace of R^N. Throughout this book we study the consequences of this fundamental piece of structural information, for it forms the basis of modern signal processing.

There are two equivalent interpretations of the linear model $\mathbf{x} = \mathbf{H}\boldsymbol{\theta}$. The first interpretation comes from writing the system matrix $\mathbf{H}$ in terms of its rows:

$$\begin{bmatrix} x_1 \\ x_2 \\ \vdots \\ x_N \end{bmatrix} = \begin{bmatrix} \mathbf{c}_1^{\mathrm{T}} \\ \mathbf{c}_2^{\mathrm{T}} \\ \vdots \\ \mathbf{c}_N^{\mathrm{T}} \end{bmatrix} \begin{bmatrix} \theta_1 \\ \vdots \\ \theta_p \end{bmatrix} \qquad\qquad 1.26$$

$$x_n = \mathbf{c}_n^{\mathrm{T}} \boldsymbol{\theta}.$$

The nth entry in the vector $\mathbf{x}$ is just a correlation of the vector $\mathbf{c}_n$ with the parameter vector $\boldsymbol{\theta}$. The system matrix may be interpreted as a set of correlators $(\mathbf{c}_n^{\mathrm{T}})_1^N$. The second interpretation comes from writing the system matrix $\mathbf{H}$ in terms of its columns:

$$\mathbf{x} = \begin{bmatrix} \mathbf{h}_1 & \mathbf{h}_2 & \cdots & \mathbf{h}_p \end{bmatrix} \begin{bmatrix} \theta_1 \\ \vdots \\ \theta_p \end{bmatrix}$$

$$= \sum_{i=1}^{p} \theta_i \mathbf{h}_i. \qquad\qquad 1.27$$

The vector $\mathbf{x}$ is a linear combination of modes $\mathbf{h}_i$, weighted by their respective coefficients θ_i. Each mode is, in turn, the (impulse) response to an impulse applied at time i:

$$\mathbf{h}_i = \begin{bmatrix} \mathbf{h}_1 & \cdots & \mathbf{h}_p \end{bmatrix} \begin{bmatrix} 0 \\ \vdots \\ 1 \\ \vdots \\ 0 \end{bmatrix} \qquad\qquad 1.28$$

$$= \mathbf{H}\boldsymbol{\delta}_i ; \qquad \boldsymbol{\delta}_i = [0 \quad \cdots \quad \underset{i}{1} \quad \cdots \quad 0]^{\mathrm{T}}.$$

The system matrix may be interpreted as a set of (time-varying) impulse responses $\mathbf{h}_i$.

Example 1.1 (Sum of Sinusoids)

In Figure 1.8, $p/2$ continuous time cosinusoids of the form $A_i \cos(\omega_i t + \phi_i)$ are summed to produce the signal $x(t)$:

$$x(t) = \sum_{i=1}^{p/2} A_i \cos(\omega_i t + \phi_i) \qquad\qquad 1.29$$

$$= \mathrm{Re} \sum_{i=1}^{p/2} A_i e^{j(\omega_i t + \phi_i)}.$$

If this signal is sampled at the sampling instants $t = nT$, then the discrete time signal is

$$x_n = \mathrm{Re} \sum_{i=1}^{p/2} A_i e^{j(\omega_i Tn + \phi_i)} = \sum_{i=1}^{p/2} A_i \cos(\omega_i Tn + \phi_i). \qquad 1.30$$

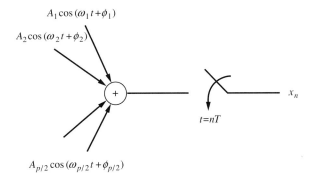

Figure 1.8 Sum of sinusoids.

Typically, such samples are taken over an interval $[0 \leq t < NT]$ to produce the samples $x_0, x_1, \ldots, x_{N-1}$. When organized into the vector $\mathbf{x} = [x_0 x_1 \cdots x_{N-1}]^T$, the result is

$$\mathbf{x} = \begin{bmatrix} x_0 \\ x_1 \\ x_2 \\ \vdots \\ x_{N-1} \end{bmatrix} = \mathrm{Re} \begin{bmatrix} 1 & 1 & \cdots & 1 \\ e^{j\omega_1 T} & e^{j\omega_2 T} & & e^{j\omega_{p/2} T} \\ (e^{j\omega_1 T})^2 & & \cdots & (e^{j\omega_{p/2} T})^2 \\ \vdots & \vdots & \vdots & \vdots \\ (e^{j\omega_1 T})^{N-1} & (e^{j\omega_2 T})^{N-1} & \cdots & (e^{j\omega_{p/2} T})^{N-1} \end{bmatrix} \begin{bmatrix} A_1 e^{j\phi_1} \\ A_2 e^{j\phi_2} \\ \vdots \\ A_{p/2} e^{j\phi_{p/2}} \end{bmatrix}.$$

$$1.31$$

The vector of samples $\mathbf{x}$ may be written as

$$\mathbf{x} = \mathrm{Re}\ \mathbf{H}\boldsymbol{\theta}$$

$$\mathbf{H} = [\mathbf{h}_1\ \mathbf{h}_2\ \cdots\ \mathbf{h}_{p/2}] \qquad \boldsymbol{\theta} = [\theta_1\ \theta_2\ \cdots\ \theta_{p/2}]^T \qquad 1.32$$

$$\mathbf{h}_i = [1\ \ e^{j\omega_i T}\ \ \cdots\ \ e^{j\omega_i T(N-1)}]^T \qquad \theta_i = A_i e^{j\phi_i}.$$

The mode $\mathbf{h}_i$ is a complex exponential mode, the matrix $\mathbf{H}$ is a *Vandermonde matrix* whose columns take the form z_i^n, and the coefficient θ_i is a complex amplitude composed of the amplitude A_i and the phase ϕ_i.

If the term $\cos(\omega_i n T + \phi_i)$ is expanded as $\cos(\omega_i n T)\cos\phi_i - \sin(\omega_i n T)\sin\phi_i$, then the vector $\mathbf{x}$ may be written as

$$\mathbf{x} = \begin{bmatrix} 1 & \cdots & 1 & 0 & \cdots & 0 \\ \cos(\omega_1 T) & & \cos(\omega_{p/2} T) & -\sin\omega_1 T & & -\sin\omega_{p/2} T \\ \cos(\omega_1 2T) & & \vdots & -\sin\omega_1 2T & & \vdots \\ \vdots & & & \vdots & & \\ \cos(\omega_1(N-1)T) & \cdots & \cos(\omega_{p/2}(N-1)T) & -\sin\omega_1(N-1)T & \cdots & -\sin\omega_{p/2}(N-1)T \end{bmatrix}$$

$$\times \begin{bmatrix} A_1 \cos\phi_1 \\ \vdots \\ A_{p/2}\cos\phi_{p/2} \\ A_1 \sin\phi_1 \\ \vdots \\ A_{p/2}\sin\phi_{p/2} \end{bmatrix}. \qquad 1.33$$

We see that $\mathbf{x}$ is an N-vector that is constructed from a linear combination of linearly independent sines and cosines, provided $\omega_i \neq \omega_j$. We say that $\mathbf{x}$ lies in a real linear vector space of dimension p. (This is explained more fully in

Chapter 2.) We also say that $\mathbf{x}$ obeys the linear model $\mathbf{x} = [\mathbf{C}|\mathbf{S}]\,[\boldsymbol{\theta}_c^T \quad \boldsymbol{\theta}_s^T]^T$. Can you identify $\mathbf{C}$, $\mathbf{S}$, $\boldsymbol{\theta}_c$, and $\boldsymbol{\theta}_s$? ∎

Example 1.2 (Array Processing)

In Figure 1.9, p radiating sources r_i transmit plane waves that are sensed by N sensors labeled $S_0, S_1, \ldots, S_{N-1}$. Sensor S_0 is located at the center of a coordinate system, and the coordinate of sensor S_n is $\mathbf{z}_n$. The source r_m transmits a propagating wave whose complex representation is

$$r_m(t, \mathbf{z}) = A_m e^{j(\omega_m t - \mathbf{k}_m^T \mathbf{z})} = A_m e^{j\omega_m t} e^{-j\mathbf{k}_m^T \mathbf{z}}.$$

The scalar ω_m is the radian frequency of the source, and the vector $\mathbf{k}_m$ is the wavenumber for the source. This wavenumber may be written as $\mathbf{k}_m = (2\pi/\lambda_m)\mathbf{d}_m$, where the vector $\mathbf{d}_m$ is a vector of direction cosines. The field propagating from radiator r_m may now be written as

$$r_m(t, \mathbf{z}) = A_m e^{j\omega_m t} e^{-j(2\pi/\lambda_m)\mathbf{d}_m^T \mathbf{z}}. \qquad 1.34$$

The scalar waveform sensed by sensor S_n is the sum of all signals $r_m(t, \mathbf{z})$, read at $\mathbf{z} = \mathbf{z}_n$:

$$x_n(t) = \sum_{m=1}^{p} r_m(t, \mathbf{z}_n).$$

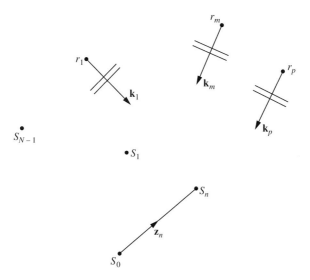

Figure 1.9 An array processing problem.

When the waveforms $x_n(t)$ are organized into the N vector $\mathbf{x} = [x_0(t)\, x_1(t) \cdots x_{N-1}(t)]^T$, we obtain the snapshot $\mathbf{x}$:

$$
\mathbf{x} = \begin{bmatrix} x_0(t) \\ x_1(t) \\ \vdots \\ x_{N-1}(t) \end{bmatrix} = \begin{bmatrix} e^{-j(2\pi/\lambda_1)\mathbf{d}_1^T\mathbf{x}_0} & \cdots & e^{-j(2\pi/\lambda_p)\mathbf{d}_p^T\mathbf{x}_0} \\ e^{-j(2\pi/\lambda_1)\mathbf{d}_1^T\mathbf{x}_1} & & \\ \vdots & & \vdots \\ e^{-j(2\pi/\lambda_1)\mathbf{d}_1^T\mathbf{x}_{N-1}} & \cdots & e^{-j(2\pi/\lambda_p)\mathbf{d}_p^T\mathbf{x}_{N-1}} \end{bmatrix} \begin{bmatrix} A_1 e^{j\omega_1 t} \\ A_2 e^{j\omega_2 t} \\ \vdots \\ A_p e^{j\omega_p t} \end{bmatrix}
$$

$$
= \begin{bmatrix} \mathbf{h}_1 & \mathbf{h}_2 & \cdots & \mathbf{h}_p \end{bmatrix} \begin{bmatrix} A_1 e^{j\omega_1 t} \\ \vdots \\ A_p e^{j\omega_p t} \end{bmatrix}.
$$

1.35

The column $\mathbf{h}_m$, $m = 1, 2, \ldots, p$, characterizes the phase delays for source r_m to each of the sensors S_n, $n = 0, 1, \ldots, N - 1$.

If samples are taken in time as well, then the time samples may be organized into row vectors to produce the equations

$$
\begin{bmatrix} x_0(0) & x_0(T) & \cdots & x_0\big[(M-1)T\big] \\ x_1(0) & x_1(T) & \cdots & x_1\big[(M-1)T\big] \\ \vdots & & & \vdots \\ x_{N-1}(0) & x_{N-1}(T) & \cdots & x_{N-1}\big[(M-1)T\big] \end{bmatrix}
$$

$$
= \begin{bmatrix} \mathbf{h}_1 & \mathbf{h}_2 & \cdots & \mathbf{h}_p \end{bmatrix} \begin{bmatrix} A_1 & & \mathbf{0} \\ & \ddots & \\ \mathbf{0} & & A_p \end{bmatrix} \begin{bmatrix} 1 & e^{j\omega_0} & e^{j\omega_0(2)} & \cdots & e^{j\omega_0(M-1)} \\ \vdots & & & & \vdots \\ 1 & e^{j\omega_p} & e^{j\omega_p(2)} & \cdots & e^{j\omega_p(M-1)} \end{bmatrix}
$$

1.36

This system of equations may be written as

$$
\mathbf{X} = \mathbf{HAT}. \qquad\qquad 1.37
$$

where the "data matrix" $\mathbf{X}$ contains all of the space-time samples that can be generated in a multisensor array, $\mathbf{H}$ is the matrix determined by the known sensor locations and the unknown source locations, $\mathbf{A}$ is a diagonal matrix containing source amplitudes, and $\mathbf{T}$ is a row Vandermonde matrix determined by the source frequencies ω_m. We say that $\mathbf{X}$ obeys the linear model $\mathbf{X} = \mathbf{HAT}$. ∎

Example 1.3 (ARMA Impulse Responses)

Consider the discrete-time system illustrated in Figure 1.10. The filter is characterized by its ARMA$(p, p - 1)$ transfer function $H(z)$:

$$H(z) = \frac{b_0 + b_1 z^{-1} + \cdots + b_{p-1} z^{-(p-1)}}{1 + a_1 z^{-1} + \cdots + a_p z^{-p}}. \qquad 1.38$$

The partial fraction expansion of this transfer function produces the linear combination

$$H(z) = \sum_{i=1}^{p} A_i \frac{1}{1 - z_i z^{-1}}$$

$$(1 + a_1 z^{-1} + \cdots + a_p z^{-p})\big|_{z = z_i} = 0 \qquad 1.39$$

$$A_i = z^{-1} H(z)(z - z_i)\big|_{z = z_i}.$$

The corresponding unit pulse response is

$$h_t = \sum_{i=1}^{p} A_i z_i^t, \qquad t = 0, 1, 2, \ldots$$

$$= 0, \qquad t < 0. \qquad 1.40$$

The first N values of the impulse response may be written out as

$$\mathbf{h} = \begin{bmatrix} h_0 \\ h_1 \\ \vdots \\ h_{N-1} \end{bmatrix} = \begin{bmatrix} 1 & 1 & \cdots & 1 \\ z_1 & z_2 & \cdots & z_p \\ \vdots & & & \vdots \\ z_1^{N-1} & z_2^{N-1} & \cdots & z_p^{N-1} \end{bmatrix} \begin{bmatrix} A_1 \\ A_2 \\ \vdots \\ A_p \end{bmatrix} = \mathbf{H\theta}. \qquad 1.41$$

The matrix $\mathbf{H}$ is a Vandermonde matrix, where each column of $\mathbf{H}$ is a mode of the impulse response. We say that $\mathbf{h}$ obeys the linear model $\mathbf{h} = \mathbf{H\theta}$, where $\boldsymbol{\theta}$ contains the mode weights θ_i. ∎

Figure 1.10 Autoregressive moving average filter.

REFERENCES AND COMMENTS

Mathematical statistics forms the basis for the study of statistical signal processing. The textbooks and reference books in mathematical statistics that I have found most illuminating are the standards in the field: Cramer [1948]; Ferguson [1967]; Hogg and Craig [1978]; Kendall and Stuart [1966–67]; Lehmann [1959]; Rao [1973]; Takeuchi, Yanai, and Mukerjee [1982]; and Zachs [1971].

The most influential and scholarly engineering texts are the books by Anderson and Moore [1979]; Mendel [1987]; Sorenson [1980]; and Van Trees [1968].

The books by Cramer, Ferguson, Hogg and Craig, Kendall and Stuart, Lehmann, and Zachs treat the mathematical foundations of hypothesis testing and parameter estimation, the two main branches of statistical inference. The books by Rao and by Takeuchi et al. treat the linear model and inference within it. The book by Van Trees applies hypothesis testing and parameter estimation to detection, estimation, and modulation of continuous-time waveforms. The book by Anderson and Moore treats the Kalman theory of prediction, filtering, and smoothing within state-space models. The books by Mendel and Sorenson treat parameter estimation and recursive filtering within linear models.

Anderson, B. D. O. and J. B. Moore [1979]. *Optimal Filtering* (Englewood Cliffs, N.J.: Prentice Hall, 1979).

Cramer, H. [1948]. *Mathematical Methods in Statistics* (Princeton, N.J.: Princeton University Press, 1948).

Ferguson, T. S. [1967]. *Mathematical Statistics: A Decision Theoretic Approach* (New York: Academic Press, 1967).

Hogg, R. V., and A. T. Craig [1978]. *Introduction to Mathematical Statistics* (New York: Macmillan, 1978).

Kendall, M. G., and A. Stuart [1966–67]. *The Advanced Theory of Statistics*, 3 vols. (New York: Hafner, 1966–1967).

Lehmann, E. L. [1959]. *Testing Statistical Hypotheses* (New York: John Wiley & Sons, 1959).

Mendel, J. M. [1987]. *Lessons in Digital Estimation Theory* (Englewood Cliffs, N.J.: Prentice Hall, 1987).

Rao, C. R. [1973]. *Linear Statistical Inference and Its Applications*, 2nd ed. (New York: John Wiley & Sons, 1973).

Sorenson, H. W. [1980]. *Parameter Estimation* (New York: Marcel Dekker, 1980).

Takeuchi, K., H. Yanai, and B. N. Mukerjee [1982]. *The Foundations of Multivariate Analysis* (New Delhi: Wiley Eastern, 1982).

Van Trees, H. L. [1968]. *Detection, Estimation, and Modulation Theory* (New York: John Wiley & Sons, 1968).

Zacks, S. [1971]. *The Theory of Statistical Inference* (New York: John Wiley & Sons, 1971).

PROBLEMS

1.1 Let's improve our understanding of the detector in Section 1.3. Assume that the symbol $\{s_t\}_1^N$ is known, with energy E_s, and assume the noises $\{n_t\}_1^N$ are drawn from a sequence of independent, identically distributed (i.i.d.) normal (or Gaussian) random variables. That is,

$$n_t = N[0, \sigma^2]$$

$$E(n_t) = 0; \qquad \text{var}(n_t) = \sigma^2.$$

a. Show that the correlation statistic is distributed as follows under $H_0 : \theta = -\mu$ and $H_1 : \theta = \mu$:

$$c_N : N[-\mu E_s, \sigma^2 E_s] \quad \text{under } H_0$$

$$c_N : N[\mu E_s, \sigma^2 E_s] \quad \text{under } H_1.$$

Plot these normal densities.

b. Define the output signal-to-noise ratio (SNR) of the correlation statistic to be the mean squared, divided by the variance. Show

$$\text{SNR} = \frac{\mu^2}{\sigma^2} E_s.$$

c. If the alternative H_1 is selected when $c_N > 0$, show that the probability of falsely choosing H_1 is

$$P[H_1 | H_0] = \int\limits_{(\mu/\sigma)\sqrt{E_s}}^{\infty} \frac{1}{\sqrt{2\pi}} e^{-x^2/2} \, dx = 1 - \int\limits_{-\infty}^{(\mu/\sigma)\sqrt{E_s}} \frac{1}{\sqrt{2\pi}} e^{-x^2/2} \, dx$$

$$= 1 - \Phi\left(\frac{\mu}{\sigma}\sqrt{E_s}\right)$$

$$\Phi(\eta) = \int\limits_{-\infty}^{\eta} \frac{1}{\sqrt{2\pi}} e^{-x^2/2} \, dx \quad : \quad \text{normal integral.}$$

1.2   Let's improve our understanding of the estimator in Section 1.4. Assume that
      θ is an unknown constant and the noises are drawn from a sequence of i.i.d.
      $N[0, \sigma^2]$ random variables.

   a.   Show that the estimator $\hat{\theta}_N$ is distributed as follows:

$$\hat{\theta}_N : N\left[\theta, \frac{\sigma^2}{N}\right].$$

   b.   Show that the estimator error is distributed as

$$\epsilon_N : N\left[0, \frac{\sigma^2}{N}\right].$$

   c.   Show that the probability that $|\epsilon_N|$ exceeds $\epsilon > 0$ is

$$P\left[|\epsilon_N| > \epsilon\right] = 2 \int_{-\infty}^{-(\epsilon/\sigma)\sqrt{N}} \frac{1}{\sqrt{2\pi}} e^{-x^2/2} \, dx$$

$$= 2\Phi\left(-\frac{\epsilon\sqrt{N}}{\sigma}\right)$$

$$\Phi(\eta) = \int_{-\infty}^{\eta} \frac{1}{\sqrt{2\pi}} e^{-x^2/2} \, dx.$$

1.3   Consider this diagram for generating the signal-plus-noise measurement
      $y = x + n$:

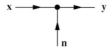

Denote the probability density function for the noise $\mathbf{n}$ by $f_N(\mathbf{n})$.

   a.   Assume the signal $\mathbf{x}$ is an unknown vector; show that the density function
        for the measurement $\mathbf{y}$ is

$$f_Y(\mathbf{y}) = f_N(\mathbf{y} - \mathbf{x}).$$

   b.   Write out the density for the case where the noise is distributed as $N[\mathbf{0}, \mathbf{R}_{nn}]$,
        a multivariate normal with mean vector $\mathbf{0}$ and covariance matrix $\mathbf{R}_{nn}$.

c. Now assume that the signal **x** is statistically independent of **n**, with a probability density function $f_X(\mathbf{x})$; show that the *conditional* density function for **y** remains

$$f_{Y|X}(\mathbf{y}|\mathbf{x}) = f_N(\mathbf{y} - \mathbf{x}).$$

d. Show that the unconditional density function for **Y** is

$$f_Y(\mathbf{y}) = \int_{-\infty}^{\infty} f_N(\mathbf{y} - \mathbf{x}) f_X(\mathbf{x}) \, d\mathbf{x}.$$

e. Write out the density function for **Y** when **X** is distributed as $N[\mathbf{0}, \mathbf{R}_{xx}]$.

1.4 Consider the autoregressive filter

$$H(z) = \frac{b_0}{1 + a_1 z^{-1} + \cdots + a_p z^{-p}}.$$

Call $\mathbf{h} = [h_0 \quad h_1 \quad \cdots \quad h_{N-1}]^T$ a vector of impulse responses for the filter.

a. Show that **h** obeys two kinds of linear models:

(1) $\quad h_t = \sum_{i=1}^{p} A_i z_i^t, \qquad t \geq 0$

(2) $\quad h_t = \begin{cases} 0, & t < 0 \\ b_0, & t = 0 \\ -\sum_{i=1}^{p} a_i h_{t-i}, & t > 0. \end{cases}$

b. Now write out these equations as

(1) $\quad \mathbf{h} = \mathbf{H}\boldsymbol{\theta}$

(2) $\quad \mathbf{G}^{-1}\mathbf{h} = b_0 \boldsymbol{\delta}; \qquad \boldsymbol{\delta} = [1 \quad 0 \quad \cdots \quad 0]^T.$

Find **H**, $\boldsymbol{\theta}$, and $\mathbf{G}^{-1}$.

c. Show that **G** is the matrix

$$\mathbf{G} = \begin{bmatrix} g_0 & & & \\ g_1 & g_0 & & \mathbf{0} \\ \vdots & \vdots & \ddots & \\ g_{N-1} & & \cdots & g_0 \end{bmatrix}$$

where $\{g_t\}$ is the impulse response of the autoregressive filter $A(z)$ $1/(1 + a_1 z^{-1} + \cdots + a_p z^{-p})$. That is, $\sum_{i=0}^{p} a_i g_{t-i} = \delta_t$.

1.5   Consider the following experimental setup.

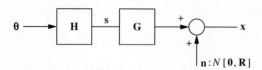

The problem is to observe **x** and estimate **s**. Redraw this diagram in the form of Figure 1.2 to show that it fits our structure of statistical reasoning. Can you describe an experiment where this diagram applies?

Rudiments of Linear Algebra and Multivariate Normal Theory

Observations typically come in strings or arrays. The strings are organized into *vectors*, and the arrays are organized into *matrices*. This means that signal processing algorithms are really procedures for efficiently manipulating vectors and matrices. Typically, these procedures may be written as elementary matrix operations from a matrix algebra. These matrix operations are designed to reveal the underlying structure of the observations and to extract relevant statistics that may be used to draw inferences.

In this chapter we review matrix theory and linear algebra. We begin with vector spaces and then proceed to matrices and linear transformations. We study the structure of Euclidean spaces and their subspaces. The study of eigenvalues and eigenvectors leads to the spectral theory of Hermitian matrices and to the study of projection operators. Along the way we encounter quadratic forms and a variety of specially structured matrices.

In Section 2.10 we endow vectors with multivariate normal distributions in order to complete our characterization of multivariate measurements.

2.1   VECTOR SPACES

Call R^n the set of all real ordered n-tuples, or n-dimensional strings, of the form
$(x_1, x_2, \ldots, x_n)$. Think of this n-tuple as a point in an n-dimensional space with
coordinates $x_1, x_2, \ldots, x_n$ as illustrated in Figure 2.1. Or think of the string as a
function defined on the integers $(1, 2, \ldots, n)$:

$$(x_1, x_2, \ldots, x_n) = f(1, 2, \ldots, n). \qquad 2.1$$

We simplify our language considerably if we organize the entries of the n-tuple into
the vector $\mathbf{x}$:

$$\mathbf{x} = [x_1 \quad x_2 \quad \cdots \quad x_n]^T. \qquad 2.2$$

We denote by R^n the set of all real n-vectors and say that $\mathbf{x} \in R^n$.

The set R^n is a linear vector space because, for any real scalar $a \in R$ and any
two vectors $\mathbf{x} \in R^n$ and $\mathbf{y} \in R^n$, the following vectors are contained in R^n:

$$\mathbf{x} \in R^n \Rightarrow a\mathbf{x} = [ax_1 \quad ax_2 \quad \cdots \quad ax_n]^T \in R^n;$$

$$\mathbf{x} \in R^n, \mathbf{y} \in R^n \Rightarrow \mathbf{x} + \mathbf{y} = [(x_1 + y_1) \quad \cdots \quad (x_n + y_n)]^T \in R^n.$$

We call $a\mathbf{x}$ *scalar multiplication* and $\mathbf{x} + \mathbf{y}$ *vector addition*.

Euclidean Space

Define the length of any vector $\mathbf{x}$ contained in R^n to be

$$\|\mathbf{x}\| = (\mathbf{x}^T\mathbf{x})^{1/2} = \left(\sum_{i=1}^{n} x_i^2 \right)^{1/2} \qquad 2.3$$

This is sometimes called the *Euclidean norm* of $\mathbf{x}$. The real linear space R^n, with the
Euclidean norm so defined, is a normed linear vector space that we call n-dimensional
Euclidean space. This norm may be used to build a metric for measuring the distance

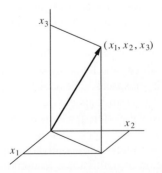

Figure 2.1 A point in an n-dimensional space.

between two vectors $\mathbf{x}$ and $\mathbf{y}$ contained in R^n:

$$d(\mathbf{x}, \mathbf{y}) = \|\mathbf{x} - \mathbf{y}\| = \left[(\mathbf{x} - \mathbf{y})^{\mathrm{T}}(\mathbf{x} - \mathbf{y})\right]^{1/2} = \left[\sum_{i=1}^{n}(x_i - y_i)^2\right]^{1/2}. \qquad 2.4$$

This metric satisfies the usual conditions required of a metric:

1. $\|\mathbf{x} - \mathbf{y}\| \geq 0$, and $\|\mathbf{x} - \mathbf{y}\| = 0 \;\rightarrow\; \mathbf{x} = \mathbf{y}$;
2. $\|\mathbf{x} - \mathbf{y}\| = \|\mathbf{y} - \mathbf{x}\|$ (symmetry);
3. $\|\mathbf{x} - \mathbf{y}\| \leq \|\mathbf{x} - \mathbf{z}\| + \|\mathbf{z} - \mathbf{y}\|$ (triangle inequality).

Now R^n is a metric space.

Hilbert Space

The inner product between two nonzero vectors in R^n is defined as follows:

$$\mathbf{x}^{\mathrm{T}}\mathbf{y} = \sum_{i=1}^{n} x_i y_i. \qquad 2.5$$

This inner product is related to the norm squared of $\mathbf{x} + \mathbf{y}$:

$$\|\mathbf{x} + \mathbf{y}\|^2 = \|\mathbf{x}\|^2 + \|\mathbf{y}\|^2 + 2\mathbf{x}^{\mathrm{T}}\mathbf{y}. \qquad 2.6$$

If $\mathbf{x}^{\mathrm{T}}\mathbf{y} = 0$, then $\|\mathbf{x} + \mathbf{y}\|^2 = \|\mathbf{x}\|^2 + \|\mathbf{y}\|^2$, and we say that $\mathbf{x}$ and $\mathbf{y}$ are orthogonal. A normed linear vector space in which every Cauchy sequence converges, and on which is defined an inner product, is called a *Hilbert space*. This is the space that underlies everything we do in statistical signal processing.

Matrices

Just as we organized real variables x_i into vectors $\mathbf{x} = [x_1 \quad x_2 \quad \cdots \quad x_n]^{\mathrm{T}}$ and said that $\mathbf{x}$ belonged to the linear vector space R^n, we may organize vectors $\mathbf{x}_i$ into matrices $\mathbf{X} = [\mathbf{x}_1 \quad \mathbf{x}_2 \quad \cdots \quad \mathbf{x}_p]$ and say that $\mathbf{X}$ belongs to the linear vector space $R^{n \times p}$. This linear vector space contains all real matrices consisting of p n-vectors or, equivalently, np scalars of the form $\mathbf{X} = \{x_{ij}\}$. It is easy to see that this space is closed under scalar multiplication and matrix addition.

2.2   LINEAR INDEPENDENCE

Our definitions of scalar multiplication and vector addition may be applied recursively to generate linear combinations of vectors of the form

$$\mathbf{x} = a_1\mathbf{x}_1 + a_2\mathbf{x}_2 + \cdots + a_p\mathbf{x}_p. \qquad 2.7$$

If one or more of the vectors in this sum is a linear combination of the others, then $\mathbf{x}$ may be generated without them. We then say that there is *redundancy* in the sum. This simple notion of redundancy motivates us to study linear independence and its implications.

If the sum $a_1\mathbf{x}_1 + a_2\mathbf{x}_2 + \cdots + a_p\mathbf{x}_p = 0$ when one or more of the a_i are nonzero, then we say that the vectors $(\mathbf{x}_1, \mathbf{x}_2, \ldots, \mathbf{x}_p)$ are *linearly dependent*. For example, if the sum is zero and $a_k \neq 0$, then the vector $\mathbf{x}_k$ is linearly dependent on the other vectors:

$$\mathbf{x}_k = -\frac{1}{a_k} \sum_{i \neq k} a_i \mathbf{x}_i. \qquad 2.8$$

If the sum $a_1\mathbf{x}_1 + a_2\mathbf{x}_2 + \cdots + a_p\mathbf{x}_p$ equals zero only when $a_1 = a_2 = \cdots = a_p = 0$, we say that the vectors $(\mathbf{x}_1, \mathbf{x}_2, \ldots, \mathbf{x}_p)$ are *linearly independent*.

The first question that naturally comes to mind is, how can we check to see whether the vectors $(\mathbf{x}_1, \mathbf{x}_2, \cdots, \mathbf{x}_p)$ are linearly independent? The answer is that we may check a *Gram matrix* or use a *Gram-Schmidt* procedure to generate orthogonal versions of the $\mathbf{x}_i$ and see how many such orthogonal vectors can be generated.

Gram Determinant

The vectors $(\mathbf{x}_1, \mathbf{x}_2, \ldots, \mathbf{x}_p)$ may be organized into the *data matrix* $\mathbf{X}$:

$$\mathbf{X} = [\mathbf{x}_1 \quad \mathbf{x}_2 \quad \cdots \quad \mathbf{x}_p]. \qquad 2.9$$

The matrix $\mathbf{X}$ is an $n \times p$ array of real numbers, or an array of p n-vectors, with $\mathbf{x}_i$ occupying the ith column of the array. The Gram matrix for the data matrix $\mathbf{X}$ is the $p \times p$ matrix

$$\mathbf{G} = \mathbf{X}^T\mathbf{X} = \{g_{ij}\} \qquad 2.10$$

$$g_{ij} = \mathbf{x}_i^T\mathbf{x}_j.$$

The (i, j) element of the Gram matrix is the inner product between vectors $\mathbf{x}_i$ and $\mathbf{x}_j$. The determinant of the Gram matrix is the Gram determinant. Its significance is summarized in the following theorem.

Theorem 2.1 The vectors $(\mathbf{x}_1, \mathbf{x}_2, \ldots, \mathbf{x}_p)$ are linearly independent iff the Gram determinant is nonzero.

Proof It is actually easier to prove the equivalent statement that the vectors $(\mathbf{x}_1, \mathbf{x}_2, \ldots, \mathbf{x}_p)$ are linearly dependent iff the Gram determinant is zero.

Only if: Assume that the vectors $(\mathbf{x}_1, \mathbf{x}_2, \ldots, \mathbf{x}_p)$ are linearly dependent. Then the kth column of the Gram matrix may be written as

$$\mathbf{g}_k = \mathbf{X}^T\mathbf{x}_k \qquad 2.11$$

$$\sum_{k=1}^{p} a_k\mathbf{g}_k = \mathbf{X}^T \sum_{k=1}^{p} a_k\mathbf{x}_k = \mathbf{0}.$$

This means that the columns of the Gram matrix are linearly dependent and that the Gram determinant is zero.

If: Assume that the Gram determinant is zero. This implies that the columns of **G** are linearly dependent. Therefore,

$$\sum_{k=1}^{p} a_k \mathbf{g}_k = \mathbf{X}^{\mathrm{T}} \sum_{k=1}^{p} a_k \mathbf{x}_k = \mathbf{0}.$$

Writing this out term by term, we obtain the result

$$\mathbf{x}_i^{\mathrm{T}} \sum_{k=1}^{p} a_k \mathbf{x}_k = 0, \qquad i = 1, 2, \ldots, p \qquad\qquad 2.12$$

$$\sum_{i=1}^{p} a_i \mathbf{x}_i^{\mathrm{T}} \sum_{k=1}^{p} a_k \mathbf{x}_k = \left\| \sum_{k=1}^{p} a_k \mathbf{x}_k \right\|^2 = 0.$$

This implies $\sum_{k=1}^{p} a_k \mathbf{x}_k = \mathbf{0}$, meaning that the vectors $(\mathbf{x}_1, \mathbf{x}_2, \ldots, \mathbf{x}_p)$ are linearly dependent. □

Sequences of Gram Determinants

Closely related to the Gram determinant is a procedure for sequentially testing each vector in a set $(\mathbf{x}_1, \mathbf{x}_2, \ldots, \mathbf{x}_p)$ to see if $\mathbf{x}_k$ is linearly independent of the preceding vectors $(\mathbf{x}_1, \mathbf{x}_2, \ldots, \mathbf{x}_{k-1})$. Our procedure will be to discard each new vector that does not pass the test, retaining only those (linearly independent) vectors that pass.

Let's suppose that the first $k - 1$ vectors have passed the test for independence. If $\mathbf{x}_k$ is linearly independent, then the following vector is nonzero for all choices of $a_i^k \neq 0$:

$$\mathbf{e}_k = a_{k-1}^k \mathbf{x}_1 + a_{k-2}^k \mathbf{x}_2 + \cdots + a_1^k \mathbf{x}_{k-1} + \mathbf{x}_k \neq \mathbf{0}. \qquad 2.13$$

In this formula, we have set the coefficient of $\mathbf{x}_k$ to unity, without loss of generality, and we have superscripted the coefficients a_i^k as a reminder that these are the coefficients that will tell us whether or not $\mathbf{x}_k$ is linearly independent of $\mathbf{x}_1, \mathbf{x}_2, \ldots, \mathbf{x}_{k-1}$. The error $\mathbf{e}_k$ is a measure of how linearly independent $\mathbf{x}_k$ is. Let's write this error as

$$\mathbf{e}_k = [\mathbf{X}_{k-1} \quad \mathbf{x}_k] \mathbf{a}_k \qquad\qquad 2.14$$

where the data matrix $\mathbf{X}_{k-1}$ and the prediction error vector $\mathbf{a}_k$ are

$$\mathbf{X}_{k-1} = [\mathbf{x}_1 \quad \mathbf{x}_2 \quad \cdots \quad \mathbf{x}_{k-1}]$$

$$\mathbf{a}_k = \begin{bmatrix} a_{k-1}^k & \cdots & a_1^k & 1 \end{bmatrix}^{\mathrm{T}}.$$

The norm squared of the error $\mathbf{e}_k$ is

$$\sigma_k^2 = \mathbf{e}_k^{\mathrm{T}} \mathbf{e}_k = (\mathbf{a}_k)^{\mathrm{T}} \begin{bmatrix} \mathbf{X}_{k-1}^{\mathrm{T}} \\ \mathbf{x}_k^{\mathrm{T}} \end{bmatrix} [\mathbf{X}_{k-1} \quad \mathbf{x}_k] \mathbf{a}_k$$

$$\qquad\qquad 2.15$$

$$= (\mathbf{a}_k)^{\mathrm{T}} \begin{bmatrix} \mathbf{G}_{k-1} & \mathbf{h}_k \\ \mathbf{h}_k^{\mathrm{T}} & g_{kk} \end{bmatrix} \mathbf{a}_k.$$

In this formula, $\mathbf{G}_{k-1}$ is the Gram matrix of $\mathbf{X}_{k-1}$, $\mathbf{h}_k$ is a vector of inner products between $\mathbf{x}_i$ and $\mathbf{x}_k$, $i = 1, 2, \ldots, k - 1$, and g_{kk} is the norm squared of $\mathbf{x}_k$:

$$\mathbf{G}_{k-1} = \mathbf{X}_{k-1}^{\mathrm{T}}\mathbf{X}_{k-1}$$

$$\mathbf{h}_k = \mathbf{X}_{k-1}^{\mathrm{T}}\mathbf{x}_k \qquad\qquad 2.16$$

$$g_{kk} = \mathbf{x}_k^{\mathrm{T}}\mathbf{x}_k.$$

The vector $\mathbf{x}_k$ is linearly independent of $(\mathbf{x}_1, \mathbf{x}_2, \ldots, \mathbf{x}_{k-1})$ iff $\sigma_k^2 > 0$. If $\sigma_k^2 = 0$, then $\mathbf{x}_k$ is linearly dependent.

Let's see what the minimum value of σ_k^2 is. Take the gradient with respect to $\mathbf{a}_k$, under the constraint that the trailing element of $\mathbf{a}_k$ is 1, and equate to zero. The resulting normal equations are

$$\frac{1}{2}\frac{\partial}{\partial \mathbf{a}_k}\left\{(\mathbf{a}_k)^{\mathrm{T}}\begin{bmatrix} \mathbf{G}_{k-1} & \mathbf{h}_k \\ \mathbf{h}_k^{\mathrm{T}} & g_{kk} \end{bmatrix}\mathbf{a}_k - \lambda\big[(\mathbf{a}_k)^{\mathrm{T}}\boldsymbol{\delta} - 1\big]\right\} = \begin{bmatrix} \mathbf{G}_{k-1} & \mathbf{h}_k \\ \mathbf{h}_k^{\mathrm{T}} & g_{kk} \end{bmatrix}\mathbf{a}_k - \lambda\boldsymbol{\delta} = \mathbf{0}$$

$$\boldsymbol{\delta} = [0 \quad \cdots \quad 0 \quad 1]^{\mathrm{T}}. \qquad\qquad 2.17$$

Multiply through by $(\mathbf{a}_k)^{\mathrm{T}}$ to see that $\sigma_k^2 = \lambda\mathbf{a}_k^{\mathrm{T}}\boldsymbol{\delta} = \lambda$. Therefore, we may organize the *normal equations* as

$$\begin{bmatrix} \mathbf{G}_{k-1} & \mathbf{h}_k \\ \mathbf{h}_k^{\mathrm{T}} & g_{kk} \end{bmatrix}\mathbf{a}_k = \sigma_k^2\boldsymbol{\delta}. \qquad\qquad 2.18$$

The explicit solution of the normal equations is

$$\mathbf{G}_{k-1}\begin{bmatrix} a_{k-1}^k \\ \vdots \\ a_1^k \end{bmatrix} + \mathbf{h}_k = \mathbf{0} \Rightarrow \begin{bmatrix} a_{k-1}^k \\ \vdots \\ a_1^k \end{bmatrix} = -\mathbf{G}_{k-1}^{-1}\mathbf{h}_k \qquad\qquad 2.19$$

$$\sigma_k^2 = g_{kk} - \mathbf{h}_k^{\mathrm{T}}\mathbf{G}_{k-1}^{-1}\mathbf{h}_k.$$

The norm squared σ_k^2 is called the *Schur complement* of the patterned Gram matrix $\mathbf{G}_k$:

$$\mathbf{G}_k = \begin{bmatrix} \mathbf{X}_{k-1}^{\mathrm{T}} \\ \mathbf{x}_k^{\mathrm{T}} \end{bmatrix}[\mathbf{X}_{k-1} \quad \mathbf{x}_k] = \begin{bmatrix} \mathbf{G}_{k-1} & \mathbf{h}_k \\ \mathbf{h}_k^{\mathrm{T}} & g_{kk} \end{bmatrix}. \qquad\qquad 2.20$$

If this Schur complement is zero, then $\mathbf{x}_k$ is discarded as a linearly dependent vector. Otherwise it is retained to build the data matrix $\mathbf{X}_k = [\mathbf{x}_1 \quad \mathbf{x}_2 \quad \cdots \quad \mathbf{x}_k]$, and the test continues for the vector $\mathbf{x}_{k+1}$.

The solution for the vector $\mathbf{a}_k$ that minimizes σ_k^2 produces some remarkable results that are of interest in their own right. In order to derive them, let's rewrite the normal equations as

$$\mathbf{G}_k\mathbf{a}_k = \sigma_k^2\boldsymbol{\delta} \Rightarrow \mathbf{a}_k = \sigma_k^2\mathbf{G}_k^{-1}\boldsymbol{\delta}. \qquad\qquad 2.21$$

(We assume here that $\mathbf{a}_k$ has passed the test for independence, meaning that the Gram matrix $\mathbf{G}_k$ is nonsingular.) If we premultiply the solution for $\mathbf{a}_k$ by $\boldsymbol{\delta}^T$, we obtain

$$\boldsymbol{\delta}^T \mathbf{a}_k = 1 = \sigma_k^2 \boldsymbol{\delta}^T \mathbf{G}_k^{-1} \boldsymbol{\delta}. \qquad 2.22$$

The inverse of $\mathbf{G}_k$ is $\mathbf{G}_k^{-1} = \mathrm{adj}\,\mathbf{G}_k / \det \mathbf{G}_k$. The pre- and postmultiplication by $\boldsymbol{\delta}^T$ and $\boldsymbol{\delta}$ picks out the element of $\mathrm{adj}\,\mathbf{G}_k$ in the lower right-hand corner, and this is just $\det \mathbf{G}_{k-1}$. Therefore,

$$1 = \sigma_k^2 \frac{\det \mathbf{G}_{k-1}}{\det \mathbf{G}_k} \qquad 2.23$$

$$\sigma_k^2 = \frac{\det \mathbf{G}_k}{\det \mathbf{G}_{k-1}}.$$

This finding may be iterated to obtain the following formula for the determinant of the Gram matrix $\mathbf{G}_k$:

$$\det \mathbf{G}_k = \sigma_k^2 \det \mathbf{G}_{k-1}$$

$$= \prod_{i=1}^{k} \sigma_i^2. \qquad 2.24$$

In summary, each correlation determinant $\det \mathbf{G}_k$ may be computed from this formula to test for linear independence. If $\det \mathbf{G}_{k-1}$ is nonzero, then $\det \mathbf{G}_k$ is nonzero iff σ_k^2 is nonzero.

Cholesky Factors of the Gram Matrix

Let's suppose that n vectors organized into a data matrix $\mathbf{X}_n$ have passed the test for linear independence. Let's define their Gram matrix to be

$$\mathbf{G}_n = \mathbf{X}_n^T \mathbf{X}_n = \begin{bmatrix} \begin{bmatrix} \mathbf{G}_{k-1} & \mathbf{h}_k \\ \mathbf{h}_k^T & g_{kk} \end{bmatrix} & \star \\ & \star \end{bmatrix}$$

$$\mathbf{X}_n = [\mathbf{x}_1 \quad \mathbf{x}_2 \quad \cdots \quad \mathbf{x}_n]. \qquad 2.25$$

The northwest block of the Gram matrix is $\mathbf{G}_k$, and the $\star$s denote the remaining elements of the Gram matrix. Let's organize the kth normal equation of Equation 2.18 as follows:

$$\mathbf{G}_n \begin{bmatrix} \mathbf{a}_k \\ \mathbf{0} \end{bmatrix} = \begin{bmatrix} \mathbf{0} \\ \sigma_k^2 \\ \star \end{bmatrix}. \qquad 2.26$$

In this equation, $\star$ denotes nonzero entries. By stacking up such equations, we see that all n normal equations may be organized into the equation

$$\mathbf{G}_n \begin{bmatrix} \mathbf{a}_1 & & & \\ & \mathbf{a}_2 & & \\ & & \ddots & \\ \mathbf{0} & & & \mathbf{a}_n \end{bmatrix} = \begin{bmatrix} \sigma_1^2 & & & \\ & \sigma_2^2 & & \mathbf{0} \\ \star & & \ddots & \\ & & & \sigma_n^2 \end{bmatrix}. \qquad 2.27$$

What we have found here is that the Gram matrix $\mathbf{G}_n$ may be taken to lower-triangular form by postmultiplying by an upper-triangular matrix that is built from the solutions to the normal equations. Let's write this finding as

$$\mathbf{G}_n \mathbf{A}_n = \mathbf{H}_n \mathbf{D}_n^2$$

$$\mathbf{A}_n = \begin{bmatrix} \mathbf{a}_1 & & & \\ & \mathbf{a}_2 & & \\ & & \ddots & \\ \mathbf{0} & & & \mathbf{a}_n \end{bmatrix} \qquad \mathbf{H}_n = \begin{bmatrix} 1 & & & \\ & 1 & & \mathbf{0} \\ \star & & \ddots & \\ & & & 1 \end{bmatrix} \qquad 2.28$$

$$\mathbf{D}_n^2 = \begin{bmatrix} \sigma_1^2 & & & \\ & \sigma_2^2 & & \mathbf{0} \\ & & \ddots & \\ \mathbf{0} & & & \sigma_n^2 \end{bmatrix}.$$

If we premultiply both sides of this equation by the lower-triangular matrix $\mathbf{A}_n^{\mathrm{T}}$, we find

$$\mathbf{A}_n^{\mathrm{T}} \mathbf{G}_n \mathbf{A}_n = \mathbf{A}_n^{\mathrm{T}} \mathbf{H}_n \mathbf{D}_n^2. \qquad 2.29$$

The matrix on the left-hand side is symmetric, and the matrix on the right-hand side is lower-triangular. The only lower-triangular matrix that is symmetric is the diagonal matrix. Therefore,

$$\mathbf{A}_n^{\mathrm{T}} \mathbf{G}_n \mathbf{A}_n = \mathbf{D}_n^2. \qquad 2.30$$

(It is easy to see that the diagonal elements of $\mathbf{A}_n^{\mathrm{T}} \mathbf{H}_n$ are 1.) We interpret this finding by rewriting it as

$$\mathbf{G}_n = (\mathbf{A}_n^{\mathrm{T}})^{-1} \mathbf{D}_n^2 \mathbf{A}_n^{-1} \qquad 2.31$$

$$\mathbf{G}_n^{-1} = \mathbf{A}_n \mathbf{D}_n^{-2} \mathbf{A}_n^{\mathrm{T}}.$$

Now the interpretation is straightforward: The order-increasing solutions to the normal equations, organized into the matrix $\mathbf{A}_n$, Cholesky-factor the inverse of the Gram matrix into an upper-diagonal-lower factorization (UDL factorization). Equivalently, the inverse of $\mathbf{A}_n$ Cholesky-factors the Gram matrix into a lower-diagonal-upper factorization (LDU factorization).

These results actually provide a proof of the following theorem, which says that *every* vector in R^n may be represented as a linear combination of n linearly independent vectors $(\mathbf{x}_1, \mathbf{x}_2, \ldots, \mathbf{x}_n)$, $\mathbf{x}_i \in R^n$. A corollary of the theorem is that there can exist no more than n linearly independent vectors in R^n.

Theorem 2.2   Given the vector $\mathbf{x} \in R^n$ and the n linearly independent vectors $\{\mathbf{x}_i\}_1^n$, then there exists a set of coefficients $\{a_i\}_1^n$, $a_i \in R$, such that

$$\mathbf{x} = -\sum_{i=1}^{n} a_{n+1-i}\mathbf{x}_i. \qquad 2.32$$

Proof   See Problem 2.6. $\square$

2.3   *QR* FACTORS

There is another way to check for linear independence, and that is to transform a set of vectors into orthogonal vectors and then check to see how many nonzero orthogonal vectors can be generated. This idea leads rather naturally to a study of schemes for orthogonalizing, or *QR*-factoring, data matrices. When we speak of the *QR* factors for a data matrix, we have in mind the following representation of the $n \times p$ matrix $\mathbf{X}$, for $n \geq p$:

$$\mathbf{X} = [\mathbf{x}_1 \quad \mathbf{x}_2 \quad \cdots \quad \mathbf{x}_p] = \mathbf{QR}$$

$$= \begin{bmatrix} \mathbf{u}_1 & \mathbf{u}_2 & \cdots & \mathbf{u}_p \end{bmatrix} \begin{bmatrix} r_{11} & r_{12} & \cdots & r_{1p} \\ & r_{22} & & \\ \mathbf{0} & & \ddots & \vdots \\ & & & r_{pp} \end{bmatrix} \qquad 2.33$$

In this representation, the vectors $(\mathbf{x}_1, \mathbf{x}_2, \ldots, \mathbf{x}_p)$ are linearly independent, and the vectors $(\mathbf{u}_1, \mathbf{u}_2, \ldots, \mathbf{u}_p)$ are orthogonal. It follows that $\mathbf{Q}^T\mathbf{Q} = \mathbf{I}$, so we call $\mathbf{Q}$ a slice of an orthogonal matrix. This factorization provides us with a representation for each $\mathbf{x}_i$,

$$\mathbf{x}_i = \sum_{j=1}^{i} \mathbf{u}_j r_{ji}, \qquad 2.34$$

and, in fact, a Cholesky (or LU) factor of the Grammian:

$$\mathbf{G} = \mathbf{X}^T\mathbf{X} = \mathbf{R}^T\mathbf{Q}^T\mathbf{QR} = \mathbf{R}^T\mathbf{R}$$

$$= \begin{bmatrix} r_{11} & & \mathbf{0} \\ \vdots & \ddots & \\ r_{1p} & \cdots & r_{pp} \end{bmatrix} \begin{bmatrix} r_{11} & \cdots & r_{1p} \\ \mathbf{0} & \ddots & \vdots \\ & & r_{pp} \end{bmatrix}. \qquad 2.35$$

Before proceeding to a study of algorithms for *QR*-factoring a data matrix, we record the following interpretations of the *QR* factor $\mathbf{X} = \mathbf{QR}$:

- $\mathbf{X} = \mathbf{QR}$:   The synthesis form of the *QR* factorization that shows how orthogonal columns of $\mathbf{Q}$ are used to "causally" construct the linearly independent columns of $\mathbf{X}$;

- $\mathbf{Q} = \mathbf{XR}^{-1}$:   The analysis form of the *QR* factorization that shows how linearly independent columns of $\mathbf{X}$ are analyzed to produce the orthogonal columns of $\mathbf{Q}$;

- $\mathbf{Q}^{\mathrm{T}}\mathbf{X} = \mathbf{R}$:   The cross-correlation form of the *QR* factorization that shows how the data matrix $\mathbf{X}$ is taken to upper-triangular form by the orthogonal matrix $\mathbf{Q}^{\mathrm{T}}$; also shows that orthogonal vectors $\mathbf{u}_i$ are causally uncorrelated with the linearly independent vectors $\mathbf{x}_j : \mathbf{u}_i^{\mathrm{T}}\mathbf{x}_j = 0$ for $i > j$;

- $\mathbf{X}^{\mathrm{T}}\mathbf{X} = \mathbf{R}^{\mathrm{T}}\mathbf{R}$:   The Gram matrix of $\mathbf{X}$ is Cholesky-factored into the LU factors $\mathbf{R}^{\mathrm{T}}\mathbf{R}$.

Gram-Schmidt Procedure

Begin the Gram-Schmidt procedure with the initialization

$$\mathbf{y}_1 = \mathbf{x}_1; \qquad \mathbf{u}_1 = \frac{\mathbf{y}_1}{\|\mathbf{y}_1\|}. \qquad\qquad 2.36$$

Construct $\mathbf{u}_2$ as follows:

$$\mathbf{y}_2 = \mathbf{x}_2 - \frac{\mathbf{x}_2^{\mathrm{T}}\mathbf{y}_1}{\|\mathbf{y}_1\|^2}\,\mathbf{y}_1 = \mathbf{x}_2 - \mathbf{x}_2^{\mathrm{T}}\mathbf{u}_1\mathbf{u}_1; \qquad \mathbf{u}_2 = \frac{\mathbf{y}_2}{\|\mathbf{y}_2\|}. \qquad 2.37$$

The vector $\mathbf{u}_2 = \mathbf{0}$ iff $\mathbf{x}_2 = a\mathbf{x}_1$, meaning that $\mathbf{x}_2$ is linearly dependent upon $\mathbf{x}_1$. Assume $\mathbf{u}_2 \neq 0$. Then $\mathbf{u}_2$ and $\mathbf{u}_1$ are linearly independent *and* orthogonal:

$$\|\mathbf{y}_2\|\,\|\mathbf{y}_1\|\,\mathbf{u}_2^{\mathrm{T}}\mathbf{u}_1 = \mathbf{y}_2^{\mathrm{T}}\mathbf{y}_1 = \mathbf{x}_2^{\mathrm{T}}\mathbf{y}_1 - \frac{\mathbf{x}_2^{\mathrm{T}}\mathbf{y}_1}{\|\mathbf{y}_1\|^2}\,\mathbf{y}_1^{\mathrm{T}}\mathbf{y}_1$$

$$= 0. \qquad\qquad 2.38$$

Continue this procedure, generating each new vector $\mathbf{u}_k$ as follows:

$$\mathbf{y}_k = \mathbf{x}_k - \sum_{i=1}^{k-1}(\mathbf{x}_k^{\mathrm{T}}\mathbf{u}_i)\mathbf{u}_i; \qquad \mathbf{u}_k = \frac{\mathbf{y}_k}{\|\mathbf{y}_k\|}. \qquad 2.39$$

The vector $\mathbf{u}_k = 0$ iff $\mathbf{x}_k$ is a linear combination of $\mathbf{u}_1, \mathbf{u}_2, \ldots, \mathbf{u}_{k-1}$. (See Problem 2.8.)

This procedure maps the vectors $(\mathbf{x}_1, \mathbf{x}_2, \ldots, \mathbf{x}_p)$ into the vectors $(\mathbf{u}_1, \mathbf{u}_2, \ldots, \mathbf{u}_p)$. In the latter set, there will be r nonzero vectors that are linearly independent and orthogonal and $(p - r)$ vectors that are zero. For each $\mathbf{u}_i$ that is zero, we conclude that the corresponding $\mathbf{x}_i$ is linearly dependent. By discarding each $\mathbf{x}_i$ for which the corresponding $\mathbf{u}_i$ is zero, we retain only those r vectors $\mathbf{x}_i$ that are linearly independent.

The discarded $(p - r)$ vectors are linearly dependent on the r linearly independent vectors.

When the Gram-Schmidt procedure terminates, we have the following representation for the vectors $(\mathbf{u}_1, \mathbf{u}_2, \ldots, \mathbf{u}_p)$:

$$\mathbf{Q} = \mathbf{XR}^{-1}$$

$$\mathbf{Q} = [\mathbf{u}_1 \quad \mathbf{u}_2 \quad \cdots \quad \mathbf{u}_p] \qquad \mathbf{X} = [\mathbf{x}_1 \quad \mathbf{x}_2 \quad \cdots \quad \mathbf{x}_p] \qquad \qquad 2.40$$

$$\mathbf{R}^{-1} = \begin{bmatrix} a_{11} & a_{12} & \cdots & a_{1p} \\ & a_{22} & & \\ & & \ddots & \vdots \\ 0 & & & a_{pp} \end{bmatrix}.$$

In this representation of what the Gram-Schmidt procedure delivers, we say that the data matrix is analyzed to produce the orthogonal vectors $\mathbf{u}_i$. Each a_{ii} is nonzero provided that $\mathbf{x}_i$ is linearly independent of $\mathbf{x}_1, \ldots, \mathbf{x}_{i-1}$.

The Gram-Schmidt procedure delivers the matrix $\mathbf{Q}$ and, with a little extra book-keeping, the matrix $\mathbf{R}^{-1}$. (See Problem 2.7.) There are many other ways to obtain $\mathbf{Q}$, and $\mathbf{R}$ or $\mathbf{R}^{-1}$. Among these, the Householder and Givens transformations are the most commonly used because they generate the matrix $\mathbf{Q}$ from sequences of very simple orthogonal transformations.

Householder Transformation

Let's assume that the data matrix $\mathbf{X} = [\mathbf{x}_1 \quad \mathbf{x}_2 \quad \cdots \quad \mathbf{x}_n]$ contains n linearly independent columns $\mathbf{x}_i$, and let's represent the *QR* factorization of the data matrix $\mathbf{X}$ as

$$\mathbf{Q}^{\mathrm{T}}\mathbf{X} = \mathbf{R}. \qquad \qquad 2.41$$

We think of the matrix $\mathbf{Q}^{\mathrm{T}}$ as an orthogonal matrix that "takes $\mathbf{X}$ to upper-triangular form." With this insight, we wonder if there is a representation of $\mathbf{Q}^{\mathrm{T}}$ that would enable us to take $\mathbf{X}$ to upper-triangular form, column-by-column. We have in mind a representation of $\mathbf{Q}^{\mathrm{T}}$ in factored form,

$$\mathbf{Q}^{\mathrm{T}} = \mathbf{Q}_n^{\mathrm{T}}\mathbf{Q}_{n-1}^{\mathrm{T}} \cdots \mathbf{Q}_1^{\mathrm{T}}, \qquad \qquad 2.42$$

where each of the transformations $\mathbf{Q}_i$ is itself orthogonal: $\mathbf{Q}_i^{\mathrm{T}}\mathbf{Q}_i = \mathbf{I}$. If we can achieve such a factorization, then the matrix $\mathbf{Q}$ will be stored implicitly in the matrices $\mathbf{Q}_i$.

Let's choose $\mathbf{Q}_1^{\mathrm{T}}$ to take just the first column of $\mathbf{X}$ to the desired form:

$$\mathbf{Q}_1^{\mathrm{T}}\mathbf{X} = \left[\begin{array}{c|ccc} r_{11} & r_{12} & \cdots & r_{1n} \\ \hline \mathbf{0} & \hat{\mathbf{x}}_2 & & \star \end{array} \right]. \qquad \qquad 2.43$$

If we repeated this kind of operation over and over again, we might deliver a sequence of approximations of the form

$$\mathbf{Q}_2^T\mathbf{Q}_1^T\mathbf{X} = \begin{bmatrix} r_{11} & r_{12} & r_{13} & \cdots & r_{1n} \\ 0 & r_{22} & r_{23} & \cdots & r_{2n} \\ & & 0 & & \\ \vdots & \vdots & & \hat{\mathbf{x}}_3 & \star \\ 0 & 0 & & & \end{bmatrix}$$

$$\mathbf{Q}_3^T\mathbf{Q}_2^T\mathbf{Q}_1^T\mathbf{X} = \begin{bmatrix} r_{11} & r_{12} & r_{13} & \cdots & r_{1n} \\ 0 & r_{22} & r_{23} & \cdots & r_{2n} \\ & 0 & r_{33} & \cdots & r_{3n} \\ & & 0 & & \\ \vdots & \vdots & \vdots & \hat{\mathbf{x}}_4 & \star \\ 0 & 0 & 0 & & \end{bmatrix} \qquad 2.44$$

$$\vdots$$

$$\mathbf{Q}_n^T \cdots \mathbf{Q}_2^T\mathbf{Q}_1^T\mathbf{X} = \begin{bmatrix} r_{11} & r_{12} & \cdots & r_{1n} \\ & r_{22} & \cdots & r_{2n} \\ & & \ddots & \vdots \\ & & & r_{nn} \\ & \mathbf{0} & & \end{bmatrix}.$$

The implication here is that, in going from approximation $k - 1$ to approximation k, all but the southeast corner of the matrix remains unchanged:

$$\mathbf{Q}_k^T\mathbf{Q}_{k-1}^T \cdots \mathbf{Q}_1^T\mathbf{X} = \mathbf{Q}_k \begin{bmatrix} r_{11} & r_{12} & \cdots & r_{1(k-1)} & \cdots & r_{1n} \\ 0 & r_{22} & & & & \\ & 0 & \ddots & & & \vdots \\ & & & r_{(k-1)(k-1)} & \cdots & r_{(k-1)n} \\ & & & 0 & & \\ \vdots & \vdots & & \vdots & \hat{\mathbf{x}}_k & \star \\ 0 & 0 & & 0 & & \end{bmatrix}$$

$$\qquad 2.45$$

$$= \begin{bmatrix} r_{11} & \cdots & r_{1(k-1)} & \cdots & & r_{1n} \\ 0 & & \vdots & & & \vdots \\ & & r_{(k-1)(k-1)} & \cdots & & r_{(k-1)n} \\ & & 0 & r_{kk} & & r_{kn} \\ & & & 0 & & \\ \vdots & & \vdots & \vdots & & \star \\ 0 & & 0 & 0 & & \end{bmatrix}.$$

This suggests that Q_k^T is a matrix of the form

$$\mathbf{Q}_k^T = \left[\begin{array}{c|c} \mathbf{I}_{k-1} & \mathbf{0} \\ \hline \mathbf{0} & \hat{\mathbf{Q}}_k^T \end{array} \right]. \qquad 2.46$$

The matrix $[\mathbf{I}_{k-1} \mid \mathbf{0}]$ leaves the northern $(k-1) \times n$ band of a matrix unchanged. The matrix $[\mathbf{0} \mid \hat{\mathbf{Q}}_k]$ leaves the southwestern $(n-k) \times (k-1)$ block filled with zeros. If we can find $\hat{\mathbf{Q}}_k^{\mathrm{T}}$ so that

$$\hat{\mathbf{Q}}_k^{\mathrm{T}} \hat{\mathbf{x}}_k = \begin{bmatrix} r_{kk} \\ \mathbf{0} \end{bmatrix}, \tag{2.47}$$

then we will have achieved our purpose by zeroing one more column and leaving previously zeroed columns untouched.

The matrix $\hat{\mathbf{Q}}_k^{\mathrm{T}}$ can be constructed from a Householder transformation. This transformation is based on the very simple geometrical idea illustrated in Figure 2.2. In the figure, $\mathbf{e}_1$ is the unit vector $\mathbf{e}_1 = [1 \quad 0 \quad \cdots \quad 0]^{\mathrm{T}}$, and the vector $\hat{\mathbf{x}}_k$ is to be "taken to one of the vectors" $\pm \|\hat{\mathbf{x}}_k\| \mathbf{e}_1$ by a transformation $\hat{\mathbf{Q}}_k^{\mathrm{T}} \hat{\mathbf{x}}_k$. Note that the vectors $\mathbf{v}_k = \hat{\mathbf{x}}_k - \|\hat{\mathbf{x}}_k\| \mathbf{e}_1$ and $\mathbf{v}_k^{\perp} = \hat{\mathbf{x}}_k + \|\hat{\mathbf{x}}_k\| \mathbf{e}_1$ are orthogonal. We can build the vector $\|\hat{\mathbf{x}}_k\| \mathbf{e}_1$ by "projecting $\hat{\mathbf{x}}_k$ through $\mathbf{v}_k^{\perp}$," and we can build the vector $-\|\hat{\mathbf{x}}_k\| \mathbf{e}_1$ by "projecting $\hat{\mathbf{x}}_k$ through the vector $\mathbf{v}_k$." The matrix that projects through $\mathbf{v}_k^{\perp}$ is

$$\hat{\mathbf{Q}}_k^{\mathrm{T}} = \mathbf{I} - \frac{2\mathbf{v}_k \mathbf{v}_k^{\mathrm{T}}}{\mathbf{v}_k^{\mathrm{T}} \mathbf{v}_k} = \hat{\mathbf{Q}}_k. \tag{2.48}$$

The transformation $\hat{\mathbf{Q}}_k^{\mathrm{T}}$ is the Householder transformation, which we summarize here:

$$\hat{\mathbf{Q}}_k^{\mathrm{T}} = \mathbf{I} - \frac{2\mathbf{v}_k \mathbf{v}_k^{\mathrm{T}}}{\mathbf{v}_k^{\mathrm{T}} \mathbf{v}_k} \tag{2.49}$$

$$\mathbf{v}_k = \hat{\mathbf{x}}_k - \|\hat{\mathbf{x}}_k\| \mathbf{e}_1.$$

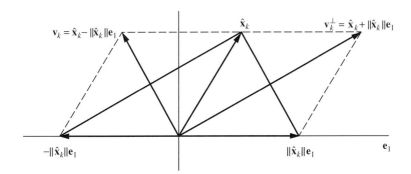

Figure 2.2 Householder transformation.

We complete our development of the Householder transformation by showing that each of the Householder transformations is itself orthogonal:

$$\mathbf{Q}_k^T = \left[\begin{array}{c|c} \mathbf{I}_{k-1} & \mathbf{0} \\ \hline \mathbf{0} & \hat{\mathbf{Q}}_k^T \end{array} \right]$$

2.50

$$\mathbf{Q}_k^T \mathbf{Q}_k = \left[\begin{array}{c|c} \mathbf{I}_{k-1} & \mathbf{0} \\ \hline \mathbf{0} & \hat{\mathbf{Q}}_k^T \end{array} \right] \left[\begin{array}{c|c} \mathbf{I}_{k-1} & \mathbf{0} \\ \hline \mathbf{0} & \hat{\mathbf{Q}}_k \end{array} \right]$$

$$= \left[\begin{array}{c|c} \mathbf{I}_{k-1} & \mathbf{0} \\ \hline \mathbf{0} & \hat{\mathbf{Q}}_k^T \hat{\mathbf{Q}}_k \end{array} \right]$$

$$\hat{\mathbf{Q}}_k^T \hat{\mathbf{Q}}_k = \left(\mathbf{I} - \frac{2\mathbf{v}_k \mathbf{v}_k^T}{\mathbf{v}_k^T \mathbf{v}_k} \right) \left(\mathbf{I} - \frac{2\mathbf{v}_k \mathbf{v}_k^T}{\mathbf{v}_k^T \mathbf{v}_k} \right)$$

$$= \mathbf{I}.$$

Givens Transformation

The sequence of Householder transformations takes the data matrix $\mathbf{X}$ to upper-triangular form, column-by-column, by introducing zeros in a column. The Givens rotations take $\mathbf{X}$ to upper-triangular form by introducing zeros element-by-element. The basic idea remains the same: Take $\mathbf{X}$ to upper-triangular form with a sequence of orthogonal transformations. Our scheme will be to introduce zeros element-by-element, completing each row before proceeding to the next:

$$\mathbf{Q}_{21}^T \mathbf{X} = \begin{bmatrix} \star & \star & \star & \star \\ 0 & \star & \star & \star \\ \star & \star & \star & \star \\ \star & \star & \star & \star \end{bmatrix}$$

$$\mathbf{Q}_{31}^T \mathbf{Q}_{21}^T \mathbf{X} = \begin{bmatrix} \star & \star & \star & \star \\ 0 & \star & \star & \star \\ 0 & \star & \star & \star \\ \star & \star & \star & \star \end{bmatrix}$$

2.51

$$\mathbf{Q}_{32}^T \mathbf{Q}_{31}^T \mathbf{Q}_{21}^T \mathbf{X} = \begin{bmatrix} \star & \star & \star & \star \\ 0 & \star & \star & \star \\ 0 & 0 & \star & \star \\ \star & \star & \star & \star \end{bmatrix}$$

$$\mathbf{Q}_{41}^T \mathbf{Q}_{32}^T \mathbf{Q}_{31}^T \mathbf{Q}_{21}^T \mathbf{X} = \begin{bmatrix} \star & \star & \star & \star \\ 0 & \star & \star & \star \\ 0 & 0 & \star & \star \\ 0 & \star & \star & \star \end{bmatrix}.$$

From this pattern, we see that $\mathbf{Q}^T\mathbf{X} = \mathbf{R}$ may be written as

$$\prod_{i=2}^{n} \prod_{j=1}^{i-1} \mathbf{Q}_{ij}^T \mathbf{X} = \mathbf{R}. \qquad 2.52$$

The transformation $\mathbf{Q}_{ij}^T$ zeros out the jth element of row i or, equivalently, the ith element of column j, $i > j$:

$$\mathbf{Q}_{ij}^T \mathbf{x}_j = \begin{bmatrix} x_{1j} \\ \vdots \\ x_{jj} \\ \vdots \\ 0 \\ \vdots \\ x_{nj} \end{bmatrix} \begin{matrix} \\ \\ \\ \\ i \\ \\ \end{matrix} . \qquad 2.53$$

The Givens rotation does it:

$$\mathbf{Q}_{ij}^T = \begin{bmatrix} \mathbf{I} & & & \\ & \cos\theta_{ij} & & \sin\theta_{ij} \\ & & \mathbf{I} & \\ & -\sin\theta_{ij} & & \cos\theta_{ij} \\ & & & & \mathbf{I} \end{bmatrix} \begin{matrix} \\ j \\ \\ i \\ \\ \end{matrix}$$

$$\begin{matrix} & j & & i \end{matrix} \qquad 2.54$$

$$\cos\theta_{ij} = \frac{x_{jj}}{(x_{jj}^2 + x_{ij}^2)^{1/2}} \qquad \sin\theta_{ij} = \frac{x_{ij}}{(x_{jj}^2 + x_{ij}^2)^{1/2}} .$$

It is easy to see that $\mathbf{Q}_{ij}^T$ leaves previously introduced zeros unchanged and that $\mathbf{Q}_{ij}$ is orthogonal:

$$\mathbf{Q}_{ij}^T \mathbf{Q}_{ij} = \mathbf{I}. \qquad 2.55$$

2.4   LINEAR SUBSPACES

Let's begin our discussion of linear subspaces by considering a set of vectors $(\mathbf{x}_1, \mathbf{x}_2, \ldots, \mathbf{x}_p)$, $\mathbf{x}_i \in R^n$. The vectors need not be linearly independent. The *span* of these vectors is the set of all vectors $\mathbf{x} \in R^n$ that may be generated from linear combinations of the set:

$$S(\mathbf{x}_1, \mathbf{x}_2, \ldots, \mathbf{x}_p) = \left\{ \mathbf{x} : \mathbf{x} = \sum_{i=1}^{p} a_i \mathbf{x}_i \right\}. \qquad 2.56$$

The span S is, itself, a linear vector space, and it is a *subspace* of R^n. We shall use the terms *span* and *subspace* interchangeably.

Basis

If the linearly dependent vectors in the set $(\mathbf{x}_1, \mathbf{x}_2, \ldots, \mathbf{x}_p)$ are removed (using Gram-Schmidt orthogonalization, for example), and all of the linearly independent vectors are retained, then these vectors form a *basis* for the subspace S. Call this basis $(\mathbf{x}_{(1)}, \mathbf{x}_{(2)}, \ldots, \mathbf{x}_{(r)})$, $r \leq p$. For example, if the vectors $\mathbf{x}_2$, $\mathbf{x}_3$, and $\mathbf{x}_7$ are linearly dependent in a set of eight vectors, then $\mathbf{x}_{(1)} = \mathbf{x}_1$, $\mathbf{x}_{(2)} = \mathbf{x}_4$, $\mathbf{x}_{(3)} = \mathbf{x}_5$, $\mathbf{x}_{(4)} = \mathbf{x}_6$, $\mathbf{x}_{(5)} = \mathbf{x}_8$. Every vector in the subspace S may be written as the sum

$$\mathbf{x} = \sum_{i=1}^{r} a_i \mathbf{x}_{(i)} = \mathbf{X}\mathbf{a}; \qquad \mathbf{X} = [\mathbf{x}_{(1)} \quad \cdots \quad \mathbf{x}_{(r)}]. \qquad 2.57$$

If the basis $\mathbf{X}$ is orthogonalized, then every vector in the subspace S may be written as

$$\mathbf{x} = \mathbf{Q}\mathbf{R}\mathbf{a} = \mathbf{Q}\mathbf{b} \qquad 2.58$$

$$\mathbf{Q} = [\mathbf{u}_1 \quad \cdots \quad \mathbf{u}_r]; \qquad \mathbf{b} = \mathbf{R}\mathbf{a}$$

where the $(\mathbf{u}_1, \mathbf{u}_2, \ldots, \mathbf{u}_r)$ are orthogonal. For this reason, we call the orthogonal vectors $(\mathbf{u}_1, \mathbf{u}_2, \ldots, \mathbf{u}_r)$ an *orthogonal basis* for the subspace S.

Direct Subspaces

The linearly independent set of vectors $\{\mathbf{x}_{(i)}\}_1^r$ may be divided into k disjoint sets of vectors. Each disjoint set spans its respective subspace S_j (or forms a basis for its span). Therefore, the subspace S may be written as a *direct sum* of disjoint (or linearly independent) subspaces:

$$S = S_1 \oplus S_2 \oplus \cdots \oplus S_k \qquad 2.59$$

$$S_j = \left\{ \mathbf{x} : \mathbf{x} = \sum_{i \in I_j} a_i \mathbf{x}_{(i)} \right\}; \qquad I_j : j\text{th disjoint subset of } \{1, 2, \ldots, r\}.$$

If the orthogonal vectors $\{\mathbf{u}_i\}_1^r$ are similarly divided into subsets, then S may be written as the direct sum of orthogonal subspaces:

$$S = U_1 \oplus U_2 \oplus \cdots \oplus U_k \qquad 2.60$$

$$U_j = \left\{ \mathbf{x} : \mathbf{x} = \sum_{i \in I_j} b_i \mathbf{u}_i \right\}.$$

These direct sums are illustrated in Figure 2.3.

Unicity

The representation of any vector in S is *unique*. That is, if there are two representations for $\mathbf{x}$ of the form

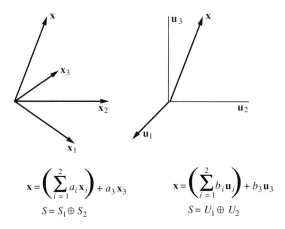

$$x = \left(\sum_{i=1}^{2} a_i x_i\right) + a_3 x_3 \qquad\qquad x = \left(\sum_{i=1}^{2} b_i u_i\right) + b_3 u_3$$

$$S = S_1 \oplus S_2 \qquad\qquad\qquad S = U_1 \oplus U_2$$

Figure 2.3 Direct sum of subspaces.

$$\mathbf{x} = \sum_{i=1}^{r} \alpha_i \mathbf{x}_{(i)} = \sum_{i=1}^{r} \beta_i \mathbf{x}_{(i)}, \qquad\qquad 2.61$$

then $\sum_{i=1}^{r}(\alpha_i - \beta_i)\mathbf{x}_{(i)} = \mathbf{0}$. As the $\{\mathbf{x}_{(i)}\}$ are linearly independent, it follows that $\alpha_i = \beta_i$. These unique coefficients may be determined as follows. Replace the linear equation $\mathbf{x} = \mathbf{Xa}$ with $\mathbf{x} = \mathbf{QRa}$ and solve for $\mathbf{a}$:

$$\mathbf{Ra} = \mathbf{Q}^\mathsf{T}\mathbf{x} = \mathbf{b} \qquad\qquad 2.62$$

Because $\mathbf{R}$ is upper-triangular, $\mathbf{a}$ may be solved from bottom to top.

Dimension, Rank, and Nullity

Let S denote the subspace spanned by the vectors $(\mathbf{x}_1, \mathbf{x}_2, \ldots, \mathbf{x}_p)$:

$$S = \left\{\mathbf{x} : \mathbf{x} = \sum_{i=1}^{p} a_i \mathbf{x}_i\right\}. \qquad\qquad 2.63$$

The spanning vectors may be organized into the data matrix $\mathbf{X} = [\mathbf{x}_1 \quad \mathbf{x}_2 \quad \cdots \quad \mathbf{x}_p]$, and the subspace may be represented as

$$R(\mathbf{X}) = \{\mathbf{x} : \mathbf{x} = \mathbf{Xa}\}. \qquad\qquad 2.64$$

We call this set the *range* of the linear transformation $\mathbf{x} = \mathbf{Xa}$ and denote it $R(\mathbf{X})$ or $\langle\mathbf{X}\rangle$. Thus, $S = R(\mathbf{X}) = \langle\mathbf{X}\rangle$. The dimension of the subspace S, denoted dim S, is just the maximum number of linearly independent vectors in the set $(\mathbf{x}_1, \mathbf{x}_2, \ldots, \mathbf{x}_p)$. But this is just the number of linearly independent columns in the data matrix $\mathbf{X}$. This number we call the *rank* of $\mathbf{X}$, and we denote it $\rho(\mathbf{X})$. Therefore, we have the identity

$$\dim S = \dim R(\mathbf{X}) = \rho(\mathbf{X}). \qquad\qquad 2.65$$

In words, "the dimension of the range of $\mathbf{X}$ equals the rank of $\mathbf{X}$."

Now consider the set of vectors $\mathbf{a}$ for which $\mathbf{Xa} = \mathbf{0}$:

$$N(\mathbf{X}) = \{\mathbf{a} : \mathbf{Xa} = \mathbf{0}\}. \tag{2.66}$$

Let's assume that the p-vector $\mathbf{a}$ lies in a subspace of dimension $p - r$. That is, there exists a basis for $N(\mathbf{X})$ of the form $(\mathbf{a}_1, \mathbf{a}_2, \ldots, \mathbf{a}_{p-r})$:

$$N(\mathbf{X}) = \left\{ \mathbf{a} : \mathbf{a} = \sum_{i=1}^{p-r} \alpha_i \mathbf{a}_i \right\} \tag{2.67}$$

$$\dim N(\mathbf{X}) = p - r.$$

To this basis we adjoin any additional set of r linearly independent vectors $(\mathbf{a}_{p-r+1}, \ldots, \mathbf{a}_p)$. Then any $\mathbf{a} \in R^p$ may be written as

$$\mathbf{a} = \sum_{i=1}^{p-r} \alpha_i \mathbf{a}_i + \sum_{i=p-r+1}^{p} \alpha_i \mathbf{a}_i. \tag{2.68}$$

The range of $\mathbf{X}$ is

$$R(\mathbf{X}) = \left\{ \mathbf{x} : \mathbf{x} = \mathbf{X} \left[\sum_{i=1}^{p-r} \alpha_i \mathbf{a}_i + \sum_{i=p-r+1}^{p} \alpha_i \mathbf{a}_i \right] \right\}$$

$$= \left\{ \mathbf{x} : \mathbf{x} = \sum_{i=p-r+1}^{p} \alpha_i \mathbf{Xa}_i \right\}. \tag{2.69}$$

The second step follows from the fact that $\mathbf{Xa}_i = \mathbf{0}$ for $i = 1, 2, \ldots, p - r$. This result says that the range of $\mathbf{X}$ is in fact the range of the vectors $(\mathbf{Xa}_{p-r+1}, \ldots, \mathbf{Xa}_p)$. These r vectors form a linearly independent basis for $R(\mathbf{X})$. Therefore the dimensions of $R(\mathbf{X})$ and $N(\mathbf{X})$ are

$$\dim R(\mathbf{X}) = r$$

$$\dim N(\mathbf{X}) = p - r \tag{2.70}$$

$$\dim R(\mathbf{X}) + \dim N(\mathbf{X}) = p.$$

This is called *Sylvester's Law of Nullity*. What it says is that rank deficiency in the data matrix $\mathbf{X}$ manifests itself in a null space whose dimension is the degree of rank deficiency, $\dim N(\mathbf{X}) = p - \rho(\mathbf{X})$. This is zero when $\mathbf{X}$ has full rank.

Linear Equations

When we characterize the range of $\mathbf{X}$ as the set of vectors $\mathbf{x} = \mathbf{Xa}$, we are saying that $\mathbf{x}$ is contained in $R(\mathbf{X})$ iff the equation $\mathbf{x} = \mathbf{Xa}$ has a solution for $\mathbf{a}$. Such a solution exists iff

$$\begin{bmatrix} \mathbf{X} & \mathbf{x} \end{bmatrix} \begin{bmatrix} -\mathbf{a} \\ 1 \end{bmatrix} = \mathbf{0}. \tag{2.71}$$

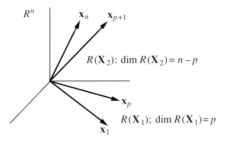

Figure 2.4 Decomposition of R^n.

But this holds iff $\rho[\mathbf{X} \quad \mathbf{x}] = \dim R[\mathbf{X} \quad \mathbf{x}] = \rho(\mathbf{X}) = \dim R(\mathbf{X})$. This is another way of saying that $\mathbf{x}$ lies in the range of $\mathbf{X}$.

Decomposition of R^n

Let $(\mathbf{x}_1, \mathbf{x}_2, \ldots, \mathbf{x}_n)$ denote n linearly independent vectors in R^n. Divide these vectors into two sets, $(\mathbf{x}_1, \ldots, \mathbf{x}_p)$ and $(\mathbf{x}_{p+1}, \ldots, \mathbf{x}_n)$. Associate the data matrices $\mathbf{X}_1 = [\mathbf{x}_1 \quad \mathbf{x}_2 \quad \cdots \quad \mathbf{x}_p]$ and $\mathbf{X}_2 = [\mathbf{x}_{p+1} \quad \cdots \quad \mathbf{x}_n]$ with the respective sets. Any vector $\mathbf{x} \in R^n$ is a linear combination of the $\mathbf{x}_i$:

$$
\begin{aligned}
\mathbf{x} &= \mathbf{x}_1 + \mathbf{x}_2 \\
&= \mathbf{X}_1 \mathbf{a}_1 + \mathbf{X}_2 \mathbf{a}_2.
\end{aligned}
\qquad 2.72
$$

We say that $R^n = R(\mathbf{X}_1) \oplus R(\mathbf{X}_2)$ is a direct sum of linearly independent subspaces and note that $\dim R^n = \dim R(\mathbf{X}_1) + \dim R(\mathbf{X}_2) = p + (n - p) = n$. This is illustrated in Figure 2.4.

If the linearly independent span $(\mathbf{x}_1, \mathbf{x}_2, \ldots, \mathbf{x}_n)$ is replaced by the orthogonal span $(\mathbf{u}_1, \mathbf{u}_2, \ldots, \mathbf{u}_n)$, then we may decompose R^n by building the data matrices $\mathbf{U}_1 = [\mathbf{u}_1 \quad \mathbf{u}_2 \quad \cdots \quad \mathbf{u}_p]$ and $\mathbf{U}_2 = [\mathbf{u}_{p+1} \quad \cdots \quad \mathbf{u}_n]$. Any $\mathbf{x} \in R^n$ has the representation

$$
\mathbf{x} = \mathbf{U}_1 \mathbf{a}_1 + \mathbf{U}_2 \mathbf{a}_2,
\qquad 2.73
$$

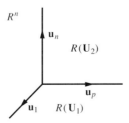

Figure 2.5 Direct sum of orthogonal subspaces.

meaning that $R^n = R(\mathbf{U}_1) \oplus R(\mathbf{U}_2)$ and $\dim R^n = \dim R(\mathbf{U}_1) + \dim R(\mathbf{U}_2) = p + (n - p) = n$. R^n is a direct sum of orthogonal subspaces, as illustrated in Figure 2.5.

2.5   HERMITIAN MATRICES

A Hermitian matrix $\mathbf{R}$ has the property that it equals the complex conjugate of its transpose: $(\mathbf{R}^T)^\star = \mathbf{R}$. We simplify our notation by writing $(\mathbf{R}^T)^\star$ as $\mathbf{R}^H$. There are a number of elegant results that accrue to Hermitian matrices. They are stated and their proofs are outlined here.

The Eigenvalues of a Hermitian Matrix Are Real

Let λ denote an eigenvalue of the Hermitian matrix $\mathbf{R} = \mathbf{R}^H$:

$$\mathbf{Ru} = \lambda\mathbf{u}. \qquad\qquad 2.74$$

Then the quadratic form $\mathbf{u}^H\mathbf{Ru}$ may be written as

$$\mathbf{u}^H\mathbf{Ru} = \lambda\mathbf{u}^H\mathbf{u}. \qquad\qquad 2.75$$

The quadratic form on the left-hand side is real, and $\mathbf{u}^H\mathbf{u}$ is real, meaning that λ is real.

The Eigenvectors of a Hermitian Matrix Are Orthogonal

Let $\mathbf{u}_i$ and $\mathbf{u}_j$ denote eigenvectors of a Hermitian matrix, corresponding to distinct eigenvalues. Then

$$\begin{aligned}\mathbf{Ru}_i &= \lambda_i\mathbf{u}_i \\ \mathbf{Ru}_j &= \lambda_j\mathbf{u}_j, \qquad \lambda_i \neq \lambda_j.\end{aligned} \qquad\qquad 2.76$$

Note that the inner product $\mathbf{u}_j^H\mathbf{u}_i$ may be written two ways:

$$\begin{aligned}\mathbf{u}_j^H\mathbf{Ru}_i &= \lambda_i\mathbf{u}_j^H\mathbf{u}_i \\ \mathbf{u}_j^H\mathbf{Ru}_i &= \lambda_j\mathbf{u}_j^H\mathbf{u}_i.\end{aligned} \qquad\qquad 2.77$$

This means that $(\lambda_j - \lambda_i)\mathbf{u}_j^H\mathbf{u}_i = 0$, implying that $\mathbf{u}_j^H\mathbf{u}_i = 0$.

Hermitian Matrices Are Diagonalizable

Let $\mathbf{R}$ denote a Hermitian matrix with distinct eigenvalues. Then organize the equations $\mathbf{Ru}_i = \lambda_i\mathbf{u}_i$, $i = 1, 2, \ldots, n$, into the following equation:

$$\mathbf{RU} = \mathbf{U\Lambda} \qquad\qquad 2.78$$

$$\mathbf{U} = [\mathbf{u}_1 \quad \mathbf{u}_2 \quad \cdots \quad \mathbf{u}_n] \qquad \mathbf{\Lambda} = \mathrm{diag}[\lambda_1 \quad \lambda_2 \quad \cdots \quad \lambda_n].$$

The vectors $\mathbf{u}_i$ are mutually orthogonal, meaning $\mathbf{u}_i^H \mathbf{u}_j = \delta_{ij}$ and making $\mathbf{U}$ a unitary matrix: $\mathbf{U}^H \mathbf{U} = \mathbf{I} = \mathbf{U}\mathbf{U}^H$. Postmultiply $\mathbf{RU} = \mathbf{U}\boldsymbol{\Lambda}$ by $\mathbf{U}^H$ to obtain

$$\mathbf{R} = \mathbf{U}\boldsymbol{\Lambda}\mathbf{U}^H$$

$$= \sum_{i=1}^{n} \lambda_i \mathbf{u}_i \mathbf{u}_i^H. \qquad 2.79$$

We call this a *unitary decomposition* of $\mathbf{R}$. We also say that $\mathbf{R}$ is *unitarily similar* to the diagonal matrix $\boldsymbol{\Lambda}$, because a unitary matrix $\mathbf{U}$ takes $\mathbf{R}$ to diagonal form: $\mathbf{U}^H \mathbf{R} \mathbf{U} = \boldsymbol{\Lambda}$.

If $\boldsymbol{\Lambda} = \mathbf{I}$, then we have the following representation of identity:

$$\mathbf{I} = \mathbf{U}\mathbf{U}^H = \sum_{i=1}^{n} \mathbf{u}_i \mathbf{u}_i^H. \qquad 2.80$$

Each of the terms $\mathbf{u}_i \mathbf{u}_i^H$ is a rank 1 projection matrix:

$$\mathbf{P}_i = \mathbf{u}_i \mathbf{u}_i^H = \mathbf{P}_i^H = \mathbf{P}_i^2. \qquad 2.81$$

Therefore we may give $\mathbf{R}$ and $\mathbf{I}$ the spectral representations

$$\mathbf{R} = \sum_{i=1}^{n} \lambda_i \mathbf{P}_i \qquad 2.82$$

$$\mathbf{I} = \sum_{i=1}^{n} \mathbf{P}_i.$$

Now we may group the matrices $\mathbf{P}_i$ in 2^n possible ways. For any grouping, the result is a projection. For example,

$$(\mathbf{P}_i + \mathbf{P}_j) = \mathbf{E}_k = \mathbf{E}_k^H = \mathbf{E}_k^2. \qquad 2.83$$

Therefore we may write $\mathbf{I}$ as

$$\mathbf{I} = \sum_{k=1}^{r} \mathbf{E}_k \qquad 2.84$$

where the ranks of the projections $\mathbf{E}_k$ sum to n.

2.6   SINGULAR VALUE DECOMPOSITION

The singular value decomposition (SVD) is a powerful representation for rectangular matrices. The representation simultaneously diagonalizes the Gram matrix $\mathbf{H}^H\mathbf{H}$ *and* the outer product $\mathbf{H}\mathbf{H}^H$. As we shall see in Section 2.8, this latter property ensures that the Gram matrix $\mathbf{H}^H\mathbf{H}$ *and* the projection $\mathbf{H}(\mathbf{H}^H\mathbf{H})^{-1}\mathbf{H}^H$ are simultaneously diagonalized. The singular value decomposition also brings additional insight into the range and null space of a matrix $\mathbf{H}$ and provides the best low-rank approximation to $\mathbf{H}$.

Consider an $N \times p$ matrix $\mathbf{H}$ and its Gram matrix $\mathbf{G} = \mathbf{H}^H\mathbf{H}$. The Gram matrix is Hermitian and nonnegative definite. Call $\lambda_1^2, \ldots, \lambda_p^2$ the eigenvalues of $\mathbf{G}$ (and, incidentally, $\lambda_1, \ldots, \lambda_p$ the singular values of $\mathbf{H}$). We know from our Gram theorem that the rank of $\mathbf{G}$ equals the rank of $\mathbf{H}$. Assume this to be $r \leq p$. Then

$$\lambda_1^2 \geq \lambda_2^2 \geq \cdots \geq \lambda_r^2 > 0; \qquad \lambda_{r+1} = \lambda_{r+2} = \cdots = \lambda_p = 0. \qquad 2.85$$

The results of Section 2.5 for Hermitian matrices may be used to represent the Gram matrix as

$$\mathbf{G} = [\, \mathbf{V}_1 \,|\, \mathbf{V}_2 \,] \left[\begin{array}{c|c} \boldsymbol{\Lambda}_1^2 & \mathbf{0} \\ \hline \mathbf{0} & \boldsymbol{\Lambda}_2^2 \end{array} \right] \left[\begin{array}{c} \mathbf{V}_1^H \\ \mathbf{V}_2^H \end{array} \right] \qquad 2.86$$

$$\mathbf{V}_1^H\mathbf{G}\mathbf{V}_1 = \boldsymbol{\Lambda}_1^2; \qquad \mathbf{V}_1^H\mathbf{V}_1 = \mathbf{I}; \qquad \boldsymbol{\Lambda}_1^2 = \text{diag}[\lambda_1^2 \quad \cdots \quad \lambda_r^2]$$

$$\mathbf{V}_2^H\mathbf{G}\mathbf{V}_2 = \boldsymbol{\Lambda}_2^2; \qquad \mathbf{V}_2^H\mathbf{V}_2 = \mathbf{I}; \qquad \boldsymbol{\Lambda}_2^2 = \text{diag}[0 \quad 0 \quad \cdots \quad 0]$$

$$\mathbf{V}_2^H\mathbf{V}_1 = \mathbf{0}.$$

These equations may be organized into the single equation

$$\left[\begin{array}{c} \mathbf{V}_1^H \\ \mathbf{V}_2^H \end{array} \right] \mathbf{H}^H\mathbf{H} [\, \mathbf{V}_1 \quad \mathbf{V}_2 \,] = \left[\begin{array}{cc} \boldsymbol{\Lambda}_1^2 & \mathbf{0} \\ \mathbf{0} & \boldsymbol{\Lambda}_2^2 \end{array} \right]. \qquad 2.87$$

It follows that $\mathbf{H}$ is a matrix of the form

$$\mathbf{H} [\, \mathbf{V}_1 \quad \mathbf{V}_2 \,] = [\, \mathbf{U}_1 \quad \mathbf{U}_2 \,] \left[\begin{array}{cc} \boldsymbol{\Lambda}_1 & \mathbf{0} \\ \mathbf{0} & \boldsymbol{\Lambda}_2 \end{array} \right] \qquad 2.88$$

where $\mathbf{U}_1$ and $\mathbf{U}_2$ are unitary matrices:

$$\left[\begin{array}{c} \mathbf{U}_1^H \\ \mathbf{U}_2^H \end{array} \right] [\, \mathbf{U}_1 \quad \mathbf{U}_2 \,] = \left[\begin{array}{cc} \mathbf{I} & \mathbf{0} \\ \mathbf{0} & \mathbf{I} \end{array} \right]. \qquad 2.89$$

We solve for $\mathbf{H}$ as follows:

$$\mathbf{H} = [\, \mathbf{U}_1 \quad \mathbf{U}_2 \,] \left[\begin{array}{cc} \boldsymbol{\Lambda}_1 & \mathbf{0} \\ \mathbf{0} & \boldsymbol{\Lambda}_2 \end{array} \right] \left[\begin{array}{c} \mathbf{V}_1^H \\ \mathbf{V}_2^H \end{array} \right] \qquad 2.90$$

$$= \mathbf{U}\boldsymbol{\Lambda}\mathbf{V}^H$$

$$\mathbf{U} = [\, \mathbf{U}_1 \quad \mathbf{U}_2 \,]; \qquad \mathbf{V} = [\, \mathbf{V}_1 \quad \mathbf{V}_2 \,].$$

This is the singular value decomposition of $\mathbf{H}$. The matrix $\mathbf{U}$ is $N \times p$ and $\mathbf{U}_1$ is $N \times r$. The matrix $\mathbf{V}^H$ is $p \times p$ and $\mathbf{V}_1$ is $r \times p$. The matrix $\boldsymbol{\Lambda}$ is $p \times p$ and $\boldsymbol{\Lambda}_1$ is an $r \times r$ diagonal matrix of nonzero singular values. Note that the matrices $\mathbf{H}^H\mathbf{H}$ and

$\mathbf{HH}^H$ may be written as

$$\mathbf{H}^H\mathbf{H} = \mathbf{V}_1\mathbf{\Lambda}_1^2\mathbf{V}_1^H + \mathbf{V}_2\mathbf{\Lambda}_2^2\mathbf{V}_2^H$$

$$= \mathbf{V}_1\mathbf{\Lambda}_1^2\mathbf{V}_1^H$$

2.91

$$\mathbf{HH}^H = \mathbf{U}_1\mathbf{\Lambda}_1\mathbf{V}_1^H\mathbf{V}_1\mathbf{\Lambda}_1\mathbf{U}_1^H + \mathbf{U}_2\mathbf{\Lambda}_2\mathbf{V}_2^H\mathbf{V}_2\mathbf{\Lambda}_2\mathbf{U}_2^H$$

$$= \mathbf{U}_1\mathbf{\Lambda}_1^2\mathbf{U}_1^H.$$

We say that the singular value decomposition *simultaneously diagonalizes* the inner product $\mathbf{H}^H\mathbf{H}$ and the outer product $\mathbf{HH}^H$.

Range and Null Space

Recall our definitions of range and null space for a matrix $\mathbf{H}$:

1. $R(\mathbf{H}) = \{\mathbf{x} : \mathbf{x} = \mathbf{H\theta}\}$ 2.92

2. $N(\mathbf{H}) = \{\mathbf{\theta} : \mathbf{x} = \mathbf{H\theta} = \mathbf{0}\}.$

From the singular value decomposition of $\mathbf{H}$, we may write these spaces as

1. $R(\mathbf{H}) = \{\mathbf{x} : \mathbf{x} = \mathbf{U\Lambda V}^H\mathbf{\theta}\}$ 2.93

         $= \{\mathbf{x} : \mathbf{x} = \mathbf{U}_1\mathbf{\phi}\}$

         $= R(\mathbf{U}_1)$

2. $N(\mathbf{H}) = \{\mathbf{\theta} : \mathbf{x} = \mathbf{U\Lambda V}^H\mathbf{\theta} = \mathbf{0}\}$

         $= \{\mathbf{\theta} : \mathbf{V}_1^H\mathbf{\theta} = \mathbf{0}\}$

         $= N(\mathbf{V}_1^H).$

We say that the range of $\mathbf{H}$ is the range of $\mathbf{U}_1$ and the null space of $\mathbf{H}$ is the null space of $\mathbf{V}_1^H$. The dimension of the range is r, and the dimension of the null space is $p - r$.

Low Rank Approximation

The SVD solves a very important approximation problem. Let $\mathbf{H}$ denote a rank p matrix and consider the error between $\mathbf{H}$ and a rank r approximation to it:

$$e^2 = \mathrm{tr}(\mathbf{H} - \mathbf{H}_r)^H(\mathbf{H} - \mathbf{H}_r).$$

2.94

This is called the *Frobenius norm* of $\mathbf{H} - \mathbf{H}_r$, the sum of squares of all elements. Let $\mathbf{H}_r = \mathbf{U\Sigma}_r\mathbf{V}^H$, where $\mathbf{\Sigma}_r = \mathrm{diag}[\lambda_1 \quad \cdots \quad \lambda_r \quad 0 \quad \cdots \quad 0]$. Then

$$e^2 = \mathrm{tr}\,\mathbf{V}(\mathbf{\Sigma} - \mathbf{\Sigma}_r)\mathbf{U}^H\mathbf{U}(\mathbf{\Sigma} - \mathbf{\Sigma}_r)\mathbf{V}^H$$

$$= \sum_{r+1}^{p} \lambda_i^2.$$

2.95

The error $\mathbf{H} - \mathbf{H}_r$ is orthogonal to $\mathbf{H}_r$:

$$
\begin{aligned}
(\mathbf{H} - \mathbf{H}_r)\mathbf{H}_r^{\mathrm{H}} &= \mathbf{U}(\mathbf{\Sigma} - \mathbf{\Sigma}_r)\mathbf{V}^{\mathrm{H}}\mathbf{V}\mathbf{\Sigma}_r\mathbf{U}^{\mathrm{H}} \\
&= \mathbf{U}(\mathbf{\Sigma} - \mathbf{\Sigma}_r)\mathbf{\Sigma}_r\mathbf{U}^{\mathrm{H}} = \mathbf{0}.
\end{aligned}
$$
2.96

This makes $\mathbf{H}_r = \mathbf{U}\mathbf{\Sigma}_r\mathbf{V}^{\mathrm{H}}$ the least-squares, rank r, approximation to $\mathbf{H}$.

Resolution (or Decomposition) of Identity

Consider the SVD of a matrix $\mathbf{S}$:

$$
\mathbf{S} = \begin{bmatrix} \mathbf{U}_1 \mid \mathbf{U}_2 \end{bmatrix} \begin{bmatrix} \mathbf{\Sigma}_1 & \mathbf{0} \\ \mathbf{0} & \mathbf{\Sigma}_2 \end{bmatrix} \begin{bmatrix} \mathbf{V}_1^{\mathrm{H}} \\ \hline \mathbf{V}_2^{\mathrm{H}} \end{bmatrix}.
$$
2.97

The Grammian of $\mathbf{S}$ is

$$
\mathbf{S}^{\mathrm{H}}\mathbf{S} = \begin{bmatrix} \mathbf{V}_1 \mid \mathbf{V}_2 \end{bmatrix} \begin{bmatrix} \mathbf{\Sigma}_1^2 & \mathbf{0} \\ \mathbf{0} & \mathbf{\Sigma}_2^2 \end{bmatrix} \begin{bmatrix} \mathbf{V}_1^{\mathrm{H}} \\ \hline \mathbf{V}_2^{\mathrm{H}} \end{bmatrix}
$$
2.98

$$
\mathbf{S}\mathbf{S}^{\mathrm{H}} = \begin{bmatrix} \mathbf{U}_1 \mid \mathbf{U}_2 \end{bmatrix} \begin{bmatrix} \mathbf{\Sigma}_1^2 & \mathbf{0} \\ \mathbf{0} & \mathbf{\Sigma}_2^2 \end{bmatrix} \begin{bmatrix} \mathbf{U}_1^{\mathrm{H}} \\ \hline \mathbf{U}_2^{\mathrm{H}} \end{bmatrix}.
$$

This gives us a fancy way to write the identity:

$$
\mathbf{I} = \mathbf{S}(\mathbf{S}^{\mathrm{H}}\mathbf{S})^{-1}\mathbf{S}^{\mathrm{H}} = \begin{bmatrix} \mathbf{U}_1 & \mathbf{U}_2 \end{bmatrix} \begin{bmatrix} \mathbf{U}_1^{\mathrm{H}} \\ \mathbf{U}_2^{\mathrm{H}} \end{bmatrix}
$$
2.99

$$
= \mathbf{U}_1\mathbf{U}_1^{\mathrm{H}} + \mathbf{U}_2\mathbf{U}_2^{\mathrm{H}}.
$$

This result is one more way of saying that the subspaces $\langle \mathbf{U}_1 \rangle$ and $\langle \mathbf{U}_2 \rangle$ span R^N.

2.7　PROJECTIONS, ROTATIONS, AND PSEUDOINVERSES

Throughout our studies of detection, estimation, and time series analysis, we encounter projections, rotations, and pseudoinverses. These three matrix transformations are intimately related.

To begin our discussion we define the $N \times p$ matrix $\mathbf{H} = [\mathbf{h}_1 \quad \mathbf{h}_2 \quad \cdots \quad \mathbf{h}_p]$ and its corresponding subspace $\langle \mathbf{H} \rangle$, which is the span of $\{\mathbf{h}_n\}_1^p$. We next define the $N \times (N-p)$ matrix $\mathbf{A} = [\mathbf{a}_1 \quad \mathbf{a}_2 \quad \cdots \quad \mathbf{a}_{N-p}]$ and its corresponding subspace $\langle \mathbf{A} \rangle$, which is the span of $\{\mathbf{a}_n\}_1^{N-p}$. We shall assume that $\mathbf{H}$ and $\mathbf{A}$ are, respectively, full rank and that $\mathbf{A}^{\mathrm{T}}\mathbf{H} = \mathbf{0}$. This means that the matrix $[\mathbf{H}\ \mathbf{A}] = [\mathbf{h}_1 \quad \cdots \quad \mathbf{h}_p \quad \mathbf{a}_1 \quad \cdots \quad \mathbf{a}_{N-p}]$ is a full-rank $N \times N$ matrix whose columns span R^N. We say that the subspaces $\langle \mathbf{H} \rangle$ and $\langle \mathbf{A} \rangle$ provide an orthogonal decomposition of R^N and illustrate this decomposition as in Figure 2.6(a). The subspace $\langle \mathbf{H} \rangle$ is often called a *signal subspace*, and the subspace $\langle \mathbf{A} \rangle$ is often called an *orthogonal subspace*.

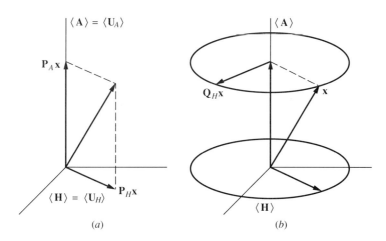

Figure 2.6 Projections and rotations. (*a*) Subspaces and projections.
(*b*) Rotations.

The real matrices $\mathbf{H}$ and $\mathbf{A}$ have SVDs and QR factors, meaning for example that

$$\mathbf{H} = \mathbf{U}_H\mathbf{\Sigma}_H\mathbf{V}_H^T; \qquad \mathbf{A} = \mathbf{U}_A\mathbf{\Sigma}_A\mathbf{V}_A^T, \qquad\qquad 2.100$$

where the $\mathbf{U}_H$, $\mathbf{U}_A$, $\mathbf{V}_H$, and $\mathbf{V}_A$ are the orthogonal matrices defined in Section 2.6. When $\mathbf{U}_H$ and $\mathbf{U}_A$ are assembled into the matrix $\mathbf{U} = [\mathbf{U}_H \quad \mathbf{U}_A]$, we obtain another orthogonal matrix:

$$\mathbf{U}^T\mathbf{U} = \begin{bmatrix} \mathbf{U}_H^T \\ \mathbf{U}_A^T \end{bmatrix} \begin{bmatrix} \mathbf{U}_H & \mathbf{U}_A \end{bmatrix} = \begin{bmatrix} \mathbf{I} & \mathbf{0} \\ \mathbf{0} & \mathbf{I} \end{bmatrix} \qquad\qquad 2.101$$

$$\mathbf{U}\mathbf{U}^T = \begin{bmatrix} \mathbf{U}_H & \mathbf{U}_A \end{bmatrix} \begin{bmatrix} \mathbf{U}_H^T \\ \mathbf{U}_A^T \end{bmatrix} = \mathbf{U}_H\mathbf{U}_H^T + \mathbf{U}_A\mathbf{U}_A^T = \mathbf{I}.$$

The subspace $\langle\mathbf{U}_H\rangle$ is identical with the subspace $\langle\mathbf{H}\rangle$ and the subspace $\langle\mathbf{U}_A\rangle$ is identical with the subspace $\langle\mathbf{A}\rangle$. This is illustrated in Figure 2.6(*a*).

Projections

Let's consider the vector $\mathbf{x}$ illustrated in Figure 2.6(*a*) and try to decompose it into two orthogonal components, one of which lies in $\langle\mathbf{H}\rangle$ and the other of which lies in $\langle\mathbf{A}\rangle$. The appropriate decomposition is

$$\mathbf{x} = \mathbf{P}_H\mathbf{x} + \mathbf{P}_A\mathbf{x}, \qquad\qquad 2.102$$

where $\mathbf{P}_H$ and $\mathbf{P}_A$ are the projections

$$\mathbf{P}_H = \mathbf{H}(\mathbf{H}^T\mathbf{H})^{-1}\mathbf{H}^T \qquad \mathbf{P}_A = \mathbf{A}(\mathbf{A}^T\mathbf{A})^{-1}\mathbf{A}^T = \mathbf{I} - \mathbf{P}_H \qquad 2.103$$

$$\mathbf{P}_H^T = \mathbf{P}_H = \mathbf{P}_H^2 \qquad \mathbf{P}_A^T = \mathbf{P}_A = \mathbf{P}_A^2 \qquad \mathbf{P}_A\mathbf{P}_H = \mathbf{P}_H\mathbf{P}_A = \mathbf{0}.$$

These projections are more illuminating in their orthogonal forms (obtained by using $\mathbf{H} = \mathbf{U}_H \boldsymbol{\Sigma}_H \mathbf{V}_H^T$, etc.):

$$\mathbf{P}_H = \mathbf{U}_H \mathbf{U}_H^T \qquad \mathbf{P}_A = \mathbf{U}_A \mathbf{U}_A^T = \mathbf{I} - \mathbf{U}_H \mathbf{U}_H^T \qquad\qquad 2.104$$

$$\mathbf{P}_H + \mathbf{P}_A = \mathbf{I}.$$

The decomposition of $\mathbf{x}$ now takes the form

$$\mathbf{x} = \mathbf{U}_H \mathbf{U}_H^T \mathbf{x} + \mathbf{U}_A \mathbf{U}_A^T \mathbf{x}. \qquad\qquad 2.105$$

The term $\mathbf{U}_H^T \mathbf{x}$ produces coefficients for the columns of $\mathbf{U}_H$, which is reminiscent of Fourier series expansions.

We often say the projection matrices $\mathbf{P}_H$ and $\mathbf{P}_A$ are *idempotent*, meaning $\mathbf{P}_H^2 = \mathbf{P}_H$ and $\mathbf{P}_A^2 = \mathbf{P}_A$. We also say that they are *unitarily similar to identity*, meaning

$$\begin{bmatrix} \mathbf{U}_H^T \\ \mathbf{U}_A^T \end{bmatrix} \mathbf{P}_H \begin{bmatrix} \mathbf{U}_H & \mathbf{U}_A \end{bmatrix} = \begin{bmatrix} \mathbf{I} & \mathbf{0} \\ \mathbf{0} & \mathbf{0} \end{bmatrix} \qquad\qquad 2.106$$

$$\begin{bmatrix} \mathbf{U}_H^T \\ \mathbf{U}_A^T \end{bmatrix} \mathbf{P}_A \begin{bmatrix} \mathbf{U}_H & \mathbf{U}_A \end{bmatrix} = \begin{bmatrix} \mathbf{0} & \mathbf{0} \\ \mathbf{0} & \mathbf{I} \end{bmatrix}.$$

This shows that the eigenvalues of $\mathbf{P}_H$ and $\mathbf{P}_A$ are 1 and 0.

Rotations

A vector $\mathbf{x} \in R^N$ may be "rotated in the subspace $\mathbf{H}$" or, equivalently, "rotated around the subspace $\mathbf{A}$," as illustrated in Figure 2.6(*b*). How do we do this algebraically? The answer is to form the matrix product

$$\mathbf{Q}_H \mathbf{x}, \qquad\qquad 2.107$$

where $\mathbf{Q}_H$ is a rotation matrix that we build from $\mathbf{P}_H$, $\mathbf{P}_A$, and an orthogonal matrix $\mathbf{Q}$:

$$\mathbf{Q}_H = \mathbf{U}_H \mathbf{Q} \mathbf{U}_H^T + \mathbf{P}_A \qquad\qquad 2.108$$

$$\mathbf{Q}^T \mathbf{Q} = \mathbf{Q} \mathbf{Q}^T = \mathbf{I}.$$

(In this construction we take $\mathbf{I} = \mathbf{P}_H + \mathbf{P}_A$ and insert $\mathbf{Q}$ between the two halves of $\mathbf{P}_H$.) The matrix $\mathbf{Q}_H$ is an orthogonal matrix that rotates the component of $\mathbf{x}$ that lies in $\langle \mathbf{H} \rangle$ and leaves the component of $\mathbf{x}$ that lies in $\langle \mathbf{A} \rangle$ unchanged:

$$(\mathbf{U}_H \mathbf{Q} \mathbf{U}_H^T + \mathbf{P}_A)\mathbf{P}_H \mathbf{x} = \mathbf{U}_H \mathbf{Q} \mathbf{U}_H^T \mathbf{x} \qquad\qquad 2.109$$

$$(\mathbf{U}_H \mathbf{Q} \mathbf{U}_H^T + \mathbf{P}_A)\mathbf{P}_A \mathbf{x} = \mathbf{P}_A \mathbf{x}.$$

Can you see that $\mathbf{P}_H \mathbf{Q}_H = \mathbf{Q}_H \mathbf{P}_H$? Can you illustrate it on Figure 2.6?

Pseudoinverse

The pseudoinverse of the real matrix $\mathbf{H}$ is defined to be the matrix

$$\begin{aligned}
\mathbf{H}^{\#} &= (\mathbf{H}^{\mathrm{T}}\mathbf{H})^{-1}\mathbf{H}^{\mathrm{T}} \\
&= \mathbf{V}_H \mathbf{\Sigma}_H^{-1} \mathbf{U}_H^{\mathrm{T}}.
\end{aligned}$$

$$2.110$$

The pseudoinverse obeys these four fundamental properties:

1. $\mathbf{H}^{\#}\mathbf{H} = \mathbf{I}$
2. $\mathbf{H}\mathbf{H}^{\#} = \mathbf{P}_H$
3. $\mathbf{H}\mathbf{H}^{\#}\mathbf{H} = \mathbf{H}$
4. $\mathbf{H}^{\#}\mathbf{H}\mathbf{H}^{\#} = \mathbf{H}^{\#}$.

In Chapter 9 we study least-squares problems in detail. We find that the pseudoinverse solves the following problem:

$$\min_{\boldsymbol{\theta}}(\mathbf{y} - \mathbf{H}\boldsymbol{\theta})^{\mathrm{T}}(\mathbf{y} - \mathbf{H}\boldsymbol{\theta})$$

$$\hat{\boldsymbol{\theta}} = \mathbf{H}^{\#}\mathbf{y}.$$

$$2.111$$

The projection $\mathbf{P}_H$ solves the related problem

$$\min_{\mathbf{x}:\mathbf{x}=\mathbf{H}\boldsymbol{\theta}} (\mathbf{y} - \mathbf{x})^{\mathrm{T}}(\mathbf{y} - \mathbf{x}) \qquad\qquad 2.112$$

$$\hat{\mathbf{x}} = \mathbf{P}_H\mathbf{y}.$$

The error between $\mathbf{y}$ and $\hat{\mathbf{x}}$ is orthogonal to $\hat{\mathbf{x}}$:

$$\begin{aligned}
(\mathbf{y} - \hat{\mathbf{x}})^{\mathrm{T}}\hat{\mathbf{x}} &= \mathbf{y}^{\mathrm{T}}(\mathbf{I} - \mathbf{P}_H)\mathbf{P}_H\mathbf{y} \\
&= 0.
\end{aligned}$$

$$2.113$$

This orthogonality is illustrated in Figure 2.6(a).

Orthogonal Representations

The SVD simultaneously diagonalizes the projection $\mathbf{P}_H$, the Gram matrix $\mathbf{G}_H = \mathbf{H}^{\mathrm{T}}\mathbf{H}$, and the pseudoinverse $\mathbf{H}^{\#}$:

$$\mathbf{P}_H = \mathbf{U}\begin{bmatrix} \mathbf{I} & \mathbf{0} \\ \mathbf{0} & \mathbf{0} \end{bmatrix}\mathbf{U}^{\mathrm{T}}$$

$$\mathbf{G}_H = \mathbf{V}\mathbf{\Sigma}^2\mathbf{V}^{\mathrm{T}}$$

$$\mathbf{H}^{\#} = \mathbf{V}\mathbf{\Sigma}^{-1}\mathbf{U}^{\mathrm{T}}; \qquad \mathbf{\Sigma}^{-1} = \begin{bmatrix} \mathbf{\Sigma}_H^{-1} & \mathbf{0} \\ \mathbf{0} & \mathbf{0} \end{bmatrix}.$$

Therefore the estimates $\hat{\boldsymbol{\theta}}$ and $\hat{\mathbf{x}}$ may also be written as

$$\hat{\boldsymbol{\theta}} = \mathbf{V}\boldsymbol{\Sigma}^{-1}\mathbf{U}^T\mathbf{y}$$

$$= \sum_{i=1}^{p} \frac{1}{\lambda_i}\mathbf{v}_i\mathbf{u}_i^T\mathbf{y} \qquad\qquad 2.114$$

$$\hat{\mathbf{x}} = \sum_{i=1}^{p} \mathbf{u}_i\mathbf{u}_i^T\mathbf{y}.$$

Can you interpret these formulas? And illustrate them?

Example 2.1 (Projections and Rotations)

Let $\mathbf{H}$ and $\mathbf{A}$ be the matrices

$$\mathbf{H} = \begin{bmatrix} 1 & 0 \\ 0 & 1 \\ 0 & 0 \end{bmatrix} \quad \text{and} \quad \mathbf{A} = \begin{bmatrix} 0 \\ 0 \\ 1 \end{bmatrix}.$$

In this case $\mathbf{U}_H = \mathbf{H}$ and $\mathbf{U}_A = \mathbf{A}$. The projections $\mathbf{P}_H$ and $\mathbf{P}_A$ are

$$\mathbf{P}_H = \begin{bmatrix} 1 & 0 \\ 0 & 1 \\ 0 & 0 \end{bmatrix}\begin{bmatrix} 1 & 0 & 0 \\ 0 & 1 & 0 \end{bmatrix} = \begin{bmatrix} 1 & 0 & 0 \\ 0 & 1 & 0 \\ 0 & 0 & 0 \end{bmatrix}$$

$$\mathbf{P}_A = \begin{bmatrix} 0 \\ 0 \\ 1 \end{bmatrix}\begin{bmatrix} 0 & 0 & 1 \end{bmatrix} = \begin{bmatrix} 0 & 0 & 0 \\ 0 & 0 & 0 \\ 0 & 0 & 1 \end{bmatrix} = \mathbf{I} - \mathbf{P}_H.$$

The rotation $\mathbf{Q}_H$ is

$$\mathbf{Q}_H = \mathbf{U}_H \begin{bmatrix} \cos\theta & -\sin\theta \\ \sin\theta & \cos\theta \end{bmatrix} \mathbf{U}_H^T + \mathbf{P}_A$$

$$= \begin{bmatrix} 1 & 0 \\ 0 & 1 \\ 0 & 0 \end{bmatrix}\begin{bmatrix} \cos\theta & -\sin\theta \\ \sin\theta & \cos\theta \end{bmatrix}\begin{bmatrix} 1 & 0 & 0 \\ 0 & 1 & 0 \end{bmatrix} + \begin{bmatrix} 0 \\ 0 \\ 1 \end{bmatrix}\begin{bmatrix} 0 & 0 & 1 \end{bmatrix}$$

$$= \begin{bmatrix} \cos\theta & -\sin\theta & 0 \\ \sin\theta & \cos\theta & 0 \\ 0 & 0 & 0 \end{bmatrix} + \begin{bmatrix} 0 & 0 & 0 \\ 0 & 0 & 0 \\ 0 & 0 & 1 \end{bmatrix} = \begin{bmatrix} \cos\theta & -\sin\theta & 0 \\ \sin\theta & \cos\theta & 0 \\ 0 & 0 & 1 \end{bmatrix}.$$

The pseudoinverse is

$$\mathbf{H}^{\#} = \left(\begin{bmatrix} 1 & 0 & 0 \\ 0 & 1 & 0 \end{bmatrix}\begin{bmatrix} 1 & 0 \\ 0 & 1 \\ 0 & 0 \end{bmatrix}\right)^{-1}\begin{bmatrix} 1 & 0 & 0 \\ 0 & 1 & 0 \end{bmatrix} = \begin{bmatrix} 1 & 0 & 0 \\ 0 & 1 & 0 \end{bmatrix}.$$

You should apply these matrices to a typical vector $\mathbf{x}$ to see how they work. Illustrate your results.   ∎

2.8   QUADRATIC FORMS

We have already encountered quadratic forms in our study of Gram determinants and linear independence in Section 2.2:

$$\sigma_k^2 = [\mathbf{a}_k]^T \begin{bmatrix} \mathbf{G}_{k-1} & \mathbf{h}_k \\ \mathbf{h}_k^T & g_{kk} \end{bmatrix} [\mathbf{a}_k]. \qquad 2.115$$

We will generalize this formula and call any real, scalar function

$$f = \mathbf{x}^T \mathbf{Q} \mathbf{x}; \qquad \mathbf{x} \in R^n, \ \mathbf{Q} \in R^{n \times n} \qquad 2.116$$

a quadratic form in the matrix $\mathbf{Q}$. Note that $f = (\mathbf{x}^T \mathbf{Q} \mathbf{x})^T = \mathbf{x}^T \mathbf{Q}^T \mathbf{x} = \mathbf{x}^T[(\mathbf{Q} + \mathbf{Q}^T)/2]\mathbf{x}$. The matrix $(\mathbf{Q} + \mathbf{Q}^T)/2$ is symmetric, so we can without loss of generality assume that $\mathbf{Q}$ is symmetric.

The symmetric matrix $\mathbf{Q}$ is nonnegative definite if $\mathbf{x}^T \mathbf{Q} \mathbf{x} \geq 0$ for all $\mathbf{x} \neq 0$ and positive definite if $\mathbf{x}^T \mathbf{Q} \mathbf{x} > 0$ for all $\mathbf{x} \neq 0$. If the symmetric matrix $\mathbf{Q}$ is partitioned, then $\mathbf{x}^T \mathbf{Q} \mathbf{x}$ may be written as

$$\mathbf{x}^T \mathbf{Q} \mathbf{x} = \begin{bmatrix} \mathbf{x}_1^T & \mathbf{x}_2^T \end{bmatrix} \begin{bmatrix} \mathbf{Q}_{11} & \mathbf{Q}_{12} \\ \mathbf{Q}_{12}^T & \mathbf{Q}_{22} \end{bmatrix} \begin{bmatrix} \mathbf{x}_1 \\ \mathbf{x}_2 \end{bmatrix}$$

$$= \mathbf{x}_1^T \mathbf{Q}_{11} \mathbf{x}_1 + \mathbf{x}_1^T \mathbf{Q}_{12} \mathbf{x}_2 + \mathbf{x}_2^T \mathbf{Q}_{12}^T \mathbf{x}_1 + \mathbf{x}_2^T \mathbf{Q}_{22} \mathbf{x}_2. \qquad 2.117$$

If $\mathbf{Q}$ is nonnegative definite, then so are the partitions $\mathbf{Q}_{11}$ and $\mathbf{Q}_{22}$. To see this, set $\mathbf{x}_2 = 0$ and then $\mathbf{x}_1 = 0$ to find

$$\begin{bmatrix} \mathbf{x}_1^T & \mathbf{0}^T \end{bmatrix} \mathbf{Q} \begin{bmatrix} \mathbf{x}_1 \\ \mathbf{0} \end{bmatrix} = \mathbf{x}_1^T \mathbf{Q}_{11} \mathbf{x}_1 \geq 0 \qquad 2.118$$

$$\begin{bmatrix} \mathbf{0}^T & \mathbf{x}_2^T \end{bmatrix} \mathbf{Q} \begin{bmatrix} \mathbf{0} \\ \mathbf{x}_2 \end{bmatrix} = \mathbf{x}_2^T \mathbf{Q}_{22} \mathbf{x}_2 \geq 0.$$

For fixed $\mathbf{x}_2$, the quadratic form $\mathbf{x}^T \mathbf{Q} \mathbf{x}$ is minimized with respect to $\mathbf{x}_1$ by choosing

$$\hat{\mathbf{x}}_1 = -\mathbf{Q}_{11}^{-1} \mathbf{Q}_{12} \mathbf{x}_2. \qquad 2.119$$

The corresponding value of $\mathbf{x}^T \mathbf{Q} \mathbf{x}$ is

$$\hat{\mathbf{x}}^T \mathbf{Q} \hat{\mathbf{x}} = \mathbf{x}_2^T \mathbf{Q}_{12}^T \mathbf{Q}_{11}^{-1} \mathbf{Q}_{11} \mathbf{Q}_{11}^{-1} \mathbf{Q}_{12} \mathbf{x}_2 - \mathbf{x}_2^T \mathbf{Q}_{12}^T \mathbf{Q}_{11}^{-1} \mathbf{Q}_{12} \mathbf{x}_2 - \mathbf{x}_1^T \mathbf{Q}_{12}^T \mathbf{Q}_{11}^{-1} \mathbf{Q}_{12} \mathbf{x}_2 + \mathbf{x}_2^T \mathbf{Q}_{22} \mathbf{x}_2$$

$$= \mathbf{x}_2^T [\mathbf{Q}_{22} - \mathbf{Q}_{12}^T \mathbf{Q}_{11}^{-1} \mathbf{Q}_{12}] \mathbf{x}_2 \geq 0. \qquad 2.120$$

We call $\mathbf{Q}_{22} - \mathbf{Q}_{12}^T \mathbf{Q}_{11}^{-1} \mathbf{Q}_{12}$ a *Schur complement* of the partitioned matrix $\mathbf{Q}$. The Schur complement is nonnegative definite. Similarly, by fixing $\mathbf{x}_1$ and minimizing with respect to $\mathbf{x}_2$, we obtain $\hat{\mathbf{x}}_2 = -\mathbf{Q}_{22}^{-1}\mathbf{Q}_{12}^{\mathsf{T}}\mathbf{x}_1$ and $\hat{\mathbf{x}}^{\mathsf{T}}\mathbf{Q}\hat{\mathbf{x}} = \mathbf{x}_1^{\mathsf{T}}[\mathbf{Q}_{11} - \mathbf{Q}_{12}\mathbf{Q}_{22}^{-1}\mathbf{Q}_{12}^{\mathsf{T}}]\mathbf{x}_1 \geq$ 0. The Schur complement $\mathbf{Q}_{11} - \mathbf{Q}_{12}\mathbf{Q}_{22}^{-1}\mathbf{Q}_{12}^{\mathsf{T}}$ is also nonnegative definite.

If $\hat{\mathbf{x}}_1 = -\mathbf{Q}_{11}^{-1}\mathbf{Q}_{12}\mathbf{x}_2$ minimizes the quadratic form with respect to $\mathbf{x}_1$ for fixed $\mathbf{x}_2$, then we suspect that the quadratic form $\mathbf{x}^{\mathsf{T}}\mathbf{Q}\mathbf{x}$ has a representation in terms of $\mathbf{x}_1 - \hat{\mathbf{x}}_1$ and $\mathbf{x}_2$. Let's find it. Build the vector

$$\begin{bmatrix} \mathbf{x}_1 - \hat{\mathbf{x}}_1 \\ \mathbf{x}_2 \end{bmatrix} = \begin{bmatrix} \mathbf{I} & \mathbf{Q}_{11}^{-1}\mathbf{Q}_{12} \\ \mathbf{0} & \mathbf{I} \end{bmatrix} \begin{bmatrix} \mathbf{x}_1 \\ \mathbf{x}_2 \end{bmatrix}. \qquad 2.121$$

This transformation has the inverse

$$\begin{bmatrix} \mathbf{x}_1 \\ \mathbf{x}_2 \end{bmatrix} = \begin{bmatrix} \mathbf{I} & -\mathbf{Q}_{11}^{-1}\mathbf{Q}_{12} \\ \mathbf{0} & \mathbf{I} \end{bmatrix} \begin{bmatrix} \mathbf{x}_1 - \hat{\mathbf{x}}_1 \\ \mathbf{x}_2 \end{bmatrix}. \qquad 2.122$$

Therefore the quadratic form $\mathbf{x}^{\mathsf{T}}\mathbf{Q}\mathbf{x}$ may be written as

$$\mathbf{x}^{\mathsf{T}}\mathbf{Q}\mathbf{x} = \begin{bmatrix} (\mathbf{x}_1 - \hat{\mathbf{x}}_1)^{\mathsf{T}} & \mathbf{x}_2^{\mathsf{T}} \end{bmatrix} \begin{bmatrix} \mathbf{I} & \mathbf{0} \\ -\mathbf{Q}_{12}^{\mathsf{T}}\mathbf{Q}_{11}^{-1} & \mathbf{I} \end{bmatrix} \begin{bmatrix} \mathbf{Q}_{11} & \mathbf{Q}_{12} \\ \mathbf{Q}_{12}^{\mathsf{T}} & \mathbf{Q}_{22} \end{bmatrix} \begin{bmatrix} \mathbf{I} & -\mathbf{Q}_{11}^{-1}\mathbf{Q}_{12} \\ \mathbf{0} & \mathbf{I} \end{bmatrix} \begin{bmatrix} \mathbf{x}_1 - \hat{\mathbf{x}}_1 \\ \mathbf{x}_2 \end{bmatrix}$$

$$= \begin{bmatrix} (\mathbf{x}_1 - \hat{\mathbf{x}}_1)^{\mathsf{T}} & \mathbf{x}_2^{\mathsf{T}} \end{bmatrix} \begin{bmatrix} \mathbf{Q}_{11} & \mathbf{0} \\ \mathbf{0} & \mathbf{Q}_{22} - \mathbf{Q}_{12}^{\mathsf{T}}\mathbf{Q}_{11}^{-1}\mathbf{Q}_{12} \end{bmatrix} \begin{bmatrix} \mathbf{x}_1 - \hat{\mathbf{x}}_1 \\ \mathbf{x}_2 \end{bmatrix}$$

$$= (\mathbf{x}_1 - \hat{\mathbf{x}}_1)^{\mathsf{T}}\mathbf{Q}_{11}(\mathbf{x}_1 - \hat{\mathbf{x}}_1) + \mathbf{x}_2^{\mathsf{T}}(\mathbf{Q}_{22} - \mathbf{Q}_{12}^{\mathsf{T}}\mathbf{Q}_{11}^{-1}\mathbf{Q}_{12})\mathbf{x}_2. \qquad 2.123$$

We have "completed the square," and we have found that $\mathbf{Q}$ is nonnegative definite iff $\mathbf{Q}_{11}$ is nonnegative definite *and* $\mathbf{Q}_{22} - \mathbf{Q}_{12}^{\mathsf{T}}\mathbf{Q}_{11}^{-1}\mathbf{Q}_{12}$ is nonnegative definite. The matrix

$$\begin{bmatrix} \mathbf{I} & -\mathbf{Q}_{11}^{-1}\mathbf{Q}_{12} \\ \mathbf{0} & \mathbf{I} \end{bmatrix} \qquad 2.124$$

takes the partitioned matrix $\mathbf{Q}$ to block diagonal form and provides a representation for $\mathbf{Q}$:

$$\begin{bmatrix} \mathbf{I} & \mathbf{0} \\ -\mathbf{Q}_{12}^{\mathsf{T}}\mathbf{Q}_{11}^{-1} & \mathbf{I} \end{bmatrix} \begin{bmatrix} \mathbf{Q}_{11} & \mathbf{Q}_{12} \\ \mathbf{Q}_{12}^{\mathsf{T}} & \mathbf{Q}_{22} \end{bmatrix} \begin{bmatrix} \mathbf{I} & -\mathbf{Q}_{11}^{-1}\mathbf{Q}_{12} \\ \mathbf{0} & \mathbf{I} \end{bmatrix} = \begin{bmatrix} \mathbf{Q}_{11} & \mathbf{0} \\ \mathbf{0} & \mathbf{Q}_{22} - \mathbf{Q}_{12}^{\mathsf{T}}\mathbf{Q}_{11}^{-1}\mathbf{Q}_{12} \end{bmatrix}$$

$$\qquad 2.125$$

$$\begin{bmatrix} \mathbf{Q}_{11} & \mathbf{Q}_{12} \\ \mathbf{Q}_{12}^{\mathsf{T}} & \mathbf{Q}_{22} \end{bmatrix} = \begin{bmatrix} \mathbf{I} & \mathbf{0} \\ \mathbf{Q}_{12}\mathbf{Q}_{11}^{-1} & \mathbf{I} \end{bmatrix} \begin{bmatrix} \mathbf{Q}_{11} & \mathbf{0} \\ \mathbf{0} & \mathbf{Q}_{22} - \mathbf{Q}_{12}^{\mathsf{T}}\mathbf{Q}_{11}^{-1}\mathbf{Q}_{12} \end{bmatrix} \begin{bmatrix} \mathbf{I} & \mathbf{Q}_{11}^{-1}\mathbf{Q}_{12} \\ \mathbf{0} & \mathbf{I} \end{bmatrix}.$$

2.9   MATRIX INVERSION FORMULAS

There are two fundamental matrix inversion lemmas that go hand-in-glove to produce a striking representation for the inverse of a partitioned matrix.

Lemma 1 (Inverse of a Partitioned Matrix)   Let $\mathbf{R}$ denote the partitioned matrix

$$\mathbf{R} = \left[\begin{array}{c|c} \mathbf{A} & \mathbf{B} \\ \hline \mathbf{C} & \mathbf{D} \end{array}\right]. \tag{2.126}$$

The inverse of $\mathbf{R}$ is

$$\mathbf{R}^{-1} = \left[\begin{array}{c|c} \mathbf{E}^{-1} & \mathbf{F}\mathbf{H}^{-1} \\ \hline \mathbf{H}^{-1}\mathbf{G} & \mathbf{H}^{-1} \end{array}\right]$$

$$\mathbf{E} = \mathbf{A} - \mathbf{B}\mathbf{D}^{-1}\mathbf{C}$$

$$\mathbf{A}\mathbf{F} = -\mathbf{B} \tag{2.127}$$

$$\mathbf{G}\mathbf{A} = -\mathbf{C}$$

$$\mathbf{H} = \mathbf{D} - \mathbf{C}\mathbf{A}^{-1}\mathbf{B}.$$

All indicated inverses are assumed to exist. The matrix $\mathbf{E}$ is called the Schur complement of $\mathbf{A}$, and the matrix $\mathbf{H}$ is called the Schur complement of $\mathbf{D}$.   □

Lemma 2 (Matrix Inversion Lemma)   Let $\mathbf{E}$ denote the Schur complement of $\mathbf{A}$:

$$\mathbf{E} = \mathbf{A} - \mathbf{B}\mathbf{D}^{-1}\mathbf{C}. \tag{2.128}$$

Then the inverse of $\mathbf{E}$ is

$$\mathbf{E}^{-1} = \mathbf{A}^{-1} + \mathbf{F}\mathbf{H}^{-1}\mathbf{G}$$

$$\mathbf{A}\mathbf{F} = -\mathbf{B} \tag{2.129}$$

$$\mathbf{G}\mathbf{A} = -\mathbf{C}$$

$$\mathbf{H} = \mathbf{D} - \mathbf{C}\mathbf{A}^{-1}\mathbf{B}.$$

Lemmas 1 and 2 combine to form the following representation for the inverse of a partitioned matrix.   □

Theorem (Partitioned Matrix Inverse)   The inverse of the partitioned matrix

$$\mathbf{R} = \left[\begin{array}{c|c} \mathbf{A} & \mathbf{B} \\ \hline \mathbf{C} & \mathbf{D} \end{array}\right] \tag{2.130}$$

is the matrix

$$\mathbf{R}^{-1} = \left[\begin{array}{c|c} \mathbf{A}^{-1} & \mathbf{0} \\ \hline \mathbf{0} & \mathbf{0} \end{array}\right] + \left[\begin{array}{c} \mathbf{F} \\ \hline \mathbf{I} \end{array}\right]\left[\mathbf{H}^{-1}\right]\left[\,\mathbf{G}\,|\,\mathbf{I}\,\right]$$

$$\mathbf{A}\mathbf{F} = -\mathbf{B} \tag{2.131}$$

$$\mathbf{G}\mathbf{A} = -\mathbf{C}$$

$$\mathbf{H} = \mathbf{D} - \mathbf{C}\mathbf{A}^{-1}\mathbf{B}.   □$$

Example 2.2 (Partitioned Covariance Matrix)

The inverse of the partitioned covariance matrix

$$\mathbf{R}_t = \left[\begin{array}{c|c} \mathbf{R}_{t-1} & \mathbf{r}_t \\ \hline \mathbf{r}_t^T & r_{tt} \end{array} \right] \qquad\qquad 2.132$$

is the matrix

$$\mathbf{R}_t^{-1} = \left[\begin{array}{c|c} \mathbf{R}_{t-1}^{-1} & \mathbf{0}^T \\ \hline \mathbf{0}^T & 0 \end{array} \right] + \gamma_t^{-1} \left[\begin{array}{c} -\mathbf{R}_{t-1}^{-1}\mathbf{r}_t \\ \hline 1 \end{array} \right] \left[-\mathbf{r}_t^T\mathbf{R}_{t-1}^{-1} \,|\, 1 \right]$$

$$\gamma_t = r_{tt} - \mathbf{r}_t^T\mathbf{R}_{t-1}^{-1}\mathbf{r}_t. \qquad\qquad (2.133)$$

∎

Example 2.3 (Partitioned Gram Matrix)

Let $\mathbf{H}_t$ denote the partitioned system matrix

$$\mathbf{H}_t = \left[\begin{array}{c} \mathbf{H}_{t-1} \\ \mathbf{c}_t^T \end{array} \right]. \qquad\qquad 2.134$$

The Grammian and its inverse are

$$\mathbf{G}_t = \mathbf{H}_t^T\mathbf{H}_t = \mathbf{G}_{t-1} + \mathbf{c}_t\mathbf{c}_t^T$$

$$\mathbf{G}_t^{-1} = \mathbf{G}_{t-1}^{-1} - \gamma_t\mathbf{G}_{t-1}^{-1}\mathbf{c}_t\mathbf{c}_t^T\mathbf{G}_{t-1}^{-1} \qquad\qquad 2.135$$

$$\gamma_t^{-1} = 1 + \mathbf{c}_t^T G_{t-1}^{-1}\mathbf{c}_t.$$

In least-squares problems involving uncorrelated errors, the inverse of the Grammian is the error covariance matrix $\mathbf{P}_t$. Therefore, these equations are often written

$$\mathbf{P}_t^{-1} = \mathbf{P}_{t-1}^{-1} + \mathbf{c}_t\mathbf{c}_t^T$$

$$\mathbf{P}_t = \mathbf{P}_{t-1} - \gamma_t\mathbf{P}_{t-1}\mathbf{c}_t\mathbf{c}_t^T\mathbf{P}_{t-1} \qquad\qquad 2.136$$

$$\gamma_t^{-1} = 1 + \mathbf{c}_t^T\mathbf{P}_{t-1}\mathbf{c}_t.$$

∎

Example 2.4 (Woodbury's Identity)

The inverse of the matrix

$$\mathbf{R} = \mathbf{R}_0 + \gamma^2\mathbf{u}\mathbf{u}^T \qquad\qquad 2.137$$

is the matrix

$$\mathbf{R}^{-1} = \mathbf{R}_0^{-1} - \frac{\gamma^2}{1 + \gamma^2\mathbf{u}^T\mathbf{R}_0^{-1}\mathbf{u}} \mathbf{R}_0^{-1}\mathbf{u}\mathbf{u}^T\mathbf{R}_0^{-1}. \qquad\qquad 2.138$$

∎

2.10   THE MULTIVARIATE NORMAL DISTRIBUTION

Without doubt, the multivariate normal distribution is the most important distribution in science and engineering. The reasons are manifold: Central limit theorems make it the limiting distribution for sums of random variables, its marginal distributions are normal, linear transformations of multivariate normals are multivariate normal, and so on. In the paragraphs to follow, we review the theory of multivariate normal distributions and of distributions that are related to it.

To proceed, we let $\mathbf{X} = [X_1 \quad X_2 \quad \cdots \quad X_N]^T$ denote an $N \times 1$ random vector. The mean value of $\mathbf{X}$ is

$$
\begin{aligned}
\mathbf{m} &= E\mathbf{X} \\
&= (m_1 \quad m_2 \quad \cdots \quad m_N)^T \\
m_i &= EX_i.
\end{aligned}
\tag{2.141}
$$

So the mean vector $\mathbf{m}$ is a vector of means.

The covariance matrix for $\mathbf{X}$ is

$$
\begin{aligned}
\mathbf{R} &= E(\mathbf{X} - \mathbf{m})(\mathbf{X} - \mathbf{m})^T \\
&= \{r_{ij}\} \\
r_{ij} &= E(X_i - m_i)(X_j - m_j).
\end{aligned}
\tag{2.142}
$$

The covariance matrix $\mathbf{R}$ is a matrix of covariances r_{ij}.

The random vector $\mathbf{X}$ is said to be multivariate normal if its density function is

$$
f(\mathbf{x}) = (2\pi)^{-N/2} (\det \mathbf{R})^{-1/2} \exp\left\{ -\frac{1}{2}(\mathbf{x} - \mathbf{m})^T \mathbf{R}^{-1}(\mathbf{x} - \mathbf{m}) \right\},
\tag{2.143}
$$

where the notations $\det \mathbf{R}$ and $|\mathbf{R}|$ are used interchangeably for the determinant of $\mathbf{R}$. We assume that the nonnegative definite matrix $\mathbf{R}$ is nonsingular. It is apparent that the scale constant $(2\pi)^{-N/2}(\det \mathbf{R})^{-1/2}$ is obtained as follows:

$$
\int \exp\left\{ -\frac{1}{2}(\mathbf{x} - \mathbf{m})^T \mathbf{R}^{-1}(\mathbf{x} - \mathbf{m}) \right\} d\mathbf{x} = (2\pi)^{N/2}(\det \mathbf{R})^{1/2}.
\tag{2.144}
$$

The quadratic form

$$
r^2 = (\mathbf{x} - \mathbf{m})^T \mathbf{R}^{-1}(\mathbf{x} - \mathbf{m})
\tag{2.145}
$$

is a weighted norm called the *Mahalanobis distance* from $\mathbf{x}$ to $\mathbf{m}$. The locus of points $\mathbf{x}$ for which r^2 is constant is also a locus of points for which the density $f(\mathbf{x})$ is constant. This locus is a hyperellipsoid that is sometimes called a *level curve for the density*. The volume enclosed by one of these ellipsoids is

$$
V = V_N (\det \mathbf{R})^{1/2} r^N
\tag{2.146}
$$

where V_N is the volume of an N-dimensional unit hypersphere:

$$V_N = \begin{cases} \pi^{N/2}/(N/2)!, & N \text{ even} \\ 2^N \pi^{(N-1)/2}[(N-1)/2]\,!/N!, & N \text{ odd}. \end{cases} \qquad 2.147$$

For $N = 2$, this volume is the area enclosed by the ellipse of constant density:

$$V = (\det \mathbf{R})^{1/2} \pi r^2. \qquad 2.148$$

This area measures the spread of points under the two-dimensional density. Points widely spread correspond to large determinants of $\mathbf{R}$.

Characteristic Function

The characteristic function of $\mathbf{X}$ is the multidimensional Fourier transform of the density:

$$\phi(\boldsymbol{\omega}) = E e^{-j\boldsymbol{\omega}^T \mathbf{X}}$$

$$= \int d\mathbf{x}(2\pi)^{-N/2}(\det \mathbf{R})^{-1/2} \exp\left\{ -j\boldsymbol{\omega}^T \mathbf{x} - \frac{1}{2}(\mathbf{x} - \mathbf{m})^T \mathbf{R}^{-1}(\mathbf{x} - \mathbf{m}) \right\}.$$

$$2.149$$

By completing the square in the exponent, we can write

$$\phi(\boldsymbol{\omega}) = \int d\mathbf{x}(2\pi)^{-N/2}(\det \mathbf{R})^{-1/2} \exp\left\{ -j\boldsymbol{\omega}^T \mathbf{x} - \frac{1}{2}(\mathbf{x} - \mathbf{m} + j\mathbf{R}\boldsymbol{\omega})^T \right.$$

$$2.150$$

$$\left. \times \mathbf{R}^{-1}(\mathbf{x} - \mathbf{m} + j\mathbf{R}\boldsymbol{\omega}) + j\boldsymbol{\omega}^T(\mathbf{x} - \mathbf{m}) - \frac{1}{2}\boldsymbol{\omega}^T \mathbf{R}\boldsymbol{\omega} \right\}$$

$$= \exp\left\{ -j\boldsymbol{\omega}^T \mathbf{m} - \frac{1}{2}\boldsymbol{\omega}^T \mathbf{R}\boldsymbol{\omega} \right\} \int d\mathbf{x}(2\pi)^{N/2}(\det \mathbf{R})^{-1/2}$$

$$\times \exp\left\{ -\frac{1}{2}(\mathbf{x} - \mathbf{m} + j\mathbf{R}\boldsymbol{\omega})^T \mathbf{R}^{-1}(\mathbf{x} - \mathbf{m} + j\mathbf{R}\boldsymbol{\omega}) \right\}$$

$$= \exp\left\{ -j\boldsymbol{\omega}^T \mathbf{m} - \frac{1}{2}\boldsymbol{\omega}^T \mathbf{R}\boldsymbol{\omega} \right\}$$

$$= \exp\left\{ -\frac{1}{2}\mathbf{m}^T \mathbf{R}^{-1}\mathbf{m} \right\} \exp\left\{ -\frac{1}{2}(\boldsymbol{\omega} + j\mathbf{R}^{-1}\mathbf{m})^T \mathbf{R}(\boldsymbol{\omega} + j\mathbf{R}^{-1}\mathbf{m}) \right\}.$$

This result shows that the characteristic function itself is a multivariate normal function of the frequency variable $\boldsymbol{\omega}$.

The volume of $\boldsymbol{\omega}$ values enclosed by the ellipsoid $\boldsymbol{\omega}^T \mathbf{R}\boldsymbol{\omega} = s^2$ is

$$V = V_N(\det \mathbf{R})^{-1/2} s^N. \qquad 2.151$$

For $r = s = 1$, the product of $\mathbf{x}$ and $\boldsymbol{\omega}$ volumes is

$$P = V_N^2.$$ 2.152

This is the well-known inverse spreading property of Fourier transforms: Widely spread densities have narrowly peaked characteristic functions and vice versa.

The characteristic function may be used to generate moments of $\mathbf{X}$:

$$j\frac{\partial}{\partial \boldsymbol{\omega}}\phi(\boldsymbol{\omega} = \mathbf{0}) = \mathbf{m}$$ 2.153

$$\frac{\partial^2}{\partial \boldsymbol{\omega}^2}\phi(\boldsymbol{\omega} = \mathbf{0}) = \mathbf{R} + \mathbf{m}\mathbf{m}^{\mathrm{T}}.$$

The Bivariate Normal Distribution

The bivariate normal distribution arises in many applications, and its study brings insight into the more general multivariate case. We obtain the bivariate normal by specializing the general case to $\mathbf{X} = (X_1 \quad X_2)^{\mathrm{T}}$. Then the density function is

$$f(\mathbf{x}) = (2\pi)^{-1}(\det \mathbf{R})^{-1/2}\exp\left\{-\frac{1}{2}(\mathbf{x} - \mathbf{m})^{\mathrm{T}}\mathbf{R}^{-1}(\mathbf{x} - \mathbf{m})\right\}$$ 2.154

where the mean value is $\mathbf{m} = (m_1 \quad m_2)^{\mathrm{T}}$, and the covariance matrix is

$$\mathbf{R} = \begin{bmatrix} r_{11} & r_{12} \\ r_{21} & r_{22} \end{bmatrix}$$ 2.155

$$r_{12} = r_{21}.$$

The inverse and determinant of $\mathbf{R}$ are

$$\mathbf{R}^{-1} = \frac{1}{r_{11}r_{22} - r_{21}r_{12}}\begin{bmatrix} r_{22} & -r_{12} \\ -r_{21} & r_{11} \end{bmatrix}$$ 2.156

$$\det \mathbf{R} = r_{11}r_{22} - r_{21}r_{12}.$$

The bivariate normal density may be written out as

$$f(x_1, x_2) = \frac{1}{2\pi(r_{11}r_{22} - r_{12}^2)^{1/2}}\exp\left\{-\frac{1}{2(r_{11}r_{22} - r_{12}^2)}\left[(x_1 - m_1)^2 r_{22}\right.\right.$$

$$\left.\left. - 2(x_1 - m_1)r_{12}(x_2 - m_2) + (x_2 - m_2)^2 r_{11}\right]\right\}$$ 2.157

The marginal distributions of X_1 and X_2 are obtained by integrating the joint density with respect to x_2 and x_1 respectively:

$$f(x_1) = \frac{1}{(2\pi r_{11})^{1/2}} \exp\left\{ -\frac{1}{2r_{11}}(x_1 - m_1)^2 \right\} : N[m_1, r_{11}]$$

$$f(x_2) = \frac{1}{(2\pi r_{22})^{1/2}} \exp\left\{ -\frac{1}{2r_{22}}(x_2 - m_2)^2 \right\} : N[m_2, r_{22}].$$

2.158

These results may be cast in a slightly different form by defining variances and normalized cross-covariances:

$$\sigma_1^2 = r_{11} \qquad \sigma_2^2 = r_{22}$$

$$\rho_{12}^2 = \frac{r_{12}^2}{\sigma_1^2 \sigma_2^2}.$$

2.159

The joint and marginal distributions are then

$$f(x_1, x_2) = \frac{1}{2\pi\sigma_1\sigma_2(1 - \rho_{12}^2)^{1/2}} \exp\left\{ -\frac{1}{2(1 - \rho_{12}^2)}\left[\frac{(x_1 - m_1)^2}{\sigma_1^2} \right.\right.$$

$$\left.\left. -\frac{2\rho_{12}}{\sigma_1\sigma_2}(x_1 - m_1)(x_2 - m_2) + \frac{(x_2 - m_2)^2}{\sigma_2^2} \right]\right\}$$

2.160

$$f(x_1) = \frac{1}{(2\pi\sigma_1^2)^{1/2}} \exp\left\{ -\frac{1}{2\sigma_1^2}(x_1 - m_1)^2 \right\}$$

$$f(x_2) = \frac{1}{(2\pi\sigma_2^2)^{1/2}} \exp\left\{ -\frac{1}{2\sigma_2^2}(x_2 - m_2)^2 \right\}.$$

The conditional density for X_1, given $X_2 = x_2$, is computed as follows:

$$f_{X_1 | X_2}(x_1 | x_2) = \frac{f(x_1, x_2)}{f(x_2)}$$

$$= \frac{1}{\left[2\pi\sigma_1^2(1 - \rho_{12}^2)\right]^{1/2}} \exp\left\{ -\frac{1}{2\sigma_1^2(1 - \rho_{12}^2)}\left[(x_1 - m_1)^2\right.\right.$$

$$- 2\rho_{12}\frac{\sigma_1}{\sigma_2}(x_1 - m_1)(x_2 - m_2)$$

$$\left.\left. +(x_2 - m_2)^2\frac{\sigma_1^2}{\sigma_2^2} - \frac{(x_2 - m_2)^2}{\sigma_2^2}\sigma_1^2(1 - \rho_{12}^2)\right]\right\}$$

2.161

$$= \frac{1}{\left[2\pi\sigma_1^2(1 - \rho_{12}^2)\right]^{1/2}} \exp\left\{ \frac{1}{2\sigma_1^2(1 - \rho_{12}^2)}\left[x_1 - m_1\right.\right.$$

$$\left.\left. - \rho_{12}\sigma_1\sigma_2(x_2 - m_2)\right]^2\right\}$$

This result may be written compactly as

$$f_{X_1|X_2}(x_1|x_2) = \frac{1}{(2\pi\sigma_{1|2}^2)^{1/2}} \exp\left\{ -\frac{1}{2\sigma_{1|2}^2}\left[x_1 - \hat{x}_{1|2}\right]^2\right\} \qquad 2.162$$

where $\hat{x}_{1|2}$ and $\sigma_{1|2}^2$ are the following conditional mean and conditional variance:

$$\hat{x}_{1|2} = m_1 + \rho_{12}\frac{\sigma_1}{\sigma_2}(x_2 - m_2)$$

$$\sigma_{1|2}^2 = \sigma_1^2(1 - \rho_{12}^2) \leq \sigma_1^2. \qquad 2.163$$

We say that X_1 is conditionally normal, with conditional mean $\hat{x}_{1|2}$ and conditional variance $\sigma_{1|2}^2$. The conditional variance is independent of x_2, and it is never larger than the unconditional variance σ_1^2.

When $r_1^2 = r_2^2 = \sigma^2$, we say that the random variables X_1 and X_2 have common variances. Then the results above specialize to

$$\hat{x}_{1|2} = m_1 + \rho_{12}(x_2 - m_2)$$

$$\sigma_{1|2}^2 = \sigma^2(1 - \rho_{12}^2). \qquad 2.164$$

The normalized correlation ρ_{12} satisfies the condition $-1 \leq \rho_{12} \leq 1$, so the conditional variance is less than the unconditional variance. When $\rho_{12} = \pm 1$, then

$$\hat{x}_{1|2} = m_1 \pm (x_2 - m_2) \qquad 2.165$$

$$\sigma_{1|2}^2 = 0. \qquad 2.166$$

This means that the probability mass for X_1 is concentrated entirely at $m_1 \pm (x_2 - m_2)$.

Linear Transformations

Let $\mathbf{Y}$ be a linear transformation of a multivariate normal random variable:

$$\mathbf{Y} = \mathbf{A}^T\mathbf{X} \qquad 2.167$$

$$\mathbf{A}^T : m \times N \quad (m \leq N).$$

The characteristic function of $\mathbf{Y}$ is

$$\phi(\boldsymbol{\omega}) = E e^{-j\boldsymbol{\omega}^T\mathbf{Y}} = E e^{-j\boldsymbol{\omega}^T\mathbf{A}^T\mathbf{X}}$$

$$= \exp\left\{ -j\boldsymbol{\omega}^T\mathbf{A}^T\mathbf{m} - \frac{1}{2}\boldsymbol{\omega}^T\mathbf{A}^T\mathbf{R}\mathbf{A}\boldsymbol{\omega}\right\}. \qquad 2.168$$

Thus, $\mathbf{Y}$ is distributed as

$$\mathbf{Y} : N\left[\mathbf{A}^T\mathbf{m}, \mathbf{A}^T\mathbf{R}\mathbf{A}\right] \qquad 2.169$$

provided the matrix $\mathbf{A}^T\mathbf{R}\mathbf{A}$ is nonsingular. If $m \leq N$, the requirement that $\mathbf{A}^T\mathbf{R}\mathbf{A}$ be nonsingular is the requirement that $\mathbf{R}$ be nonsingular and that $\mathbf{A}$ have rank m.

If $m > N$, then $\mathbf{A}^T\mathbf{R}\mathbf{A}$ is singular, indicating that at least $m - N$ components of $\mathbf{Y}$ are linearly dependent. When $\mathbf{A}^T = \mathbf{a}^T$ is $1 \times N$, then the scalar $Y = \mathbf{a}^T\mathbf{X}$ is distributed as

$$Y : N\left[\mathbf{a}^T\mathbf{m}, \mathbf{a}^T\mathbf{R}\mathbf{a}\right]. \qquad 2.170$$

For the special case

$$\mathbf{a}^T = \begin{bmatrix} 0 \cdots 1 & 0 \cdots 0 \end{bmatrix}, \qquad 2.171$$
$$\underset{i}{\phantom{\mathbf{a}^T = [0\cdots 1}}$$

the random variable $Y = \mathbf{a}^T\mathbf{X}$ is distributed as

$$Y : N\left[m_i, r_{ii}\right], \qquad 2.172$$

indicating that the variables Y_i have normal marginal distributions with means m_i and variances r_{ii}.

Analysis and Synthesis

If the transformation matrix $\mathbf{A}$ is $N \times N$, then we can consider two variations on our discussion of linear transformations. In the first we try to synthesize an $N[\mathbf{0}, \mathbf{R}]$ random vector from an $N[\mathbf{0}, \mathbf{I}]$ random vector, and in the second we try to transform (or analyze) an $N[\mathbf{0}, \mathbf{R}]$ random vector into an $N[\mathbf{0}, \mathbf{I}]$ random vector. This means that we need to find transformation matrices that solve these two problems:

1. synthesis

$$\mathbf{B}\mathbf{I}\mathbf{B}^T = \mathbf{R} \qquad 2.173$$

2. analysis

$$\mathbf{A}^T\mathbf{R}\mathbf{A} = \mathbf{I}. \qquad 2.174$$

We assume that $\mathbf{B}$ and $\mathbf{A}^T$ are lower triangular matrices:

$$\mathbf{B} = \begin{bmatrix} b_{11} & & & \\ b_{21} & b_{22} & & \mathbf{0} \\ \vdots & & \ddots & \\ b_{N1} & \cdots & & b_{NN} \end{bmatrix} \qquad \mathbf{A} = \begin{bmatrix} a_{11} & a_{12} & \cdots & a_{1N} \\ & a_{22} & & \vdots \\ & \mathbf{0} & & \ddots & \\ & & & & a_{NN} \end{bmatrix}. \qquad 2.175$$

When $\mathbf{B}\mathbf{B}^T$ is substituted for $\mathbf{R}$ in the analysis equation, we get the identity

$$\mathbf{A}^T\mathbf{B}(\mathbf{A}^T\mathbf{B})^T = \mathbf{I} \quad \Rightarrow \quad \mathbf{A}^T\mathbf{B} = (\mathbf{A}^T\mathbf{B})^{-T}. \qquad 2.176$$

The matrix $\mathbf{A}^T\mathbf{B}$ is a product of lower triangular matrices and is therefore lower triangular. The matrix $(\mathbf{A}^T\mathbf{B})^T$ is the upper triangular transpose of $(\mathbf{A}^T\mathbf{B})$. For $(\mathbf{A}^T\mathbf{B})(\mathbf{A}^T\mathbf{B})^T$ to equal identity, we require

$$\mathbf{A}^T\mathbf{B} = \mathbf{I}, \qquad 2.177$$

meaning that $\mathbf{A}^T$ is the inverse of $\mathbf{B}$ and vice versa:

$$\mathbf{A}^T = \mathbf{B}^{-1}. \qquad 2.178$$

Note that the inverse of the lower triangular matrix $\mathbf{B}$ is lower triangular.

We may solve the synthesis problem by noting that $\mathbf{BB}^T = \mathbf{R}$ is a Cholesky factorization of $\mathbf{R}$. Write $\mathbf{R}$ and $\mathbf{B}$ as

$$\mathbf{R} = [\mathbf{r}_1 \quad \cdots \quad \mathbf{r}_N]$$

$$\mathbf{B} = [\mathbf{b}_1 \quad \cdots \quad \mathbf{b}_N] = \begin{bmatrix} b_{11} & & & \\ b_{21} & b_{22} & & \mathbf{0} \\ \vdots & & \ddots & \\ b_{N1} & \cdots & & b_{NN} \end{bmatrix}. \qquad 2.179$$

The factoring equations are then

$$\begin{aligned} \mathbf{r}_1 &= \mathbf{b}_1 b_{11} \\ \mathbf{r}_2 &= \mathbf{b}_1 b_{21} + \mathbf{b}_2 b_{22} \\ &\vdots \\ \mathbf{r}_N &= \mathbf{b}_1 b_{N1} + \cdots + \mathbf{b}_N b_{NN}. \end{aligned} \qquad 2.180$$

These are solved for the columns $\mathbf{b}_i$ and the diagonal terms b_{ii} as follows:

$$\begin{aligned} b_{11}^2 &= r_{11} \\ b_{11}\mathbf{b}_1 &= \mathbf{r}_1 \\ &\vdots \\ b_{NN}^2 &= r_{NN} - \sum_{i=1}^{N-1} b_{Ni}^2 \\ b_{NN}\mathbf{b}_N &= \mathbf{r}_N - \sum_{i=1}^{N-1} b_{Ni}\mathbf{b}_i = \begin{bmatrix} \mathbf{0} \\ b_{NN} \end{bmatrix}. \end{aligned} \qquad 2.181$$

The analysis equation may be solved by inverting $\mathbf{B}$ or by using the procedures of Section 2.2 to factor $\mathbf{R}$ as $\mathbf{A}^T\mathbf{RA} = \mathbf{D}^2$ and then replacing $\mathbf{A}$ by $\mathbf{AD}^{-1}$:

$$\mathbf{RA} = \mathbf{A}^{-T}\mathbf{D}^2 \quad \Rightarrow \quad \mathbf{R}(\mathbf{AD}^{-1}) = (\mathbf{AD}^{-1})^{-T}. \qquad 2.182$$

Diagonalizing Transformations

The correlation matrix is symmetric and nonnegative definite. Therefore, there exists an orthogonal matrix $\mathbf{U}$ such that

$$\mathbf{U}^T\mathbf{RU} = \text{diag}[\lambda_1^2 \quad \cdots \quad \lambda_N^2]. \qquad 2.183$$

It follows that the vector $\mathbf{Y} = \mathbf{U}^T\mathbf{X}$ is distributed as follows:

$$\mathbf{Y} = \mathbf{U}^T\mathbf{X} : N\big[\mathbf{U}^T\mathbf{m}, \operatorname{diag}[\lambda_1^2 \quad \cdots \quad \lambda_N^2]\big]. \qquad 2.184$$

We say that the random variables $Y_1, Y_2, \ldots, Y_N$ are uncorrelated because

$$E(\mathbf{Y} - \mathbf{U}^T\mathbf{m})(\mathbf{Y} - \mathbf{U}^T\mathbf{m})^T = \mathbf{U}^T\mathbf{R}\mathbf{U} = \operatorname{diag}[\lambda_1^2 \quad \cdots \quad \lambda_N^2]. \qquad 2.185$$

Writing out the density for $\mathbf{Y}$, we see that the Y_n are independent normal random variables with means $(\mathbf{U}^T\mathbf{m})_n$ and variances λ_n^2:

$$f(\mathbf{y}) = \prod_{n=1}^{N}(2\pi\lambda_n^2)^{-1/2}\exp\left\{-\frac{1}{2\lambda_n^2}\big[y_n - (\mathbf{U}^T\mathbf{m})_n\big]^2\right\}. \qquad 2.186$$

This approach to diagonalization provides an alternative to the analysis procedure discussed previously for generating independent random variables. The transformation $\mathbf{Y} = \mathbf{U}^T\mathbf{X}$ is called a *Karhunen-Loeve* or *Hotelling* transform. It simply diagonalizes the covariance matrix $\mathbf{R} : \mathbf{U}^T\mathbf{R}\mathbf{U} = \mathbf{\Lambda}^2$.

2.11   QUADRATIC FORMS IN MVN RANDOM VARIABLES

Linear functions of MVN random vectors remain MVN. But what about quadratic functions of MVNs? In some very important cases the quadratic functions have χ^2 distributions, as we now illustrate. These results are used extensively in Chapters 4 and 9.

Let $\mathbf{X}$ denote an $N[\mathbf{m}, \mathbf{R}]$ random variable. The distribution of the quadratic form

$$Q = (\mathbf{X} - \mathbf{m})^T\mathbf{R}^{-1}(\mathbf{X} - \mathbf{m}) \qquad 2.187$$

is χ_N^2 distributed, as the following calculation shows. The characteristic function of Q is

$$
\begin{aligned}
\phi(\omega) = E e^{-j\omega Q} &= \int d\mathbf{x} e^{-j\omega(\mathbf{x}-\mathbf{m})^T\mathbf{R}^{-1}(\mathbf{x}-\mathbf{m})}(2\pi)^{-N/2}|\mathbf{R}|^{-1/2} \\
&\quad \times \exp\left\{-\frac{1}{2}(\mathbf{x}-\mathbf{m})^T\mathbf{R}^{-1}(\mathbf{x}-\mathbf{m})\right\} \\
&= \int d\mathbf{x}\,\frac{1}{(1+2j\omega)^{N/2}}(2\pi)^{-N/2}|\mathbf{R}|^{-1/2}(1+2j\omega)^{N/2} \qquad 2.188 \\
&\quad \times \exp\left\{-\frac{1}{2}(\mathbf{x}-\mathbf{m})^T\mathbf{R}^{-1}(\mathbf{I}+2j\omega\mathbf{I})(\mathbf{x}-\mathbf{m})\right\} \\
&= \frac{1}{(1+2j\omega)^{N/2}}.
\end{aligned}
$$

This is the characteristic function of a chi-squared distribution with N degrees of freedom, denoted χ_N^2. The density function for Q is the inverse transform

$$f(q) = \frac{1}{\Gamma(N/2)2^{N/2}}q^{(N/2)-1}e^{-q/2}; \qquad q \geq 0. \qquad 2.189$$

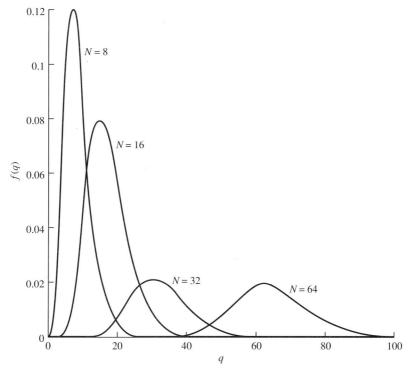

Figure 2.7 Chi-squared distribution, χ_N^2.

Think of the density as the impulse response of a linear system with $N/2$ repeated roots at $s = j\omega = -\frac{1}{2}$. It is plotted for numerous values of N in Figure 2.7.

The mean and variance of Q are obtained from the characteristic function:

$$EQ = N$$
$$\text{var}\, Q = 2N.$$

2.190

Example 2.5 (Chi-Squared)

As a special case of the result derived above, note that $Q = (X - \mu)^2/\sigma^2$ is χ_1^2 (chi-squared with one degree of freedom) when $X : N[\mu, \sigma^2]$. To prove it in a slightly different way, note

$$F(q) = P[Q \le q] = P\left[-q^{1/2} \le \frac{X - \mu}{\sigma} \le q^{1/2}\right]$$

$$= 2\int_0^{q^{1/2}} \frac{1}{\sqrt{2\pi}} \exp\{-x^2/2\}\, dx$$

$$= \int_0^{q} \frac{1}{\sqrt{2\pi}} e^{-y/2} y^{-1/2}\, dy.$$

So the density is

$$f(q) = \frac{1}{\sqrt{2\pi}} q^{-1/2} e^{-q/2}; \qquad q \geq 0$$

and the characteristic function is

$$\phi(\omega) = \int_0^\infty dq \, e^{-j\omega q} \frac{1}{2\pi} q^{-1/2} e^{-q/2}$$

$$= \int_0^\infty dq \, \frac{1}{2\pi} q^{-1/2} e^{-(q/2)(1+2j\omega)}$$

$$= \int_0^\infty \frac{dv}{2\pi} \frac{2}{1+2j\omega} \left[\frac{2v}{1+2j\omega}\right]^{-1/2} e^{-v}; \qquad v = \frac{q}{2}(1+2j\omega)$$

$$= \left[\frac{2}{\pi^2(1+2j\omega)}\right]^{1/2} \int_0^\infty dv \, v^{-1/2} e^{-v}$$

$$= \frac{1}{(1+2j\omega)^{1/2}}.$$

The characteristic function of the random variable $Q = \sum_{n=1}^N (X_n - \mu)^2/\sigma^2$, where the X_n are i.i.d. $N[\mu, \sigma^2]$ random variables is

$$\phi(\omega) = \frac{1}{(1+2j\omega)^{N/2}}, \qquad\qquad 2.191$$

making the sum chi-squared distributed with N degrees of freedom:

$$Q : \chi_N^2. \qquad\qquad 2.192$$

∎

Quadratic Forms Using Projection Matrices

What about more general quadratic forms in the symmetric matrix $\mathbf{P}$:

$$Q = (\mathbf{X} - \mathbf{m})^\mathsf{T} \mathbf{P}(\mathbf{X} - \mathbf{m})$$

$$\mathbf{X} : N[\mathbf{m}, \mathbf{R}]? \qquad\qquad 2.193$$

The characteristic function of Q is easily computed and from it the mean and variance of Q. The results are

$$EQ = \operatorname{tr} \mathbf{PR}$$

$$\operatorname{var} Q = 2 \operatorname{tr}(\mathbf{PR})^2 \qquad\qquad 2.194$$

(where tr denotes trace). The characteristic function of Q is then

$$\phi(\omega) = \int d\mathbf{x}(2\pi)^{-N/2}(\det \mathbf{R})^{-1/2} \exp\left\{-\frac{1}{2}(\mathbf{x} - \mathbf{m})^{\mathrm{T}}(\mathbf{I} + 2j\omega\mathbf{PR})\mathbf{R}^{-1}(\mathbf{x} - \mathbf{m})\right\}$$

$$= \int d\mathbf{x}(2\pi)^{-N/2} \frac{1}{|\mathbf{R}(\mathbf{I} + 2j\omega\mathbf{PR})^{-1}|^{1/2}} \frac{1}{|\mathbf{I} + 2j\omega\mathbf{PR}|^{1/2}}$$

$$\times \exp\left\{-\frac{1}{2}(\mathbf{x} - \mathbf{m})^{\mathrm{T}}(\mathbf{I} + 2j\omega\mathbf{PR})\mathbf{R}^{-1}(\mathbf{x} - \mathbf{m})\right\}$$

$$= \frac{1}{|\mathbf{I} + 2j\omega\mathbf{PR}|^{1/2}}.$$

2.195

If $\mathbf{PR}$ is symmetric (in which case $\mathbf{RP} = \mathbf{PR}$), the characteristic function may be written as

$$\phi(\omega) = \frac{1}{\prod_{n=1}^{N}(1 + 2j\omega\lambda_n)^{1/2}}$$

2.196

where the λ_n are eigenvalues of $\mathbf{PR}$:

$$\mathbf{U}^{\mathrm{T}}\mathbf{PRU} = \mathrm{diag}[\lambda_1 \quad \cdots \quad \lambda_N].$$

2.197

This is the characteristic function of a χ_r^2 random variable iff

$$\lambda_n = \begin{cases} 1, & n = 1, 2, \ldots, r \\ 0, & n = r + 1, \ldots, N. \end{cases}$$

2.198

This in turn implies that $\mathbf{PR}$ is orthogonally similar to a rank-deficient identity and that $\mathbf{PR}$ is idempotent. That is,

$$\mathbf{U}^{\mathrm{T}}\mathbf{PRU} = \mathrm{diag}[1 \quad \cdots \quad 1 \quad 0 \quad \cdots \quad 0] = \mathbf{I}_r$$

$$\mathbf{PR} = \mathbf{UI}_r\mathbf{U}^{\mathrm{T}}$$

2.199

$$(\mathbf{PR})(\mathbf{PR}) = \mathbf{UI}_r\mathbf{U}^{\mathrm{T}}\mathbf{UI}_r\mathbf{U}^{\mathrm{T}} = \mathbf{UI}_r\mathbf{U}^{\mathrm{T}} = \mathbf{PR}.$$

If $\mathbf{R} = \mathbf{I}$, implying that $\mathbf{X}$ consists of independent components, then the quadratic form Q is χ_r^2 iff $\mathbf{P}$ is idempotent:

$$\mathbf{P}^2 = \mathbf{P}.$$

2.200

Such a matrix is said to be a projection matrix. Therefore we have the important result that the linear transformation $\mathbf{Y} = \mathbf{PX}$ is $N[\mathbf{0}, \mathbf{P}]$ and the quadratic form $\mathbf{Y}^{\mathrm{T}}\mathbf{Y} = \mathbf{X}^{\mathrm{T}}\mathbf{PX}$ is χ_r^2 whenever $\mathbf{X}$ is $N[\mathbf{0}, \mathbf{I}]$ and $\mathbf{P}$ is a rank r projection. More generally, if $\mathbf{X} : N[\mathbf{0}, \mathbf{R}]$, $\mathbf{R} = \mathbf{U}\mathbf{\Lambda}^2\mathbf{U}^{\mathrm{T}}$, with $\mathbf{\Lambda}^2 = \mathrm{diag}[\lambda_1^2 \quad \lambda_2^2 \quad \cdots \quad \lambda_N^2]$ and $\mathbf{P}$ is chosen to be $\mathbf{U}\mathbf{\Lambda}_r^{-1}\mathbf{U}^{\mathrm{T}}$ with $\mathbf{\Lambda}_r^{-2} = \mathrm{diag}[\lambda_1^{-2} \quad \lambda_2^{-2} \quad \cdots \quad \lambda_r^{-2} \quad 0 \quad \cdots \quad 0]$, then $\mathbf{PRP} = \mathbf{UI}_r\mathbf{U}^{\mathrm{T}}$ and the quadratic form $\mathbf{Y}^{\mathrm{T}}\mathbf{Y}$ is χ_r^2.

Asymptotics

Let $Q = (\mathbf{X} - \mathbf{m})^{\mathrm{T}} \mathbf{R}^{-1} (\mathbf{X} - \mathbf{m})$ be a quadratic form in the $N[\mathbf{m}, \mathbf{R}]$ random vector $\mathbf{X}$. Equivalently, let $Q = \sum_{n=1}^{N} (X_n - \mu)^2/\sigma^2$ be a quadratic form in the i.i.d. $N[\mu, \sigma^2]$ random variables $X_1, X_2, \ldots, X_N$. We have shown that Q is χ_N^2. Form the random variable

$$V = \frac{Q - N}{\sqrt{2N}}. \qquad\qquad 2.201$$

The random variable V has mean 0 and variance 1:

$$EV = 0$$

$$\operatorname{var} V = \frac{1}{2N} 2N = 1. \qquad\qquad 2.202$$

The random variable V is asymptotically $N[0, 1]$, as the following calculation shows. The characteristic function of V is

$$
\begin{aligned}
\phi(\omega) = E e^{-j\omega V} &= E e^{-j\omega(Q-N)/(2N)^{1/2}} \\[2mm]
&= e^{+j\omega(N/2)^{1/2}} E e^{-j\omega Q/(2N)^{1/2}} \\[2mm]
&= e^{-j\omega(2/N)^{1/2}(-N/2)} \frac{1}{\left[1 + 2j\,\omega/(2N)^{1/2}\right]^{N/2}} \\[2mm]
&= \left[e^{-j\omega(2/N)^{1/2}} \left(1 + 2j \frac{\omega}{(2N)^{1/2}}\right) \right]^{-N/2} \\[2mm]
&= \left[1 + \frac{\omega^2}{N} + \frac{0(N)}{N} \right]^{-N/2}.
\end{aligned}
\qquad 2.203
$$

The term $0(N)$ includes terms that converge to zero as N diverges to ∞. The limit of this polynomial is

$$\lim_{N \to \infty} \left[1 + \frac{\omega^2}{N} + \frac{0(N)}{N} \right]^{-N/2} = \exp(-\omega^2/2). \qquad 2.204$$

This result says that the random variable V is asymptotically $N[0, 1]$ as the sample size $N \to \infty$.

REFERENCES AND COMMENTS

In a short chapter on linear algebra, one must pick and choose among relevant topics. I have been influenced by the early chapters of Rao [1973]; and Takeuchi, Yanai, and Mukerjee [1982]. For a systematic treatment of linear algebra and matrix theory, I recommend Gantmacher [1959]; Golub and Van Loan [1989]; Marcus and Minc [1965]; and Shilov [1977]. The book by Golub and Van Loan has become the standard for numerical linear algebra. I have borrowed the section on quadratic forms from Roberts and Mullis [1987].

Gantmacher, F. R. [1959]. *The Theory of Matrices*, vols. I and II (New York: Chelsea, 1959).

Golub, G. H., and C. F. Van Loan [1989]. *Matrix Computations*, 2nd ed. (Baltimore: Johns Hopkins, 1989).

Marcus, M., and H. Minc [1965]. *Introduction to Linear Algebra* (New York: Dover, 1965).

Rao, C. R. [1973]. *Linear Statistical Inference and Its Applications*, 2nd ed. (New York: John Wiley & Sons, 1973).

Roberts, R. A., and C. T. Mullis [1987]. *Digital Signal Processing*, Appendix 7A (Reading, MA: Addison-Wesley Publishing Co., 1987).

Shilov, G. E. [1977]. *Linear Algebra*, R. A. Silverman, translator (New York: Dover, 1977).

Takeuchi, K., H. Yanai, and B. N. Mukerjee [1982]. *The Foundations of Multivariate Analysis* (New Delhi: Wiley Eastern, 1982).

PROBLEMS

2.1 Let $\mathbf{y} = f(\mathbf{x})$ denote a linear mapping from R^p to R^n:

$$\mathbf{y} = f(\mathbf{x})$$

$$f(a\mathbf{x}) = a f(\mathbf{x})$$

$$f(\mathbf{x}_1 + \mathbf{x}_2) = f(\mathbf{x}_1) + f(\mathbf{x}_2).$$

Show that this mapping may always be written

$$\mathbf{y} = \mathbf{A}\mathbf{x}.$$

Determine $\mathbf{A} = [\mathbf{a}_1 \quad \cdots \quad \mathbf{a}_p]$.

2.2 Begin with the data matrix

$$\mathbf{X} = \begin{bmatrix} \mathbf{x}_1 & \mathbf{x}_2 & \mathbf{x}_3 \end{bmatrix} = \begin{bmatrix} 3 & 2 & 1 \\ 1 & 3 & 2 \\ 2 & 1 & 3 \end{bmatrix}.$$

The *QR* factor of *X* is

$$\mathbf{X} = \mathbf{QR}; \qquad \mathbf{Q} = \begin{bmatrix} \mathbf{u}_1 & \mathbf{u}_2 & \mathbf{u}_3 \end{bmatrix}; \qquad \mathbf{R} = \begin{bmatrix} r_{11} & r_{12} & r_{13} \\ 0 & r_{22} & r_{23} \\ 0 & 0 & r_{33} \end{bmatrix}$$

$$\mathbf{u}_i^T \mathbf{u}_j = \delta_{ij}.$$

Find $\mathbf{Q}$ using

a. Gram-Schmidt;

b. Householder (show typical transformation $\mathbf{Q}_k$);

c. Givens (show typical transformation $\mathbf{Q}_{ij}$).

Use $(\mathbf{u}_1, \mathbf{u}_2, \mathbf{u}_3)$ as your new orthogonal basis and sketch $\mathbf{x}_1, \mathbf{x}_2, \mathbf{x}_3$ with respect to it. Can you see how the QR factorization has orthogonalized the $\mathbf{x}_i$?

2.3 Let $\mathbf{R} = \{r_{ij}\}$ be a positive-definite, symmetric, matrix and let $\mathbf{R}^{-1} = \{s_{ij}\}$ be the inverse of $\mathbf{R}$. Consider quadratic forms like $r_n^2 = \mathbf{a}_n^T \mathbf{R} \mathbf{a}$, with $\mathbf{a}_n^T = (a_n^n \ldots a_1^n 1, \mathbf{0})$. Minimize r_n^2 with respect to a_n and use your result to show $(1/s_{nn}) \le r_{nn}$. Interpret this result.

2.4 Let $\mathbf{R} = \{\mathbf{r}_1 \quad \cdots \quad \mathbf{r}_N\} = \{r_{ij}\}$, with $r_{ii} = 1$ for all i, be a (possibly singular)$_m$ matrix whose mth column $\mathbf{r}$ non-negative-definite is linearly independent of all others. Call $\mathbf{R}^{\#} = \{s_{ij}\}$ a pseudoinverse of $\mathbf{R}$. Show $s_{ii} \ge 1$.

2.5 Begin with the data matrix

$$\mathbf{X} = \begin{bmatrix} 1 & 4 & 7 \\ 2 & 5 & 8 \\ 3 & 6 & 0 \end{bmatrix}.$$

Compute the Gram matrix $\mathbf{G} = \mathbf{X}^T \mathbf{X}$. Then compute the matrices

$$\mathbf{A} = \begin{bmatrix} \mathbf{a}_1 & & \\ & \mathbf{a}_2 & \\ & & \mathbf{a}_3 \end{bmatrix} \quad \text{and} \quad \mathbf{D}^2 = \begin{bmatrix} \sigma_1^2 & & \\ & \sigma_2^2 & \\ & & \sigma_3^2 \end{bmatrix}$$

in the representation $\mathbf{A}^T \mathbf{G} \mathbf{A} = \mathbf{D}^2$. Verify that $\mathbf{G}\mathbf{A} = \mathbf{H}\mathbf{D}^2$, with $\mathbf{A}^T\mathbf{H} = \mathbf{I}$.

2.6 Prove Theorem 2.2: Every vector $\mathbf{x} \in R^n$ is a linear combination of n linearly independent vectors $(\mathbf{x}_1, \mathbf{x}_2, \ldots, \mathbf{x}_n)$, $\mathbf{x}_i \in R^n$.

2.7 The Gram-Schmidt procedure produces this representation for the orthogonal matrix $\mathbf{Q}$:

$$\mathbf{Q} = \mathbf{X}\mathbf{R}^{-1}; \qquad \mathbf{R}^{-1} = \begin{bmatrix} a_{11} & a_{12} & \cdots & a_{1n} \\ & a_{22} & \cdots & a_{2n} \\ \mathbf{0} & & \ddots & \vdots \\ & & & a_{nn} \end{bmatrix}.$$

From the Gram-Schmidt procedure, determine the entries in $\mathbf{R}^{-1}$.

2.8 In the Gram-Schmidt procedure for QR-factoring a data matrix $\mathbf{X}$, show that the kth step of the procedure may be written as $\mathbf{y}_k = (\mathbf{I} - \mathbf{P}_{k-1})\mathbf{x}_k$, where $\mathbf{P}_{k-1}$ is a projection matrix. Then prove the orthogonal vector $\mathbf{u}_k = \mathbf{0}$ iff $\mathbf{x}_k$ is a linear combination of the $\mathbf{x}_1, \mathbf{x}_2, \ldots, \mathbf{x}_{k-1}$.

2.9 Prove that the matrix $\mathbf{Q} = \mathbf{I} - 2(\mathbf{v}\mathbf{v}^T/\mathbf{v}^T\mathbf{v})$ is orthogonal ($\mathbf{Q}^T\mathbf{Q} = \mathbf{I}$) and that it "projects through any vector that is perpendicular to $\mathbf{v}$."

2.10 Begin with the data matrix

$$\mathbf{X} = \begin{bmatrix} 1 & 4 & 7 \\ 2 & 5 & 8 \\ 3 & 6 & 0 \end{bmatrix}.$$

Determine the *QR* factorization of $\mathbf{X}$, using these three techniques:

a. Gram-Schmidt,

b. Householder transformations,

c. Givens rotations.

Plot your original vectors $\mathbf{x}_1, \mathbf{x}_2, \mathbf{x}_3$ in the data matrix $\mathbf{X} = [\mathbf{x}_1 \quad \mathbf{x}_2 \quad \mathbf{x}_3]$ versus your new coordinate system $\mathbf{Q} = [\mathbf{u}_1 \quad \mathbf{u}_2 \quad \mathbf{u}_3]$.

2.11 How many different direct sums $S_1 \oplus S_2 \oplus \cdots \oplus S_k$ can you build from the span $S(\mathbf{x}_1, \mathbf{x}_2, \ldots, \mathbf{x}_p)$?

2.12 Find the vector $\hat{\mathbf{x}} \in R^N$ that minimizes

$$(\mathbf{x} - \hat{\mathbf{x}})^T(\mathbf{x} - \hat{\mathbf{x}})$$

under the constraint that $\hat{\mathbf{x}} = \mathbf{H}\boldsymbol{\theta}$, with $\mathbf{H}$ an $N \times p$ matrix that has rank $p \leq N$.

2.13 Find the vector $\hat{\mathbf{x}} \in R^N$ that minimizes

$$(\mathbf{x} - \hat{\mathbf{x}})^T(\mathbf{x} - \hat{\mathbf{x}})$$

under the constraint that $\mathbf{A}^T\hat{\mathbf{x}} = \mathbf{0}$, with $\mathbf{A}^T$ a $p \times N$ matrix with rank $p \leq N$.

2.14 Define the "cyclic shift matrix" $\mathbf{R}$ and the "DFT matrix" $\mathbf{V}$:

$$\mathbf{R} = \begin{bmatrix} 0 & \cdots & 0 & 1 \\ & & & 0 \\ & \mathbf{I} & & \vdots \\ & & & 0 \end{bmatrix}$$

$$\mathbf{V} = \frac{1}{\sqrt{N}} \begin{bmatrix} 1 & 1 & \cdots & 1 \\ 1 & e^{-j(2\pi/N)} & & e^{-j(2\pi/N)(N-1)} \\ \vdots & \vdots & & \vdots \\ 1 & e^{-j(2\pi/N)(N-1)} & \cdots & e^{-j(2\pi/N)(N-1)^2} \end{bmatrix}$$

a. Show $\mathbf{V}^H\mathbf{V} = \mathbf{V}\mathbf{V}^H = \mathbf{I}$.

b. Show $\mathbf{V}\mathbf{R} = \mathbf{W}\mathbf{V}$ and $\mathbf{R} = \mathbf{V}^H\mathbf{W}\mathbf{V}$, where $\mathbf{W}$ is the diagonal matrix $\mathbf{W} = \text{diag}(1 \quad W_N^{-1} \quad \cdots \quad W_N^{-(N-1)})$, with $W_N = e^{j(2\pi/N)}$.

c.　Show that every "circulant matrix" $\mathbf{C}$ may be written as

$$\mathbf{C} = \begin{bmatrix} c_0 & c_{N-1} & \cdots & c_1 \\ c_1 & c_0 & & c_2 \\ \vdots & \vdots & & \vdots \\ c_{N-1} & c_{N-2} & \cdots & c_0 \end{bmatrix} = c_0\mathbf{I} + c_1\mathbf{R} + \cdots + c_{N-1}\mathbf{R}^{N-1}.$$

d.　Show that every circulant matrix $\mathbf{C}$ is diagonalized by the DFT matrix:

$$\mathbf{C} = \mathbf{V}^H\mathbf{DV}$$

$$\mathbf{D} = c_0\mathbf{I} + c_1\mathbf{W} + \cdots + c_{N-1}\mathbf{W}^{N-1}.$$

e.　Interpret the terms of the diagonal matrix $\mathbf{D}$ as a "DFT of the first column of $\mathbf{C}$."

2.15 *Kumaresan.* Let $\mathbf{R}$ denote an $N \times N$ banded Toeplitz matrix of the form

$$\mathbf{R} = \begin{bmatrix} r_0 & r_1 & \cdots & r_p & & & \\ r_1 & r_0 & & & & \mathbf{0} & \\ r_2 & r_1 & & & & & \\ \vdots & & & & & \ddots & \\ r_p & & & & & & r_p \\ & & & & & \ddots & \ddots & \vdots \\ \mathbf{0} & \ddots & & & \ddots & & \\ & & & r_p & \cdots & r_1 & r_0 \end{bmatrix}$$

Assume $N > 2p$.

a.　Let $\mathbf{J}_p$ denote a $p \times p$ exchange matrix (ones on antidiagonal). Show that the matrix $\mathbf{S} = \mathbf{R} + \mathbf{V}^T\mathbf{J}_{2p}\mathbf{V}$ is circulant, where

$$\mathbf{V} = \left[\begin{array}{cc|c} & & \begin{matrix} r_p & \cdots & r_1 \\ & \ddots & \end{matrix} \\ \mathbf{0} & \mathbf{0} & \mathbf{0} \\ & & \quad r_p \\ \hline \mathbf{J}_p & \mathbf{0} & \mathbf{0} \end{array} \right]$$

b.　Use a matrix inversion lemma on $\mathbf{R} = \mathbf{S} - \mathbf{V}^T\mathbf{J}_{2p}\mathbf{V}$ and the DFT to find a fast algorithm for inverting $\mathbf{R}$.

2.16 Prove Sylvester's Theorem: if $f(\lambda)$ is the polynomial

$$f(\lambda) = 1 + a_1\lambda + \cdots + a_N\lambda^N$$

and $\mathbf{A} = \mathbf{V\Lambda V}^{-1}$, with $\mathbf{\Lambda} = \text{diag}[\lambda_1 \quad \lambda_2 \quad \cdots \quad \lambda_N]$, then the matrix polynomial $f(A)$ may be written as

$$f(\mathbf{A}) = \mathbf{V}f(\mathbf{\Lambda})\mathbf{V}^{-1}$$

$$f(\mathbf{\Lambda}) = \mathbf{I} + a_1\mathbf{\Lambda} + \cdots + a_N\mathbf{\Lambda}^N.$$

2.17 Consider the Cholesky factor of the Gram matrix $\mathbf{G} = \mathbf{X}^T\mathbf{X}$:

$$\mathbf{A}^T\mathbf{G}\mathbf{A} = \mathbf{D}^2.$$

Show that the matrix $\mathbf{U} = \mathbf{X}\mathbf{A}\mathbf{D}^{-1}$ is orthogonal ($\mathbf{U}^T\mathbf{U} = \mathbf{I}$).

2.18 Prove that the Vandermonde matrix $\mathbf{V}$ satisfies the eigenvalue problem

$$\mathbf{A}\mathbf{V} = \mathbf{V}\mathbf{\Lambda}$$

when $\mathbf{A}$ is a companion matrix. Show that $\mathbf{V}$ is nonsingular iff $\lambda_1 \neq \lambda_2 \neq \cdots \neq \lambda_N$.

2.19 Derive a formula for the determinant of a Vandermonde matrix. Verify that the determinant is nonzero iff the roots of the Vandermonde matrix, $z_1, z_2, \ldots, z_N$, are distinct.

2.20 Begin with a real, nonnegative definite symmetric matrix $\mathbf{R}$. There exists an orthogonal matrix $\mathbf{U}$ such that

$$\mathbf{U}^T\mathbf{R}\mathbf{U} = \mathbf{\Lambda}^2 = \mathrm{diag}[\lambda_1^2 \quad \cdots \quad \lambda_N^2].$$

a. Show that $\lambda_n^2 \geq 0$ for all n.

b. Find a symmetric, nonnegative definite matrix $\mathbf{H}$ with the property $\mathbf{H}\mathbf{H} = \mathbf{R}$.

2.21 Let $\mathbf{H}_{p-1}\mathbf{H}_{p-1}^T$ be a lower-upper Cholesky factor of the $(p-1) \times (p-1)$ nonnegative definite, symmetric matrix $\mathbf{R}_{p-1}$:

$$\mathbf{R}_{p-1} = \mathbf{H}_{p-1}\mathbf{H}_{p-1}^T; \qquad \mathbf{R}_{p-1}^{-1} = (\mathbf{H}_{p-1}^T)^{-1}(\mathbf{H}_{p-1})^{-1}.$$

Define the $p \times p$, nonnegative definite, symmetric matrix $\mathbf{R}_p$:

$$\mathbf{R}_p = \begin{bmatrix} \mathbf{R}_{p-1} & \mathbf{r}_p \\ \mathbf{r}_p^T & r_{pp} \end{bmatrix}.$$

Show that

a. $\mathbf{R}_p$ has the lower-upper Cholesky factorization

$$\mathbf{R}_p = \begin{bmatrix} \mathbf{H}_{p-1} & \mathbf{0} \\ \mathbf{h}_p^T & h_{pp} \end{bmatrix} \begin{bmatrix} \mathbf{H}_{p-1}^T & \mathbf{h}_p \\ \mathbf{0}^T & h_{pp} \end{bmatrix}; \qquad \begin{aligned} \mathbf{H}_{p-1}\mathbf{h}_p &= -\mathbf{r}_p \\ \mathbf{h}_p^T\mathbf{h}_p + h_{pp}^2 &= r_{pp}; \end{aligned}$$

b. $\mathbf{R}_p^{-1}$ has the upper-lower Cholesky factorization

$$\mathbf{R}_p^{-1} = \begin{bmatrix} (\mathbf{H}_{p-1}^T)^{-1} & \mathbf{b}_p \\ \mathbf{0}^T & b_{pp} \end{bmatrix} \begin{bmatrix} \mathbf{H}_{p-1}^{-1} & \mathbf{0} \\ \mathbf{b}_p^T & b_{pp} \end{bmatrix}; \qquad \begin{aligned} \mathbf{H}_{p-1}^T\mathbf{b}_p &= -b_{pp}\mathbf{h}_p \\ h_{pp}b_{pp} &= 1. \end{aligned}$$

2.22 Begin with $\mathbf{x} = [x_1 \quad \cdots \quad x_n]^T$, $x_i = |x_i|e^{j\theta_i}$. Define $\mathbf{u} = \mathbf{x} + e^{j\theta_i}\|\mathbf{x}\|\mathbf{e}_1$, $\mathbf{e}_1 = [1 \quad 0 \quad \cdots \quad 0]$. Show that $\mathbf{P} = \mathbf{I} - (2/\mathbf{u}^H\mathbf{u})\mathbf{u}\mathbf{u}^H$ is unitary and that $\mathbf{P}\mathbf{x} = -e^{j\theta_1}\|\mathbf{x}\|\mathbf{e}_1$. Sketch.

2.23 Begin with $\mathbf{x} = [x_1 \quad x_2 \quad \cdots \quad x_n]^{\mathrm{T}}$, $x_i \in R$. Build $\mathbf{M}_k = \mathbf{I} - \boldsymbol{\alpha}\mathbf{e}_k^{\mathrm{T}}$ with $\boldsymbol{\alpha}^{\mathrm{T}} = [0 \quad \cdots \quad 0 \quad \alpha_{k+1} \quad \cdots \quad \alpha_n]$ and $\mathbf{e}_k^{\mathrm{T}} = [0 \quad \cdots \quad 1 \quad 0 \quad \cdots \quad 0]$. Show how to choose $(\alpha_i)_{k+1}^n$ so that $\mathbf{M}_k\mathbf{x} = [x_1 \quad \cdots \quad x_k \quad 0 \quad \cdots \quad 0]^{\mathrm{T}}$. This is called the Gauss transformation.

2.24 *Leverrier's Algorithm.* Denote by $p(z)$ the characteristic polynomial of the $n \times n$ matrix $\mathbf{A}$:

$$p(z) = \det(z\mathbf{I} - \mathbf{A}) = z^n + a_1 z^{n-1} + \cdots + a_n.$$

The matrix resolvent is

$$(z\mathbf{I} - \mathbf{A})^{-1} = \frac{1}{p(z)}(z^{n-1}\mathbf{I} + z^{n-2}\mathbf{B}_2 + \cdots + \mathbf{B}_n),$$

where each $\mathbf{B}_i$ is an $n \times n$ matrix.

a. Prove

$$\frac{d}{dz}p(z) = \mathrm{tr}(z^{n-1}\mathbf{I} + z^{n-2}\mathbf{B}_2 + \cdots + \mathbf{B}_n).$$

b. Derive Leverrier's algorithm [Roberts and Mullis, 1987]:

$$a_0 = 1$$

$$\mathbf{B}_1 = \mathbf{I}$$

loop on $i = 1, \ldots, n - 1$

$$\begin{bmatrix} a_i = -\frac{1}{i}\,\mathrm{tr}(\mathbf{A}\mathbf{B}_i) \\ \mathbf{B}_{i+1} = \mathbf{A}\mathbf{B}_i + a_i\mathbf{I}. \end{bmatrix}$$

c. Derive the Cayley-Hamilton Theorem:

$$p(\mathbf{A}) = \mathbf{0}.$$

2.25 The "spectral" decomposition of a matrix is defined as

$$\mathbf{A} = \sum_{i=1}^n \lambda_i \mathbf{E}_i + \mathbf{N}_i,$$

with $\{\mathbf{E}_i\}$ a set of orthogonal projections ($\mathbf{E}_i^2 = \mathbf{E}_i$, $\mathbf{E}_i\mathbf{E}_j = 0$ for $i \neq j$) summing to the identity $\mathbf{I} = \sum_1^n \mathbf{E}_i$. If m_i is the multiplicity of the eigenvalue λ_i, then rank $(\mathbf{E}_i) = m_i$ and $\mathbf{N}_i^{m_i} = \mathbf{0}$ (nilpotent). Furthermore,

$$\mathbf{N}_i\mathbf{E}_i = \mathbf{E}_i\mathbf{N}_i = \mathbf{N}_i$$

$$\mathbf{N}_i\mathbf{E}_j = \mathbf{E}_j\mathbf{N}_i = \mathbf{0}, \qquad i \neq j$$

$$\mathbf{N}_i\mathbf{N}_j = \mathbf{0}, \qquad i \neq j.$$

For a function f we have

$$f(\mathbf{A}) = \sum_{i=1}^{m}\left[f(\lambda_i)\mathbf{E}_i + \sum_{j=1}^{m_i-1} \frac{f^{(j)}(\lambda_i)\mathbf{N}_i^j}{j!} \right]$$

where $f^{(j)}$ is the jth derivative of f. Use this equation with $f(\mathbf{A}) = (z\mathbf{I}-\mathbf{A})^{-1}$, together with Leverrier's algorithm, to find the "spectral" decomposition for

$$\mathbf{A} = \begin{bmatrix} 1 & -1 \\ 2 & 3 \end{bmatrix}.$$

2.26 Follow the procedure in problem 2.25 to find the "spectral" decomposition for

$$\mathbf{A} = \begin{bmatrix} 6 & 9 & 0 \\ -4 & -6 & 1 \\ 0 & 0 & 2 \end{bmatrix}.$$

2.27 Let $\mathbf{A}$ be an $N \times p$ matrix of rank r with real coefficients. The matrix $\mathbf{A}^{\#}$ is called a pseudoinverse of $\mathbf{A}$ if

$$\mathbf{A}\mathbf{A}^{\#} = (\mathbf{A}\mathbf{A}^{\#})^{\mathrm{T}}, \qquad \mathbf{A}\mathbf{A}^{\#}\mathbf{A} = \mathbf{A}$$
$$\mathbf{A}^{\#}\mathbf{A} = (\mathbf{A}^{\#}\mathbf{A})^{\mathrm{T}}, \qquad \mathbf{A}^{\#}\mathbf{A}\mathbf{A}^{\#} = \mathbf{A}^{\#}.$$

a. Prove the existence of the matrices $\mathbf{B}$ ($N \times r$) and $\mathbf{C}$ ($r \times p$) for which

$$\mathbf{A} = \mathbf{B}\mathbf{C}$$
$$\det(\mathbf{C}\mathbf{C}^{\mathrm{T}}) \neq 0$$
$$\det(\mathbf{B}^{\mathrm{T}}\mathbf{B}) \neq 0.$$

Hint: Use the SVD of $\mathbf{A}$.

b. Find pseudoinverses for $\mathbf{B}$ and $\mathbf{C}$.

c. Show that $\mathbf{A}^{\#} = \mathbf{C}^{\#}\mathbf{B}^{\#}$ is a pseudoinverse for $\mathbf{A}$.

d. Prove that the pseudoinverse for $\mathbf{A}$ is unique.

2.28 Consider the quadratic form

$$\mathbf{x}^{\mathrm{T}}\mathbf{Q}\mathbf{x} = \begin{bmatrix} \mathbf{x}_1^{\mathrm{T}} & \mathbf{x}_2^{\mathrm{T}} \end{bmatrix}\begin{bmatrix} \mathbf{Q}_{11} & \mathbf{Q}_{12} \\ \mathbf{Q}_{21} & \mathbf{Q}_{22} \end{bmatrix}\begin{bmatrix} \mathbf{x}_1 \\ \mathbf{x}_2 \end{bmatrix}.$$

Show that for fixed $\mathbf{x}_2$, $\mathbf{x}^{\mathrm{T}}\mathbf{Q}\mathbf{x}$ is minimized when $\hat{\mathbf{x}}_1 = -\mathbf{Q}_{11}^{-1}\mathbf{Q}_{12}\mathbf{x}_2$. Show that $\hat{\mathbf{x}}_2 = -\mathbf{Q}_{22}^{-1}\mathbf{Q}_{21}\mathbf{x}_1$ for $\mathbf{x}_1$ fixed.

2.29 Prove Woodbury's identity (Example 2.4). Generalize it by showing that the inverse of $\mathbf{R} = \mathbf{R}_0 + \mathbf{u}\mathbf{v}^{\mathrm{T}}$ is

$$\mathbf{R}^{-1} = \mathbf{R}_0^{-1} - \frac{1}{1 + \mathbf{v}^{\mathrm{T}}\mathbf{R}_0^{-1}\mathbf{u}} \mathbf{R}_0^{-1}\mathbf{u}\mathbf{v}^{\mathrm{T}}\mathbf{R}_0^{-1}.$$

2.30 Begin with the identity

$$I(a) = \int_{-\infty}^{\infty} e^{-ax^2} \, dx = (\pi/a)^{1/2}.$$

Differentiate n times with respect to a to derive this identity:

$$\int_{-\infty}^{\infty} x^{2n} e^{-ax^2} \, dx = \frac{1 \quad 3 \quad \cdots \quad (2n-1)}{2n} = (\pi/a^{2n+1})^{1/2}.$$

Use this result to find the even moments EX^{2n} for an $N[0, \sigma^2]$ random variable. Now find the even and odd moments for an $N[\theta, \sigma^2]$ random variable.

2.31 Select an $n \times N$ transformation A^T that selects the variables $X_{i_1} \quad \cdots \quad X_{i_n}$ $(1 \le i_1 < i_2 < \cdots < i_n \le N)$ out of the vector $\mathbf{X}$. Find the joint distribution of these variables when $\mathbf{X} : N[\mathbf{m}, R]$.

2.32 Use the characteristic function of a χ_N^2 random variable to show that the mean and variance of $Q : \chi_N^2$ are

$$EQ = N$$

$$\text{var } Q = 2N.$$

2.33 What is the characteristic function for the general bivariate normal distribution?

2.34 Let (U_1, U_2) denote independent normal random variables with respective means of zero and respective variances of (σ_1^2, σ_2^2). Define (X_1, X_2) as the following rotation of (U_1, U_2):

$$\begin{bmatrix} X_1 \\ X_2 \end{bmatrix} = \begin{bmatrix} \cos\theta & -\sin\theta \\ \sin\theta & \cos\theta \end{bmatrix} \begin{bmatrix} U_1 \\ U_2 \end{bmatrix}.$$

Draw the contours of constant probability for (X_1, X_2) and compare them with those of (U_1, U_2).

2.35 Let $\mathbf{X} = \begin{bmatrix} X_1 & X_2 \end{bmatrix}^T$ denote a bivariate normal random vector. Assume

$$E\mathbf{X} = \mathbf{0} \quad \text{and} \quad E\mathbf{X}\mathbf{X}^T = \begin{bmatrix} 1 & \rho \\ \rho & 1 \end{bmatrix}.$$

Define $Y_1 = X_1 + X_2$ and $Y_2 = -X_1 + X_2$.

a.   Find the joint distributions of Y_1 and Y_2; find the marginal distributions of Y_1 and Y_2.

b.   Find the conditional density of X_1, given Y_1; find the conditional density of X_1, given Y_2.

    c.   Find the conditional mean and variance of X_1, given Y_1; find the conditional mean and variance of X_1, given Y_2.

2.36 Begin with $\mathbf{X} : N[\mathbf{0}, \mathbf{U}\mathbf{\Lambda}^2\mathbf{U}^T]$ with $\mathbf{U}$ orthogonal and $\mathbf{\Lambda}^2$ diagonal. Define $\mathbf{Y} = \mathbf{\Gamma}\mathbf{U}^T\mathbf{X}$. Choose $\mathbf{\Gamma}$ to make $\mathbf{Y}^T\mathbf{Y}$ chi-squared distributed with r degrees of freedom.

2.37 Prove $EQ = \mathrm{tr}\ \mathbf{PR}$ and $\mathrm{var}\ Q = 2\ \mathrm{tr}\ (\mathbf{PR})^2$ when $Q = (\mathbf{X} - \mathbf{m})^T\mathbf{P}(\mathbf{X} - \mathbf{m})$ and $\mathbf{X} : N[\mathbf{m}, \mathbf{R}]$.

Sufficiency and MVUB Estimators

Sufficiency is one of those concepts that leaves its imprint on every aspect of statistical signal processing. Throughout this book we shall use sufficiency to solve problems in detection, estimation, and time-series analysis. In this chapter we will show that any estimator that hopes to win a competition for the "minimum variance unbiased (MVUB) estimator" of a parameter must be a function of a sufficient statistic.

Among the sufficient statistics for a particular problem there may be one that is *minimal* in the sense that it is a function of all others. Such a statistic is valuable because it typically compresses the raw measurements into a low-dimensional statistic. The raw measurements are then discarded and only the minimal sufficient statistic is stored. We shall find that the only practical way to establish that a sufficient statistic is minimal is to establish that it is *complete*, a technical characterization that we shall develop in our treatment of sufficiency.

Throughout this chapter we shall consider measurements $\mathbf{x}$ that are drawn from the multivariate normal distribution $N[\mathbf{m}, \mathbf{R}]$. We will encounter sufficient statistics that are linear and quadratic forms in the measurements. The resulting statistics are distributed as normal and chi-squared random variables. The rudiments of multivariate normal (MVN) theory, and quadratic forms in MVN random variables, are summarized in Section 2.10 of Chapter 2.

3.1   SUFFICIENCY

Let's suppose we have designed an experiment to take a suite of measurements $\mathbf{x}$. The measurements are described by the probability law $F_\theta(\mathbf{x})$, and $\boldsymbol{\theta}$ is a vector of parameters to be estimated. We say that $\boldsymbol{\theta}$ parameterizes the distribution of $\mathbf{x}$ or, conversely, that the measurement $\mathbf{x}$ brings information about $\boldsymbol{\theta}$ through the probability law $F_\theta(\mathbf{x})$. Is it possible that some function of the measurements could carry all of the useful information about the underlying parameters? If so, we could discard the raw measurements themselves and retain only the appropriate function of them. In this way, we would dramatically compress the data into a sufficient statistic that would be used to draw inference about parameters $\boldsymbol{\theta}$. The sufficient statistic is more memory-efficient than is the original data. The key ideas behind sufficiency are best illustrated with a commonplace example associated with a binary information source.

Example 3.1 (Binary Information Source)

Let S denote a binary source that produces independent binary symbols $x_0, x_1, \ldots, x_{N-1}$. Each symbol (or bit) may be regarded as a realization of a Bernoulli trial. The probability mass function for the random variable X_n is

$$P_\theta(x_n) = \theta^{x_n}(1 - \theta)^{1-x_n}, \qquad x_n = 0, 1. \qquad\qquad 3.1$$

That is, the random variable X_n equals 1 with probability θ and 0 with probability $(1 - \theta)$. A sequence of N independent Bernoulli trials produces the random sample $\mathbf{x} = (x_0, x_1, \ldots, x_{N-1})$ consisting of N independent source symbols. The distribution of the random sample is the product

$$P_\theta(\mathbf{x}) = \prod_{n=0}^{N-1} \theta^{x_n}(1 - \theta)^{1-x_n} = \theta^k(1 - \theta)^{N-k} \qquad\qquad 3.2$$

$$k = \sum_{n=0}^{N-1} x_n \quad : \text{ number of ones} \in \mathbf{x} = (x_0, x_1, \ldots, x_{N-1}).$$

The distribution of the number of ones in N independent Bernoulli trials is binomial:

$$P_\theta(k) = \binom{N}{k} \theta^k(1 - \theta)^{N-k}. \qquad\qquad 3.3$$

Now consider the joint probability of observing the sample $\mathbf{x}$ and the number k. The probability is zero unless k equals the number of ones in $\mathbf{x}$. When it does, then the joint probability is just the marginal probability of $\mathbf{x}$:

$$P_\theta(\mathbf{x}, k) = \begin{cases} P_\theta(\mathbf{x}), & k = \text{number of ones} \in \mathbf{x} \\ 0, & \text{otherwise.} \end{cases} \qquad\qquad 3.4$$

The conditional probability of $\mathbf{x}$, given the number of ones in $\mathbf{x}$, may be written

$$P_\theta(\mathbf{x}|k) = \frac{P_\theta(\mathbf{x}, k)}{P_\theta(k)}$$

$$= \frac{\theta^k (1 - \theta)^{N-k}}{\binom{N}{k} \theta^k (1 - \theta)^{N-k}} = \frac{1}{\binom{N}{k}}. \qquad 3.5$$

The conditional probability is distributed uniformly over the $\binom{N}{k}$ sequences $\mathbf{x}$ that have k ones in them—independently of θ! We say that the conditional distribution of the random sample, given the number of ones in the sample, is independent of the parameter θ. As a consequence, the probability mass function for $\mathbf{x}$ is just a scaled version of the probability mass function for k:

$$P_\theta(\mathbf{x}) = P_\theta(\mathbf{x}|k)P_\theta(k)$$

$$= \frac{1}{\binom{N}{k}} P_\theta(k). \qquad 3.6$$

$P(x) = P(x|N+)P(N+)$

$= \frac{1}{\binom{N}{N+}} P(N+)$

This result suggests that the dependence of the random sample on the parameter θ is carried by k. In fact, as the binomial distribution for k shows, this is the case. ■

Discrete Random Variables

To expand our discussion, let $\mathbf{X}$ denote an N-dimensional random vector with probability mass function $P_\theta(\mathbf{x})$. The parameter $\boldsymbol{\theta}$ is p-dimensional. Define a statistic $\mathbf{T}(\mathbf{X})$, which is k-dimensional and whose probability mass function is $P_\theta(\mathbf{t})$. Consider the joint probability of $\mathbf{x}$ and $\mathbf{t}$:

$$P_\theta(\mathbf{x}, \mathbf{t}) = P_\theta(\mathbf{X} = \mathbf{x}, \mathbf{T}(\mathbf{X}) = \mathbf{t}). \qquad 3.7$$

This is zero unless $\mathbf{t} = \mathbf{T}(\mathbf{x})$, where it equals the marginal probability of $\mathbf{x}$:

$$P_\theta(\mathbf{x}, \mathbf{t}) = \begin{cases} P_\theta(\mathbf{x}), & \mathbf{t} = \mathbf{T}(\mathbf{x}) \\ 0, & \text{otherwise.} \end{cases}$$

But the joint probability on the left-hand side may always be written as the product of a conditional probability and a marginal probability. Therefore,

$$P_\theta(\mathbf{x}, \mathbf{t}) = P_\theta(\mathbf{x}) = \begin{cases} P_\theta(\mathbf{x}|\mathbf{t})P_\theta(\mathbf{t}), & \mathbf{t} = \mathbf{T}(\mathbf{x}) \\ 0, & \text{otherwise} \end{cases} \qquad 3.8$$

where $P_\theta(\mathbf{x}|\mathbf{t})$ is the conditional probability of $\mathbf{x}$, given $\mathbf{t}$.

Suppose the statistic $\mathbf{T}(\mathbf{X})$ has been chosen so that the conditional probability is independent of $\boldsymbol{\theta}$:

$$P_\theta(\mathbf{x}|\mathbf{t}) = P(\mathbf{x}|\mathbf{t}). \qquad 3.9$$

Then the probability of $\mathbf{x}$ may be written as the probability of $\mathbf{t}$, scaled by a function that is dependent only on $\mathbf{x}$ and not on $\boldsymbol{\theta}$:

$$P_{\boldsymbol{\theta}}(\mathbf{x}) = \begin{cases} P(\mathbf{x}|\mathbf{t})P_{\boldsymbol{\theta}}(\mathbf{t}), & \mathbf{t} = \mathbf{T}(\mathbf{x}) \\ 0, & \text{otherwise.} \end{cases} \qquad 3.10$$

What does this mean? It means that

1. the parameterization of the probability law for the measurement $\mathbf{x}$ is really manifested in the parameterization of the probability law for the statistic $\mathbf{t}$,
2. any inference strategy based on the probability of $\mathbf{x}$, namely $P_{\boldsymbol{\theta}}(\mathbf{x})$, may be replaced with a strategy based on $\mathbf{T}(\mathbf{x})$ and its probability $P_{\boldsymbol{\theta}}(\mathbf{t})$, and
3. the information that $\mathbf{x}$ brings about $\boldsymbol{\theta}$ is also brought by the statistic $\mathbf{T}(\mathbf{x})$.

These heuristic beginnings lead us to the formal definition of a sufficient statistic.

Definition (Sufficient Statistic)   Let $\mathbf{X}$ denote an N-dimensional random vector or, more generally, a random sample. Let $\boldsymbol{\theta}$ denote a p-dimensional parameter, and let $P_{\boldsymbol{\theta}}(\mathbf{x})$ denote the density of $\mathbf{x}$. The statistic $\mathbf{T}(\mathbf{X})$ is sufficient for $\boldsymbol{\theta}$ if the density of $\mathbf{x}$, given $\mathbf{T}(\mathbf{x}) = \mathbf{t}$, is independent of $\boldsymbol{\theta}$.   □

   In the discrete Bernoulli case, we have had to guess at a sufficient statistic and then show that a conditional distribution is independent of $\boldsymbol{\theta}$. This is a lot of work. The following factorization theorem helps us spot sufficient statistics from the form of the distribution and gives an if-and-only-if condition for a statistic to be sufficient.

Fisher-Neyman Factorization Theorem   Let $\mathbf{X}$ be a discrete random vector with probability mass function $P_{\boldsymbol{\theta}}(\mathbf{x})$. The statistic $\mathbf{T}(\mathbf{X})$ is sufficient for the parameter $\boldsymbol{\theta}$ iff the probability mass function for $\mathbf{x}$ factors into a product:

$$P_{\boldsymbol{\theta}}(\mathbf{x}) = a(\mathbf{x})b_{\boldsymbol{\theta}}(\mathbf{t}); \qquad \mathbf{t} = \mathbf{T}(\mathbf{x}). \qquad 3.11$$

Proof (if)   Assume $P_{\boldsymbol{\theta}}(\mathbf{x}) = a(\mathbf{x})b_{\boldsymbol{\theta}}(\mathbf{t})$. The probability mass function for $\mathbf{t}$ is obtained by summing over all $\mathbf{x}$ such that $\mathbf{T}(\mathbf{x}) = \mathbf{t}$:

$$P_{\boldsymbol{\theta}}(\mathbf{t}) = \sum_{\mathbf{x}:\mathbf{T}(\mathbf{x})=\mathbf{t}} P_{\boldsymbol{\theta}}(\mathbf{x}) = \left(\sum_{\mathbf{x}:\mathbf{T}(\mathbf{x})=\mathbf{t}} a(\mathbf{x}) \right) b_{\boldsymbol{\theta}}(\mathbf{t}). \qquad 3.12$$

The conditional density for $\mathbf{x}$, given $\mathbf{t}$, is therefore

$$P_{\boldsymbol{\theta}}(\mathbf{x}|\mathbf{t}) = \frac{P_{\boldsymbol{\theta}}(\mathbf{x}, \mathbf{t})}{P_{\boldsymbol{\theta}}(\mathbf{t})} = \frac{P_{\boldsymbol{\theta}}(\mathbf{x})}{P_{\boldsymbol{\theta}}(\mathbf{t})}$$

$$= \frac{a(\mathbf{x})}{\displaystyle\sum_{\mathbf{x}:\mathbf{T}(\mathbf{x})=\mathbf{t}} a(\mathbf{x})}. \qquad 3.13$$

This is independent of $\boldsymbol{\theta}$, meaning that $\mathbf{t}$ is sufficient and proving the *if* part of the theorem. The *only if* part is proven by assuming that $P_{\boldsymbol{\theta}}(\mathbf{x}|\mathbf{t}) = P(\mathbf{x}|\mathbf{t})$ and writing $P_{\boldsymbol{\theta}}(\mathbf{x}) = P(\mathbf{x}|\mathbf{t})P_{\boldsymbol{\theta}}(\mathbf{t})$, which is the desired factorization. $\square$

Comments

When a factorization exists, then the probability mass function for $\mathbf{x}$ is

$$P_{\boldsymbol{\theta}}(\mathbf{x}) = \frac{a(\mathbf{x}) \displaystyle\sum_{\mathbf{x}:T(\mathbf{x})=\mathbf{t}} a(\mathbf{x})b_{\boldsymbol{\theta}}(\mathbf{t})}{\displaystyle\sum_{\mathbf{x}:T(\mathbf{x})=\mathbf{t}} a(\mathbf{x})} = P(\mathbf{x}|\mathbf{t})P_{\boldsymbol{\theta}}(\mathbf{t}). \qquad 3.14$$

The factorization theorem says that $\mathbf{T}(\mathbf{X})$ is sufficient for $\boldsymbol{\theta}$ if and only if the probability mass function for $\mathbf{x}$ may be written as a scale constant, dependent on $\mathbf{x}$ and $\mathbf{t}$ but independent of $\boldsymbol{\theta}$, times the probability mass function for $\mathbf{t}$. The theorem also shows that the scale constant is the conditional density of $\mathbf{x}$, given $\mathbf{t}$, and it shows how to compute the probability mass function for the sufficient statistic $\mathbf{t}$.

Example 3.2 (Bernoulli Trials Revisited)

Write the probability mass function for $\mathbf{x}$ as

$$P_{\theta}(\mathbf{x}) = a(\mathbf{x})b_{\theta}(k) \qquad 3.15$$

$$a(\mathbf{x}) = 1; \qquad b_{\theta}(k) = \theta^{k}(1-\theta)^{N-k}.$$

The conditional probability of $\mathbf{x}$, given k, is

$$P_{\theta}(\mathbf{x}|k) = \frac{a(\mathbf{x})}{\displaystyle\sum_{\mathbf{x}:\sum x_n = k} a(\mathbf{x})} = \frac{1}{\displaystyle\sum_{\mathbf{x}:\sum x_n = k} 1}$$

$$= \frac{1}{\binom{N}{k}},$$

$$\qquad\qquad 3.16$$

and the probability mass function for $\mathbf{x}$ is

$$P_{\theta}(\mathbf{x}) = \frac{1}{\binom{N}{k}} P_{\theta}(k). \qquad 3.17$$

$\blacksquare$

Example 3.3 (Poisson Example)

Let $\mathbf{X}$ denote a random sample of N Poisson random variables X_n, each with probability mass function

$$P_{\theta}(x_n) = e^{-\theta} \frac{1}{x_n!} \theta^{x_n} \qquad (x_n = 0, 1, \ldots). \qquad 3.18$$

The probability of the random sample is

$$P_\theta(\mathbf{x}) = a(\mathbf{x})b_\theta(k) \qquad\qquad 3.19$$

$$k = \sum_{n=0}^{N-1} x_n, \qquad a(\mathbf{x}) = \frac{1}{\prod_{n=0}^{N-1} x_n!}, \qquad b_\theta(k) = e^{-N\theta}\theta^k.$$

This makes k, the total number of Poisson counts, sufficient for θ. The conditional density for $\mathbf{x}$, given k, is

$$P_\theta(\mathbf{x}|k) = \begin{cases} \dfrac{P_\theta(\mathbf{x})}{P_\theta(k)}, & k = \sum_{n=0}^{N-1} x_n \\ 0, & \text{otherwise.} \end{cases} \qquad\qquad 3.20$$

But the total number of Poisson counts is Poisson with parameter $N\theta$, so we may write

$$P_\theta(\mathbf{x}|k) = \frac{a(\mathbf{x})e^{-N\theta}\theta^k}{e^{-N\theta}(1/k!)(N\theta)^k}$$

$$= \frac{k!}{N^k} \frac{1}{\prod_{n=0}^{N-1} x_n!}. \qquad\qquad 3.21$$

This is independent of θ.   ∎

Continuous Random Variables

The factorization for discrete random variables may be extended to continuous random variables in the following way. Assume that the density function for the continuous random vector $\mathbf{x}$ factors:

$$f_\theta(\mathbf{x}) = a(\mathbf{x})b_\theta(\mathbf{t}); \qquad \mathbf{t} = \mathbf{T}(\mathbf{x}). \qquad\qquad 3.22$$

Adjoin to the statistic $\mathbf{T}(\mathbf{x})$ the auxiliary statistic $\mathbf{U}(\mathbf{x})$ so that the mapping from $\mathbf{x}$ to $\mathbf{y} = (\mathbf{t}, \mathbf{u})$ is $1:1$ onto and therefore invertible. Call the mapping from $\mathbf{x}$ to $\mathbf{y}$, $\mathbf{y} = \mathbf{W}(\mathbf{x})$, and call the inverse mapping $\mathbf{x} = \mathbf{W}^{-1}(\mathbf{y})$. The density function for the random vector $\mathbf{y}$ is

$$f_\theta(\mathbf{y}) = f_\theta[\mathbf{x} = \mathbf{W}^{-1}(\mathbf{y})]\,|\mathbf{J}| \qquad\qquad 3.23$$

where $\mathbf{J} = \det \frac{\partial}{\partial \mathbf{y}}[\mathbf{W}^{-1}(\mathbf{y})]^T$ is the Jacobian of the transformation $\mathbf{W}^{-1}$. Because the density function for $\mathbf{x}$ factors, the density function for $\mathbf{y}$ factors as well:

$$f_\theta(\mathbf{y}) = a[\mathbf{x} = \mathbf{W}^{-1}(\mathbf{y})]\,|\mathbf{J}|\,b_\theta(\mathbf{t}); \qquad \mathbf{t} = \mathbf{T}[\mathbf{W}^{-1}(\mathbf{y})]. \qquad\qquad 3.24$$

The density function for $\mathbf{t}$ is obtained by integrating the density for $\mathbf{y} = (\mathbf{t}, \mathbf{u})$ over $\mathbf{u}$:

$$f_{\boldsymbol{\theta}}(\mathbf{t}) = \left(\int d\mathbf{u}\, a\big[\mathbf{x} = \mathbf{W}^{-1}(\mathbf{y})\big] |\mathbf{J}| \right) b_{\boldsymbol{\theta}}(\mathbf{t}). \qquad 3.25$$

The conditional density for $\mathbf{x}$, given $\mathbf{t}$, is therefore independent of $\boldsymbol{\theta}$:

$$f(\mathbf{x}|\mathbf{t}) = \frac{f_{\boldsymbol{\theta}}(\mathbf{x}, \mathbf{t})}{f_{\boldsymbol{\theta}}(\mathbf{t})}$$

$$= \frac{f_{\boldsymbol{\theta}}(\mathbf{x})}{f_{\boldsymbol{\theta}}(\mathbf{t})} \qquad 3.26$$

$$= \frac{a(\mathbf{x})}{\int d\mathbf{u}\, a\big[\mathbf{x} = \mathbf{W}^{-1}(\mathbf{y})\big] |\mathbf{J}|}.$$

Finally, the density function for $\mathbf{x}$ may be written as

$$f_{\boldsymbol{\theta}}(\mathbf{x}) = f(\mathbf{x}|\mathbf{t}) f_{\boldsymbol{\theta}}(\mathbf{t}), \qquad 3.27$$

showing that the density is a scaled version of the density for $\mathbf{t}$. These results also show that the scale constant is the conditional density of $\mathbf{x}$, given $\mathbf{t}$, and they show how to compute the density function for the sufficient statistic $\mathbf{T}(\mathbf{x})$ from Equation 3.25.

Example 3.4 (Normal)

Let $\mathbf{X} = (\mathbf{X}_0, \mathbf{X}_1, \ldots, \mathbf{X}_{M-1})$ denote a random sample of N-dimensional normal random vectors $\mathbf{X}_n$, each of which has mean value $\mathbf{m}$ and covariance matrix $\mathbf{R}$:

$$\mathbf{X}_n : N[\mathbf{m}, \mathbf{R}]. \qquad 3.28$$

The density function for $\mathbf{x}$ is (see Section 2.10 of Chapter 2 for a discussion of the multivariate normal distribution)

$$f_{\boldsymbol{\theta}}(\mathbf{x}) = \prod_{n=0}^{M-1} f_{\boldsymbol{\theta}}(\mathbf{x}_n)$$

$$= (2\pi)^{-MN/2} |\mathbf{R}|^{-M/2} \exp\left\{ -\frac{1}{2} \sum_{n=0}^{M-1} (\mathbf{x}_n - \mathbf{m})^{\mathrm{T}} \mathbf{R}^{-1} (\mathbf{x}_n - \mathbf{m}) \right\} \qquad 3.29$$

$$= (2\pi)^{-MN/2} |\mathbf{R}|^{-M/2} \exp\left\{ -\frac{M}{2} \operatorname{tr} \mathbf{R}^{-1} \mathbf{S}(\mathbf{m}) \right\}$$

where $\mathbf{S}(\mathbf{m})$ is the scatter matrix

$$\mathbf{S}(\mathbf{m}) = \frac{1}{M} \sum_{n=0}^{M-1} (\mathbf{x}_n - \mathbf{m})(\mathbf{x}_n - \mathbf{m})^{\mathrm{T}}. \qquad 3.30$$

Define the sample mean and sample covariance:

$$\hat{\mathbf{m}} = \frac{1}{M} \sum_{n=0}^{M-1} \mathbf{x}_n \;:\; \text{sample mean} \qquad\qquad 3.31$$

$$\mathbf{S}(\hat{\mathbf{m}}) = \frac{1}{M} \sum_{n=0}^{M-1} (\mathbf{x}_n - \hat{\mathbf{m}})(\mathbf{x}_n - \hat{\mathbf{m}})^{\mathrm{T}} \;:\; \text{sample covariance.}$$

Add and subtract terms in $\hat{\mathbf{m}}$ in the exponent of the density function, and manipulate the quadratic form to obtain the formula

$$f_{\boldsymbol{\theta}}(\mathbf{x}) = (2\pi)^{-MN/2} |\mathbf{R}|^{-M/2} \exp\left\{ -\frac{1}{2} \sum_{n=0}^{M-1} (\mathbf{x}_n - \hat{\mathbf{m}} + \hat{\mathbf{m}} - \mathbf{m})^{\mathrm{T}} \mathbf{R}^{-1} (\mathbf{x}_n - \hat{\mathbf{m}} + \hat{\mathbf{m}} - \mathbf{m}) \right\}$$

$$= (2\pi)^{-MN/2} |\mathbf{R}|^{-M/2} \exp\left\{ -\frac{1}{2} \sum_{n=0}^{M-1} (\hat{\mathbf{m}} - \mathbf{m})^{\mathrm{T}} \mathbf{R}^{-1} (\hat{\mathbf{m}} - \mathbf{m}) \right\}$$

$$\times \exp\left\{ -\frac{M}{2} \frac{1}{M} \sum_{n=0}^{M-1} (\mathbf{x}_n - \hat{\mathbf{m}})^{\mathrm{T}} \mathbf{R}^{-1} (\mathbf{x}_n - \hat{\mathbf{m}}) \right\}$$

$$= (2\pi)^{-MN/2} |\mathbf{R}|^{-M/2} \exp\left\{ -\frac{M}{2} \operatorname{tr} \left[\mathbf{R}^{-1}(\hat{\mathbf{m}} - \mathbf{m})(\hat{\mathbf{m}} - \mathbf{m})^{\mathrm{T}} \right] \right\}$$

$$\times \exp\left\{ -\frac{M}{2} \operatorname{tr} \left[\mathbf{R}^{-1} \mathbf{S}(\hat{\mathbf{m}}) \right] \right\}. \qquad\qquad 3.32$$

This makes the statistics $(\hat{\mathbf{m}}, \mathbf{S}(\hat{\mathbf{m}}))$ sufficient for the parameter $\boldsymbol{\theta} = (\mathbf{m}, \mathbf{R})$, meaning that the raw data $\mathbf{x} = (\mathbf{x}_0, \ldots, \mathbf{x}_{M-1})$ may be discarded and only the sample mean and sample covariance retained. ∎

Example 3.5 (Uniform)

Let $\mathbf{X}$ denote a random sample of N uniform random variables X_n, each with density

$$f_{\boldsymbol{\theta}}(x_n) = \frac{1}{\theta_2 - \theta_1} I_{[\theta_1, \theta_2]}(x_n) \qquad\qquad 3.33$$

$$I_{[\theta_1, \theta_2]}(x_n) = \begin{cases} 1, & \theta_1 \leq x_n \leq \theta_2 \\ 0, & \text{otherwise.} \end{cases}$$

The probability density function for the random sample is

$$f_{\boldsymbol{\theta}}(\mathbf{x}) = \frac{1}{(\theta_2 - \theta_1)^N} I_{[\theta_1, \infty)}(\min x_n) I_{(-\infty, \theta_2]}(\max x_n). \qquad\qquad 3.34$$

This makes the minimum and maximum of the sample, namely $(\min X_n, \max X_n)$, sufficient for the parameter $\boldsymbol{\theta} = (\theta_1, \theta_2)$. ∎

Nonsingular Transformations and Sufficiency

Suppose you have a random vector $\mathbf{x}$ whose density function is $f_\theta(\mathbf{x})$. Even with the help of the Neyman-Pearson factorization theorem, you are having trouble spotting a sufficient statistic for θ. Perhaps you can find a nonsingular transformation $\mathbf{y} = \mathbf{W}(\mathbf{x})$ whose density $f_\theta(\mathbf{y})$ factors to produce a sufficient statistic $\mathbf{S}(\mathbf{y})$:

$$f_\theta(\mathbf{y}) = a(\mathbf{y})b_\theta(\mathbf{s}); \qquad \mathbf{s} = \mathbf{S}(\mathbf{y}). \qquad\qquad 3.35$$

[Caution: $f_\theta(\mathbf{y})$ is a different function than $f_\theta(\mathbf{x})$.] Can we steal a sufficient statistic $\mathbf{T}(\mathbf{x})$ from the statistic $\mathbf{S}(\mathbf{y})$? The answer is yes, and the argument proceeds as follows. Write the density for $\mathbf{x}$ as

$$f_\theta(\mathbf{x}) = f_\theta\big[\mathbf{y} = \mathbf{W}(\mathbf{x})\big]\,|\mathbf{J}|, \qquad\qquad 3.36$$

where $|\mathbf{J}|$ is the Jacobian of the transformation $\mathbf{y} = \mathbf{W}(\mathbf{x})$. Substitute the factorization of $f_\theta(\mathbf{y})$ to obtain the following Neyman-Pearson factorization of $f_\theta(\mathbf{x})$:

$$f_\theta(\mathbf{x}) = a\big[\mathbf{W}(\mathbf{x})\big]\,|\mathbf{J}|\,b_\theta(\mathbf{s}); \qquad \mathbf{s} = \mathbf{S}\big[\mathbf{W}(\mathbf{x})\big]. \qquad\qquad 3.37$$

The statistic $\mathbf{S}\big[\mathbf{W}(\mathbf{x})\big]$ is sufficient for θ.

Example 3.6 (Variance Components)

The random vector $\mathbf{X}$ is distributed as $N[\mathbf{0}, \mathbf{H}\Lambda^2\mathbf{H}^\mathsf{T}]$, where $\mathbf{H}$ is a known nonsingular matrix, and $\Lambda^2 = \operatorname{diag}[\lambda_1^2 \quad \lambda_2^2 \quad \cdots \quad \lambda_N^2]$ is a diagonal matrix of unknown "variance components" λ_n^2. The random vector $\mathbf{Y} = \mathbf{H}^{-1}\mathbf{X}$ is a linear transformation of $\mathbf{X}$ that is distributed as $N[\mathbf{0}, \Lambda^2]$. The density function for $\mathbf{y}$ is

$$
\begin{aligned}
f_\theta(\mathbf{y}) &= (2\pi)^{-N/2}\left(\prod_{n=1}^{N}\lambda_n^2\right)^{-1/2}\exp\left\{-\frac{1}{2}\,\mathbf{y}^\mathsf{T}\Lambda^{-2}\mathbf{y}\right\} \\
&= (2\pi)^{-N/2}\left(\prod_{n=1}^{N}\lambda_n^2\right)^{-1/2}\exp\left\{-\frac{1}{2}\sum_{n=1}^{N}\frac{y_n^2}{\lambda_n^2}\right\}.
\end{aligned}
\qquad 3.38
$$

The $\{y_n^2\}_1^N$ are sufficient statistics for the $\{\lambda_n^2\}_1^N$. Each y_n is $\big[\mathbf{H}^{-1}\mathbf{x}\big]_n$. Can you see how to write the vector of sufficient statistics as $\mathbf{s} = \mathbf{S}[\mathbf{W}(\mathbf{x})]$? ∎

3.2 MINIMAL AND COMPLETE SUFFICIENT STATISTICS

Our original motivation for studying sufficient statistics was to find functions of the measurements that were memory-efficient and that were sufficient in the sense that they carried all useful information about the underlying parameter θ. The factorization theorem provides a way to find a sufficient statistic, but how do we know whether it is the most memory-efficient one that can be found?

To illustrate the problem, let's consider just two Bernoulli trials (or two transmissions from the binary source in Example 3.1). The measurement is $\mathbf{x} = (x_1, x_2)$ where $x_i \in \{0, 1\}$. The four possible outcomes are illustrated in Figure 3.1(a). The measurement $\mathbf{x}$ is, itself, sufficient for this problem because the conditional probability mass function for $\mathbf{x}$, given $\mathbf{I(x)} = \mathbf{i}$, is independent of θ:

$$P_\theta(\mathbf{x}|\mathbf{i}) = \begin{cases} 1, & \mathbf{i} = \mathbf{x} \\ 0, & \text{otherwise.} \end{cases} \qquad 3.39$$

But the order statistic $\mathbf{U(x)} = (u_1, u_2) = (\max x_i, \min x_i)$, illustrated in Figure 3.1(*b*), is also sufficient:

$$P_\theta(\mathbf{x}|\mathbf{u}) = \begin{cases} 1, & \mathbf{x} = (0, 0) \text{ and } \mathbf{u} = (0, 0) \\ \frac{1}{2}, & \mathbf{x} = (0, 1) \text{ and } \mathbf{u} = (1, 0) \\ \frac{1}{2}, & \mathbf{x} = (1, 0) \text{ and } \mathbf{u} = (1, 0) \\ 1, & \mathbf{x} = (1, 1) \text{ and } \mathbf{u} = (1, 1). \end{cases} \qquad 3.40$$

We know from our previous study of this problem that the sum, $\mathbf{T(x)} = k = x_1 + x_2$, is sufficient and that the conditional density $P_\theta(\mathbf{x}|k)$ is

$$P_\theta(\mathbf{x}|k) = \begin{cases} \frac{1}{\binom{2}{k}}, & k = x_1 + x_2 \\ 0, & \text{otherwise.} \end{cases} \qquad 3.41$$

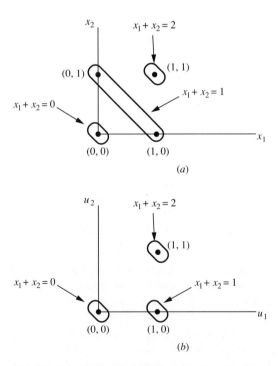

Figure 3.1 Sufficient statistics for Bernoulli trials.

The sufficient statistic $\mathbf{T}(\mathbf{x}) = x_1 + x_2$ is illustrated in Figures 3.1(a) and (b). It subsumes the other two sufficient statistics in the sense that it is a function of both. Any value of $\mathbf{x}$ or $\mathbf{u}$ produces a unique value of $\mathbf{T}(\mathbf{x}) = x_1 + x_2 = u_1 + u_2$, but the converse is not true. We think $\mathbf{T}(\mathbf{x})$ is a "better" sufficient statistic than $\mathbf{U}(\mathbf{x})$, but we would like to make this idea precise.

Minimality

A sufficient statistic $\mathbf{T}(\mathbf{x})$ that is a function of all other sufficient statistics is said to be minimal. A minimal sufficient statistic must have minimal dimensionality (or maximum memory efficiency) because its dimensionality can be no larger than the dimensionalities of the sufficient statistics it subsumes. This is illustrated in Figure 3.2, where the minimal sufficient statistic $\mathbf{t}$ is a function of all other sufficient statistics $\mathbf{t}_n$. That is, $\mathbf{t} = \mathbf{W}_n(\mathbf{t}_n)$.

Example 3.7 (Binary Communication Source)

For Example 3.1, all of the statistics

$$\mathbf{T}_n(\mathbf{x}) = \left(\sum_{i=0}^{n} x_i, x_{n+1}, \ldots, x_{N-1} \right), \qquad n = 0, 1, \ldots, N-1 \qquad 3.42$$

are sufficient. The respective dimensions are $N - n$, with $\mathbf{T}_{N-1}(\mathbf{x}) = \sum_{i=0}^{N-1} x_i$ being the minimal sufficient statistic of dimension 1. It is a function of every other $\mathbf{T}_n(\mathbf{x})$, but no other $\mathbf{T}_n(\mathbf{x})$ is a function of it.   ■

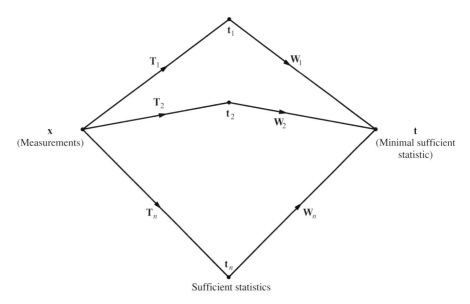

Figure 3.2 Concept of minimality.

Completeness

A sufficient statistic $\mathbf{t} = \mathbf{T}(\mathbf{x})$ is said to be complete (or unique) if the condition

$$E_\theta W(\mathbf{t}) = \mathbf{0} \ \forall \ \boldsymbol{\theta} \qquad\qquad 3.43$$

implies $W(\mathbf{t}) = \mathbf{0}$, with probability 1, for all $\boldsymbol{\theta}$. In this equation E_θ denotes expectation with respect to the distribution $F_\theta(\mathbf{x})$. This is an abstract property of sufficient statistics that will enable us to identify minimal sufficient statistics and to construct minimum variance unbiased estimators. The following example clarifies the notion of a complete sufficient statistic.

Example 3.8 (Binary Communication Source)

Each of the statistics $\mathbf{T}_n(\mathbf{x})$ in Example 3.7 is sufficient. Furthermore, the scalar function

$$W\big[\mathbf{T}_n(\mathbf{x})\big] = \left(\sum_{i=0}^{n} x_i\right) - \frac{n+1}{N-1-n} \sum_{j=n+1}^{N-1} x_j \qquad\qquad 3.44$$

has mean value 0 for all $\boldsymbol{\theta}$. But, for $n = 0, 1, \ldots, N-2$, we cannot say W is 0 with probability 1. Therefore, none of the sufficient statistics $\mathbf{T}_n(\mathbf{x})$, $n = 0, 1, \ldots, N-2$, is complete. For $n = N-1$, the sufficient statistic $\mathbf{T}_{N-1}(\mathbf{x})$ is the statistic $k = x_0 + x_1 + \cdots + x_{N-1}$. This statistic has probability mass function $P_\theta(k) = \binom{N}{k} \theta^k (1-\theta)^{N-k}$. The expected value of any function $W(k)$ is

$$E_\theta W(k) = \sum_{k=0}^{N} \binom{N}{k} W(k)\theta^k (1-\theta)^{N-k}. \qquad\qquad 3.45$$

This is an Nth order polynomial in θ that can equal 0 iff $W(k) = 0$. Therefore the sufficient statistic $k = \mathbf{T}_{N-1}(\mathbf{x})$ is complete. ∎

Unbiasedness

The most remarkable property of a complete sufficient statistic is that there is just one function $\mathbf{W}$ of a complete sufficient statistic $\mathbf{T}(\mathbf{x})$ that produces an unbiased estimate of the parameter $\boldsymbol{\theta}$. To show this, we assume that there are two functions, $\mathbf{W}_1$ and $\mathbf{W}_2$, such that

$$E_\theta \mathbf{W}_1(\mathbf{t}) = E_\theta \mathbf{W}_2(\mathbf{t}) = \boldsymbol{\theta} \ \forall \ \boldsymbol{\theta}. \qquad\qquad 3.46$$

Then

$$E_\theta\big[\mathbf{W}_1(\mathbf{t}) - \mathbf{W}_2(\mathbf{t})\big] = \mathbf{0} \ \forall \ \boldsymbol{\theta}. \qquad\qquad 3.47$$

But the function inside the brackets [ ] is a function of the complete sufficient statistic $\mathbf{T}$. Therefore $\mathbf{W}_1(\mathbf{t}) = \mathbf{W}_2(\mathbf{t})$ with probability 1.

Completeness Ensures Minimality

If you can show that a sufficient statistic $T(x)$ is complete, then you have shown that it is minimal. To prove this result, we assume that $S(x)$ is another sufficient statistic and that $S(x)$ is minimal. From the fundamental property of conditional expectation, we know that $ET = EE[T|S]$. The conditional expectation of T, given S, is a function of S, but because S is minimal it is also a function of T. Therefore $T - E[T|S]$ is a function of T whose mean is zero for all θ. As T is complete, it follows that $T = E[T|S]$. This makes T a function of S and of all other sufficient statistics. Therefore T is minimal.

Summary

Our strategy will be to use the Fisher-Neyman factorization theorem to spot a sufficient statistic and then check to see if the sufficient statistic is complete. If i is, then it is minimal, and we have terminated our search for the most memory-efficient sufficient statistic. Furthermore, there will be only one function of this complete sufficient statistic that is unbiased. This strategy is very easy to implement in the case of the exponential family of distributions, our next topic.

3.3   SUFFICIENCY AND COMPLETENESS IN THE EXPONENTIAL FAMILY

A family of distributions is said to be a k-parameter exponential family if the probability mass function or probability density function for the measurement x may be written as the following exponential form:

$$f_\theta(x) = c(\theta)a(x)\exp\left\{\sum_{i=1}^{k}\pi_i(\theta)t_i(x)\right\} \qquad 3.48$$

$$c^{-1}(\theta) = \int dx\, a(x)\exp\left\{\sum_{i=1}^{k}\pi_i(\theta)t_i(x)\right\}.$$

The parameter $\pi = (\pi_1, \pi_2, \ldots, \pi_k)$ contains k functions of the parameter vector θ, and the statistic $t = (t_1(x), t_2(x), \ldots, t_k(x))$ contains k functions of the measurement vector x.

Sufficiency

Let X denote an N-dimensional random vector whose distribution belongs to the k-parameter exponential family:

$$f_\theta(x) = c(\theta)a(x)\exp\left\{\sum_{i=1}^{k}\pi_i(\theta)t_i(x)\right\}. \qquad 3.49$$

Then the k-dimensional statistic $\mathbf{t}$ is sufficient for the parameter vector $\boldsymbol{\theta}$. This follows directly from the factorization theorem:

$$f_{\boldsymbol{\theta}}(\mathbf{x}) = a(\mathbf{x})b_{\boldsymbol{\theta}}(\mathbf{t}) \qquad\qquad 3.50$$

$$b_{\boldsymbol{\theta}}(\mathbf{t}) = c(\boldsymbol{\theta})\exp\left\{\sum_{i=1}^{k}\pi_i(\boldsymbol{\theta})t_i(\mathbf{x})\right\}.$$

Can you see that the statistic $t(\mathbf{x}) = (t_1(\mathbf{x}), \ldots, t_k(\mathbf{x}))$, with

$$t_i(\mathbf{x}) = \sum_{n=0}^{M-1} t_i(\mathbf{x}_n), \qquad\qquad 3.51$$

is sufficient for $\boldsymbol{\theta}$ when a random sample $\mathbf{X} = (\mathbf{X}_0, \mathbf{X}_1, \ldots, \mathbf{X}_{M-1})$ is drawn from a k-parameter exponential family? This result is important because it says that the dimension of the sufficient statistic is k, independent of the multiplicity of the random sample.

Recall from the comments advanced in the proof of the factorization theorem that the density for the sufficient statistic may be written

$$f_{\boldsymbol{\theta}}(\mathbf{t}) = \left(\int d\mathbf{u}\,a\left[\mathbf{x} = \mathbf{W}^{-1}(\mathbf{y})\right]|\mathbf{J}|\right)b_{\boldsymbol{\theta}}(\mathbf{t}) = \frac{\int d\mathbf{u}\,a\left[\mathbf{x} = \mathbf{W}^{-1}(\mathbf{y})\right]|\mathbf{J}|}{a(\mathbf{x})}f_{\boldsymbol{\theta}}(\mathbf{x}).$$

In the case of the k-parameter exponential family, this produces the following density for the sufficient statistic:

$$f_{\boldsymbol{\theta}}(\mathbf{t}) = c(\boldsymbol{\theta})\left[\int d\mathbf{u}\,a\left[\mathbf{x} = \mathbf{W}^{-1}(\mathbf{y})\right]|\mathbf{J}|\right]\exp\left\{\sum_{i=1}^{k}\pi_i(\boldsymbol{\theta})t_i\right\}$$

$$\qquad\qquad 3.52$$

$$= c(\boldsymbol{\theta})a_0(\mathbf{t})\exp\left\{\sum_{i=1}^{k}\pi_i(\boldsymbol{\theta})t_i\right\}$$

$$a_0(\mathbf{t}) = \int d\mathbf{u}\,a\left[\mathbf{x} = \mathbf{W}^{-1}(\mathbf{y})\right]|\mathbf{J}|.$$

The sufficient statistic itself belongs to a k-parameter exponential family with $\mathbf{t}$ sufficient for $\boldsymbol{\theta}$.

Completeness

In the k-parameter exponential family, the sufficient statistic $\mathbf{t}$ is complete. The proof goes as follows. If $E_{\boldsymbol{\theta}}W(\mathbf{t}) = 0$ for all $\boldsymbol{\theta}$, then we may write this condition as

$$\int d\mathbf{t}\,W(\mathbf{t})a_0(\mathbf{t})\exp\left\{\sum_{i=1}^{k}\pi_i(\boldsymbol{\theta})t_i\right\} = 0\ \forall\ \boldsymbol{\theta}. \qquad\qquad 3.53$$

This is a Laplace transform of a function of the vector variable $\mathbf{t}$. From the unicity of the Laplace transform, $\mathbf{W(t)} = \mathbf{0}$ for almost all $\mathbf{t}$. This means that a sufficient statistic spotted from an exponential family is complete and minimal.

3.4   SUFFICIENCY IN THE LINEAR STATISTICAL MODEL

Let $\mathbf{X}$ denote a normal random vector with mean $\mathbf{m}$ and covariance matrix $\mathbf{R}$. If $\mathbf{m} = \mathbf{H}\boldsymbol{\theta}$, where $\boldsymbol{\theta}$ is $p \times 1$ and $\mathbf{H}$ is $N \times p$ with $N > p$, then we say that the mean obeys a linear model. The distribution of $\mathbf{X}$ is then

$$\mathbf{X} : N(\mathbf{H}\boldsymbol{\theta}, \mathbf{R}). \qquad 3.54$$

Assume that $\mathbf{H}$ and $\mathbf{R}$ are known but that $\boldsymbol{\theta}$ is unknown. The density function for $\mathbf{X}$ may be written as

$$f_{\boldsymbol{\theta}}(\mathbf{x}) = (2\pi)^{-N/2}|\mathbf{R}|^{-1/2}\exp\left\{-\frac{1}{2}(\mathbf{x} - \mathbf{H}\boldsymbol{\theta})^{\mathrm{T}}R^{-1}(\mathbf{x} - \mathbf{H}\boldsymbol{\theta})\right\}$$

$$= (2\pi)^{-N/2}|\mathbf{R}|^{-1/2}\exp\left\{-\frac{1}{2}\boldsymbol{\theta}^{\mathrm{T}}\mathbf{H}^{\mathrm{T}}\mathbf{R}^{-1}\mathbf{H}\boldsymbol{\theta}\right\}\exp\left\{-\frac{1}{2}\mathbf{x}^{\mathrm{T}}\mathbf{R}^{-1}\mathbf{x}\right\} \qquad 3.55$$

$$\times \exp\left\{\boldsymbol{\theta}^{\mathrm{T}}\mathbf{H}^{\mathrm{T}}\mathbf{R}^{-1}\mathbf{x}\right\}.$$

This makes the vector $\mathbf{H}^{\mathrm{T}}\mathbf{R}^{-1}\mathbf{x}$ sufficient for $\boldsymbol{\theta}$. The distribution for this sufficient statistic is normal (see Section 2.10 of Chapter 2 for a discussion of the multivariate normal distribution):

$$\mathbf{H}^{\mathrm{T}}\mathbf{R}^{-1}\mathbf{X} : N(\mathbf{H}^{\mathrm{T}}\mathbf{R}^{-1}\mathbf{H}\boldsymbol{\theta}, \mathbf{H}^{\mathrm{T}}\mathbf{R}^{-1}\mathbf{H}). \qquad 3.56$$

The sufficient statistic $\mathbf{H}^{\mathrm{T}}\mathbf{R}^{-1}\mathbf{X}$ is also complete and minimal. Why?

Recursive Computation of the Sufficient Statistic in the Linear Statistical Model

Suppose that scalar measurements are generated in the following way:

$$x_t = \mathbf{c}_t^{\mathrm{T}}\boldsymbol{\theta} + n_t \qquad (t = 0, 1, \ldots). \qquad 3.57$$

The $p \times 1$ parameter $\boldsymbol{\theta}$ is fixed, but the $p \times 1$ vector $\mathbf{c}_t$ is time varying. A sequence of measurements may be organized into a linear statistical model:

$$\mathbf{x}_t = \mathbf{H}_t\boldsymbol{\theta} + \mathbf{n}_t \qquad 3.58$$

$$\mathbf{H}_t = \begin{bmatrix} \mathbf{H}_{t-1} \\ \mathbf{c}_t^{\mathrm{T}} \end{bmatrix}$$

$$\mathbf{x}_t = [x_0, x_1, \ldots, x_t]^{\mathrm{T}}$$

$$\mathbf{n}_t = [n_0, n_1, \ldots, n_t]^{\mathrm{T}}.$$

Assume that the noise vector $\mathbf{n}_t$ is multivariate normal with mean $\mathbf{0}$ and covariance $\mathbf{R}_t$:

$$\mathbf{n}_t : N\left[\mathbf{0}, \mathbf{R}_t\right]$$

$$\mathbf{R}_t = E\mathbf{n}_t\mathbf{n}_t^T = \begin{bmatrix} \mathbf{R}_{t-1} & \mathbf{r}_t \\ \mathbf{r}_t^T & r_{tt} \end{bmatrix} \qquad\qquad 3.59$$

$$\mathbf{r}_t = E\mathbf{n}_{t-1}n_t$$

$$r_{tt} = E{n_t}^2.$$

The linear statistical model for $\mathbf{x}_t$ says that $\mathbf{x}_t$ is distributed as follows:

$$\mathbf{x}_t : N[\mathbf{H}_t\boldsymbol{\theta}, \mathbf{R}_t] \qquad\qquad 3.60$$

For a fixed value of t, the sufficient statistic for $\boldsymbol{\theta}$ is

$$T(\mathbf{x}_t) = \mathbf{H}_t^T\mathbf{R}_t^{-1}\mathbf{x}_t. \qquad\qquad 3.61$$

The question we ask is, "can the sufficient statistic be computed recursively and, if so, how?" To answer the question, we need to recall a result from Chapter 2 on partitioned matrices and their inverses.

Partitioned Matrix Inverse

The formula for inverting a symmetric partitioned matrix is

$$\mathbf{R}_t^{-1} = \begin{bmatrix} \mathbf{R}_{t-1}^{-1} & \mathbf{0} \\ \mathbf{0}^T & 0 \end{bmatrix} + \gamma_t^{-1}\begin{bmatrix} \mathbf{b}_t \\ 1 \end{bmatrix}[\mathbf{b}_t^T \quad 1]$$

$$\mathbf{b}_t = -\mathbf{R}_{t-1}^{-1}\mathbf{r}_t \qquad\qquad 3.62$$

$$\gamma_t = r_{tt} - \mathbf{r}_t^T\mathbf{R}_{t-1}^{-1}\mathbf{r}_t = r_{tt} + \mathbf{r}_t^T\mathbf{b}_t.$$

Sufficient Statistic

The sufficient statistic may now be written

$$T(\mathbf{x}_t) = \mathbf{H}_t^T\mathbf{R}_t^{-1}\mathbf{x}_t$$

$$= [\mathbf{H}_{t-1}^T \quad \mathbf{c}_t]\left(\begin{bmatrix} \mathbf{R}_{t-1}^{-1} & \mathbf{0} \\ \mathbf{0}^T & 0 \end{bmatrix} + \gamma_t^{-1}\begin{bmatrix} \mathbf{b}_t \\ 1 \end{bmatrix}[\mathbf{b}_t^T \quad 1]\right)\begin{bmatrix} \mathbf{x}_{t-1} \\ x_t \end{bmatrix}$$

$$\qquad\qquad 3.63$$

$$= [\mathbf{H}_{t-1}^T\mathbf{R}_{t-1}^{-1} \quad \mathbf{0}]\begin{bmatrix} \mathbf{x}_{t-1} \\ x_t \end{bmatrix} + \gamma_t^{-1}[\mathbf{H}_{t-1}^T\mathbf{b}_t + \mathbf{c}_t](\mathbf{b}_t^T\mathbf{x}_{t-1} + x_t)$$

$$= T(\mathbf{x}_{t-1}) + \gamma_t^{-1}[\mathbf{c}_t + \mathbf{H}_{t-1}^T\mathbf{b}_t](x_t + \mathbf{b}_t^T\mathbf{x}_{t-1}).$$

This general result says that $\mathbf{T}$ may be recursively updated. However, except in special cases, the update uses the entire past of the observations, $\mathbf{x}_{t-1}$. Let's examine a special case.

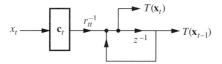

Figure 3.3 Updating the sufficient statistic
in the white noise case.

White Noise

If the additive noises n_t are independent, then the covariance matrix $\mathbf{R}_t$ is diagonal, meaning that $\mathbf{r}_t = \mathbf{0}$ and $\mathbf{b}_t = \mathbf{0}$. Then the sufficient statistic may be updated as follows:

$$\mathbf{T}(\mathbf{x}_t) = \mathbf{T}(\mathbf{x}_{t-1}) + r_{tt}^{-1}\mathbf{c}_t x_t. \qquad 3.64$$

In this simple recursion, the sufficient statistic is updated by using only the new scalar measurement x_t. A structure is illustrated in Figure 3.3. When a measurement is very noisy, meaning that r_{tt} is large, then the update to the sufficient statistic is small.

3.5   SUFFICIENCY IN THE COMPONENTS OF VARIANCE MODEL

Let $\mathbf{x} = [x_1 \ x_2 \ \cdots \ x_N]^T$ denote an $N \times 1$ measurement vector. Assume $\mathbf{x}$ is a linear combination of p linearly independent modes $\mathbf{h}_i$:

$$\mathbf{x} = \sum_{i=1}^{p} \theta_i \mathbf{h}_i = \mathbf{H}\boldsymbol{\theta}. \qquad 3.65$$

The modes are known, but the unknown mode weights θ_i are drawn from the normal distribution $\boldsymbol{\theta} : N[\mathbf{0}, \boldsymbol{\Lambda}^2]$, with $\boldsymbol{\Lambda}^2 = \mathrm{diag}[\lambda_1^2 \ \lambda_2^2 \ \cdots \ \lambda_p^2]$. The λ_n^2 are called variance components. The random vector $\mathbf{x}$ is distributed as $\mathbf{x} : N[\mathbf{0}, \mathbf{H}\boldsymbol{\Lambda}^2\mathbf{H}^T]$. This is the distribution of randomly weighted and summed modes when the mode weights are independent and normal.

Let's construct a full-rank $N \times (N - p)$ matrix $\mathbf{A}$ whose columns are orthogonal to the modes $\mathbf{h}_i$:

$$\mathbf{A}^T\mathbf{H} = \mathbf{0}. \qquad 3.66$$

The linear transformation

$$\mathbf{y} = \begin{bmatrix} (\mathbf{H}^T\mathbf{H})^{-1}\mathbf{H}^T \\ \mathbf{A}^T \end{bmatrix} \mathbf{x} \qquad 3.67$$

produces an $N \times 1$ vector whose distribution is

$$\mathbf{y} : N\left[\mathbf{0}, \ \left[\begin{array}{c|c} \boldsymbol{\Lambda}^2 & \mathbf{0} \\ \hline \mathbf{0} & \mathbf{0} \end{array}\right]\right]. \qquad 3.68$$

The trailing $N - p$ components of $\mathbf{y}$ are identically zero. The first p components of $\mathbf{y}$ have density

$$f_{\boldsymbol{\theta}}(y_1 \cdots y_p) = (2\pi)^{-p/2} \left(\prod_{i=1}^{p} \lambda_i^2 \right)^{-1/2} \exp \left\{ -\frac{1}{2} \sum_{n=1}^{p} \frac{y_n^2}{\lambda_n^2} \right\}. \qquad 3.69$$

From Example 3.6 we know that the $\{y_n^2\}_1^p$ are sufficient for the $\{\lambda_n^2\}_1^p$. Therefore the $([(\mathbf{H}^{\mathrm{T}}\mathbf{H})^{-1}\mathbf{H}^{\mathrm{T}}\mathbf{x}]_n)^2$ are sufficient. Recall from Section 2.7 of Chapter 2 that

$$\hat{\boldsymbol{\theta}} = (\mathbf{H}^{\mathrm{T}}\mathbf{H})^{-1}\mathbf{H}^{\mathrm{T}}\mathbf{x} \qquad 3.70$$

is a least-squares estimate of $\boldsymbol{\theta}$. The squared elements of $\hat{\boldsymbol{\theta}}$, namely the $\hat{\theta}_n^2$, are sufficient for the variance components λ_n^2. From our discussion of completeness in the exponential family, we know that the sufficient statistics $\hat{\theta}_n^2$ are also complete and minimal.

3.6 MINIMUM VARIANCE UNBIASED (MVUB) ESTIMATORS

Suppose the measurement $\mathbf{x}$ is distributed as $f_{\boldsymbol{\theta}}(\mathbf{x})$ and $\boldsymbol{\theta}$ is an unknown parameter to be estimated. Call $\hat{\boldsymbol{\theta}}(\mathbf{x})$ an estimate of $\boldsymbol{\theta}$ and denote it by $\hat{\boldsymbol{\theta}}$. The estimator $\hat{\boldsymbol{\theta}}$ is said to be *unbiased* if

$$E\hat{\boldsymbol{\theta}} = \boldsymbol{\theta}. \qquad 3.71$$

The estimator $\hat{\boldsymbol{\theta}}$ is minimum variance unbiased if the variance

$$\sigma^2 = E[\hat{\boldsymbol{\theta}} - \boldsymbol{\theta}]^{\mathrm{T}}[\hat{\boldsymbol{\theta}} - \boldsymbol{\theta}] \qquad 3.72$$

cannot be decreased by any other *unbiased* estimator of $\boldsymbol{\theta}$. The Rao-Blackwell theorem, correctly interpreted, tells us that in our search for MVUB estimators we may restrict attention to functions of a sufficient statistic. That is, if the statistic $\hat{\boldsymbol{\theta}}$ is an arbitrary unbiased estimator of $\boldsymbol{\theta}$ that has variance σ^2,

$$E\hat{\boldsymbol{\theta}} = \boldsymbol{\theta} \qquad \text{and} \qquad E[\hat{\boldsymbol{\theta}} - \boldsymbol{\theta}]^{\mathrm{T}}[\hat{\boldsymbol{\theta}} - \boldsymbol{\theta}] = \sigma^2, \qquad 3.73$$

then this estimator may be improved upon with an estimator that is a function of a sufficient statistic for the parameter.

Theorem (Rao-Blackwell) Let $\mathbf{Y}$ and $\mathbf{Z}$ denote random vectors such that $\mathbf{Y}$ has mean $\boldsymbol{\theta}$ and variance $E(\mathbf{Y} - \boldsymbol{\theta})^{\mathrm{T}}(\mathbf{Y} - \boldsymbol{\theta}) = \operatorname{var} \mathbf{Y}$. Define the function $\mathbf{g}(\mathbf{z})$ to be the following conditional expectation:

$$\mathbf{g}(\mathbf{z}) = E[\mathbf{Y}|\mathbf{Z} = \mathbf{z}]. \qquad 3.74$$

The statistic $\mathbf{g}(\mathbf{Z})$ improves on $\mathbf{Y}$ in the following way:

1. $E\mathbf{g}(\mathbf{Z}) = \boldsymbol{\theta}$
2. $\operatorname{var} \mathbf{g}(\mathbf{Z}) \le \operatorname{var} \mathbf{Y}$. \qquad 3.75

The variables $\mathbf{Y}$, $\mathbf{Z}$, and $\mathbf{g}(\mathbf{z})$ are illustrated in Figure 3.4(*a*).

Proof The proof of (1) follows from the definition of conditional expectation:

$$E\mathbf{g}(\mathbf{Z}) = EE[\mathbf{Y}|\mathbf{Z}] = E\mathbf{Y} = \boldsymbol{0}. \qquad 3.76$$

Written out explicitly in terms of densities, the result is

$$E\mathbf{g}(\mathbf{Z}) = \int d\mathbf{z}\, f(\mathbf{z})\mathbf{g}(\mathbf{z}) = \int d\mathbf{z}\, f(\mathbf{z}) \int d\mathbf{y}\, f(\mathbf{y}|\mathbf{z})\mathbf{y}$$
$$= \int d\mathbf{y}\, f(\mathbf{y})\mathbf{y} = E\mathbf{Y} = \boldsymbol{0}. \qquad 3.77$$

The proof of (2) goes like this:

$$\text{var } \mathbf{Y} = E[\mathbf{Y} - \boldsymbol{\theta}]^T[\mathbf{Y} - \boldsymbol{\theta}] = E\big[\mathbf{Y} - \mathbf{g}(\mathbf{Z}) + \mathbf{g}(\mathbf{Z}) - \boldsymbol{\theta}\big]^T\big[\mathbf{Y} - \mathbf{g}(\mathbf{Z}) + \mathbf{g}(\mathbf{Z}) - \boldsymbol{\theta}\big]$$
$$= E\big[\mathbf{Y} - \mathbf{g}(\mathbf{Z})\big]^T\big[\mathbf{Y} - \mathbf{g}(\mathbf{Z})\big] + \text{var } \mathbf{g}(\mathbf{Z}) + 2E\big[\mathbf{Y} - \mathbf{g}(\mathbf{Z})\big]^T\big[\mathbf{g}(\mathbf{Z}) - \boldsymbol{\theta}\big]$$
$$= \gamma^2 + \text{var } \mathbf{g}(\mathbf{Z}) + 2E\big[\mathbf{Y} - g(\mathbf{Z})\big]^T\big[g(\mathbf{Z}) - \boldsymbol{\theta}\big]. \qquad 3.78$$

The expectation in this equation may be written as

$$E\big[\mathbf{Y} - \mathbf{g}(\mathbf{Z})\big]^T\big[\mathbf{g}(\mathbf{Z}) - \boldsymbol{\theta}\big] = EE\{\big[\mathbf{Y} - \mathbf{g}(\mathbf{Z})\big]^T\big[\mathbf{g}(\mathbf{Z}) - \boldsymbol{\theta}\big]|\mathbf{Z}\} = 0. \qquad 3.79$$

Therefore var $\mathbf{Y} = \text{var } \mathbf{g}(\mathbf{Z}) + \gamma^2.$ □

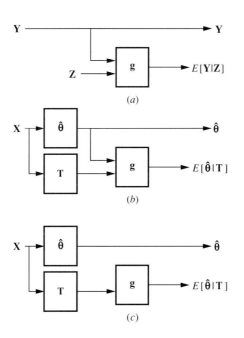

Figure 3.4 The Rao-Blackwell theorem illustrated.

Interpretation

Here is what this theorem tells us about sufficient statistics. Let $\mathbf{Y} = \hat{\boldsymbol{\theta}}$ be an unbiased estimator of $\boldsymbol{\theta}$. Let $\mathbf{Z} = \mathbf{T}(\mathbf{X})$ be a sufficient statistic for $\boldsymbol{\theta}$. Then the statistic $E[\hat{\boldsymbol{\theta}}|\mathbf{T}(\mathbf{X})]$ is a function of the sufficient statistic that is unbiased and has smaller variance than $\hat{\boldsymbol{\theta}}$. Refer to Figure 3.4($b$) for an illustration. Finally, as illustrated in Figure 3.4(c), $E[\hat{\boldsymbol{\theta}}|\mathbf{T}(\mathbf{X})]$ is really just a computed function of $\mathbf{T}(\mathbf{X})$, the sufficient statistic.

This is a startling result. If any unbiased estimator may be improved upon with an estimator that is a function of a sufficient statistic, why not restrict attention to only sufficient statistics and functions of them? More than this, why not restrict attention to a function of a complete (and minimal) sufficient statistic? Then there is one and only one function of the complete (and minimal) sufficient statistic that is unbiased. We are stuck with the variance it brings us, and it is misleading to talk about a minimum variance unbiased estimator. So what about the Rao-Blackwell theorem? It has told us where to look for good unbiased estimators—among functions of a (minimal) complete sufficient statistic. When there is only one such function to be found, it is the minimum variance unbiased estimator in a class of just one such estimator. As Ferguson puts it, "... being best in a class of one is no optimum property at all." On the other hand, the estimator will have smaller variance than any unbiased estimator that is not a function of the complete sufficient statistic, and this is why we value it.

Example 3.9 (The Normal Example)
Let $\mathbf{X}$ denote a random sample of N-dimensional random variables $\mathbf{X}_n : N[\mathbf{m}, \mathbf{R}]$. The following estimators are unbiased estimators of $\mathbf{m}$:

1. $\hat{\mathbf{m}}_1 = \mathbf{X}_n$ (some n)

2. $\hat{\mathbf{m}}_2 = \text{median } \mathbf{X}_n$ (vector of medians)   3.80

3. $\hat{\mathbf{m}}_3 = \dfrac{1}{M} \sum\limits_{n=0}^{M-1} \mathbf{X}_n$.

Their respective variances are

1. $\text{tr } \mathbf{R}$

2. ?   3.81

3. $\dfrac{1}{M} \text{tr } \mathbf{R}$.

The sample mean $\hat{\mathbf{m}}_3$ is the unique unbiased function of the complete sufficient statistic, and therefore it is the unique MVUB estimator. ■

Example 3.10 (The Linear Statistical Model)
Recall the following sufficient statistic in the linear statistical model:

$$\mathbf{T}(\mathbf{X}) = \mathbf{H}^\mathsf{T}\mathbf{R}^{-1}\mathbf{X}.$$ 3.82

It is normally distributed and therefore complete. The statistic

$$\hat{\boldsymbol{\theta}} = (\mathbf{H}^T \mathbf{R}^{-1} \mathbf{H})^{-1} \mathbf{H}^T \mathbf{R}^{-1} \mathbf{X} \qquad 3.83$$

is an unbiased function of the complete sufficient statistic. Therefore, it is the minimum variance unbiased estimator of $\boldsymbol{\theta}$. ∎

Example 3.11 (Variance Components)

Recall the following sufficient statistic $\{\hat{\theta}_n^2\}_1^p$ for the parameter $\{\lambda_n^2\}_1^p$ in the components of variance model. The vector

$$\hat{\boldsymbol{\theta}} = \begin{bmatrix} \hat{\theta}_1 \\ \vdots \\ \hat{\theta}_p \end{bmatrix} = (\mathbf{H}^T \mathbf{H})^{-1} \mathbf{H}^T \mathbf{x} \qquad 3.84$$

is distributed as $N[\mathbf{0}, \boldsymbol{\Lambda}^2]$, so $E\hat{\theta}_n^2 = \lambda_n^2$. This means that $\hat{\theta}_n^2$ is an unbiased function of a complete sufficient statistic, making it a minimum variance unbiased estimator of λ_n^2. ∎

REFERENCES AND COMMENTS

For discussions of sufficiency, I recommend these books by Ferguson and Zachs. I have drawn liberally from both of these gentlemen and added several examples and insights of my own.

Ferguson, T. S. [1967]. *Mathematical Statistics: A Decision Theoretic Approach* (New York: Academic Press, 1967).

Zacks, S. [1971]. *The Theory of Statistical Inference* (New York: John Wiley & Sons, 1971).

PROBLEMS

3.1 For Bernoulli trials, the statistic

$$k = \sum_{n=0}^{N-1} x_n, \qquad x_n \in (0, 1)$$

is sufficient. Compare the number of bits required to code k with the number of bits required to code the sequence $\mathbf{x} = (x_0, x_1, \ldots, x_{N-1})$.

3.2 Let $\mathbf{X} = (X_1, X_2, \ldots, X_N)$ denote a random sample of $U[0, \theta]$ random variables:

$$f_\theta(x_n) = \frac{1}{\theta} I_{[0,\theta]}(x_n).$$

a.   Show that $t = \max(x_n)$ is sufficient for θ.

b.   Find $f_\theta(t)$.

c.   Find $f_\theta(\mathbf{x}|t)$ and show that it is independent of θ.

d.   Write $f_\theta(\mathbf{x})$ as $f(\mathbf{x}|t)f_\theta(t)$.

e.   Is t complete? Unbiased?

f.   Find the minimum variance unbiased estimator of θ.

3.3 Return to the normal example of Example 3.4. Show that

a.   for $\mathbf{R}$ known, $\hat{\mathbf{m}}$ is sufficient for $\mathbf{m}$;

b.   for $\mathbf{m}$ known, $\mathbf{S}(\mathbf{m})$ is sufficient for $\mathbf{R}$.

Explain your findings in words.

3.4 Let $\mathbf{x} = (x_1, x_2)$ be independent Bernoulli random variables with $P(x_n = 1) = \theta$ and $P(x_n = 0) = (1 - \theta)$. Define the order statistic $\mathbf{u} = (u_1, u_2) = (\max(x_1, x_2), \min(x_1, x_2))$. Find $P_\theta(\mathbf{x}|\mathbf{u})$ and show that it is independent of θ.

3.5 In Example 3.7, show that $f(\mathbf{x}|\mathbf{t}_n)$ is independent of $\boldsymbol{\theta}$ for each $\mathbf{T}_n(\mathbf{x})$.

3.6 Let $\mathbf{X}$ denote an N-dimensional random vector with mean $\theta \mathbf{1}$ and covariance matrix $\mathbf{I}$:

$$\mathbf{X} : N[\theta \mathbf{1}, \mathbf{I}]$$

$$\mathbf{1}^{\mathrm{T}} = (1 \quad 1 \quad \cdots \quad 1) \qquad \mathbf{I} = \mathrm{diag}[1 \quad 1 \quad \cdots \quad 1].$$

Define the following transformation on $\mathbf{X}$:

$$\mathbf{Y} = \mathbf{AX}$$

$$A = \left[\begin{array}{c|c} \begin{matrix} 1 \\ \hline -1 \\ \vdots \\ -1 \end{matrix} & \begin{matrix} 1 \cdots 1 \\ \hline \\ \mathbf{I} \\ \\ \end{matrix} \end{array}\right].$$

a.   Show that $\mathbf{T}(\mathbf{x})$, the first element of $\mathbf{Y}$, is distributed as $N[N\theta, N]$.

b.   Show that $\mathbf{Z}$, the last $(N - 1)$ elements of $\mathbf{Y}$, is distributed as $N[\mathbf{0}, \mathbf{Q}]$; find $\mathbf{Q}$.

c.   Show that $\mathbf{Y}$ is $N[[N\theta \quad \mathbf{0}^{\mathrm{T}}]^{\mathrm{T}}, \mathbf{R}]$; find $\mathbf{R}$ and show that $\mathbf{R}$ and $\mathbf{R}^{-1}$ may be written

$$\mathbf{R} = \left[\begin{array}{c|c} \begin{matrix} N \\ 0 \\ \vdots \\ 0 \end{matrix} & \begin{matrix} 0 \cdots 0 \\ \\ \mathbf{Q} \\ \\ \end{matrix} \end{array}\right] \qquad \mathbf{R}^{-1} = \left[\begin{array}{c|c} \begin{matrix} N^{-1} \\ 0 \\ \vdots \\ 0 \end{matrix} & \begin{matrix} 0 \cdots 0 \\ \\ \mathbf{Q}^{-1} \\ \\ \end{matrix} \end{array}\right].$$

    d.   Show that the conditional distribution of $\mathbf{Z}$, given $\mathbf{T}$, is $N[\mathbf{0}, \mathbf{Q}]$, independent of θ.

    e.   Show that the conditional distribution of $\mathbf{X}$, given $\mathbf{T}$, is independent of θ.

    f.   What can you say about the statistic $\mathbf{T}$?

3.7 Let $\mathbf{X} = (\mathbf{X}_0, \ldots, \mathbf{X}_{M-1})$ denote a random sample of random variables $\mathbf{X}_n :$ $N[\mathbf{H}\theta, \mathbf{R}]$. Find the sufficient statistic for θ and the minimum variance unbiased estimator of θ. What is its variance?

3.8 In the recursive computation of the sufficient statistic in the linear statistical model, show how $\mathbf{T}(\mathbf{x}_0)$ is initialized. See Equation 3.63.

3.9 Consider the recursion for updating the sufficient statistic in the linear statistical model:

$$\mathbf{T}(\mathbf{x}_t) = \mathbf{T}(\mathbf{x}_{t-1}) + \gamma_t^{-1}[\mathbf{c}_t + \mathbf{H}_{t-1}^{\mathrm{T}}\mathbf{b}_t](x_t + \mathbf{b}_t^{\mathrm{T}}\mathbf{x}_{t-1})$$

$$\gamma_t = r_{tt} + \mathbf{r}_t^{\mathrm{T}}\mathbf{b}_t$$

$$\mathbf{b}_t = -\mathbf{R}_{t-1}^{-1}\mathbf{r}_t.$$

Suppose $\mathbf{b}_t^{\mathrm{T}} = [0 \cdots 0\, \mathbf{b}_p^{\mathrm{T}}]$ for all $t \geq p$.

    a.   Show that the sufficient statistic may be updated as follows for $t \geq p$:

$$\mathbf{T}(\mathbf{x}_t) = \mathbf{T}(\mathbf{x}_{t-1}) + \gamma_p\left[\sum_{n=0}^{p} b_n \mathbf{c}_{t-n}\right]\left(\sum_{n=0}^{p} b_n x_{t-n}\right)$$

$$\gamma_p = r_0 + \mathbf{r}_p^{\mathrm{T}}\mathbf{b}_p$$

$$b_0 = 1.$$

    b.   Draw a block diagram that illustrates how this solution would be implemented in hardware.

    c.   How would you initialize $\mathbf{T}(\mathbf{x}_{p-1})$?

3.10 In an identification experiment, a known signal sequence $(s_0, s_1, \ldots, s_t)$ excites a channel with unknown impulse response $(h_0, h_1, \ldots, h_p)$. The measurement is noisy:

$$x_t = \sum_{i=0}^{p} h_i s_{t-i} + n_t$$

$$= \mathbf{c}_t^{\mathrm{T}}\theta + n_t$$

$$\mathbf{c}_t = [s_t \quad \cdots \quad s_{t-p}]^{\mathrm{T}}$$

$$\theta = [h_0 \quad \cdots \quad h_p]^{\mathrm{T}}.$$

Find a recursion for the sufficient statistic for $\boldsymbol{\theta}$ when the noises n_t are i.i.d. $N[0, \sigma^2]$.

3.11 In a deconvolution experiment, an unknown sequence $(s_0, s_1, \ldots, s_p)$ excites a channel or filter with known impulse response $(h_0, h_1, \ldots, h_t)$. The measurements are noisy:

$$x_t = \sum_{i=0}^{p} s_i h_{t-i} + n_t$$

Set up the linear statistical model for this problem. Show how to compute the sufficient statistic for $\boldsymbol{\theta}^{\mathrm{T}} = (s_0 \quad \cdots \quad s_p)$ and how to update it recursively when the noises n_t are i.i.d. $N[0, \sigma^2]$.

3.12 Prove that $\mathbf{T}(\mathbf{X}) = \sum_{n=0}^{M-1} \mathbf{T}(\mathbf{X}_n)$ is sufficient for the parameter vector $\boldsymbol{\theta}$ when $\mathbf{X} = (\mathbf{X}_0, \mathbf{X}_1, \ldots, \mathbf{X}_{M-1})$ is a random sample, and $f_{\boldsymbol{\theta}}(\mathbf{x}_n) = c(\boldsymbol{\theta})a(\mathbf{x}_n)\exp\{\sum_{i=1}^{k} \pi_i(\boldsymbol{\theta})t_i(\mathbf{x}_n)\}$. Interpret your finding.

3.13 Let $\mathbf{X} = (X_0, X_1, \ldots, X_{M-1})$ denote a random sample of scalar i.i.d. random variables. For each of the following cases, find a sufficient statistic for $\boldsymbol{\theta}$ and find the distribution of the sufficient statistic.

a.　$X_n : N[\theta_1, 1]$

b.　$X_n : N[0, \theta_2], \quad \theta_2 > 0$

c.　$X_n : N[\theta_1, \theta_2], \quad \theta_2 > 0$

d.　$P[X_n = x] = \binom{N+x-1}{x}(1-p)^N p^x; \quad x = 0, 1, \ldots$

e.　$P[X_n = x] = e^{-\lambda}\frac{\lambda^x}{x!}, \quad x = 0, 1, 2, \ldots$　　(Poisson)

f.　$P[X_n \le x] = \int_0^x [1/\Gamma(\theta_2)\theta_1^{\theta_2}]e^{-y/\theta_1}y^{(\theta_2-1)}\, dy; \quad 0 \le x < \infty$

g.　$P[X_n \le x] = \frac{\Gamma(\theta_1+\theta_2)}{\Gamma(\theta_1)\Gamma(\theta_2)}\int_0^x y^{\theta_1-1}(1-y)^{\theta_2-1}\, dy; \quad 0 \le x \le 1$

3.14 For each of the cases in Problem 3.13, find the minimum variance unbiased estimate of θ and compute the variance of the estimate.

3.15 Let $\mathbf{X} = (X_0, X_1, \ldots, X_{N-1})$ denote a random sample of i.i.d. random variables X_n that are distributed as

$$P[X_n = x] = C\theta_1^{x-\theta_0} \quad (x = \theta_0, \theta_0 + 1, \ldots); \quad 0 < \theta_1 < 1.$$

a.　Choose C to make $P[\ ]$ a probability mass function.

b.　Find the distribution for the random sample.

c.　Find the sufficient statistic for θ_1, assuming θ_0 known.

d.　Find the sufficient statistic for θ_0, assuming θ_1 known.

e.　What is the distribution of $\sum_{n=0}^{N-1} X_n$?

3.16 Show that the sample covariance of Example 3.4 may be written

$$\mathbf{S}(\hat{\mathbf{m}}) = \frac{1}{M} \sum_{n=0}^{M-1} \mathbf{x}_n \mathbf{x}_n^\mathrm{T} - \hat{\mathbf{m}}\hat{\mathbf{m}}^\mathrm{T}$$

$$\hat{\mathbf{m}} = \frac{1}{M} \sum_{n=0}^{M-1} \mathbf{x}_n .$$

3.17 In Example 3.4, show that the sample mean

$$\hat{\mathbf{m}}_t = \frac{1}{t+1} \sum_{n=0}^{t} \mathbf{x}_n \mathbf{x}_n^\mathrm{T}$$

and the sample covariance

$$S_t(\hat{\mathbf{m}}_t) = \frac{1}{t+1} \sum_{n=0}^{t} (\mathbf{x}_n - \hat{\mathbf{m}}_t)(\mathbf{x}_n - \hat{\mathbf{m}}_t)^\mathrm{T}$$

may be written recursively as follows:

$$\hat{\mathbf{m}}_t = \frac{t}{t+1} \hat{\mathbf{m}}_{t-1} + \frac{1}{t+1} \mathbf{x}_t; \qquad \hat{\mathbf{m}}_0 = \mathbf{x}_0$$

$$Q_t = \frac{t}{t+1} Q_{t-1} + \frac{1}{t+1} \mathbf{x}_t \mathbf{x}_t^\mathrm{T}; \qquad Q_0 = \mathbf{x}_0 \mathbf{x}_0^\mathrm{T}$$

$$S_t(\hat{\mathbf{m}}_t) = Q_t - \hat{\mathbf{m}}_t \hat{\mathbf{m}}_t^\mathrm{T} .$$

3.18 Consider the sample mean and sample variance in a random sample of scalar random variables $X_n : N[m, \sigma^2]$

$$\hat{m} = \frac{1}{M} \sum_{n=0}^{M-1} x_n$$

$$\hat{\sigma}^2 = \frac{1}{M} \sum_{n=0}^{M-1} (x_n - \hat{m})^2 .$$

Find the distribution of $\hat{m}$ and of $\hat{\sigma}^2$. Is $\hat{m}$ unbiased? Is $\hat{\sigma}^2$ unbiased? Find minimum variance unbiased estimates of m and σ^2.

3.19 (*Just a Curiosity.*) Prove

$$\sum_{\mathbf{x}: \sum_{n=0}^{N-1} x_n = k} \left(\frac{1}{\prod_{n=0}^{N-1} x_n!} \right) = \frac{N^k}{k!} .$$

This produces a very complicated way of writing 1:

$$\frac{1}{N} \ln \sum_{k=0}^{\infty} \sum_{\mathbf{x}: \sum_{n=0}^{N-1} x_n = k} \left(\frac{1}{\prod_{n=0}^{N-1} x_n!} \right) = 1.$$

3.20 Extend the Rao-Blackwell theorem by showing that

$$\mathbf{R} = \mathbf{Q} + \mathbf{P}$$

where $\mathbf{R} = E[\mathbf{Y} - \boldsymbol{\theta}][\mathbf{Y} - \boldsymbol{\theta}]^{\mathrm{T}}$ and $\mathbf{Q} = E[\mathbf{g}(\mathbf{Z}) - \boldsymbol{\theta}][\mathbf{g}(\mathbf{Z}) - \boldsymbol{\theta}]^{\mathrm{T}}$ are the covariance matrices for $\mathbf{Y}$ and $\mathbf{g}(\mathbf{Z})$, and $\mathbf{P}$ is the nonnegative definite matrix $\mathbf{P} = E[\mathbf{Y} - \mathbf{g}(\mathbf{Z})][\mathbf{Y} - \mathbf{g}(\mathbf{Z})]^{\mathrm{T}}$. Use this result to show

$$E[\mathbf{a}^{\mathrm{T}}(\hat{\boldsymbol{\theta}}_1 - \boldsymbol{\theta})]^2 \geq E[\mathbf{a}^{\mathrm{T}}(\hat{\boldsymbol{\theta}}_2 - \boldsymbol{\theta})]^2$$

for $\hat{\boldsymbol{\theta}}_1$ an unbiased estimator of $\boldsymbol{\theta}$ and $\hat{\boldsymbol{\theta}}_2 = E[\hat{\boldsymbol{\theta}}_1 | \mathbf{T}(\mathbf{X})]$ a "Rao-Blackwellized" version of $\hat{\boldsymbol{\theta}}_1$. Interpret this result.

3.21 Describe and contrast experiments that, respectively, produce the measurement models $\mathbf{x} : N[\mathbf{H}\boldsymbol{\theta}, \mathbf{R}]$ and $\mathbf{x} : N[\mathbf{0}, \mathbf{H}\boldsymbol{\Lambda}^2\mathbf{H}^{\mathrm{T}}]$.

Neyman-Pearson Detectors

To motivate interest in a formal treatment of hypothesis testing or detection theory, the following problem is offered. You, as an experimentalist, have two candidate models for a physical system. System S_0 generates measurements $\mathbf{X}$ that are distributed as $N[\mathbf{m}_0, \mathbf{R}_0]$, and system S_1 generates measurements that are distributed as $N[\mathbf{m}_1, \mathbf{R}_1]$. If system S_0 is the appropriate model, we say that hypothesis H_0 is in force. If S_1 is the appropriate model, we say that alternative H_1 is in force. These we denote as follows:

$$H_0 : \mathbf{X} : N[\mathbf{m}_0, \mathbf{R}_0] \qquad\qquad 4.1$$

$$H_1 : \mathbf{X} : N[\mathbf{m}_1, \mathbf{R}_1].$$

103

The problem is to test the hypothesis H_0 against the alternative H_1 and to select the one that is most consistent with our prior model and the observation $\mathbf{X} = \mathbf{x}$. We would then like to know the probability that our selection is the correct one.

The normal density $N[\mathbf{m}_i, \mathbf{R}_i]$ distributes probability mass according to the density (see Section 2.10 of Chapter 2).

$$f_{\boldsymbol{\theta}_i}(\mathbf{x}) = (2\pi)^{-N/2} |\mathbf{R}_i|^{-1/2} \exp\left\{ -\frac{1}{2}(\mathbf{x} - \mathbf{m}_i)^{\mathrm{T}} \mathbf{R}_i^{-1}(\mathbf{x} - \mathbf{m}_i) \right\} \qquad 4.2$$

$$\boldsymbol{\theta}_i = (\mathbf{m}_i, \mathbf{R}_i).$$

The probability of observing $\mathbf{x}$ is $f_{\boldsymbol{\theta}_i}(\mathbf{x})\, d\mathbf{x}$. If $f_{\boldsymbol{\theta}_1}(\mathbf{x}) > f_{\boldsymbol{\theta}_0}(\mathbf{x})$ at the observation $\mathbf{x}$, then we expect that H_1 is the alternative in force. Why not look at the likelihood ratio

$$l(\mathbf{x}) = \frac{f_{\boldsymbol{\theta}_1}(\mathbf{x})}{f_{\boldsymbol{\theta}_0}(\mathbf{x})} \qquad 4.3$$

and select H_1 when this is greater than 1 or H_0 when it is less than 1? As we shall see, this idea, suitably refined, is not a bad one.

4.1   CLASSIFYING TESTS

Let $\mathbf{X}$ denote a random vector with distribution $F_{\boldsymbol{\theta}}(\mathbf{x})$. The parameter $\boldsymbol{\theta}$ belongs to the parameter space Θ. Let $\Theta = \Theta_0 \cup \Theta_1$ be a disjoint covering of the parameter space, and let H_i denote the hypothesis that $\boldsymbol{\theta} \in \Theta_i$. The test

$$H_0 : \boldsymbol{\theta} \in \Theta_0 \text{ versus } H_1 : \boldsymbol{\theta} \in \Theta_1$$

is said to be a binary test of hypotheses. We are testing to see which of two disjoint subsets contains the unknown parameter $\boldsymbol{\theta}$. If $\Theta = \Theta_0 \cup \Theta_1 \cup \cdots \cup \Theta_{M-1}$ is a disjoint covering of the parameter space, we can construct the hypothesis $H_i : \boldsymbol{\theta} \in \Theta_i$ and test

$$H_0 \text{ versus } H_1 \text{ versus } \cdots H_{M-1}.$$

This is called a multiple, or M-ary, hypothesis test.

If Θ_i contains a single element $\boldsymbol{\theta}_i$, the hypothesis H_i is said to be simple. Otherwise it is composite. If H_0 is simple and H_1 is composite, and we are testing H_0 versus H_1, we say we are testing a simple hypothesis versus a composite alternative. A typical example is

$$H_0 : \boldsymbol{\theta} = \mathbf{0} \text{ versus } H_1 : \boldsymbol{\theta} \neq \mathbf{0}.$$

This binary hypothesis test is said to be two-sided because the alternative H_1 "lies on both sides of H_0." When $\boldsymbol{\theta}$ is scalar, the binary test

$$H_0 : \boldsymbol{\theta} < \boldsymbol{\theta}_0 \text{ versus } H_1 : \boldsymbol{\theta} > \boldsymbol{\theta}_0$$

is said to be a one-sided test of a composite hypothesis versus a composite alternative.

Example 4.1 (*M*-ary Communication)

Suppose binary strings of length b are mapped into the $M = 2^b$ symbols $\boldsymbol{\theta}_n$, $n = 0, 1, \ldots, M - 1$. These symbols are used to modulate a signal according to the rule $\mathbf{m}(\boldsymbol{\theta}_n)$. This signal is communicated over a channel to a receiver that observes $\mathbf{x}$. Assume that the density $f_{\boldsymbol{\theta}_m}(\mathbf{x})$ is known for each $\boldsymbol{\theta}_m$. The problem of decoding symbols (and, ultimately, information) is one of testing $H_0 : \boldsymbol{\theta}_0$ versus $H_1 : \boldsymbol{\theta}_1$ versus $\cdots H_{M-1} : \boldsymbol{\theta}_{M-1}$. This is a multiple test of simple hypotheses. ■

4.2   THE TESTING OF BINARY HYPOTHESES

Let $\mathbf{X}$ denote a random vector with distribution $F_{\boldsymbol{\theta}}(\mathbf{x})$, $\boldsymbol{\theta} \in \Theta$. A binary test of $H_0 : \boldsymbol{\theta} \in \Theta_0$ versus $H_1 : \boldsymbol{\theta} \in \Theta_1$ takes the form

$$\phi(\mathbf{x}) = \begin{cases} 1 \sim H_1, & \mathbf{x} \in R \\ 0 \sim H_0, & \mathbf{x} \in A. \end{cases} \qquad 4.4$$

This equation is read as, "the test function $\phi(\mathbf{x})$ equals 1, and hypothesis H_1 is accepted (H_0 is rejected), if the measurement $\mathbf{x}$ lies in the rejection region R. If the measurement lies in acceptance region A, then the test function equals zero and hypothesis H_0 is accepted." The trick is to find the regions R and A. The test function $\phi(\mathbf{x})$ represents the ultimate in data compression—it compresses the measurement $\mathbf{x}$ into a scalar that assumes the value 0 or 1.

Size or Probability of False Alarm

If the parameter $\boldsymbol{\theta} \in \Theta_0$ but $\mathbf{x} \in R$, then H_0 is rejected when in fact it is in force. This is called a *type I error* or a *false alarm*. If H_0 is simple—that is, $\Theta_0 = \{\boldsymbol{\theta}_0\}$—then the size or probability of false alarm is

$$\alpha = P_{\boldsymbol{\theta}_0}\big[\phi(\mathbf{X}) = 1\big]$$
$$= E_{\boldsymbol{\theta}_0}\,\phi(\mathbf{X}) = P_{\text{FA}}. \qquad 4.5$$

The notation $E_{\boldsymbol{\theta}_0}\,\phi(\mathbf{X})$ indicates that $\phi(\mathbf{X})$ is averaged under the density function $f_{\boldsymbol{\theta}_0}(\mathbf{x})$. If H_0 is composite, then the size is defined to be the following supremum:

$$\alpha = \sup_{\boldsymbol{\theta} \in \Theta_0} E_{\boldsymbol{\theta}}\,\phi(\mathbf{X}). \qquad 4.6$$

The size measures the worst-case probability of accepting H_1 when, in fact, H_0 is in force. We use the notations α and P_{FA} interchangeably.

Power or Detection Probability

If $\boldsymbol{\theta} \in \Theta_1$ and $\mathbf{x} \in R$, then H_1 is accepted when, in fact, H_1 is in force. If H_1 is simple, then the probability of correctly accepting H_1 is the power or detection probability

$$\beta = P_{\boldsymbol{\theta}_1}[\phi(\mathbf{X}) = 1]$$
$$= E_{\boldsymbol{\theta}_1} \phi(\mathbf{X}) = P_D. \qquad 4.7$$

It is desirable to have powerful tests of small size. We use the notations β and P_D interchangeably. One minus the power is the probability of accepting H_0 when, in fact, H_1 is in force. This is also called the probability of a *type II error* or the miss probability:

$$P_M = 1 - P_D. \qquad 4.8$$

If H_1 is composite, then the power is defined for each $\boldsymbol{\theta} \in \Theta_1$ and given the notation $\beta(\boldsymbol{\theta})$. If $\boldsymbol{\theta}$ is scalar and Θ_1 is a continuous set, then the power might look like the curve in Figure 4.1.

Bias

A test $\phi(\mathbf{x})$ is said to be unbiased if its power is never smaller than its size:

$$\beta(\boldsymbol{\theta}) \geq \alpha \text{ for all } \boldsymbol{\theta} \in \Theta_1. \qquad 4.9$$

Best Test

If H_0 and H_1 are simple, then test ϕ is the best test of size α if it has the most power among all tests of size α. That is, if $\phi(\mathbf{x})$ and $\phi'(\mathbf{x})$ are two competing tests, each of which has size α, then

$$\beta \geq \beta'. \qquad 4.10$$

The best test maximizes detection probability P_D for a fixed false alarm probability P_{FA}.

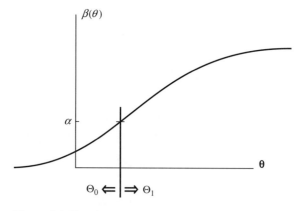

Figure 4.1 Power curve.

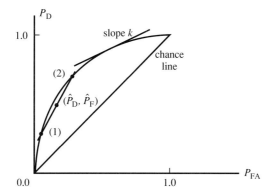

Figure 4.2 Receiver operating characteristics.

Receiver Operating Characteristics (ROC)

It is customary to plot β versus α or, equivalently, the detection probability P_D versus the false alarm probability P_{FA}. What results is a curve like that of Figure 4.2. For some choice of size α, the power β results. If the test is a good one, then the curve should be above the "chance line" that characterizes the performance of a pure guess. If P_{FA} equals zero, then H_0 is always selected, meaning that H_1 is never selected and that $P_D = 0$. Conversely, if P_{FA} equals one, then H_1 is always selected, meaning that P_D equals one. Other important characteristics of the ROC curve will be discussed shortly. For the time being, ignore the details of Figure 4.2.

Usually, in delicate experiments conducted in pharmaceutical labs, physics labs, war gamerooms, or communication systems, P_{FA} is very small—on the order of 10^{-2} to 10^{-9}. For these small values of P_{FA}, the trick is to design an experiment and a test such that P_D is on the order of $(1 - 10^{-1})$ to $(1 - 10^{-6})$.

4.3   THE NEYMAN-PEARSON LEMMA

The celebrated Neyman-Pearson lemma tells us how to find the most powerful test of size α for testing the simple hypothesis H_0 versus the simple alternative H_1. The test is a slight generalization of the test defined in Equation 4.4. The lemma tells us exactly how to find the acceptance and rejection regions for the test.

Neyman-Pearson Lemma   Let $\Theta = \{\theta_0, \theta_1\}$ and denote the distribution of $\mathbf{X}$ by $F_\theta(\mathbf{x})$ for $\theta = \theta_0$ or θ_1. Assume that the distribution has a density or probability mass function $f_{\theta_i}(\mathbf{x})$. A test ϕ of the form

$$\phi(\mathbf{x}) = \begin{cases} 1, & f_{\theta_1}(\mathbf{x}) > k f_{\theta_0}(\mathbf{x}) \\ \gamma, & f_{\theta_1}(\mathbf{x}) = k f_{\theta_0}(\mathbf{x}) \\ 0, & f_{\theta_1}(\mathbf{x}) < k f_{\theta_0}(\mathbf{x}) \end{cases} \qquad 4.11$$

for some $k \geq 0$ and some $0 \leq \gamma \leq 1$ is the most powerful test of size $\alpha > 0$ for testing $H_0 : \boldsymbol{\theta} = \boldsymbol{\theta}_0$ versus $H_1 : \boldsymbol{\theta} = \boldsymbol{\theta}_1$. If $\alpha = 0$, then the test

$$\phi(\mathbf{x}) = \begin{cases} 1, & f_{\boldsymbol{\theta}_0}(\mathbf{x}) = 0 \\ 0, & f_{\boldsymbol{\theta}_0}(\mathbf{x}) > 0 \end{cases} \qquad 4.12$$

is the most powerful test. These tests are unique except perhaps on $\mathbf{x}$ sets of probability 0 under H_0 and H_1. When $\phi(\mathbf{x}) = 1$ we choose H_1, and when $\phi(\mathbf{x}) = 0$ we choose H_0. When $\phi(\mathbf{x}) = \gamma$ we "flip a γ coin" to select H_1 with probability γ (the probability that the coin turns up heads).

Proof Let $\phi'(\mathbf{x})$ denote any test such that $0 \leq \phi'(\mathbf{x}) \leq 1$ and such that its size is less than or equal to the size of $\phi(\mathbf{x})$:

$$\alpha' = E_{\boldsymbol{\theta}_0} \phi'(\mathbf{X}) \leq E_{\boldsymbol{\theta}_0} \phi(\mathbf{X}) = \alpha.$$

Consider this inequality:

$$\int d\mathbf{x} [\phi(\mathbf{x}) - \phi'(\mathbf{x})][f_{\boldsymbol{\theta}_1}(\mathbf{x}) - k f_{\boldsymbol{\theta}_0}(\mathbf{x})] \geq 0.$$

This proves that ϕ is more powerful than ϕ':

$$\beta - \beta' \geq k(\alpha - \alpha') \geq 0.$$

A test of size $\alpha = 0$ must be zero on the set $\{\mathbf{x} : f_{\boldsymbol{\theta}_0}(\mathbf{x}) > 0\}$. Thus, the previous integral becomes the integral

$$\int\limits_{\{\mathbf{x}: f_{\boldsymbol{\theta}_0}(\mathbf{x}) = 0\}} d\mathbf{x} [\phi(\mathbf{x}) - \phi'(\mathbf{x})] f_{\boldsymbol{\theta}_1}(\mathbf{x}).$$

For the Neyman-Pearson test ϕ of size $\alpha = 0$ given in the theorem statement, this integral produces the result

$$\int\limits_{\{\mathbf{x}: \}} d\mathbf{x} [1 - \phi'(\mathbf{x})] f_{\boldsymbol{\theta}_1}(\mathbf{x}) \, d\mathbf{x} \geq 0$$

or

$$\beta - \beta' \geq 0. \qquad \blacksquare$$

Choosing the Threshold

There remains the problem of choosing the threshold k to produce the desired size α. For $0 < \alpha \leq 1$, the size may be written

$$\alpha = E_{\boldsymbol{\theta}_0} \phi(\mathbf{X}) = 1 - P_{\boldsymbol{\theta}_0}[f_{\boldsymbol{\theta}_1}(\mathbf{X}) \leq k f_{\boldsymbol{\theta}_0}(\mathbf{X})]$$
$$+ \gamma P_{\boldsymbol{\theta}_0}[f_{\boldsymbol{\theta}_1}(\mathbf{X}) = k f_{\boldsymbol{\theta}_0}(\mathbf{X})]. \qquad 4.13$$

If there exists a k_0 such that

$$P_{\boldsymbol{\theta}_0}[f_{\boldsymbol{\theta}_1}(\mathbf{X}) \le k_0 f_{\boldsymbol{\theta}_0}(\mathbf{X})] = 1 - \alpha, \qquad 4.14$$

then we take $\gamma = 0$. Otherwise there exists a k_0 such that

$$P_{\boldsymbol{\theta}_0}[f_{\boldsymbol{\theta}_1}(\mathbf{X}) < k_0 f_{\boldsymbol{\theta}_0}(\mathbf{X})] < 1 - \alpha \le P_{\boldsymbol{\theta}_0}[f_{\boldsymbol{\theta}_1}(\mathbf{X}) \le k_0 f_{\boldsymbol{\theta}_0}(\mathbf{X})]. \qquad 4.15$$

We may use k_0 and solve for γ as follows:

$$\gamma P_{\boldsymbol{\theta}_0}[f_{\boldsymbol{\theta}_1}(\mathbf{X}) = k_0 f_{\boldsymbol{\theta}_0}(\mathbf{X})] = P_{\boldsymbol{\theta}_0}[f_{\boldsymbol{\theta}_1}(\mathbf{X}) \le k_0 f_{\boldsymbol{\theta}_0}(\mathbf{X})] - (1 - \alpha). \qquad 4.16$$

Interpretation

We often summarize the Neyman-Pearson lemma by saying that the most powerful test of size α for testing $H_0 : \boldsymbol{\theta} = \boldsymbol{\theta}_0$ versus $H_1 : \boldsymbol{\theta} = \boldsymbol{\theta}_1$ is a *likelihood ratio* test of the form

$$\phi(\mathbf{x}) = \begin{cases} 1, & l(\mathbf{x}) > k \\ \gamma, & l(\mathbf{x}) = k \\ 0, & l(\mathbf{x}) < k \end{cases} \qquad 4.17$$

where $l(\mathbf{x})$ is the likelihood ratio

$$l(\mathbf{x}) = \frac{f_{\boldsymbol{\theta}_1}(\mathbf{x})}{f_{\boldsymbol{\theta}_0}(\mathbf{x})}. \qquad 4.18$$

If $l(\mathbf{x}) = k$ with probability zero, then $\gamma = 0$ and the *threshold k* is found as follows:

$$\alpha = P_{\boldsymbol{\theta}_0}\big[l(\mathbf{X}) > k\big] = \int_k^\infty f_{\boldsymbol{\theta}_0}(l)\, dl. \qquad 4.19$$

Here $f_{\boldsymbol{\theta}_0}(l)$ is the density function for $l(\mathbf{X})$ under H_0. We shall illustrate the choosing of thresholds with several examples.

Geometrical Properties of the ROC Curve

From the definition of likelihood we obtain the fundamental identity

$$E_{\boldsymbol{\theta}_1} l^n(\mathbf{x}) = \int d\mathbf{x}\, f_{\boldsymbol{\theta}_1}(\mathbf{x}) \frac{f_{\boldsymbol{\theta}_1}^n(\mathbf{x})}{f_{\boldsymbol{\theta}_0}^n(\mathbf{x})}$$

$$= \int d\mathbf{x}\, f_{\boldsymbol{\theta}_0}(\mathbf{x}) \frac{f_{\boldsymbol{\theta}_1}^{n+1}(\mathbf{x})}{f_{\boldsymbol{\theta}_0}^{n+1}(\mathbf{x})} = E_{\boldsymbol{\theta}_0} l^{n+1}(\mathbf{x}). \qquad 4.20$$

But this identity may also be written as

$$\int dl\, f_{\boldsymbol{\theta}_1}(l)l^n = \int dl\, f_{\boldsymbol{\theta}_0}(l)l^{n+1} \qquad 4.21$$

for all n, meaning $f_{\boldsymbol{\theta}_1}(l) = l f_{\boldsymbol{\theta}_0}(l)$, where $f_{\boldsymbol{\theta}_i}(l)$ is the density function for the likelihood ratio l, under H_i. This identity produces an important geometrical property of the ROC curve. Write the detection probability and the false alarm probability as

$$P_{\mathrm{D}} = \int_{k}^{\infty} f_{\boldsymbol{\theta}_1}(l) \, dl \qquad\qquad 4.22$$

$$P_{\mathrm{FA}} = \int_{k}^{\infty} f_{\boldsymbol{\theta}_0}(l) \, dl$$

where k is the threshold in the likelihood ratio test. The likelihood $l(\mathbf{x})$ is nonnegative and, therefore, so is the threshold k. The derivative of P_{D} with respect to P_{FA} is

$$\frac{dP_{\mathrm{D}}}{dP_{\mathrm{FA}}} = \frac{\partial P_{\mathrm{D}}}{\partial k} \frac{\partial k}{\partial P_{\mathrm{FA}}} = \frac{f_{\boldsymbol{\theta}_1}(k)}{f_{\boldsymbol{\theta}_0}(k)} = k \geq 0. \qquad\qquad 4.23$$

This result says that the ROC curve is nondecreasing and that the slope of the curve at any operating point $(P_{\mathrm{D}}, P_{\mathrm{FA}})$ is just the threshold required to achieve that point. This is illustrated in Figure 4.2.

The ROC curve is also convex, as the following argument shows. Let (1) and (2) denote any two points on the ROC curve of Figure 4.2, and define $(\hat{P}_{\mathrm{D}}, \hat{P}_{\mathrm{FA}})$ to be any point on the straight line that connects them. The Neyman-Pearson lemma says that, for the value of $\hat{P}_{\mathrm{FA}}$, there is a value of P_{D} that lies on the ROC curve for the likelihood ratio test that cannot be smaller than $\hat{P}_{\mathrm{D}}$. This argument holds for every pair of points (1) and (2), and makes the ROC curve convex.

North-by-Northwest: Birdsall's Insight

The Neyman-Pearson lemma establishes that the likelihood ratio test maximizes detection probability for any value of false alarm probability in a test of the simple

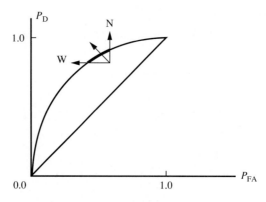

Figure 4.3   North-by-Northwest.

hypothesis H_0 versus the simple alternative H_1. As a consequence, the resulting ROC curve is convex. This convexity may be used to show that the Neyman-Pearson lemma "does even more than advertised." To illustrate, let Q denote a measure of quality for a test of hypotheses that favors small P_{FA} and large P_D. That is, $Q(P_{FA} - x, P_D + y) \geq Q(P_{FA}, P_D)$ for nonnegative x and y. Any test whose operating point lies below the ROC curve of Figure 4.3 can be improved on with a likelihood ratio test whose operating point lies on the segment of the ROC curve subtended by the wedge (of Figure 4.3), because all of these operating points have smaller P_{FA} and/or larger P_D. Therefore, all candidate tests for maximizing Q must be likelihood ratio tests whose operating points lie on the ROC curve. In other words, the likelihood ratio test maximizes any performance criterion that values low false alarm probabilities and/or high detection probabilities. Can you explain why we do not consider points above the ROC curve?

4.4   THE MULTIVARIATE NORMAL MODEL

Binary testing of simple hypotheses in the multivariate normal model forms the basis for most work in communication theory. The results are elegant, and the geometrical insights are illuminating. We proceed as follows. Under H_0, $\mathbf{X}$ is distributed as $N[\mathbf{m}_0, \mathbf{R}_0]$, and under H_1, $\mathbf{X}$ is distributed as $N[\mathbf{m}_1, \mathbf{R}_1]$. The problem is to test H_0 versus H_1. If $\mathbf{R}_0 = \mathbf{R}_1$ but $\mathbf{m}_1 \neq \mathbf{m}_0$, we have the following special case.

Uncommon Means and Common Covariance

In this case, the covariance structure is common under both hypotheses. The likelihood ratio is then

$$l(\mathbf{x}) = \exp\left\{ -\frac{1}{2}(\mathbf{x} - \mathbf{m}_1)^T \mathbf{R}^{-1}(\mathbf{x} - \mathbf{m}_1) + \frac{1}{2}(\mathbf{x} - \mathbf{m}_0)^T \mathbf{R}^{-1}(\mathbf{x} - \mathbf{m}_0) \right\}$$

$$= \exp\left\{ (\mathbf{m}_1 - \mathbf{m}_0)^T \mathbf{R}^{-1}\mathbf{x} + \frac{1}{2}(\mathbf{m}_0 + \mathbf{m}_1)^T \mathbf{R}^{-1}(\mathbf{m}_0 - \mathbf{m}_1) \right\}.$$

$\qquad$ 4.24

The natural logarithm is a monotone function that may be used in place of $l(\mathbf{x})$. After some rearrangement, the log likelihood $L(\mathbf{x}) = \ln l(\mathbf{x})$ may be written as

$$L(\mathbf{x}) = (\mathbf{m}_1 - \mathbf{m}_0)^T \mathbf{R}^{-1}\mathbf{x} + \frac{1}{2}(\mathbf{m}_0 + \mathbf{m}_1)^T \mathbf{R}^{-1}(\mathbf{m}_0 - \mathbf{m}_1)$$

$$= (\mathbf{m}_1 - \mathbf{m}_0)^T \mathbf{R}^{-1}(\mathbf{x} - \mathbf{x}_0)$$

$\qquad$ 4.25

$$\mathbf{x}_0 = \frac{1}{2}(\mathbf{m}_1 + \mathbf{m}_0).$$

The log likelihood ratio $L(\mathbf{x})$ may therefore be written as the inner product

$$L(\mathbf{x}) = \mathbf{w}^T(\mathbf{x} - \mathbf{x}_0)$$

$\qquad$ 4.26

where $\mathbf{w}$ is a "weight vector" that solves the equation

$$R\mathbf{w} = \mathbf{m}_1 - \mathbf{m}_0.$$ 4.27

We say that the weight vector $\mathbf{w}$ satisfies a linear regression equation. The Neyman-Pearson lemma says to compare $L(\mathbf{x})$ to a threshold $\eta = \ln k$ and select H_1 when the threshold is exceeded:

$$\phi(\mathbf{x}) = \begin{cases} 1, & L(\mathbf{x}) > \eta \\ 0, & L(\mathbf{x}) \leq \eta. \end{cases}$$ 4.28

This test is illustrated in Figure 4.4 for the case where $\mathbf{x}$ is two-dimensional and $\mathbf{R} = \mathbf{I}$.

How is log likelihood distributed and how do we choose the threshold to get the false alarm probability P_{FA}? Note that $L(\mathbf{X})$ is a linear transformation of the variable $\mathbf{X}$. Thus, $L(\mathbf{X})$ is normal:

$$H_0 : L(\mathbf{X}) : N\left(\frac{-d^2}{2}, d^2\right)$$ 4.29

$$H_1 : L(\mathbf{X}) : N\left(\frac{d^2}{2}, d^2\right).$$

The parameter d^2 is a "signal-to-noise ratio":

$$d^2 = \mathbf{w}^T \mathbf{R} \mathbf{w}$$
$$= (\mathbf{m}_1 - \mathbf{m}_0)^T \mathbf{R}^{-1} (\mathbf{m}_1 - \mathbf{m}_0).$$ 4.30

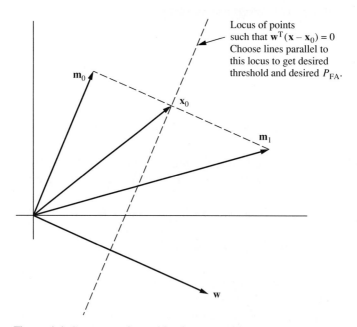

Locus of points such that $\mathbf{w}^T(\mathbf{x} - \mathbf{x}_0) = 0$ Choose lines parallel to this locus to get desired threshold and desired P_{FA}.

Figure 4.4 Geometry of a multivariate normal hypothesis test.

All we need to do is to select the threshold η so that we get the right value for P_{FA}:

$$
\begin{aligned}
P_{FA} &= \int_{\eta}^{\infty} (2\pi d^2)^{-1/2} \exp\left\{ -\frac{1}{2}(y + d^2/2)d^{-2}(y + d^2/2) \right\} dy \\
&= \int_{[\eta+d^2/2]/d}^{\infty} (2\pi)^{-1/2} \exp\{-z^2/2\} dz \\
&= 1 - \Phi\left(\frac{\eta + d^2/2}{d} \right).
\end{aligned}
\tag{4.31}
$$

The function $\Phi(z)$ is the normal integral

$$
\Phi(z) = \int_{-\infty}^{z} (2\pi)^{-1} \exp\{-x^2/2\} dx
\tag{4.32}
$$

$$
\Phi(-z) = 1 - \Phi(z).
$$

The formula for the threshold η is therefore

$$
\frac{\eta + d^2/2}{d} = z
\tag{4.33}
$$

where z is the parameter of the normal integral $\Phi(z)$ that makes the false alarm probability equal P_{FA}:

$$
1 - \Phi(z) = \alpha.
\tag{4.34}
$$

The power β is computed in a similar way:

$$
\begin{aligned}
\beta &= \int_{\eta}^{\infty} (2\pi d^2)^{-1/2} \exp\left\{ -\frac{1}{2}(y - d^2/2)d^{-2}(y - d^2/2) \right\} \\
&= \int_{[\eta-d^2/2]/d}^{\infty} (2\pi)^{-1/2} \exp\{-x^2/2\} dx \\
&= 1 - \Phi\left(\frac{\eta - d^2/2}{d} \right) \\
&= 1 - \Phi(z - d).
\end{aligned}
\tag{4.35}
$$

The power is the monotone function of "voltage signal-to-noise ratio" d illustrated in Figure 4.5.

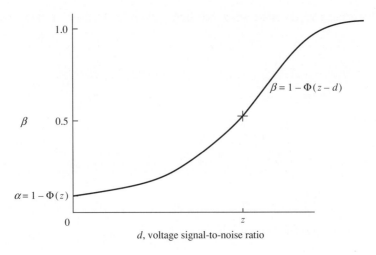

Figure 4.5 The power curve $\beta = 1 - \Phi(z - d)$.

Let's summarize our findings. The size and power are coupled through the formulas

$$\alpha = 1 - \Phi(z) \qquad\qquad 4.36$$

$$\beta = 1 - \Phi(z - d)$$

where d is the voltage signal-to-noise ratio $d = [(\mathbf{m}_1 - \mathbf{m}_0)^{\mathrm{T}}\mathbf{R}^{-1}(\mathbf{m}_1 - \mathbf{m}_0)]^{1/2}$. In

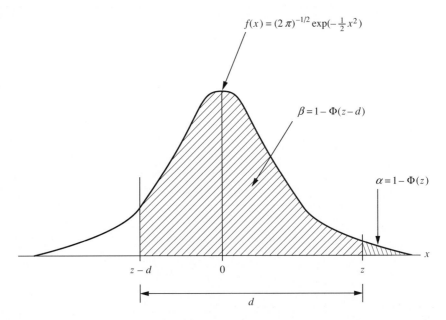

Figure 4.6 The interplay between SNR, detection probability, and false alarm probability.

Figure 4.6, the computations of α and β are illustrated by shaded areas under the $N[0, 1]$ density. You can see that, as α gets small, the parameter z must get large. This means that d (the voltage signal-to-noise ratio) must get large for β to be large. For fixed voltage signal-to-noise ratio d, decreasing the size α increases z (and the threshold η) and thereby decreases the power! In summary, the trade-off between α and β (P_{FA} and P_D) goes like this. The $N[0, 1]$ density is drawn. You have a piece of string of length d. You may slide the right end of the string to the right of Figure 4.6 to get a desired value of P_{FA}. The left end of the string tells you what value of P_D you get. Conversely, you may slide the left end of the string to the left for a desired P_D and take the P_{FA} that the right end gives you. In Problem 4.8, you are asked to compute and plot the receiver operating characteristic β versus α, and the threshold η versus the false alarm probability α.

Linear Statistical Model

When the mean value vectors $\mathbf{m}_0$ and $\mathbf{m}_1$ are composed of linear combinations of modes, then we represent them as

$$\mathbf{m}_0 = \mathbf{H}\boldsymbol{\theta}_0 \qquad\qquad 4.37$$

$$\mathbf{m}_1 = \mathbf{H}\boldsymbol{\theta}_1$$

where $\boldsymbol{\theta}_0$ and $\boldsymbol{\theta}_1$ contain mode weights and the matrix $\mathbf{H}$ organizes the modes into columns:

$$\mathbf{H} = \begin{bmatrix} \mathbf{h}_1 & \mathbf{h}_2 & \cdots & \mathbf{h}_p \end{bmatrix} \qquad\qquad 4.38$$

$$\mathbf{h}_i : i\text{th mode.}$$

The test $H_0 : \mathbf{X} : N[\mathbf{m}_0, \mathbf{R}]$ versus $H_1 : \mathbf{X} : N[\mathbf{m}_1, \mathbf{R}]$ is really a test of $H_0 : \boldsymbol{\theta} = \boldsymbol{\theta}_0$ versus $H_1 : \boldsymbol{\theta} = \boldsymbol{\theta}_1$ in the linear statistical model $\mathbf{X} : N[\mathbf{H}\boldsymbol{\theta}, \mathbf{R}]$. In words, we are testing to determine which mode weights are present. The Neyman-Pearson test of H_0 versus H_1 is

$$L(\mathbf{x}) = \mathbf{w}^T(\mathbf{x} - \mathbf{x}_0) \qquad\qquad 4.39$$

where the weight vector $\mathbf{w}$ solves the linear equation

$$\mathbf{R}\mathbf{w} = \mathbf{H}(\boldsymbol{\theta}_1 - \boldsymbol{\theta}_0) \qquad\qquad 4.40$$

$$\mathbf{x}_0 = \frac{1}{2}\mathbf{H}(\boldsymbol{\theta}_1 + \boldsymbol{\theta}_0).$$

The term $\mathbf{w}^T\mathbf{x}_0$ is independent of the data, meaning that it may be absorbed into the threshold term η. The test statistic $\mathbf{w}^T\mathbf{x}$ may be written as

$$\mathbf{w}^T\mathbf{x} = (\boldsymbol{\theta}_1 - \boldsymbol{\theta}_0)^T\mathbf{H}^T\mathbf{R}^{-1}\mathbf{x}. \qquad\qquad 4.41$$

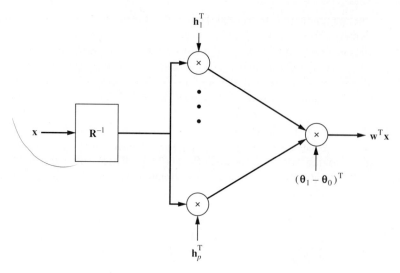

Figure 4.7 Implementation of Neyman-Pearson detector in the linear statistical model.

A structure for implementing $\mathbf{w}^T\mathbf{x}$ is illustrated in Figure 4.7. Each branch is a correlator of transformed data with stored modes. The correlator outputs are weighted and
accumulated to obtain the output $\mathbf{w}^T\mathbf{x}$.

Common Means and Uncommon Covariances

In this case, the means are common at $\mathbf{m}_0 = \mathbf{m}_1 = \mathbf{m}$, and the two hypotheses are
characterized by uncommon covariances. The log likelihood ratio is then

$$L(\mathbf{x}) = \log \frac{|\mathbf{R}_0|^{1/2}}{|\mathbf{R}_1|^{1/2}} - \frac{1}{2}(\mathbf{x} - \mathbf{m})^T(\mathbf{R}_1^{-1} - \mathbf{R}_0^{-1})(\mathbf{x} - \mathbf{m}). \qquad 4.42$$

The first term is independent of the data, so we may ignore it and use only the
quadratic form

$$L(\mathbf{x}) = (\mathbf{x} - \mathbf{m})^T(\mathbf{R}_0^{-1} - \mathbf{R}_1^{-1})(\mathbf{x} - \mathbf{m}).$$

This quadratic form may be written as

$$L(\mathbf{x}) = (\mathbf{x} - \mathbf{m})^T\mathbf{R}_0^{-T/2}(\mathbf{I} - \mathbf{S}^{-1})\mathbf{R}_0^{-1/2}(\mathbf{x} - \mathbf{m}) \qquad 4.43$$

where $\mathbf{R}_0 = \mathbf{R}_0^{1/2}\mathbf{R}_0^{T/2}$ is a symmetric factorization of $\mathbf{R}_0$, and $\mathbf{S}$ is the signal-to-noise
matrix

$$\mathbf{S} = \mathbf{R}_0^{-1/2}\mathbf{R}_1\mathbf{R}_0^{-T/2}. \qquad 4.44$$

The quadratic form must be compared with a threshold η:

$$\phi(\mathbf{x}) = \begin{cases} 1, & L(\mathbf{x}) > \eta \\ 0, & L(\mathbf{x}) \le \eta. \end{cases} \qquad 4.45$$

We shall have more to say about this result in Section 4.14, Detectors for Gaussian Random Signals.

Uncommon Means and Uncommon Variances

When neither the means nor the covariances are common, the log likelihood is

$$L(\mathbf{x}) = \log \frac{|\mathbf{R}_0|^{1/2}}{|\mathbf{R}_1|^{1/2}} - \frac{1}{2}\left[(\mathbf{x} - \mathbf{m}_1)^{\mathrm{T}}\mathbf{R}_1^{-1}(\mathbf{x} - \mathbf{m}_1) - (\mathbf{x} - \mathbf{m}_0)^{\mathrm{T}}\mathbf{R}_0^{-1}(\mathbf{x} - \mathbf{m}_0)\right]. \quad 4.46$$

This is a complicated quadratic form in the data that we would prefer not to compute. In Section 4.15, we study linear discriminant functions as a way of simplifying the computations.

4.5   BINARY COMMUNICATION

The Neyman-Pearson theory of detection applies directly to binary communication. As illustrated in Figure 4.8(a), a binary communication source S transmits a binary digit $i \in \{0, 1\}$ that is used to modulate a sequence $\{m_t(i)\}_0^{N-1}$. The modulated sequence is transmitted through a channel, where noise $\{n_t\}_0^{N-1}$ is added. The problem is to observe the signal-plus-noise sequence $\{x_t\}_0^{N-1}$ and decide whether $i = 0$ or 1.

An equivalent communication diagram is illustrated in Figure 4.8(b), where the

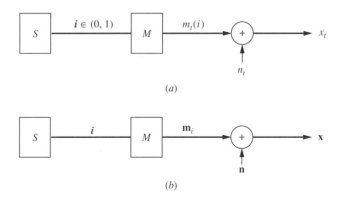

(a)

(b)

Figure 4.8 Binary communication (a). Equivalent diagram (b).

vectors $\mathbf{m}_i$, $\mathbf{n}$, and $\mathbf{x}$ are defined as follows:

$$\mathbf{m}_i = \begin{bmatrix} m_0(i) & m_1(i) & \cdots & m_{N-1}(i) \end{bmatrix}^T$$

$$\mathbf{n} = \begin{bmatrix} n_0 & n_1 & \cdots & n_{N-1} \end{bmatrix}^T \qquad 4.47$$

$$\mathbf{x} = \begin{bmatrix} x_0 & x_1 & \cdots & x_{N-1} \end{bmatrix}^T.$$

The vector $\mathbf{m}_i$ is called a symbol, and the interval $0 \le t \le N - 1$ is called a baud interval. The communication diagram of Figure 4.8(b) is called a waveform, or vector, diagram.

We shall assume that the noises n_t are zero-mean and normal, with cross-correlations $E n_t n_s = \sigma^2 r_{|t-s|}$ and $r_0 = 1$. This means that the noise vector $\mathbf{n}$ is $N[\mathbf{0}, \sigma^2 \mathbf{R}]$, with $\sigma^2 \mathbf{R} = E \mathbf{n} \mathbf{n}^T = \sigma^2 \{ r_{|t-s|} \}$. Under hypothesis H_0, the measurement vector $\mathbf{x}$ is $N[\mathbf{m}_0, \sigma^2 \mathbf{R}]$, and under alternative H_1, $\mathbf{x}$ is $N[\mathbf{m}_1, \sigma^2 \mathbf{R}]$. From the results of Section 4.4, we know that the Neyman-Pearson detector is the log likelihood ratio test

$$\phi(\mathbf{x}) = \begin{cases} 1, & L(\mathbf{x}) > \eta \\ 0, & L(\mathbf{x}) \le \eta \end{cases} \qquad 4.48$$

$$L(\mathbf{x}) = (\mathbf{m}_1 - \mathbf{m}_0)^T (\sigma^2 \mathbf{R})^{-1} \left(\mathbf{x} - \frac{1}{2}(\mathbf{m}_1 + \mathbf{m}_0) \right).$$

Typically, the symbols $\mathbf{m}_1$ and $\mathbf{m}_0$ are chosen so that their energies with respect to $\mathbf{R}^{-1}$ are equal. We call this energy E_s and write

$$E_s = \mathbf{m}_0^T \mathbf{R}^{-1} \mathbf{m}_0 = \mathbf{m}_1^T \mathbf{R}^{-1} \mathbf{m}_1.$$

Consequently, the inner product $(\mathbf{m}_1 - \mathbf{m}_0)^T (\sigma^2 \mathbf{R})^{-1}(\mathbf{m}_1 + \mathbf{m}_0)$ is zero (you should check it), and we may write the log likelihood $L(\mathbf{x})$ as

$$L(\mathbf{x}) = (\mathbf{m}_1 - \mathbf{m}_0)^T (\sigma^2 \mathbf{R})^{-1} \mathbf{x}. \qquad 4.49$$

The statistic $L(\mathbf{x})$ is distributed as follows under H_0 and H_1:

$$H_0 : L(\mathbf{x}) : N \left[-\frac{d^2}{2}, d^2 \right] \qquad 4.50$$

$$H_1 : L(\mathbf{x}) : N \left[\frac{d^2}{2}, d^2 \right].$$

The signal-to-noise ratio is

$$d^2 = (\mathbf{m}_1 - \mathbf{m}_0)^T (\sigma^2 \mathbf{R})^{-1}(\mathbf{m}_1 - \mathbf{m}_0)$$

$$= \frac{2E_s}{\sigma^2}(1 - \rho) \qquad 4.51$$

where ρ is the cosine of the angle between $\mathbf{m}_0$ and $\mathbf{m}_1$:

$$\rho = \frac{\mathbf{m}_0^T \mathbf{R}^{-1} \mathbf{m}_1}{(\mathbf{m}_0^T \mathbf{R}^{-1} \mathbf{m}_0)^{1/2}(\mathbf{m}_1^T \mathbf{R}^{-1} \mathbf{m}_1)^{1/2}} = \frac{\mathbf{m}_0^T \mathbf{R}^{-1} \mathbf{m}_1}{E_s}. \qquad 4.52$$

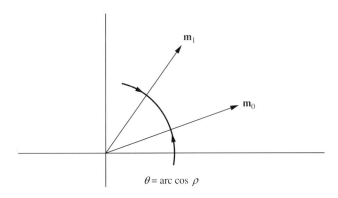

Figure 4.9 The symbols $\mathbf{m}_0$ and $\mathbf{m}_1$.

With this definition of ρ, the symbols $\mathbf{m}_0$ and $\mathbf{m}_1$ may be depicted as in Figure 4.9 for the case $\mathbf{R} = \sigma^2 I$.

It is actually a little more illuminating to normalize the statistic $L(\mathbf{x})$ by the voltage signal-to-noise ratio d. Then, under H_0 and H_1, the statistic $L(\mathbf{x})/d$ is distributed as follows:

$$H_0 : \frac{L(\mathbf{x})}{d} : N\left[-\frac{d}{2}, 1\right]$$

$$H_1 : \frac{L(\mathbf{x})}{d} : N\left[\frac{d}{2}, 1\right] \qquad\qquad 4.53$$

$$\frac{d}{2} = \sqrt{\frac{E_s}{2\sigma^2}(1-\rho)}.$$

The normally distributed statistic $L(\mathbf{x})/d$ is illustrated in Figure 4.10. It is clear from the figure that the false alarm probability α will equal the miss probability when the threshold η is chosen to be $\eta = 0$. Then $z = d/2$ and the formulae

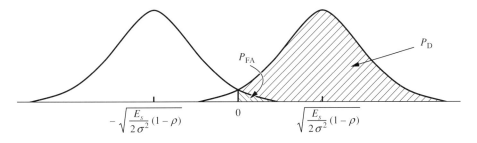

Figure 4.10 Distribution of the detector statistic.

for α and $1 - \beta$ are

$$\alpha = 1 - \Phi(d/2)$$

$$1 - \beta = 1 - \big(1 - \Phi(-d/2)\big)$$

$$= 1 - \Phi(d/2).$$

4.54

The size α (or false alarm probability P_{FA}) and power β (or detection probability P_D) are also illustrated in Figure 4.10.

The detector statistic $L(\mathbf{x})$, illustrated in Figure 4.11(a) for $\sigma^2 = 1$, is called a *correlation detector*. Each of the inner products $\mathbf{m}_i^T\mathbf{y}$, with $\mathbf{y} = \mathbf{R}^{-1}\mathbf{x}$, may also be written as

$$\mathbf{m}_i^T\mathbf{y} = \sum_{t=0}^{N-1} m_t(i)y_t$$

$$= \sum_{t=0}^{N-1} h_{N-1-t}(i)y_t,$$

4.55

where the coefficients h_{N-1-t} are related to the symbol coefficients m_t as follows:

$$h_{N-1-t}(i) = m_t(i)$$

4.56

or

$$h_t(i) = m_{N-1-t}(i).$$

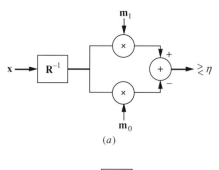

(a)

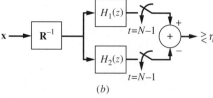

(b)

Figure 4.11 Binary detectors (a) Correlation detector. (b) Matched filter detector.

This means that the inner product may be obtained by passing the sequence $\{x_t\}_0^{N-1}$ through a digital filter whose impulse response is $m_{N-1-t}(i)$ and sampling the output at $t = N - 1$. This observation leads to the implementation of Figure 4.11(b), where $H_i(z)$ denotes the filter whose impulse response is $h_t(i) = m_{N-1-t}(i)$.

4.6   SUFFICIENCY

We have argued in Chapter 3 that a sufficient statistic summarizes the useful information that a measurement brings about a parameter. This statement can be made explicit for the problem of testing the simple hypothesis $H_0 : \boldsymbol{\theta} = \boldsymbol{\theta}_0$ versus the simple alternative $H_1 : \boldsymbol{\theta} = \boldsymbol{\theta}_1$. The Neyman-Pearson lemma says that we must compute the likelihood ratio $l(\mathbf{x}) = f_{\boldsymbol{\theta}_1}(\mathbf{x})/f_{\boldsymbol{\theta}_0}(\mathbf{x})$ where $f_{\boldsymbol{\theta}_i}(\mathbf{x})$ is the density or mass function for the measurement $\mathbf{x}$ under H_i. If $\mathbf{t}(\mathbf{x})$ is sufficient for $\boldsymbol{\theta}$, then the density function $f_{\boldsymbol{\theta}_i}(\mathbf{x})$ factors as $a(\mathbf{x})b_{\boldsymbol{\theta}_i}(\mathbf{t})$. Therefore the likelihood ratio may be written as

$$l(\mathbf{x}) = \frac{b_{\boldsymbol{\theta}_1}(\mathbf{t})}{b_{\boldsymbol{\theta}_0}(\mathbf{t})}. \qquad 4.57$$

But recall from Chapter 3 that the density function for the sufficient statistic $\mathbf{t}$ is just a scaled version of $b_{\boldsymbol{\theta}_i}(\mathbf{t})$. Therefore the likelihood ratio $l(\mathbf{x})$ may be written as the likelihood ratio for the sufficient statistic:

$$l(\mathbf{x}) = \frac{f_{\boldsymbol{\theta}_1}(\mathbf{t})}{f_{\boldsymbol{\theta}_0}(\mathbf{t})} = l(\mathbf{t}). \qquad 4.58$$

This is clearly just a function of the sufficient statistic $\mathbf{t}$.

What are the implications of this discussion for the practical application of sufficiency to hypothesis testing? Begin with a random sample $\mathbf{x}$ whose distribution is $F_{\boldsymbol{\theta}_1}(\mathbf{x})$ under H_1 and $F_{\boldsymbol{\theta}_0}(\mathbf{x})$ under H_0. Assume that $\mathbf{t}(\mathbf{x})$ is a sufficient statistic for $\boldsymbol{\theta}$. Under H_1, $\mathbf{t}$ is distributed as $F_{\boldsymbol{\theta}_1}(\mathbf{t})$ and under H_0, $\mathbf{t}$ is distributed as $F_{\boldsymbol{\theta}_0}(\mathbf{t})$. The Neyman-Pearson lemma says that the test of H_1 versus H_0 takes the form

$$\phi(\mathbf{t}) = \begin{cases} 1, & l(\mathbf{t}) > k \\ \gamma, & l(\mathbf{t}) = k \\ 0, & l(\mathbf{t}) < k \end{cases} \qquad 4.59$$

where $l(\mathbf{t})$ is the likelihood ratio in $\mathbf{t}$:

$$l(\mathbf{t}) = \frac{f_{\boldsymbol{\theta}_1}(\mathbf{t})}{f_{\boldsymbol{\theta}_0}(\mathbf{t})}. \qquad 4.60$$

The threshold is selected to produce the appropriate size. In summary, the sufficient statistic replaces the random sample and Neyman-Pearson takes over from there.

Example 4.2

Let $\mathbf{X} = (\mathbf{X}_0, \mathbf{X}_1, \ldots, \mathbf{X}_{N-1})$ denote a random sample of normal random variables $\mathbf{X}_n : N[\mathbf{m}, \mathbf{R}]$. Under H_1, $\mathbf{m} = \mathbf{m}_1$ and, under H_0, $\mathbf{m} = \mathbf{m}_0$. The sufficient statistic for $\mathbf{m}$ is

$$\mathbf{t(x)} = \frac{1}{N} \sum_{n=0}^{N-1} \mathbf{x}_n. \qquad 4.61$$

This sufficient statistic has distribution $N\left[\mathbf{m}_1, (1/N)\mathbf{R}\right]$ under H_1 and distribution $N\left[\mathbf{m}_0, (1/N)\mathbf{R}\right]$ under H_0. The likelihood ratio in $\mathbf{t}$ is

$$l(\mathbf{t}) = \exp\left\{ -\frac{1}{2}(\mathbf{t} - \mathbf{m}_1)^T N \mathbf{R}^{-1}(\mathbf{t} - \mathbf{m}_1) + \frac{1}{2}(\mathbf{t} - \mathbf{m}_0)^T N \mathbf{R}^{-1}(\mathbf{t} - \mathbf{m}_0) \right\}$$

$$= \exp\left\{ (\mathbf{m}_1 - \mathbf{m}_0)^T N \mathbf{R}^{-1}\mathbf{t} + \frac{1}{2}(\mathbf{m}_0 - \mathbf{m}_1)^T N \mathbf{R}^{-1}(\mathbf{m}_0 + \mathbf{m}_1) \right\}. \qquad 4.62$$

The natural logarithm is a monotone function that may be used in place of $l(\mathbf{t})$ to produce the log likelihood ratio:

$$\begin{aligned}
L(\mathbf{t}) &= \ln l(\mathbf{t}) \\
&= (\mathbf{m}_1 - \mathbf{m}_0)^T N \mathbf{R}^{-1}(\mathbf{t} - \mathbf{m}) \\
&= (\mathbf{m}_1 - \mathbf{m}_0)^T \mathbf{R}^{-1}\left(\sum_{n=0}^{N-1} \mathbf{x}_n - N\mathbf{m} \right)
\end{aligned} \qquad 4.63$$

$$\mathbf{m} = \frac{1}{2}(\mathbf{m}_0 + \mathbf{m}_1).$$

From here on, the story is just as in Section 4.4. ∎

Example 4.3

Let K denote a Poisson random variable with rate parameter λ:

$$P[k] = e^{-\lambda}\lambda^k/k! \qquad 4.64$$

Under hypothesis $H_0 : \lambda = \lambda_0$. The alternative is $H_1 : \lambda = \lambda_1$. A random sample of size N has probability mass function

$$\begin{aligned}
P[k_1 \quad \cdots \quad k_N] &= \frac{e^{-N\lambda}\lambda^{\sum_{n=1}^N k_n}}{\prod_{n=1}^N k_n!} \\
&= \frac{1}{\prod_{n=1}^N k_n!} \exp(-N\lambda)\exp\left\{ \ln\lambda^{\sum_{n=1}^N k_n} \right\} \\
&= \frac{1}{\prod_{n=1}^N k_n!} \exp(-N\lambda)\exp\left\{ (\ln\lambda)\sum_{n=1}^N k_n \right\}.
\end{aligned} \qquad 4.65$$

This is an exponential family in which the sum of the Poisson counts,

$$k = \sum_{n=1}^{N} k_n \qquad\qquad 4.66$$

is sufficient for the parameter λ. The Neyman-Pearson test of H_0 versus H_1 is

$$\phi(k) = \begin{cases} 1, & k > \eta \\ \gamma, & k = \eta \\ 0, & k < \eta \end{cases} \qquad\qquad 4.67$$

$$k = \sum_{n=1}^{N} k_n.$$

The random variable k is distributed as a Poisson random variable with rate parameter $N\lambda$. The false alarm probability is therefore

$$\alpha = \sum_{k>\eta}^{\infty} \frac{e^{-N\lambda_0}(N\lambda_0)^k}{k!} + \gamma \frac{e^{-N\lambda_0}(N\lambda_0)^\eta}{\eta!}. \qquad\qquad 4.68$$

By selecting the threshold η and the parameter γ, we can achieve false alarm probability α. The detection probability is

$$\beta = \sum_{k>\eta}^{\infty} \frac{e^{-N\lambda_1}(N\lambda_1)^k}{k!} + \gamma \frac{e^{-N\lambda_1}(N\lambda_1)^\eta}{\eta!}. \qquad\qquad 4.69$$

∎

4.7   THE TESTING OF COMPOSITE BINARY HYPOTHESES

We now generalize our treatment of hypothesis testing by considering binary tests of composite hypotheses. We are interested in tests that are optimum in the sense of Neyman-Pearson, even when H_0 and/or H_1 is composite.

Let's review what we mean by a composite test. A binary or multiple hypothesis-testing problem is said to be composite if one or more of the hypotheses H_i contain more than one element. When this is the case, then the definitions of size and power generalize as follows for binary hypothesis testing:

$$\alpha = \sup_{\theta \in \Theta_0} E_\theta \, \phi(\mathbf{x}) \qquad\qquad 4.70$$

$$\beta(\theta) = E_\theta \, \phi(\mathbf{x}), \, \theta \in \Theta_1.$$

We would like to design one test $\phi(\mathbf{x})$ that is optimum in the sense of Neyman-Pearson for every pair $\theta_0 \in \Theta_0$ and $\theta_1 \in \Theta_1$. This leads to the concept of a uniformly most powerful (UMP) test.

Uniformly Most Powerful Test

A test $\phi(\mathbf{x})$ of H_0 versus H_1 is uniformly most powerful (UMP) of size α if it has size α and its power is uniformly greater than the power of any other test $\phi'(\mathbf{x})$ whose size is less than or equal to α:

(size) $\displaystyle \sup_{\boldsymbol{\theta} \in \Theta_0} E_{\boldsymbol{\theta}}\, \phi(\mathbf{X}) = \alpha; \quad \sup_{\boldsymbol{\theta} \in \Theta_0} E_{\boldsymbol{\theta}}\, \phi'(\mathbf{X}) \le \alpha;$

(power) $E_{\boldsymbol{\theta}}\, \phi(\mathbf{X}) \ge E_{\boldsymbol{\theta}}\, \phi'(\mathbf{X})$ for all $\boldsymbol{\theta} \in \Theta_1$.

In order to characterize tests that can be UMP, we consider *scalar* random variables X whose density functions $f_\theta(x)$ are parameterized by scalar parameters θ. This seems very restrictive. However, as we shall see, many signal detection problems reduce to the consideration of *scalar* sufficient statistics whose densities are parameterized by a scalar parameter. For this reason, what follows is fundamentally important.

Karlin-Rubin Theorem (for UMP One-sided Tests)   Let X denote a scalar random variable whose density function is parameterized by the scalar parameter θ. Assume that the likelihood ratio

$$l_{\theta_1}(x) = \frac{f_{\theta_1}(x)}{f_{\theta_0}(x)}$$

is a nondecreasing function of x for every pair $[\theta_1 > \theta_0, \theta_0]$. We say that X has a monotone-likelihood ratio. The interpretation is that the larger x is, the more probable the alternative H_1 looks. Then the *threshold test*

$$\phi(x) = \begin{cases} 1, & x > x_0 \\ \gamma, & x = x_0 \\ 0, & x < x_0 \end{cases} \qquad\qquad 4.71$$

$$E_{\theta_0}\, \phi(X) = P_{\theta_0}[X > x_0] + \gamma P_{\theta_0}[X = x_0] = \alpha$$

is the UMP test of size α for testing $H_0 : \theta \le \theta_0$ versus $H_1 : \theta > \theta_0$. (Such a hypothesis test is said to be "one-sided.") Notice that the size is computed at the boundary θ_0.

Proof   Begin with fixed values θ_0 and $\theta_1 > \theta_0$. By the Neyman-Pearson lemma, the test

$$\phi(x) = \begin{cases} 1, & l(x) > k \\ \gamma, & l(x) = k \\ 0, & l(x) < k \end{cases} \qquad\qquad 4.72$$

$$E_{\theta_0}\, \phi(X) = \alpha$$

is most powerful for testing the *simple* hypothesis $H_0' : \theta = \theta_0$ versus the simple alternative $H_1' : \theta = \theta_1$. As likelihood is monotone, we may replace this test with the *threshold* test

$$\phi(x) = \begin{cases} 1, & x > x_0 \\ \gamma, & x = x_0 \\ 0, & x < x_0 \end{cases} \qquad\qquad 4.73$$

$$E_{\theta_0}\,\phi(X) = P_{\theta_0}[X > x_0] + \gamma P_{\theta_0}[X = x_0] = \alpha.$$

See Figure 4.12(*a*) for an illustration of the likelihood ratio test and the corresponding threshold test. This test is independent of θ_1, so the argument holds for every $\theta_1 > \theta_0$, making $\phi(x)$ uniformly most powerful among all tests of size $\leq \alpha$ at $\theta = \theta_0$ for testing the *composite alternative* H_1 versus the *simple hypothesis* H_0'. But now consider the power function for the test. At $\theta = \theta_0$, $\beta(\theta_0) = \alpha$. For any $\theta_1 > \theta_0$, $\beta(\theta_1) \geq \alpha$ because $\phi(x)$ is more powerful than the test $\phi'(x) \equiv \alpha$. Now consider any $\theta_2 < \theta_0$. As a test of $H_0'' : \theta = \theta_2$ versus $H_1'' : \theta > \theta_2$, the test $\phi(x)$ has some size α'' and power $\beta(\theta)$, with $\beta(\theta_0) = \alpha$. Again, $\phi(x)$ is more powerful than the test $\phi''(x) \equiv \alpha''$, so $\alpha'' < \alpha$. We conclude that the power function $\beta(\theta)$ is nondecreasing, as illustrated in Figure 4.12(*b*). Consequently, $\phi(x)$ is also a test whose size satisfies

$$\sup_{\theta \in \theta_\Theta 0} E_{\theta}\,\phi(X) = \alpha. \qquad\qquad 4.74$$

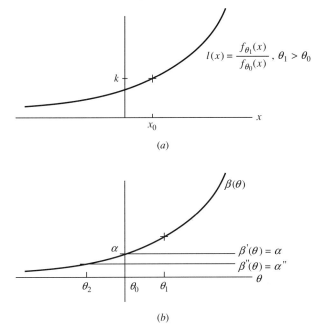

(*a*)

(*b*)

Figure 4.12 Monotone likelihood ratio (*a*) and monotone power function (*b*).

Finally, there can be no test $\phi'(x)$ whose size is $\leq \alpha$ and whose power exceeds $\beta(\theta)$, because such a test would have to have size $\leq \alpha$ at θ_0, and any such test must have power less than $\phi(x)$. This makes $\phi(x)$ UMP for testing $H_0 : \theta \leq \theta_0$ versus $H_1 : \theta > \theta_1$. $\square$

Example 4.4

Let k denote the number of events in an experiment described by a Poisson law:

$$P_\theta[k] = \frac{e^{-\theta}\theta^k}{k!}; \qquad k = 0, 1, .: . \tag{4.75}$$

The likelihood ratio is monotone (nondecreasing in k for $\theta_1 > \theta_0$):

$$\frac{P_{\theta_1}[k]}{P_{\theta_0}[k]} = \left(\frac{\theta_1}{\theta_0}\right)^k e^{-(\theta_1-\theta_0)}. \tag{4.76}$$

Therefore by the Karlin-Rubin theorem, the test

$$\phi(k) = \begin{cases} 1, & k > k_0 \\ \gamma, & k = k_0 \\ 0, & k < k_0 \end{cases} \tag{4.77}$$

$$E_{\theta_0}\phi(k) = \alpha$$

is UMP of size α for testing $H_0 : \theta \leq \theta_0$ versus $H_1 : \theta > \theta_0$.   ∎

The following theorem shows that uniformly most powerful one-sided tests exist for a host of problems.

Theorem   The one-parameter exponential family of distributions with density or probability mass function

$$f_\theta(\mathbf{x}) = c(\theta)h(\mathbf{x})\exp\{\pi(\theta)t(\mathbf{x})\} \tag{4.78}$$

has a monotone likelihood ratio in the sufficient statistic t, provided π is a *nondecreasing* function of its argument. A random sample $\mathbf{X} = (\mathbf{X}_0, \mathbf{X}_1, \dots, \mathbf{X}_{N-1})$ drawn from this distribution produces a sufficient statistic with a one-parameter distribution and a monotone-likelihood ratio. Therefore, if a random sample is drawn from such a one-parameter exponential family or the sufficient statistic is drawn from such a one-parameter exponential family, then there exists a uniformly most powerful test of the one-sided hypothesis test $H_0 : \theta \leq \theta_0$ versus $H_1 : \theta > \theta_0$.

Proof.   See Problem 4.9. $\square$

Example 4.5

Let $\mathbf{X} = (\mathbf{X}_0, \mathbf{X}_1, \dots, \mathbf{X}_{M-1})$ denote a random sample of $N[\theta\mathbf{m}, \mathbf{R}]$ random variables. The mean value vector $\mathbf{m}$ and the covariance matrix $\mathbf{R}$ are known, but the scale constant θ for the mean value vector is unknown. The distribution of the random sample is

$$f_\theta(\mathbf{x}) = (2\pi)^{-MN/2} |\mathbf{R}|^{-M/2} \exp\left\{ -\frac{1}{2} \sum_{n=0}^{M-1} (\mathbf{x}_n - \theta\mathbf{m})^\mathsf{T} \mathbf{R}^{-1}(\mathbf{x}_n - \theta\mathbf{m}) \right\}$$

$$= (2\pi)^{-MN/2} |\mathbf{R}|^{-M/2} \exp\left\{ -\frac{1}{2} \sum_{n=0}^{M-1} \mathbf{x}_n^\mathsf{T} \mathbf{R}^{-1} \mathbf{x}_n \right\} \exp\left\{ -\frac{M}{2} \theta^2 \mathbf{m}^\mathsf{T} \mathbf{R}^{-1} \mathbf{m} \right\}$$

$$\times \exp\left\{ \theta\mathbf{m}^\mathsf{T} \mathbf{R}^{-1} \sum_{n=0}^{M-1} \mathbf{x}_n \right\}. \tag{4.79}$$

This makes the statistic

$$t(\mathbf{x}) = \mathbf{m}^\mathsf{T} \mathbf{R}^{-1} \sum_{n=0}^{M-1} \mathbf{x}_n \tag{4.80}$$

a scalar sufficient statistic for the scalar parameter θ. The distribution of t is $t : N[M\theta\mathbf{m}^\mathsf{T}\mathbf{R}^{-1}\mathbf{m}, M\mathbf{m}^\mathsf{T} \mathbf{R}^{-1}\mathbf{m}]$. The likelihood ratio $l(t) = f_{\theta_1}(t)/f_{\theta_0}(t)$ is a nondecreasing function of t for $\theta_1 > \theta_0$. Therefore the sufficient statistic t has monotone-likelihood ratio. By the Karlin-Rubin theorem, the test

$$\phi(t) = \begin{cases} 1, & t > t_0 \\ 0, & t \le t_0 \end{cases} \tag{4.81}$$

$$E_{\theta_0} \phi(t) = P_{\theta_0}[t > t_0] = \alpha$$

is the uniformly most powerful test of size α for testing $H_0 : \theta \le \theta_0$ versus $H_1 : \theta > \theta_0$. ∎

4.8   INVARIANCE

The idea behind invariance in the Neyman-Pearson theory of hypothesis testing is the following. Suppose H_0 and H_1 are binary hypotheses. In general, both are composite. Ideally, we would like to have a test that is uniformly most powerful (UMP) of size α for testing H_0 versus H_1. When no such UMP test exists for a decision problem, then it is sometimes possible to impose additional constraints on the class of decision rules of interest and then find a rule within this constrained class that is UMP of size α. For example, suppose $\boldsymbol{\theta}$ is a $p \times 1$ parameter vector for the distribution $F_{\boldsymbol{\theta}}(\mathbf{x})$. Then H_0 might be a hypothesis of the form $H_0 : (\theta_1, \dots, \theta_r) \in \Theta_0$ and H_1 might be an alternative of the form $H_1 : (\theta_1, \dots, \theta_r) \in \Theta_1$. The parameters $(\theta_{r+1}, \dots, \theta_p)$ are nuisance parameters that do not enter into the hypothesis testing problem and, worse, they preclude us from finding a UMP test with the techniques we have so far developed.

If $(\theta_{r+1}, \dots, \theta_p)$ are nuisance parameters, then it may be reasonable to look for decision rules that are, in some well defined sense, invariant to them. This leads to

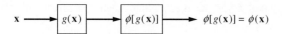

Figure 4.13 Decision rule is invariant to transformation $g(\mathbf{x})$, $g \in G$.

the key idea behind invariance in hypothesis testing: When presented with nuisance parameters that are extraneous to the hypothesis test, look for transformations of the measured data that would introduce these nuisance parameters and then look for a decision rule that is invariant to these transformations. This idea is illustrated in Figure 4.13, where $g(\mathbf{x})$ is a transformation of the measurement $\mathbf{x}$ and ϕ is a test that is invariant to g. The trick is to decide on an appropriate set of transformations $g(\mathbf{x})$ that the decision rule $\phi(\mathbf{x})$ should be invariant to, and then restrict attention to only those rules that *are* invariant to $g(\mathbf{x})$.

To turn these ideas into a workable mathematical formalism, we need to define exactly what we mean by an invariant hypothesis-testing problem and an invariant decision rule.

Invariance of an Hypothesis Testing Problem

The basic problem in binary hypothesis testing is to observe the random vector $\mathbf{X}$: $F_{\boldsymbol{\theta}}(\mathbf{x})$ and test the hypothesis $H_0 : \boldsymbol{\theta} \in \Theta_0$ versus the alternative $H_1 : \boldsymbol{\theta} \in \Theta_1$, with $\Theta_0 \cap \Theta_1 = \Phi$ and $\Theta_0 \cup \Theta_1 = \Theta$. If $g(\mathbf{x})$ is a transformation drawn from the group G (the group operation is composition: $g_2 g_1(\mathbf{x}) = g_2[g_1(\mathbf{x})]$), then the transformed vector $\mathbf{y} = g(\mathbf{x})$ is distributed as follows:

$$F'_{\boldsymbol{\theta}}(\mathbf{y}) = P_{\boldsymbol{\theta}}\big[g(\mathbf{X}) \leq \mathbf{y}\big] \qquad\qquad 4.82$$

where $F'_{\boldsymbol{\theta}}(\mathbf{y})$ is the distribution function for $\mathbf{y}$. If $P_{\boldsymbol{\theta}}\big[g(\mathbf{X}) \leq \mathbf{y}\big]$ may be written as $P_{\bar{g}(\boldsymbol{\theta})}[\mathbf{X} \leq \mathbf{y}]$, then we may write

$$F'_{\boldsymbol{\theta}}(\mathbf{y}) = F_{\bar{g}(\boldsymbol{\theta})}(\mathbf{y}), \qquad\qquad 4.83$$

meaning that the transformation $g(\mathbf{x})$ leaves the distribution invariant in form while changing the parameter from $\boldsymbol{\theta}$ to $\bar{g}(\boldsymbol{\theta})$. When this happens, we say that the family of distributions $P_{\boldsymbol{\theta}} = \big\{F_{\boldsymbol{\theta}}(\mathbf{x}), \boldsymbol{\theta} \in \Theta\big\}$ is invariant to G. If, furthermore, $\bar{g}(\Theta_0) = \Theta_0$ and $\bar{g}(\Theta_1) = \Theta_1$, meaning that the induced transformation on the parameter space preserves the original dichotomy of the parameter space, then we say that the hypothesis testing problem $H_0 : \boldsymbol{\theta} \in \Theta_0$ versus $H_1 : \boldsymbol{\theta} \in \Theta_1$ is invariant to G.

Example 4.6

Let $\mathbf{X}$ denote an $N[\mathbf{m}, \mathbf{I}]$ random vector. We would like to test $H_0 : \mathbf{m} = \mathbf{m}_0$ versus $H_1 : \mathbf{m} = \mathbf{m}_1$. However, suppose a channel adds an unknown bias of $\gamma\mathbf{1}$, $-\infty < \gamma < \infty$, as illustrated in Figure 4.14(a). The measurement $\mathbf{y} = \mathbf{x} + \gamma\mathbf{1}$ is now distributed as $\mathbf{Y} : N[\mathbf{m} + \gamma\mathbf{1}, \mathbf{I}]$, so it is only reasonable to test $H_0 : \mathbf{m} =$

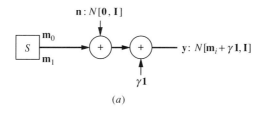

(a)

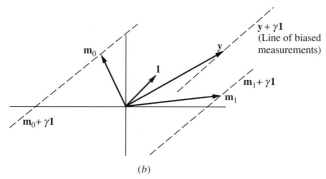

(b)

Figure 4.14 Channel bias. (a) Channel adds unknown bias. (b) Corresponding geometry of measurements.

$\mathbf{m}_0 + \gamma\mathbf{1}$ versus $H_1 : \mathbf{m} = \mathbf{m}_1 + \gamma\mathbf{1}$ in the distribution for $\mathbf{Y}$. The transformed measurements $\mathbf{y}$ are illustrated in Figure 4.14(b), together with the parameter sets $\mathbf{m}_i + \gamma\mathbf{1}$. If we do not know the bias γ, then we should ask for invariance to it. This suggests that we look for invariance to the group of transformations $G = \{g : g(\mathbf{y}) = \mathbf{y} + c\mathbf{1}\}$, with $-\infty < c < \infty$. The transformation $g(\mathbf{y})$ induces the distribution $N[\mathbf{m} + (\gamma + c)\mathbf{1}, \mathbf{I}]$ for $g(\mathbf{y})$, meaning that the distribution for $\mathbf{y}$ is invariant to G and showing that $\bar{g}(\mathbf{m}_i + \gamma\mathbf{1}) = \mathbf{m}_i + (\gamma + c)\mathbf{1}$. The induced sets $\bar{g}(\Theta_0)$ and $\bar{g}(\Theta_1)$ are therefore

$$\bar{g}(\Theta_0) = \mathbf{m}_0 + (\gamma + c)\mathbf{1} = \Theta_0$$
$$\bar{g}(\Theta_1) = \mathbf{m}_1 + (\gamma + c)\mathbf{1} = \Theta_1. \qquad 4.84$$

Therefore the hypothesis testing problem is invariant to G, the transformation group that adds an unknown bias of c to each measurement. ■

Example 4.7

Let $\mathbf{X} : N[\mu\mathbf{m}, \mathbf{I}]$ denote a normal measurement vector that is produced when a source produces $\mu\mathbf{m}$ and a channel adds $N[\mathbf{0}, \mathbf{I}]$ noise. We would like to test $H_0 : \mu \leq 0$ versus $H_1 : \mu > 0$. But, as illustrated in Figure 4.15(a), a channel also introduces the unknown gain $\gamma > 0$, and the unknown rotation $\mathbf{Q}_A$ in the subspace $\langle \mathbf{A} \rangle$ that is orthogonal to $\mathbf{m}$. The rotation is represented as $\mathbf{Q}_A = \mathbf{U}_A\mathbf{Q}\mathbf{U}_A^T + \mathbf{P_m}$, where $\mathbf{P}_A = \mathbf{U}_A\mathbf{U}_A^T$ is the projection onto

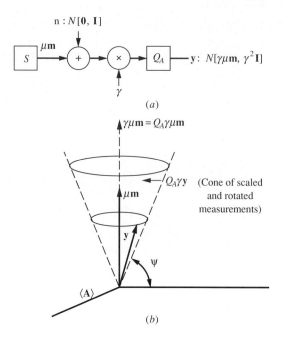

Figure 4.15 Channel gain and rotation. (*a*) Channel
introduces unknown gain and rotation in
subspace orthogonal to **m**. (*b*) Corresponding geometry of measurements.

$\langle\mathbf{A}\rangle$ and $\mathbf{P_m} = \mathbf{m}(1/(\mathbf{m^Tm}))\mathbf{m^T}$ is the projection onto $\langle\mathbf{m}\rangle$. The rotation
is orthogonal, meaning $\mathbf{Q}_A\mathbf{Q}_A^T = \mathbf{Q}_A^T\mathbf{Q}_A = \mathbf{I}$, and it leaves $\mathbf{m}$ unaffected:
$\mathbf{Q}_A\mathbf{m} = \mathbf{P_m}\mathbf{m} = \mathbf{m}$. The transformed measurements $\mathbf{y} = \mathbf{Q}_A\gamma\mathbf{x} = \gamma\mathbf{Q}_A\mathbf{x}$
are now distributed as $N[\gamma\mu\mathbf{m}, \gamma^2\mathbf{Q}_A\mathbf{I}\mathbf{Q}_A^T] = N[\gamma\mu\mathbf{m}, \gamma^2\mathbf{I}]$, so it is only reasonable to test $H_0 : (\gamma\mu \leq 0, \gamma^2 > 0)$ versus $H_1 : (\gamma\mu > 0, \gamma^2 > 0)$.
The transformed measurements $\mathbf{y}$ are illustrated in Figure 4.15(*b*). If we do
not know the gain γ and the rotation Q_A, then we should ask for invariance
to it. This suggests that we look for invariance to the group of transformations
$G = \{g : g(\mathbf{y}) = \mathbf{Q}_A c\mathbf{y}$, with $\mathbf{Q}_A = \mathbf{U}_A\mathbf{Q}\mathbf{U}_A^T + \mathbf{P_m}$ and $c > 0\}$. The transformation G induces the distribution $N[c\gamma\mu\mathbf{m}, c^2\gamma^2\mathbf{I}]$ for $g(\mathbf{y})$, meaning that the
distribution of $\mathbf{y}$ is invariant to G and showing that the transformation of the pa-
rameters $(\gamma\mu, \gamma^2\mathbf{I})$ is $\bar{g}(\gamma\mu, \gamma^2\mathbf{I}) = (c\gamma\mu, c^2\gamma^2\mathbf{I})$. The induced sets $\bar{g}(\Theta_0)$ and
$\bar{g}(\Theta_1)$ are

$$\bar{g}(\Theta_0) = (c\gamma\mu \leq 0, c^2\gamma^2 > 0) = \Theta_0$$

$$\bar{g}(\Theta_1) = (c\gamma\mu > 0, c^2\gamma^2 > 0) = \Theta_1.$$

Therefore the hypothesis testing problem is invariant to G, the transformation that introduces unknown gain and rotation in the subspace orthogonal to $\langle \mathbf{m} \rangle$.   ∎

Example 4.8

Let $\mathbf{X} = \mu \mathbf{H}\boldsymbol{\theta} + \mathbf{N}$ be generated by a source that transmits $\mu\mathbf{H}\boldsymbol{\theta}$ and a channel that adds $N[\mathbf{0}, \sigma^2\mathbf{I}]$ noise. The distribution of $\mathbf{X}$ is $\mathbf{X} : N[\mu\mathbf{H}\boldsymbol{\theta}, \sigma^2\mathbf{I}]$. We would like to test $H_0 : \mu = 0$ versus $H_1 : \mu > 0$. Under hypothesis H_0 the mean of $\mathbf{X}$ is zero and under the alternative H_1 the mean of $\mathbf{X}$ lies somewhere in the subspace $\langle \mathbf{H} \rangle$. But a channel introduces a bias $\mathbf{v}$ that lies in $\langle \mathbf{A} \rangle$, the subspace orthogonal to $\langle \mathbf{H} \rangle$, and it also rotates $\mathbf{x} + \mathbf{v}$ in the subspace $\langle \mathbf{H} \rangle$. This is illustrated in Figure 4.16(a). The rotation in $\langle \mathbf{H} \rangle$ is represented as $\mathbf{Q}_H = \mathbf{U}_H\mathbf{Q}\mathbf{U}_H^T + \mathbf{P}_A$, where $\mathbf{P}_H = \mathbf{U}_H\mathbf{U}_H^T$ is the projection onto $\langle \mathbf{H} \rangle$ and $\mathbf{P}_A$ is the projection onto $\langle \mathbf{A} \rangle$. The rotation of $\mathbf{v}$ leaves $\mathbf{v}$ unaffected ($\mathbf{Q}_H\mathbf{v} = \mathbf{P}_A\mathbf{v} = \mathbf{v}$), and the rotation of $\mu\mathbf{H}\boldsymbol{\theta}$ just produces a rotated

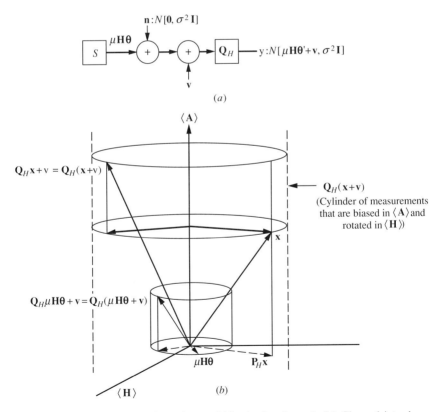

Figure 4.16 Parameter sets for orthogonal bias in the channel. (*a*) Channel introduces orthogonal subspace bias and subspace rotation. (*b*) Corresponding geometry of measurements.

version of $\mu\mathbf{H}\boldsymbol{\theta}$ ($\mathbf{Q}_H\mu\mathbf{H}\boldsymbol{\theta} = \mu\mathbf{H}\boldsymbol{\theta}'$, with $(\boldsymbol{\theta}')^\mathsf{T}\mathbf{H}^\mathsf{T}\mathbf{H}\boldsymbol{\theta}' = \boldsymbol{\theta}^\mathsf{T}\mathbf{H}^\mathsf{T}\mathbf{H}\boldsymbol{\theta}$). Because $\mathbf{Q}_H$ is an orthogonal matrix ($\mathbf{Q}_H\mathbf{Q}_H^\mathsf{T} = \mathbf{Q}_H^\mathsf{T}\mathbf{Q}_H = \mathbf{I}$), the transformed measurements $\mathbf{y} = \mathbf{Q}_H(\mathbf{x} + \mathbf{v}) = \mathbf{Q}_H\mathbf{x} + \mathbf{v}$ are distributed as $\mathbf{Y} : N[\mu\mathbf{H}\boldsymbol{\theta}' + \mathbf{v}, \sigma^2\mathbf{I}]$. It is only reasonable to test $H_0 : \mathbf{Y} : N[\mathbf{v}, \sigma^2\mathbf{I}]$ versus $H_1 : \mathbf{Y} : N[\mu\mathbf{H}\boldsymbol{\theta}' + \mathbf{v}, \sigma^2\mathbf{I}]$. The transformed measurements are illustrated in Figure 4.16(b). If we do not know the orthogonal bias $\mathbf{v}$ or the subspace rotation $\mathbf{Q}_H$, then we should ask for invariance to it. This suggests that we look for invariance to the group of transformations

$$G = \left\{ g : g(\mathbf{y}) = \mathbf{Q}_H(\mathbf{y} + \mathbf{w}), \quad \mathbf{Q}_H = \mathbf{U}_H\mathbf{Q}\mathbf{U}_H^\mathsf{T} + \mathbf{P}_A \quad \text{and} \quad \mathbf{P}_A\mathbf{w} = \mathbf{w} \right\}.$$

Because $\mathbf{Q}_H$ is an orthogonal matrix, the distribution of the transformed measurement $\mathbf{Q}_H(\mathbf{y} + \mathbf{w})$ is $N[\mu\mathbf{H}\boldsymbol{\theta}' + \mathbf{v} + \mathbf{w}, \sigma^2\mathbf{I}]$. This means that the distribution of $\mathbf{y}$ is invariant to G and it shows that the transformation of the parameter $\mu\mathbf{H}\boldsymbol{\theta} + \mathbf{v}$ is $\bar{g}(\mu\mathbf{H}\boldsymbol{\theta} + \mathbf{v}) = \mu\mathbf{H}\boldsymbol{\theta}' + (\mathbf{v} + \mathbf{w})$. The parameter sets $\bar{g}(\Theta_0)$ and $\bar{g}(\Theta_1)$ are

$$\bar{g}(\Theta_0) = \mathbf{v} + \mathbf{w} = \Theta_0$$

$$\bar{g}(\Theta_1) = \mu\mathbf{H}\boldsymbol{\theta}' + \mathbf{v} + \mathbf{w} = \Theta_1.$$

Therefore the hypothesis testing problem is invariant to G, the transformation group that adds orthogonal subspace bias and rotates within the signal subspace. ∎

These examples illustrate that for some hypothesis testing problems there are natural transformations to which a problem should be invariant. So, when applying invariance principles, the first problem is to find a natural group of transformations to which the hypothesis testing problem should remain invariant. Then find an hypothesis test that is invariant to the same group. This brings us to the topic of invariant tests.

Invariant Tests and Maximal Invariant Statistics

The hypothesis test $\phi(\mathbf{x})$ is said to be invariant to G if

$$\phi[g(\mathbf{x})] = \phi(\mathbf{x}) \qquad\qquad 4.85$$

for all $g \in G$. A *statistic* $M(\mathbf{x})$ is said to be a maximal invariant statistic if

1. (invariant) $M[g(\mathbf{x})] = M[\mathbf{x}]$ for all $g \in G$, and
2. (maximal) $M[\mathbf{x}_1] = M[\mathbf{x}_2]$ implies $\mathbf{x}_2 = g(\mathbf{x}_1)$ for some $g \in G$.

Maximality is a form of uniqueness. It says that $M[\mathbf{x}_1] = M[\mathbf{x}_2]$ iff $\mathbf{x}_2 = g(\mathbf{x}_1)$ for some $g \in G$. We say that $M[\mathbf{x}_1] = M[\mathbf{x}_2]$ iff $\mathbf{x}_1$ and $\mathbf{x}_2$ lie on the same orbit. This is illustrated in Figure 4.17. The maximal invariant statistic simply organizes measurements $\mathbf{x}$ into distinguishable equivalence classes where $M(\mathbf{x})$ is constant and where every $\mathbf{x}$ is related to every other through a transformation $g(\mathbf{x})$. The only requirement of an invariant test $\phi(\mathbf{x})$ is that it remain constant on these equivalence classes. Therefore every invariant test may be written as a function of a maximal invariant statistic:

$$\phi(\mathbf{x}) = \phi[M(\mathbf{x})]. \qquad\qquad 4.86$$

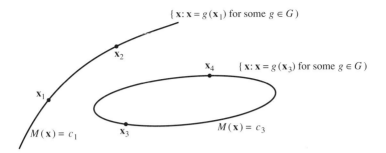

Figure 4.17 The maximal invariant statistic $M(\mathbf{x})$ is constant on orbits.

The interpretation of this result is that, when solving invariant hypothesis-testing problems, it is permissible to restrict attention to decision rules that are functions of a maximal invariant statistic. The application of invariance principles to physical problems must be justified, as in Examples 4.6 through 4.8, by the existence of a natural invariance requirement for the problem. The following examples illustrate how maximal invariant statistics are found.

Example 4.9 (Invariance to Bias)

Let's return to Example 4.6, where the problem was to test $H_0 : \mathbf{Y} : N[\mathbf{m}_0 + \gamma\mathbf{1}, \mathbf{I}]$ versus $H_1 : \mathbf{Y} : N[\mathbf{m}_1 + \gamma\mathbf{1}, \mathbf{I}]$. The statistic

$$\mathbf{z} = \mathbf{Py} \qquad\qquad 4.87$$

$$\mathbf{P} = \left(\mathbf{I} - \frac{1}{N}\mathbf{11}^{\mathrm{T}}\right) = \mathbf{P}^2$$

is a rank $N - 1$ projection onto the subspace that is orthogonal to the vector $\mathbf{1}$. This projection is illustrated in Figure 4.18. We claim that $\mathbf{Py}$ is a maximal invariant to the group $G = \{g : g(\mathbf{y}) = \mathbf{y} + c\mathbf{1}\}$. Is it? Let's see.

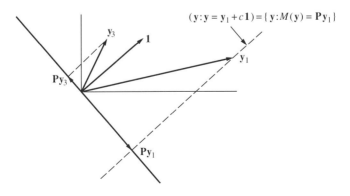

Figure 4.18 The maximal invariant statistic $\mathbf{Py}$.

1. $P(y + c1) = Py$

2. $Py_1 = Py_2 \Rightarrow \left(I - \dfrac{1}{N} 11^T\right) y_1 = \left(I - \dfrac{1}{N} 11^T\right) y_2$

$$\Rightarrow y_2 = y_1 + \dfrac{1}{N} 1^T(y_2 - y_1)1$$

$$= y_1 + c1.$$

The maximal invariant statistic Py is illustrated in Figure 4.18.   ■

Example 4.10 (Invariance to Gain and Rotation)

Let's return to Example 4.7, where the problem was to test $H_0 : (\gamma\mu \le 0, \gamma^2 > 0)$ versus $H_1 : (\gamma\mu > 0, \gamma^2 > 0)$ in the model $Y : N[\gamma\mu m, \gamma^2 I]$. We want to have invariance to the gain and rotation transformation $cQ_A y = Q_A cy = (U_A Q U_A^T + P_m)cy$. We claim that the statistic

$$t = \dfrac{m^T y}{(m^T m)^{1/2}(y^T y)^{1/2}} \qquad\qquad 4.88$$

is a maximal invariant statistic to the group $G = \{g : g(y) = cQ_A y\}$. From Figure 4.15 we see that this statistic is just the cosine of the angle that the measurement y makes with the subspace $\langle m \rangle$. It is also the sine of the angle that the side of the cone makes with the horizontal in Figure 4.15. From the figure we see that this angle is clearly invariant to gain and rotation of y, and the following calculations confirm it:

1. $\dfrac{m^T(Q_A cy)}{(m^T m)^{1/2}[(Q_A cy)^T(Q_A cy)]^{1/2}} = \dfrac{m^T y}{(m^T m)^{1/2}(y^T y)^{1/2}} \qquad 4.89$

2. $\dfrac{m^T y_1}{(m^T m)^{1/2}(y_1^T y_1)^{1/2}} = \dfrac{m^T y_2}{(m^T m)^{1/2}(y_2^T y_2)^{1/2}};$

$$\Rightarrow m^T(y_2 - cy_1) = 0, \quad c = \sqrt{\dfrac{y_2^T y_2}{y_1^T y_1}}$$

$$\Rightarrow y_2 - cy_1 \perp m$$

$$\Rightarrow y_2 = Q_A cy_1.$$

This choice for a maximal invariant statistic is not as inventive as it seems. It follows from the geometry of Figure 4.15 or from the observation that $t_1 = m^T y$ and $t_2 = y^T y$ are sufficient for the pair $(\gamma\mu, \gamma^2)$. The statistic

$$t = \dfrac{t_1}{t_2^{1/2}}, \quad \text{or} \quad \dfrac{1}{(m^T m)^{1/2}} t,$$

is an obvious function to try as a maximal invariant.   ■

Example 4.11 (Invariance to Orthogonal Bias and Subspace Rotation)

Let's return to Example 4.8, where the problem was to test $H_0 : \mu = 0$ versus $\mu > 0$ in the model $\mathbf{Y} : N[\mu \mathbf{H} \boldsymbol{\theta}, \sigma^2 \mathbf{I}]$. The quadratic form $\mathbf{y}^T P_H \mathbf{y} = (P_H \mathbf{y})^T (P_H \mathbf{y})$ is a maximal invariant to the group of transformations $G = \{ g : g(\mathbf{y}) = \mathbf{Q}_H (\mathbf{y} + \mathbf{v}) \}$ that adds a bias from the orthogonal subspace $\langle \mathbf{A} \rangle$ and rotates in the subspace $\langle \mathbf{H} \rangle$. This is obvious from Figure 4.16, and these calculations confirm it:

1. $(\mathbf{Q}_H (\mathbf{y} + \mathbf{v}))^T \mathbf{P}_H \mathbf{Q}_H (\mathbf{y} + \mathbf{v}) = (\mathbf{y} + \mathbf{v})^T \mathbf{P}_H (\mathbf{y} + \mathbf{v}) = \mathbf{y}^T \mathbf{P}_H \mathbf{y}$ 4.90

2. $\mathbf{y}_1^T \mathbf{P}_H \mathbf{y}_1 = \mathbf{y}_2^T \mathbf{P}_H \mathbf{y}_2 \Rightarrow \mathbf{y}_2 = \mathbf{Q}_H (\mathbf{y}_1 + \mathbf{v})$.   ∎

Suppose we have established the invariance of a hypothesis testing problem and found a maximal invariant statistic $M(\mathbf{x})$. What kind of optimality statements might we make about the invariant decision rule $\phi(\mathbf{x}) = \phi[M(\mathbf{x})]$? Here is one answer.

Uniformly Most Powerful Invariant Test

The invariant test $\phi(\mathbf{x}) = \phi[M(\mathbf{x})]$ is uniformly most powerful of size α for testing $H_0 : \boldsymbol{\theta} \in \Theta_0$ versus $H_1 : \boldsymbol{\theta} \in \Theta_1$ in the model $\mathbf{X} : F_{\boldsymbol{\theta}}(\mathbf{x})$ if, for the test ϕ and for every competing test ϕ' that is also invariant-G,

1. (size) $\sup_{\boldsymbol{\theta} \in \Theta_0} E_{\boldsymbol{\theta}} \phi[M(\mathbf{X})] = \alpha; \ \sup_{\boldsymbol{\theta} \in \Theta_0} E_{\boldsymbol{\theta}} \phi'(\mathbf{X}) \leq \alpha$ 4.91

2. (power) $E_{\boldsymbol{\theta}} \phi[M(\mathbf{X})] \geq E_{\boldsymbol{\theta}} \phi'[\mathbf{X}]$ for all $\boldsymbol{\theta} \in \Theta_1$.

To establish that a test $\phi[M(\mathbf{x})]$ is UMP-invariant, we will have to establish that the maximal invariant statistic has a monotone likelihood ratio. This we shall do in the next several sections of this chapter where we work out several applications of invariance in the theory of signal detection.

Reduction by Sufficiency and Invariance

The principles of sufficiency and invariance may be used to compress measurements into statistics of low dimensionality that satisfy invariance conditions. Then within a restricted class of invariant tests, it is often possible to find a uniformly most powerful test. Often the invariance condition is so basic to the problem that a decision rule that does not exhibit the invariant condition would never be accepted. Then the restricted class of invariant tests is just the class of interest to us, and optimality statements within this class are meaningful.

The steps when applying the principles of sufficiency and invariance are always the following:

1. begin with the hypothesis testing problem $H_0 : \boldsymbol{\theta} \in \Theta_0$ versus $H_1 : \boldsymbol{\theta} \in \Theta_1$ for the measurements $\mathbf{X} : F_{\boldsymbol{\theta}}(\mathbf{x})$;

2. if possible, replace the data $\mathbf{X}$ by the sufficient statistic $\mathbf{T}(\mathbf{X})$;

3. rephrase the hypothesis testing problem as one of testing $H_0 : \boldsymbol{\theta} \in \Theta_0$ versus $H_1 : \boldsymbol{\theta} \in \Theta_1$ for the sufficient statistic $\mathbf{T}(\mathbf{X})$;

4. establish that the rephrased hypothesis testing problem is invariant to a group of transformations G that is meaningful for the hypothesis testing problem;

5. find a maximal invariant statistic;

6. construct a likelihood ratio test of H_0 versus H_1 that has size α;

7. check to see whether the statistic M has a monotone-likelihood ratio. If it does, then the likelihood ratio test is uniformly most powerful invariant for testing one-sided hypotheses of the form $H_0 : \theta \le \theta_0$ versus $H_1 : \theta > \theta_0$ or of the form $H_0 : \theta = \theta_0$ versus $H_1 : \theta > \theta_0$.

In the next four sections, we study four problems from the theory of signal detection. For each problem, we apply the principles of sufficiency and invariance in order to find a uniformly most powerful test within a class of invariant tests. As we proceed from known signal and noise models to partially known models, we proceed from the normally distributed matched filter to the t-distributed, normalized, matched filter, to the χ^2-distributed energy detector, and to the F-distributed, normalized energy detector. By comparing detector performance for each case, we are able to assess the performance penalties that accrue to model uncertainties.

4.9   MATCHED FILTERS (NORMAL)

In signal detection problems, we assume that a sequence of measurements is read. Each measurement is a sum of a signal component and a noise component:

$$x_n = \mu s_n + \sigma w_n; \qquad n = 0, 1, \ldots, N - 1. \qquad\qquad 4.92$$

The measurements are organized into an N-dimensional measurement vector

$$\mathbf{x} = \mu\mathbf{s} + \sigma\mathbf{w}$$

where the vector $\mu\mathbf{s}$ contains samples of the signal to be detected and the vector $\sigma\mathbf{w}$ contains samples of the added noise. It is useful to think of $\mathbf{s}$ as a signal originating from a source and μ as an attenuation constant. Similarly, think of $\mathbf{w}$ as a noise vector that is scaled by the level constant σ and added in a channel between the source and the receiver. This is illustrated in Figure 4.19.

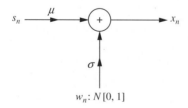

Figure 4.19 Signal plus noise model.

We assume that the noise vector $\mathbf{w}$ is drawn from a multivariate normal distribution $\mathbf{w} : N[\mathbf{0}, \mathbf{I}]$. This means that the measurement $\mathbf{x}$ is drawn from a multivariate normal distribution $\mathbf{x} : N[\mu\mathbf{s}, \sigma^2\mathbf{I}]$. In any communication system where the source transmits a signal to a receiver through a distortionless channel, the signal vector $\mathbf{s}$ is known. The receiver sees the vector $\mathbf{x} = \mu\mathbf{s} + \sigma\mathbf{w}$, and the question is whether $\mu = 0$ or $\mu > 0$. The signal detection problem is phrased as the following test of hypotheses: $H_0 : \mu = 0$ versus $H_1 : \mu > 0$ in the model $\mathbf{x} : N[\mu\mathbf{s}, \sigma^2\mathbf{I}]$. We emphasize that the signal $\mathbf{s}$, the noise correlation matrix $\mathbf{I}$, and the variance parameter σ^2 are known. Only μ is unknown, and we wish to determine whether it is zero or greater than zero.

There is no loss in generality by assuming that the noise correlation matrix is $\mathbf{I}$. If the noise has correlation matrix $\mathbf{R}$, then we can replace the measurement $\mathbf{x}$ by $\mathbf{R}^{-T/2}\mathbf{x}$, which is then distributed as $\mathbf{R}^{-T/2}\mathbf{x} : N[\mu\mathbf{R}^{-T/2}\mathbf{s}, \sigma^2\mathbf{I}]$. We say that the measurements $\mathbf{R}^{-T/2}\mathbf{x}$ are *prewhitened* because their elements are uncorrelated with each other. The multivariate normal distribution is invariant to the transformation $\mathbf{R}^{-T/2}\mathbf{x}$. We know this because the transformation is a linear one that leaves the distribution multivariate normal and induces the following transformation on the parameter space $(\mu\mathbf{s}, \sigma^2\mathbf{R})$:

$$\overline{g}(\mu\mathbf{s}, \sigma^2\mathbf{R}) = (\mu\mathbf{R}^{-T/2}\mathbf{s}, \sigma^2\mathbf{I}). \qquad 4.93$$

This transformation on the parameter space leaves the hypothesis testing problem invariant because $\mathbf{R}^{-T/2}\mathbf{x}$ is $N[\mathbf{0}, \sigma^2\mathbf{I}]$ if and only if $\mathbf{x}$ is $N[\mathbf{0}, \sigma^2\mathbf{R}]$, and $\mathbf{R}^{-T/2}\mathbf{x}$ is $N[\mu\mathbf{R}^{-T/2}\mathbf{s}, \sigma^2\mathbf{I}]$ if and only if $\mathbf{x}$ is $N[\mu\mathbf{s}, \sigma^2\mathbf{R}]$, with $\mu > 0$.

The importance of this result is that there is really only one signal detection problem when the signal correlation matrix is known, and that is the problem of detecting signals in *uncorrelated* or *white* noise. All other problems involving non-white noise with known correlation matrices may be transformed into white noise problems. The transformation $\mathbf{R}^{-T/2}\mathbf{x}$ is called a whitening transformation.

The density function for the observation vector $\mathbf{x}$ is

$$f_\mu(\mathbf{x}) = \frac{1}{(2\pi\sigma^2)^{N/2}} \exp\left\{ -\frac{1}{2\sigma^2} (\mathbf{x} - \mu\mathbf{s})^T(\mathbf{x} - \mu\mathbf{s}) \right\}. \qquad 4.94$$

This density belongs to a one-parameter exponential family. From the Fisher-Neyman factorization theorem, we know that a sufficient statistic for the parameter μ is

$$m = \frac{\mathbf{s}^T\mathbf{x}}{\sigma(\mathbf{s}^T\mathbf{s})^{1/2}}. \qquad 4.95$$

The normalizing constant $\sigma(\mathbf{s}^T\mathbf{s})^{1/2}$ merely simplifies subsequent calculations. The distribution of the sufficient statistic is normal, with mean $\mu\mathbf{s}^T\mathbf{s}/\sigma$ and variance 1:

$$m : N\left[\frac{\mu\sqrt{E_s}}{\sigma}, 1 \right] \qquad 4.96$$

$$E_s = \mathbf{s}^T\mathbf{s}.$$

We say that the *noncentrality* parameter of the normal distribution is $(\mu/\sigma)\sqrt{E_s}$. The corresponding signal-to-noise ratio is

$$\text{SNR} = \frac{\mu^2}{\sigma^2} E_s.$$ (4.97)

The principle of sufficiency applied to hypothesis testing says that we may replace the N-dimensional measurement with the one-dimensional sufficient statistic m and replace the hypothesis testing problem with the following test: $H_0 : \mu = 0$ versus $H_1 : \mu > 0$ in the model $m : N[(\mu/\sigma)\sqrt{E_s}, 1]$. The statistic m is called a *matched filter* because it is matched to the signal $\mathbf{s}$. The matched filter is illustrated in Figure 4.20. In the figure we have introduced the projector $\mathbf{P_s} = \mathbf{ss}^{\mathrm{T}}/\mathbf{s}^{\mathrm{T}}\mathbf{s}$ for convenience in comparing the matched filter with the matched subspace filter that we shall derive in Section 4.11. We are allowed to introduce it because $\mathbf{s}^{\mathrm{T}}\mathbf{P_s} = \mathbf{s}^{\mathrm{T}}$.

The matched filter has monotone likelihood because its distribution is $N[(\mu/\sigma)\sqrt{E_s}, 1]$. By the Karlin-Rubin theorem, the following one-sided threshold test is uniformly most powerful for testing H_0 versus H_1:

$$\phi(\mathbf{x}) = \begin{cases} 1 \sim H_1, & m > m_0 \\ 0 \sim H_0, & m \le m_0 \end{cases}$$ (4.98)

$$m = \frac{\mathbf{s}^{\mathrm{T}} P_s \mathbf{x}}{\sigma(\mathbf{s}^{\mathrm{T}}\mathbf{s})^{1/2}}.$$

The performance of the matched filter is summarized by its size and power or, equivalently, its false alarm probability and detection probability. The false alarm probability is given by

$$P_{\text{FA}} = \int_{m_0}^{\infty} (2\pi)^{-1/2} \exp\{-x^2/2\}\, dx$$

$$= 1 - \Phi(m_0).$$ (4.99)

The threshold m_0 is determined from this formula to produce the desired size. The detection probability is then determined from the formula

$$P_{\text{D}} = \int_{m_0}^{\infty} (2\pi)^{-1/2} \exp\left\{-\left(x - \frac{\mu\sqrt{E_s}}{\sigma}\right)^2/2\right\}\, dx$$

$$= \int_{m_0-(\mu/\sigma)\sqrt{E_s}}^{\infty} (2\pi)^{-1/2} \exp\{-z^2/2\}\, dz$$ (4.100)

$$= 1 - \Phi\left(m_0 - (\mu/\sigma)\sqrt{E_s}\right).$$

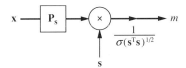

Figure 4.20 Matched filter.

When the threshold has been set to achieve the size α, then the power is computed for given values of $(\mu/\sigma)\sqrt{E_s}$ from this formula. The power is only a function of the signal-to-noise ratio SNR $= (\mu^2/\sigma^2)E_s$. Receiver operating characteristics for the matched filter are illustrated in Figure 4.21. In the figure, each curve shows the detection probability plotted versus the voltage signal-to-noise ratio $(\mu/\sigma)\sqrt{E_s}$, for some fixed false alarm probability. Then, for example, at $P_{FA} = 0.1$ and $(\mu/\sigma)\sqrt{E_s} = 3$, $P_D = 0.95$. The curves of Figure 4.21 are straight lines because they are plotted on normal probability paper. (Normal probability paper is defined in Problem 4.3.)

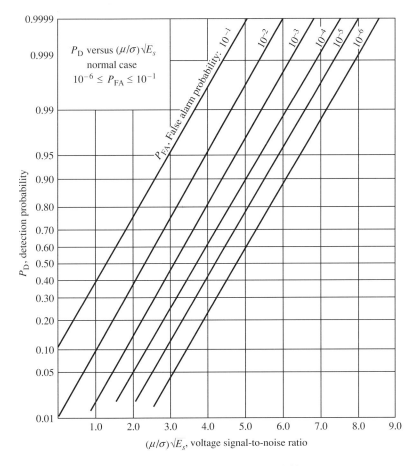

Figure 4.21 Receiver operating characteristics: matched filter.

4.10   CFAR MATCHED FILTERS (t)

When the noise variance σ^2 is unknown in the signal detection problem, then it is unknown in the distribution $\mathbf{x} : N[\mu\mathbf{s}, \sigma^2\mathbf{I}]$. There is no way to construct the matched filter m that has the false alarm probability P_{FA} (or constant false alarm rate [CFAR]) and the uniformly most powerful detection probability P_D, because the variance of m cannot be set to 1 by dividing by $1/[\sigma(\mathbf{s}^T\mathbf{s})^{1/2}]$. Consequently, the threshold m_0 cannot be set to achieve false alarm probability α. There is no UMP test of H_0 versus H_1. Perhaps there is a way to proceed, reducing the problem by sufficiency and invariance, in order to find a detector that is UMP within a restricted class of detectors that are invariant to a group of transformations.

The measurements are distributed as $\mathbf{x} : N[\mu\mathbf{s}, \sigma^2\mathbf{I}]$. The signal $\mathbf{s}$ is known, but the variance σ^2 is unknown. The detection problem is one of testing $H_0 : \mu = 0$ versus $H_1 : \mu > 0$ in the model $\mathbf{x} : N[\mu\mathbf{s}, \sigma^2\mathbf{I}]$ when the variance σ^2 is unknown. Think of this as a problem of testing $H_0 : \mu = 0$ versus $H_1 : \mu > 0$ in the model $\sigma\mathbf{x}$ when $\mathbf{x}$ is distributed as $N[(\mu/\sigma)\mathbf{s}, \mathbf{I}]$. This problem is essentially Example 4.7.

What kind of invariance requirement would be appropriate for this problem? Here is an answer. Ignorance of σ^2 in the distribution for $\mathbf{x} : N[\mu\mathbf{s}, \sigma^2\mathbf{I}]$ is equivalent to ignorance of the scale constant σ in the measurement $\sigma\mathbf{x}$ when $\mathbf{x} : N[(\mu/\sigma)\mathbf{s}, \mathbf{I}]$. If the scale constant is unknown, then we should value a test that is invariant to scaling of the measurements and rotation of the measurements in the subspace $\langle \mathbf{A} \rangle$ orthogonal to $\langle \mathbf{s} \rangle$. (Refer to the discussion in Examples 4.7 and 4.10.) Therefore, we consider the group of transformations $G = \{ g : g(\mathbf{y}) = \mathbf{Q}_A c\mathbf{y} \}$ that scale the measurements by a positive constant $c > 0$ and rotate in the subspace $\langle \mathbf{A} \rangle$. This transformation leaves the distribution of $\mathbf{x}$ multivariate normal and induces a transformation on the parameter space:

$$\mathbf{Q}_A c\mathbf{x} : N[c\mu\mathbf{s}, c^2\sigma^2\mathbf{I}]. \qquad 4.101$$

This transformation on the parameter space leaves the hypothesis testing problem invariant because $\mathbf{x} : N[\mathbf{0}, \sigma^2\mathbf{I}]$ if and only if $\mathbf{Q}_A c\mathbf{x} : N[\mathbf{0}, c^2\sigma^2\mathbf{I}]$, and $\mathbf{x} : N[\mu\mathbf{s}, \sigma^2\mathbf{I}]$ with $\mu > 0$, if and only if $\mathbf{Q}_A c\mathbf{x} : N[c\mu\mathbf{s}, c^2\sigma^2\mathbf{I}]$, with $c\mu > 0$.

The density function for the measurements is

$$f_\theta(\mathbf{x}) = (2\pi\sigma^2)^{-N/2} \exp\left\{ -\frac{1}{2\sigma^2}(\mathbf{x} - \mu\mathbf{s})^T(\mathbf{x} - \mu\mathbf{s}) \right\}. \qquad 4.102$$

This density belongs to a two-parameter exponential family of distributions, with $\theta = (\mu, \sigma^2)$. From the Fisher-Neyman factorization theorem, we know that the pair of statistics (t_1, t_2) is sufficient for the parameter pair (μ, σ^2):

$$t_1 = \frac{\mathbf{s}^T\mathbf{x}}{(\mathbf{s}^T\mathbf{s})^{1/2}} \qquad 4.103$$

$$t_2 = \mathbf{x}^T\mathbf{x}.$$

From the principle of sufficiency for hypothesis testing, we may replace the N-dimensional measurements $\mathbf{x}$ with the two-dimensional sufficient statistic (t_1, t_2) and look for a function of these sufficient statistics that is a maximal invariant to the transformation G. We know from Examples 4.7 and 4.10 that the statistic $s^T x/(\sqrt{s^T s}\ \sqrt{x^T x})$ is a maximal invariant to gain and rotation in the subspace $\langle A\rangle$ that is orthogonal to the one-dimensional subspace $\langle s\rangle$. Furthermore, as illustrated in Figure 4.15, this statistic measures the cosine of the angle that the measurement x makes with the subspace $\langle s\rangle$. This angle is a maximal invariant to gain and phase, but so is the tangent of the angle ψ that the cone of Figure 4.15 makes with the subspace $\langle A\rangle$:

$$\tan\psi = \frac{s^T P_s x}{\sqrt{s^T s}\ \sqrt{x^T (I - P_s)x}}.\qquad\qquad 4.104$$

In this formula, $P_s x$ is the projection of the measurement x onto the subspace $\langle s\rangle$ and $(I - P_s)x$ is its projection onto the orthogonal subspace $\langle A\rangle$. The term $\sqrt{x^T (I - P_s)x}$ is the length of the latter projection. Of course, $s^T x/\sqrt{s^T s}$ is the signed magnitude of x projected onto the subspace $\langle s\rangle$. To see this, note that $s^T P_s x/\sqrt{s^T s}$ is the inner product of $P_s x$ and the unit vector $s^T/\sqrt{s^T s}$.) We shall actually slightly modify the statistic $\tan\psi$ as follows:

$$t = \frac{s^T P_s x/\sqrt{\sigma^2 s^T s}}{\sqrt{x^T (I - P_s)x/(N - 1)\sigma^2}}.\qquad\qquad 4.105$$

The division by σ in the numerator and denominator is an artifact that cancels out — that is, it is not really used in the computation of t. But it simplifies our study of t.

The numerator of the statistic t is an $N[(\mu/\sigma)\sqrt{E_s}, 1]$ random variable. The random variable $(1/\sigma)(I - P_s)x$ is an $N[0, (I - P_s)]$ random vector that is *independent* of $P_s x$ and therefore independent of the numerator. The random variable $(1/\sigma^2)x^T(I - P_s)x$ is therefore a χ^2_{N-1} random variable, as explained in Section 2.11 of Chapter 2. In the Appendix of this chapter it is shown that the ratio of a normal random variable to the scaled square-root of an independent chi-squared random variable is "t-distributed." Therefore we write

$$t : \begin{cases} t_{N-1}, & \text{under } H_0 \\ t_{N-1}\left(\dfrac{\mu\sqrt{E_s}}{\sigma}\right), & \text{under } H_1, \end{cases}\qquad\qquad 4.106$$

meaning that the statistic t is distributed as a "central t statistic" under H_0 and as a "noncentral t statistic" under H_1. The noncentrality parameter $(\mu/\sigma)\sqrt{E_s}$ is identical to the noncentrality parameter discovered in the case of the matched filter. What is so important here is that the distribution of t is *completely* characterized under H_0, even though the noise variance σ^2 is unknown. That is, under H_0, t is invariant to σ^2. This means that we can set a threshold t_0 in the test

$$\phi(t) = \begin{cases} 1, & t > t_0 \\ 0, & t \le t_0 \end{cases}\qquad\qquad 4.107$$

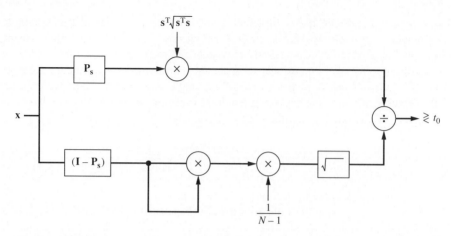

Figure 4.22 CFAR matched filter for detecting known form signal in noise of unknown
level.

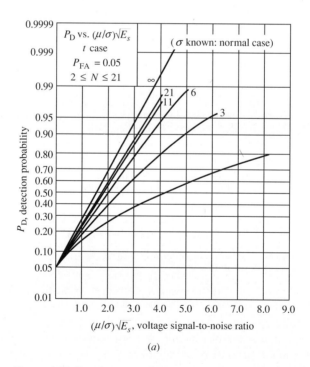

(a)

Figure 4.23 Receiver operating characteristics: CFAR matched
filter.

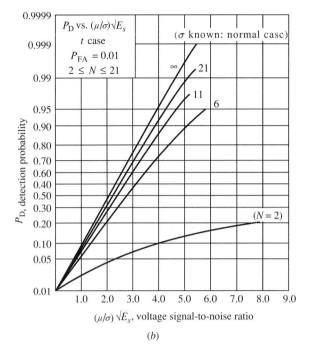

(*b*)

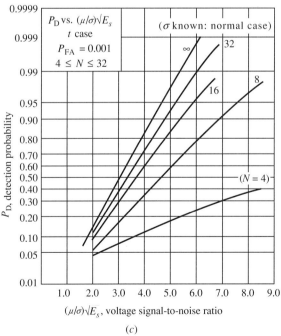

(*c*)

Figure 4.23 (continued)

in order to obtain a false alarm probability P_{FA}. Furthermore, because the likelihood ratio of a noncentral t is monotone, this threshold test is UMP invariant for testing $H_0 : \mu = 0$ versus $H_1 : \mu > 0$ in the distribution $\mathbf{x} : N[\mu\mathbf{s}, \sigma^2\mathbf{I}]$ when σ^2 is unknown! The actual probability of detection depends on the actual value of SNR.

The detector $\phi(t)$ is illustrated in Figure 4.22. The measurement is first projected onto orthogonal signal and noise subspaces. In the signal subspace, the projection is correlated with $\mathbf{s}^T/\sqrt{\mathbf{s}^T\mathbf{s}}$ to produce a matched filter output. In the noise subspace, the projection $(I - P_\mathbf{s})\mathbf{x}$ is squared, divided by $(N - 1)$, and rooted to produce an estimate of σ. This estimate is divided into the matched filter output to produce the maximal invariant statistic t.

The performance of the t detector is obtained by integrating the tails of the central and noncentral t-distribution. The results of P_D versus the voltage signal-to-noise ratio $(\mu/\sigma)\sqrt{E_s}$ are plotted in Figure 4.23 for several values of P_F and for several values of N. The curves are plotted on normal probability paper to illustrate their deviation from normality.

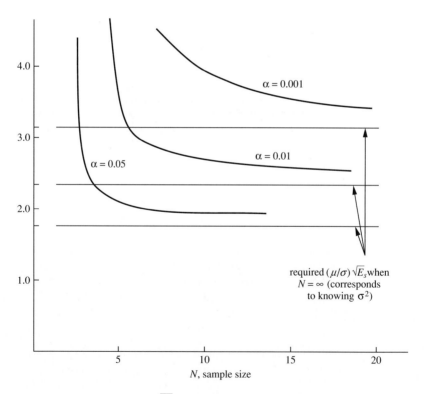

Figure 4.24 Required $(\mu/\sigma)\sqrt{E_s}$ for detection probability of 0.5 when noise level is unknown and sample size is small.

One way to assess the effects of not knowing the variance σ^2 is to plot the required voltage signal-to-noise ratio $(\mu/\sigma)\sqrt{E_s}$ to achieve a detection probability of $\beta = 0.5$ when the false alarm probability is fixed and the sample size is varied. The results are given in Figure 4.24. For small sample sizes N the effect is dramatic, and for large sample sizes it is not. As $N \to \infty$, the denominator of the t-statistic converges in probability to σ, and the performance of the normalized matched filter is the same as the matched filter. This is another way of saying that the t-distribution converges to the normal distribution or that the t-statistic converges in distribution to a normal random variable as $N \to \infty$.

4.11   MATCHED SUBSPACE FILTERS (χ^2)

It sometimes happens that the signal $\mathbf{s}$ in the measurement model $\mathbf{x} : N[\mu\mathbf{s}, \sigma^2\mathbf{I}]$ is a linear combination of modes or basis vectors, in which case it may be represented as

$$\mathbf{s} = \sum_{n=1}^{p} \theta_n \mathbf{h}_n$$

$$= \mathbf{H}\boldsymbol{\theta}.$$

4.108

Here $\mathbf{H}$ is a known $N \times p$ matrix with columns $\mathbf{h}_n$ and $\boldsymbol{\theta}$ is a $p \times 1$ vector with elements θ_n. Such a signal is represented in Figure 4.25. All of the results for detecting known signals, in noise of known or unknown variance, apply. The signal $\mathbf{s}$ is replaced by $\mathbf{H}\boldsymbol{\theta}$ in all of the formulas. It is more interesting to ask what happens when the mode matrix $\mathbf{H}$ is known but the mode weights are *unknown*. In this case, the signal is known to lie in the linear subspace $\langle\mathbf{H}\rangle$ spanned by the columns of $\mathbf{H}$, but its exact location is unknown because $\boldsymbol{\theta}$ is unknown. This is illustrated in Figure 4.25. We would like to test $H_0 : \mu = 0$ versus $H_1 : \mu > 0$ when $\mathbf{x}$ is distributed as $N[\mu\mathbf{H}\boldsymbol{\theta}, \sigma^2\mathbf{I}]$ and $\boldsymbol{\theta}$ is unknown. This is essentially the problem studied in Examples 4.8 and 4.11. In those examples we argued that we should have invariance to the group of transformations that adds bias in the subspace $\langle\mathbf{A}\rangle$ that is orthogonal to

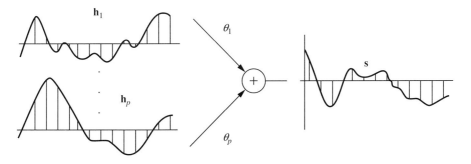

Figure 4.25 Linear combination of modes.

$\langle H \rangle$ and that rotates in the subspace $\langle H \rangle$. This group is written $G = \{ g : g(\mathbf{x}) = \mathbf{Q}_H(\mathbf{x} + \mathbf{v}), \ \mathbf{Q}_H = \mathbf{U}_H \mathbf{Q} \mathbf{U}_H^T + \mathbf{P}_A, \ \mathbf{P}_A \mathbf{v} = \mathbf{v} \}$. In Example 4.11 we established that the statistic

$$\chi^2 = \mathbf{x}^T \mathbf{P}_H \mathbf{x} \qquad\qquad 4.109$$

is a maximal invariant to G. As illustrated in Figure 4.16, $\mathbf{P}_H \mathbf{x}$ is the component of $\mathbf{x}$ that lies in the signal subspace $\langle H \rangle$ and $\mathbf{x}^T \mathbf{P}_H \mathbf{x}$ is the "power in the subspace $\langle H \rangle$." This power is a maximal invariant statistic because $(\mathbf{x} + \mathbf{v})^T \mathbf{Q}_H^T \mathbf{P}_H^T \mathbf{P}_H \mathbf{Q}_H (\mathbf{x} + \mathbf{v}) = \mathbf{x}^T \mathbf{Q}_H^T \mathbf{P}_H \mathbf{Q}_H \mathbf{x} = \mathbf{x}^T \mathbf{P}_H \mathbf{x}$ and $\mathbf{x}_1^T \mathbf{P}_H \mathbf{x}_1 = \mathbf{x}_2^T \mathbf{P}_H \mathbf{x}_2 \rightarrow \mathbf{x}_2 = \mathbf{Q}_H(\mathbf{x}_1 + \mathbf{v})$.

The statistic χ^2 is a quadratic form in the normal random vector $\mathbf{x} : N[\mu \mathbf{H}\boldsymbol{\theta}, \sigma^2 \mathbf{I}]$ and the projector $\mathbf{P}_H$. The vector $\mathbf{P}_H \mathbf{x}$ is distributed as $N[\mu \mathbf{H}\boldsymbol{\theta}, \sigma^2 \mathbf{P}_H]$. In Section 2.11 of Chapter 2 and in the Appendix of this chapter, it is shown that χ^2/σ^2 is chi-squared distributed with p degrees of freedom and noncentrality parameter $(\mu^2/\sigma^2) E_s$, $E_s = \boldsymbol{\theta}^T \mathbf{H}^T \mathbf{H} \boldsymbol{\theta}$:

$$\frac{\chi^2}{\sigma^2} : \chi_p^2 \left(\frac{\mu^2 E_s}{\sigma^2} \right). \qquad\qquad 4.110$$

The chi-squared distribution has a monotone likelihood ratio. Therefore by the Karlin-Rubin theorem, the test

$$\phi\left(\frac{\chi^2}{\sigma^2} \right) = \begin{cases} 1, & \chi^2/\sigma^2 > \chi_0^2 \quad (\chi^2 > \sigma^2 \chi_0^2) \\ 0, & \chi^2/\sigma^2 \le \chi_0^2 \quad (\chi^2 \le \sigma^2 \chi_0^2) \end{cases} \qquad 4.111$$

is the UMP invariant detector for testing $H_0 : \mu = 0$ versus $H_1 : \mu > 0$ in the measurement $\mathbf{x} : N[\mu \mathbf{H}\boldsymbol{\theta}, \sigma^2 \mathbf{I}]$. We call this a *matched subspace filter* because $\mathbf{P}_H \mathbf{x}$ is "matched" to the subspace $\langle H \rangle$ and $\mathbf{x}^T \mathbf{P}_H \mathbf{x}$ is simply the energy in $\mathbf{P}_H \mathbf{x}$. We might also call it a *generalized energy detector*, because it is a quadratic form in $\mathbf{x}$.

The detector $\phi(\chi^2/\sigma^2)$ is illustrated in Figure 4.26. The operating characteristics of the detector are illustrated in Figure 4.27, where P_D is plotted versus $(\mu/\sigma)\sqrt{E_s}$, for several subspace dimensions p, and compared with the normal case. These curves may be used to compute the effective loss in signal-to-noise ratio that results from not knowing exactly where in a subspace $\langle H \rangle$ the signal $\mathbf{s} = \mathbf{H}\boldsymbol{\theta}$ lies. The noncentrality parameter of the chi-squared distribution is just the square of the noncentrality parameter for the normal and t-distributions.

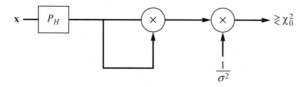

Figure 4.26 Matched subspace detector for detecting signal in a subspace.

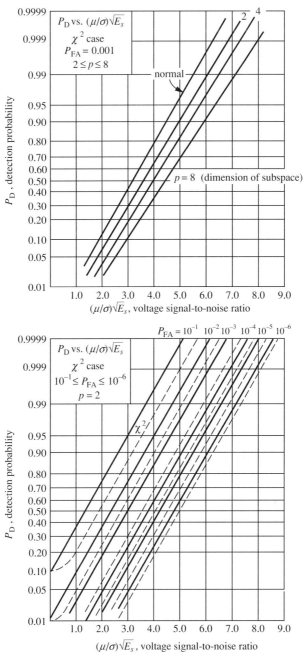

Figure 4.27 Receiver operating characteristics: matched subspace detector.

4.12   CFAR MATCHED SUBSPACE FILTERS (F)

If the variance σ^2 is unknown in the model $\mathbf{x} : N[\mu\mathbf{H}\boldsymbol{\theta}, \sigma^2\mathbf{I}]$, then there is no way to compute the statistic $\mathbf{x}^T\mathbf{P}_H\mathbf{x}/\sigma^2$. Therefore, there is no UMP test that is invariant to rotation and orthogonal bias. Perhaps there is a way to proceed, reducing the problem by sufficiency and invariance, in order to find a test that is UMP within a more restricted class.

When the location of $\mu\mathbf{H}\boldsymbol{\theta}$ in the subspace $\langle\mathbf{H}\rangle$ is unknown and the noise variance σ^2 is unknown, then we may imagine that the measurement is $\mathbf{Q}_H\sigma[(\mu/\sigma)\mathbf{H}\boldsymbol{\theta}' + \mathbf{n}]$, where $\mathbf{Q}_H\sigma(\mu/\sigma)\mathbf{H}\boldsymbol{\theta}' = \mu\mathbf{H}\boldsymbol{\theta}$ and $\mathbf{Q}_H\sigma\mathbf{n}$ is $N[\mathbf{0}, \sigma^2\mathbf{I}]$. Therefore we should ask for invariance to the group of transformations that scales the measurements and rotates them in the signal subspace $\langle\mathbf{H}\rangle$. We characterize this group as $G = \big\{ g : g(\mathbf{x}) = \mathbf{Q}_H c\mathbf{x}, \ \mathbf{Q}_H = \mathbf{U}_H\mathbf{Q}\mathbf{U}_H^T + \mathbf{P}_A\big\}$ and illustrate it in Figure 4.28. Figure 4.28 is *not* the same figure as Figure 4.15, because the rotation is within the subspace $\langle\mathbf{H}\rangle$ and *not* within the orthogonal subspace $\langle\mathbf{A}\rangle$.

From Figure 4.28 it is clear that the ratio

$$\frac{\mathbf{x}^T\mathbf{P}_H\mathbf{x}}{\mathbf{x}^T(\mathbf{I} - \mathbf{P}_H)\mathbf{x}} \qquad\qquad 4.112$$

is a maximal invariant to G. In Problem 4.24 you are asked to prove it. Actually, we shall modify the statistic as follows:

$$F = \frac{\mathbf{x}^T\mathbf{P}_H\mathbf{x}/\sigma^2 p}{\mathbf{x}^T(\mathbf{I} - \mathbf{P}_H)\mathbf{x}/\sigma^2(N - p)}. \qquad\qquad 4.113$$

This statistic is a ratio of quadratic forms in projection matrices. It measures the ratio of the energy of $\mathbf{x}$, per dimension, that lies in the subspace $\langle\mathbf{H}\rangle$ to the energy of $\mathbf{x}$, per dimension, that lies in the subspace $\langle\mathbf{A}\rangle$. This is illustrated in Figure 4.29.

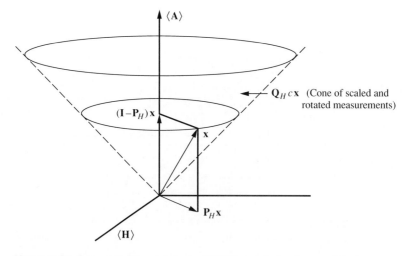

Figure 4.28 Geometric interpretation of CFAR matched subspace detector.

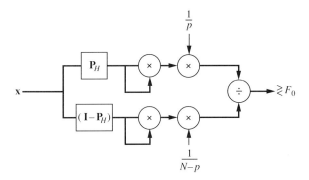

Figure 4.29 CFAR matched subspace detector.

The ratio F is a ratio of independent χ^2 random variables, because the quadratic forms in the numerator and denominator are constructed from independent normal random variables. In Section 2.11 of Chapter 2 and the Appendix of this chapter, it is shown that F is distributed as an F-statistic with p and $N - p$ degrees of freedom and noncentrality parameter $(\mu^2/\sigma^2)E_s$:

$$F : F_{(p,N-p)}\big[(\mu^2/\sigma^2)E_s\big] \qquad\qquad 4.114$$

$$E_s = \boldsymbol{\theta}^{\mathrm{T}}\mathbf{H}^{\mathrm{T}}\mathbf{H}\boldsymbol{\theta}.$$

The F-distribution has a monotone likelihood ratio. By the Karlin-Rubin theorem, the test

$$\phi(F) = \begin{cases} 1, & F > F_0 \\ 0, & F \le F_0 \end{cases} \qquad\qquad 4.115$$

is UMP invariant for detecting the signal $\mathbf{s} = \mu\mathbf{H}\boldsymbol{\theta}$ in noise of unknown variance σ^2. We call this a CFAR *matched subspace detector* because the detector has constant false alarm rate α, independent of σ^2. The detector is illustrated in Figure 4.29 and its operating characteristics are illustrated in Figure 4.30 and compared with the ROCs for the χ^2 case. Note that the division by σ^2 in Equation 4.113 cancels out and is therefore not required in the implementation of the detector.

A Comparative Summary and a Partial Ordering of Performance

When detecting signals in noise, the performance in the normal case exceeds performance in the t case which, in turn, exceeds performance in the F case:

$$N > t > F. \qquad\qquad 4.116$$

Similarly, performance in the normal case exceeds performance in the chi-squared case, which exceeds performance in the F case:

$$N > \chi^2 > F. \qquad\qquad 4.117$$

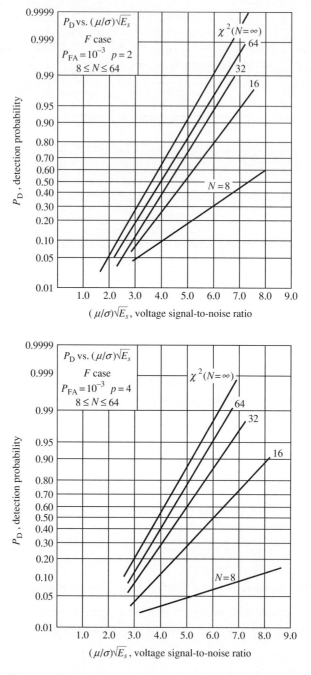

Figure 4.30 Receiver operating characteristics: CFAR matched
subspace detector.

These orderings are clear from the mathematics and from the performance curves. They comport with our intuition because the orderings correspond to increasing levels of certainty about the nuisance parameters σ^2 and $\boldsymbol{\theta}$. There is no ordering between the t and χ^2 cases because in one case it is σ^2 that is unknown and in the other it is $\boldsymbol{\theta}$ that is unknown. There is no general way to rate the relative importance of these two nuisance parameters, so we say

$$t <> \chi^2, \qquad\qquad 4.118$$

meaning that there is no ordering. This fact is illustrated in the relative performance curves of Figure 4.31, where the orderings $N > t > F$ and $N > \chi^2 > F$ are evident and the ordering between t and χ^2 switches from $t > \chi^2$ at low SNR to $\chi^2 > t$ at high SNR. It is clear that, as $N \to \infty$, the performance of the t equals the performance of the normal and the performance of the F equals the performance of the χ^2:

$$N = t > F \qquad\qquad 4.119$$

$$N > \chi^2 = F.$$

This is evident from the ROC curves. It happens because the denominators of the t- and F-statistics converge in probability to σ.

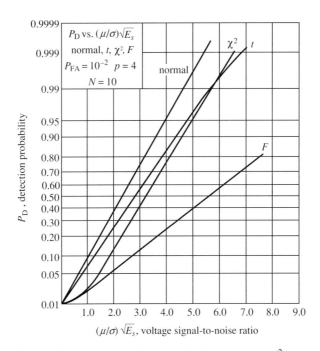

Figure 4.31 Relative performance in the normal, t, χ^2, and F problems.

Signal Model

Noise Model $\sigma\mathbf{w} : N[0,\sigma^2\mathbf{I}]$	**s** known	$\mathbf{s} = \mathbf{H}\boldsymbol{\theta}$; subspace $\langle\mathbf{H}\rangle$ known, $\boldsymbol{\theta}$ unknown
σ^2 known	Invariance: $g(\mathbf{x}) = (\mathbf{U}_A\mathbf{Q}\mathbf{U}_A^T + \mathbf{P}_s)\mathbf{x}$ $\qquad m = \dfrac{\mathbf{s}^T\mathbf{P}_s\mathbf{x}}{\sqrt{\sigma^2\mathbf{s}^T\mathbf{s}}}$ Detector: $\phi(m) = \begin{cases} 1, & m > m_0 \\ 0, & m \le m_0 \end{cases}$ Distribution: $m : N\!\left[\dfrac{\mu}{\sigma}\sqrt{E_s}, 1\right]$ SNR: $\dfrac{\mu^2}{\sigma^2}E_s$	$\mathbf{P}_A\mathbf{v} = \mathbf{v}$ Invariance: $g(\mathbf{x}) = (\mathbf{U}_H\mathbf{Q}\mathbf{U}_H^T + \mathbf{P}_A)(\mathbf{x}+\mathbf{v})$; $\qquad \chi^2/\sigma^2 = \dfrac{\mathbf{x}^T\mathbf{P}_H\mathbf{x}}{\sigma^2}$ Detector: $\phi(\chi^2/\sigma^2) = \begin{cases} 1, & \chi^2/\sigma^2 > \chi_0^2 \\ 0, & \chi^2/\sigma^2 \le \chi_0^2 \end{cases}$ Distribution: $\chi^2/\sigma^2 : \chi_p^2\!\left[\dfrac{\mu^2}{r^2}E_s\right]$ SNR: $\dfrac{\mu^2}{\sigma^2}E_s$
σ^2 unknown	Invariance: $g(\mathbf{x}) = (\mathbf{U}_A\mathbf{Q}\mathbf{U}_A^T + \mathbf{P}_s)\,c\mathbf{x}$ $\qquad t = \dfrac{\mathbf{s}^T\mathbf{P}_s\mathbf{x}}{\sqrt{\frac{1}{N-1}\mathbf{x}^T(\mathbf{I}-\mathbf{P}_s)\mathbf{x}}}$ Detector: $\phi(t) = \begin{cases} 1, & t > t_0 \\ 0, & t \le t_0 \end{cases}$ Distribution: $t : t_{N-1}\!\left[\dfrac{\mu}{\sigma}\sqrt{E_s}\right]$ SNR: $\dfrac{\mu^2}{\sigma^2}E_s$	Invariance: $g(\mathbf{x}) = (\mathbf{U}_H\mathbf{Q}\mathbf{U}_H^T + \mathbf{P}_A)\,c\mathbf{x}$ $\qquad F = \dfrac{\mathbf{x}^T\mathbf{P}_H\mathbf{x}/p}{\mathbf{x}^T(\mathbf{I}-\mathbf{P}_H)\mathbf{x}/(N-p)}$ Detector: $\phi(F) = \begin{cases} 1, & F > F_0 \\ 0, & F \le F_0 \end{cases}$ Distribution: $F : F_{p,\,N-p}\!\left[\dfrac{\mu}{\sigma}\sqrt{E_s}\right]$ SNR: $\dfrac{\mu^2}{\sigma^2}E_s$

Table 4.1 UMP invariant signal detectors for testing $H_0 : \mathbf{x} : N[\mu\mathbf{s}, \sigma^2\mathbf{I}]$, $\mu = 0$ versus $H_1 : \mathbf{x} : N[\mu\mathbf{s}, \sigma^2\mathbf{I}]$, $\mu > 0$.

The results of this section are summarized in Table 4.1, where invariances, detector statistics, detector distributions, signal-to-noise ratios, and detector structures are summarized for the four detection problems we have studied here in Sections 4.9 through 4.12.

4.13   SIGNAL DESIGN

We often speak of the signal-to-noise ratio gain of a signal detector. Here is what we mean. Each sample of the measurement vector $\mathbf{x}$ is distributed as an $N[\mu s_n, \sigma^2]$ random variable. The signal-to-noise ratio per sample (or input signal-to-noise ratio) is the average of the mean squared to the variance for each of these samples:

$$\text{SNR}_{\text{in}} = \frac{1}{N} \sum_{n=1}^{N} \mu^2 s_n^2 / \sigma^2 = \frac{\mu^2}{N\sigma^2} \mathbf{s}^{\text{T}}\mathbf{s} = \frac{1}{N} \frac{\mu^2}{\sigma^2} E_s. \qquad 4.120$$

The output signal-to-noise ratio is the quantity that determines the noncentrality parameter of each of the four detectors that we have derived:

$$\text{SNR} = \frac{\mu^2}{\sigma^2} \mathbf{s}^{\text{T}}\mathbf{s} = \frac{\mu^2}{\sigma^2} E_s = N\text{SNR}_{\text{in}}. \qquad 4.121$$

The gain provided by the signal detector is N, the ratio of the output signal-to-noise to the input signal-to-noise ratio:

$$G = \frac{\text{SNR}}{\text{SNR}_{\text{in}}} = N.$$

We sometimes write this result as

$$10 \log_{10} \text{SNR} = 10 \log_{10} \text{SNR}_{\text{in}} + 10 \log_{10} N \qquad 4.122$$

and say that a measurement of dimension N brings a gain in signal-to-noise ratio of $10 \log N$ in units of decibels (db). The signal-to-noise ratio (SNR) is a scalar quantity that brings insight to the performance of a signal detector. However, it is important to emphasize that it is the SNR, together with the distribution of the detector statistic, which determines detector performance. An output SNR of 20 db in the normal case is very high, but in the F case it can be very low.

Signal Design for Detection

The output signal-to-noise ratio for all of the signal detectors we have studied is $\text{SNR} = (\mu^2/\sigma^2) E_s$. But, if the measurements have been prewhitened, then $\mathbf{s}$ is replaced by $\mathbf{R}^{-\text{T}/2}\mathbf{s}$ and the SNR is replaced by

$$\text{SNR} = \frac{\mu^2}{\sigma^2} \mathbf{s}^{\text{T}}\mathbf{R}^{-1/2}\mathbf{R}^{-\text{T}/2}\mathbf{s} = \frac{\mu^2}{\sigma^2} \mathbf{s}^{\text{T}}\mathbf{R}^{-1}\mathbf{s}. \qquad 4.123$$

The problem of signal design is one of designing the signal $\mathbf{s}$ that maximizes this signal-to-noise ratio under the constraint that $\mathbf{s}^T\mathbf{s} = 1$:

$$\max_{\mathbf{s}} \frac{\mathbf{s}^T\mathbf{R}^{-1}\mathbf{s}}{\mathbf{s}^T\mathbf{s}}. \qquad 4.124$$

In Problem 4.14, you are asked to show that the solution is

$$\mathbf{s} = \mathbf{u}, \qquad 4.125$$

where $\mathbf{u}$ is the eigenvector corresponding to the maximum eigenvalue of $\mathbf{R}^{-1}$, or the minimum eigenvalue of $\mathbf{R}$:

$$\mathbf{R}\mathbf{u} = \lambda_{\min}\mathbf{u}. \qquad 4.126$$

This means that the maximum achievable signal-to-noise ratio is

$$\text{SNR} = \frac{\mu^2}{\sigma^2}\frac{1}{\lambda_{\min}}. \qquad 4.127$$

Constrained Signal Design

If we do not have complete freedom in the choice of $\mathbf{s}$, then we may proceed as follows. Assume that the signal is constrained to lie in the subspace spanned by the columns of a matrix $\mathbf{H}$:

$$\mathbf{s} = \mathbf{H}\boldsymbol{\theta}. \qquad 4.128$$

Then the output SNR is

$$\text{SNR} = \frac{\mu^2}{\sigma^2}\boldsymbol{\theta}^T\mathbf{H}^T\mathbf{R}^{-1}\mathbf{H}\boldsymbol{\theta}. \qquad 4.129$$

This is maximized under the constraint $\boldsymbol{\theta}^T\boldsymbol{\theta} = 1$ by choosing the vector $\boldsymbol{\theta}$ to be the maximum eigenvector of the matrix $\mathbf{H}^T\mathbf{R}^{-1}\mathbf{H}$. The maximum achievable SNR is then

$$\text{SNR} = \frac{\mu^2}{\sigma^2}\lambda_{\max} \qquad 4.130$$

where $\lambda_{\max}$ is the maximum eigenvector of $\mathbf{H}^T\mathbf{R}^{-1}\mathbf{H}$.

There is an interesting variation on this problem. Suppose the vector $\boldsymbol{\theta}$ is to be chosen to maximize the SNR, but the constraint is not that $\boldsymbol{\theta}^T\boldsymbol{\theta} = 1$ but rather that $\mathbf{s}^T\mathbf{s} = 1$. Then the problem is to

$$\max_{\boldsymbol{\theta}} \frac{\boldsymbol{\theta}^T\mathbf{H}^T\mathbf{R}^{-1}\mathbf{H}\boldsymbol{\theta}}{\boldsymbol{\theta}^T\mathbf{H}^T\mathbf{H}\boldsymbol{\theta}}.$$

In Problem 4.15, you are asked to show that the solution for $\boldsymbol{\theta}$ is the eigenvector corresponding to the maximum eigenvalue in the following generalized eigenvalue problem:

$$\mathbf{H}^T\mathbf{R}^{-1}\mathbf{H}\mathbf{u} = \lambda\mathbf{H}^T\mathbf{H}\mathbf{u}.$$

The next several examples illustrate the application of these results to practical problems.

Example 4.12 (Signals in White Noise)
When the covariance matrix $\mathbf{R}$ is the identity matrix $\mathbf{I}$, then the additive noise is said to be white because the elements of the noise vector are uncorrelated random variables. The SNR is

$$\text{SNR} = \frac{\mu^2}{\sigma^2} \mathbf{s}^T \mathbf{s} = \frac{\mu^2}{\sigma^2} E_s. \qquad 4.131$$

There is no signal design problem here. Any signal $\mathbf{s}$ for which $\mathbf{s}^T\mathbf{s} = E_s$ is as good as any other. ∎

Example 4.13 (Signals in Autoregressive Noise)
When the correlation matrix $\mathbf{R}$ is the Toeplitz matrix

$$\mathbf{R} = \begin{bmatrix} 1 & a & a^2 & \cdots & a^{N-1} \\ a & 1 & a & & \\ \vdots & & & & a \\ a^{N-1} & \cdots & & a & 1 \end{bmatrix} \qquad 4.132$$

then we say that the additive noise is autoregressive because the correlation between noise samples w_n and w_{n+i} is a^i. The inverse of $\mathbf{R}$ is

$$\mathbf{R}^{-1} = \frac{1}{1-a^2} \begin{bmatrix} 1 & -a & & & \\ -a & 1 & -a & & \mathbf{0} \\ & -a & & \ddots & \\ \mathbf{0} & & \ddots & \ddots & -a \\ & & & -a & 1 \end{bmatrix}. \qquad 4.133$$

The best designed signal is the eigenvector corresponding to the maximum eigenvalue of this matrix or, equivalently, corresponding to the minimum eigenvalue of $\mathbf{R}$. But what if $\mathbf{x}$ is simply chosen to be the constant vector $\mathbf{x} = \mathbf{1}$? How does the correlation between noise samples influence the SNR? The signal-to-noise ratio is simply the sum of the elements in $\mathbf{R}^{-1}$:

$$\text{SNR} = \frac{\mu^2}{\sigma^2} \mathbf{1}^T \mathbf{R}^{-1} \mathbf{1} = \frac{\mu^2}{\sigma^2} \frac{1}{(1-a^2)} \left[(N-2)(1+a^2) + 2 - (N-1)a \right]. \qquad 4.134$$

When $a = 0$, then $\text{SNR} = (\mu^2/\sigma^2)N$. When $a = \pm 1$, this expression is indeterminate. In Problem 4.16, you are asked to find its limit as $a \to \pm 1$. ∎

Example 4.14 (Signal Design in the Bivariate Normal Model)
Consider the bivariate normal model in which the two-dimensional random vector $\mathbf{x}$ is distributed as $N[\mathbf{0}, \sigma^2\mathbf{R}]$ under H_0 and as $N[\mu\mathbf{s}, \sigma^2\mathbf{R}]$ under H_1. The

covariance matrix $\sigma^2 \mathbf{R}$ has the bivariate form

$$\sigma^2 \mathbf{R} = \sigma^2 \begin{bmatrix} 1 & \rho \\ \rho & 1 \end{bmatrix}. \tag{4.135}$$

The signal-to-noise ratio is

$$\mathrm{SNR} = \frac{\mu^2}{\sigma^2} \mathbf{s}^{\mathrm{T}} \mathbf{R}^{-1} \mathbf{s}, \tag{4.136}$$

where $\mathbf{R}^{-1}$ is the matrix

$$\frac{1}{(1-\rho^2)} \begin{bmatrix} 1 & -\rho \\ -\rho & 1 \end{bmatrix}, \tag{4.137}$$

The signal-to-noise ratio is maximized by choosing $\mathbf{s}$ to be the eigenvector corresponding to the minimum eigenvalue of $\mathbf{R}$. The eigenvalues of $\mathbf{R}$ are $\lambda_1 = (1+\rho)$ and $\lambda_2 = (1-\rho)$. The correlation coefficient ρ can be positive or negative. Therefore, the minimum eigenvalue is

$$\lambda_{\min} = 1 - |\rho|. \tag{4.138}$$

The eigenvector corresponding to the minimum eigenvalue is

$$\mathbf{u} = \frac{1}{\sqrt{2}} \begin{bmatrix} 1 \\ \mathrm{sgn}(\rho) \end{bmatrix}. \tag{4.139}$$

The maximum achievable signal-to-noise ratio is

$$\begin{aligned} \mathrm{SNR} &= \frac{\mu^2}{\sigma^2} \lambda_{\min}^{-1} \\ &= \frac{\mu^2}{\sigma^2} \frac{1}{1-|\rho|}. \end{aligned} \tag{4.140}$$

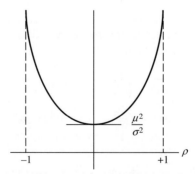

Figure 4.32 Signal-to-noise ratio versus correlation coefficient.

This function is illustrated in Figure 4.32. Note that when ρ is near ± 1, the signal-to-noise ratio for an optimally designed signal is arbitrarily large. Why is this? ■

4.14   DETECTORS FOR GAUSSIAN RANDOM SIGNALS

The Neyman-Pearson detector for testing between two known covariance matrices is generally a full-rank quadratic form. In practice, the detector may be replaced by a low-rank detector constructed from dominant eigenvectors. In the context of hypothesis testing, a dominant eigenvector will be shown to be an eigenvector associated with a dominant eigenvalue of a "signal-to-noise ratio" matrix. A dominant eigenvalue is one for which $\lambda_n + \lambda_n^{-1}$ is large. Thus in pairwise discrimination of models, it may be small eigenvalues or spectral nulls, rather than large eigenvalues or spectral peaks, that provide the most information for discrimination.

In this section, we study Neyman-Pearson detectors for Gaussian random vectors and offer a constructive procedure for designing low-rank detectors that maximize the divergence between hypotheses about covariance structure. The procedure extends the classical theory of principal component analysis. We also indicate when the rank of the quadratic detector may be reduced with no loss in performance.

Likelihood Ratios and Quadratic Detectors

Let $\mathbf{x} : N[\mathbf{0}, \mathbf{R}]$ denote an $N \times 1$ normal random vector with mean zero and covariance matrix $\mathbf{R}$. The problem considered here is the test of the hypothesis $H_0 : \mathbf{R} = \mathbf{R}_0$ versus the alternative $H_1 : \mathbf{R} = \mathbf{R}_1$. The log likelihood ratio test is

$$\phi(\mathbf{x}) = \begin{cases} 1 \sim H_1, & L(\mathbf{x}) > l_0 \\ 0 \sim H_0, & L(\mathbf{x}) \le l_0 \end{cases}$$

$$L(\mathbf{x}) = \mathbf{x}^\mathrm{T} \mathbf{Q} \mathbf{x} \qquad\qquad 4.141$$

$$\mathbf{Q} = \mathbf{R}_0^{-1} - \mathbf{R}_1^{-1}.$$

The matrix $\mathbf{Q}$ may be rewritten as

$$\mathbf{Q} = \mathbf{R}_0^{-\mathrm{T}/2}(\mathbf{I} - \mathbf{S}^{-1})\mathbf{R}_0^{-1/2} \qquad\qquad 4.142$$

where $\mathbf{R}_0^{1/2}\mathbf{R}_0^{\mathrm{T}/2}$ is a symmetric factor of $\mathbf{R}_0$ and $\mathbf{S}$ is a signal-to-noise ratio matrix:

$$\mathbf{R}_0 = \mathbf{R}_0^{1/2}\mathbf{R}_0^{\mathrm{T}/2}; \qquad \mathbf{R}_0^{\mathrm{T}/2} = \left(\mathbf{R}_0^{1/2}\right)^\mathrm{T} \qquad\qquad 4.143$$

$$\mathbf{S} = \mathbf{R}_0^{-1/2}\mathbf{R}_1\mathbf{R}_0^{-\mathrm{T}/2}; \qquad \mathbf{R}_0^{-\mathrm{T}/2} = \left(\mathbf{R}_0^{-1/2}\right)^\mathrm{T}.$$

This produces the following representation of the log likelihood ratio:

$$L(\mathbf{x}) = \mathbf{y}^\mathrm{T}(\mathbf{I} - \mathbf{S}^{-1})\mathbf{y} \qquad\qquad 4.144$$

$$\mathbf{y} = \mathbf{R}_0^{-1/2}\mathbf{x}.$$

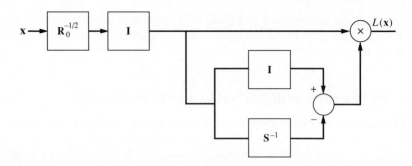

Figure 4.33 Structure for computing likelihood.

The transformed vector $\mathbf{y}$ is distributed as $N[\mathbf{0}, \mathbf{R}]$, with $\mathbf{R} = \mathbf{I}$ under H_0 and $\mathbf{R} = \mathbf{S}$ under H_1:

$$E_{H_0}\mathbf{y}\mathbf{y}^{\mathrm{T}} = \mathbf{I} \qquad\qquad 4.145$$

$$E_{H_1}\mathbf{y}\mathbf{y}^{\mathrm{T}} = \mathbf{S}.$$

An implementation of the likelihood ratio computation is illustrated in Figure 4.33.

Orthogonal Decomposition

The matrix $\mathbf{S}$ has an orthogonal decomposition

$$\mathbf{S} = \mathbf{R}_0^{-1/2}\mathbf{R}_1\mathbf{R}_0^{-\mathrm{T}/2} = \mathbf{U}\boldsymbol{\Lambda}\mathbf{U}^{\mathrm{T}} \qquad\qquad 4.146$$

$$\mathbf{S}\mathbf{U} = \mathbf{U}\boldsymbol{\Lambda}$$

where $\boldsymbol{\Lambda}$ is a diagonal matrix with diagonal elements λ_i and $\mathbf{U}$ satisfies $\mathbf{U}^{\mathrm{T}}\mathbf{U} = \mathbf{I}$. This means that $(\mathbf{R}_0^{-\mathrm{T}/2}\mathbf{U}, \boldsymbol{\Lambda})$ solves the generalized eigenvalue problem

$$\mathbf{R}_1\left(\mathbf{R}_0^{-\mathrm{T}/2}\mathbf{U}\right) - \mathbf{R}_0\left(\mathbf{R}_0^{-\mathrm{T}/2}\mathbf{U}\right)\boldsymbol{\Lambda} = \mathbf{0}. \qquad\qquad 4.147$$

With this representation for $\mathbf{S}$, the log likelihood ratio may be written

$$L(\mathbf{x}) = \mathbf{y}^{\mathrm{T}}\mathbf{U}(\mathbf{I} - \boldsymbol{\Lambda}^{-1})\mathbf{U}^{\mathrm{T}}\mathbf{y}, \qquad\qquad 4.148$$

where the random vector $\mathbf{y}$ has covariance matrix $\mathbf{U}\boldsymbol{\Lambda}\mathbf{U}^{\mathrm{T}}$ under H_1 and $\mathbf{I}$ under H_0. This implementation of likelihood is illustrated in Figure 4.34.

Distribution of Log Likelihood

The log likelihood $L(\mathbf{x})$ is a quadratic form in the normal random vector $\mathbf{y}$. Under H_0, $\mathbf{y}$ is distributed as $N[\mathbf{0}, \mathbf{U}\mathbf{U}^{\mathrm{T}}]$, and under H_1 it is distributed as $N[\mathbf{0}, \mathbf{U}\boldsymbol{\Lambda}\mathbf{U}^{\mathrm{T}}]$. In

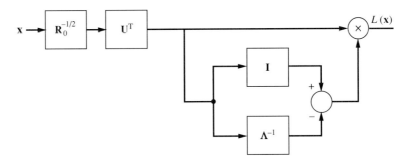

Figure 4.34 Orthogonal structure for computing likelihood.

Section 2.11 of Chapter 2 we establish that the characteristic function for a quadratic form $\mathbf{y}^T\mathbf{P}\mathbf{y}$ is $\phi(\omega) = [\det(\mathbf{I} + 2j\omega\mathbf{P}\mathbf{R})]^{-1/2}$ when $\mathbf{y} : N[\mathbf{0}, \mathbf{R}]$. Applying this result to $l(\mathbf{x})$, we obtain the following characteristic functions for $l(\mathbf{x})$:

$$H_0 : \phi(\omega) = \frac{1}{\left[\det(\mathbf{I} + 2j\omega\mathbf{U}(\mathbf{I} - \boldsymbol{\Lambda}^{-1})\mathbf{U}^T\mathbf{U}\mathbf{U}^T)\right]^{1/2}}$$

$$= \frac{1}{\left[\det[\mathbf{I} + 2j\omega(\mathbf{I} - \boldsymbol{\Lambda}^{-1})]\right]^{1/2}}$$

$$H_1 : \phi(\omega) = \frac{1}{\left[\det(\mathbf{I} + 2j\omega\mathbf{U}(\mathbf{I} - \boldsymbol{\Lambda}^{-1})\mathbf{U}^T\mathbf{U}\boldsymbol{\Lambda}\mathbf{U}^T)\right]^{1/2}}$$

$$= \frac{1}{\left[\det[\mathbf{I} + 2j\omega(\boldsymbol{\Lambda} - \mathbf{I})]\right]^{1/2}} \cdot$$

$$4.149$$

There is a fundamental result buried in here: unity eigenvalues of $\mathbf{S}$ contribute nothing to log likelihood! That is, unity eigenvalues on the diagonal of $\boldsymbol{\Lambda}$ are cancelled in the differences $\mathbf{I} - \boldsymbol{\Lambda}^{-1}$ and $\boldsymbol{\Lambda} - \mathbf{I}$. This means that the characteristic function $\phi(\omega)$ may be written as follows whenever the signal-noise matrix $\mathbf{S}$ has $N - r$ unity eigenvalues and r non-unity eigenvalues:

$$H_0 : \phi(\omega) = \frac{1}{\left[\prod_{n=1}^{r} [1 + j\omega(1 - (1/\lambda_n))]\right]^{1/2}}$$

$$4.150$$

$$H_1 : \phi(\omega) = \frac{1}{\left[\prod_{n=1}^{r} [1 + j\omega(\lambda_n - 1)]\right]^{1/2}} \cdot$$

These are χ^2-like characteristic functions that may be inverted numerically to produce the density function for $L(\mathbf{x})$ under H_0 and H_1 and to compute P_{FA} and P_{D}.

Rank Reduction

A reduced-rank version of the log likelihood ratio is

$$L_r(\mathbf{x}) = \mathbf{y}^T \mathbf{U}(\mathbf{I}_r - \mathbf{\Lambda}_r^{-1})\mathbf{U}^T \mathbf{y} \qquad 4.151$$

where $\mathbf{I}_r$ and $\mathbf{\Lambda}_r^{-1}$ are reduced-rank versions of $\mathbf{I}$ and $\mathbf{\Lambda}^{-1}$ that contain r nonzero terms and $N - r$ zero terms. An implementation of log likelihood in this form is illustrated in Figure 4.35.

This rank reduction may be carried out, without penalty, whenever the discarded eigenvalues λ_n are unity. However, sometimes non-unity eigenvalues may be discarded without *much* penalty. To illustrate this, we introduce *divergence*, a coarse measure of how the log likelihood distinguishes between H_0 and H_1:

$$J = E_{H_1}L(\mathbf{x}) - E_{H_0}L(\mathbf{x}) = \operatorname{tr}\mathbf{U}(\mathbf{I} - \mathbf{\Lambda}^{-1})\mathbf{U}^T\mathbf{U}\mathbf{\Lambda}\mathbf{U}^T - \operatorname{tr}\mathbf{U}(\mathbf{I} - \mathbf{\Lambda}^{-1})\mathbf{U}^T\mathbf{U}\mathbf{U}^T$$

$$= \operatorname{tr}(\mathbf{\Lambda} + \mathbf{\Lambda}^{-1} - 2\mathbf{I})$$

$$= \sum_{n=1}^{N}(\lambda_n + \lambda_n^{-1} - 2) = \operatorname{tr}(\mathbf{S} + \mathbf{S}^{-1} - 2\mathbf{I}). \qquad 4.152$$

Notice that it is the sum $\lambda_n + \lambda_n^{-1} - 2$ that determines the contribution of an eigenvalue to divergence, *not* λ_n itself.

When log likelihood is replaced by its reduced-rank version, then divergence is

$$J_r = \sum_{n \in G}(\lambda_n + \lambda_n^{-1} - 2), \qquad 4.153$$

where G is the set of retained eigenvalues. The goal is to select the set $G = (n_1, n_2, \ldots, n_r)$ so that divergence is maximized. Thus, $\lambda_{n_i} + (1/\lambda_{n_i}) - 2$ must be larger for every n_i included in G than for every n_j excluded. This implies $\lambda_{n_i} > 1/\lambda_{n_j}$. Assuming that the λ_i are ordered $(\lambda_1 > \lambda_2 > \cdots > \lambda_N)$, the following algorithm

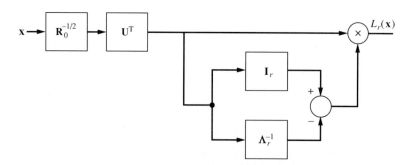

Figure 4.35 Reduced-rank orthogonal structure for computing likelihood.

selects the "dominant λs" for inclusion in the set G.

> Let $i = 1$, $j = N$
> For $k = 1$ to r, do
>   If $(\lambda_i \geq \lambda_j^{-1})$ then
> $i \in G$
> $i = i + 1$
>   Else if $(\lambda_i < \lambda_j^{-1})$ then
> $j \in G$
> $j = j - 1$
>   End if
>   Increment k

Typically, very large and very small eigenvalues are selected in this procedure.

The rank r divergence is identical to the full-rank divergence when $N - r$ of the eigenvalues in the original diagonal matrix Λ are unity. From the generalized eigenvalue problem, we note that each eigenvalue satisfies the generalized eigenproblem

$$(\mathbf{R}_1 - \lambda \mathbf{R}_0)\mathbf{u} = \mathbf{0} \qquad 4.154$$

where $\mathbf{u}$ is the corresponding generalized eigenvector. A unity eigenvalue implies

$$(\mathbf{R}_1 - \mathbf{R}_0)\mathbf{u} = \mathbf{0}, \qquad 4.155$$

so the number of unity eigenvalues equals the dimension of the null space of $\mathbf{R}_1 - \mathbf{R}_0$, namely N minus rank $(\mathbf{R}_1 - \mathbf{R}_0)$. To illustrate, consider the case when $\mathbf{R}_1$ and $\mathbf{R}_0$ are the sum of outer products and a nonsingular matrix $\mathbf{Q}$:

$$H_0 : \mathbf{R}_0 = \mathbf{Q} + \sum_{i=1}^{p} \sigma_i^2 \mathbf{u}_i \mathbf{u}_i^{\mathsf{T}} \qquad 4.156$$

$$H_1 : \mathbf{R}_1 = \mathbf{Q} + \sum_{i=p+1}^{p+q} \sigma_i^2 \mathbf{v}_i \mathbf{v}_i^{\mathsf{T}}.$$

The difference between $\mathbf{R}_1$ and $\mathbf{R}_0$ is

$$\mathbf{R}_1 - \mathbf{R}_0 = \sum_{i=p+1}^{p+q} \sigma_i^2 \mathbf{v}_i \mathbf{v}_i^{\mathsf{T}} - \sum_{i=1}^{p} \sigma_i^2 \mathbf{u}_i \mathbf{u}_i^{\mathsf{T}} \qquad 4.157$$

$$= \mathbf{LDL}^{\mathsf{T}}$$

$$\mathbf{L} = [\mathbf{v}_{p+1}, \mathbf{v}_{p+2}, \ldots, \mathbf{v}_{p+1}, \mathbf{u}_1, \ldots, \mathbf{u}_p]$$

$$\mathbf{D} = \mathrm{diag}\big[\sigma_{p+1}^2, \sigma_{p+2}^2, \ldots, \sigma_{p+q}^2, -\sigma_1^2, -\sigma_2^2, \ldots, -\sigma_p^2\big].$$

Assuming $\sigma_i^2 > 0$, then rank$(\mathbf{R}_1 - \mathbf{R}_0)$ = rank$(\mathbf{L})$ and a rank$(\mathbf{L})$ detector has the same divergence as a full-rank detector.

Example 4.15 (An Optimum Rank-One Detector)

Consider building a low-rank detector for H_0 versus H_1 when the observed data is distributed as $N[\mathbf{0}, \mathbf{R}_0]$ under H_0 and $N[\mathbf{0}, \mathbf{R}_1]$ under H_1:

$$H_0 : \mathbf{R}_0 = \sigma^2\mathbf{I} + \beta^2\mathbf{w}\mathbf{w}^T$$
$$H_1 : \mathbf{R}_1 = \sigma^2\mathbf{I} + \beta^2\mathbf{w}\mathbf{w}^T + \mathbf{v}\mathbf{v}^T. \qquad 4.158$$

After some algebra, we obtain this result for the signal-to-noise matrix $\mathbf{S}$:

$$\mathbf{S} = \mathbf{I} + \mathbf{U}\mathbf{\Sigma}\mathbf{U}^T$$
$$\mathbf{\Sigma} = \mathrm{diag}\big[\mathbf{v}^T R_0 \mathbf{v}\ 0\ 0\ \cdots\ 0\big]. \qquad 4.159$$

The eigenvalues of $\mathbf{S}$ are

$$\lambda_1 = 1 + \mathbf{v}^T\mathbf{R}_0\mathbf{v},\ \lambda_2 = 1,\ \lambda_3 = 1, \ldots, \lambda_N = 1. \qquad 4.160$$

A rank-one detector is optimum for this problem. It is constructed using the eigenvector corresponding to λ_1. ■

Example 4.16 (Gaussian Signal Plus Gaussian Noise)

The previous example is a special case of the more general signal-plus-noise problem where $\mathbf{R}_1 = \mathbf{R}_0 + \mathbf{R}_S$, resulting in $\mathbf{S} = \mathbf{I} + \mathbf{R}_0^{-1/2}\mathbf{R}_S\mathbf{R}_0^{-T/2}$. Clearly, all the eigenvalues of $\mathbf{S}$ are greater than one. This does not mean, however, that eigenvalues near 1 cannot be discarded in order to approximate log likelihood with a low-rank detector. ■

4.15   LINEAR DISCRIMINANT FUNCTIONS

Linear discriminant functions may be used to approximate a quadratic likelihood ratio when testing $\mathbf{x} : N[\mathbf{0}, \mathbf{R}_i]$ under hypothesis H_i. The Neyman-Pearson test of H_0 versus H_1 will have us compare the log likelihood ratio to a threshold:

$$L(\mathbf{x}) = \ln \frac{f_{\theta_1}(\mathbf{x})}{f_{\theta_0}(\mathbf{x})} \lessgtr \eta. \qquad 4.161$$

We hope that, on average, $L(\mathbf{x})$ will be larger than η under H_1 and smaller than η under H_0. An incomplete measure of how the test of H_0 versus H_1 will perform is the difference in means of $L(\mathbf{x})$ under the two hypotheses:

$$J = E_{\theta_1}L(\mathbf{X}) - E_{\theta_0}L(\mathbf{X})$$
$$= E_{\theta_1} \ln \frac{f_{\theta_1}(\mathbf{X})}{f_{\theta_0}(\mathbf{X})} - E_{\theta_0} \ln \frac{f_{\theta_1}(\mathbf{X})}{f_{\theta_0}(\mathbf{X})}. \qquad 4.162$$

This function is the J-divergence between H_0 and H_1 introduced in the previous section. It is related to information that a random sample brings about the hypothsis H_i.

The J-divergence for the multivariate normal problem $H_i : \mathbf{x} : N[\mathbf{0}, \mathbf{R}_i]$ is computed by carrying out the expectations:

$$J = \frac{1}{2} \operatorname{tr}(\mathbf{R}_1 - \mathbf{R}_0)(\mathbf{R}_0^{-1} - \mathbf{R}_1^{-1})$$

$$= \frac{1}{2} \operatorname{tr}\left[\mathbf{R}_1\mathbf{R}_0^{-1} + \mathbf{R}_0\mathbf{R}_1^{-1} - 2\mathbf{I}\right] \geq 0. \qquad 4.163$$

This expression does *not* completely characterize the performance of a likelihood ratio statistic, but it does bring useful information about the "distance" between H_1 and H_0.

Linear Discrimination

Assume that the data $\mathbf{x}$ is used to form the linear discriminant function (or statistic)

$$y = \mathbf{w}^T\mathbf{x}. \qquad 4.164$$

This statistic is distributed as $N[0, \mathbf{w}^T\mathbf{R}_i\mathbf{w}]$ under hypothesis H_i. If a log likelihood ratio is formed using the variable y, then the divergence between H_1 and H_0 is

$$J = \frac{1}{2}\left(\frac{\mathbf{w}^T\mathbf{R}_1\mathbf{w}}{\mathbf{w}^T\mathbf{R}_0\mathbf{w}} + \frac{\mathbf{w}^T\mathbf{R}_0\mathbf{w}}{\mathbf{w}^T\mathbf{R}_1\mathbf{w}} - 2\right). \qquad 4.165$$

Let's define the following ratio of quadratic forms,

$$\lambda[\mathbf{Q}] = \frac{\mathbf{w}^T\mathbf{Q}\mathbf{w}}{\mathbf{w}^T\mathbf{R}_0\mathbf{w}}. \qquad 4.166$$

Then we may write the divergence formula as follows:

$$J = \frac{1}{2}\left[\lambda[\mathbf{R}_1] + \frac{1}{\lambda[\mathbf{R}_1]} - 2\right]. \qquad 4.167$$

These formulas show that the choice of the discriminant $\mathbf{w}$ that maximizes divergence is also the choice of $\mathbf{w}$ that maximizes a function of a quadratic form. We study the maximization of quadratic forms in the next several paragraphs.

An Extremization Problem

Consider the extremization of the quadratic form $\lambda[\mathbf{Q}]$ with respect to the vector $\mathbf{w}$:

$$\frac{\partial}{\partial\mathbf{w}}\lambda[\mathbf{Q}] = \mathbf{0}$$

$$\frac{\mathbf{Q}\mathbf{w} - \lambda[\mathbf{Q}]\mathbf{R}_0\mathbf{w}}{\mathbf{w}^T\mathbf{R}_0\mathbf{w}} = \mathbf{0}. \qquad 4.168$$

We say that the vector $\mathbf{w}$ solves the generalized eigenvalue problem

$$[\mathbf{Q} - \lambda[\mathbf{Q}]\mathbf{R}_0]\mathbf{w} = \mathbf{0}. \qquad\qquad 4.169$$

Equivalently, we say that the vector $\mathbf{R}_0\mathbf{w}$ is an eigenvector of the matrix $\mathbf{Q}\mathbf{R}_0^{-1}$:

$$[\mathbf{Q}\mathbf{R}_0^{-1} - \lambda[\mathbf{Q}]\mathbf{I}]\mathbf{R}_0\mathbf{w} = \mathbf{0}. \qquad\qquad 4.170$$

So, while there are an infinity of values for $\lambda[\mathbf{Q}]$ that may be achieved for different choices of $\mathbf{w}$, there are just N values that can compete for the title of maximum or minimum for $\lambda[\mathbf{Q}]$. These are the eigenvalues $\lambda[\mathbf{Q}]$ of $\mathbf{Q}\mathbf{R}_0^{-1}$. Call the nth such eigenvalue $\lambda_n[\mathbf{Q}]$:

$$\mathbf{Q}\mathbf{w}_n = \lambda_n[\mathbf{Q}]\mathbf{R}_0\mathbf{w}_n. \qquad\qquad 4.171$$

To find the maximum value of $\lambda[\mathbf{Q}]$, we choose the maximum eigenvalue $\lambda_{\max}[\mathbf{Q}]$ and the corresponding maximum eigenvector $\mathbf{w}_{\max}$. To find the minimum value of $\lambda[\mathbf{Q}]$, we choose the minimum eigenvalue $\lambda_{\min}[\mathbf{Q}]$ and the corresponding minimum eigenvector $\mathbf{w}_{\min}$.

Maximizing Divergence

Let's rewrite divergence as

$$\begin{aligned}
J &= \frac{1}{2}\left[\lambda[\mathbf{R}_1] + \frac{1}{\lambda[\mathbf{R}_1]} - 2\right] \\
&= \frac{1}{2}\left[\lambda^{1/2}[\mathbf{R}_1] - \frac{1}{\lambda^{1/2}[\mathbf{R}_1]}\right]^2.
\end{aligned} \qquad\qquad 4.172$$

The function J is convex in λ as illustrated in Figure 4.36. It achieves its maximum either at $\lambda_{\max}$ or at $\lambda_{\min}$, where $\lambda_{\max}$ and $\lambda_{\min}$ are maximum and minimum eigenvalues of $\mathbf{R}_1\mathbf{R}_0^{-1}$, depending upon whether or not $\lambda_{\max} > (1/\lambda_{\min})$ or $\lambda_{\max} < (1/\lambda_{\min})$. In fact, J-divergence is maximized as follows:

$$\mathbf{w} = \begin{cases} \mathbf{w}_{\max} & \text{if } \lambda_{\max} > 1/\lambda_{\min} \\ \mathbf{w}_{\min} & \text{if } \lambda_{\max} < 1/\lambda_{\min}. \end{cases} \qquad\qquad 4.173$$

If $\lambda_{\min} > 1$, then $\lambda_{\max}$ always exceeds $1/\lambda_{\min}$. If $\lambda_{\max} < 1$, then $\lambda_{\max}$ is always less than $1/\lambda_{\min}$. So, the only interesting case is $\lambda_{\min} < 1$ and $\lambda_{\max} > 1$ as illustrated in Figure 4.36.

The linear discriminant function is either the maximum eigenvector of $\mathbf{R}_1\mathbf{R}_0^{-1}$ *or* the minimum eigenvector of $\mathbf{R}_1\mathbf{R}_0^{-1}$, depending on the nature of the maximum and minimum eigenvalues! It is *not* always the maximum eigenvector. The choice $\mathbf{w} = \mathbf{w}_{\max}$ or $\mathbf{w}_{\min}$ is also called a principal component. Without loss of generality,

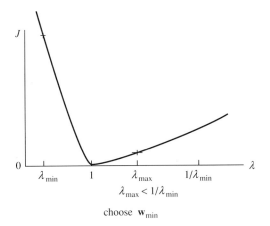

Figure 4.36 Maximizing divergence when means are common.

$\mathbf{w}$ may be normalized so that $\mathbf{w}^T\mathbf{w} = 1$. Then, if $\mathbf{R}_0 = \mathbf{I}$, the linear discriminant function is distributed as follows:

$$y = \mathbf{w}^T\mathbf{x} : \begin{cases} N[0, 1] & \text{under } H_0 \\ N[0, \lambda_{\min}] \text{ or } N[0, \lambda_{\max}] & \text{under } H_1. \end{cases} \qquad 4.174$$

This distribution is plotted in Figure 4.37.

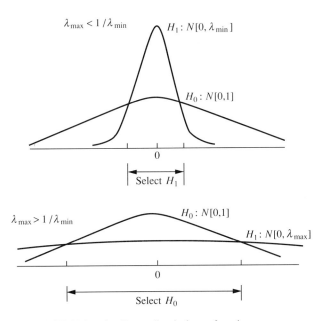

Figure 4.37 Using the linear discriminant function.

REFERENCES AND COMMENTS

For my treatment of the Neyman-Pearson lemma, I have been influenced by Ferguson [1967] and by Lehmann [1959]. In my treatment of the multivariate normal problem, I have borrowed geometrical insights from Duda and Hart [1973]. Invariance in hypothesis testing is an important topic in Ferguson's book, and I have generally followed his development, but the application of invariance to signal detection brings some new wrinkles. As far as I know, the first applications of invariance to signal detection were published in Scharf and Lytle [1971]. Then in 1984 Steve Kay and I published an application of invariance to the problem of detecting autoregressive moving average (ARMA) signals in Gaussian noise of unknown level [Kay and Scharf, 1984]. A similar problem of detecting phase incoherent signals in noise of unknown level had already been treated from the point of view of generalized likelihood ratios by Nuttall and Cable [1973].

I began to realize that there were actually four fundamental and interrelated problems in signal detection theory—matched filtering, CFAR matched filtering, matched subspace filtering, and CFAR matched subspace filtering. These problems were worked and reworked in collaboration with Michael Dunn and presented in Dunn [1986] and Scharf [1987]. Harry Cox offered the term "matched subspace filtering."

At the 1987 Rhode Island meeting I discovered that Irving Reed and Larry Stotts had also been working on a version of the matched subspace problem, using generalized likelihood ratios to study the detection of multiple, linearly independent signals for communication. Their work was reported at the Rhode Island meeting and subsequently published [Reed and Stotts, 1990].

The example on bias invariance is from Neil Endsley, and the treatment of reduced-rank Gauss-Gauss detectors is a variation on Scharf and Van Veen [1987].

Duda, R. O., and P. E. Hart [1973]. *Pattern Classification and Scene Analysis*, Chapter 2 (New York: Wiley-Interscience, 1973).

Dunn, M. J. [1986]. "Sufficiency and Invariance Principles Applied to Four Detection Problems." M.S. Thesis, University of Colorado, Boulder (1986).

Ferguson, T. S. [1967]. *Mathematical Statistics*, Chapter 5 (New York: Academic Press, 1967).

Lehmann, E. L. [1959]. *Testing Statistical Hypotheses* (New York: John Wiley & Sons, 1959).

Kay, S. M., and L. L. Scharf [1984]. "Invariant Detection of Transient ARMA Signals with Unknown Initial Conditions," *Proc IEEE Conf on ASSP*, San Diego (March 1984).

Nuttall, A., and P. G. Cable [1973]. "Operating Characteristics for Maximum Likelihood Detection of Signals in Gaussian Noise of Unknown Level–II: Phase Incoherent Signals of Unknown Level," NUSC Technical Report 4683 (April 1974).

Reed, I. S., and L. B. Stotts [1990]. "A Simultaneous *M*-ary Channel Hypothesis Test with Least Mean Square Signal Amplitude Estimation," *IEEE Trans AES*, to be published (1990).

Scharf, L. L. [1987]. "Signal Processing in the Linear Statistical Model," *Proc IEEE Workshop on Underwater Acoustic Signal Processing*, University of Rhode Island (September 1987).

Scharf, L. L., and D. W. Lytle [1971]. "Signal Detection in Gaussian Noise of Unknown Level: An Invariance Application," *IEEE Trans IT*, **IT-17**, pp. 404–411 (July 1971).

Scharf, L. L., and B. D. Van Veen [1987]. "Low-Rank Detectors for Gaussian Random Vectors," *IEEE Trans ASSP*, **ASSP-35**, pp. 1579–1582 (November 1987).

PROBLEMS

4.1   What is the verbal description of an unbiased test? Why would a biased test be a poor test?

4.2   Explain how you would implement the test

$$\phi(\mathbf{x}) = \alpha; \qquad 0 \le \alpha \le 1.$$

How would you implement the test

$$\phi(\mathbf{x}) = \begin{cases} 1, & l(\mathbf{x}) > k \\ \gamma, & l(\mathbf{x}) = k \\ 0, & l(\mathbf{x}) < k \ ? \end{cases}$$

4.3   *Normal Probability Paper.* On normal probability paper, the power function for the normally distributed matched filter is a linear function of $(\mu/\sigma)\sqrt{E_s}$. Define "normal probability paper" by describing coordinates y versus x.

4.4   *Unknown Phase.* Suppose the signal **s** contains the elements $s_n = A\cos(n\omega - \phi)$ where ω is a known angular frequency, A is an unknown amplitude, and ϕ is an unknown phase. Show that $\mathbf{s} = (s_0 \quad s_1 \quad \cdots \quad s_{N-1})^{\mathrm{T}}$ has the representation

$$\mathbf{s} = \mathbf{H\theta}$$

$$\mathbf{H} = \begin{bmatrix} 1 & 0 \\ \cos\omega & \sin\omega \\ \vdots & \vdots \\ \cos(N-1)\omega & \sin(N-1)\omega \end{bmatrix}; \ \mathbf{\theta} = \begin{bmatrix} A\cos\phi \\ A\sin\phi \end{bmatrix}.$$

Thus a sinusoidal signal with unknown amplitude and phase lies in a two-dimensional linear subspace spanned by the cosinusoidal and sinusoidal components of the matrix $\mathbf{H}$. These are often called inphase and quadrature components. Let $\mathbf{x} = \mu \mathbf{s} + \mathbf{n}$ and assume $\mathbf{n} : N[\mathbf{0}, \sigma^2 \mathbf{I}]$. Find the uniformly most powerful tests of $H_0 : \mu = 0$ versus $H_1 : \mu > 0$ when σ^2 is known and when σ^2 is unknown. Defend any invariance requirements and write out all of your quadratic forms to illustrate what is going on. Characterize the performance for each case (σ^2 known and σ^2 unknown).

4.5   *Unknown Polarity.* Let $\delta \mathbf{s}$ denote a known signal whose polarity is switched by $\delta = \pm 1$. Consider the test $H_0 : \mathbf{x} : N[\mathbf{0}, \sigma^2 \mathbf{I}]$ versus $H_1 : \mathbf{x} : N[\mu \delta \mathbf{s}, \sigma^2 \mathbf{I}]$, $\mu > 0$. For each of the following cases, determine an appropriate invariance condition and derive a UMP-invariant detector:

   a. σ^2 known, $\delta = 1$;

   b. σ^2 known, $\delta = \pm 1$ (unknown);

   c. σ^2 unknown, $\delta = 1$;

   d. σ^2 unknown, $\delta = \pm 1$ (unknown).

4.6   *Unknown Bias.* Derive the UMP-invariant test of $H_0 : \mathbf{x} : N[\gamma \mathbf{1}, \sigma^2 \mathbf{I}]$ versus $H_1 : \mathbf{x} : N[\mu \mathbf{s} + \gamma \mathbf{1}, \sigma^2 \mathbf{I}]$, $\mu > 0$, when $\mathbf{s}$ is known and the channel introduces an unknown bias $\gamma \mathbf{1}$. Treat the case where σ^2 is known and the case where σ^2 is unknown.

4.7   Prove that, if $Z = X + Y$ and $X : \chi_1^2$ and $Z : \chi_N^2$, then X and Y are independent and $Y : \chi_{N-1}^2$.

4.8   *ROC Curves.* Consider the false alarm probability $P_{FA} = 1 - \Phi(z)$ and the detection probability $P_D = 1 - \Phi(z - d)$. Compute and plot P_D versus P_{FA} for several representative values of d. What value of d is required to achieve ($P_D = 0.99$, $P_{FA} = 0.01$)? Compute and plot P_D versus d for several representative values of P_{FA}. Fix P_D at 0.5 and compute and plot the threshold η versus the false alarm probability P_{FA}.

4.9   Prove the theorem in Section 4.7 that claims that certain one-parameter exponential families have monotone likelihood ratios.

4.10 Show that normal, gamma, beta, binomial, negative binomial, and Poisson distributions have monotone likelihood ratios.

4.11 Call $\bar{G} = \{\bar{g}(\boldsymbol{\theta})\}$ the set of parameters induced by the transformation g contained in the group $G = \{g\}$. Show that $\bar{G}$ is itself a group. What is the group operation?

4.12 Derive a general expression for the output SNR $= (\mu^2/\sigma^2) \mathbf{s}^T \mathbf{R}^{-1} \mathbf{s}$ when the noise is autoregressive and the signal $\mathbf{s}$ is arbitrary.

4.13 Let Q_i be a chi-squared random variable with one degree of freedom and non-centrality parameter d_i, denoted $Q_i : \chi_1^2(d_i)$. Show that, if $Q = \sum_{i=1}^n Q_i$ is distributed as $Q : \chi_n^2(\sum_{i=1}^n d_i)$, then the Q_i are independent; show that every Q distributed as $Q : \chi_n^2(\sum_{i=1}^n d_i)$ may be represented as a sum of n independent $\chi_1^2(d_i)$ random variables.

4.14 Show that the maximum of $\mathbf{s}^T\mathbf{R}^{-1}\mathbf{s}/\mathbf{s}^T\mathbf{s}$ is $\lambda_{max}[\mathbf{R}^{-1}] = 1/\lambda_{min}[\mathbf{R}]$ and that this value is achieved for $\mathbf{s}$ the maximum eigenvector of $\mathbf{R}^{-1}$ or the minimum eigenvector of $\mathbf{R}$:

$$\mathbf{R}^{-1}\mathbf{s} = \lambda_{max}\mathbf{s}$$

$$\mathbf{R}\mathbf{s} = \lambda_{min}\mathbf{s}.$$

4.15 Show that the maximum of $\boldsymbol{\theta}^T\mathbf{H}^T\mathbf{R}^{-1}\mathbf{H}\boldsymbol{\theta}/\boldsymbol{\theta}^T\mathbf{H}^T\mathbf{H}\boldsymbol{\theta}$ is λ_{max} in the eigenvalue equation

$$(\mathbf{H}^T\mathbf{R}^{-1}\mathbf{H} - \lambda\mathbf{H}^T\mathbf{H})\mathbf{u} = \mathbf{0}.$$

4.16 Find the limit of the SNR

$$\text{SNR} = \frac{\mu^2}{\sigma^2} \frac{1}{1 - a^2}\left[(N - 2)(1 + a^2) + 2 - (N - 1)a\right]$$

as $a \to \pm 1$.

4.17 *Linear Discriminants.* Consider the problem of testing H_0 versus H_1 when $\mathbf{x}$ is distributed as $N[\mathbf{0}, \sigma^2\mathbf{I}]$ under H_0 and as $N[\mathbf{0}, \mathbf{U}\boldsymbol{\Lambda}^2\mathbf{U}^T]$ under H_1. Assume that $\mathbf{U}$ is orthogonal and $\boldsymbol{\Lambda}^2 = \text{diag}[\lambda_1^2 \quad \lambda_2^2 \quad \cdots \quad \lambda_N^2]$, $\lambda_1^2 \geq \lambda_2^2 \geq \cdots \geq \lambda_N^2$. Derive the linear discriminant function for testing H_1 versus H_0 when

a. $\lambda_1^2/\sigma^2 > \sigma^2/\lambda_N^2$;

b. $\lambda_1^2/\sigma^2 < \sigma^2/\lambda_N^2$.

For each of the cases (a) and (b), compute the distribution of your linear discriminant function under H_0 and H_1 and show how to set decision thresholds to get a false alarm probability of α. Compute the detection probability, β.

4.18 Let $\mathbf{R}_0 = \sigma^2\mathbf{I} + \beta^2\mathbf{w}\mathbf{w}^T$ and $\mathbf{R}_1 = \sigma^2\mathbf{I} + \beta^2\mathbf{w}\mathbf{w}^T + \mathbf{v}\mathbf{v}^T$. Show

$$\mathbf{S} = \mathbf{R}_0^{-1/2}\mathbf{R}_1\mathbf{R}_0^{-T/2} = \mathbf{I} + \mathbf{U}\boldsymbol{\Sigma}^2\mathbf{U}^T$$

$$\boldsymbol{\Sigma}^2 = \text{diag}[\mathbf{v}^T\mathbf{R}_0\mathbf{v} \quad 0 \quad 0 \quad \cdots \quad 0].$$

4.19 Consider the matrix $\mathbf{PRP}$ where $\mathbf{P}$ is a rank r projection and $\mathbf{R}$ is a full-rank, $N \times N$ covariance matrix. Show that the pseudoinverse of $\mathbf{PRP}$ is

$$(\mathbf{PRP})^{\#} = \mathbf{U}(\mathbf{U}^T\mathbf{RU})^{-1}\mathbf{U}^T$$

$$\mathbf{UU}^T = \mathbf{P}.$$

4.20 *Sufficiency and Composite Tests.* Suppose the random vector $\mathbf{x}$ is drawn from a one-parameter exponential family

$$f_\theta(\mathbf{x}) = c(\theta)h(\mathbf{x})\exp\left\{Q(\theta)t(\mathbf{x})\right\}$$

where θ is scalar, $Q(\theta)$ is a scalar nondecreasing function of θ, and t is a scalar sufficient statistic for $Q(\theta)$. The distribution of a random sample $\mathbf{X} = (\mathbf{X}_0, \mathbf{X}_1, \ldots, \mathbf{X}_{N-1})$ is

$$f_\theta(\mathbf{x}) = c^N(\theta) \prod_{n=0}^{N-1} h(\mathbf{x}_n) \exp\left\{ Q(\theta) \sum_{n=0}^{N-1} t(\mathbf{x}_n) \right\}.$$

This makes the statistic

$$t = \sum_{n=0}^{N-1} t(\mathbf{x}_n)$$

sufficient for $Q(\theta)$. Find a UMP test of $H_0 : \theta \leq \theta_0$ versus $H_1 : \theta > \theta_0$.

4.21 *Symbol Constellations.* In an M-ary communication system, the plot of $\mathbf{m}_0, \mathbf{m}_1, \ldots, \mathbf{m}_{M-1}$ on C^N is called the symbol constellation:

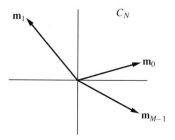

For each of the following cases, compute pairwise values of $d_{ij}^2 = (\mathbf{m}_i - \mathbf{m}_j)^{\mathrm{T}}(1/\sigma^2)(\mathbf{m}_i - \mathbf{m}_j) = 2(E_s/\sigma^2)(1 - \rho_{ij})$. (We assume $\mathbf{R} = \sigma^2 \mathbf{I}$.) Illustrate d_{ij}^2 on an appropriate constellation for the following cases:

a. Binary Antipodal ($M = 2$): $\mathbf{m}_1 = -\mathbf{m}_0$

b. Binary Orthogonal ($M = 2$): $\mathbf{m}_1^{\mathrm{T}} \mathbf{m}_0 = 0$; $\mathbf{m}_1^{\mathrm{T}} \mathbf{m}_1 = \mathbf{m}_0^{\mathrm{T}} \mathbf{m}_0$

c. M-ary Phase: $\mathbf{m}_i = \mathbf{m}_0 e^{j 2\pi i/M}$, $i = 0, 1, \ldots, M - 1$

d. M-ary Phase-Amplitude ($M = KL$):

$$\mathbf{m}_i = \mathbf{s}_0 e^{j 2\pi i/K}, \quad i = 0, 1, \ldots, K - 1$$

$$= \mathbf{s}_1 e^{j 2\pi i/K}, \quad i = K, \ldots, 2K - 1$$

$$\vdots$$

$$= \mathbf{s}_{L-1} e^{j 2\pi i/K}, \quad i = (L-1)K, \ldots, LK - 1.$$

4.22 *ASK.* Digital information is transmitted in an amplitude-shift-keyed (ASK) communication system by the symbols

$$m_t(i) = (-1)^i A \cos 2\pi \frac{k}{N} t; \qquad t = 0, 1, \ldots, N - 1.$$

a. Define $\mathbf{m}(0)$ and $\mathbf{m}(1)$ and show the symbol constellation.

b. Compute $d^2 = (2E_s/\sigma^2)(1 - \rho)$.

4.23 There is another way to derive the matched subspace filter of Section 4.11. Carry out the following sequence of steps to do it.

a. Write identity as $\mathbf{I} = \mathbf{Q}\mathbf{Q}^{\mathrm{T}}$ where

$$\mathbf{Q} = \begin{bmatrix} (\mathbf{H}^{\mathrm{T}}\mathbf{H})^{-1/2} & 0 \\ 0 & (\mathbf{A}^{\mathrm{T}}\mathbf{A})^{-1/2} \end{bmatrix} \begin{bmatrix} \mathbf{H}^{\mathrm{T}} \\ \mathbf{A}^{\mathrm{T}} \end{bmatrix}; \qquad \mathbf{A}^{\mathrm{T}}\mathbf{H} = \mathbf{0}.$$

b. Use $\mathbf{Q}$ to transform $\mathbf{x}$:

$$\mathbf{Q}\mathbf{x} = \begin{bmatrix} \mathbf{x}_1 \\ \mathbf{x}_2 \end{bmatrix} : N\left[\begin{bmatrix} (\mathbf{H}^{\mathrm{T}}\mathbf{H})^{\mathrm{T}/2}\boldsymbol{\theta} \\ \mathbf{0} \end{bmatrix}, \sigma^2\mathbf{I} \right].$$

c. Show

$$\mathbf{x} : N[\mu\mathbf{H}\boldsymbol{\theta}, \sigma^2\mathbf{I}] \qquad \text{iff } \mathbf{Q}\mathbf{x} : N\left[\mu \begin{bmatrix} (\mathbf{H}^{\mathrm{T}}\mathbf{H})^{\mathrm{T}/2}\boldsymbol{\theta} \\ \mathbf{0} \end{bmatrix}, \sigma^2\mathbf{I} \right].$$

d. Show that $\mathbf{x}_1^{\mathrm{T}}\mathbf{x}_1 = \mathbf{x}^{\mathrm{T}}\mathbf{P}_H\mathbf{x}$ is a maximal invariant to rotation of $\mathbf{x}_1$, and show that such an invariance is appropriate for this problem.

4.24 Show that the ratio of quadratic forms,

$$\frac{\mathbf{y}^{\mathrm{T}}\mathbf{P}_H\mathbf{y}}{\mathbf{y}^{\mathrm{T}}(\mathbf{I} - \mathbf{P}_H)\mathbf{y}},$$

is a maximal invariant to subspace rotation $\mathbf{Q}_H$ and gain σ.

4.25 Redraw the block diagrams for the matched subspace detector and the CFAR matched subspace detector when the projector $\mathbf{P}_H = \mathbf{U}_H\mathbf{U}_H^{\mathrm{T}}$ is built from a $\mathbf{U}_H$ that is a slice of a DFT matrix. Interpret your diagrams in terms of bandpass filters.

4.26 *Rotation.* Show that the transformation $\mathbf{y} = (\mathbf{U}_H\mathbf{Q}\mathbf{U}_H^{\mathrm{T}} + \mathbf{U}_A\mathbf{U}_A^{\mathrm{T}})\mathbf{x}$ rotates the vector $\mathbf{x}$ in the plane of $\langle\mathbf{H}\rangle$ when $\mathbf{P}_H = \mathbf{U}_H\mathbf{U}_H^{\mathrm{T}}$, $\mathbf{P}_A = \mathbf{U}_A\mathbf{U}_A^{\mathrm{T}}$, and $\mathbf{Q}$ is an orthogonal matrix. Hint: Write $\mathbf{x}$ as $(\mathbf{P}_H + \mathbf{P}_A)\mathbf{x}$.

4.27 *An Optimum Rank-Two Detector and its Approximants.* Let $\mathbf{x}$ be distributed as $N[\mathbf{0}, \mathbf{R}_i]$ and let $\mathbf{R}_1$ and $\mathbf{R}_0$ be defined as

$$H_0 : \mathbf{R}_0 = \mathbf{I} + \sigma_1^2\mathbf{w}\mathbf{w}^{\mathrm{T}}$$

$$H_1 : \mathbf{R}_1 = \mathbf{I} + \sigma_2^2\mathbf{v}\mathbf{v}^{\mathrm{T}}.$$

a. Show that the signal-to-noise ratio matrix $\mathbf{S}$ has $N - 2$ unity eigenvalues and two that satisfy the quadratic equation

$$\lambda^2 - \{1 + \sigma + \sigma_2^2\mathbf{v}^{\mathrm{T}}\mathbf{R}_0\mathbf{v}\}\lambda + \sigma\{1 + \sigma_1^2\sigma_2^2\sigma(\mathbf{w}^{\mathrm{T}}\mathbf{v})^2 + \sigma_2^2\mathbf{v}^{\mathrm{T}}\mathbf{R}_0\mathbf{v}\} = 0$$

with $\sigma = (1 + \sigma_1^2)^{-1}$. This means that the rank 2 detector is optimum for this problem.

b.   Show that, if both roots of the quadratic equation are greater than unity, then a rank-one detector is constructed by selecting the eigenvector corresponding to the largest eigenvalue, and if both roots are less than unity, the eigenvector corresponding to the smallest eigenvalue is used.

c.   Show that, in the case where one root is greater than unity and one is smaller, the eigenvector corresponding to the smallest eigenvector is chosen if and only if

$$\sigma\{1 + \sigma_1^2\sigma_2^2\sigma(\mathbf{w}^T\mathbf{v})^2 + \sigma_2^2\mathbf{v}^T\mathbf{R}_0\mathbf{v}\} < 1.$$

4.28 *Van Trees.* The likelihood ratio $l(\mathbf{X})$ is a random variable when the realization $\mathbf{x}$ is replaced by the random variable $\mathbf{X}$:

$$l(\mathbf{X}) = \frac{f_{\boldsymbol{\theta}_1}(\mathbf{X})}{f_{\boldsymbol{\theta}_0}(\mathbf{X})}.$$

This random variable has a number of interesting properties:

a. $E_{\boldsymbol{\theta}_0} l^{n+1}(\mathbf{X}) = E_{\boldsymbol{\theta}_1} l^n(\mathbf{X})$

b. $E_{\boldsymbol{\theta}_0} l(\mathbf{X}) = 1$

c. $\mathrm{var}_{\boldsymbol{\theta}_0} l(\mathbf{X}) = E_{\boldsymbol{\theta}_1} l(\mathbf{X}) - E_{\boldsymbol{\theta}_0} l(\mathbf{X})$

d. $f_{\boldsymbol{\theta}_1}(l) = l\, f_{\boldsymbol{\theta}_0}(l).$

Prove properties (a) through (d).

4.29 *Optical Communications.* Here is an optical communication system:

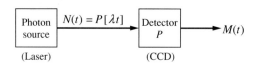

A laser transmits n photons with probability $P[N(t) = n] = e^{-\lambda t}(\lambda t)^n/n!$. Each photon is detected with probability p by the imperfect charge-coupled device (CCD).

a.   Find $P[M(t) = m]$.

b.   Find the NP (Neyman-Pearson) test of $H_0 : \lambda = \lambda_0$ versus $H_1 : \lambda = \lambda_1$. (If $\lambda = \lambda_0$ is used for 0 and $\lambda = \lambda_1$ is used for 1, then this is a scheme for transmitting binary digits.)

c.   Compute the ROC curve for representative values of λ_0 and λ_1.

4.30 Consider the problem of testing $H_0 : \mathbf{x} : N[\mathbf{m}_0, \mathbf{R}]$ versus $H_1 : \mathbf{x} : N[\mathbf{m}_1, \mathbf{R}]$. The SNR is $d^2 = (\mathbf{m}_1 - \mathbf{m}_0^T)\mathbf{R}^{-1}(\mathbf{m}_1 - \mathbf{m}_0)$. Find the value of η in the log likelihood ratio

$$\phi(\mathbf{x}) = \begin{cases} 1, & L(\mathbf{x}) > \eta \\ 0, & L(\mathbf{x}) \le \eta \end{cases}$$

that makes $\alpha = 1 - \beta$. (False alarm probability equals the miss probability.) Redraw Figure 4.6 for this case and show how the position of the interval of length d is chosen.

4.31 Let's send the binary digit $i \in (0, 1)$ by transmitting this signal on the baud interval $n = 0, 1, \ldots, N - 1$:

$$s_n = \cos 2\pi \frac{m_i}{N} n; \qquad m_i \quad \text{(integer)}.$$

This is called a frequency-shift-keyed (FSK) symboling scheme. Observations are $y_n = s_n + n_n$ with the n_n i.i.d. $N[0, \sigma^2]$ random variables.

a. Derive an optimum detector for this problem. Illustrate the detector and completely characterize its performance when α is required to equal $1 - \beta$.

b. How would you extend this signalling scheme to the M-ary problem of transmitting binary strings of length $r = \log_2 M$? What role could you see the DFT playing?

4.32 Let's send the binary digit 0 by transmitting

$$s_t = \cos 2\pi \frac{m}{N} t; \qquad t = 0, 1, \ldots, N - 1$$

and send the binary digit 1 by transmitting

$$s_t = \cos \left(2\pi \frac{m}{N} t + \theta \right); \qquad t = 0, 1, \ldots, N - 1.$$

The received signal is $x_t = s_t + n_t$ with $\{n_t\}$ a sequence of i.i.d. $N[0, \sigma^2]$ random variables.

a. Derive an optimum detector for detecting 1s and 0s.

b. Draw a block diagram.

c. How would you pick a decision threshold to get size α?

d. Write down the expression for power, β.

e. Compute and plot β versus θ when $\alpha = 1 - \beta$.

4.33 *The DFT and a Special Projector.* Show that the rank $N - 1$ projection $\mathbf{P} = \mathbf{I} - (1/N) \mathbf{1}\mathbf{1}^{\mathrm{T}}$ may be represented as

$$\mathbf{P} = \mathbf{U} \mathbf{\Lambda} \mathbf{U}^H$$

where $\mathbf{\Lambda}$ is the diagonal matrix $\mathbf{\Lambda} = [0 \quad 1 \quad 1 \quad \cdots \quad 1]$ and $\mathbf{U}$ is the *DFT* (discrete Fourier transform) matrix

$$\mathbf{U} = \frac{1}{\sqrt{N}} \begin{bmatrix} 1 & 1 & \cdots & 1 \\ 1 & e^{j2\pi/N} & & e^{j2\pi(N-1)/N} \\ \vdots & \vdots & & \vdots \\ 1 & e^{j2\pi(N-1)/N} & & e^{j2\pi(N-1)^2/N} \end{bmatrix}.$$

4.34 Find the distribution of the maximal invariant statistic $\mathbf{z} = \mathbf{Py}$ in Example 4.9. Then derive the best invariant detector for $H_0 : \mathbf{x} : N[\mathbf{m}_0 + \gamma\mathbf{1}, \sigma^2\mathbf{I}]$ versus $H_1 : \mathbf{x} : N[\mathbf{m}_1 + \gamma\mathbf{1}, \sigma^2\mathbf{I}]$. Treat the case where σ^2 is known and the case where σ^2 is unknown. Completely characterize performance in each case. Use the result of Problem 4.33 to show how the DFT could be used in your detector. Sketch your detector structure.

4.35 Find the UMP invariant detector for $H_0 : \mathbf{x} : N[\mathbf{0} + \gamma\mathbf{1}, \sigma^2\mathbf{I}]$ versus $H_1 : \mathbf{x} : N[\mu\mathbf{s} + \gamma\mathbf{1}, \sigma^2\mathbf{I}]$. Treat the case where σ^2 is known and where σ^2 is unknown. Completely characterize performance in each case. Show how the DFT may be used. (Cross-reference Problems 4.33 and 4.34.)

4.36 *Signal Design.* Extend the previous problem by allowing $\mathbf{x}$ to be distributed as $N[\mu\mathbf{s} + \gamma\mathbf{1}, \sigma^2\mathbf{R}]$ with $\mathbf{R}$ known. How would you choose $\mathbf{s}$ to maximize SNR?

4.37 Let $X = (X_0 \quad X_1 \quad \cdots \quad X_{N-1})$ denote a random sample of Poisson random variables. That is, $X_n = P[\lambda]$.

    a.   Show that $\sum_{n=0}^{N-1} X_n$ is sufficient for λ.

    b.   Show that $\sum_{n=0}^{N-1} X_n$ is distributed as $P[N\lambda]$.

    c.   Find the likelihood ratio for some $\lambda_1 > \lambda_0$.

    d.   Find the NP detector of size α for testing $H_0 : \lambda \leq 1$ versus $H_1 : \lambda > 1$; show how the threshold is chosen.

    e.   Show that your detector is UMP of size α.

APPENDIX: THE t, χ^2, AND F DISTRIBUTIONS

In our study of signal detectors in Sections 4.10 through 4.12, we have encountered complicated functions of normal random variables. These functions produce three of the most important distributions in science and engineering: the t, χ^2, and F distributions. In this appendix we study these distributions and outline how they may be used to compute the performance curves in Figures 4.23, 4.27, and 4.30. We begin with the central χ^2 distribution, because it may be used to generate the central t- and F-distributions.

Central χ^2

In Section 2.11 of Chapter 2 we have established that the distribution of the sum $Y = \sum_{n=1}^{N} X_n^2$ is central χ_N^2 when the X_n are i.i.d. $N[0, 1]$ random variables. The density for y is

$$f_Y(y) = \frac{1}{\Gamma(N/2)\, 2^{N/2}}\, y^{(N/2)-1} e^{-y/2}; \qquad y \geq 0.$$

We denote this property by writing $Y : \chi_N^2$ and saying that Y is a central χ^2 random variable with N degrees of freedom.

Central t

If $X : N[0, 1]$ and $Y : \chi^2_N$ are independent random variables, then the ratio

$$T = \frac{X}{\sqrt{Y/N}}$$

is called a t-statistic, and its distribution is called a t-distribution. In order to compute the t-distribution, we define the invertible transformations $(t, u) = W(x, y)$ and $(x, y) = W^{-1}(t, u)$:

1. $(t, u) = W(x, y)$

   $$t = x/\sqrt{y/N}, \quad u = y$$

2. $(x, y) = W^{-1}(t, u)$

   $$x = t\sqrt{u/N}, \quad y = u.$$

The Jacobian of the transformation W^{-1} is

$$|\mathbf{J}| = \det \begin{bmatrix} \dfrac{\partial x}{\partial t} & \dfrac{\partial x}{\partial u} \\ \dfrac{\partial y}{\partial t} & \dfrac{\partial y}{\partial u} \end{bmatrix} = \sqrt{u/N}.$$

The joint density function for x and y is a product distribution:

$$f_{XY}(x, y) = f_X(x) f_Y(y)$$

$$= \frac{1}{\sqrt{2\pi}} e^{-x^2/2} \frac{1}{\Gamma(N/2)\, 2^{N/2}} y^{(N/2)-1} e^{-y/2}.$$

Therefore the joint density for t and u is

$$f_{TU}(t, u) = f_X(t\sqrt{u/N}) f_Y(u) |\mathbf{J}|$$

$$= \frac{1}{\sqrt{2\pi}} e^{-t^2 u/2N} \frac{1}{\Gamma(N/2)\, 2^{N/2}} u^{(N/2)-1} e^{-u/2} \sqrt{\frac{u}{N}}.$$

By making the change of variable $w = (u/2)[1 + (t^2/N)]$ and integrating over w, we obtain the t density:

$$f_T(t) = \frac{\Gamma[(N+1)/2]}{\Gamma(N/2)\sqrt{\pi N}} \frac{1}{(1 + t^2/N)^{(N+1)/2}}; \qquad \infty < t < \infty.$$

Central F

Let $Y : \chi^2_p$ and $Z : \chi^2_{N-p}$ denote independent central χ^2 random variables with respective degrees of freedom p and $N - p$. The ratio

$$F = \frac{Y/p}{Z/(N-p)}$$

is called an F-statistic, and the distribution of F is called an F-distribution. In order to compute the F-distribution, we define the invertible transformations $(f, g) = W(y, z)$ and $(y, z) = W^{-1}(f, g)$:

1. $(f, g) = W(y, z)$

$$f = \frac{y/p}{z/(N-p)}, \quad g = z$$

2. $(y, z) = W^{-1}(f, g)$

$$y = \frac{p}{N-p} fg, \quad z = g.$$

The Jacobian of the transformation W^{-1} is

$$|\mathbf{J}| = \det \begin{bmatrix} \dfrac{\partial y}{\partial f} & \dfrac{\partial y}{\partial g} \\[2mm] \dfrac{\partial z}{\partial f} & \dfrac{\partial z}{\partial g} \end{bmatrix} = \frac{p}{N-p} g.$$

The joint density of y and z is a product density, so the joint density of f and g is

$$f_{FG}(f, g) = f_Y\left(\frac{p}{N-p} fg\right) f_Z(g) |\mathbf{J}|$$

$$= \frac{1}{\Gamma(p/2) 2^{p/2}} \left(\frac{p}{N-p} fg\right)^{(p/2)-1}$$

$$\times e^{(-p/(N-p)) fg/2} \frac{1}{\Gamma[(N-p)/2] 2^{(N-p)/2}} g^{(N-p)/2-1} e^{-g/2} \frac{p}{N-p} g$$

$$= \frac{1}{\Gamma(p/2) \Gamma[(N-p)/2] 2^{N/2}} \left(\frac{p}{N-p}\right)^{p/2} f^{(p/2)-1} g^{(N/2)-1} e^{-(g/2)[1+(p/(N-p))f]}$$

By making the change of variable $h = (g/2)[1 + (p/(N-p))f]$ and integrating over h, we obtain the density function for f:

$$f_F(f) = \frac{\Gamma(N/2) [p/(N-p)]^{p/2}}{\Gamma(p/2) \Gamma[(N-p)/2]} \frac{f^{p/2-1}}{[1+(p/(N-p))f]^{N/2}}; \quad f \geq 0.$$

(The notation $f_F(f)$ is natural, but admittedly awkward, for the density of f.)

Noncentral χ^2

When the independent random variables X_n in the sum $Y = \sum_{n=1}^{N} X_n^2$ are distributed as $X_n : N[\mu_n, 1]$, then the distribution of Y is noncentral χ^2 with noncentrality parameter $d^2 = \sum_{n=1}^{N} \mu_n^2$. This we denote $Y : \chi_N^2(d^2)$. In order to determine the density function for y, we begin with the characteristic function for one of the X_n^2:

$$\phi_{X_n^2}(\omega) = \int_{-\infty}^{\infty} e^{-j\omega x^2} \frac{1}{\sqrt{2\pi}} e^{(-1/2)(x-\mu_n)^2} dx$$

$$= \int_{-\infty}^{\infty} \frac{1}{\sqrt{2\pi}} \exp\left[\frac{-1}{2}\mu_n^2\left(\frac{j2\omega}{1+2j\omega}\right)\right]$$

$$\times \exp\left\{\frac{-1}{2}\left[x(1+2j\omega)^{1/2} - \frac{\mu_n}{(1+2j\omega)^{1/2}}\right]^2\right\} dx.$$

Make the change of variable $z = x(1 + 2j\omega)^{1/2} - \mu_n/[(1 + 2j\omega)^{1/2}]$ and integrate over z to obtain the characteristic function

$$\phi_{X_n^2}(\omega) = \exp\left[\frac{-1}{2}\mu_n^2\left(\frac{2j\omega}{1+2j\omega}\right)\right] \frac{1}{(1+2j\omega)^{1/2}}.$$

The characteristic function for the sum of the X_n^2 is then the product of these characteristic functions:

$$\phi_Y(\omega) = \exp\left[\frac{-1}{2}d^2\left(\frac{2j\omega}{1+2j\omega}\right)\right] \frac{1}{(1+2j\omega)^{N/2}}.$$

The density function for y is the inverse transform of $\phi_Y(\omega)$:

$$f(y) = \sum_{n=0}^{\infty} e^{-d^2/2} \frac{(d^2/2)^n}{n!} \frac{1}{\Gamma[(N+2n)/2]\, 2^{(N+2n)/2}} e^{-y/2} y^{[(N+2n)/2]-1}.$$

Noncentral t

The noncentral t-distribution is obtained by replacing $X : N[0, 1]$ with $X : N[\mu, 1]$ in the density for x. The joint density function for t and u is then

$$f_{TU}(t, u) = \frac{1}{\sqrt{2\pi}} \exp\left\{\frac{-1}{2}\left(t\sqrt{\frac{u}{N}} - \mu\right)^2\right\} \frac{1}{\Gamma(N/2)\, 2^{N/2}} u^{(N/2)-1} e^{-u/2} \sqrt{\frac{u}{N}}.$$

When this density is integrated over u, the result is the density function for the noncentral t-distribution:

$$f_T(t) = \frac{N^{N/2}}{\sqrt{\pi}\, \Gamma(N/2)\, 2^{(N-1)/2}} e^{(-N\mu^2)/2(t^2+N)} \frac{1}{(t^2+N)^{(N+1)/2}}$$

$$\times \int_0^{\infty} \exp\left\{\frac{-1}{2}\left(x - \frac{\mu t}{\sqrt{t^2+N}}\right)^2\right\} x^N dx.$$

Noncentral F

The noncentral F-distribution is obtained by replacing $Y : \chi_p^2$ with the noncentral χ^2 random variable $Y : \chi_p^2(d^2)$ in the ratio that defines F. The steps for finding the density for f remain unchanged, but the noncentral χ^2 density is used for the density of y. The result is

$$f_F(f) = \sum_{n=0}^{\infty} e^{-d^2/2} \frac{(d^2/2)^n}{n!} \frac{\Gamma(N/2 + n)}{\Gamma(p/2 + n)\Gamma[(N-p)/2]} \left(\frac{p}{N-p}\right)^{p/2+n}$$

$$\times \frac{f^{(p/2)-1+n}}{[1 + (p/(N-p))\, f]^{N/2+n}} \, .$$

Size and Power

In Sections 4.10 through 4.12, we have constructed detectors that select hypothesis H_1 whenever the test statistic (t, χ^2, or F) exceeds a threshold. This means that the size α (or detection probability P_{FA}) is a tail probability of a central t, χ^2, or F. The power β (or detection probability P_D) is a tail probability of a noncentral t, χ^2, or F. The size and power may be obtained by numerically integrating the densities derived in this appendix, or they may be determined by finding transformations of the t, χ^2, or F random variables that are approximately normal and then computing tail probabilities for these normals.

CHAPTER **5**

Bayes Detectors

In this chapter, we return to the simple hypothesis testing problems studied in Chapter 4. That is, given a measurement $\mathbf{x}$ drawn from the distribution $F_{\boldsymbol{\theta}}(\mathbf{x})$, how do we decide whether $\boldsymbol{\theta} = \boldsymbol{\theta}_0$ or $\boldsymbol{\theta} = \boldsymbol{\theta}_1$? As before, we define the hypothesis $H_0 : \boldsymbol{\theta} = \boldsymbol{\theta}_0$ and the alternative $H_1 : \boldsymbol{\theta} = \boldsymbol{\theta}_1$. We look for a test of H_1 versus H_0 that is optimum with respect to a criterion. In the Bayes theory of hypothesis testing, the criterion is minimum Bayes risk, where Bayes risk is defined to be the average of a loss function with respect to the joint distribution of the measurement $\mathbf{x}$ and the parameter $\boldsymbol{\theta}$.

The conceptual framework for Bayes detection is this. Mother Nature selects the parameter $\boldsymbol{\theta} = \boldsymbol{\theta}_0$ with probability p_0 and the parameter $\boldsymbol{\theta} = \boldsymbol{\theta}_1$ with probability $p_1 = 1 - p_0$. Her selection determines from which of the distributions, $F_{\boldsymbol{\theta}_0}(\mathbf{x})$ or $F_{\boldsymbol{\theta}_1}(\mathbf{x})$, Father Nature draws his measurement. The experimenter uses the measurement to decide between H_0 and H_1. Each run of this experiment produces a parameter and a decision. A loss function is defined on this pair, and the loss function is then averaged over the joint distribution of $\boldsymbol{\theta}$ and $\mathbf{x}$. This joint distribution describes the experiment consisting of Mother Nature's draw of $\boldsymbol{\theta} = \boldsymbol{\theta}_i$ followed by Father Nature's draw of $\mathbf{x}$ from distribution $F_{\boldsymbol{\theta}_i}(\mathbf{x})$. See Figure 5.1.

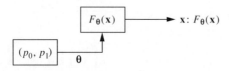

Figure 5.1 Generating measurements in the
Bayesian experiment.

The joint distribution of $\boldsymbol{\theta}$ and $\mathbf{x}$ is

$$F(\boldsymbol{\theta}, \mathbf{x}) = \begin{cases} p_0\, F_{\boldsymbol{\theta}_0}(\mathbf{x}), & \boldsymbol{\theta} = \boldsymbol{\theta}_0 \\ p_1\, F_{\boldsymbol{\theta}_1}(\mathbf{x}), & \boldsymbol{\theta} = \boldsymbol{\theta}_1. \end{cases} \qquad 5.1$$

The marginal distributions of $\boldsymbol{\theta}$ and $\mathbf{x}$ are

$$P(\boldsymbol{\theta}) = \int_{\mathbf{x}} dF(\boldsymbol{\theta}, \mathbf{x})$$

$$\qquad 5.2$$

$$= \begin{cases} p_0, & \boldsymbol{\theta} = \boldsymbol{\theta}_0 \\ p_1, & \boldsymbol{\theta} = \boldsymbol{\theta}_1 \end{cases}$$

$$F(\mathbf{x}) = p_0 F_{\boldsymbol{\theta}_0}(\mathbf{x}) + p_1 F_{\boldsymbol{\theta}_1}(\mathbf{x}).$$

The conditional distribution of $\mathbf{x}$, given $\boldsymbol{\theta}$, is

$$F(\mathbf{x}|\boldsymbol{\theta}) = \begin{cases} F_{\boldsymbol{\theta}_0}(\mathbf{x}), & \boldsymbol{\theta} = \boldsymbol{\theta}_0 \\ F_{\boldsymbol{\theta}_1}(\mathbf{x}), & \boldsymbol{\theta} = \boldsymbol{\theta}_1. \end{cases} \qquad 5.3$$

We shall use the notation $F_{\boldsymbol{\theta}}(\mathbf{x})$ rather than $F(\mathbf{x}|\boldsymbol{\theta})$ to denote the conditional distribution.

5.1 BAYES RISK FOR HYPOTHESIS TESTING

We begin with the parameter $\boldsymbol{\theta}$ drawn from the binary distribution

$$P(\boldsymbol{\theta}) = \begin{cases} p_0, & \boldsymbol{\theta} = \boldsymbol{\theta}_0 \\ p_1, & \boldsymbol{\theta} = \boldsymbol{\theta}_1 \end{cases} \qquad 5.4$$

and the measurement $\mathbf{x}$ drawn from the distribution $F_{\boldsymbol{\theta}}(\mathbf{x})$. The decision rule ϕ maps each $\mathbf{x}$ into 0 or 1, corresponding to hypothesis H_0 or alternative H_1:

$$\phi(\mathbf{x}) = \begin{cases} 1\ (\sim H_1), & \mathbf{x} \in S_1 \\ 0\ (\sim H_0), & \mathbf{x} \in S_0. \end{cases} \qquad 5.5$$

The sets S_0 and S_1 are disjoint sets that span the outcome space for the measurements. We call S_0 the acceptance region (for H_0) and S_1 the rejection region (for H_0). These regions, and the map ϕ, are illustrated in Figure 5.2.

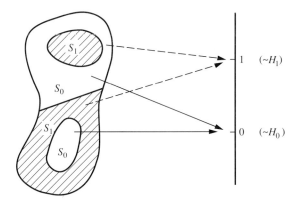

Figure 5.2 The decision rule ϕ and the regions S_0
and S_1.

Loss

For each parameter $\boldsymbol{\theta} = \boldsymbol{\theta}_i$ and decision $\phi(\mathbf{x}) = j$, we form the pair $(\boldsymbol{\theta}_i, j)$ and assign the nonnegative loss $L[\boldsymbol{\theta}, \phi(\mathbf{x})] = L_{ij}$ to $(\boldsymbol{\theta}_i, j)$. We say that L_{ij} is the loss sustained when Mother Nature selects hypothesis H_i (or parameter $\boldsymbol{\theta} = \boldsymbol{\theta}_i$) and we, as decision makers, select hypothesis H_j (or parameter $\boldsymbol{\theta}_j$). For example, L_{01} is the loss sustained when H_0 is in force and we decide H_1. The loss is illustrated in Figure 5.3(a).

Risk

The risk associated with the test of H_0 versus H_1 is the average of the loss over the distribution of the measurements. So, $R(\boldsymbol{\theta}_i, \phi)$ is the risk associated with the decision ϕ when measurements are distributed as $F_{\boldsymbol{\theta}_i}(\mathbf{x})$:

$$R(\boldsymbol{\theta}, \phi) = E_{\boldsymbol{\theta}}L[\boldsymbol{\theta}, \phi(\mathbf{x})] = \int L[\boldsymbol{\theta}, \phi(\mathbf{x})] \, dF_{\boldsymbol{\theta}}(\mathbf{x})$$

$$= \begin{cases} L_{00}P_{\boldsymbol{\theta}_0}\big[\phi(\mathbf{X}) = 0\big] + L_{01}P_{\boldsymbol{\theta}_0}\big[\phi(\mathbf{X}) = 1\big], & \boldsymbol{\theta} = \boldsymbol{\theta}_0 \\ L_{10}P_{\boldsymbol{\theta}_1}\big[\phi(\mathbf{X}) = 0\big] + L_{11}P_{\boldsymbol{\theta}_1}\big[\phi(\mathbf{X}) = 1\big], & \boldsymbol{\theta} = \boldsymbol{\theta}_1. \end{cases}$$

5.6

In these expressions, $P_{\boldsymbol{\theta}_i}[\phi(\mathbf{X}) = j]$ is the probability that hypothesis H_j is selected when hypothesis H_i is in force. So we may rewrite the risk as

$$R(\boldsymbol{\theta}, \phi) = \begin{cases} L_{00}P_{00} + L_{01}P_{01}, & \boldsymbol{\theta} = \boldsymbol{\theta}_0 \\ L_{10}P_{10} + L_{11}P_{11}, & \boldsymbol{\theta} = \boldsymbol{\theta}_1 \end{cases}$$

5.7

where the P_{ij} have these interpretations:

P_{00} probability of (correct) rejection

P_{01} probability of false alarm

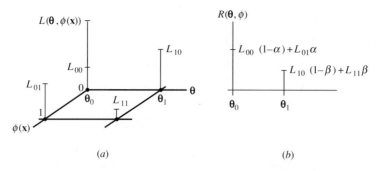

Figure 5.3 Loss and risk. (a) Loss $L(\boldsymbol{\theta}, \phi(\mathbf{x})) = L_{ij}$ versus $(\boldsymbol{\theta}, \phi(\mathbf{x}))$.
(b) Risk $R(\boldsymbol{\theta}, \phi)$ versus $\boldsymbol{\theta}$.

P_{10} probability of miss
P_{11} probability of detection.

Actually, $P_{00} = 1 - P_{01}$ and $P_{10} = 1 - P_{11}$. Furthermore, in Chapter 4 we agreed to denote the probability of false alarm as $P_{01} = \alpha$ and the probability of detection as $P_{11} = \beta$. With these conventions, we have

$$P_{00} = 1 - \alpha$$
$$P_{01} = \alpha$$
$$P_{10} = 1 - \beta$$
$$P_{11} = \beta.$$

$\qquad$ 5.8

The risk function is now written as

$$R(\boldsymbol{\theta}, \phi) = \begin{cases} L_{00}(1 - \alpha) + L_{01}\alpha, & \boldsymbol{\theta} = \boldsymbol{\theta}_0 \\ L_{10}(1 - \beta) + L_{11}\beta, & \boldsymbol{\theta} = \boldsymbol{\theta}_1. \end{cases}$$

$$= \begin{cases} (L_{01} - L_{00})\alpha + L_{00}, & \boldsymbol{\theta} = \boldsymbol{\theta}_0 \\ (L_{10} - L_{11})(1 - \beta) + L_{11}, & \boldsymbol{\theta} = \boldsymbol{\theta}_1. \end{cases}$$

$\qquad$ 5.9

The risk is illustrated in Figure 5.3(b). It is no longer a function of the data because this dependence has been averaged out. The second expression for risk in Equation 5.9 shows two of the components of risk, namely the constant terms L_{00} and L_{11}, to be independent of the decision ϕ. Therefore, in our search for decision rules that will minimize Bayes risk, we may ignore the constant terms L_{00} and L_{11}. Without loss of generality, then, we shall assume $L_{00} = 0 = L_{11}$ and assume for the remainder of our discussion that the risk is

$$R(\boldsymbol{\theta}, \phi) = \begin{cases} L_{01}\alpha, & \boldsymbol{\theta} = \boldsymbol{\theta}_0 \\ L_{10}(1 - \beta), & \boldsymbol{\theta} = \boldsymbol{\theta}_1. \end{cases}$$

$\qquad$ 5.10

The risk function is nothing more nor less than a scaled version of the false alarm probability α and the miss probability $(1 - \beta)$. In fact, when $L_{01} = L_{10} = 1$,

$R(\boldsymbol{\theta}_0, \phi) = \alpha$ and $R(\boldsymbol{\theta}_1, \phi) = 1 - \beta$. One suspects that this scaling will bring only minor modifications to the results we found in Chapter 4 and not change the basic character of our solutions.

Bayes Risk

Bayes risk is the average of risk over the distribution of the parameter $\boldsymbol{\theta}$ that Mother Nature has drawn from her distribution $P(\boldsymbol{\theta})$:

$$R(\mathbf{p}, \phi) = ER(\boldsymbol{\theta}, \phi) = p_0 R(\boldsymbol{\theta}_0, \phi) + p_1 R(\boldsymbol{\theta}_1, \phi)$$
$$= p_0 L_{01}\alpha + p_1 L_{10}(1 - \beta). \tag{5.11}$$

Bayes risk actually describes the average loss when loss is averaged over the joint distribution of the parameter $\boldsymbol{\theta}$ and the measurement $\mathbf{x}$:

$$R(\mathbf{p}, \phi) = EL[\boldsymbol{\theta}, \phi(\mathbf{x})] = \int dF(\boldsymbol{\theta}, \mathbf{x})L[\boldsymbol{\theta}, \phi(\mathbf{x})]$$

$$= \int dF(\boldsymbol{\theta}) \int dF_{\boldsymbol{\theta}}(\mathbf{x})L[\boldsymbol{\theta}, \phi(\mathbf{x})] \tag{5.12}$$

$$= \int dF(\boldsymbol{\theta})R(\boldsymbol{\theta}, \phi) = p_0 R(\boldsymbol{\theta}_0, \phi) + p_1 R(\boldsymbol{\theta}_1, \phi).$$

It is this average loss, or Bayes risk, that we wish to minimize with the clever choice of a detector ϕ. From Birdsall's North-by-Northwest insight, we know that the detector will be a likelihood ratio detector. This we prove in the next section.

5.2   BAYES TESTS OF SIMPLE HYPOTHESES

The Bayes risk for testing $H_0 : \boldsymbol{\theta} = \boldsymbol{\theta}_0$ versus $H_1 : \boldsymbol{\theta} = \boldsymbol{\theta}_1$ is

$$R(\mathbf{p}, \phi) = p_0 L_{01}\alpha + p_1 L_{10}(1 - \beta). \tag{5.13}$$

But α and $(1 - \beta)$ are, respectively, false alarm and miss probabilities, so they can be written

$$\alpha = P_{\boldsymbol{\theta}_0}\big[\phi(\mathbf{X}) = 1\big]$$

$$= E_{\boldsymbol{\theta}_0}\phi(\mathbf{X}) = \int_{S_1} dF_{\boldsymbol{\theta}_0}(\mathbf{x}) \tag{5.14}$$

$$1 - \beta = 1 - P_{\boldsymbol{\theta}_1}\big[\phi(\mathbf{X}) = 1\big]$$

$$= 1 - E_{\boldsymbol{\theta}_1}\big[\phi(\mathbf{X})\big] = 1 - \int_{S_1} dF_{\boldsymbol{\theta}_1}(\mathbf{x})$$

where $S_1 = \{\mathbf{x} : \phi(\mathbf{x}) = 1\}$ is the set of measurements $\mathbf{x}$ that produce $\phi(\mathbf{x}) = 1$ ($\sim H_1$). When these results are substituted into the risk formula, we obtain the result

$$R(\mathbf{p}, \phi) = p_0 L_{01} \int_{S_1} dF_{\theta_0}(\mathbf{x}) + p_1 L_{10} \left[1 - \int_{S_1} dF_{\theta_1}(\mathbf{x}) \right]$$

$$= p_1 L_{10} + \int_{S_1} \left[p_0 L_{01} \, dF_{\theta_0}(\mathbf{x}) - p_1 L_{10} \, dF_{\theta_1}(\mathbf{x}) \right].$$

5.15

How do we select the set S_1 or, equivalently, the test $\phi(\mathbf{x})$ to minimize the Bayes risk? Here is a procedure:

1. Begin with S_1 empty—that is, $\phi(\mathbf{x}) = 0$ for all $\mathbf{x}$. The Bayes risk is then the risk associated with a test that always chooses H_0:

$$R = p_1 L_{10}.$$

2. Now examine each $\mathbf{x}$ to see if it can be included in S_1 to decrease the Bayes risk. It can, whenever the integrand of equation 5.15 is negative:

$$p_1 L_{10} \, dF_{\theta_1}(\mathbf{x}) > p_0 L_{01} \, dF_{\theta_0}(\mathbf{x}).$$

3. Proceed in this way to generate S_1 as the set

$$S_1 = \left\{ \mathbf{x} : \frac{dF_{\theta_1}(\mathbf{x})}{dF_{\theta_0}(\mathbf{x})} > \frac{p_0 L_{01}}{p_1 L_{10}} \right\}.$$

The corresponding test is

$$\phi(\mathbf{x}) = \begin{cases} 1, & dF_{\theta_1}(\mathbf{x})/dF_{\theta_0}(\mathbf{x}) > p_0 L_{01}/p_1 L_{10} \\ 0, & \text{otherwise.} \end{cases}$$

5.16

This is the familiar likelihood ratio test. The term $dF_{\theta_1}(\mathbf{x})/dF_{\theta_0}(\mathbf{x})$ is the likelihood ratio and the term $p_0 L_{01}/p_1 L_{10}$ is the threshold. If the likelihood ratio exceeds the threshold, then H_1 is accepted. Otherwise, H_0 is accepted. This is illustrated in Figure 5.4 for the case where $dF_{\theta_i}(\mathbf{x}) = f_{\theta_i}(\mathbf{x}) \, d\mathbf{x}$, with $f_{\theta_i}(\mathbf{x})$ a continuous density; α is the false alarm probability and $1 - \beta$ is the miss probability.

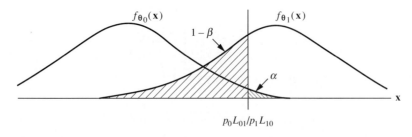

Figure 5.4 Likelihood and threshold.

5.3 MINIMAX TESTS

If Mother Nature is perverse, she will select a prior distribution $P(\boldsymbol{\theta})$ that makes (even) a Bayesian look bad. Anticipating this, we might try to design a detector that a perverse Mother Nature cannot defeat. The game is this: We know that Mother Nature is trying to make us look bad, so we are going to try to neutralize her attempts. In mathematical language, Mother Nature will try to maximize risk for each detector we choose:

$$\max_{\mathbf{p}} \min_{\phi} R(\mathbf{p}, \phi). \qquad 5.17$$

Conversely, we will try to minimize risk for the worst case distribution that Mother Nature chooses:

$$\min_{\phi} \max_{\mathbf{p}} R(\mathbf{p}, \phi). \qquad 5.18$$

Our detector is called a minimax detector. We shall show that it is a Bayes detector against Mother Nature's least favorable (or maximizing) prior, meaning that

$$\min_{\phi} \max_{\mathbf{p}} R(\mathbf{p}, \phi) = \max_{\mathbf{p}} \min_{\phi} R(\mathbf{p}, \phi). \qquad 5.19$$

In order to show this, we must undertake a study of risk sets for Bayes detectors.

Risk Set

The risk set consists of all values of risk that may be achieved with some randomized decision rule ϕ:

$$R = \{ R(\boldsymbol{\theta}, \phi) : \phi \text{ randomized decision rule} \}. \qquad 5.20$$

Here is what we mean by a randomized decision rule. You choose any two rules ϕ_1 and ϕ_2. Then the rule

$$\phi = \begin{cases} \phi_1 \text{ with probability } p \\ \phi_2 \text{ with probability } (1 - p) \end{cases} \qquad 5.21$$

is a randomized decision rule. The risk associated with ϕ is simply

$$R(\boldsymbol{\theta}, \phi) = pR(\boldsymbol{\theta}, \phi_1) + (1 - p)R(\boldsymbol{\theta}, \phi_2). \qquad 5.22$$

If $R(\boldsymbol{\theta}, \phi_1)$ and $R(\boldsymbol{\theta}, \phi_2)$ are two points in the risk set, then $R(\boldsymbol{\theta}, \phi)$ is in the risk set. This makes the risk set convex. But we know a lot more about it than this.

From the Neyman-Pearson lemma in Chapter 4, we know that the receiver operating characteristic (ROC), a plot of detection probability β versus false alarm probability α, looks like the curve of Figure 5.5(a). The value of β on the ROC is the maximum achievable β for the corresponding α, and the maximum is achieved by the Neyman-Pearson test. This we know from the Neyman-Pearson lemma. We call the rule $\hat{\phi} = 1 - \phi$ the rule that is conjugate to ϕ. It is obviously a decision rule. If the Neyman-Pearson rule ϕ achieves detection probability β and false alarm probability α, then the conjugate rule $\hat{\phi} = 1 - \phi$ achieves detection probability $(1 - \beta)$ and false

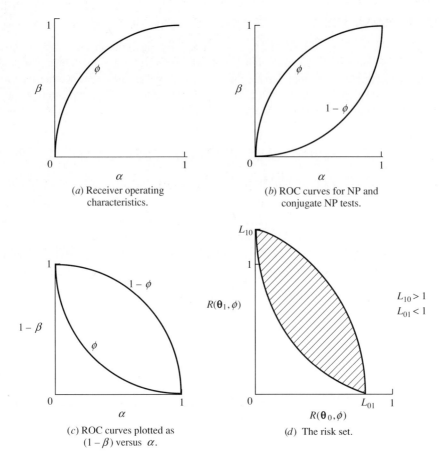

(a) Receiver operating
characteristics.

(b) ROC curves for NP and
conjugate NP tests.

(c) ROC curves plotted as
$(1 - \beta)$ versus α.

(d) The risk set.

Figure 5.5 Development of the risk set.

alarm probability $(1 - \alpha)$. (If this is not obvious, work Problem 5.3.) The ROC for
these conjugate rules is illustrated in Figure 5.5(b). Each point on the conjugate ROC
curve is the minimum β that can be achieved for the corresponding α.

The risk is just a scaled version of the false alarm probability α and the miss
probability $(1 - \beta)$. So let's replot the ROC curves for the Neyman-Pearson rules and
the conjugate Neyman-Pearson rules, as in Figure 5.5(c), with $(1-\beta)$ plotted versus α.
Then the risk $R(\boldsymbol{\theta}_1, \phi) = L_{10}(1 - \beta)$ may be plotted versus the risk $R(\boldsymbol{\theta}_0, \phi) = L_{01}\alpha$
as in Figure 5.5(d) for $L_{01} < 1$ and $L_{10} > 1$. There are no risk points $R(\boldsymbol{\theta}, \phi)$ outside
the "Brazil nut" of Figure 5.5(d), because such points would correspond to (α, β)
pairs that violate the Neyman-Pearson lemma. Finally, by choosing decision rules ϕ_1
and ϕ_2, which produce risks on the boundary of the Brazil nut, we can construct
randomized decision rules whose risks lie in the interior:

$$R(\boldsymbol{\theta}, \phi) = pR(\boldsymbol{\theta}, \phi_1) + (1 - p)R(\boldsymbol{\theta}, \phi_2).$$ 5.23

The risk set R consisting of all such risks is shaded in Figure 5.5(d). This is the risk set for Bayes hypothesis testing. We are able to characterize it because its boundary is completely characterized by the Neyman-Pearson lemma.

Bayes Tests

Bayes risk is the average of risk over the distribution of the parameter $\boldsymbol{\theta}$:

$$R(\mathbf{p}, \phi) = p_0 R(\boldsymbol{\theta}_0, \phi) + p_1 R(\boldsymbol{\theta}_1, \phi)$$
$$= p_0 L_{01} \alpha + p_1 L_{10}(1 - \beta). \qquad 5.24$$

We can rewrite this as the inner product of the probability vector $\mathbf{p} = [p_0 \quad p_1]^{\mathrm{T}}$ and the risk vector $\mathbf{R}(\phi) = \left[R(\boldsymbol{\theta}_0, \phi) \quad R(\boldsymbol{\theta}_1, \phi)\right]^{\mathrm{T}}$:

$$R(\mathbf{p}, \phi) = \mathbf{R}^{\mathrm{T}}(\phi)\mathbf{p}. \qquad 5.25$$

This means that the points in the risk set that produce the same value of Bayes risk lie on a line in the risk set that is perpendicular to the probability vector $\mathbf{p}$. Even though Bayes risk is constant on this line, the components of risk, $R(\boldsymbol{\theta}_1, \phi)$ and $R(\boldsymbol{\theta}_0, \phi)$, are unequal. (The actual value of Bayes risk produced may be read from the intersection of this line with the 45° line of Figure 5.6.) The minimum Bayes risk detector is found by sliding this locus line perpendicular to $\mathbf{p}$ and toward the origin until the locus is tangent to the boundary of the risk set. The intersection with the risk set determines the Bayes decision rule, and the intersection with the 45° line determines the corresponding value of Bayes risk. The components of risk, $\min R(\boldsymbol{\theta}_0, \phi)$ and $\min R(\boldsymbol{\theta}_1, \phi)$, are illustrated in Figure 5.6. The Bayes rule is

$$\phi(\mathbf{p}) = \arg \min_{\phi} R(\mathbf{p}, \phi). \qquad 5.26$$

This procedure produces a Bayes detector for any prior distribution $\mathbf{p}$.

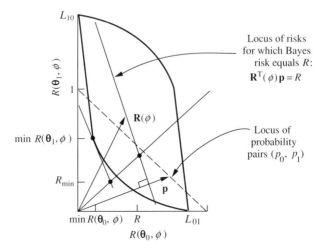

Figure 5.6 Bayes risk and Bayes decision rules.

Minimax and Maximin Tests

Let's set up the game in which the Bayes decision maker makes the first move and
Mother Nature makes the last move. The Bayes decision maker uses the Bayes decision
rule ϕ for each prior distribution $\mathbf{p}$, in order to minimize the Bayes risk $R(\mathbf{p}, \phi)$.
Mother Nature then chooses the prior distribution $\mathbf{p}$ to maximize the Bayes risk. The
maximum value of the game played this way is the maximin value

$$\max_{\mathbf{p}} \min_{\phi} R(\mathbf{p}, \phi). \qquad 5.27$$

How should Mother Nature compute this least favorable distribution? The answer
is that she should choose her prior distribution so that the minimum Bayes risk line
intersects the $45°$ line at the boundary of the risk set, as illustrated in Figure 5.7(a).
Her least favorable prior will always force the Bayes rule to be an equalizer rule that
produces equal values for the components of risk, $R(\boldsymbol{\theta}_0, \phi)$ and $R(\boldsymbol{\theta}_1, \phi)$. This also
is illustrated in Figure 5.7(a).

There is another way to play the game. Mother Nature makes the first move by
choosing a worst case prior for each decision rule ϕ (or equivalently for each risk
$R(\boldsymbol{\theta}, \phi)$). The decision maker then chooses the decision rule to minimize the risk
$R(\mathbf{p}, \phi)$. The minimum value of the game played this way is the minimax value

$$\min_{\phi} \max_{\mathbf{p}} R(\mathbf{p}, \phi). \qquad 5.28$$

The decision rule that produces the minimax value is called the minimax decision
rule:

$$\phi_{mM} = \arg \min_{\phi} \max_{\mathbf{p}} R(\mathbf{p}, \phi). \qquad 5.29$$

If Mother Nature goes first, her strategy will be to assign probability 1 to $\boldsymbol{\theta}_0$ if
$R(\boldsymbol{\theta}_0, \phi) > R(\boldsymbol{\theta}_1, \phi)$ and to assign probability 1 to $\boldsymbol{\theta}_1$ if $R(\boldsymbol{\theta}_1, \phi) > R(\boldsymbol{\theta}_0, \phi)$. How

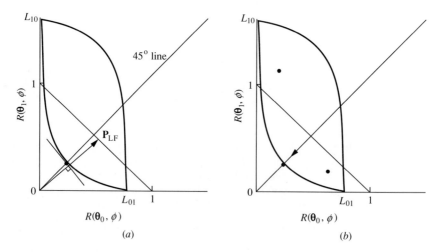

Figure 5.7 Maximin (a) and minimax (b) tests.

should the decision maker choose the minimax decision rule? The answer is to crawl along the 45° line of Figure 5.7(b) to find the intersection of the 45° line and the boundary of the risk set. Why is this? Because for any choice of ϕ that produces a risk $R(\boldsymbol{\theta}, \phi)$, that lies off the 45° line, Mother Nature will maximize the average risk by choosing the worst case prior $p_0 = 1$ if $R(\boldsymbol{\theta}_0, \phi) > R(\boldsymbol{\theta}_1, \phi)$ and $p_1 = 1$ if $R(\boldsymbol{\theta}_1, \phi) > R(\boldsymbol{\theta}_0, \phi)$. The resulting risks are illustrated by the dots in Figure 5.7(b). If this is Mother Nature's strategy, then the best a decision maker can do is to crawl along the 45° line where the two risk components are equal. The minimum average risk is achieved at the intersection of the 45° line and the boundary of the risk set. This means that the minimax value of the risk equals the maximin value:

$$\min_{\phi} \max_{\mathbf{p}} R(\mathbf{p}, \phi) = \max_{\mathbf{p}} \min_{\phi} R(\mathbf{p}, \phi). \qquad 5.30$$

But the result means more than this; it means that the minimax decision rule ϕ_{mM} may be computed by computing the Bayes decision rule ϕ for the least favorable prior distribution $\mathbf{p}_{LF}$.

Bayes and minimax tests produce risks that lie on the boundary of the risk set. These boundaries are determined by the ROC for the Neyman-Pearson test of H_1 versus H_0. This means that Bayes and minimax tests are also likelihood ratio, or Neyman-Pearson, tests whose false alarm probability and detection probability lie on the ROC. Let's see how to compute them.

Computing Minimax Tests

We have established that the minimax test of H_1 versus H_0 is a Bayes test against a least favorable prior and that this least favorable prior forces the Bayes test to be an equalizer rule that equalizes the risks $R(\boldsymbol{\theta}_0, \phi)$ and $R(\boldsymbol{\theta}_1, \phi)$. To see the implications of these facts, consider the following formulas for risk and Bayes risk:

$$R(\boldsymbol{\theta}_0, \phi) = L_{01}\alpha$$

$$R(\boldsymbol{\theta}_1, \phi) = L_{10}(1 - \beta)$$

$$R(\mathbf{p}, \phi) = p_0 L_{01}\alpha + (1 - p_0)L_{10}(1 - \beta) \qquad 5.31$$

$$= L_{10}(1 - \beta) + p_0\big[L_{01}\alpha - L_{10}(1 - \beta)\big].$$

When the risks are equalized, we have the *minimax condition*

$$L_{01}\alpha = L_{10}(1 - \beta). \qquad 5.32$$

This condition makes the minimax risk independent of the prior probability p_0:

$$R(\mathbf{p}, \phi) = L_{10}(1 - \beta). \qquad 5.33$$

So, evidently, the Bayes decision rule neutralizes Mother Nature by making the minimax risk independent of her choice for p_0! No matter what she does, the Bayesian can achieve a minimax risk of $L_{10}(1 - \beta)$.

The decision rule is determined from the minimax condition as follows. Plot the minimax condition $L_{01}\alpha = L_{10}(1 - \beta)$ as the minimax curve

$$\beta = 1 - \frac{L_{01}}{L_{10}}\alpha \qquad\qquad 5.34$$

and overlay it on the ROC as illustrated in Figure 5.8. Where the curves intersect is where a minimax test operates. So, for a given problem, proceed as follows:

1. Compute the ROC for a Neyman-Pearson detector.
2. Find the point on the ROC where $\beta = 1 - (L_{01}/L_{10})\alpha$. Call the solution β^*.
3. Solve for $\alpha^* = (L_{10}/L_{01})(1 - \beta^*)$.
4. Find the threshold k^* for the likelihood ratio test from the formula for size,

$$\int_{k^*} dF_{\theta_0}(l),$$

where $dF_{\theta_0}(l)$ is the distribution of the likelihood ratio statistic l under H_0.

In Example 5.1 we shall illustrate this procedure.

Least Favorable Prior

The minimax detector is a likelihood ratio detector whose threshold k^* produces the size and power pair (α^*, β^*). But it is also a Bayes detector against the least favorable prior. This means that k^* is the Bayes threshold

$$k^* = \frac{p_0 L_{01}}{p_1 L_{10}} = \frac{p_0 L_{01}}{(1 - p_0)L_{10}} \qquad\qquad 5.35$$

where p_0 is the corresponding least favorable prior. That is, $\mathbf{p}_{\mathrm{LF}} = (p_0^*, p_1^*)$ where

$$p_0^* = \frac{k^*}{k^* + (L_{01}/L_{10})} \qquad\qquad 5.36$$

$$p_1^* = 1 - p_0^*.$$

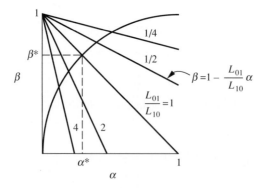

Figure 5.8 Computing minimax tests.

Example 5.1

In order to illustrate the computation of a minimax test, we return to the binary communication problem of testing $H_0 : \mathbf{x} : N[\mathbf{s}_0, \sigma^2\mathbf{I}]$ versus $H_1 : \mathbf{x} : N[\mathbf{s}_1, \sigma^2\mathbf{I}]$. From our treatment of this problem in Chapter 4, we know that the log-likelihood ratio is

$$L(\mathbf{x}) = \log \frac{f_{\theta_1}(\mathbf{x})}{f_{\theta_0}(\mathbf{x})} = \frac{1}{\sigma^2}(\mathbf{s}_1 - \mathbf{s}_0)^{\mathsf{T}}\left[\mathbf{x} - \frac{1}{2}(\mathbf{s}_1 + \mathbf{s}_0)\right]. \qquad 5.37$$

The statistic L is distributed as $L : N[-d^2/2, d^2]$ under hypothesis H_0 and as $L : N[d^2/2, d^2]$ under alternative H_1. Recall that d^2 is the signal-to-noise ratio $d^2 = (\mathbf{s}_1 - \mathbf{s}_0)^{\mathsf{T}}(\mathbf{s}_1 - \mathbf{s}_0)/\sigma^2$. The performance of any log-likelihood ratio test of the form

$$\phi(\mathbf{x}) = \begin{cases} 1, & L(\mathbf{x}) > \eta \\ 0, & L(\mathbf{x}) \le \eta \end{cases} \qquad 5.38$$

is characterized by the size and power that we computed in Chapter 4:

$$\alpha = 1 - \Phi(z)$$
$$\beta = 1 - \Phi(z - d).$$

Recall that the threshold η required to achieve the pair (α, β) is $\eta = zd - d^2/2$. For the losses $L_{00} = L_{11} = 0$ and $L_{01} = cL_{10}$, the minimax condition is

$$\beta = 1 - c\alpha. \qquad 5.39$$

When we substitute the normal integrals that determine α and β into this equation, we obtain the equation

$$1 - \Phi(z - d) = 1 - c\big[1 - \Phi(z)\big] \qquad 5.40$$

or

$$c\big[1 - \Phi(z)\big] = \Phi(z - d). \qquad 5.41$$

The value of z that satisfies this equation is determined from a table of normal integrals, and the corresponding threshold η is determined from the equation $\eta = zd - d^2/2$. This completes the specification of the minimax test. In order to determine the least favorable prior, we note that the threshold in a likelihood ratio test (*not* a log-likelihood ratio test) is $k = e^\eta$ (i.e., $\eta = \log k$). Therefore, to find the least favorable prior, we equate

$$k = e^\eta = \frac{p_0 L_{01}}{p_1 L_{10}} = \frac{p_0 c}{(1 - p_0)} \qquad 5.42$$

to determine p_0 and $p_1 = 1 - p_0$. ∎

5.4   BAYES TESTS OF MULTIPLE HYPOTHESES

Let $\Theta = \Theta_0 \cup \Theta_1 \cup \cdots \cup \Theta_{M-1}$, $\Theta_i \cap \Theta_j = \Theta_i \delta_{ij}$ denote a disjoint covering of a parameter space Θ. Assume that each subset Θ_i contains the single element $\theta_i : \Theta_i =$

$\{\boldsymbol{\theta}_i\}$. We seek a decision rule $\phi(\mathbf{x})$ that takes on values in the set $\{0, 1, \ldots, M - 1\}$ where $\phi(\mathbf{x}) = m$ corresponds to selection of hypothesis $H_m : \boldsymbol{\theta} = \boldsymbol{\theta}_m$. Define the loss function for the multiple hypothesis testing problem to be one for an incorrect classification and zero for a correct one:

$$L\big[\boldsymbol{\theta}_n, \phi(\mathbf{x})\big] = \begin{cases} 1, & \phi \neq n \\ 0, & \phi = n. \end{cases} \qquad 5.43$$

The risk is then

$$\begin{aligned} R(\boldsymbol{\theta}_n, \phi) &= E_{\boldsymbol{\theta}_n} L\big[\boldsymbol{\theta}_n, \phi(\mathbf{x})\big] \\ &= 1 - P_{\boldsymbol{\theta}_n}(\mathbf{X} \in S_n). \end{aligned} \qquad 5.44$$

The average (or Bayes) risk is just the probability of misclassification:

$$\begin{aligned} P(E) = R(\mathbf{p}, \phi) &= 1 - \sum_{n=0}^{M-1} p_n P_{\boldsymbol{\theta}_n}(\mathbf{X} \in S_n) \\ &= 1 - \sum_{n=0}^{M-1} p_n \int_{S_n} dF_{\boldsymbol{\theta}_n}(\mathbf{x}) \end{aligned} \qquad 5.45$$

where p_n is the a priori probability of H_n.

The optimum solution is to select $\phi(\mathbf{x}) = n$—that is, to place $\mathbf{x}$ in S_n—whenever the joint probability of $\boldsymbol{\theta}_n$ and $\mathbf{x}$ exceeds the joint probability of $\boldsymbol{\theta}_m$ and $\mathbf{x}$ $\forall\ m \neq n$:

$$\phi(\mathbf{x}) = n, \qquad \text{whenever } p_n\, dF_{\boldsymbol{\theta}_n}(\mathbf{x}) > p_m\, dF_{\boldsymbol{\theta}_m}(\mathbf{x}). \qquad 5.46$$

When this is the case, the a posteriori probability of $\boldsymbol{\theta}_n$, given $\mathbf{x}$, exceeds the a posteriori probability of $\boldsymbol{\theta}_m$. Can you see why?

When $p_m = 1/M$ for all m, then $\phi(\mathbf{x}) = m$ whenever the likelihood of $\boldsymbol{\theta}_m$ exceeds the likelihood for any other hypothesis:

$$\phi(\mathbf{x}) = n \qquad \text{whenever } f_{\boldsymbol{\theta}_n}(\mathbf{x}) > \max_{m \neq n} f_{\boldsymbol{\theta}_m}(\mathbf{x}). \qquad 5.47$$

The probability of misclassification is then

$$P(E) = 1 - \frac{1}{M} \sum_{n=0}^{M-1} P_{\boldsymbol{\theta}_n}[\mathbf{X} \in S_n]. \qquad 5.48$$

Example 5.2 (Normal)

Let $\mathbf{X}$ denote a normal random vector with mean $\mathbf{m}_n$ and covariance $\mathbf{R}$:

$$\mathbf{X} : N[\mathbf{m}_n, \mathbf{R}], \qquad \mathbf{R} \text{ known.} \qquad 5.49$$

We wish to test $H_0 : \boldsymbol{\theta}_0$ versus $H_1 : \boldsymbol{\theta}_1$ versus $\cdots H_{M-1} : \boldsymbol{\theta}_{M-1}$. Consider the inequality

$$p_n \exp\left\{ -\frac{1}{2} (\mathbf{x} - \mathbf{m}_n)^{\mathrm{T}} \mathbf{R}^{-1} (\mathbf{x} - \mathbf{m}_n) \right\} > p_m \exp\left\{ -\frac{1}{2} (\mathbf{x} - \mathbf{m}_m)^{\mathrm{T}} \mathbf{R}^{-1} (\mathbf{x} - \mathbf{m}_m) \right\}.$$

$$5.50$$

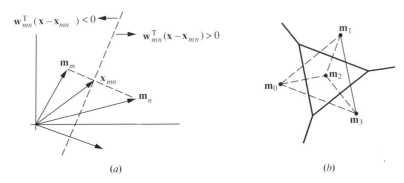

Figure 5.9 Decision regions for *M*-ary hypothesis testing. (*a*) Typical decision boundary. (*b*) Intersecting boundaries and decision regions.

This implies

$$\mathbf{w}_{mn}^{T}(\mathbf{x} - \mathbf{x}_{mn}) > \log p_m / p_n$$

$$\mathbf{R}\mathbf{w}_{mn} = \mathbf{m}_n - \mathbf{m}_m \qquad\qquad 5.51$$

$$2\mathbf{x}_{mn} = \mathbf{m}_n + \mathbf{m}_m.$$

When $p_m = p_n$ and $\mathbf{R} = \mathbf{I}$, this condition may be illustrated as in Figure 5.9(*a*).

The $(N - 1)$-dimensional hyperplane separating the region where H_m is favored over H_n is the perpendicular bisector of the line between $\mathbf{m}_n$ and $\mathbf{m}_m$. The set where $\phi(\mathbf{x}) = m$ is the intersection of half spaces like the one illustrated in Figure 5.9(*a*). This is illustrated in Figure 5.9(*b*). The decision boundaries form a convex polyhedron with at *most M* $-$ 1 faces. ∎

5.5   *M*-ORTHOGONAL SIGNALS

There is a version of the multiple hypothesis testing problem that has great importance in communication theory. To set up this version, we consider the following problem. Binary digits from a digital information source are blocked into binary words, or strings, of length b. There are 2^b such words that can be constructed, and to each we assign a symbol $\mathbf{s}_n$. The symbol $\mathbf{s}_n$ is transmitted (or stored) when the appropriate binary word is generated. There are $M = 2^b$ such signals, which we denote $\{\mathbf{s}_n\}_0^{M-1}$. If the signals are mutually orthogonal, then the following Gram matrix is diagonal:

$$G = \begin{bmatrix} \mathbf{s}_0^T \\ \mathbf{s}_1^T \\ \vdots \\ \mathbf{s}_{M-1}^T \end{bmatrix} \begin{bmatrix} \mathbf{s}_0 \ \mathbf{s}_1 \cdots \ \mathbf{s}_{M-1} \end{bmatrix} = E_s \mathbf{I} \qquad\qquad 5.52$$

$$E_s = \mathbf{s}_n^T \mathbf{s}_n \ \forall \ n.$$

If normal noise $\mathbf{n} : N[\mathbf{0}, \sigma^2 \mathbf{I}]$ is added to $\mathbf{s}_n$ to produce the measurement (or received signal) $\mathbf{x} = \mathbf{s}_n + \mathbf{n}$, then the measurement is distributed as $N[\mathbf{s}_n, \sigma^2 \mathbf{I}]$. The problem of deciding which binary word was transmitted is one of testing $H_0 : \mathbf{s} = \mathbf{s}_0$ versus $H_1 : \mathbf{s} = \mathbf{s}_1$ versus $\cdots$ versus $H_{M-1} : \mathbf{s} = \mathbf{s}_{M-1}$ in the model $\mathbf{x} : N[\mathbf{s}, \sigma^2 \mathbf{I}]$. If the hypotheses are equally likely ($p_m = 1/M$), then the Bayes detector that minimizes the probability of misclassification (or probability of error) is the detector

$$\phi = n \quad (\sim H_n : \mathbf{s} = \mathbf{s}_n), \qquad \text{whenever } g_n > \max_{m \neq n} g_m \qquad 5.53$$

where $g_n = \mathbf{s}_n^T \mathbf{x}$ is the nth matched filter output. This is just a special case of Example 5.2. The resulting error probability may be written as

$$P(E) = R(\mathbf{p}, \phi) = 1 - \frac{1}{M} \sum_{n=0}^{M-1} P_{H_n}[g_n > \max_{m \neq n} g_m]. \qquad 5.54$$

Figure 5.10 illustrates the problem and its solution.

The matched filter outputs $\{g_m\}_0^{N-1}$ are normal and statistically independent by virtue of the orthogonality of the symbols. Under hypothesis H_n, g_m is distributed as

$$g_m = \mathbf{s}_m^T \mathbf{x} : N[E_s \delta_{mn}, \sigma^2 E_s] \qquad \text{under } H_n. \qquad 5.55$$

The probability under H_n that $g_n > \max_{m \neq n} g_m$ is independent of n. The random variable $\max_{m \neq n} g_m$ is distributed as follows:

$$P_{H_n}\left[\max_{m \neq n} g_m \leq g \right] = \left[P_{H_n}[g_m \leq g] \right]^{M-1}$$

$$= \left[\int_{-\infty}^{g} \frac{1}{\sqrt{2\pi\sigma^2 E_s}} \exp\left\{ -\frac{1}{2\sigma^2 E_s} y^2 \right\} dy \right]^{M-1} \qquad 5.56$$

$$= \left[\Phi\left(\frac{g}{\sqrt{\sigma^2 E_s}} \right) \right]^{M-1}$$

Therefore, the probability of error for M-orthogonal communication is

$$P(E) = 1 - P_{H_n}\left[\max_{m \neq n} g_m \leq g_n \right]$$

$$= 1 - \int_{-\infty}^{\infty} \frac{1}{\sqrt{2\pi\sigma^2 E_s}} \exp\left\{ -\frac{1}{2\sigma^2 E_s} (g - E_s)^2 \right\} \left[\Phi\left(\frac{g}{\sqrt{\sigma^2 E_s}} \right) \right]^{M-1} dg \qquad 5.57$$

$$= 1 - \int_{-\infty}^{\infty} \frac{1}{\sqrt{2\pi}} \exp\left\{ -\frac{z^2}{2} \right\} \left[\Phi\left(z + \sqrt{\frac{E_s}{\sigma^2}} \right) \right]^{M-1} dz.$$

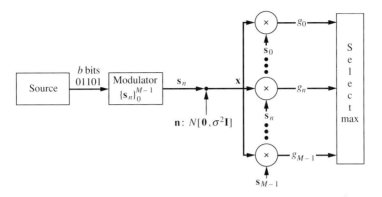

Figure 5.10 *M*-orthogonal communication.

This integral has been tabulated numerically. The results are plotted in Figure 5.11, together with the bound

$$P(E) \le (M - 1)P_{H_n}[g_m > g_n]$$

$$= (M - 1)P_{H_n}[g_n - g_m < 0]$$

$$= (M - 1) \int_{-\infty}^{0} \frac{1}{\sqrt{2\pi 2\sigma^2 E_s}} \exp\left\{ - \frac{1}{2(2\sigma^2 E_s)} (y - E_s)^2 \right\} dy$$

$$\text{5.58}$$

$$= (M - 1) \int_{-\infty}^{-\sqrt{E_s/(2\sigma^2)}} \frac{1}{\sqrt{2\pi}} \exp\{-z^2/2\} \, dz$$

$$= (M - 1)\Phi\left(-\sqrt{\frac{E_s}{2\sigma^2}}\right). \qquad\qquad \square$$

Example 5.3 (Binary Orthogonal Communication)

When $M = 2$, then the signals s_0 and s_1 may be depicted geometrically as in Figure 5.12. The decision boundary $\{x : s_1^T x > s_0^T x\}$ is also illustrated. When $p_0 = p_1 = 1/2$, the probability of error is just

$$P(E) = \frac{1}{2} P_{H_0}[s_1^T x > s_0^T x] + \frac{1}{2} P_{H_1}[s_0^T x > s_1^T x]$$

$$\text{5.59}$$

$$= P_{H_0}\left[s_0^T x - s_1^T x \le 0\right].$$

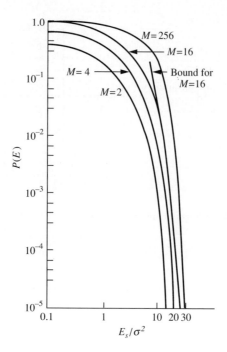

Figure 5.11 Error probability for M-orthogonal
signals [see Van Trees, 1968].

But $\mathbf{s}_0^T\mathbf{x} - \mathbf{s}_1^T\mathbf{x}$ is distributed as $N[E_s, 2\sigma^2 E_s]$ under H_0 (the difference between
two independent normals, each of variance $\sigma^2 E_s$). So,

$$
P(E) = \int\limits_{-\infty}^{0} \frac{1}{\sqrt{2\pi 2\sigma^2 E_s}} \exp\left\{ -\frac{1}{2(2\sigma^2 E_s)}(y - E_s)^2\right\}
$$

$$
= \int\limits_{-\infty}^{-\sqrt{E_s/(2\sigma)^2}} \frac{1}{\sqrt{2\pi}} \exp\left\{ -\frac{z^2}{2}\right\} dz = \Phi\left(-\sqrt{\frac{E_s}{2\sigma^2}}\right). \quad\blacksquare
$$

5.60

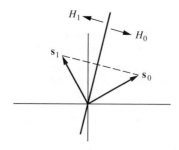

Figure 5.12 Binary orthogonal communication.

Example 5.4 (Channel Capacity)

The results for *M*-orthogonal communication bring quite a bit of insight into the problem of transmitting binary digits over a white noise channel. To develop this insight, we introduce the notion of bit rate. If bits are transmitted at rate R (bits/second), then each b-bit binary word is transmitted at the rate of R/b (words/sec). This may be written as $R/\log_2 M$ (words/sec). Each word is represented by the symbol $\mathbf{s}_n$. So, signals are transmitted at rate $R/\log_2 M$ signals per second. The energy in symbol $\mathbf{s}_n$ is

$$E_s = PT_s = \frac{P \log_2 M}{R} \qquad 5.61$$

where P is the average power of a transmitter and $T_s = (R/\log_2 M)^{-1}$ is the period of time occupied by symbol $\mathbf{s}_n$. The exact expression for probability of error is then

$$1 - P(E) = \int_{-\infty}^{\infty} \frac{1}{\sqrt{2\pi}} \exp\{-z^2/2\} \left[\Phi\left(z + \sqrt{\frac{P \log_2 M}{\sigma^2 R}} \right) \right]^{M-1} dz. \qquad 5.62$$

If we block the bits into longer and longer words ($M \rightarrow \infty$), then we have a larger and larger set of orthogonal signals $\{\mathbf{s}_n\}_0^{M-1}$. In the limit as $M \rightarrow \infty$, we have the result that

$$\lim_{M \to \infty} \ln\left[\Phi\left(z + \sqrt{\frac{P \log_2 M}{\sigma^2 R}} \right) \right]^{M-1} = \begin{cases} -\infty, & P/\sigma^2 R < 2\ln 2 \\ 0, & P/\sigma^2 R > 2\ln 2. \end{cases} \qquad 5.63$$

This means that the limit of $1 - P(E)$ is

$$\lim_{M \to \infty} \left[1 - P(E) \right] = \begin{cases} 0, & P/\sigma^2 R < 2\ln 2 \\ 1, & P/\sigma^2 R > 2\ln 2. \end{cases} \qquad 5.64$$

In words, the error probability $P(E)$ goes to zero if the bit rate is

$$R < \frac{P}{2\sigma^2 \ln 2}. \qquad 5.65$$

This remarkable result says that there is a constraint on the *rate* at which bits may be transmitted, if error-free transmission is to be achieved. The corresponding bound on signal-to-noise ratio is

$$E_s = \frac{P \log_2 M}{R}$$

$$\qquad\qquad\qquad\qquad\qquad\qquad\qquad\qquad\qquad\qquad 5.66$$

$$> \frac{P \log_2 M}{P/2\sigma^2 \ln 2} \quad \Rightarrow \quad \frac{E_s}{\sigma^2 \log_2 M} > 2\ln 2.$$

But $E_s / \log_2 M$ is the energy per bit, so we say that the minimum per bit signal-to-noise ratio for error-free communication is

$$\frac{E_b}{\sigma^2} > 2 \ln 2; \qquad E_b : \text{ energy per bit.} \qquad 5.67$$

∎

5.6 COMPOSITE MATCHED FILTERS AND ASSOCIATIVE MEMORIES

Suppose that you are testing the multiple hypotheses $\{H_n\}_0^{M-1}$ where $M = 2^b$ is a power of 2. Assume that hypothesis H_n corresponds to the hypothesis that the data is distributed as $\mathbf{x} : N [\mathbf{s}_n, \sigma^2 \mathbf{I}]$, with $\mathbf{s}_n^T \mathbf{s}_n = E_s$ for all n. Assume that the prior probability of H_n is $P[H_n] = 1/M$. Then the Bayes detector decides in favor of hypothesis H_n whenever

$$\mathbf{s}_n^T \mathbf{x} > \mathbf{s}_m^T \mathbf{x} \qquad \text{for all } m \neq n. \qquad 5.68$$

The detector requires the implementation of M matched filters of the form $\mathbf{s}_n^T \mathbf{x}$ and the finding of 1 maximum. In a communication system, the signal vectors $\mathbf{s}_n$ are typically discrete-time sinusoids whose amplitudes, frequencies, and phases depend on n. It should be obvious that the set of signals $\{\mathbf{s}_n\}_0^{M-1}$ are used for waveform coding of the 2^b different binary words of length $b = \log_2 M$.

The Bayes detector of Section 5.5 may be described as follows:

$$\text{compute} \qquad \mathbf{g} = \mathbf{S}^T \mathbf{x}$$

$$\mathbf{g} = [g_0 \quad g_1 \quad \cdots \quad g_{M-1}]^T$$

$$\mathbf{S} = [\mathbf{s}_0 \quad \mathbf{s}_1 \quad \cdots \quad \mathbf{s}_{M-1}] \qquad 5.69$$

$$\text{set} \qquad \mathbf{d} = \mathbf{1}_n \quad \text{if } g_n > g_m \text{ for all } m \neq n$$

$$\mathbf{1}_n = [0 \quad \cdots \quad 0 \underset{n}{\;1\;} 0 \quad \cdots \quad 0]^T.$$

The M-vector $\mathbf{d} = \mathbf{1}_n$, containing a single 1 in its nth position, codes or represents the selection of hypothesis H_n; that is, $\mathbf{d} = \mathbf{1}_n$ when the matched filter output g_n is maximum.

The idea behind composite matched filtering is to replace the M matched filters with $b = \log_2 M$ composite filters whose outputs may be used to select the appropriate hypothesis. The implementation is hardware-efficient, but it is sub-optimum. To illustrate the idea, we consider the case where the noise variance σ^2 goes to zero. When the signal set is orthogonal and the noise variance is *zero*, the vector of matched filter outputs may be written

$$\mathbf{g} = \mathbf{S}^T \mathbf{x} = E_s \mathbf{1}_n \qquad \text{when } H_n \text{ is in force.} \qquad 5.70$$

This means that

$$\mathbf{d} = \mathbf{1}_n \qquad \text{when } H_n \text{ is in force.} \qquad 5.71$$

Now this is a very inefficient scheme for coding hypothesis H_n because we have used M different vectors of the form $\mathbf{1}_n$, each of dimension M. This is called unary coding. It ought to take only M different vectors, each of dimension $b = \log_2 M$. That is, the vector

$$\mathbf{b}_n = \mathbf{B1}_n$$

$$\mathbf{B} = \begin{bmatrix} \mathbf{b}_0 & \cdots & \mathbf{b}_{M-1} \end{bmatrix} = \begin{bmatrix} 0 & 1 & 0 & 1 & 0 & 1 & 0 & 1 \\ 0 & 0 & 1 & 1 & 0 & 0 & 1 & 1 \\ 0 & 0 & 0 & 0 & 1 & 1 & 1 & 1 \end{bmatrix} \quad (M = 8) \qquad 5.72$$

represents $\mathbf{1}_n$ with a binary representation of its index. By decoding $\mathbf{b}_n$, we decode the index n of $\mathbf{1}_n$, and this is an efficient way to code hypothesis H_n. In the noise-free case, $\mathbf{g}$ is a scalar multiple of $\mathbf{1}_n$, so we may try the composite vector

$$\mathbf{b} = \mathbf{Bg} = \mathbf{BS}^\mathsf{T}\mathbf{x}$$
$$= \mathbf{Cx} \qquad\qquad 5.73$$

$$\mathbf{C} = \mathbf{BS}^\mathsf{T} = \begin{bmatrix} \mathbf{c}_0^\mathsf{T} \\ \vdots \\ \mathbf{c}_{b-1}^\mathsf{T} \end{bmatrix} = \begin{bmatrix} \mathbf{s}_1^\mathsf{T} + \mathbf{s}_3^\mathsf{T} + \mathbf{s}_5^\mathsf{T} + \mathbf{s}_7^\mathsf{T} \\ \mathbf{s}_2^\mathsf{T} + \mathbf{s}_3^\mathsf{T} + \mathbf{s}_6^\mathsf{T} + \mathbf{s}_7^\mathsf{T} \\ \mathbf{s}_4^\mathsf{T} + \mathbf{s}_5^\mathsf{T} + \mathbf{s}_6^\mathsf{T} + \mathbf{s}_7^\mathsf{T} \end{bmatrix} \quad (M = 8).$$

The matrix $\mathbf{C}$ contains the composite vectors $\mathbf{c}_i^\mathsf{T}$.

The filter

$$\mathbf{b} = \mathbf{Cx}$$
$$\mathbf{C} = \mathbf{BS}^\mathsf{T} \qquad\qquad 5.74$$

is called a composite matched filter. When the noise variance is zero, then the filter $\mathbf{b}$ delivers perfect binary vectors $E_s\mathbf{b}_n$ that may be decoded to determine H_n. However, when the noise covariance matrix is $\sigma^2\mathbf{I}$ and $\sigma^2 > 0$, then $\mathbf{b}$ is distributed as follows under hypothesis H_n:

$$\mathbf{b} = N[E_s\mathbf{b}_n, \mathbf{R}]$$
$$\mathbf{R} = \sigma^2\mathbf{CC}^\mathsf{T}$$

$$= \frac{M\sigma^2 E_s}{2} \begin{bmatrix} 1 & \frac{1}{2} & \cdots & \frac{1}{2} \\ \frac{1}{2} & 1 & & \vdots \\ \frac{1}{2} & & \ddots & \frac{1}{2} \\ \frac{1}{2} & \frac{1}{2} & \cdots & \frac{1}{2} & 1 \end{bmatrix}. \qquad 5.75$$

The vectors $\mathbf{b}_n$ are points on the cube illustrated in Figure 5.13. The noise components of $\mathbf{b}$ are highly correlated.

Now we have the problem of designing the decision boundaries for assignment of composite filter outputs to appropriate binary vectors $\mathbf{b}_n$. The answer has been

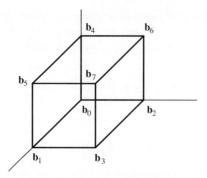

Figure 5.13 Geometry of the composite matched filter outputs ($M = 8$).

derived in our discussion of multiple hypothesis testing: choose $\mathbf{b}_n$ whenever, for all $m \neq n$,

$$-(\mathbf{b} - \mathbf{b}_n)^{\mathrm{T}}\, \mathbf{R}^{-1}(\mathbf{b} - \mathbf{b}_n) > -(\mathbf{b} - \mathbf{b}_m^{\mathrm{T}})\mathbf{R}^{-1}(\mathbf{b} - \mathbf{b}_m). \qquad 5.76$$

This determines $(b - 1)$-dimensional hyperplanes that are characterized by

$$\mathbf{w}_{mn}^{\mathrm{T}}(\mathbf{b} - \mathbf{b}_{mn}) > 0$$

$$\mathbf{R}\mathbf{w}_{mn} = \mathbf{b}_n - \mathbf{b}_m \qquad 5.77$$

$$2\mathbf{b}_{mn} = \mathbf{b}_n + \mathbf{b}_m.$$

The hyperplanes are not the simple separating hyperplanes of Figure 5.9, because the correlation matrix is *not* diagonal!

The analysis of the composite matched filter may be simplified by replacing the usual binary addresses $\mathbf{b}_n$ by binary addresses in which zeros are replaced by -1s. Then

$$\mathbf{b}_n = \mathbf{B}\mathbf{1}_n$$

$$\mathbf{B} = \begin{bmatrix} \mathbf{b}_0 & \cdots & \mathbf{b}_{p-1} \end{bmatrix} = \begin{bmatrix} -1 & +1 & -1 & +1 & -1 & +1 & -1 & +1 \\ -1 & -1 & +1 & +1 & -1 & -1 & +1 & +1 \\ -1 & -1 & -1 & -1 & +1 & +1 & +1 & +1 \end{bmatrix}. \qquad 5.78$$

The composite matched filter is then

$$\mathbf{b} = \mathbf{Cy}$$

$$\mathbf{C} = \mathbf{B}\mathbf{S}^{\mathrm{T}} = \begin{bmatrix} -\mathbf{s}_0^{\mathrm{T}} & +\mathbf{s}_1^{\mathrm{T}} & -\mathbf{s}_2^{\mathrm{T}} & +\mathbf{s}_3^{\mathrm{T}} & -\mathbf{s}_4^{\mathrm{T}} & +\mathbf{s}_5^{\mathrm{T}} & -\mathbf{s}_6^{\mathrm{T}} & +\mathbf{s}_7^{\mathrm{T}} \\ -\mathbf{s}_0^{\mathrm{T}} & -\mathbf{s}_1^{\mathrm{T}} & +\mathbf{s}_2^{\mathrm{T}} & +\mathbf{s}_3^{\mathrm{T}} & -\mathbf{s}_4^{\mathrm{T}} & -\mathbf{s}_5^{\mathrm{T}} & +\mathbf{s}_6^{\mathrm{T}} & +\mathbf{s}_7^{\mathrm{T}} \\ -\mathbf{s}_0^{\mathrm{T}} & -\mathbf{s}_1^{\mathrm{T}} & -\mathbf{s}_2^{\mathrm{T}} & -\mathbf{s}_3^{\mathrm{T}} & +\mathbf{s}_4^{\mathrm{T}} & +\mathbf{s}_5^{\mathrm{T}} & +\mathbf{s}_6^{\mathrm{T}} & +\mathbf{s}_7^{\mathrm{T}} \end{bmatrix}. \qquad 5.79$$

This filter delivers an output $\mathbf{b}$ that is distributed as follows under H_n:

$$N[\mathbf{C}\mathbf{s}_n, \sigma^2 \mathbf{C}\mathbf{C}^{\mathrm{T}}] = N[E_s \mathbf{b}_n, M\sigma^2 E_s \mathbf{I}].\qquad 5.80$$

The Bayes decision boundaries correspond to comparing each element of $\mathbf{b}$ to zero and decoding as -1 if it is less than zero and $+1$ if it is larger. The probability of misclassification if $p_m = 1/M$ is

$$P(E) = R(\mathbf{p}, \mathbf{d}) = 1 - \frac{1}{M} \sum_{n=0}^{M-1} P_{H_n}[\mathbf{b} \in S_n]$$

$$= 1 - P_{H_0}[\mathbf{b} \in S_0]$$

$$= 1 - \left[\int_{-\infty}^{0} \frac{1}{\sqrt{2\pi M\sigma^2 E_s}} \exp\left\{ -\frac{1}{2\sigma^2 M E_s} (y + E_s)^2 \right\} dy \right]^b$$

$$= 1 - \left[\int_{-\infty}^{\sqrt{E_s/(M\sigma^2)}} \frac{1}{\sqrt{2\pi}} \exp\left\{ -\frac{z^2}{2} \right\} dz \right]^b$$

$$= 1 - \left[\Phi\left(\sqrt{E_s/M\sigma^2} \right) \right]^{\log_2 M}.$$

$$5.81$$

The term $1 - P(E)$ may be written as

$$1 - P(E) = \left[\Phi\left(\sqrt{\frac{E_s}{M\sigma^2}} \right) \right]^{\log_2 M}.\qquad 5.82$$

If we replace E_s by $(P \log_2 M)/R$, as we did in our study of matched filters for M-orthogonal signals, then we obtain the formula

$$1 - P(E) = \left[\Phi\left(\frac{P \log_2 M}{M\sigma^2 R} \right) \right]^{\log_2 M} \longrightarrow 0 \text{ as } M \to \infty.\qquad 5.83$$

Regardless of rate R, the $P(E)$ goes to 1 as $M \to \infty$.

Application to Associative Memories

Suppose that the orthogonal signals $\{\mathbf{s}_n\}_0^{M-1}$ represent an arbitrary multidimensional data set, stored in memory locations $\{\mathbf{b}_n\}_0^{M-1}$. A memory produces the data $\mathbf{s}_n$ in response to a memory address:

$$\mathbf{s}_n = f(\mathbf{b}_n).\qquad 5.84$$

An associative memory produces the address $\mathbf{b}_n$ in response to the data $\mathbf{s}_n$:

$$\mathbf{b}_n = f^{-1}(\mathbf{s}_n).\qquad 5.85$$

This is identical to the functionality of the composite matched filter. It is also identical to the functionality of a vector quantizer.

Summary

The composite matched filter may be used to reduce the complexity of matched filtering or to implement associative memories. The reduction brings hardware gains, but performance penalties.

5.7 LIKELIHOOD RATIOS, POSTERIOR PROBABILITIES, AND ODDS

There are numerous connections between likelihood ratios, posterior probabilities, and odds. By exploring them, we develop insight into the nature of Bayes tests.

Let H_0 and H_1 denote two hypotheses with prior probabilities

$$p_0 = p[H_0] \qquad\qquad 5.86$$

$$p_1 = p[H_1] = 1 - p_0.$$

The ratio of p_1 to p_0 is called the prior odds on H_1:

$$0(H_1) = p_1/p_0. \qquad\qquad 5.87$$

The measurement $\mathbf{x}$ is assumed to have conditional probability $p(\mathbf{x}|H_i)$ under hypothesis H_i. We denote this conditional probability by $p_{H_i}(\mathbf{x})$ and call it the measurement probability:

$$p(\mathbf{x}|H_i) = p_{H_i}(\mathbf{x}). \qquad\qquad 5.88$$

If $\mathbf{x}$ is a continuous random vector, then the measurement probability is

$$p_{H_i}(\mathbf{x}) = f_{H_i}(\mathbf{x})\,d\mathbf{x}. \qquad\qquad 5.89$$

From Bayes rule, we may calculate the posterior probability of hypothesis H_i, given $\mathbf{x}$:

$$p(H_i|\mathbf{x}) = \frac{p_{H_i}(\mathbf{x})p_i}{p(\mathbf{x})} \qquad\qquad 5.90$$

$$p(\mathbf{x}) = p_{H_1}(\mathbf{x})p_1 + p_{H_0}(\mathbf{x})p_0.$$

The ratio of the posterior probabilities is called the posterior odds on H_1, given $\mathbf{x}$:

$$0(H_1|\mathbf{x}) = \frac{p(H_1|\mathbf{x})}{p(H_0|\mathbf{x})}$$

$$= l(\mathbf{x})0(H_1) \qquad\qquad 5.91$$

$$l(\mathbf{x}) = \frac{p_{H_1}(\mathbf{x})}{p_{H_0}(\mathbf{x})}.$$

This formula shows that likelihood is just a function of the data that scales prior odds to produce posterior odds. The log odds are called evidence, so log likelihood adds to prior evidence to produce posterior evidence:

$$\epsilon(H_1|\mathbf{x}) = \log 0(H_1|\mathbf{x})$$

$$= \log l(\mathbf{x}) + \epsilon(H_1).$$

5.92

Bayes Tests

With these interpretations, we can write the Bayes test of H_1 versus H_0 in the following two ways:

$$\phi(\mathbf{x}) = \begin{cases} 1(\sim H_1), & 0(H_1|\mathbf{x}) > L_{01}/L_{10} \\ 0(\sim H_0), & \text{otherwise} \end{cases}$$

$$= \begin{cases} 1, & p(H_1|\mathbf{x}) > (L_{01}/L_{10})p(H_0|\mathbf{x}) \\ 0, & \text{otherwise.} \end{cases}$$

5.93

The ratio of losses L_{01}/L_{10} just tells us how high the posterior odds must be for us to select H_1. Equivalently, L_{01}/L_{10} tells us how much more probable H_1 must look than H_0 after observing $\mathbf{x}$.

Example 5.5

We have five signal sources. Four of them transmit the binary digits 0 and 1 independently, with equal probabilities $p(0) = 1/2 = p(1)$. The fifth is "stuck at 1," such that it transmits 1 with probability 1. A source is selected at random: $P[S_i] = 1/5$, $i = 1, 2, \ldots, 5$. The output of the selected source is monitored, and it is found that k out of n digits are 1. What is the posterior probability that the defective source is transmitting? How many samples n must be observed until the posterior odds exceed the prior odds by a factor of 128?

Call H_1 the hypothesis that the defective source is selected and H_0 the hypothesis that a good source is selected. The prior odds are $0(H_1) = (1/5)/(4/5) = 1/4$. The conditional probability of observing k ones in n transmissions is

$$p_{H_0}(k) = \binom{n}{k} 2^{-n}$$

5.94

$$p_{H_1}(k) = \begin{cases} 1, & k = n \\ 0, & \text{otherwise.} \end{cases}$$

The posterior probability that H_1 is transmitting is 0 if $k \neq n$. If $k = n$, then

$$P[H_1|k = n] = \frac{1P[H_1]}{1P[H_1] + 2^{-n}P[H_0]}$$

5.95

$$= \frac{1}{1 + 4(2^{-n})}.$$

The posterior odds are

$$0(H_1|k = n) = 2^n 0(H_1).$$ 5.96

So $n = 7$. If $n = 7$, the a posteriori probability of H_1 is

$$P(H_1|k = n = 7) = \frac{1}{1 + 4/128} = \frac{1}{1 + 1/32}$$ 5.97

$$\cong 1 - 1/32.$$

In Bayes theory, the prior probabilities are assumed known. The losses L_{01} and L_{10} are, to a certain extent, arbitrary. Some of this arbitrariness is resolved when we realize that the ratio L_{01}/L_{10} determines the required scaling of the prior odds in order to produce enough posterior odds for a detection of H_1. ∎

5.8 BALANCED TESTS

In the Neyman-Pearson theory of hypothesis testing, it is the selection of the false alarm probability that is somewhat arbitrary. In Bayes theory, it is the selection of losses that is arbitrary. We can use our insights about Bayes tests to remove some of the arbitrariness. To do so, we pose the following question: What is the posterior probability that H_1 is true, given that we have selected H_1? This is the *probabilistic inverse* of the question: What is the probability of selecting H_1, given that H_1 is true? The answer to the first of these questions may be given in terms of the answer to the second:

$$P[H_1|\phi(\mathbf{x}) = 1] = \frac{P[\phi(\mathbf{x}) = 1|H_1]p_1}{P[\phi(\mathbf{x}) = 1|H_1]p_1 + P[\phi(\mathbf{x}) = 1|H_0]p_0}$$

$$= \frac{1}{1 + (\alpha/\beta)(p_0/p_1)}.$$ 5.98

Here α and β are the false alarm and detection probabilities of Neyman-Pearson theory.

Let's define a test to be balanced when the posterior probability $P[H_1|\phi(\mathbf{x}) = 1]$ equals the detection probability $P[\phi(\mathbf{x}) = 1|H_1]$. Then we have the balancing equation

$$\beta = \frac{1}{1 + (\alpha/\beta)(p_0/p_1)}$$

or

$$\beta = 1 - \frac{p_0}{p_1}\alpha.$$ 5.99

This equation bears a striking resemblance to the minimax equation

$$\beta = 1 - \frac{L_{01}}{L_{10}}\alpha.$$ 5.100

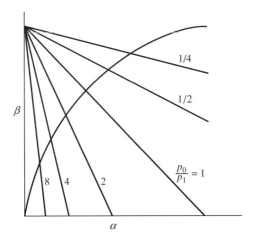

Figure 5.14 Computing balanced tests.

When the balancing equation is plotted on the ROC of Figure 5.14, we find inter-sections between the ROC and the balancing equation. These intersections produce operating points of Neyman-Pearson detectors that are balanced. They also show that a minimax test is balanced when $L_{01}/L_{10} = p_0/p_1$. The threshold in the corresponding Bayes or Neyman-Pearson test is $k = (p_0/p_1)^2$. When $(p_0/p_1) \gg 1$, indicating that H_0 is very likely a priori, then α is forced to be very small, justifying small false alarm probabilities in any problem where the H_0 hypothesis is very likely. (See Figure 5.14.)

REFERENCES AND COMMENTS

In my treatment of Bayes detectors I have been influenced by R. O. Duda and P. E. Hart, T. S. Ferguson, and H. L. Van Trees. Ferguson's analysis of risk sets is partic-ularly illuminating, and Van Trees' graphical analysis of minimax tests is insightful. The discussion of M-orthogonal signaling follows Van Trees fairly closely, and the discussion of composite filters is an adaptation of ideas I picked up from R. J. Marks and L. E. Atlas. The notion of a *balanced detector* appears to be new.

Duda, R. O., and P. E. Hart [1973]. *Pattern Classification and Scene Analysis* (New York: John Wiley & Sons, 1973).

Ferguson, T. S. [1967]. *Mathematical Statistics* (New York: Academic Press, 1967).

Marks, R. J., and L. E. Atlas [1987]. "Composite Matched Filtering with Error Correction," *Optics Letters* **12**, p. 135 (February 1987).

Van Trees, H. L. [1968]. *Detection, Estimation, and Modulation Theory*, Part I (New York: John Wiley & Sons, 1968).

PROBLEMS

5.1   Call $f(\boldsymbol{\theta}, \mathbf{x})$ the joint probability mass function-probability density function (pmf-pdf) of $\boldsymbol{\theta}$ and $\mathbf{x}$:

$$f(\boldsymbol{\theta}, \mathbf{x}) = \begin{cases} p_0 f_{\boldsymbol{\theta}_0}(\mathbf{x}), & \boldsymbol{\theta} = \boldsymbol{\theta}_0 \\ p_1 f_{\boldsymbol{\theta}_1}(\mathbf{x}), & \boldsymbol{\theta} = \boldsymbol{\theta}_1 \end{cases}$$

Find the conditional pmf of $\boldsymbol{\theta}$, given $\mathbf{x}$. This conditional pmf is also called the posterior pmf of $\boldsymbol{\theta}$.

5.2   One criterion for testing H_1 versus H_0 would be to evaluate the posterior pmf for $\boldsymbol{\theta}$ (see preceding problem) at the measurement, call the result the posterior likelihood, and select H_1 whenever the posterior likelihood for H_1 exceeds the posterior likelihood for H_0. Show that the resulting test is the likelihood ratio test

$$\phi(\mathbf{x}) = \begin{cases} 1, & \dfrac{f_{\boldsymbol{\theta}_1}(\mathbf{x})}{f_{\boldsymbol{\theta}_0}(\mathbf{x})} > \dfrac{p_0}{p_1} \\ 0, & \text{otherwise} \end{cases}$$

5.3   Suppose the decision rule ϕ produces false alarm probability α and detection probability β. Show that the decision rule $\hat{\phi} = 1 - \phi$ produces false alarm probability $1 - \alpha$ and detection probability $1 - \beta$.

5.4   You are given M signals $\{\mathbf{s}_n\}_0^{M-1}$ where each $\mathbf{s}_n$ is N-dimensional and $M \le N$. Describe how you would orthogonalize this set of signals. Can you do this when $M > N$? Why or why not?

5.5   Verify that $\mathbf{s}_n^T \mathbf{x} : N[E_s \delta_{nm}, \sigma^2 E_s]$ when $\mathbf{x}$ is $N[\mathbf{s}_m, \sigma^2 \mathbf{I}]$ and $\mathbf{s}_n^T \mathbf{s}_m = E_s \delta_{nm}$. Verify that $\mathbf{s}_n^T \mathbf{x}$ is independent of $\mathbf{s}_m^T \mathbf{x}$ for $m \ne n$.

5.6   Consider the hypothesis test $H_0 : \mathbf{x} : N[\mathbf{s}_0, \sigma^2 \mathbf{I}]$ versus $H_1 : \mathbf{x} : N[\mathbf{s}_1, \sigma^2 \mathbf{I}]$. Assume $L_{00} = L_{11} = 0$ and $L_{01} = 2L_{10}$. Use the *analytical* results of Example 5.1 and a table of normal integrals to determine (α, β), η, and k for a minimax test. Carry out your computations for $d^2 = (\mathbf{s}_1 - \mathbf{s}_0)^T (\mathbf{s}_1 - \mathbf{s}_0)/\sigma^2 = 0.1$, 1.0, and 10. For each case, determine Mother Nature's least favorable prior.

5.7   In Problem 5.6, *analytically* determine (α, β), η, and k for a Bayes test of H_0 versus H_1 when $p_0 = 3/4$, $L_{00} = L_{11} = 0$, and $L_{01} = 2L_{10}$; let $d^2 = 0.1$, 1.0, 10.

5.8   Consider the hypothesis $H_0 : \mathbf{x} : N[\mathbf{s}_0, \sigma^2 \mathbf{I}]$ and the alternative $H_1 : \mathbf{x} : N[\mathbf{s}_1, \sigma^2 \mathbf{I}]$. Call the signal-to-noise ratio $d^2 = (\mathbf{s}_1 - \mathbf{s}_0)^T (\mathbf{s}_1 - \mathbf{s}_0)/\sigma^2$. Assume $L_{00} = L_{11} = 0$ and $L_{01} = 2L_{10}$. Plot the ROC curves β versus α for $d^2 = 0.1$, 1.0, and 10. From each ROC curve, graphically determine (α, β) and the thresholds k and $\eta = \ln k$ for a minimax test. Determine the least favorable prior in each case. Compare your findings with the analytical results of Problem 5.6.

5.9   From the ROC curves of Problem 5.8, graphically determine (α, β) and the thresholds k and $\eta = \ln k$ for a Bayes test when $p_0 = 3/4$ and $L_{01} = 2L_{10}$.

5.10 In Example 5.1, prove that $L_{00} = L_{11} = 0$ and $L_{01} = L_{10}$ produce the thresholds $\eta = 0$ and $k = e^{\eta} = 1$ for a minimax test and force Mother Nature into the least favorable prior $p_0 = p_1 = 1/2$.

5.11 Generalize the discussion of balanced tests by deriving a test for which $P[H_1|\phi(\mathbf{x}) = 1] = q$. Explain how the threshold would be determined.

5.12 Prove that the Bayes test of $H_0 : \mathbf{x} : F_{\boldsymbol{\theta}_0}(\mathbf{x})$ versus $H_1 : \mathbf{x} : F_{\boldsymbol{\theta}_1}(\mathbf{x})$ versus $\cdots$ versus $H_{M-1} : \mathbf{x} : F_{\boldsymbol{\theta}_{M-1}}(\mathbf{x})$ is

$$\phi(\mathbf{x}) = n \qquad \text{whenever } p_n \, dF_{\boldsymbol{\theta}_n}(\mathbf{x}) > p_m \, dF_{\boldsymbol{\theta}_m}(\mathbf{x}).$$

5.13 Prove the results of Equations 5.63 and 5.64:

$$\lim_{M \to \infty} \ln \left[\Phi \left(z + \sqrt{\frac{P \log_2 M}{\sigma^2 R}} \right) \right]^{M-1} = \begin{cases} -\infty, & \dfrac{P}{\sigma^2 R} < 2 \ln 2 \\[2mm] 0, & \dfrac{P}{\sigma^2 R} > 2 \ln 2 \end{cases}$$

$$\lim_{M \to \infty} \left[1 - P(E) \right] = \begin{cases} 0, & \dfrac{P}{\sigma^2 R} < 2 \ln 2 \\[2mm] 1, & \dfrac{P}{\sigma^2 R} > 2 \ln 2. \end{cases}$$

5.14 *Binary Communication.* A communication source generates binary digits 0 and 1 and transmits the respective signals $\{s_t(i)\}_{t=0}^{N-1}$ for $i = 0, 1$. A sequence of independent, identically distributed noises $\{n_t\}_0^{N-1}$ is added to produce the measurements $\{x_t\}_0^{N-1}$, with $x_t : N[s_t(i), \sigma^2]$. You are to test $H_0 : i = 0$ versus $H_1 : i = 1$. Assume $\sum_{t=0}^{N-1} s_t^2(i) = E_s$ and $\sum_{t=0}^{N-1} s_t(0)s_t(1) = \rho E_s$, $-1 < \rho < 1$. Then

   a. vectorize the problem and sketch the vector source and channel;

   b. find the likelihood ratio detector;

   c. compute and plot the ROC curve (β versus α) for $E_s/\sigma^2 = 10$ and $\rho = 0$; repeat for $\rho = 1/2$;

   d. show operating points on the ROC curves for (i) NP detector with $\alpha = 0.01$, (ii) Bayes detector when $p_0 = 7/16$, $p_1 = 9/16$, and $L_{01} = L_{10} = 1$, (iii) minimax detector when $L_{01} = 4L_{10}$; find least favorable prior, (iv) balanced detector when $p_0 = 15/16$;

   e. for each of the operating points in (d), determine the thresholds η and $k = e^{\eta}$;

   f. discuss how your results apply to ASK, PSK, and FSK communication.

5.15 In our discussion of Bayes tests, we argued that the value of Bayes risk could be read from the intersection of the line $\mathbf{R}^{\mathsf{T}}(\phi)\mathbf{p} = R$ with the 45° line of Figure 5.6. Prove this.

5.16 Here is an optical communication system:

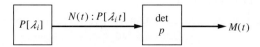

A laser transmits n photons on the baud interval $[0, t)$ with Poisson probability $P[N(t) = n] = e^{-\lambda_i t}(\lambda_i t)^n / n!$. The rate parameter λ_i is λ_0 for a binary digit 0 and λ_1 for a binary digit 1. Each photon is detected with probability p. You are to observe $M(t)$, the number of photons detected on $[0, t)$, and test $H_0 : i = 0$ versus $H_1 : i = 1$.

a.   Compute the distribution of $M(t)$.

b.   Find the likelihood ratio detector for testing H_0 versus H_1.

c.   Compute and plot the ROC curve (β versus α) for $\lambda_0 p = 1$ and $\lambda_1 p = 2$, and for t chosen so that $\beta = 0.5$ at $\alpha = 0.01$.

d.   Show operating points on the ROC curve for (i) NP detector with $\alpha = 0.1$, (ii) Bayes detector when $p_0 = 1/4$, and $L_{01} = 4L_{10}$, (iii) minimax detector when $L_{01} = 4L_{10}$; find least favorable prior, (iv) balanced detector when $p_0 = 1/4$.

Maximum Likelihood Estimators

The principle of maximum likelihood is deceptively simple. Application of this principle to inference problems is straightforward and, by and large, the results are useful. In many cases, maximum likelihood estimators correspond to least squares and minimum variance unbiased estimators.

In spite of its simplicity, the theory of maximum likelihood is fraught with potential dangers for the incautious user. To begin, it is important to distinguish between *likelihood functions* and *probability density functions*. Next, it is important to understand how random samples bring information about underlying parameters. This information is carried by the *score function*. Finally, it is essential to understand how the Cramer-Rao bound places limits on the performance of maximum likelihood estimators.

In this chapter we apply the theory of maximum likelihood (ML) to a number of interesting problems in signal processing. We study the invariance theorem of ML theory and the Cramer-Rao (CR) bounds on unbiased estimators. The chapter appendix contains a number of results for vector and matrix gradients that are used throughout the chapter.

6.1   MAXIMUM LIKELIHOOD PRINCIPLE

Let $\mathbf{X}$ denote a random variable whose probability density function or probability mass function $f_\theta(\mathbf{x})$ is parameterized by the unknown parameter θ. A typical density function is illustrated in Figure 6.1. In the figure, we have shown two densities, one for parameter θ_1 and one for parameter θ_2.

Suppose the value $\hat{\mathbf{x}}$ is observed. Based on the prior model $f_\theta(\mathbf{x})$ illustrated in Figure 6.1, we can say that $\hat{\mathbf{x}}$ is more probably observed when $\theta = \theta_2$ than when $\theta = \theta_1$. More generally, there may be a unique value of θ for which the value $\hat{\mathbf{x}}$ is more probably observed than for any other. We call this value of θ that makes $\hat{\mathbf{x}}$ most probable, or most likely, the *maximum likelihood estimate* and denote it $\hat{\theta}$:

$$\hat{\theta} = \arg\left[\max_\theta f_\theta(\hat{\mathbf{x}})\right]. \qquad 6.1$$

We obtain the maximum likelihood estimate by evaluating the density $f_\theta(\mathbf{x})$ at the value of the observation $\hat{\mathbf{x}}$ and then searching for the value of θ that maximizes $f_\theta(\hat{\mathbf{x}})$. The function $l(\theta, \hat{\mathbf{x}}) = f_\theta(\hat{\mathbf{x}})$ is called the *likelihood function*, and the logarithm

$$L(\theta, \hat{\mathbf{x}}) = \ln f_\theta(\hat{\mathbf{x}}) \qquad 6.2$$

is called the *log-likelihood function*. When this function is continuously differentiable in θ, then the maximum likelihood estimate may be determined by differentiating the

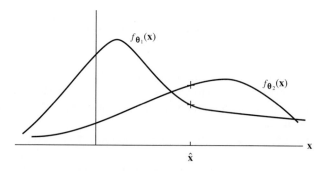

Figure 6.1 Typical density functions.

log-likelihood function. The maximum likelihood estimate $\hat{\boldsymbol{\theta}}$ is then called the root of the maximum likelihood cquation

$$\frac{\partial}{\partial \boldsymbol{\theta}} L(\boldsymbol{\theta}, \hat{\mathbf{x}}) = \frac{\partial}{\partial \boldsymbol{\theta}} \ln f_{\boldsymbol{\theta}}(\hat{\mathbf{x}}) = \mathbf{0}. \qquad 6.3$$

We shall assume that there is only one value of $\boldsymbol{\theta}$ for which this derivative is zero.
The gradient of the log-likelihood function is called the *score function*:

$$\mathbf{s}(\boldsymbol{\theta}, \mathbf{x}) = \frac{\partial}{\partial \boldsymbol{\theta}} L(\boldsymbol{\theta}, \mathbf{x}). \qquad 6.4$$

So we can say that the maximum likelihood estimate $\hat{\boldsymbol{\theta}}$ is the zero of the score function:

$$\mathbf{s}(\hat{\boldsymbol{\theta}}, \mathbf{x}) = \mathbf{0}. \qquad 6.5$$

In this formula and throughout most of this chapter, we drop the notation $\hat{\mathbf{x}}$ for the observation and use, simply, $\mathbf{x}$.

The likelihood, log-likelihood, and score function are all related to the underlying density function. When we evaluate the performance of the maximum likelihood estimator, it will be essential to remind ourselves that the log-likelihood function $L(\boldsymbol{\theta}, \mathbf{x})$ becomes a random variable when $\mathbf{x}$ is replaced by the random variable $\mathbf{X}$. Similarly, the score function becomes a random variable, and it is precisely the behavior of the score function random variable that determines how the maximum likelihood estimator will perform.

Example 6.1

Let $\mathbf{X} = (X_0, X_1)$ denote a random sample of size 2 from a uniform $U[0, \theta]$ distribution. The density function for the sample is

$$f_{\theta}(x_0, x_1) = \prod_{n=0}^{1} \frac{1}{\theta} I_{[0,\theta]}(x_n)$$

$$= \frac{1}{\theta^2} I_{[0,\max(x_0,x_1)]}\big[\min(x_0, x_1)\big] I_{[\min(x_0,x_1),\theta]}\big[\max(x_0, x_1)\big]. \qquad 6.6$$

This function is illustrated in Figure 6.2.

Now suppose the measurement $\mathbf{x} = (x_0, x_1)$ is observed. The likelihood function is now

$$l(\theta, \mathbf{x}) = f_{\theta}(x_0, x_1) = \frac{1}{\theta^2} I_{[\min(x_0,x_1),\theta]}\big[\max(x_0, x_1)\big]$$

$$= \frac{1}{\theta^2} I_{[\max(x_0,x_1),\infty)}(\theta). \qquad 6.7$$

This likelihood function is illustrated in Figure 6.3. It is maximum if θ is as small as possible, but θ must be at least as large as $\max(x_0, x_1)$. Therefore, the

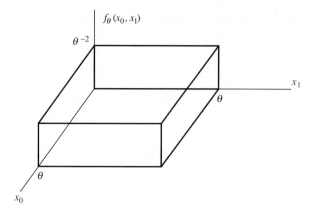

Figure 6.2 Distribution of uniform random sample.

likelihood function achieves its maximum at the maximum likelihood estimate:

$$\hat{\theta} = \max(x_0, x_1). \qquad\qquad 6.8$$

This example serves many purposes. It illustrates the distinction between densities
and likelihoods and, generalized to random samples of size M, it enables us to
sample dinner prices in Paris and estimate the amount of money it takes to taste
la bonne vie de Paris. If it helps, think of the density $f_{\boldsymbol{\theta}}(\mathbf{x})$ as a map from $\mathbf{x}$ to
$f_{\boldsymbol{\theta}}(\mathbf{x})$ and think of $l(\boldsymbol{\theta}, \mathbf{x})$ as a map from $\boldsymbol{\theta}$ to $l(\boldsymbol{\theta}, \mathbf{x})$. In the first case, the map
is parametrically described by the parameter $\boldsymbol{\theta}$, and in the second case the map
is parametrically described by the measurement $\mathbf{x}$.   ■

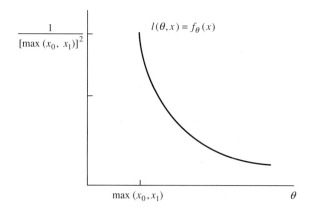

Figure 6.3 Likelihood function for a uniform random
sample.

Example 6.2

Let $\mathbf{X} = (X_0, X_1, \ldots, X_{M-1})$ denote a random sample of size M from an $N[\mu,\sigma^2]$ distribution. Define the parameter $\boldsymbol{\theta}$ to be $\boldsymbol{\theta} = (\mu,\sigma^2)$. The distribution of the random sample is then

$$f_{\boldsymbol{\theta}}(\mathbf{x}) = \prod_{n=0}^{M-1} (2\pi\sigma^2)^{-1/2} \exp\left\{ -\frac{1}{2\sigma^2} (x_n - \mu)^2 \right\}$$

$$= (2\pi\sigma^2)^{-M/2} \exp\left\{ -\frac{1}{2\sigma^2} \sum_{n=0}^{M-1} (x_n - \mu)^2 \right\}. \qquad 6.9$$

Can you show that the maximum likelihood estimates of μ and σ^2 are

$$\hat{\mu} = \frac{1}{M} \sum_{n=0}^{M-1} x_n \qquad 6.10$$

$$\hat{\sigma}^2 = \frac{1}{M} \sum_{n=0}^{M-1} (x_n - \hat{\mu})^2 \; ?$$

These are called the *sample mean* and the *sample variance*. ∎

Random Parameters

The principle of maximum likelihood extends essentially unchanged to the estimation of random parameters. The random variable $\mathbf{x}$ and the *random* parameter $\boldsymbol{\theta}$ are assumed to be jointly distributed according to the joint density function $f(\mathbf{x},\boldsymbol{\theta})$. The contours of constant probability for such a density are illustrated in Figure 6.4. Once the value

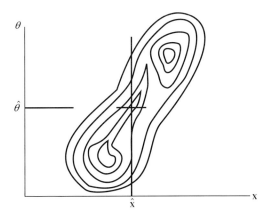

Figure 6.4 Contours of constant probability for observations and parameters.

$\hat{\mathbf{x}}$ is observed, then the vertical line $\mathbf{x} = \hat{\mathbf{x}}$ is searched to find the value $\boldsymbol{\theta} = \hat{\boldsymbol{\theta}}$ for which the likelihood function $l(\boldsymbol{\theta},\hat{\mathbf{x}}) = f(\hat{\mathbf{x}},\boldsymbol{\theta})$ is maximum. The cross hair in Figure 6.4 determines the maximum likelihood estimate:

$$\hat{\boldsymbol{\theta}} = \arg\left[\max_{\boldsymbol{\theta}} f(\hat{\mathbf{x}}, \boldsymbol{\theta})\right]. \qquad 6.11$$

This estimate can actually be computed in two different ways. To illustrate these computations, we assume that the joint density function can be written as

$$f(\mathbf{x}, \boldsymbol{\theta}) = f(\mathbf{x}|\boldsymbol{\theta})f(\boldsymbol{\theta}), \qquad 6.12$$

where $f(\mathbf{x}|\boldsymbol{\theta})$ is the conditional density of $\mathbf{x}$, given $\boldsymbol{\theta}$, and $f(\boldsymbol{\theta})$ is the marginal (or prior) density of $\boldsymbol{\theta}$. Then the log-likelihood function is

$$\begin{aligned} L(\boldsymbol{\theta}, \mathbf{x}) &= \ln f(\mathbf{x}, \boldsymbol{\theta}) \\ &= \ln f(\mathbf{x}|\boldsymbol{\theta}) + \ln f(\boldsymbol{\theta}) \end{aligned} \qquad 6.13$$

and the maximum likelihood estimate of $\hat{\boldsymbol{\theta}}$ is determined from the equation

$$\frac{\partial}{\partial\boldsymbol{\theta}} [\ln f(\mathbf{x}|\boldsymbol{\theta}) + \ln f(\boldsymbol{\theta})] = \mathbf{0}. \qquad 6.14$$

The principle of maximum likelihood is unchanged, but the details are different. The log-likelihood function now includes a term $f(\boldsymbol{\theta})$ that describes the prior, or marginal, distribution of the random parameter $\boldsymbol{\theta}$. The function $\ln f(\mathbf{x}|\boldsymbol{\theta})$ plays the same role as the function $\ln f_{\boldsymbol{\theta}}(\mathbf{x})$ played in the case where $\boldsymbol{\theta}$ was an unknown parameter. However, when a prior distribution for $\boldsymbol{\theta}$ is known, this distribution enters the computation of log likelihood.

There is one more way to interpret maximum likelihood for random parameters. Write the joint density $f(\mathbf{x}, \boldsymbol{\theta})$ as follows:

$$f(\mathbf{x}, \boldsymbol{\theta}) = f(\boldsymbol{\theta}|\mathbf{x})f(\mathbf{x}) \qquad 6.15$$

where $f(\boldsymbol{\theta}|\mathbf{x})$ is the conditional (or posterior) density of $\boldsymbol{\theta}$, given $\mathbf{x}$, and $f(\mathbf{x})$ is the marginal density of $\mathbf{x}$:

$$f(\boldsymbol{\theta}|\mathbf{x}) = \frac{f(\mathbf{x}, \boldsymbol{\theta})}{f(\mathbf{x})} \qquad 6.16$$

$$f(\mathbf{x}) = \int f(\mathbf{x}, \boldsymbol{\theta}) \, d\boldsymbol{\theta}.$$

The log-likelihood function is

$$L(\boldsymbol{\theta}, \mathbf{x}) = \ln f(\mathbf{x}, \boldsymbol{\theta}) = \ln f(\boldsymbol{\theta}|\mathbf{x}) + \ln f(\mathbf{x}) \qquad 6.17$$

and the maximum likelihood estimate of $\boldsymbol{\theta}$ is determined from the equation

$$\frac{\partial}{\partial\boldsymbol{\theta}} L(\boldsymbol{\theta}, \mathbf{x}) = \frac{\partial}{\partial\boldsymbol{\theta}} \ln f(\boldsymbol{\theta}|\mathbf{x}) = \mathbf{0}. \qquad 6.18$$

The function $f(\theta|x)$ is the posterior probability of θ, given x. The function $\ln f(\theta|x)$ is the posterior log likelihood of θ, given x. The maximum likelihood estimate $\hat{\theta}$ is often called the *maximum a posteriori probability* (MAP) estimate of θ. (It should be called the *maximum a posteriori likelihood* (MAL) estimate of θ, even though the acronym MAL might not please some readers.)

Example 6.3

Let X denote a Poisson random variable with probability mass function

$$f(x|\lambda) = \frac{\lambda^x}{x!} e^{-\lambda}; \qquad x = 0, 1, \dots. \qquad 6.19$$

Assume that the rate parameter λ is exponentially distributed:

$$f(\lambda) = \frac{1}{\lambda_0} e^{-\lambda/\lambda_0}; \qquad \lambda > 0. \qquad 6.20$$

The joint density of x and λ is

$$f(x, \lambda) = f(x|\lambda)f(\lambda)$$

$$= \frac{\lambda^x}{\lambda_0 x!} e^{-\lambda(1+1/\lambda_0)}. \qquad 6.21$$

The log likelihood function is

$$L(\lambda, x) = x \log \lambda - \log(\lambda_0 x!) - \lambda(1 + 1/\lambda_0) \qquad 6.22$$

and the maximum likelihood, or MAP, estimate is obtained by finding the value of λ for which $(\partial/\partial\lambda)L(\lambda,x) = 0$:

$$\frac{\partial}{\partial\lambda} L(\lambda,x) = \frac{x}{\lambda} - (1 + 1/\lambda_0) = 0 \qquad 6.23$$

$$\hat{\lambda} = \frac{\lambda_0}{1 + \lambda_0} x.$$

When λ_0 is much less than 1, then the maximum likelihood estimate is simply λ_0 scaled by x:

$$\hat{\lambda} \cong x\lambda_0. \qquad 6.24$$

The variable x will assume the values 0 or 1 with high probability, so $\hat{\lambda}$ will assume the values 0 and λ_0 with high probability. When λ_0 is much greater than one, the maximum likelihood estimate is approximately x, indicating that the prior is so diffuse that it should be ignored:

$$\hat{\lambda} \cong x. \qquad 6.25$$

This result enables us to estimate rate parameters in counting experiments that are described by Poisson's Law. ∎

6.2   SUFFICIENCY

Recall the definition of a sufficient statistic: $t(x)$ is a *sufficient statistic* for the parameter
θ if and only if the density function for x factors as $a(x)b_\theta(t)$, in which case the density
function for the sufficient statistic t is a scaled version of the density function for t:

$$f_\theta(x) = \frac{a(x)}{\int d u\, a[x = W^{-1}(y)]\,|J|}\, f_\theta(t). \qquad 6.26$$

(See Equations 3.11 through 3.13.) When this is the case, the log likelihood is (ignoring constants that are independent of θ)

$$L(\theta, x) \sim \log f_\theta(t). \qquad 6.27$$

In maximum likelihood theory, this log likelihood is differentiated with respect to θ.
Therefore we may replace the log likelihood $L(\theta, x)$ by the log likelihood $L(\theta, t)$:

$$L(\theta, t) = \log f_\theta(t). \qquad 6.28$$

The function $L(\theta, t)$ is simply the log likelihood of θ in an experiment where the
sufficient statistic t is the measurement. It follows, then, that the score function is
just the score function for the sufficient statistic and that the maximum likelihood
estimate will be a function of the sufficient statistic. The implication of this result is
the following. When solving maximum likelihood problems, you may always begin
with the sufficient statistic for the problem and maximize the log-likelihood function
for θ using the sufficient statistic.

Example 6.4

Let X denote a Bernoulli random variable and let $X = (X_0, \ldots, X_{M-1})$ denote a
random sample of size M from the Bernoulli distribution. The density function
for X is

$$f_\theta(x) = \prod_{n=0}^{M-1} f_\theta(x_n) = \prod_{n=0}^{M-1} \theta^{x_n}(1 - \theta)^{1-x_n}; \qquad x_n = 0, 1. \qquad 6.29$$

The sufficient statistic for this problem is

$$k = \sum_{n=0}^{M-1} x_n. \qquad 6.30$$

The distribution of k is binomial:

$$f_\theta(k) = \binom{M}{k} \theta^k (1 - \theta)^{M-k}. \qquad 6.31$$

The log-likelihood and score functions are

$$L(\theta, k) = \log \binom{M}{k} + k \log \theta + (M - k) \log(1 - \theta) \qquad 6.32$$

$$s(\theta, k) = \frac{k}{\theta} - (M - k) \frac{1}{1 - \theta}.$$

The maximum likelihood estimate of θ is the value of θ for which the score function is zero:

$$\hat{\theta} = \frac{k}{M}.$$ 6.33

This function of the sufficient statistic is just the fraction of ones in a sequence of ones and zeros. ∎

Example 6.5

Let $\mathbf{X}$ denote a random sample of M random vectors $\mathbf{X}_n$. Assume that each vector $\mathbf{X}_n$ is an N-dimensional normal random vector with mean $\mathbf{m}$ and covariance matrix $\mathbf{R}$. That is, $\mathbf{X}_n : N[\mathbf{m}, \mathbf{R}]$. Assume that $\mathbf{R}$ is known and that $\boldsymbol{\theta} = \mathbf{m}$ is unknown. The density for $\mathbf{x}$ is

$$f_{\boldsymbol{\theta}}(\mathbf{x}) = (2\pi)^{-MN/2}|\mathbf{R}|^{-M/2}\exp\left\{ -\frac{1}{2}\sum_{n=0}^{M-1}(\mathbf{x}_n - \mathbf{m})^{\mathrm{T}}\mathbf{R}^{-1}(\mathbf{x}_n - \mathbf{m})\right\},   \quad 6.34$$

and the sufficient statistic for $\mathbf{m}$ is

$$\mathbf{t} = \frac{1}{M}\sum_{n=0}^{M-1}\mathbf{x}_n.   \quad 6.35$$

The sufficient statistic is normally distributed:

$$\mathbf{t} : N\left[\mathbf{m}, \frac{1}{M}\mathbf{R}\right].   \quad 6.36$$

The likelihood and score functions are

$$L(\boldsymbol{\theta}, \mathbf{t}) = -\frac{N}{2}\log(2\pi) - \frac{1}{2}\log\left|\frac{1}{M}\mathbf{R}\right| - \frac{1}{2}(\mathbf{t} - \mathbf{m})^{\mathrm{T}}M\mathbf{R}^{-1}(\mathbf{t} - \mathbf{m})   \quad 6.37$$

$$s(\boldsymbol{\theta}, \mathbf{t}) = \frac{\partial}{\partial\mathbf{m}}L(\boldsymbol{\theta}, \mathbf{t}) = M\mathbf{R}^{-1}(\mathbf{t} - \mathbf{m}).$$

The maximum likelihood estimate is the sample mean:

$$\hat{\mathbf{m}} = \mathbf{t}.   \quad 6.38$$

∎

6.3   INVARIANCE

The maximum likelihood estimate of the parameter $\boldsymbol{\theta}$ maximizes the log-likelihood function $L(\boldsymbol{\theta}, \mathbf{x}) = \ln f_{\boldsymbol{\theta}}(\mathbf{x})$. But suppose we wish to estimate the function $\mathbf{w} = W(\boldsymbol{\theta})$ and not $\boldsymbol{\theta}$. Our impulse is to construct the estimate $\hat{\mathbf{w}} = W(\hat{\boldsymbol{\theta}})$ from the maximum likelihood estimate $\hat{\boldsymbol{\theta}}$ and call $\hat{\mathbf{w}}$ the maximum likelihood estimate of $\mathbf{w}$. Is this justified? The answer is yes, and the argument goes as follows.

Let's denote the parameter space for $\boldsymbol{\theta}$ by Θ, a subset of R^p, and the parameter space for $\mathbf{w}$ by Ω, a subset of R^n. Let W denote the mapping from Θ to Ω illustrated

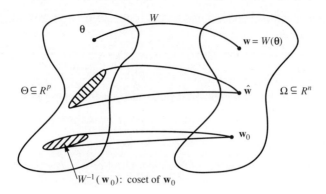

Figure 6.5 The mapping $\mathbf{w} = W(\boldsymbol{\theta})$.

in Figure 6.5. The domain and range of W are, respectively, Θ and Ω. Choose any point $\mathbf{w}_0$ contained in Ω and call $W^{-1}(\mathbf{w}_0)$ its coset or inverse image:

$$W^{-1}(\mathbf{w}_0) = \left\{ \boldsymbol{\theta} : W(\boldsymbol{\theta}) = \mathbf{w}_0 \right\}. \qquad\qquad 6.39$$

Such a coset is illustrated in Figure 6.5. The set of cosets $\left\{ W^{-1}(\mathbf{w}), \mathbf{w} \in \Omega \right\}$ is a disjoint covering of the parameter space Θ, meaning that $\hat{\boldsymbol{\theta}}$ is contained in just one coset, say $W^{-1}(\hat{\mathbf{w}})$.

Define $L(\mathbf{w}, \mathbf{x})$ to be the following "induced log likelihood for $\mathbf{w}$":

$$L(\mathbf{w}, \mathbf{x}) = \max_{\boldsymbol{\theta} \in W^{-1}(\mathbf{w})} L(\boldsymbol{\theta}, \mathbf{x}). \qquad\qquad 6.40$$

This is the maximum log likelihood that can be obtained for $\mathbf{w}$. If we search Ω for the maximum of these induced likelihoods, then we obtain

$$\max_{\mathbf{w}} L(\mathbf{w}, \mathbf{x}) = \max_{\mathbf{w}} \max_{\boldsymbol{\theta} \in W^{-1}(\mathbf{w})} L(\boldsymbol{\theta}, \mathbf{x}). \qquad\qquad 6.41$$

This is what we mean by the maximum likelihood for $\mathbf{w}$. The two maximizing steps on the right-hand side determine $\hat{\mathbf{w}} = W(\hat{\boldsymbol{\theta}})$ where $\hat{\boldsymbol{\theta}}$ is just the maximizing value of $\hat{\boldsymbol{\theta}}$ that would have been returned from maximizing $L(\boldsymbol{\theta}, \mathbf{x})$ with respect to $\boldsymbol{\theta}$. Therefore,

$$L(\hat{\mathbf{w}}, \mathbf{x}) = L(\hat{\boldsymbol{\theta}}, \mathbf{x}); \qquad \hat{\mathbf{w}} = W(\hat{\boldsymbol{\theta}}). \qquad\qquad 6.42$$

The application of this fundamental result is illustrated in the following examples. The first of the examples illustrates how the two maximizing steps on the right-hand side of (6.41) determine $\hat{\mathbf{w}} = W(\hat{\boldsymbol{\theta}})$, where $\hat{\boldsymbol{\theta}}$ is the maximum likelihood estimate of $\boldsymbol{\theta}$.

Example 6.6

Let $\mathbf{x} : N[\boldsymbol{\theta}, \mathbf{I}]$ denote a two-dimensional normal random vector with mean $\boldsymbol{\theta} = (\theta_1, \theta_2)$. The log-likelihood function is (ignoring constants)

$$L(\boldsymbol{\theta}, \mathbf{x}) = -\frac{1}{2} \|\mathbf{x} - \boldsymbol{\theta}\|^2. \qquad\qquad 6.43$$

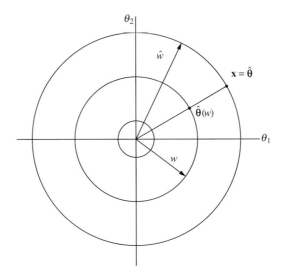

Figure 6.6 The action of computing $\hat{\mathbf{w}} = W(\hat{\boldsymbol{\theta}})$.

The induced log likelihood for the parameter $w = W(\boldsymbol{\theta}) = (\boldsymbol{\theta}^T\boldsymbol{\theta})^{1/2}$ is

$$L(w, \mathbf{x}) = \max_{\boldsymbol{\theta}:\boldsymbol{\theta}^T\boldsymbol{\theta}=w} L(\boldsymbol{\theta}, \mathbf{x}). \qquad 6.44$$

As illustrated in Figure 6.6, the induced log likelihood $L(w, \mathbf{x})$ is determined by searching for the minimum of $\|\mathbf{x} - \boldsymbol{\theta}\|^2$ on the circle $(\boldsymbol{\theta}^T\boldsymbol{\theta})^{1/2} = w$. This minimum is achieved for $\hat{\boldsymbol{\theta}}(w)$ lying at the intersection of the circle of radius w with the ray to $\mathbf{x}$. When these minima are computed for each circle of radius w, the minimum is found at $\hat{w} = (\hat{\boldsymbol{\theta}}^T\hat{\boldsymbol{\theta}})^{1/2}$, where $\hat{\boldsymbol{\theta}} = \mathbf{x}$ is the maximum likelihood estimate of $\boldsymbol{\theta}$. ■

Example 6.7

Let X denote a Poisson random variable with rate parameter θ:

$$f_\theta(x) = e^{-\theta}\frac{\theta^x}{x!}; \qquad x = 0, 1, \ldots, \qquad \theta \geq 0 \quad \text{(unknown)}. \qquad 6.45$$

From the random sample $\mathbf{X} = (X_0, X_1, \ldots, X_{M-1})$, find the maximum likelihood estimate of the probability that X will exceed its mean θ.

We note that $w(\theta)$ may be written as

$$w(\theta) = P[X > \theta] = \sum_{x=\lceil\theta\rceil}^{\infty} e^{-\theta}\frac{\theta^x}{x!}; \qquad \lceil z \rceil : \text{smallest integer} \geq z. \qquad 6.46$$

The maximum likelihood estimate of w is

$$\hat{w} = \sum_{\lceil\hat\theta\rceil}^{\infty} e^{-\hat\theta}\frac{\hat\theta^x}{x!} \qquad 6.47$$

where $\hat{\boldsymbol{\theta}}$ is the maximum likelihood estimate of θ:

$$\hat{\theta} = \frac{1}{M} \sum_{n=0}^{M-1} x_n. \qquad 6.48$$

■

Example 6.8

Let $\mathbf{X}$ denote a normal random vector with unknown mean $\boldsymbol{\theta}$ and known covariance $\mathbf{R}$. That is, $\mathbf{X} : N[\boldsymbol{\theta}, \mathbf{R}]$. From the random sample $\mathbf{X} = (\mathbf{X}_0, \mathbf{X}_1, \ldots, \mathbf{X}_{M-1})$, find the maximum likelihood estimate of the norm squared of $\boldsymbol{\theta}$, namely $w = \boldsymbol{\theta}^T \boldsymbol{\theta}$. The maximum likelihood estimate of w is

$$\hat{w} = \hat{\boldsymbol{\theta}}^T \hat{\boldsymbol{\theta}} \qquad 6.49$$

where $\hat{\boldsymbol{\theta}}$ is the maximum likelihood estimate

$$\hat{\boldsymbol{\theta}} = \frac{1}{M} \sum_{n=0}^{M-1} \mathbf{x}_n. \qquad 6.50$$

■

Example 6.9

Let $\mathbf{X} = (X_0, X_1, \ldots, X_{M-1})$ denote a random sample of size M from a univariate normal distribution $N[\mu, \sigma^2]$. Find the maximum likelihood estimates of

1. $w(\mu) = \mu^2$ $\qquad\qquad$ 6.51
2. $w(\sigma^2) = \sigma$.

These are, respectively, the mean-squared and the standard deviation. Apply the invariance principle to find

1. $\hat{w} = w(\hat{\mu}) = (\hat{\mu})^2$ $\qquad\qquad$ 6.52

$$\hat{\mu} = \frac{1}{M} \sum_{n=0}^{M-1} x_n$$

2. $\hat{w} = w(\hat{\sigma}^2) = (\hat{\sigma}^2)^{1/2}$

$$\hat{\sigma}^2 = \frac{1}{M} \sum_{n=0}^{M-1} (x_n - \hat{\mu})^2.$$

No surprises here. ■

Example 6.10

Let $\mathbf{X} = (X_0, X_1, \ldots, X_{M-1})$ denote a random sample of size M from a univariate normal distribution $N[\mu, \sigma^2]$. Find the maximum likelihood estimates of $w(\mu, \sigma^2)$ in each of these problems:

1. $w(\mu, \sigma^2) = P[X \le c]$; c known and $-\infty < c < \infty$;
2. $P[X \le w(\mu, \sigma^2)] = c$; c known and $0 \le c \le 1$.

In the first problem, we are using the random sample to estimate the probability w that another realization of $X : N[\mu, \sigma^2]$ will be less than or equal to c. In the second, we are using the random sample to estimate the value w that another realization of X will be less than or equal to with probability c. Apply the invariance principle to find

1. $\hat{w} = w(\hat{\mu}, \hat{\sigma}^2) = \displaystyle\int_{-\infty}^{c} (2\pi\hat{\sigma}^2)^{-1/2} \exp\left\{ -\frac{1}{2}(x - \hat{\mu})^2/\hat{\sigma}^2 \right\}$ 6.53

$$= \int_{-\infty}^{(c-\hat{\mu})/\hat{\sigma}} (2\pi)^{-1/2} \exp\left\{ -\frac{1}{2}y^2 \right\} dy$$

$$\hat{w} = \Phi\left(\frac{c - \hat{\mu}}{\hat{\sigma}} \right)$$

2. $\hat{w} = w(\hat{\mu}, \hat{\sigma}^2) = \displaystyle\int_{-\infty}^{(\hat{w}-\hat{\mu})/\hat{\sigma}} (2\pi)^{-1/2} \exp\{-y^2/2\} \, dy = c$ 6.54

$$\Phi\left(\frac{\hat{w} - \hat{\mu}}{\hat{\sigma}} \right) = c.$$

Problems (1) and (2) are inverses of each other. In (1) we compute $\hat{w} = \Phi\big((c - \hat{\mu})/\hat{\sigma}\big)$, and in (2) we look for the value of $\hat{w}$ that makes $\Phi\big[(\hat{w} - \hat{\mu})/\hat{\sigma}\big] = c$. This example illustrates the power of maximum likelihood for estimating rather complicated functions of the parameters μ and σ^2. ∎

6.4   THE FISHER MATRIX AND THE CRAMER-RAO BOUND

After we estimate a parameter θ, we are usually interested in at least two properties of the estimator. First, what is its mean, and second, what is its covariance? These we denote by

$$E\hat{\theta}$$ 6.55

$$E[\hat{\theta} - E\hat{\theta}][\hat{\theta} - E\hat{\theta}]^{\mathrm{T}}.$$

The *error covariance matrix* is the matrix

$$\mathbf{C} = E[\hat{\theta} - \theta][\hat{\theta} - \theta]^{\mathrm{T}}.$$ 6.56

Each entry in $\mathbf{C}$ is the cross-covariance between an error $(\hat{\theta}_n - \theta_n)$ and an error $(\hat{\theta}_m - \theta_m)$. The diagonal term $E(\hat{\theta}_n - \theta_n)^2$ is the mean-squared error between the estimator $\hat{\theta}_n$ and the true parameter θ_n. The error covariance matrix may be rewritten as

$$\mathbf{C} = E[\hat{\theta} - E\hat{\theta} + E\hat{\theta} - \theta][\hat{\theta} - E\hat{\theta} + E\hat{\theta} - \theta]^{\mathrm{T}}$$
$$= E[\hat{\theta} - E\hat{\theta}][\hat{\theta} - E\hat{\theta}]^{\mathrm{T}} + [E\hat{\theta} - \theta][E\hat{\theta} - \theta]^{\mathrm{T}}.$$ 6.57

The first term on the right-hand side is the *covariance matrix* for the estimator $\hat{\boldsymbol{\theta}}$, and the second term is the *bias-squared matrix* for the estimator. The nth diagonal term of C is the mean-squared error (MSE) of the estimator $\hat{\theta}_n$. It may be written as

$$\mathrm{MSE}(\hat{\theta}_n) = E(\hat{\theta}_n - \theta_n)^2 = E(\hat{\theta}_n - E\hat{\theta}_n)^2 + (E\hat{\theta}_n - \theta_n)^2$$
$$= \mathrm{var}(\hat{\theta}_n) + b^2(\hat{\theta}_n). \tag{6.58}$$

We say that the mean-squared error of the estimator $\hat{\theta}_n$ is the variance of the estimator plus its bias-squared. If the estimator $\hat{\boldsymbol{\theta}}$ is unbiased, then the error covariance matrix is the estimator covariance matrix, and the mean-squared error is the variance of the estimator:

$$E\hat{\boldsymbol{\theta}} = \boldsymbol{\theta} \ \Rightarrow \ E[\hat{\boldsymbol{\theta}} - \boldsymbol{\theta}][\hat{\boldsymbol{\theta}} - \boldsymbol{\theta}]^{\mathrm{T}} = E[\hat{\boldsymbol{\theta}} - E\hat{\boldsymbol{\theta}}][\hat{\boldsymbol{\theta}} - E\hat{\boldsymbol{\theta}}]^{\mathrm{T}}$$
$$\Rightarrow \ \mathrm{MSE}(\hat{\theta}_n) = \mathrm{var}(\hat{\theta}_n). \tag{6.59}$$

Cramer-Rao Bound

The Cramer-Rao bound establishes a lower bound on the error covariance matrix for any unbiased estimator of a parameter $\boldsymbol{\theta}$. There are several results that are collateral to the Cramer-Rao bound and numerous interpretations that give insight into its meaning. To set up the Cramer-Rao theorem, we need to review the definition of the score function, interpret it, and establish its statistical properties.

Recall that the score function is the gradient of the log-likelihood function:

$$\mathbf{s}(\boldsymbol{\theta}, \mathbf{x}) = \frac{\partial}{\partial \boldsymbol{\theta}} L(\boldsymbol{\theta}, \mathbf{x}) = \frac{\partial}{\partial \boldsymbol{\theta}} \ln f_{\boldsymbol{\theta}}(\mathbf{x}). \tag{6.60}$$

When the realization $\mathbf{x}$ is replaced by the random variable $\mathbf{X}$, then the log-likelihood and score functions become random variables:

$$\mathbf{s}(\boldsymbol{\theta}, \mathbf{X}) = \frac{\partial}{\partial \boldsymbol{\theta}} L(\boldsymbol{\theta}, \mathbf{X}) = \frac{\partial}{\partial \boldsymbol{\theta}} \ln f_{\boldsymbol{\theta}}(\mathbf{X}). \tag{6.61}$$

The score function "scores" values of $\boldsymbol{\theta}$ as the random vector $\mathbf{X}$ assumes values from the distribution of $f_{\boldsymbol{\theta}}(\mathbf{x})$. Scores near zero are "good scores," and scores different from zero are "bad scores." The score function has mean zero:

$$E\mathbf{s}(\theta, \mathbf{X}) = E \frac{\partial}{\partial \boldsymbol{\theta}} \ln f_{\boldsymbol{\theta}}(\mathbf{X})$$

$$= \int d\mathbf{x} f_{\boldsymbol{\theta}}(\mathbf{x}) \frac{\partial}{\partial \boldsymbol{\theta}} \ln f_{\boldsymbol{\theta}}(\mathbf{x}) \tag{6.62}$$

$$= \int d\mathbf{x} \frac{\partial}{\partial \boldsymbol{\theta}} f_{\boldsymbol{\theta}}(\mathbf{x}) = \frac{\partial}{\partial \boldsymbol{\theta}} \int d\mathbf{x} f_{\boldsymbol{\theta}}(\mathbf{x}) = \mathbf{0}.$$

The nonnegative definite covariance matrix $\mathbf{Q}$ may be diagonalized as follows:

$$\begin{bmatrix} \mathbf{I} & -\mathbf{J}^{-1} \\ \mathbf{0} & \mathbf{I} \end{bmatrix} \begin{bmatrix} \mathbf{C} & \mathbf{I} \\ \mathbf{I} & \mathbf{J} \end{bmatrix} \begin{bmatrix} \mathbf{I} & \mathbf{0} \\ -\mathbf{J}^{-1} & \mathbf{I} \end{bmatrix} = \begin{bmatrix} \mathbf{C} - \mathbf{J}^{-1} & \mathbf{0} \\ \mathbf{0} & \mathbf{J} \end{bmatrix}. \qquad 6.76$$

So the covariance matrix $\mathbf{Q}$ is similar to the matrix on the right-hand side. Thus, $\mathbf{C} - \mathbf{J}^{-1}$ is nonnegative definite, meaning

$$\mathbf{C} \geq \mathbf{J}^{-1} \quad \text{or} \quad \mathbf{J} \geq \mathbf{C}^{-1}. \qquad 6.77$$

The ii element of $\mathbf{C}$ is the mean-squared error of the estimator of θ_i:

$$C_{ii} = E(\hat{\theta}_i - \theta_i)^2 \geq (\mathbf{J}^{-1})_{ii}. \qquad 6.78$$

So, the ii element of the inverse of the Fisher information matrix lower bounds the mean-squared error of any unbiased estimator of θ_i. $\square$

Concentration Ellipses

Suppose the unbiased estimator $\hat{\boldsymbol{\theta}}$, with covariance matrix $\mathbf{C}$, were distributed as $N[\boldsymbol{\theta}, \mathbf{C}]$:

$$f_C(\hat{\boldsymbol{\theta}}) = \frac{1}{(2\pi)^{p/2}(\det \mathbf{C})^{1/2}} \exp\left\{ -\frac{1}{2}(\hat{\boldsymbol{\theta}} - \boldsymbol{\theta})^{\mathrm{T}}\mathbf{C}^{-1}(\hat{\boldsymbol{\theta}} - \boldsymbol{\theta}) \right\}. \qquad 6.79$$

The random variable $r_C^2 = (\hat{\boldsymbol{\theta}} - \boldsymbol{\theta})^{\mathrm{T}}\mathbf{C}^{-1}(\hat{\boldsymbol{\theta}} - \boldsymbol{\theta})$ would be distributed as χ_p^2, a chi-squared random variable with p degrees of freedom. (See Section 2.11 in Chapter 2.) The probability that $r_C^2 \leq r^2$ would be determined by integrating the χ_p^2 density from 0 to r^2. The volume of Euclidean p space for which $r_C^2 \leq r^2$ is also the volume enclosed by the concentration ellipse $r_C^2 = r^2$ (see Section 2.10 of Chapter 2):

$$V_C = V_p(\det \mathbf{C})^{1/2} r^p; \qquad V_p = \frac{(\pi)^{p/2}}{(p/2)!}, \quad p \text{ even}. \qquad 6.80$$

This concentration ellipse and its enclosed volume are illustrated in Figure 6.7.

If the estimator $\hat{\boldsymbol{\theta}}$ were normal with mean $\boldsymbol{\theta}$ and covariance matrix $\mathbf{J}^{-1}$, then the density function for $\hat{\boldsymbol{\theta}}$ would be

$$f_{J^{-1}}(\hat{\boldsymbol{\theta}}) = \frac{1}{(2\pi)^{p/2}(\det \mathbf{J}^{-1})^{1/2}} \exp\left\{ -\frac{1}{2}(\hat{\boldsymbol{\theta}} - \boldsymbol{\theta})^{\mathrm{T}}\mathbf{J}(\hat{\boldsymbol{\theta}} - \boldsymbol{\theta}) \right\} \qquad 6.81$$

The random variable $r_{J^{-1}}^2 = (\hat{\boldsymbol{\theta}} - \boldsymbol{\theta})^{\mathrm{T}}\mathbf{J}(\hat{\boldsymbol{\theta}} - \boldsymbol{\theta})$ would be distributed as χ_p^2, and the probability that $r_{J^{-1}}^2 \leq r^2$ would be computed as before. The probability that $r_{J^{-1}}^2 \leq r^2$ equals the probability that $r_C^2 \leq r^2$! However, the *volume* enclosed by the concentration ellipse $r_{J^{-1}}^2 = r^2$ is smaller than the *volume* enclosed by the ellipse $r_C^2 = r^2$:

$$V_{J^{-1}} = V_p(\det \mathbf{J}^{-1})^{1/2} r^p \leq V_C. \qquad 6.82$$

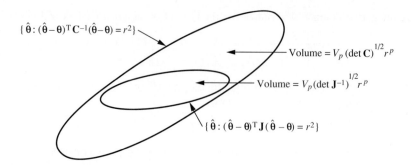

Figure 6.7 Concentration ellipses.

See Figure 6.7. The Fisher information matrix defines the smallest concentration ellipse that can be achieved for any unbiased estimator.

Efficiency

An efficient estimator is an unbiased estimator whose error covariance matrix $\mathbf{C}$ meets the Cramer-Rao (CR) bound, meaning that $\mathbf{C} = \mathbf{J}^{-1}$. The following theorem establishes that this happens if and only if the estimator is a linear transformation of the score function.

Theorem (Efficiency)   An unbiased estimator $\hat{\boldsymbol{\theta}}$ is efficient, that is,

$$E\hat{\boldsymbol{\theta}} = \boldsymbol{\theta} \qquad\qquad 6.83$$

$$E[\hat{\boldsymbol{\theta}} - \boldsymbol{\theta}][\hat{\boldsymbol{\theta}} - \boldsymbol{\theta}]^{\mathrm{T}} = \mathbf{J}^{-1}(\boldsymbol{\theta}),$$

if and only if

$$\mathbf{J}(\boldsymbol{\theta})(\hat{\boldsymbol{\theta}} - \boldsymbol{\theta}) = \mathbf{s}(\boldsymbol{\theta}, \mathbf{X}); \qquad\qquad 6.84$$

Proof (If)   If $\mathbf{J}(\boldsymbol{\theta})(\hat{\boldsymbol{\theta}} - \boldsymbol{\theta}) = \mathbf{s}(\boldsymbol{\theta}, \mathbf{X})$, then

$$E\mathbf{J}(\boldsymbol{\theta})(\hat{\boldsymbol{\theta}} - \boldsymbol{\theta})(\hat{\boldsymbol{\theta}} - \boldsymbol{\theta})^{\mathrm{T}}\mathbf{J}^{\mathrm{T}}(\boldsymbol{\theta}) = \mathbf{J}(\boldsymbol{\theta})\mathbf{C}\mathbf{J}(\boldsymbol{\theta}) = E\mathbf{s}(\boldsymbol{\theta}, \mathbf{X})\mathbf{s}^{\mathrm{T}}(\boldsymbol{\theta}, \mathbf{X}) = \mathbf{J}(\boldsymbol{\theta}), \qquad 6.85$$

meaning $\mathbf{C} = \mathbf{J}^{-1}(\boldsymbol{\theta})$.

Only if :   Recall the cross-covariance between $\mathbf{s}(\boldsymbol{\theta}, \mathbf{X})$ and $(\hat{\boldsymbol{\theta}} - \boldsymbol{\theta})$:

$$\left[E\mathbf{s}(\boldsymbol{\theta}, \mathbf{X})(\hat{\boldsymbol{\theta}} - \boldsymbol{\theta})^{\mathrm{T}}\right]^2 = \mathbf{I}^2 = \mathbf{I}. \qquad\qquad 6.86$$

Schwartz' inequality for random variables says

$$\mathbf{I} = \left[E\mathbf{s}(\boldsymbol{\theta}, \mathbf{X})(\hat{\boldsymbol{\theta}} - \boldsymbol{\theta})^{\mathrm{T}}\right]^2 \leq E\mathbf{s}(\boldsymbol{\theta}, \mathbf{X})\mathbf{s}^{\mathrm{T}}(\boldsymbol{\theta}, \mathbf{X})E(\hat{\boldsymbol{\theta}} - \boldsymbol{\theta})(\hat{\boldsymbol{\theta}} - \boldsymbol{\theta})^{\mathrm{T}}$$
$$= \mathbf{JC}$$

6.87

with equality iff $\mathbf{J}(\boldsymbol{\theta})(\hat{\boldsymbol{\theta}} - \boldsymbol{\theta}) = \mathbf{s}(\boldsymbol{\theta}, \mathbf{X})$.

Every efficient estimator is minimum variance unbiased (MVUB), but the converse is not true. That is, an estimator may be MVUB but not achieve the CR bound. □

Example 6.11

Let $\mathbf{X} = (X_0, X_1, \ldots, X_{M-1})$ denote a random sample of size M from a Poisson distribution. The distribution of the random sample is

$$f_\theta(\mathbf{x}) = \frac{e^{-M\theta}}{\Pi x_n!} \theta^k$$

6.88

$$k = \sum_{n=0}^{M-1} x_n.$$

The maximum likelihood estimate of θ is

$$\hat{\theta} = \frac{1}{M} \sum_{n=0}^{M-1} x_n,$$

6.89

and the mean and variance of $\hat{\theta}$ are

$$E\hat{\theta} = \theta$$

6.90

$$E(\hat{\theta} - \theta)^2 = \frac{\theta}{M}.$$

The score function is

$$s(\theta, \mathbf{x}) = \frac{\partial}{\partial \theta} \ln f_\theta(\mathbf{x}) = -M + \frac{1}{\theta} \sum_{n=0}^{M-1} x_n.$$

6.91

The Fisher information matrix (a scalar in this case) is

$$J = -E \frac{\partial^2}{\partial \theta^2} \ln f_\theta(\mathbf{x}) = -E\left(-\frac{1}{\theta^2} \sum_{n=0}^{M-1} x_n\right)$$

6.92

$$= \frac{1}{\theta^2} M\theta = \frac{M}{\theta}.$$

The inverse of J is $J^{-1} = \theta/M$. This means that the unbiased estimator $\hat{\theta}$ is efficient and therefore MVUB. Can we write $\hat{\theta}$ as $J(\theta)(\hat{\theta} - \theta) = s(\theta, \mathbf{x})$? Let's try:

$$J(\theta)(\hat{\theta} - \theta) = \frac{M}{\theta}\left(\frac{1}{M}\sum_{n=0}^{M-1}(x_n - \theta)\right) = -M + \frac{1}{\theta}\sum_{n=0}^{M-1}x_n \qquad 6.93$$

$$= s(\theta, \mathbf{x}).$$

This checks.   ∎

Example 6.12

Let $\mathbf{X} = (X_1, X_2, \ldots, X_M)$ denote a random sample of size M from the exponential distribution

$$f_\theta(x_n) = \frac{1}{\theta}e^{-x_n/\theta}; \qquad 0 \le x_n < \infty, \qquad 0 \le \theta < \infty \qquad 6.94$$

$$f_\theta(\mathbf{x}) = \frac{1}{\theta^M}e^{-k/\theta}$$

$$k = \sum_{n=1}^{M}x_n$$

The score function is

$$s(\theta, \mathbf{x}) = -\frac{M}{\theta} + \frac{1}{\theta^2}\sum_{n=1}^{M}x_n, \qquad 6.95$$

and the maximum likelihood estimate is

$$\hat{\theta} = \frac{1}{M}\sum_{n=1}^{M}x_n. \qquad 6.96$$

The characteristic function of the estimator $\hat{\theta}$ is

$$\phi(\omega) = \prod_{n=1}^{M}\phi_n\left(\frac{\omega}{M}\right)$$

$$\phi_n\left(\frac{\omega}{M}\right) = E\exp\left(-j\frac{\omega}{M}X_n\right)$$

$$= \int_0^\infty \frac{1}{\theta}\exp\left\{-\left(j\frac{\omega}{M} + \frac{1}{\theta}\right)x\right\}dx \qquad 6.97$$

$$= \frac{1}{1 + j(\omega\theta/M)}.$$

The first three moments of $\hat{\theta}$ are

$$\phi(0) = 1$$

$$j \frac{\partial \phi(0)}{\partial \omega} = E\hat{\theta} = \theta \qquad\qquad 6.98$$

$$-\frac{\partial^2}{\partial \omega^2} \phi(0) = E\hat{\theta}^2 = \frac{\theta^2}{M} + \theta^2.$$

The mean and variance of the maximum likelihood estimate are

$$E\hat{\theta} = \theta \qquad\qquad 6.99$$

$$E(\hat{\theta} - \theta)^2 = \frac{\theta^2}{M}.$$

The maximum likelihood estimator is unbiased. Is it minimum variance and/or efficient? (See Problem 6.23.) ■

Cramer-Rao Bound for Functions of Parameters

The invariance principle for maximum likelihood tells us that we may construct the maximum likelihood estimate for the parameter $\mathbf{w} = W(\boldsymbol{\theta})$ by computing $\hat{\mathbf{w}} = W(\hat{\boldsymbol{\theta}})$. What then can we say about the Cramer-Rao bound for the estimator $\hat{\mathbf{w}}$ that is constructed in this way? In order to answer this question we shall assume that the function W is a continuous, nonsingular mapping from R^p to R^p, meaning $\boldsymbol{\theta} \in R^p$ and $\mathbf{w} \in R^p$.

From the original log-likelihood function $L(\boldsymbol{\theta}, \mathbf{x})$ we construct the induced log-likelihood function $L(\mathbf{w}, \boldsymbol{\theta})$ and vice versa:

$$L(\mathbf{w}, \mathbf{x}) = L\big[\boldsymbol{\theta} = W^{-1}(\mathbf{w}), \mathbf{x}\big] \qquad\qquad 6.100$$

$$L(\boldsymbol{\theta}, \mathbf{x}) = L\big[\mathbf{w} = W(\boldsymbol{\theta}), \mathbf{x}\big] .$$

The gradients of these log likelihoods may be written as

$$\frac{\partial}{\partial w_i} L(\mathbf{w}, \mathbf{x}) = \sum_{j=1}^{p} \frac{\partial}{\partial \theta_j} L(\boldsymbol{\theta}, \mathbf{x}) \frac{\partial \theta_j}{\partial w_i}$$

$$\qquad\qquad 6.101$$

$$= \frac{\partial \boldsymbol{\theta}^T}{\partial w_i} \frac{\partial}{\partial \boldsymbol{\theta}} L(\boldsymbol{\theta}, \mathbf{x})$$

$$\frac{\partial}{\partial \theta_i} L(\boldsymbol{\theta}, \mathbf{x}) = \sum_{j=1}^{p} \frac{\partial}{\partial w_j} L(\mathbf{w}, \mathbf{x}) \frac{\partial w_j}{\partial \theta_i}$$

$$= \frac{\partial \mathbf{w}^T}{\partial \theta_i} \frac{\partial}{\partial \mathbf{w}} L(\mathbf{w}, \mathbf{x}).$$

When written out for $i = 1, 2, \ldots, p$, these equations produce the following connections between the score functions for $\mathbf{w}$ and $\boldsymbol{\theta}$:

$$\frac{\partial}{\partial \mathbf{w}} L(\mathbf{w}, \mathbf{x}) = \mathbf{G} \frac{\partial}{\partial \boldsymbol{\theta}} L(\boldsymbol{\theta}, \mathbf{x})$$

$$\frac{\partial}{\partial \boldsymbol{\theta}} L(\boldsymbol{\theta}, \mathbf{x}) = \mathbf{H} \frac{\partial}{\partial \mathbf{w}} L(\mathbf{w}, \mathbf{x}) \qquad \qquad 6.102$$

$$\mathbf{G} = \left\{ \frac{\partial \theta_j}{\partial w_i} \right\}; \qquad \mathbf{H} = \left\{ \frac{\partial w_j}{\partial \theta_i} \right\}$$

$$\mathbf{GH} = \mathbf{I}.$$

The connection between Fisher information matrices is

$$\mathbf{J}(\mathbf{w}) = \mathbf{G}\mathbf{J}(\boldsymbol{\theta})\mathbf{G}^{\mathrm{T}}; \qquad \boldsymbol{\theta} = W^{-1}(\mathbf{w}) \qquad \qquad 6.103$$

$$\mathbf{J}(\boldsymbol{\theta}) = \mathbf{H}\mathbf{J}(\mathbf{w})\mathbf{H}^{\mathrm{T}}; \qquad \mathbf{w} = W(\boldsymbol{\theta}),$$

and the corresponding connection between Cramer-Rao bounds is

$$\mathbf{J}^{-1}(\boldsymbol{\theta}) = \mathbf{H}^{-\mathrm{T}}\mathbf{J}^{-1}(\mathbf{w})\mathbf{H}^{-1} = \mathbf{G}^{\mathrm{T}}\mathbf{J}^{-1}(\mathbf{w})\mathbf{G} \qquad \qquad 6.104$$

$$\mathbf{J}^{-1}(\mathbf{w}) = \mathbf{G}^{-\mathrm{T}}\mathbf{J}^{-1}(\boldsymbol{\theta})\mathbf{G}^{-1} = \mathbf{H}^{\mathrm{T}}\mathbf{J}^{-1}(\boldsymbol{\theta})\mathbf{H}.$$

This result actually extends to maps from R^p to R^m, $m \leq p$, provided the induced likelihood $L(\mathbf{w}, \mathbf{x}) = \max_{\boldsymbol{\theta} \in W^{-1}(\mathbf{w})} L(\boldsymbol{\theta}, \mathbf{x})$ defines a continuous, bounded mapping. Then $\mathbf{GH} = \mathbf{I}$, but $\mathbf{HG} \neq \mathbf{I}$.

Numerical Maximum Likelihood and the Stochastic Fisher Matrix

The stochastic Fisher matrix plays an important role in numerical procedures for solving the maximum likelihood equation $\mathbf{s}(\hat{\boldsymbol{\theta}}, \mathbf{x}) = \mathbf{0}$. To illustrate this point, we consider the log-likelihood function $L(\boldsymbol{\theta}, \mathbf{x}) = \ln f_{\boldsymbol{\theta}}(\mathbf{x})$ and its Taylor series expansion about some estimate of $\boldsymbol{\theta}$, denoted $\boldsymbol{\theta}_n$:

$$L(\boldsymbol{\theta}, \mathbf{x}) = L(\boldsymbol{\theta}_n, \mathbf{x}) + [\boldsymbol{\theta} - \boldsymbol{\theta}_n]^{\mathrm{T}} \left[\frac{\partial}{\partial \boldsymbol{\theta}} L(\boldsymbol{\theta}_n, \mathbf{x}) \right]$$

$$+ \frac{1}{2}[\boldsymbol{\theta} - \boldsymbol{\theta}_n]^{\mathrm{T}} \frac{\partial}{\partial \boldsymbol{\theta}} \left[\frac{\partial}{\partial \boldsymbol{\theta}} L(\boldsymbol{\theta}_n, \mathbf{x}) \right]^{\mathrm{T}} [\boldsymbol{\theta} - \boldsymbol{\theta}_n] + \text{higher order terms}$$

$$\cong L(\boldsymbol{\theta}_n, \mathbf{x}) + [\boldsymbol{\theta} - \boldsymbol{\theta}_n]^{\mathrm{T}}\mathbf{s}(\boldsymbol{\theta}_n, \mathbf{x}) - \frac{1}{2}[\boldsymbol{\theta} - \boldsymbol{\theta}_n]^{\mathrm{T}}\mathbf{J}(\boldsymbol{\theta}_n, \mathbf{x})[\boldsymbol{\theta} - \boldsymbol{\theta}_n]. \qquad 6.105$$

This expression shows that log likelihood is quadratic in the neighborhood of the maximum likelihood estimate $\boldsymbol{\theta}_n = \hat{\boldsymbol{\theta}}$, where $\mathbf{s}(\hat{\boldsymbol{\theta}}, \mathbf{x}) = \mathbf{0}$. This suggests that a Newton-Raphson procedure for finding the maximum of $L(\boldsymbol{\theta}, \mathbf{x})$, or the zero of $\mathbf{s}(\boldsymbol{\theta}, \mathbf{x}) = (\partial/\partial \boldsymbol{\theta}) L(\boldsymbol{\theta}, \mathbf{x})$, is appropriate.

The Taylor series for log likelihood produces the following linear model for the score function:

$$\mathbf{s}(\boldsymbol{\theta}, \mathbf{x}) = \mathbf{s}(\boldsymbol{\theta}_n, \mathbf{x}) - \mathbf{J}(\boldsymbol{\theta}_n, \mathbf{x})[\boldsymbol{\theta} - \boldsymbol{\theta}_n]. \qquad 6.106$$

The Newton-Raphson map equates $\mathbf{s}(\boldsymbol{\theta}_{n+1}, \mathbf{x})$ to zero to produce this equation for $\boldsymbol{\theta}_{n+1}$:

$$\mathbf{J}(\boldsymbol{\theta}_n, \mathbf{x})[\boldsymbol{\theta}_{n+1} - \boldsymbol{\theta}_n] = \mathbf{s}(\boldsymbol{\theta}_n, \mathbf{x})$$

or

$$\boldsymbol{\theta}_{n+1} = \boldsymbol{\theta}_n + \mathbf{J}^{-1}(\boldsymbol{\theta}_n, \mathbf{x})\mathbf{s}(\boldsymbol{\theta}_n, \mathbf{x}). \qquad 6.107$$

In summary, the stochastic Fisher matrix $\mathbf{J}(\boldsymbol{\theta}, \mathbf{x})$ plays the role of the Hessian in a Newton-Raphson iteration for the zero of the score function.

6.5 NUISANCE PARAMETERS

It sometimes happens that only a subset of the parameters in the parameter vector $\boldsymbol{\theta}$ are of real interest. The parameter θ_n may be an important parameter, and all others may be (nuisance) parameters with unknown values of no particular interest to us. For example, θ_3 may be the frequency of a discrete time sinusoid, and θ_1 and θ_2 may be amplitude and phase. We may be interested only in the frequency. The natural question is, "how does an unknown nuisance parameter influence our ability to estimate an important parameter?" In order to answer this question, we let $\boldsymbol{\theta}_p^T = (\theta_1 \quad \cdots \quad \theta_k \quad \cdots \quad \theta_p) = (\boldsymbol{\theta}_{p-1}^T, \theta_p)$ denote a parameter vector and $\mathbf{J}$ its corresponding Fisher information matrix with elements $(\mathbf{J})_{ij}$:

$$(\mathbf{J})_{ij} = -E \frac{\partial^2}{\partial \theta_i \partial \theta_j} \ln f_{\boldsymbol{\theta}}(\mathbf{X}). \qquad 6.108$$

Denote the inverse of $\mathbf{J}$ as $\mathbf{J}^{-1}$ with elements $(\mathbf{J}^{-1})_{ij}$. The variance of any unbiased estimator of θ_k is bounded by $(J^{-1})_{kk}$:

$$E(\hat{\theta}_k - \theta_k)^2 \geq (\mathbf{J}^{-1})_{kk}. \qquad 6.109$$

This is the Cramer-Rao bound. But what if the parameters θ_1 through θ_p, excluding θ_k, are known. Then the Fisher information matrix is the scalar matrix $(\mathbf{J})_{kk}$, and the Cramer-Rao bound on an unbiased estimator of θ_k is

$$E(\hat{\theta}_k - \theta_k)^2 \geq 1/(\mathbf{J})_{kk}. \qquad 6.110$$

The bound $(\mathbf{J}^{-1})_{kk}$ represents a variance bound on the estimator $\hat{\theta}_k$ when the unknown nuisance parameters θ_1 through θ_p are also estimated. The bound $1/(\mathbf{J})_{kk}$ represents a variance bound on the estimator $\hat{\theta}_k$ when all potential nuisance parameters are known. Now consider Schwartz' inequality for the positive definite, Fisher information matrix $\mathbf{J}$:

$$(\mathbf{y}^T\mathbf{J}\mathbf{y})(\mathbf{x}^T\mathbf{J}\mathbf{x}) \geq (\mathbf{y}^T\mathbf{J}\mathbf{x})^2. \qquad 6.111$$

Choose $\mathbf{y} = \mathbf{J}^{-1}\mathbf{x}$ and $\mathbf{x} = \mathbf{e}_k$, the unit vector with a 1 in its kth position and zero elsewhere. Then

$$(\mathbf{e}_k^T \mathbf{J}^{-1} \mathbf{e}_k)(\mathbf{e}_k^T \mathbf{J} \mathbf{e}_k) \geq \mathbf{e}_k^T \mathbf{e}_k$$

$$(\mathbf{J}^{-1})_{kk}(\mathbf{J})_{kk} \geq 1 \qquad\qquad 6.112$$

$$(\mathbf{J}^{-1})_{kk} \geq 1/(\mathbf{J})_{kk}$$

meaning that the variance bound can only increase when unknown nuisance parameters are estimated.

This result can actually be generalized by considering a different argument. Again, let $\mathbf{J}$ denote the Fisher information matrix for the parameter vector $\boldsymbol{\theta}$. Write it as the patterned matrix

$$\mathbf{J}_p = \left[\begin{array}{c|c} \mathbf{J}_{p-1} & \mathbf{j}_p \\ \hline \mathbf{j}_p^T & \gamma_p \end{array}\right] \qquad\qquad 6.113$$

where $\mathbf{J}_p$ is the Fisher information matrix for $\boldsymbol{\theta}_p$ and $\mathbf{J}_{p-1}$ is the Fisher information matrix for $\boldsymbol{\theta}_{p-1}$, the first $(p-1)$ elements of $\boldsymbol{\theta}$, when θ_p is known. The matrix $\mathbf{J}_{p-1}$ has the Cholesky factor $\mathbf{J}_{p-1} = \mathbf{A}_{p-1}\mathbf{A}_{p-1}^T$ with inverse $\mathbf{J}_{p-1}^{-1} = \mathbf{A}_{p-1}^{-T}\mathbf{A}_{p-1}^{-1}$. Similarly, the matrix $\mathbf{J}_p$ has Cholesky factor

$$\mathbf{J}_p = \left[\begin{array}{c|c} \mathbf{A}_{p-1} & \mathbf{0} \\ \hline \mathbf{a}_p^T & \alpha_p \end{array}\right] \left[\begin{array}{c|c} \mathbf{A}_{p-1}^T & \mathbf{a}_p \\ \hline \mathbf{0}^T & \alpha_p \end{array}\right] = \left[\begin{array}{c|c} \mathbf{J}_{p-1} & \mathbf{j}_p \\ \hline \mathbf{j}_p^T & \gamma_p \end{array}\right]. \qquad 6.114$$

The solutions for $\mathbf{a}_p$ and α_p are

$$\mathbf{A}_{p-1}\mathbf{a}_p = \mathbf{j}_p \qquad\qquad 6.115$$

$$\mathbf{a}_p^T \mathbf{a}_p + \alpha_p^2 = \gamma_p.$$

These equations update the Cholesky factors of $\mathbf{J}_p$ from the Cholesky factors of $\mathbf{J}_{p-1}$.

The Cholesky factor of $\mathbf{J}_p$ may be inverted to obtain the following upper-lower factorization of $\mathbf{J}_p^{-1}$:

$$\mathbf{J}_p^{-1} = \left[\begin{array}{c|c} \mathbf{A}_{p-1}^{-T} & \mathbf{b}_p \\ \hline \mathbf{0}^T & \beta_p \end{array}\right] \left[\begin{array}{c|c} \mathbf{A}_{p-1}^{-1} & \mathbf{0} \\ \hline \mathbf{b}_p^T & \beta_p \end{array}\right] = \left[\begin{array}{c|c} \mathbf{J}_{p-1}^{-1} + \mathbf{b}_p\mathbf{b}_p^T & \beta_p\mathbf{b}_p \\ \hline \beta_p\mathbf{b}_p^T & \beta_p^2 \end{array}\right]. \qquad 6.116$$

The solutions for $\mathbf{b}_p$ and β_p are obtained from the constraint that $\mathbf{J}_p\mathbf{J}_p^{-1} = \mathbf{I}$:

$$\left[\begin{array}{c|c} \mathbf{A}_{p-1}^T & \mathbf{a}_p \\ \hline \mathbf{0}^T & \alpha_p \end{array}\right] \left[\begin{array}{c|c} \mathbf{A}_{p-1}^{-T} & \mathbf{b}_p \\ \hline \mathbf{0}^T & \beta_p \end{array}\right] = \mathbf{I} \qquad\qquad 6.117$$

$$\mathbf{A}_{p-1}^T \mathbf{b}_p = -\beta_p \mathbf{a}_p$$

$$\alpha_p \beta_p = 1.$$

These equations update the Cholesky factors of $\mathbf{J}_p^{-1}$. From the solution for $\mathbf{J}_p^{-1}$, we note that its upper $(p-1) \times (p-1)$ corner — call it $[\mathbf{J}_p^{-1}]_{p-1}$ — may be written as

$$\left[\mathbf{J}_p^{-1}\right]_{p-1} = \mathbf{J}_{p-1}^{-1} + \mathbf{b}_p\mathbf{b}_p^{\mathsf{T}}. \qquad 6.118$$

The upper left-hand corner of $\mathbf{J}_p^{-1}$ is the covariance bound for $\hat{\boldsymbol{\theta}}_{p-1}$ when θ_p is unknown, and $\mathbf{J}_{p-1}^{-1}$ is the covariance bound for $\hat{\boldsymbol{\theta}}_{p-1}$ when θ_p is known. The difference is $\mathbf{b}_p\mathbf{b}_p^{\mathsf{T}}$, a nonnegative definite matrix. Therefore, the difference is a nonnegative definite rank-one matrix:

$$\left[\mathbf{J}_p^{-1}\right]_{p-1} - \mathbf{J}_{p-1}^{-1} = \mathbf{b}_p\mathbf{b}_p^{\mathsf{T}} \geq 0. \qquad 6.119$$

So error ellipses can only increase as more nuisance parameters are unknown.

6.6   ENTROPY, LIKELIHOOD, AND NONLINEAR LEAST SQUARES

In this section we study the connection between entropy and likelihood. We show that when one of the unknown parameters in a multivariate normal distribution is a variance parameter, maximization of likelihood with respect to the variance produces a "compressed likelihood" formula that is identical with the negative of estimated entropy. Further maximization of likelihood with respect to other parameters is then equivalent to minimization of estimated entropy. When the covariance matrix in the multivariate normal model is known, then maximization of compressed likelihood and minimization of estimated entropy are equivalent to nonlinear least squares fitting of a parametric mean value vector to observed data.

Why should a maximum likelihood principle produce the same results as a minimum entropy principle? Because estimating parameters of a distribution to make measurements look probable is equivalent to estimating parameters to make the distribution as peaky as it can possibly be around the measurements.

Entropy

The entropy of a distribution measures the spread of the distribution or, equivalently, the uncertainty in random variables drawn from it. Entropy is defined to be

$$\mathscr{E} = -E \ln f_{\boldsymbol{\theta}}(\mathbf{X}) \qquad 6.120$$

where $\ln f_{\boldsymbol{\theta}}(\mathbf{X})$ is the log-likelihood random variable. So entropy is just the negative expected value of log likelihood. If $\mathbf{X}$ is the random sample $(\mathbf{X}_1 \quad \mathbf{X}_2 \quad \cdots \quad \mathbf{X}_M)$ and $\mathbf{X}_i$ is distributed as $\mathbf{X}_i : N[\mathbf{m}, \sigma^2\mathbf{R}]$, then entropy is

$$\mathscr{E} = M \ln\left[(2\pi\sigma^2)^{N/2}|\mathbf{R}|^{1/2}\right] + E\left[\frac{M}{2\sigma^2}v^2\right]$$

$$v^2 = \frac{1}{M}\sum_{i=1}^{M}(\mathbf{X}_i - \mathbf{m})^{\mathsf{T}}\mathbf{R}^{-1}(\mathbf{X}_i - \mathbf{m}) \qquad \text{sample variance.} \qquad 6.121$$

The random variable $(M/\sigma^2)\, v^2$ is χ^2_{MN} distributed, with mean value MN. Therefore, entropy is

$$\mathcal{E} = M \ln\left[(2\pi\sigma^2)^{N/2}|\mathbf{R}|^{1/2}\right] + MN/2$$

$$= M \ln\left[(2\pi\sigma^2 e)^{N/2}|\mathbf{R}|^{1/2}\right]. \qquad\qquad 6.122$$

Likelihood

The log likelihood for the sample $\mathbf{x}$ may be written

$$L = \ln f_{\boldsymbol{\theta}}(\mathbf{x}) = -M \ln\left[(2\pi\sigma^2)^{N/2}|\mathbf{R}|^{1/2}\right] - \frac{M}{2\sigma^2}\frac{1}{M}\sum_{i=1}^{M}(\mathbf{x}_i - \mathbf{m})^{\mathrm{T}}\mathbf{R}^{-1}(\mathbf{x}_i - \mathbf{m}).$$

This function may be maximized by differentiating with respect to σ^2 and equating the result to zero. The maximum likelihood estimate of σ^2 is then

$$\hat{\sigma}^2 = \frac{1}{MN}\sum_{i=1}^{M}(\mathbf{x}_i - \mathbf{m})^{\mathrm{T}}\mathbf{R}^{-1}(\mathbf{x}_i - \mathbf{m}) = \frac{1}{N}v^2. \qquad\qquad 6.123$$

When $\hat{\sigma}^2$ is substituted into the likelihood, we obtain "compressed likelihood":

$$\hat{L} = -M \ln\left[(2\pi\hat{\sigma}^2)^{N/2}|\mathbf{R}|^{1/2}\right] - MN/2$$

$$= -M \ln\left[(2\pi\hat{\sigma}^2 e)^{N/2}|\mathbf{R}|^{1/2}\right] \qquad\qquad 6.124$$

$$= -\hat{\mathcal{E}}.$$

So, compressed likelihood is the negative of estimated entropy. If the mean value $\mathbf{m}$ and the covariance $\mathbf{R}$ are also unknown, then maximum likelihood estimation of $\mathbf{m}$ and $\mathbf{R}$ is equivalent to maximization of compressed likelihood or minimization of estimated entropy.

Nonlinear Least Squares

If the covariance matrix $\mathbf{R}$ is known and $\mathbf{m} = \mathbf{m}(\boldsymbol{\theta})$ is an unknown mean value vector that is parameterized by $\boldsymbol{\theta}$, then maximization of compressed likelihood or minimization of estimated entropy becomes

$$\min_{\boldsymbol{\theta}} \hat{\sigma}^2 \qquad\qquad 6.125$$

$$\hat{\sigma}^2 = \frac{1}{N}v^2.$$

This amounts to nonlinear, weighted, least-squares fitting of the mean value vector $\mathbf{m}(\boldsymbol{\theta})$ to the data $[\mathbf{x}_1 \quad \mathbf{x}_2 \quad \cdots \quad \mathbf{x}_M]$. In summary, for this case,

$$\max_{\boldsymbol{\theta}} \hat{L} \leftrightarrow \min_{\boldsymbol{\theta}} \hat{\mathscr{E}} \leftrightarrow \min_{\boldsymbol{\theta}} \hat{\sigma}^2 \qquad\qquad 6.126$$

max compressed likelihood $\leftrightarrow$ min estimated entropy $\leftrightarrow$ NL least squares.

Comments

The results of this section apply whenever $\mathbf{X}_i$ is distributed as $\mathbf{X}_i : N[\mathbf{m}, \mathbf{W}]$ and the covariance matrix $\mathbf{W}$ has the representation $\mathbf{W} = \sigma^2 \mathbf{R}$. Some explicit forms for $\mathbf{W}$ where these results apply are

1. $\mathbf{W} = \sigma^2 \mathbf{I}$ (uncorrelated random variables of variance σ^2)

2. $\mathbf{W} = \sigma^2 \begin{bmatrix} 1 & & \star \\ & \ddots & \\ \star & & 1 \end{bmatrix}$ (correlated random variables of variance σ^2)

3. $\mathbf{W} = \{r_{|i-j|}\} = \sigma^2\{\rho_{|i-j|}\}$ (stationary correlated random variables of variance σ^2); $\sigma^2 = r_0$; $\rho_{|i-j|} = r_{|i-j|}/r_0$.

6.7   THE MULTIVARIATE NORMAL MODEL

When a random sample $\mathbf{X} = (\mathbf{X}_1, \mathbf{X}_2, \dots, \mathbf{X}_M)$ consists of independent, N-dimensional, normal random vectors, then we may derive generally applicable results for the score function and the Fisher information matrix. These general results may then be applied to special cases, including the linear statistical model to be covered in Section 6.8 and the structured covariance matrices to be covered in Section 6.13.

Let $\mathbf{X} = (\mathbf{X}_1, \mathbf{X}_2, \dots, \mathbf{X}_M)$ denote a random sample of normal random vectors $\mathbf{X}_i : N[\mathbf{m}(\boldsymbol{\theta}), R(\boldsymbol{\theta})]$. To simplify notation, let's denote $\mathbf{m}(\boldsymbol{\theta})$ as $\mathbf{m}$ and $R(\boldsymbol{\theta})$ as $\mathbf{R}$. The joint distribution of the random sample has density

$$f_{\boldsymbol{\theta}}(\mathbf{x}) = (2\pi)^{-MN/2}|\mathbf{R}|^{-M/2} \exp\left\{ -\frac{1}{2}\sum_{i=1}^{M}(\mathbf{x}_i - \mathbf{m})^{\mathrm{T}}\mathbf{R}^{-1}(\mathbf{x}_i - \mathbf{m}) \right\}$$

$$6.127$$

$$= (2\pi)^{-MN/2}|\mathbf{R}|^{-M/2} \exp\left\{ -\frac{M}{2}\operatorname{tr}\left[\mathbf{R}^{-1}\mathbf{S}\right] \right\}$$

where $\mathbf{S}$ is the sample covariance matrix

$$\mathbf{S} = \frac{1}{M}\sum_{i=1}^{M}(\mathbf{x}_i - \mathbf{m})(\mathbf{x}_i - \mathbf{m})^{\mathrm{T}}.$$

The score function is the following $p \times 1$ vector of derivatives:

$$s(\boldsymbol{\theta}, \mathbf{x}) = \frac{\partial}{\partial \boldsymbol{\theta}} \ln f_{\boldsymbol{\theta}}(\mathbf{x})$$

$$= \frac{\partial}{\partial \boldsymbol{\theta}} \left\{ -\frac{M}{2} \ln |\mathbf{R}| - \frac{M}{2} \operatorname{tr}\left[\mathbf{R}^{-1}\mathbf{S} \right] \right\}.$$

6.128

The kth element of the score function is

$$s_k = \frac{\partial}{\partial \theta_k} \left\{ -\frac{M}{2} \ln |\mathbf{R}| - \frac{M}{2} \operatorname{tr}\left[\mathbf{R}^{-1}\mathbf{S} \right] \right\}.$$

6.129

Use the identities for matrix gradients in the Appendix of this chapter to write s_k as

$$s_k = -\frac{M}{2} \operatorname{tr}\left[\mathbf{R}^{-1} \frac{\partial \mathbf{R}}{\partial \theta_k} \right] + \frac{M}{2} \operatorname{tr}\left[\mathbf{R}^{-1} \frac{\partial \mathbf{R}}{\partial \theta_k} \mathbf{R}^{-1}\mathbf{S} \right] - \frac{M}{2} \operatorname{tr}\left[\mathbf{R}^{-1} \frac{\partial \mathbf{S}}{\partial \theta_k} \right]$$

6.130

$$= -\frac{M}{2} \operatorname{tr}[\mathbf{A}_k] + \frac{M}{2} \operatorname{tr}\left[\mathbf{A}_k \mathbf{R}^{-1}\mathbf{S} \right] - \frac{M}{2} \operatorname{tr}\left[\mathbf{R}^{-1} \frac{\partial \mathbf{S}}{\partial \theta_k} \right]$$

where $\mathbf{A}_k$ is the matrix

$$\mathbf{A}_k = \mathbf{R}^{-1} \frac{\partial \mathbf{R}}{\partial \theta_k}.$$

6.131

When equated to zero, this formula provides a prescription for finding the ML estimate of θ_k.

In order to compute the CR bound, we need the stochastic Fisher matrix, which is the second derivative of the log-likelihood function, and the first derivative of the score function:

$$J_{ij}(\boldsymbol{\theta}, \mathbf{x}) = -\frac{\partial^2}{\partial \theta_i \partial \theta_j} \ln f_{\boldsymbol{\theta}}(\mathbf{x}) = -\frac{\partial}{\partial \theta_i} s_j$$

$$= \frac{M}{2} \operatorname{tr}\left[\frac{\partial \mathbf{A}_j}{\partial \theta_i} \right] - \frac{M}{2} \operatorname{tr}\left[\frac{\partial \mathbf{A}_j}{\partial \theta_i} \mathbf{R}^{-1}\mathbf{S} \right] + \frac{M}{2} \operatorname{tr}\left[\mathbf{A}_j \mathbf{R}^{-1} \frac{\partial \mathbf{R}}{\partial \theta_i} \mathbf{R}^{-1}\mathbf{S} \right]$$

6.132

$$- \frac{M}{2} \operatorname{tr}\left[\mathbf{A}_j \mathbf{R}^{-1} \frac{\partial \mathbf{S}}{\partial \theta_i} \right] - \frac{M}{2} \operatorname{tr}\left[\mathbf{R}^{-1} \frac{\partial \mathbf{R}}{\partial \theta_i} \mathbf{R}^{-1} \frac{\partial \mathbf{S}}{\partial \theta_j} \right] + \frac{M}{2} \operatorname{tr}\left[\mathbf{R}^{-1} \frac{\partial^2 \mathbf{S}}{\partial \theta_i \partial \theta_j} \right].$$

The expectation of the random variable $J_{ij}(\boldsymbol{\theta}, \mathbf{x})$ is the ij element in the Fisher information matrix. But note the following expectations of the sample covariance matrix and its derivatives:

$$E\mathbf{S} = \mathbf{R}$$

$$E \frac{\partial \mathbf{S}}{\partial \theta_n} = [\mathbf{0}]$$

6.133

$$E \frac{\partial^2 \mathbf{S}}{\partial \theta_i \partial \theta_j} = \frac{\partial \mathbf{m}}{\partial \theta_i} \frac{\partial \mathbf{m}^{\mathsf{T}}}{\partial \theta_j} + \frac{\partial \mathbf{m}}{\partial \theta_j} \frac{\partial \mathbf{m}^{\mathsf{T}}}{\partial \theta_i}$$

You are asked to prove these identities in Problem 6.6. With them, we may write the Fisher information matrix as follows:

$$\mathbf{J} = \{J_{ij}\} \qquad\qquad 6.134$$

$$J_{ij} = E J_{ij}(\boldsymbol{\theta}, \mathbf{x})$$

$$= \frac{M}{2} \operatorname{tr}\left[\mathbf{R}^{-1} \frac{\partial \mathbf{R}}{\partial \theta_i} \mathbf{R}^{-1} \frac{\partial \mathbf{R}}{\partial \theta_j}\right] + M \frac{\partial \mathbf{m}^{\mathrm{T}}}{\partial \theta_i} \mathbf{R}^{-1} \frac{\partial \mathbf{m}}{\partial \theta_j}.$$

This is the general result for the Fisher information matrix in the normal case. Special cases are

- $J_{ij} = M \dfrac{\partial \mathbf{m}^{\mathrm{T}}}{\partial \theta_i} \mathbf{R}^{-1} \dfrac{\partial \mathbf{m}}{\partial \theta_j}, \quad \mathbf{R}$ known, $\qquad\qquad 6.135$

- $J_{ij} = \dfrac{M}{2} \operatorname{tr}\left[\mathbf{R}^{-1} \dfrac{\partial \mathbf{R}}{\partial \theta_i} \mathbf{R}^{-1} \dfrac{\partial \mathbf{R}}{\partial \theta_j}\right], \quad \mathbf{m}$ known.

Example 6.13 (Mean Value Parameters in the Normal Model)

When $\mathbf{R}$ is known and $\mathbf{m} = \mathbf{m}(\boldsymbol{\theta})$ is unknown in the multivariate normal model, then the jth element of the score function is

$$s_j = -\frac{M}{2} \operatorname{tr}\left[\mathbf{R}^{-1} \frac{\partial \mathbf{S}}{\partial \theta_j}\right]$$

$$= \operatorname{tr}\left[\mathbf{R}^{-1} \sum_{i=1}^{M} [\mathbf{x}_i - \mathbf{m}] \frac{\partial \mathbf{m}^{\mathrm{T}}}{\partial \theta_j}\right] = \frac{\partial \mathbf{m}^{\mathrm{T}}}{\partial \theta_j} \mathbf{R}^{-1} \sum_{i=1}^{M} [\mathbf{x}_i - \mathbf{m}].$$

$\qquad\qquad 6.136$

This result may be organized into the vector of scores:

$$\mathbf{s}(\boldsymbol{\theta}, \mathbf{x}) = \frac{\partial}{\partial \boldsymbol{\theta}} \ln f_{\boldsymbol{\theta}}(\mathbf{x}) = \frac{\partial}{\partial \boldsymbol{\theta}} \mathbf{m}^{\mathrm{T}} \mathbf{R}^{-1} \sum_{i=1}^{M} [\mathbf{x}_i - \mathbf{m}]. \qquad 6.137$$

Note that the expectation of $\mathbf{s}(\boldsymbol{\theta}, \mathbf{x})$ is zero because $E\mathbf{x}_i = \mathbf{m}$, and the covariance of the score function is the Fisher information matrix:

$$\mathbf{J}(\boldsymbol{\theta}) = E\mathbf{s}(\boldsymbol{\theta}, \mathbf{X})\mathbf{s}^{\mathrm{T}}(\boldsymbol{\theta}, \mathbf{X}) = \left(\frac{\partial}{\partial \boldsymbol{\theta}} \mathbf{m}^{\mathrm{T}}\right) \mathbf{R}^{-1} M \mathbf{R} \mathbf{R}^{-1} \left(\frac{\partial}{\partial \boldsymbol{\theta}} \mathbf{m}^{\mathrm{T}}\right)^{\mathrm{T}}$$

$\qquad\qquad 6.138$

$$= M \frac{\partial}{\partial \boldsymbol{\theta}} \mathbf{m}^{\mathrm{T}} \mathbf{R}^{-1} \left(\frac{\partial}{\partial \boldsymbol{\theta}} \mathbf{m}^{\mathrm{T}}\right)^{\mathrm{T}}.$$

This result is general for the estimation of mean value parameters in the multivariate normal model. ∎

6.8   THE LINEAR STATISTICAL MODEL

The general results of the multivariate normal model produce illuminating results when the mean value vector $\mathbf{m}$ obeys the linear model $\mathbf{m} = \mathbf{H}\boldsymbol{\theta}$. In this case, $\mathbf{X}_n$ is an N-dimensional normal random vector distributed as $N[\mathbf{H}\boldsymbol{\theta}, \mathbf{R}]$. The matrix $\mathbf{H}$ is $N \times p$ with full-column rank $p < N$, and $\boldsymbol{\theta}$ is an unknown $p \times 1$ vector. The covariance matrix is known. Think of $\mathbf{H}\boldsymbol{\theta}$ as a structured model for the mean value vector $\mathbf{m}$ with known structure $\mathbf{H}$ and unknown parameters $\boldsymbol{\theta}$.

The distribution of a random sample $\mathbf{x} = (\mathbf{x}_1, \ldots, \mathbf{x}_M)$ is

$$f_{\boldsymbol{\theta}}(\mathbf{x}) = (2\pi)^{-MN/2}|\mathbf{R}|^{-M/2} \exp\left\{ -\frac{1}{2} \sum_{n=1}^{M} (\mathbf{x}_n - \mathbf{H}\boldsymbol{\theta})^{\mathrm{T}} \mathbf{R}^{-1}(\mathbf{x}_n - \mathbf{H}\boldsymbol{\theta}) \right\}. \qquad 6.139$$

The score function is

$$s(\boldsymbol{\theta}, \mathbf{x}) = \frac{\partial}{\partial \boldsymbol{\theta}} \ln f_{\boldsymbol{\theta}}(\mathbf{x})$$

$$= \mathbf{H}^{\mathrm{T}} \mathbf{R}^{-1} \sum_{n=1}^{M} (\mathbf{x}_n - \mathbf{H}\boldsymbol{\theta}). \qquad 6.140$$

The maximum likelihood estimate is

$$\hat{\boldsymbol{\theta}} = (\mathbf{H}^{\mathrm{T}} \mathbf{R}^{-1} \mathbf{H})^{-1} \mathbf{H}^{\mathrm{T}} \mathbf{R}^{-1} \hat{\mathbf{m}} \qquad 6.141$$

$$\hat{\mathbf{m}} = \frac{1}{M} \sum_{n=1}^{M} \mathbf{x}_n \qquad \text{sample mean.}$$

The Fisher information matrix is

$$\mathbf{J} = M \left(\frac{\partial}{\partial \boldsymbol{\theta}} \mathbf{m}^{\mathrm{T}} \right)^{\mathrm{T}} \mathbf{R}^{-1} \frac{\partial}{\partial \boldsymbol{\theta}} \mathbf{m}^{\mathrm{T}} = M \mathbf{H}^{\mathrm{T}} \mathbf{R}^{-1} \mathbf{H}. \qquad 6.142$$

The score function may be written as

$$s(\boldsymbol{\theta}, \mathbf{x}) = \mathbf{J}[\hat{\boldsymbol{\theta}} - \boldsymbol{\theta}], \qquad 6.143$$

so we know that $\hat{\boldsymbol{\theta}}$ in the linear statistical model is efficient. The estimate $\hat{\boldsymbol{\theta}}$ is a function of the complete sufficient statistic $\mathbf{H}^{\mathrm{T}} \mathbf{R}^{-1} \hat{\mathbf{m}}$. It is unbiased with variance $(1/M)(\mathbf{H}^{\mathrm{T}} \mathbf{R}^{-1} \mathbf{H})^{-1}$:

$$E\hat{\boldsymbol{\theta}} = (\mathbf{H}^{\mathrm{T}} \mathbf{R}^{-1} \mathbf{H})^{-1} \mathbf{H}^{\mathrm{T}} \mathbf{R}^{-1} \mathbf{H}\boldsymbol{\theta} = \boldsymbol{\theta}$$

$$\mathbf{C} = E[\hat{\boldsymbol{\theta}} - \boldsymbol{\theta}][\hat{\boldsymbol{\theta}} - \boldsymbol{\theta}]^{\mathrm{T}} = (\mathbf{H}^{\mathrm{T}} \mathbf{R}^{-1} \mathbf{H})^{-1} \mathbf{H}^{\mathrm{T}} \mathbf{R}^{-1} \frac{1}{M} \mathbf{R} \mathbf{R}^{-1} \mathbf{H}(\mathbf{H}^{\mathrm{T}} \mathbf{R}^{-1} \mathbf{H})^{-1}$$

$$\qquad\qquad 6.144$$

$$= \frac{1}{M} (\mathbf{H}^{\mathrm{T}} \mathbf{R}^{-1} \mathbf{H})^{-1}.$$

Therefore, $\hat{\boldsymbol{\theta}}$ is MVUB, efficient, and consistent.

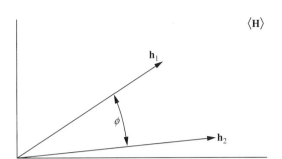

Figure 6.8 Two-dimensional subspace.

Example 6.14

Suppose $\mathbf{x} = (\mathbf{x}_1, \mathbf{x}_2, \ldots, \mathbf{x}_M)$ is a random sample with $\mathbf{x}_i : N[\mathbf{H}\boldsymbol{\theta}, \sigma^2\mathbf{I}]$ and $\mathbf{H}$ is the $N \times 2$ matrix $\mathbf{H} = [\mathbf{h}_1 \ \mathbf{h}_2]$. Then the error covariance matrix for the maximum likelihood estimator $\hat{\boldsymbol{\theta}}$ is

$$\mathbf{C} = E[\hat{\boldsymbol{\theta}} - \boldsymbol{\theta}][\hat{\boldsymbol{\theta}} - \boldsymbol{\theta}]^{\mathrm{T}} = \frac{\sigma^2}{M} \begin{bmatrix} \mathbf{h}_1^{\mathrm{T}}\mathbf{h}_1 & \mathbf{h}_1^{\mathrm{T}}\mathbf{h}_2 \\ \mathbf{h}_1^{\mathrm{T}}\mathbf{h}_2 & \mathbf{h}_2^{\mathrm{T}}\mathbf{h}_2 \end{bmatrix}^{-1}. \qquad 6.145$$

This is also $\mathbf{J}^{-1}$, the inverse of the Fisher matrix. Let $\mathbf{h}_i^{\mathrm{T}}\mathbf{h}_i = E_s$ for $i = 1, 2$, and let $\mathbf{h}_i^{\mathrm{T}}\mathbf{h}_j = \rho E_s$ for $i \neq j$. Then the error covariance matrix is

$$\mathbf{C} = \frac{\sigma^2}{ME_s} \begin{bmatrix} 1 & \rho \\ \rho & 1 \end{bmatrix}^{-1} = \frac{\sigma^2}{ME_s(1 - \rho^2)} \begin{bmatrix} 1 & -\rho \\ -\rho & 1 \end{bmatrix}. \qquad 6.146$$

This result is illustrated in Figure 6.8. The cosine of the angle ϕ between the mode vectors $\mathbf{h}_1$ and $\mathbf{h}_2$ is

$$\rho = \cos\phi = \frac{\mathbf{h}_1^{\mathrm{T}}\mathbf{h}_2}{\|\mathbf{h}_1\| \|\mathbf{h}_2\|}. \qquad 6.147$$

When $\mathbf{h}_1$ and $\mathbf{h}_2$ are orthogonal, then $\rho = 0$ and $\mathbf{C} = (\sigma^2/ME_s)\mathbf{I}$. As $|\rho| \to 1$, meaning that $\mathbf{h}_1$ and $\mathbf{h}_2$ are nearly colinear, then

$$(1 - \rho^2)\mathbf{C} \to \frac{T^2}{ME_s} \begin{bmatrix} 1 & -1 \\ -1 & 1 \end{bmatrix}.$$

Can you give a physical interpretation of this finding?   ■

6.9   MODE IDENTIFICATION IN THE LINEAR STATISTICAL MODEL

Recall from our discussion of the linear model $\mathbf{H}\boldsymbol{\theta}$ that we have called $\boldsymbol{\theta}$ the "mode weights" and the columns of $\mathbf{H} = [\mathbf{h}_1 \quad \mathbf{h}_2 \quad \cdots \quad \mathbf{h}_p]$ the "modes." It often turns out in time-series analysis and in array processing that the modes are incompletely known, in which case we write $\mathbf{H}(\boldsymbol{\omega}) = [\mathbf{h}_1(\boldsymbol{\omega})\ \ \mathbf{h}_2(\boldsymbol{\omega})\ \ \cdots\ \ \mathbf{h}_p(\boldsymbol{\omega})]$ to indicate that the mode structure $\mathbf{h}_i(\boldsymbol{\omega})$ is finitely parameterized by the vector of unknowns $\boldsymbol{\omega}$.

Maximum Likelihood Equations

If $\mathbf{x} = (\mathbf{x}_1, \mathbf{x}_2, \ldots, \mathbf{x}_M)$ is a random sample of $N[\mathbf{H}(\boldsymbol{\omega})\boldsymbol{\theta}, \mathbf{R}]$ random variables, then the log likelihood is (ignoring terms independent of $\boldsymbol{\omega}$ and $\boldsymbol{\theta}$)

$$L(\boldsymbol{\omega}, \boldsymbol{\theta}, \mathbf{x}) = -\frac{1}{2}\sum_{i=1}^{M}\left[\mathbf{x}_i - \mathbf{H}(\boldsymbol{\omega})\boldsymbol{\theta}\right]^{\mathrm{T}}\mathbf{R}^{-1}\left[\mathbf{x}_i - \mathbf{H}(\boldsymbol{\omega})\boldsymbol{\theta}\right]. \qquad 6.148$$

For any $\boldsymbol{\omega}$, the maximum likelihood estimate of $\boldsymbol{\theta}$ is

$$\hat{\boldsymbol{\theta}} = \left[\mathbf{H}^{\mathrm{T}}(\boldsymbol{\omega})\mathbf{R}^{-1}\mathbf{H}(\boldsymbol{\omega})\right]^{-1}\mathbf{H}^{\mathrm{T}}(\boldsymbol{\omega})\mathbf{R}^{-1}\mathbf{t} \qquad 6.149$$

$$\mathbf{t} = \frac{1}{M}\sum_{j=1}^{M}\mathbf{x}_j : N\left[\mathbf{H}(\boldsymbol{\omega})\boldsymbol{\theta},\ \frac{1}{M}\mathbf{R}\right].$$

When this estimate is substituted into log likelihood, the compressed log likelihood becomes

$$L(\boldsymbol{\omega}, \hat{\boldsymbol{\theta}}, \mathbf{x}) = -\frac{1}{2}\sum_{i=1}^{M}[\mathbf{x}_i - \mathbf{P}(\boldsymbol{\omega})\mathbf{t}]^{\mathrm{T}}\mathbf{R}^{-1}[\mathbf{x}_i - \mathbf{P}(\boldsymbol{\omega})\mathbf{t}], \qquad 6.150$$

where $\mathbf{P}(\boldsymbol{\omega})$ is the projector $\mathbf{P}(\boldsymbol{\omega}) = \mathbf{H}(\boldsymbol{\omega})\left[\mathbf{H}^{\mathrm{T}}(\boldsymbol{\omega})\mathbf{R}^{-1}\mathbf{H}(\boldsymbol{\omega})\right]^{-1}\mathbf{H}^{\mathrm{T}}(\boldsymbol{\omega})\mathbf{R}^{-1}$. We may give and take $\mathbf{t}$, to rewrite this expression as

$$L(\boldsymbol{\omega}, \hat{\boldsymbol{\theta}}, \mathbf{x}) = -\frac{1}{2}\sum_{i=1}^{M}[\mathbf{t} - \mathbf{P}(\boldsymbol{\omega})\mathbf{t} + \mathbf{x}_i - \mathbf{t}]^{\mathrm{T}}\mathbf{R}^{-1}[\mathbf{t} - \mathbf{P}(\boldsymbol{\omega})\mathbf{t} + \mathbf{x}_i - \mathbf{t}]$$

$$= -\frac{M}{2}\mathbf{t}^{\mathrm{T}}\left[\mathbf{I} - \mathbf{P}(\boldsymbol{\omega})\right]\mathbf{R}^{-1}\left[\mathbf{I} - \mathbf{P}(\boldsymbol{\omega})\right]\mathbf{t} - \frac{1}{2}\sum_{i=1}^{M}(\mathbf{x}_i - \mathbf{t})^{\mathrm{T}}\mathbf{R}^{-1}(\mathbf{x}_i - \mathbf{t}).$$

$$6.151$$

This compressed likelihood is maximized with respect to $\boldsymbol{\omega}$ by

$$\min_{\boldsymbol{\omega}}\mathbf{t}^{\mathrm{T}}\left[\mathbf{I} - \mathbf{P}(\boldsymbol{\omega})\right]\mathbf{R}^{-1}\left[\mathbf{I} - \mathbf{P}(\boldsymbol{\omega})\right]\mathbf{t} \qquad 6.152$$

This is the general result. When $\mathbf{R} = \sigma^2\mathbf{I}$, then the problem is to find the value of $\hat{\boldsymbol{\omega}}$ that minimizes $\mathbf{t}^{\mathrm{T}}\left[\mathbf{I} - \mathbf{P}(\boldsymbol{\omega})\right]\mathbf{t}$. That is,

$$\min_{\boldsymbol{\omega}}\mathbf{t}^{\mathrm{T}}\left[\mathbf{I} - \mathbf{P}(\boldsymbol{\omega})\right]\mathbf{t}\quad\leftrightarrow\quad\max_{\boldsymbol{\omega}}\mathbf{t}^{\mathrm{T}}\mathbf{P}(\boldsymbol{\omega})\mathbf{t}, \qquad 6.153$$

where, now, $\mathbf{P}(\boldsymbol{\omega})$ is the projector $\mathbf{P}(\boldsymbol{\omega}) = \mathbf{H}(\boldsymbol{\omega})\left[\mathbf{H}^{\mathsf{T}}(\boldsymbol{\omega})\mathbf{H}(\boldsymbol{\omega})\right]^{-1}\mathbf{H}^{\mathsf{T}}(\boldsymbol{\omega})$. That is, we choose the parameter $\boldsymbol{\omega}$ so that the projection of the sample mean $\mathbf{t}$ onto the subspace $\langle\mathbf{H}(\boldsymbol{\omega})\rangle$ has maximum norm-squared. We can think of this as constrained subspace identification, because the subspace is constrained to look like $\mathbf{H}(\boldsymbol{\omega})$, where $\boldsymbol{\omega}$ is its parametric description. This result applies to mode identification in time series and direction of arrival estimation in multisensor arrays. We treat these problems in Section 6.10 and in Chapters 9 and 11, where we discuss algorithms for identifying $\mathbf{H}(\boldsymbol{\omega})$.

The Fisher Information Matrix

The Fisher information matrix for simultaneous estimation of mode weights $\boldsymbol{\theta}$ and mode parameters $\boldsymbol{\omega}$ is characterized by partial derivatives of the form $\partial^2/\partial\theta_i\partial\theta_j$, $\partial^2/\partial\theta_i\partial\omega_j$, and $\partial^2/\partial\omega_i\partial\omega_j$. Let's organize the Fisher matrix as follows:

$$\mathbf{J}(\boldsymbol{\theta}, \boldsymbol{\omega}) = \begin{bmatrix} -E\dfrac{\partial^2}{\partial\theta_i\partial\theta_j}\ln f_{\boldsymbol{\theta},\boldsymbol{\omega}}(\mathbf{x}) & -E\dfrac{\partial^2}{\partial\theta_i\partial\omega_j}\ln f_{\boldsymbol{\theta},\boldsymbol{\omega}}(\mathbf{x}) \\[2mm] -E\dfrac{\partial^2}{\partial\omega_i\partial\theta_j}\ln f_{\boldsymbol{\theta},\boldsymbol{\omega}}(\mathbf{x}) & -E\dfrac{\partial^2}{\partial\omega_i\partial\omega_j}\ln f_{\boldsymbol{\theta},\boldsymbol{\omega}}(\mathbf{x}) \end{bmatrix}$$

$$= \begin{bmatrix} \mathbf{A} & \mathbf{B} \\ \mathbf{B}^{\mathsf{T}} & \mathbf{D} \end{bmatrix}. \tag{6.154}$$

Thus, $\mathbf{A}$ contains the Fisher matrix for $\boldsymbol{\theta}$, $\mathbf{D}$ contains the Fisher matrix for $\boldsymbol{\omega}$, and $\mathbf{B}$ contains the cross-terms between $\boldsymbol{\theta}$ and $\boldsymbol{\omega}$.

We shall assume $\mathbf{R} = \sigma^2\mathbf{I}$ and proceed to compute the elements of $\mathbf{J}$. In the formulas to follow, the vector $\mathbf{e}_i$ is a coordinate vector of zeros, with a one in the ith coordinate:

1. $$-\frac{\partial^2}{\partial\theta_i\partial\theta_j}\ln f_{\boldsymbol{\theta},\boldsymbol{\omega}}(\mathbf{x}) = \frac{1}{2\sigma^2}\frac{\partial}{\partial\theta_i\partial\theta_j}(\mathbf{x}-\mathbf{H}\boldsymbol{\theta})^{\mathsf{T}}(\mathbf{x}-\mathbf{H}\boldsymbol{\theta})$$

    $$= \frac{1}{2\sigma^2}\frac{\partial}{\partial\theta_i}\left(-\mathbf{e}_j^{\mathsf{T}}\mathbf{H}^{\mathsf{T}}(\mathbf{x}-\mathbf{H}\boldsymbol{\theta}) - (\mathbf{x}-\mathbf{H}\boldsymbol{\theta})^{\mathsf{T}}\mathbf{H}\mathbf{e}_j\right)$$

    $$= \frac{1}{2\sigma^2}\mathbf{e}_j^{\mathsf{T}}\mathbf{H}^{\mathsf{T}}\mathbf{H}\mathbf{e}_i + \frac{1}{2\sigma^2}\mathbf{e}_i^{\mathsf{T}}\mathbf{H}^{\mathsf{T}}\mathbf{H}\mathbf{e}_j \tag{6.155}$$

    $$= \frac{1}{\sigma^2}\mathbf{e}_i^{\mathsf{T}}\mathbf{H}^{\mathsf{T}}\mathbf{H}\mathbf{e}_j$$

    $$-E\frac{\partial^2}{\partial\theta_i\partial\theta_j}\ln f_{\boldsymbol{\theta},\boldsymbol{\omega}}(\mathbf{x}) = \frac{1}{\sigma^2}\mathbf{e}_i^{\mathsf{T}}\mathbf{H}^{\mathsf{T}}\mathbf{H}\mathbf{e}_j$$

2. $\quad -\dfrac{\partial^2}{\partial\theta_i\partial\omega_j}\ln f_{\theta,\omega}(\mathbf{x}) = \dfrac{1}{2\sigma^2}\dfrac{\partial}{\partial\theta_i}\dfrac{\partial}{\partial\omega_j}(\mathbf{x}-\mathbf{H}\theta)^\mathrm{T}(\mathbf{x}-\mathbf{H}\theta)$

$$= -\dfrac{1}{2\sigma^2}\dfrac{\partial}{\partial\theta_i}\,\theta^\mathrm{T}\dfrac{\partial\mathbf{H}^\mathrm{T}}{\partial\omega_j}(\mathbf{x}-\mathbf{H}\theta)$$

$$-\dfrac{1}{2\sigma^2}\dfrac{\partial}{\partial\theta_i}(\mathbf{x}-\mathbf{H}\theta)^\mathrm{T}\dfrac{\partial\mathbf{H}}{\partial\omega_j}\theta$$

$\qquad\qquad\qquad\qquad\qquad\qquad\qquad\qquad\qquad\qquad\qquad\qquad$ 6.156

$$= -\dfrac{1}{\sigma^2}\dfrac{\partial}{\partial\theta_i}\,\theta^\mathrm{T}\dfrac{\partial\mathbf{H}^\mathrm{T}}{\partial\omega_j}(\mathbf{x}-\mathbf{H}\theta)$$

$$= -\dfrac{1}{\sigma^2}\,\mathbf{e}_i^\mathrm{T}\dfrac{\partial\mathbf{H}^\mathrm{T}}{\partial\omega_j}(\mathbf{x}-\mathbf{H}\theta)+\dfrac{1}{\sigma^2}\,\theta^\mathrm{T}\dfrac{\partial\mathbf{H}^\mathrm{T}}{\partial\omega_j}\mathbf{H}\mathbf{e}_i$$

$\quad -E\,\dfrac{\partial^2}{\partial\theta_i\partial\omega_j}\ln f_{\theta,\omega}(\mathbf{x}) = \dfrac{1}{\sigma^2}\,\mathbf{e}_i^\mathrm{T}\mathbf{H}^\mathrm{T}\dfrac{\partial\mathbf{H}}{\partial\omega_j}\theta$

3. $\quad -\dfrac{\partial^2}{\partial\omega_i\partial\omega_j}\ln f_{\theta,\omega}(\mathbf{x}) = -\dfrac{1}{\sigma^2}\dfrac{\partial}{\partial\omega_i}\,\theta^\mathrm{T}\dfrac{\partial\mathbf{H}^\mathrm{T}}{\partial\omega_j}(\mathbf{x}-\mathbf{H}\theta)$

$$= -\dfrac{1}{\sigma^2}\dfrac{\partial^2}{\partial\omega_i\partial\omega_j}\mathbf{H}^\mathrm{T}(\mathbf{x}-\mathbf{H}\theta)+\dfrac{1}{\sigma^2}\,\theta^\mathrm{T}\dfrac{\partial\mathbf{H}^\mathrm{T}}{\partial\omega_j}\dfrac{\partial\mathbf{H}}{\partial\omega_i}\theta$$

$\quad -E\,\dfrac{\partial^2}{\partial\omega_i\partial\omega_j}\ln f_{\theta,\omega}(\mathbf{x}) = \dfrac{1}{\sigma^2}\,\theta^\mathrm{T}\dfrac{\partial\mathbf{H}^\mathrm{T}}{\partial\omega_i}\dfrac{\partial\mathbf{H}}{\partial\omega_j}\theta.$

$\qquad\qquad\qquad\qquad\qquad\qquad\qquad\qquad\qquad\qquad\qquad\qquad$ 6.157

In summary, then, the Fisher information matrix for simultaneous estimation of modes and mode weights is

$$\mathbf{J}(\theta,\omega)=\dfrac{1}{\sigma^2}\begin{bmatrix}\{\mathbf{e}_i^\mathrm{T}\mathbf{H}^\mathrm{T}\mathbf{H}\mathbf{e}_j\} & \left\{\mathbf{e}_i^\mathrm{T}\mathbf{H}^\mathrm{T}\dfrac{\partial\mathbf{H}}{\partial\omega_j}\theta\right\}\\[2ex] \left\{\theta^\mathrm{T}\dfrac{\partial\mathbf{H}^\mathrm{T}}{\partial\omega_j}\mathbf{H}\mathbf{e}_i\right\} & \left\{\theta^\mathrm{T}\dfrac{\partial\mathbf{H}^\mathrm{T}}{\partial\omega_i}\dfrac{\partial\mathbf{H}}{\partial\omega_j}\theta\right\}\end{bmatrix}.$$

$\qquad\qquad\qquad\qquad\qquad\qquad\qquad\qquad\qquad\qquad\qquad\qquad$ 6.158

In the next section we apply these results to the identification of autoregressive moving average parameters.

6.10   MAXIMUM LIKELIHOOD IDENTIFICATION OF ARMA PARAMETERS

The theoretical results of the previous section produce important findings when applied to the problem of identifying the parameters of an autoregressive moving average (ARMA) system. The problem is illustrated in Figure 6.9: a unit pulse sequence $\{\delta_t\}$

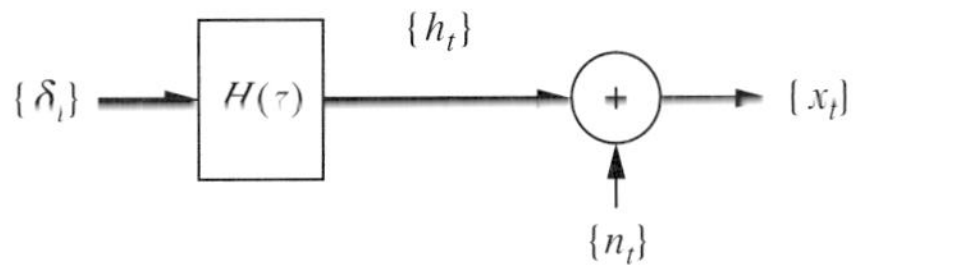

$$H(z) = \frac{B(z)}{A(z)}; \quad B(z) = \sum_{i=0}^{p-1} b_i z^{-i}; \quad A(z) = \sum_{i=0}^{p} a_i z^{-1}; \quad a_0 = 1$$

Figure 6.9 Noisy observation of ARMA impulse response.

excites the digital filter, or discrete-time system $H(z)$, to produce the impulse response sequence $\{h_t\}$. This sequence obeys the ARMA recursion

$$A(z)\{h_t\} = B(z)\{\delta_t\} \qquad\qquad 6.159$$

or

$$h_t = \begin{cases} 0, & t < 0 \\ -\sum_{n=1}^{p} a_n h_{t-n} + b_t, & 0 \le t < p \\ -\sum_{n=1}^{p} a_n h_{t-n}, & t \ge p. \end{cases} \qquad 6.160$$

$$H(z) = \frac{B(z)}{A(z)}; \quad B(z) = \sum_{i=0}^{p-1} b_i z^{-i}; \quad A(z) = \sum_{i=0}^{p} a_i z^{-i}; \quad a_0 = 1$$

These recursions may be written as

$$\sum_{n=0}^{p} a_n h_{t-n} = b_t; \qquad a_0 = 1 \qquad\qquad 6.161$$

and organized into the matrix equations

$$\begin{bmatrix} 1 & & & & & \\ a_1 & 1 & & & & \\ & & \ddots & & \mathbf{0} & \\ a_{p-1} & \cdots & a_1 & 1 & & \\ a_p & & \cdots & a_1 & 1 & \\ \mathbf{0} & \ddots & & & \ddots & \ddots \\ & & a_p & & \cdots & a_1 & 1 \end{bmatrix} \begin{bmatrix} h_0 \\ h_1 \\ \vdots \\ h_{p-1} \\ h_p \\ \vdots \\ h_{N-1} \end{bmatrix} = \begin{bmatrix} b_0 \\ b_1 \\ \vdots \\ b_{p-1} \\ 0 \\ \vdots \\ 0 \end{bmatrix}. \qquad 6.162$$

We shall write these equations as

$$\mathbf{K}^{-1}\mathbf{h} = \begin{bmatrix} \mathbf{b} \\ \mathbf{0} \end{bmatrix}; \qquad \mathbf{h} = \begin{bmatrix} h_0 \\ \vdots \\ h_{N-1} \end{bmatrix}; \qquad \mathbf{b} = \begin{bmatrix} b_0 \\ \vdots \\ b_{p-1} \end{bmatrix}$$

$$\mathbf{K}^{-1} = \begin{bmatrix} 1 & & & \\ a_1 & 1 & & \mathbf{0} \\ & & \ddots & \\ a_{p-1} & \cdots & a_1 & 1 \\ & & \mathbf{A}^{\mathrm{T}} & \end{bmatrix}; \qquad \mathbf{A}^{\mathrm{T}} = \begin{bmatrix} a_p & \cdots & a_1 & 1 & & & \\ & \ddots & & & & \mathbf{0} & \\ & \mathbf{0} & & & \ddots & & \\ & & & a_p & \cdots & a_1 & 1 \end{bmatrix}.$$

$$6.163$$

We say that the impulse response vector $\mathbf{h}$ obeys the "analysis model" $\mathbf{K}^{-1}\mathbf{h} = [\mathbf{b}^{\mathrm{T}} \quad \mathbf{0}^{\mathrm{T}}]^{\mathrm{T}}$. In this model the "prediction error" matrix $\mathbf{A}^{\mathrm{T}}$ annihilates $\mathbf{h}$ to produce zeros in the equation $\mathbf{A}^{\mathrm{T}}\mathbf{h} = \mathbf{0}$.

Corresponding to the analysis model for $\mathbf{h}$ is the "synthesis model"

$$\mathbf{h} = \mathbf{K}\begin{bmatrix} \mathbf{b} \\ \mathbf{0} \end{bmatrix}. \qquad 6.164$$

We may rewrite this model as

$$\mathbf{h} = \mathbf{H}\mathbf{b} \qquad 6.165$$

where $\mathbf{H}$ is the model matrix that comprises the first p columns of $\mathbf{K}$:

$$\mathbf{K} = [\mathbf{H} | \star]. \qquad 6.166$$

The synthesis model shows very clearly that the unit pulse response $\mathbf{h}$ obeys a linear model in which the autoregressive (AR) parameters $\mathbf{a}^{\mathrm{T}} = (a_p \quad \cdots \quad a_1 \quad 1)$ determine $\mathbf{H}$ (through $\mathbf{K}$) and the moving average (MA) parameters $\mathbf{b}^{\mathrm{T}} = (b_0 \quad b_1 \quad \cdots \quad b_{p-1})$ determine $\mathbf{b}$. We say that the model is *separable* and write

$$\mathbf{h} = \mathbf{H}(\mathbf{a})\mathbf{b}$$

$$\mathbf{H}(\mathbf{a}) \qquad \text{function of AR parameters} \qquad 6.167$$

$$\mathbf{b} \qquad \text{function of MA parameters.}$$

Now, as illustrated in Figure 6.9, noise $\{n_t\}$ is added to the unit pulse response $\{h_t\}$ to produce the noisy measurements

$$x_t = h_t + n_t; \qquad 0 \leq t \leq N - 1. \qquad 6.168$$

These measurements are organized into the linear statistical model

$$\mathbf{x} = \mathbf{h} + \mathbf{n}$$
$$= \mathbf{H}(\mathbf{a})\mathbf{b} + \mathbf{n}. \qquad 6.169$$

Our problem is to find maximum likelihood estimates of $\mathbf{a}$ and $\mathbf{b}$ in the model. The MA coefficients are mode weights, and the AR parameters $\mathbf{a}$ are mode parameters.

Maximum Likelihood Equations

We shall assume that the noises $\{n_t\}$ are i.i.d. $N[0, \sigma^2]$ random variables, meaning that the noise vector $\mathbf{n}$ is distributed as $\mathbf{n} : N[\mathbf{0}, \sigma^2\mathbf{I}]$. The log-likelihood function is then

$$\ln f_{(\mathbf{a},\mathbf{b})}(\mathbf{x}) = -\ln\left[(2\pi\sigma^2)^{-N/2}\right] - \frac{1}{2\sigma^2}[\mathbf{x} - \mathbf{H}(\mathbf{a})\mathbf{b}]^T[\mathbf{x} - \mathbf{H}(\mathbf{a})\mathbf{b}]. \qquad 6.170$$

The maximum likelihood estimates for $\mathbf{a}$ and $\mathbf{b}$ minimize the quadratic form

$$(\hat{\mathbf{a}}, \hat{\mathbf{b}}) = \arg\min_{(\mathbf{a},\mathbf{b})}[\mathbf{x} - \mathbf{H}(\mathbf{a})\mathbf{b}]^T[\mathbf{x} - \mathbf{H}(\mathbf{a})\mathbf{b}]. \qquad 6.171$$

The maximum likelihood estimate of $\mathbf{b}$ is

$$\hat{\mathbf{b}} = \left[\mathbf{H}^T(\mathbf{a})\mathbf{H}(\mathbf{a})\right]^{-1}\mathbf{H}^T(\mathbf{a})\mathbf{x}. \qquad 6.172$$

When this solution is substituted into log likelihood, the compressed likelihood is (ignoring constants)

$$\log f_{\mathbf{a},\hat{\mathbf{b}}}(\mathbf{x}) = -\mathbf{x}^T[\mathbf{I} - \mathbf{P}(\mathbf{a})]\mathbf{x} \qquad 6.173$$

where $\mathbf{P}(\mathbf{a})$ is the projector

$$\mathbf{P}(\mathbf{a}) = \mathbf{H}(\mathbf{a})\left[\mathbf{H}^T(\mathbf{a})\mathbf{H}(\mathbf{a})\right]^{-1}\mathbf{H}^T(\mathbf{a}). \qquad 6.174$$

Instead of maximizing log likelihood, we may minimize its negative. Therefore the maximum likelihood estimate of $\mathbf{a}$ is

$$\hat{\mathbf{a}} = \arg\min_{\mathbf{a}} \mathbf{x}^T[\mathbf{I} - \mathbf{P}(\mathbf{a})]\mathbf{x} . \qquad 6.175$$

That is, $\hat{\mathbf{a}}$ is the value of $\mathbf{a}$ that accounts for the most energy in $\left\|\mathbf{P}(\mathbf{a})\mathbf{x}\right\|^2$, under the constraint that the projector $\mathbf{P}(\mathbf{a})$ must be constructed from the model $\mathbf{H}(\mathbf{a})$.

The Projector P(a)

The projector $\mathbf{P}(\mathbf{a})$ and the orthogonal projector $\mathbf{I} - \mathbf{P}(\mathbf{a})$ have alternative representations to those given in Equation (6.174). We obtain them by noting from the analysis and synthesis models for $\mathbf{h}$ that

$$\begin{bmatrix} 1 & & & \\ a_1 & 1 & & \mathbf{0} \\ \vdots & \ddots & \ddots & \\ a_{p-1} & \cdots & a_1 & 1 \\ \hline & & \mathbf{A}^T & \end{bmatrix} \mathbf{H}(\mathbf{a}) \begin{bmatrix} \mathbf{b} \\ \\ \mathbf{0} \end{bmatrix} = \begin{bmatrix} \mathbf{b} \\ \\ \mathbf{0} \end{bmatrix}. \qquad 6.176$$

This result shows two things:

1.
$$\begin{bmatrix} 1 & & & \\ a_1 & 1 & & \mathbf{0} \\ \vdots & \ddots & \ddots & \\ a_{p-1} & \cdots & a_1 & 1 \end{bmatrix} \mathbf{H(a)} = \mathbf{I};$$

2.
$$\mathbf{A}^{\mathrm{T}}\mathbf{H(a)} = \mathbf{0}.$$

The second of these equations shows that the $N - p$ linearly independent rows of $\mathbf{A}^{\mathrm{T}}$ are orthogonal to the p linearly independent columns of $\mathbf{H(a)}$. Therefore the columns of $\mathbf{A}$ span the subspace that is orthogonal to the subspace $\langle \mathbf{H(a)} \rangle$. The projector onto the orthogonal subspace $\langle \mathbf{A} \rangle$ is

$$\begin{aligned} \mathbf{P}_A &= \mathbf{I} - \mathbf{P(a)} \\ &= \mathbf{A(A^TA)^{-1}A^T}. \end{aligned} \qquad 6.177$$

This fundamental characterization of $\mathbf{P}_A$ (and of $\mathbf{P(a)} = \mathbf{I} - \mathbf{P}_A$) permits us to say that the maximum likelihood estimate of $\mathbf{a}$ is

$$\hat{\mathbf{a}} = \arg\min_{\mathbf{a}} \mathbf{x}^{\mathrm{T}}\mathbf{P}_A\mathbf{x}$$

$$\mathbf{P}_A = \mathbf{A(A^TA)^{-1}A^T} \qquad 6.178$$

$$\mathbf{A} = \begin{bmatrix} a_p & \cdots & a_1 & 1 & & \mathbf{0} \\ & \ddots & & \ddots & \ddots & \\ \mathbf{0} & & a_0 & \cdots & a_1 & 1 \end{bmatrix} \qquad \text{prediction error matrix.}$$

Interpretations

Before we study ways to solve this nonlinear minimization problem, we observe several things about the solution. First, the term $\mathbf{A}^{\mathrm{T}}\mathbf{x}$ may be written as

$$\mathbf{A}^{\mathrm{T}}\mathbf{x} = \mathbf{A}^{\mathrm{T}}\mathbf{h} + \mathbf{A}^{\mathrm{T}}\mathbf{n} = \mathbf{A}^{\mathrm{T}}\mathbf{n}. \qquad 6.179$$

The tth term in this equation is

$$\sum_{n=0}^{p} a_n x_{t-n} = \sum_{n=0}^{p} a_n n_{t-n}. \qquad 6.180$$

The terms on the left are residuals that would be zero if there were no noise. The terms on the right are colored noises whose covariance matrix is

$$E\mathbf{A}^{\mathrm{T}}\mathbf{n}\mathbf{n}^{\mathrm{T}}\mathbf{A} = \sigma^2\mathbf{A}^{\mathrm{T}}\mathbf{A}. \qquad 6.181$$

So the solution to the maximum likelihood equation is equivalent to the maximum likelihood estimate of **a** in the linear statistical model

$$\mathbf{A}^T\mathbf{x} = \mathbf{0}. \tag{6.182}$$

As $\mathbf{A}^T\mathbf{x}$ is distributed as $\mathbf{A}^T\mathbf{x} : N[\mathbf{0}, \sigma^2\mathbf{A}^T\mathbf{A}]$, the negative of log likelihood is (ignoring constants)

$$\begin{aligned}
-\ln f_\mathbf{a}(\mathbf{A}^T\mathbf{x}) &= (\mathbf{A}^T\mathbf{x})^T(\mathbf{A}^T\mathbf{A})^{-1}(\mathbf{A}^T\mathbf{x}) \\
&= \mathbf{x}^T\mathbf{A}(\mathbf{A}^T\mathbf{A})^{-1}\mathbf{A}^T\mathbf{x} = \mathbf{x}^T\mathbf{P}_A\mathbf{x}.
\end{aligned} \tag{6.183}$$

This shows the equivalence of the two problems.

KiSS

The "prediction errors"

$$\mathbf{A}^T\mathbf{x} = \begin{bmatrix} a_p & \cdots & a_1 & 1 & & \mathbf{0} \\ & \ddots & & \ddots & \ddots & \\ \mathbf{0} & & a_p & \cdots & a_1 & 1 \end{bmatrix} \begin{bmatrix} x_0 \\ x_1 \\ \vdots \\ x_{N-1} \end{bmatrix} \tag{6.184}$$

may actually be rewritten as the matrix equation

$$\begin{aligned}
\mathbf{A}^T\mathbf{x} &= \begin{bmatrix} x_0 & x_1 & \cdots & x_p \\ x_1 & \cdots & x_p & x_{p+1} \\ \vdots & & \ddots & \\ x_p & & & \\ \vdots & & & \vdots \\ x_{N-1-p} & & \cdots & x_{N-1} \end{bmatrix} \begin{bmatrix} a_p \\ \vdots \\ a_1 \\ 1 \end{bmatrix} \\
&= \mathbf{X}^T\mathbf{a}.
\end{aligned} \tag{6.185}$$

In the absence of noise, the matrix $\mathbf{X}$ would be a Hankel matrix of impulse responses. With this identity we can write negative log likelihood as

$$\mathbf{a}^T\mathbf{X}(\mathbf{A}^T\mathbf{A})^{-1}\mathbf{X}^T\mathbf{a}. \tag{6.186}$$

Kumaresan, Scharf, and Shaw [1986] proposed the KiSS algorithm for minimizing this quadratic form using the iteration

$$\min_{\mathbf{a}_{n+1}} \mathbf{a}_{n+1}^T\mathbf{X}\left[\mathbf{A}^T(\mathbf{a}_n)\mathbf{A}(\mathbf{a}_n)\right]^{-1}\mathbf{X}^T\mathbf{a}_{n+1}. \tag{6.187}$$

In this recursion, the solution $\mathbf{a}_n$ is used to build $\mathbf{A}(\mathbf{a}_n)$, and the resulting Grammian $\mathbf{A}^T(\mathbf{a})\mathbf{A}(\mathbf{a})$ is inverted using the fast algorithm published in the same article. The data is stored in $\mathbf{X}$, and the resulting quadratic minimization problem is

$$\frac{\partial}{\partial \mathbf{a}_{n+1}} \left[\mathbf{a}_{n+1}^T \mathbf{Q}_n \mathbf{a}_{n+1} + 2\lambda(\mathbf{a}_{n+1}^T \boldsymbol{\delta} - 1) \right] = \mathbf{0}. \tag{6.188}$$

When this equation is solved for $\mathbf{a}_{n+1}$, the result is

$$\mathbf{a}_{n+1} = \frac{1}{\boldsymbol{\delta}^T \mathbf{Q}_n^{-1} \boldsymbol{\delta}} \mathbf{Q}_n^{-1} \boldsymbol{\delta} \tag{6.189}$$

$$\mathbf{Q}_n = \mathbf{X}\left[\mathbf{A}^T(\mathbf{a}_n)\mathbf{A}(\mathbf{a}_n) \right]^{-1} \mathbf{X}^T; \qquad \boldsymbol{\delta} = \begin{bmatrix} 0 \\ \vdots \\ 0 \\ 1 \end{bmatrix}$$

This system of linear equations is solved at each step of the iteration. In McClellan [1990] it is shown that the KiSS algorithm, and the IQML algorithm of Bressler and Macovski [1986], are equivalent to the Steiglitz-McBride [1965] algorithm.

The Fisher Information Matrix

The Fisher information matrix for ARMA identification is obtained by replacing $\boldsymbol{\theta}$ with $\mathbf{b}$ and $\boldsymbol{\omega}$ with $\mathbf{a}$ in the Fisher information matrix of Equation 6.158:

$$\mathbf{J}(\mathbf{b}, \mathbf{a}) = \begin{bmatrix} \{ \mathbf{e}_i^T \mathbf{H}^T \mathbf{H} \mathbf{e}_j \} & \left\{ \mathbf{e}_i^T \mathbf{H}^T \dfrac{\partial \mathbf{H}}{\partial a_j} \mathbf{b} \right\} \\ \left\{ \mathbf{b}^T \dfrac{\partial \mathbf{H}^T}{\partial a_j} \mathbf{H} \mathbf{e}_i \right\} & \left\{ \mathbf{b}^T \dfrac{\partial \mathbf{H}^T}{\partial a_i} \dfrac{\partial \mathbf{H}}{\partial a_j} \mathbf{b} \right\} \end{bmatrix}. \tag{6.190}$$

From the identity $\mathbf{KH} = [\mathbf{I}^T \quad \mathbf{0}^T]^T$, we derive the following formula for $\partial \mathbf{H}/\partial a_i$:

$$\frac{\partial}{\partial a_i} \mathbf{KH} = \frac{\partial \mathbf{K}}{\partial a_i} \mathbf{H} + \mathbf{K} \frac{\partial \mathbf{H}}{\partial a_i}$$

$$= \begin{bmatrix} \mathbf{0} \\ \mathbf{0} \end{bmatrix} \longrightarrow \frac{\partial \mathbf{H}}{\partial a_i} = -\mathbf{K}^{-1} \frac{\partial \mathbf{K}}{\partial a_i} \mathbf{H} = -\mathbf{K}^{-1} \frac{\partial \mathbf{K}}{\partial a_i} \mathbf{K}^{-1} \begin{bmatrix} \mathbf{I} \\ \mathbf{0} \end{bmatrix}. \tag{6.191}$$

The terms of the Fisher matrix involving $\partial \mathbf{H}/\partial a_j$ may be written

$$\mathbf{e}_i^T \mathbf{H}^T \frac{\partial \mathbf{H}}{\partial a_j} \mathbf{b} = -\mathbf{e}_i^T \left[\mathbf{I}^T \mathbf{0}^T \right] \mathbf{K}^{-T} \mathbf{K}^{-1} \frac{\partial \mathbf{K}}{\partial a_j} \mathbf{K}^{-1} \begin{bmatrix} \mathbf{b} \\ \mathbf{0} \end{bmatrix};$$

$$\mathbf{b}^T \frac{\partial \mathbf{H}^T}{\partial a_i} \frac{\partial \mathbf{H}}{\partial a_j} \mathbf{b} = \left[\mathbf{b}^T \mathbf{0}^T \right] \mathbf{K}^{-T} \frac{\partial \mathbf{K}^T}{\partial a_i} \mathbf{K}^{-T} \mathbf{K}^{-1} \frac{\partial \mathbf{K}}{\partial a_j} \mathbf{K}^{-1} \begin{bmatrix} \mathbf{b} \\ \mathbf{0} \end{bmatrix}.$$

In these formulas, the matrix $\mathbf{K}^{-1}$ is the Toeplitz impulse response matrix

$$\mathbf{K}^{-1} = \begin{bmatrix} 1 & & & & \\ g_1 & 1 & & \mathbf{0} & \\ g_2 & g_1 & 1 & & \\ \vdots & & & \ddots & \\ g_{N-1} & \cdots & g_2 & g_1 & 1 \end{bmatrix}, \qquad 6.192$$

where $\{g_n\}_0^{N-1}$ is the impulse response of the purely autoregressive filter $1/A(z)$:

$$\sum_{n=0}^{p} a_n g_{t-n} = \delta_t; \qquad a_0 = 1. \qquad 6.193$$

Check it:

$$\begin{bmatrix} 1 & & & & & \\ a_1 & 1 & & & \mathbf{0} & \\ \vdots & & \ddots & & & \\ a_{p-1} & & & 1 & & \\ a_p & & & a_1 & 1 & \\ & \ddots & & & \ddots & \\ \mathbf{0} & & a_p & \cdots & a_1 & 1 \end{bmatrix} \begin{bmatrix} 1 & & & & \\ g_1 & 1 & & \mathbf{0} & \\ & \ddots & \ddots & & \\ & & & & \\ g_{N-1} & & \cdots & g_1 & 1 \end{bmatrix}$$

$$= \begin{bmatrix} 1 & & & & \\ & 1 & & \mathbf{0} & \\ & & 1 & & \\ & & & \ddots & \\ \mathbf{0} & & & 1 & \\ & & & & 1 \end{bmatrix}. \qquad 6.194$$

Therefore, our algorithm for generating the Fisher information matrix is

1. compute $\{g_t\}$ from the recursion $A(z)\{g_t\} = \{\delta_t\}$;
2. compute

$$\partial \mathbf{K}/\partial a_j = \begin{bmatrix} & & & & \mathbf{0} \\ 1 & & & & \\ & 1 & & & \\ \mathbf{0} & & \ddots & & \\ & & & & 1 \end{bmatrix},$$

a matrix with a single diagonal of ones on the ith diagonal below the main diagonal;

3. compute $\mathbf{J}(\mathbf{b}, \mathbf{a})$:

$$\mathbf{J}(\mathbf{b}, \mathbf{a}) = \begin{bmatrix} \left\{ [\mathbf{e}_i^T \ \mathbf{0}^T] \mathbf{K}^{-T} \mathbf{K}^{-1} \begin{bmatrix} \mathbf{e}_j \\ \mathbf{0} \end{bmatrix} \right\} & \left\{ [-\mathbf{e}_i^T \ \mathbf{0}^T] \mathbf{K}^{-T} \mathbf{K}^{-1} \frac{\partial \mathbf{K}}{\partial a_j} \mathbf{K}^{-1} \begin{bmatrix} \mathbf{b} \\ \mathbf{0} \end{bmatrix} \right\} \\ \left\{ [\mathbf{b}^T \ \mathbf{0}^T] \mathbf{K}^{-T} \frac{\partial \mathbf{K}^T}{\partial a_j} \mathbf{K}^{-T} \mathbf{K}^{-1} \begin{bmatrix} \mathbf{e}_i \\ \mathbf{0} \end{bmatrix} \right\} & \left\{ [\mathbf{b}^T \ \mathbf{0}^T] \mathbf{K}^{-T} \frac{\partial \mathbf{K}}{\partial a_i} \mathbf{K}^{-T} \mathbf{K}^{-1} \frac{\partial \mathbf{K}}{\partial a_j} \mathbf{K}^{-1} \begin{bmatrix} \mathbf{b} \\ \mathbf{0} \end{bmatrix} \right\} \end{bmatrix}.$$

Mode Identification

In many cases it is the modes of the unit pulse response that are of interest—not the ARMA difference equation. That is, the solution for h_t may be written as

$$h_t = \sum_{n=1}^{p} c_n z_n^t, \qquad\qquad 6.195$$

where the modes z_n^t are constructed from the roots of $A(z)$, and the residues c_n are constructed from a partial fraction expansion:

$$A(z_n) = 0 \qquad\qquad 6.196$$

$$\frac{B(z)}{A(z)} = \sum_{n=1}^{p} \frac{c_n}{1 - z_n z^{-1}}; \qquad c_n = (1 - z_n z^{-1}) \frac{B(z)}{A(z)} \Big|_{z=z_n}.$$

The corresponding model for the unit pulse response vector is

$$\mathbf{h} = \mathbf{V}\mathbf{c} \qquad\qquad 6.197$$

$$\mathbf{V} = \begin{bmatrix} 1 & \cdots & 1 \\ z_1 & & z_p \\ \vdots & & \vdots \\ z_1^{N-1} & \cdots & z_p^{N-1} \end{bmatrix}; \qquad \mathbf{c} = \begin{bmatrix} c_1 \\ c_2 \\ \vdots \\ c_p \end{bmatrix}.$$

From the equation $\mathbf{K}\mathbf{h} = [\mathbf{b}^T \ \ \mathbf{0}^T]^T$ we establish this connection between the ARMA model and the modal model:

$$\mathbf{K}\mathbf{V}\mathbf{c} = \begin{bmatrix} \mathbf{b} \\ \mathbf{0} \end{bmatrix}. \qquad\qquad 6.198$$

The second connection we shall need is the connection between the roots z_n and the AR coefficients $\mathbf{a} = (a_p \ \ \cdots \ \ a_1 \ \ 1)$:

$$\prod_{n=1}^{p} (1 - z_n z^{-1}) = \sum_{n=0}^{p} a_n z^{-n}; \qquad a_0 = 1. \qquad\qquad 6.199$$

The Fisher information matrix for the roots $\mathbf{z} = (z_1 \quad z_2 \quad \cdots \quad z_p)$ and the residues $\mathbf{c} = [c_1 \quad \cdots \quad c_p]^T$ is denoted $J(\mathbf{c}, \mathbf{z})$. From our results for the CR bound, we may write this matrix as

$$\mathbf{J(c, z)} = \mathbf{GJ(b, a)G^T}, \qquad\qquad 6.200$$

where $\mathbf{G}$ is the matrix

$$\mathbf{G} = \begin{bmatrix} \dfrac{\partial \mathbf{b}^T}{\partial \mathbf{c}} & \dfrac{\partial \mathbf{a}^T}{\partial \mathbf{c}} \\[2ex] \dfrac{\partial \mathbf{b}^T}{\partial \mathbf{z}} & \dfrac{\partial \mathbf{a}^T}{\partial \mathbf{z}} \end{bmatrix}. \qquad\qquad 6.201$$

The corresponding dependence of $\mathbf{J(b, a)}$ on $\mathbf{J(c, z)}$ is

$$\mathbf{J(b, a)} = \mathbf{HJ(c, z)H^T}, \qquad\qquad 6.202$$

where

$$\mathbf{H} = \begin{bmatrix} \dfrac{\partial \mathbf{c}^T}{\partial \mathbf{b}} & \dfrac{\partial \mathbf{z}^T}{\partial \mathbf{b}} \\[2ex] \dfrac{\partial \mathbf{c}^T}{\partial \mathbf{a}} & \dfrac{\partial \mathbf{z}^T}{\partial \mathbf{a}} \end{bmatrix}. \qquad\qquad 6.203$$

Let's evaluate the *sensitivity* matrix $\mathbf{G}$. The northeast term $\partial \mathbf{a}^T/\partial \mathbf{c}$ is $\mathbf{0}$. The northwest term $\partial \mathbf{b}^T/\partial \mathbf{c}$ is obtained from Equation 6.196:

$$\frac{\partial B(z)}{\partial c_j} = \frac{\partial}{\partial c_j} A(z) \sum_{n=1}^{p} \frac{c_n}{1 - z_n z^{-1}} = \frac{A(z)}{1 - z_j z^{-1}} \qquad\qquad 6.204$$

$$\Rightarrow \frac{\partial b_i}{\partial c_j} : i\text{th coefficient of } \frac{A(z)}{1 - z_j z^{-1}}.$$

The southeast term is obtained from the equation for $A(z)$ in Equation 6.199:

$$\frac{\partial}{\partial z_j} A(z) = -z^{-1} \frac{A(z)}{1 - z_j z^{-1}}$$

$$\Rightarrow \frac{\partial a_{i+1}}{\partial z_j} = -\frac{\partial b_i}{\partial c_j} \qquad\qquad 6.205$$

The southwest term is obtained from the equation for $B(z)$ given in Equation 6.196:

$$\frac{\partial}{\partial z_j} B(z) = \frac{\partial}{\partial z_j} \cdot \sum_{n=1}^{p} c_n \prod_{i \neq n} (1 - z_i z^{-1})$$

$$= \sum_{n \neq j} c_n \left(-z^{-1} \frac{A(z)}{1 - z_n z^{-1}} \right) \qquad\qquad 6.206$$

$$\Rightarrow \frac{\partial b_i}{\partial z_j} = \sum_{n \neq j} c_n \frac{\partial a_i}{\partial z_n} = -\sum_{n \neq j} c_n \frac{\partial b_{i-1}}{\partial z_n}$$

These formulas characterize the Fisher information matrix for mode identification. It remains only to program them for any particular problem. When the modes z_i, and the mode weights c_i, appear in complex conjugate pairs, then these formal partials with respect to complex numbers may not be used. Rather we must differentiate with respect to real and imaginary parts.

6.11   MAXIMUM LIKELIHOOD IDENTIFICATION OF A SIGNAL SUBSPACE

What if the subspace $\mathbf{H}$ in the linear statistical model is unknown and there is no parametric description to guide us as in Sections 6.9 and 6.10? Is there a maximum likelihood principle that would allow us to proceed? The answer is yes. To illustrate, suppose we take a sequence of measurements of the form $\{\mathbf{x}_t\}_0^{M-1}$. Each measurement $\mathbf{x}_t$ is an $N \times 1$ vector of the form

$$\mathbf{x}_t = \mathbf{s}_t + \mathbf{n}_t. \qquad\qquad 6.207$$

The signal vectors $\mathbf{s}_t$ are unknown, but they are known to lie in a p-dimensional subspace, meaning $\mathbf{s}_t = \mathbf{H}\boldsymbol{\theta}_t$. That is, there exist $N - p$ linearly independent vectors $\mathbf{a}_i$ such that

$$\mathbf{a}_i^{\mathsf{T}} \mathbf{s}_t = 0; \qquad i = p+1, \ldots, N, \quad t = 0, 1, \ldots, M - 1. \qquad 6.208$$

This condition is illustrated in Figure 6.10.

The question we pose is, "is there a maximum likelihood criterion for estimating the signal subspace $\mathbf{H}$ or, equivalently, the orthogonal subspace $\mathbf{A} = [\mathbf{a}_{p+1} \ldots \mathbf{a}_N]$?" The answer is yes, and to obtain it we proceed as follows.

Assume that the observation noises $\{\mathbf{n}_t\}_0^{M-1}$ are drawn independently from an $N[\mathbf{0}, \sigma^2 \mathbf{I}]$ distribution. Then the log-likelihood function for $\{\mathbf{x}_t\}_0^{M-1}$ is (ignoring constants)

$$L = -\frac{1}{2\sigma^2} \sum_{t=0}^{M-1} (\mathbf{x}_t - \mathbf{s}_t)^{\mathsf{T}} (\mathbf{x}_t - \mathbf{s}_t). \qquad\qquad 6.209$$

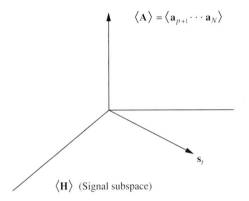

Figure 6.10 Signal subspace.

A maximum likelihood principle for identifying the orthogonal subspace $\mathbf{A}$ will maximize log likelihood under the constraints

$$\mathbf{a}_i^{\mathrm{T}} \mathbf{s}_t = 0; \qquad i = p + 1, \ldots, N, \quad t = 0, 1, \ldots, M - 1. \qquad 6.210$$

Therefore, let's form a Lagrangian for minimizing $-2\sigma^2 L$ under these constraints:

$$\mathcal{L} = \sum_{t=0}^{M-1} (\mathbf{x}_t - \mathbf{s}_t)^{\mathrm{T}} (\mathbf{x}_t - \mathbf{s}_t) + 2 \sum_{i=p+1}^{N} \sum_{t=0}^{M-1} \lambda_{ti} \mathbf{a}_i^{\mathrm{T}} \mathbf{s}_t. \qquad 6.211$$

Let's assume for the moment that $\mathbf{A} = [\mathbf{a}_{p+1} \cdots \mathbf{a}_N]$ is known. Then the gradient of $\mathcal{L}$ with respect to $\mathbf{s}_t$ is

$$\nabla_{\mathbf{s}_t} \mathcal{L} = -2(\mathbf{x}_t - \mathbf{s}_t) + 2 \sum_{i=p+1}^{N} \lambda_{ti} \mathbf{a}_i. \qquad 6.212$$

The candidate solution for $\mathbf{s}_t$ is

$$\mathbf{s}_t = \mathbf{x}_t - \sum_{i=p+1}^{N} \lambda_{ti} \mathbf{a}_i \qquad 6.213$$

$$= \mathbf{x}_t - \mathbf{A}\boldsymbol{\lambda}_t$$

$$\mathbf{A} = [\mathbf{a}_{p+1} \quad \cdots \quad \mathbf{a}_N]$$

$$\boldsymbol{\lambda}_t = [\lambda_{t(p+1)} \quad \cdots \quad \lambda_{tN}]^{\mathrm{T}}.$$

The matrix $\mathbf{A}$ contains the vectors $\mathbf{a}_i$ for its columns, and the vector $\boldsymbol{\lambda}_t$ contains the Lagrangians λ_{ti} for its elements.

The constraints are enforced by writing

$$\mathbf{A}^{\mathrm{T}}(\mathbf{x}_t - \mathbf{A}\boldsymbol{\lambda}_t) = \mathbf{0} \qquad 6.214$$

and solving for $\boldsymbol{\lambda}_t$:

$$\boldsymbol{\lambda}_t = (\mathbf{A}^T\mathbf{A})^{-1}\mathbf{A}^T\mathbf{x}_t. \qquad 6.215$$

The corresponding solution for $\mathbf{s}_t$ is

$$\begin{aligned} \mathbf{s}_t &= \left[\mathbf{I} - \mathbf{A}(\mathbf{A}^T\mathbf{A})^{-1}\mathbf{A}^T\right]\mathbf{x}_t \\ &= (\mathbf{I} - \mathbf{P}_A)\mathbf{x}_t. \end{aligned} \qquad 6.216$$

The resulting maximum likelihood is

$$\begin{aligned} L &= -\frac{1}{2\sigma^2}\sum_{t=0}^{M-1}\mathbf{x}_t^T\mathbf{P}_A\mathbf{x}_t \\ &= -\frac{M}{2\sigma^2}\,\mathrm{tr}\,[\mathbf{P}_A\mathbf{S}] \end{aligned} \qquad 6.217$$

where $\mathbf{S}$ is the sample correlation matrix:

$$\mathbf{S} = \frac{1}{M}\sum_{t=0}^{M-1}\mathbf{x}_t\mathbf{x}_t^T. \qquad 6.218$$

At this point in our development, $\hat{\mathbf{s}}_t = (\mathbf{I} - \mathbf{P}_A)\mathbf{x}_t$ represents the maximum likelihood solution for $\mathbf{s}_t$ when the subspace $\mathbf{A}$, or equivalently a subspace $\mathbf{H} = [\mathbf{h}_1 \quad \cdots \quad \mathbf{h}_p]$, is known. This is of great interest in its own right. However, what if $\mathbf{A}$ is unknown? Then we must further maximize likelihood with respect to $\mathbf{A}$. Let the sample covariance matrix $\mathbf{S}$ have the orthogonal decomposition

$$\mathbf{S} = \mathbf{U}\mathbf{\Lambda}^2\mathbf{U}^T$$

$$\mathbf{U} = [\mathbf{u}_1 \quad \cdots \quad \mathbf{u}_{p+1} \quad \cdots \quad \mathbf{u}_N] \qquad 6.219$$

$$\mathbf{\Lambda}^2 = \mathrm{diag}[\lambda_1^2 \quad \cdots \quad \lambda_{p+1}^2 \quad \cdots \quad \lambda_N^2]; \quad \lambda_1^2 \geq \lambda_2^2 \geq \cdots \geq \lambda_N^2.$$

Then it is a simple matter to show that L is bounded as follows (see Problem 6.14):

$$\begin{aligned} L &= -\frac{M}{2\sigma^2}\,\mathrm{tr}\,[\mathbf{P}_A\mathbf{S}] \\ &\leq -\frac{M}{2\sigma^2}\sum_{i=p+1}^{N}\lambda_i^2 \end{aligned} \qquad 6.220$$

for any rank $N - p$ projector $\mathbf{P}_A$. This bound is achieved for a projector $\mathbf{P}_A$ onto the subspace $\langle\mathbf{U}_{p+1}\rangle$:

$$\mathbf{P}_A = \mathbf{U}_{p+1}\mathbf{U}_{p+1}^T = \sum_{i=p+1}^{N}\mathbf{u}_i\mathbf{u}_i^T \qquad 6.221$$

$$\mathbf{U}_{p+1} = [\mathbf{u}_{p+1} \quad \cdots \quad \mathbf{u}_N].$$

In summary, maximum likelihood theory builds a rank p subspace $\langle \mathbf{H} \rangle$ from the dominant p eigenvectors of the sample covariance matrix, and it builds a rank $N - p$ subspace $\langle \mathbf{A} \rangle$ from the subdominant eigenvectors. This is illustrated in Figure 6.11. It is interesting to note that maximum likelihood theory has constructed the following decomposition of the sample covariance matrix:

$$
\begin{aligned}
\mathbf{S} &= \frac{1}{M} \sum_{t=0}^{M-1} \mathbf{x}_t \mathbf{x}_t^{\mathrm{T}} \\
&= \frac{1}{M} \sum_{t=0}^{M-1} (\mathbf{I} - \mathbf{P}_A)\mathbf{x}_t \mathbf{x}_t^{\mathrm{T}}(\mathbf{I} - \mathbf{P}_A) + \frac{1}{M} \sum_{t=0}^{M-1} \mathbf{P}_A \mathbf{x}_t \mathbf{x}_t^{\mathrm{T}} \mathbf{P}_A \\
&= (\mathbf{I} - \mathbf{P}_A)\mathbf{S}(\mathbf{I} - \mathbf{P}_A) + \mathbf{P}_A \mathbf{S} \mathbf{P}_A .
\end{aligned}
\tag{6.222}
$$

In this decomposition, $\mathbf{P}_A \mathbf{S} \mathbf{P}_A$ is the estimated "noise" covariance matrix and $(\mathbf{I} - \mathbf{P}_A)\mathbf{S}(\mathbf{I} - \mathbf{P}_A)$ is the estimated "signal" covariance matrix.

Example 6.15 (Rank $N - 1$ Subspace)

When the signal subspace is known to have rank $N - 1$, meaning that the orthogonal subspace $\mathbf{A} = \mathbf{a}_N$ has rank 1, then $\mathbf{a}_N$ is the eigenvector of $\mathbf{S}$ with minimum eigenvalue λ_N. The maximum value of log-likelihood is

$$
L = -\frac{M}{2\sigma^2} \lambda_N^2 .
\tag{6.223}
$$

∎

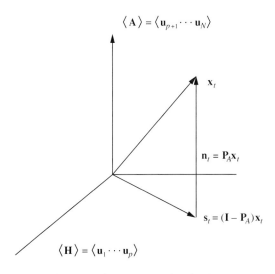

Figure 6.11 Building a rank p signal subspace.

6.12   ESTIMATION OF PARAMETERS IN SINUSOIDAL MODELS

We now turn to an important problem in signal processing: Estimation of amplitude, phase, and frequency in a sinusoidal signal model. We assume that the following measurements are taken:

$$x_t = s_t + n_t; \qquad t = 0, 1, \ldots, N - 1 \qquad\qquad 6.224$$

$$s_t = A\cos(\omega t - \phi); \qquad A > 0.$$

The signal parameters (A, ϕ, ω) are unknown. The noise terms are a sequence of independent, identically distributed $N[0, \sigma^2]$ random variables, so the measurements x_t are a sequence of i.i.d. $N[s_t, \sigma^2]$ random variables. The noise variance σ^2 is assumed unknown.

The input, or per sample, signal-to-noise ratio is

$$\mathrm{SNR_{in}} = \frac{\frac{1}{N}\sum\limits_{t=0}^{N-1} s_t^2}{\sigma^2} \cong \frac{A^2}{2\sigma^2} \quad \text{(large } N\text{)}. \qquad\qquad 6.225$$

The output signal-to-noise ratio is $\mathrm{SNR} = N(\mathrm{SNR})_{\mathrm{in}}$. We shall use the approximation $\mathrm{SNR_{in}} = A^2/(2\sigma^2)$ throughout our development.

Joint Density

The joint density function for the random sample $\mathbf{x} = (x_0, x_1, \ldots, x_{N-1})$ is the product density

$$f_{\boldsymbol{\theta}}(\mathbf{x}) = \prod_{t=0}^{N-1} f_{\boldsymbol{\theta}}(x_t) = (2\pi\sigma^2)^{-N/2} \exp\left\{ -\frac{1}{2\sigma^2} \sum_{t=0}^{N-1} (x_t - s_t)^2 \right\}$$

$$s_t = A\cos(\omega t - \phi) \qquad\qquad 6.226$$

$$\boldsymbol{\theta} = (A, \phi, \omega, \sigma^2).$$

The problem is to find the maximum likelihood estimate of $\boldsymbol{\theta}$.

Likelihood

The log-likelihood of $\boldsymbol{\theta}$ is the function $ln f_{\boldsymbol{\theta}}(\mathbf{x})$ evaluated at the measurement, or sample, $\mathbf{x}$:

$$L(\boldsymbol{\theta}, \mathbf{x}) = ln f_{\boldsymbol{\theta}}(\mathbf{x}); \qquad \mathbf{x} = (x_0, x_1, \ldots, x_{N-1}). \qquad\qquad 6.227$$

Let's maximize log-likelihood first with respect to σ^2:

$$\frac{\partial}{\partial\sigma^2} L(\boldsymbol{\theta}, \mathbf{x}) = -\frac{N}{2\sigma^2} + \frac{1}{2\sigma^4} \sum_{t=0}^{N-1} (x_t - s_t)^2 = 0.$$

So the maximum likelihood estimate of σ^2 is

$$\hat{\sigma}^2 = \frac{1}{N} v^2 \qquad\qquad 6.228$$

$$v^2 = \sum_{t=0}^{N-1} (x_t - s_t)^2.$$

The result says that the maximum likelihood estimate of the noise variance is the average squared residual between the observations x_t and the model s_t. When likelihood is jointly maximized with respect to (A, ϕ, ω) as well, then $\hat{\sigma}^2$ becomes the minimum average squared residuals, as we now show.

Substitute the expression for $\hat{\sigma}^2$ into likelihood:

$$\hat{l}(\boldsymbol{\theta}, \mathbf{x}) = (2\pi v^2 / N)^{-N/2} \exp\{-N/2\}. \qquad\qquad 6.229$$

This is the compressed likelihood we studied in Section 6.7. It shows that maximization of likelihood with respect to $(A, \phi, \omega, \sigma^2)$ reduces to minimization of the squared residuals v^2. From here on, we consider minimization of v^2 with respect to (A, ϕ, ω).

Approximation of v^2

The squared residuals may be approximated as follows:

$$v^2 = \sum_{t=0}^{N-1} x_t^2 - 2A \sum_{t=0}^{N-1} x_t \cos(\omega t - \phi) + A^2 \sum_{t=0}^{N-1} \cos^2(\omega t - \phi)$$

$$\cong \sum_{t=0}^{N-1} x_t^2 - 2A \sum_{t=0}^{N-1} x_t \cos(\omega t - \phi) + \frac{A^2 N}{2} \qquad\qquad 6.230$$

$$= \sum_{t=0}^{N-1} x_t^2 - 2 \operatorname{Re} A e^{j\phi} \sum_{t=0}^{N-1} x_t e^{-j\omega t} + \frac{A^2 N}{2}.$$

Minimization with respect to ϕ

Differentiate with respect to phase to obtain the regression equation

$$\frac{\partial}{\partial \phi} \operatorname{Re}\left\{ e^{j\phi} \sum_{t=0}^{N-1} x_t e^{-j\omega t} \right\} = \operatorname{Re}\left\{ j e^{j\phi} \sum_{t=0}^{N-1} x_t e^{-j\omega t} \right\} = 0. \qquad 6.231$$

This means that everything to the right of j is real. So the maximum likelihood estimate of ϕ is

$$\hat{\phi} = -\arg X(\omega) \qquad\qquad 6.232$$

$$X(\omega) = \sum_{t=0}^{N-1} x_t e^{-j\omega t}.$$

Of course, $X(\omega)$ is just the discrete-time Fourier transform of the measurements $(x_0, \ldots, x_{N-1})$.

Minimization with Respect to A

Substitute the solution for $\hat{\phi}$ to further compress likelihood and differentiate to produce the regression equation

$$AN - 2|X(\omega)| = 0 \qquad\qquad 6.233$$

which produces the following maximum likelihood estimate of A:

$$\hat{A} = \frac{2}{N}|X(\omega)|. \qquad\qquad 6.234$$

Minimization with Respect to ω

Substitute the solution for $\hat{A}$ to obtain

$$v^2 = \sum_{t=0}^{N-1} x_t^2 - 2\frac{2}{N}|X(\omega)|^2 + \frac{4}{2N^2}|X(\omega)|^2 N$$

$$\qquad\qquad 6.235$$

$$= \sum_{t=0}^{N-1} x_t^2 - \frac{2}{N}|X(\omega)|^2.$$

This is minimized by maximizing $|X(\omega)|$. Therefore, the maximum likelihood estimate of ω is

$$\hat{\omega} = \arg\max_{\omega} |X(\omega)|^2. \qquad\qquad 6.236$$

In 1897, Schuster named $(1/N)|X(\omega)|^2$ the periodogram.

Summary and Algorithm

In summary, the maximum likelihood estimates of $(A, \phi, \omega, \sigma^2)$ are

$$\hat{\omega} = \arg\max_{\omega} |X(\omega)|^2$$

$$\hat{A} = \frac{2}{N}|X(\hat{\omega})| \qquad\qquad 6.237$$

$$\hat{\phi} = -\arg X(\hat{\omega})$$

$$\hat{\sigma}^2 = \frac{1}{N}\sum_{t=0}^{N-1}[x_t - \hat{A}\cos(\hat{\omega}t - \hat{\phi})]^2.$$

Here is an algorithm: Decide on an acceptable "bias" in your ability to resolve the maximum of $\left|X(\omega)\right|$. Call it $\Delta\omega$:

$$\Delta\omega = \frac{2\pi}{M} \qquad\qquad 6.238$$

$$M = \frac{2\pi}{\Delta\omega}.$$

Zeropad $(x_0 \quad \cdots \quad x_{N-1})$ to produce $(x_0, x_1, \ldots, x_{N-1}, x_N, \ldots, x_{M-1})$. Fast-Fourier transform it to produce the DFT variables

$$X_m = \sum_{t=0}^{M-1} x_t e^{-j(2\pi mt)/M} = \sum_{t=0}^{N-1} x_t e^{-j2\pi(m/M)t}. \qquad\qquad 6.239$$

Then the approximate maximum likelihood estimates are

$$\hat{m} = \arg\max_m \left|X_m\right|$$

$$\hat{\omega} = \frac{2\pi\hat{m}}{M}$$

$$\hat{A} = \frac{2}{N}\left|X_{\hat{m}}\right| \qquad\qquad 6.240$$

$$\hat{\phi} = -\arg X_{\hat{m}}$$

$$\hat{\sigma}^2 = \frac{1}{N}\sum_{t=0}^{N-1}[x_t - \hat{A}\cos(\hat{\omega}t - \hat{\phi})]^2.$$

Caution

Zeropadding only allows us to get within $2\pi/M$ of the maximum of $\left|X(\omega)\right|$. It does not improve the accuracy, or variance, of our estimator $\hat{\omega}$. The variance can only be improved by increasing N, the size of the measurement set. This brings us to the Cramer-Rao bound for this example.

Cramer-Rao Bounds

Let $\hat{\boldsymbol{\theta}}$ be any unbiased estimator of $\boldsymbol{\theta}$. The Cramer-Rao bound says that the error covariance matrix for $\hat{\boldsymbol{\theta}}$ is bounded as

$$\mathbf{C} = E(\hat{\boldsymbol{\theta}} - \boldsymbol{\theta})(\hat{\boldsymbol{\theta}} - \boldsymbol{\theta})^{\mathrm{T}} \geq \mathbf{J}^{-1}, \qquad\qquad 6.241$$

where $\mathbf{J}$ is the Fisher information matrix:

$$\mathbf{J} = [J_{ij}] \qquad\qquad 6.242$$

$$J_{ij} = -E\frac{\partial^2}{\partial\theta_i\partial\theta_j}\ln f_{\boldsymbol{\theta}}(\mathbf{x}).$$

In our case, $\boldsymbol{\theta} = (A, \phi, \omega, \sigma^2)$, so, for example, $\theta_3 = \omega$.

The natural logarithm of the random variable $f_\theta(\mathbf{x})$ is

$$\ln f_\theta(\mathbf{x}) = -\frac{N}{2} \ln \sigma^2 - \frac{1}{2\sigma^2} \sum_{t=0}^{N-1} (x_t - s_t)^2. \qquad 6.243$$

From this formula for log likelihood we may differentiate with respect to A, ϕ, ω, and σ^2 to compute the Fisher information matrix. Its inverse is the CR bound. See Problem 6.8.

6.13   STRUCTURED COVARIANCE MATRICES

In this section we study the problem of estimating parameters in a structured correlation matrix. We assume that the observed data consists of realizations from the sequence of i.i.d. normal random vectors $\{\mathbf{X}_m\}_1^M$. For each m, the N-vector $\mathbf{X}_m$ is distributed as $N[\mathbf{0}, \mathbf{R}(\boldsymbol{\theta})]$ where the covariance matrix $\mathbf{R}(\boldsymbol{\theta})$ is a function of the parameters $\boldsymbol{\theta}$.

The maximum likelihood estimate of $\boldsymbol{\theta}$ is obtained by maximizing the likelihood function for the realization $\mathbf{x} = \{\mathbf{x}_m\}_1^M$ with respect to $\boldsymbol{\theta}$:

$$\max_{\boldsymbol{\theta}} f_\theta(\mathbf{x})$$

$$f_\theta(\mathbf{x}) = (2\pi)^{-MN/2} (\det \mathbf{R})^{-M/2} \exp\left\{ -\frac{M}{2} \operatorname{tr}\left[\mathbf{R}^{-1}\mathbf{S}\right] \right\} \qquad 6.244$$

$$\mathbf{S} = \frac{1}{M} \sum_{m=1}^M \mathbf{x}_m \mathbf{x}_m^T \qquad \text{sample covariance.}$$

The sample covariance would be the maximum likelihood estimate of $\mathbf{R}$ if the maximum likelihood estimate were not constrained to be the structured estimate $\mathbf{R}(\hat{\boldsymbol{\theta}})$.

To find the maximum likelihood estimate of $\boldsymbol{\theta}$, we construct the derivative of log likelihood with respect to θ_n. The details are delicate, but the procedure is straightforward:

$$\begin{aligned}
\frac{\partial}{\partial \theta_n} \ln f_\theta(\mathbf{x}) &= -\frac{M}{2} \frac{\partial}{\partial \theta_n} \ln \det \mathbf{R} - \frac{M}{2} \frac{\partial}{\partial \theta_n} \operatorname{tr}\left[\mathbf{R}^{-1}\mathbf{S}\right] \\
&= -\frac{M}{2} \sum_{ij} \frac{\partial}{\partial r_{ij}} \ln \det \mathbf{R} \frac{\partial r_{ij}}{\partial \theta_n} - \frac{M}{2} \sum_{ij} \frac{\partial}{\partial r_{ij}} \operatorname{tr}\left[\mathbf{R}^{-1}\mathbf{S} \frac{\partial r_{ij}}{\partial \theta_n}\right] \\
&= -\frac{M}{2} \operatorname{tr}\left[\mathbf{R}^{-1} \frac{\partial \mathbf{R}}{\partial \theta_n}\right] + \frac{M}{2} \operatorname{tr}\left[\mathbf{R}^{-1}\mathbf{S}\mathbf{R}^{-1} \frac{\partial \mathbf{R}}{\partial \theta_n}\right].
\end{aligned}$$

When this is equated to zero, we obtain the regression equation for the maximum likelihood estimate of θ_n:

$$\operatorname{tr}\left[\mathbf{R}^{-1}(\mathbf{R} - \mathbf{S})\mathbf{R}^{-1} \frac{\partial \mathbf{R}}{\partial \theta_n}\right] = 0. \qquad 6.245$$

Recall from our discussion of Cramer-Rao bounds that the inverse of the Fisher information matrix lower bounds the covariance matrix for any unbiased estimator. The Fisher information matrix for this problem is

$$\mathbf{J} = [J_{ij}]$$

$$J_{ij} = -E \frac{\partial^2}{\partial \theta_i \partial \theta_j} \ln f_{\boldsymbol{\theta}}(\mathbf{X})$$

$$= \frac{M}{2} \operatorname{tr}\left[\mathbf{R}^{-1} \frac{\partial \mathbf{R}}{\partial \theta_i} \mathbf{R}^{-1} \frac{\partial \mathbf{R}}{\partial \theta_j} \right].$$

6.246

Linear Structure

When the correlation matrix $\mathbf{R}(\boldsymbol{\theta})$ has linear structure, then these general results specialize rather nicely. To illustrate, we consider the model

$$\mathbf{R} = \sum_{i=1}^{p} \theta_i \mathbf{Q}_i,$$

6.247

with $\{\mathbf{Q}_i\}_1^p$ known matrices and $\{\theta_i\}_1^p$ unknown coefficients. The regression equation for θ_i is

$$\operatorname{tr}\left[\mathbf{R}^{-1}(\mathbf{R} - \mathbf{S})\mathbf{R}^{-1}\mathbf{Q}_i \right] = 0,$$

6.248

and the ij term of the Fisher information matrix is

$$J_{ij} = \frac{M}{2} \operatorname{tr}\left[\mathbf{R}^{-1}\mathbf{Q}_i\mathbf{R}^{-1}\mathbf{Q}_j \right].$$

6.249

We now focus our attention on concrete applications of these results. In each application, the data is a random sample $\{\mathbf{X}_m\}_1^M$ of $N[\mathbf{0}, \mathbf{R}(\boldsymbol{\theta})]$ random vectors.

Low-Rank Orthogonal Correlation Matrix

Let $\mathbf{U}$ denote an orthogonal matrix with columns $\mathbf{u}_n$:

$$\mathbf{U} = [\mathbf{u}_1 \quad \cdots \quad \mathbf{u}_p \mathbf{u}_{p+1} \quad \cdots \quad \mathbf{u}_N] = [\mathbf{U}_p | \mathbf{U}_{N-p}]$$

6.250

$$\mathbf{U}^{\mathrm{T}}\mathbf{U} = \mathbf{U}\mathbf{U}^{\mathrm{T}} = \mathbf{I}.$$

The first p vectors of $\mathbf{U}$ span a p-dimensional subspace, and the remaining $N - p$

vectors span the orthogonal $(N-p)$-dimensional subspace. The projections onto these respective subspaces are

$$\mathbf{P} = \mathbf{U}_p\mathbf{U}_p^{\mathrm{T}} = \sum_{i=1}^{p}\mathbf{u}_i\mathbf{u}_i^{\mathrm{T}} = \sum_{i=1}^{p}\mathbf{P}_i$$

$$\mathbf{P}_i = \mathbf{u}_i\mathbf{u}_i^{\mathrm{T}} \tag{6.251}$$

$$\mathbf{I} - \mathbf{P} = \mathbf{U}_{N-p}\mathbf{U}_{N-p}^{\mathrm{T}} = \sum_{i=p+1}^{N}\mathbf{u}_i\mathbf{u}_i^{\mathrm{T}}.$$

From the orthogonal matrix $\mathbf{U}$, we build the covariance matrix

$$\mathbf{R}(\boldsymbol{\theta}) = \mathbf{U}(\boldsymbol{\Sigma}^2 + \sigma^2\mathbf{I})\mathbf{U}^{\mathrm{T}} \tag{6.252}$$

$$\boldsymbol{\Sigma}^2 = \mathrm{diag}[\sigma_1^2 \quad \cdots \quad \sigma_p^2 \quad 0 \quad \cdots \quad 0].$$

This is the correlation structure of a multivariate normal signal that is composed of p uncorrelated and orthogonal "modes" and observed in uncorrelated multivariate noise.

The beauty of the orthogonal representation for $\mathbf{R}$ is that $\mathbf{R}^{-1}$ has the same representation:

$$\mathbf{R}^{-1} = \mathbf{U}(\boldsymbol{\Sigma}^2 + \sigma^2\mathbf{I})^{-1}\mathbf{U}^{\mathrm{T}}. \tag{6.253}$$

Furthermore, treating the parameters $\{\sigma_i^2\}_1^p$ and σ^2 as the unknown parameters in the characterization of $\mathbf{R}$, we have

$$\frac{\partial\mathbf{R}}{\partial\sigma_i^2} = \mathbf{U}\frac{\partial}{\partial\sigma_i^2}(\boldsymbol{\Sigma}^2 + \sigma^2\mathbf{I})\mathbf{U}^{\mathrm{T}} = \mathbf{U}\begin{bmatrix} 0 & & & & & & \\ & \ddots & & & & \mathbf{0} & \\ & & 0 & & & & \\ & & & 1 & & & \\ & & & & 0 & & \\ & \mathbf{0} & & & & \ddots & \\ & & & & & & 0 \end{bmatrix}\mathbf{U}^{\mathrm{T}} = \mathbf{u}_i\mathbf{u}_i^{\mathrm{T}} = \mathbf{P}_i \tag{6.254}$$

$$\frac{\partial\mathbf{R}}{\partial\sigma^2} = \mathbf{U}\frac{\partial}{\partial\sigma^2}(\boldsymbol{\Sigma}^2 + \sigma^2\mathbf{I})\mathbf{U}^{\mathrm{T}} = \mathbf{I}.$$

These results may be substituted into the regression equation for maximum likelihood estimates to obtain the regressions

$$(\sigma_i^2 + \sigma^2 - \mathrm{tr}[\mathbf{P}_i\mathbf{S}])\frac{1}{(\sigma_i^2 + \sigma^2)^2} = 0; \quad i = 1, 2, \ldots, p \tag{6.255}$$

$$\sum_{i=1}^{N}[\sigma_i^2 + \sigma^2 - \mathrm{tr}[\mathbf{P}_i\mathbf{S}]]\frac{1}{\sigma_i^2 + \sigma^2} = 0.$$

The first of these equations produces the solution

$$\hat{\sigma}_i^2 + \hat{\sigma}^2 = \text{tr}\,[\mathbf{P}_i\mathbf{S}]\,;\quad i = 1, 2, \ldots, p. \tag{6.256}$$

When this solution is substituted into the second equation, the result is

$$(N - p)\hat{\sigma}^2 = \sum_{i=p+1}^{N} \text{tr}\,[\mathbf{P}_i\mathbf{S}] = \text{tr}\,[(\mathbf{I} - \mathbf{P})\mathbf{S}]\,. \tag{6.257}$$

In summary, the maximum likelihood estimates of the variance parameters $\{\hat{\sigma}_i^2\}_1^p$ and the noise variance $\hat{\sigma}^2$ in the multivariate normal model with a low-rank plus identity correlation matrix are

$$\hat{\sigma}_i^2 = \text{tr}\,[\mathbf{P}_i\mathbf{S}] - \hat{\sigma}^2 = \mathbf{u}_i^{\mathsf{T}}\mathbf{S}\mathbf{u}_i - \hat{\sigma}^2$$

$$\hat{\sigma}^2 = \frac{1}{N - p}\,\text{tr}\,(\mathbf{I} - \mathbf{P})\mathbf{S} \tag{6.258}$$

$$\mathbf{S} = \frac{1}{M}\sum_{m=1}^{M}\mathbf{x}_m\mathbf{x}_m^{\mathsf{T}}.$$

The maximum likelihood estimate of σ_i^2 is the trace of the sample correlation matrix after it has been projected onto the ith subspace. The maximum likelihood estimate of σ^2 is the average of the trace of $\mathbf{S}$ after it has been projected onto the space that is orthogonal to the signal subspace. These results may be rewritten by noting that

$$\text{tr}\,[\mathbf{P}_i\mathbf{S}] = \text{tr}\left[\mathbf{P}_i\frac{1}{M}\sum_{m=1}^{M}\mathbf{x}_m\mathbf{x}_m^{\mathsf{T}}\right] = \frac{1}{M}\sum_{m=1}^{M}\|\mathbf{P}_i\mathbf{x}_m\|^2$$

$$\text{tr}\,[(\mathbf{I} - \mathbf{P})\mathbf{S}] = \text{tr}\left[(\mathbf{I} - \mathbf{P})\frac{1}{M}\sum_{m=1}^{M}\mathbf{x}_m\mathbf{x}_m^{\mathsf{T}}\right] = \frac{1}{M}\sum_{m=1}^{M}\|(\mathbf{I} - \mathbf{P})\mathbf{x}_m\|^2 \tag{6.259}$$

$$= \frac{1}{M}\sum_{m=1}^{M}\left\|\mathbf{x}_m - \sum_{i=1}^{p}\mathbf{P}_i\mathbf{x}_m\right\|^2.$$

The maximum likelihood estimates may now be written as

$$\hat{\sigma}_i^2 = \frac{1}{M}\sum_{m=1}^{M}\|\mathbf{P}_i\mathbf{x}_m\|^2 - \hat{\sigma}^2 \tag{6.260}$$

$$\hat{\sigma}^2 = \frac{1}{N - p}\frac{1}{M}\sum_{m=1}^{M}\left\|\mathbf{x}_m - \sum_{i=1}^{p}\mathbf{P}_i\mathbf{x}_m\right\|^2.$$

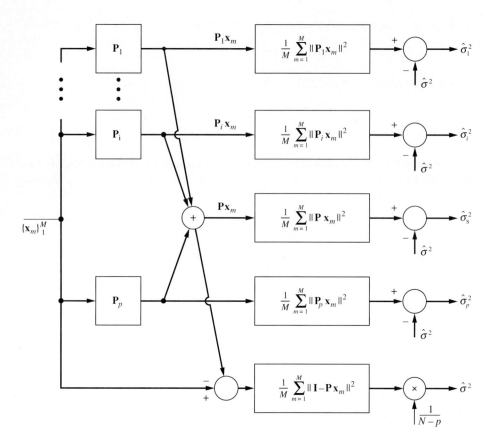

Figure 6.12 Block diagram for estimating variance parameters in low-rank orthogonal correlation matrix.

These solutions are illustrated in Figure 6.12. The estimates $\hat{\sigma}_i^2$ are powers in the respective rank-one subspaces $\langle \mathbf{u}_i \rangle$, $\hat{\sigma}_s^2$ is the estimated power in the signal subspace $\langle \mathbf{U}_p \rangle$, and $(N - p)\hat{\sigma}^2$ is the estimated power in the noise subspace $\langle \mathbf{U}_{N-p} \rangle$. These estimates are unbiased:

$$E\hat{\sigma}_i^2 = \text{tr}\,[\mathbf{P}_i \mathbf{R}] - \sigma^2 = \sigma_i^2 \qquad 6.261$$

$$E\hat{\sigma}^2 = \frac{1}{N - p}(N - p)\sigma^2 = \sigma^2.$$

The Fisher information matrix for this problem is diagonal with iith elements J_{ii}:

$$J_{ii} = \frac{M}{2} \operatorname{tr}\left\{ \mathbf{U}(\mathbf{\Sigma}^2 + \sigma^2\mathbf{I})^{-1}\mathbf{U}^{\mathrm{T}}\mathbf{U} \begin{bmatrix} & & 0 \\ & 1 & \\ 0 & & \end{bmatrix} \mathbf{U}^{\mathrm{T}}\mathbf{U}(\mathbf{\Sigma}^2 + \sigma^2\mathbf{I})^{-1}\mathbf{U}^{\mathrm{T}}\mathbf{U} \begin{bmatrix} & & 0 \\ & 1 & \\ 0 & & \end{bmatrix} \mathbf{U}^{\mathrm{T}} \right\}$$

$$= \frac{M}{2}(\sigma_i^2 + \sigma^2)^{-2}; \qquad i = 1, 2, \ldots, p$$

6.262

$$J_{(p+1)(p+1)} = \frac{M}{2}\sum_{i=1}^{p}(\sigma_i^2 + \sigma^2)^{-2} + \frac{M}{2}\sum_{i=p+1}^{N}(\sigma^2)^{-2}$$

$$= \sum_{i=1}^{p} J_{ii} + (N-p)\frac{M}{2}(\sigma^2)^{-2}.$$

As the maximum likelihood estimates of σ_i^2 and σ^2 are unbiased, the inverses of these diagonal elements lower bound the variances of the maximum likelihood estimates:

$$E(\hat{\sigma}_i^2 - \sigma_i^2)^2 \geq \frac{2}{M}(\sigma_i^2 + \sigma^2)^2 = (J^{-1})_{ii}$$

6.263

$$E(\hat{\sigma}^2 - \sigma^2)^2 \geq J_{(p+1)(p+1)}^{-1} = \left(\sum_{i=1}^{p} J_{ii} + (N-p)\frac{M}{2}(\sigma^2)^{-2} \right)^{-1}.$$

When $\sigma^2 = 0$, the variables $\mathbf{P}_i\mathbf{x}_m$ are distributed as $N[\mathbf{0}, \mathbf{P}_i\mathbf{R}\mathbf{P}_i]$. But $\mathbf{P}_i\mathbf{R}\mathbf{P}_i$ is $\sigma_i^2\mathbf{u}_i\mathbf{u}_i^{\mathrm{T}}$, so the variable $\mathbf{P}_i\mathbf{x}_m/\sigma_i$ is distributed as $N[\mathbf{0}, \mathbf{u}_i\mathbf{u}_i^{\mathrm{T}}]$. This means that $\|P_i\mathbf{x}_m\|^2$ is χ_1^2 distributed with one degree of freedom, and the estimator

$$M\frac{\hat{\sigma}_i^2}{\sigma_i^2} = \sum_{m=1}^{M}\left\| \frac{P_i\mathbf{x}_m}{\sigma_i} \right\|^2$$

6.264

is χ_M^2 distributed with M degrees of freedom. The mean and variance of such a distribution are M and $2M$, so the estimator $\hat{\sigma}_i^2$ has mean and variance

$$E\hat{\sigma}_i^2 = \sigma_i^2$$

6.265

$$\operatorname{var}\hat{\sigma}_i^2 = \frac{2}{M}\sigma_i^4 = J_{ii} \quad \text{when } \sigma^2 = 0.$$

So, we now know that the ML estimates of σ_i^2 are MVUB and efficient. (Can you justify the claim for MVUB?)

Example 6.14

These results extend directly to the complex case

$$\mathbf{R} = \sigma_1^2 \mathbf{e}_1 \mathbf{e}_1^H + \sigma^2 \mathbf{I} = \sigma_1^2 (\mathbf{e}_1^H \mathbf{e}_1) \mathbf{u}_1 \mathbf{u}_1^H + \sigma^2 \mathbf{I}$$

$$\mathbf{u}_1 = \mathbf{e}_1 / (\mathbf{e}_1^H \mathbf{e}_1)^{1/2} \qquad\qquad 6.266$$

$$\mathbf{e}_1^H = (1 \quad e^{j\mathbf{k}\cdot\mathbf{r}_1} \quad \cdots \quad e^{j\mathbf{k}\cdot\mathbf{r}_{N-1}}).$$

This is the correlation matrix for a complex wavefront of wavenumber $\mathbf{k}$ that is sampled by an array of elements located at geometric points $\{\mathbf{r}_i\}_0^{N-1}$. The resulting estimates of σ_1^2 and σ^2 are

$$(\mathbf{e}_1^H \mathbf{e}_1)\hat{\sigma}_1^2 = \frac{\mathbf{e}_1^H \mathbf{S} \mathbf{e}_1}{\mathbf{e}_1^H \mathbf{e}_1} - \hat{\sigma}^2 \qquad\qquad 6.267$$

$$\hat{\sigma}^2 = \frac{1}{N-1} \left[\text{tr } \mathbf{S} - \frac{\mathbf{e}_1^H \mathbf{S} \mathbf{e}_1}{\mathbf{e}_1^H \mathbf{e}_1} \right].$$

The correlation matrix $\mathbf{R}$ may also be written as

$$\mathbf{R} = (\sigma_1^2 \mathbf{e}_1^H \mathbf{e}_1 + \sigma^2) \mathbf{u}_1 \mathbf{u}_1^H + \sigma^2 (\mathbf{I} - \mathbf{u}_1 \mathbf{u}_1^H). \qquad\qquad 6.268$$

We call $\sigma_1^2 \mathbf{e}_1^H \mathbf{e}_1 + \sigma^2$ the total power in mode $\mathbf{u}_1$ and estimate it as follows:

$$(\mathbf{e}_1^H \mathbf{e}_1)\hat{\sigma}_1^2 + \hat{\sigma}^2 = \frac{\mathbf{e}_1^H \mathbf{S} \mathbf{e}_1}{\mathbf{e}_1^H \mathbf{e}_1}. \qquad\qquad 6.269$$

This is certainly an intuitive result that may also be written as

$$(\mathbf{e}_1^H \mathbf{e}_1)\hat{\sigma}_1^2 + \hat{\sigma}^2 = \frac{1}{M} \sum_{m=1}^{M} \frac{|\mathbf{e}_1^H \mathbf{x}_m|^2}{\mathbf{e}_1^H \mathbf{e}_1}. \qquad\qquad 6.270$$

This is just the average power that $\mathbf{x}$ has in mode $\mathbf{e}_1$, normalized by the power of mode $\mathbf{e}_1$. ∎

REFERENCES AND COMMENTS

The theory of maximum likelihood is standard fare in most books on mathematical statistics and estimation theory. (Reference Hogg and Craig [1965], Mendel [1987], Sorenson [1980], or Zachs [1971].) The derivation of the Cramer-Rao bound is a modest generalization of the derivation in Van Trees [1965]. The treatment of nuisance parameters appears to be original and more complete than the usual treatments. The three approaches we have used to study Cramer-Rao bounds were first used in Phillippe Tourtier's M.S. thesis. The special Cramer-Rao bounds for structured means and covariances in the multivariate normal model are well known. They have been used to study a number of specialized problems in the identification of autoregressive moving-average time series. See, for example, the paper by Porat and Friedlander [1986]. The derivation of Cramer-Rao bounds for ARMA impulse responses observed in additive noise appears to be new.

Maximum likelihood theory has been applied to a number of fundamental problems in signal processing. It is now an active area of research for time series identification, spectrum analysis, and multisensor array processing. Some of these applications we cover in Chapters 10 and 11. Many authors have applied maximum likelihood theory (or its equivalent, least-squares theory, for deterministic signals observed in additive Gaussian noise) to the problems treated here in Sections 6.9 and 6.10. Kumaresan, Scharf, and Shaw [1986] have presented a fast algorithm for use in the recursion step of Equation 6.189. Bressler and Macovski [1986] present an equivalent algorithm. In fact, J. McClellan in a paper to appear in 1990 in the IEEE's *Transactions ASSP* has shown that the recursions of these authors just mentioned are equivalent to the famous Steiglitz-McBride algorithm [1965].

Examples of maximum likelihood theory applied to time series analysis can be found in Akaike [1973], Gueguen and Scharf [1980], or Kay [1983]; these are just three representative papers out of hundreds. For a fairly complete set of references, see the paper by Dugre, Scharf, and Gueguen [1986].

The estimation of structured covariance matrices dates at least to the early work of Anderson [1969]. But it was the paper by Burg, Luenberger, and Wenger [1982] that generated most of the enthusiasm in the engineering literature for parametric estimators of structured covariance matrices. In this paper, the authors studied the maximum likelihood estimation of Toeplitz matrices and the use of such estimators in autoregressive, or all-pole, spectrum estimators. The properties of maximum likelihood estimators of covariance matrices were further studied by Fuhrmann and Miller [1988].

It turns out that there are other ways to connect spectrum analysis with the maximum likelihood estimation of structured covariance matrices. These connections are treated in the following papers: Tourtier and Scharf [1987]; Van Veen and Scharf [1988 and 1990].

Akaike, H. [1973]. "Maximum Likelihood Identification of Gaussian Autoregressive Moving Average Models," *Biometrika* **60**, pp. 255–265 (1973).

Anderson, T. W. [1969]. "Statistical Inference for Covariance Matrices with Linear Structure," *Multivariate Analysis—II*, P. R. Krishnaiah, Ed. (New York: Academic Press, 1969).

Bressler, Y., and A. Macovski [1986]. "Exact Maximum Likelihood Estimation of Superimposed Exponential Signals in Noise," *IEEE Trans ASSP* **ASSP:34**, pp. 1081–1089 (October 1986).

Burg, J. P., D. G. Luenberger, and D. L. Wenger [1982]. "Estimation of Structured Covariance Matrices," *IEEE Proc* **70**, pp. 963–974 (September 1982).

Dugre, J. P., L. L. Scharf, and C. Gueguen [1986]. "Exact Likelihood for Stationary Vector Autoregressive Moving Average Processes," *Signal Processing* **11**, pp. 105–118 (September 1986).

Evans, A. G., and Fischl [1973]. "Optimal Least Squares Time-Domain Synthesis of Recursive Digital Filters," *IEEE Trans Audio Electroacoust* **AE:27**, pp. 61–65 (February 1973).

Fuhrmann, D. R., and M. I. Miller [1988]. "On the Existence of Positive Definite Maximum Likelihood Estimates of Structured Covariance Matrices," *IEEE Trans IT* **IT-34**, pp. 722–729 (July 1988).

Gueguen, C. J., and L. L. Scharf [1980]. "Exact Maximum Likelihood Identification of ARMA Models: A Signal Processing Perspective," *Proc EUSIPCO*, pp. 759–769 (September 1980).

Hogg, R. V., and A. T. Craig [1965]. *Introduction to Mathematical Statistics*, Second Ed., Chapter 9 (New York: MacMillan Publishing Company, 1965).

Kay, S. [1983]. "Recursive Maximum Likelihood Estimation of Autoregressive Processes," *IEEE Trans ASSP* **ASSP-31**, pp. 56–65 (February 1983).

Kumaresan, R., L. L. Scharf, and A. K. Shaw [1986]. "An Algorithm for Pole-Zero Modeling and Spectral Analysis," *IEEE Trans ASSP* **ASSP:34**, pp. 637–640 (June 1986).

McClellan, J. "Exact Equivalence of the Steiglitz-McBride Iteration and IQML," *IEEE Trans ASSP* (to appear 1990).

Mendel, J. M. [1987]. *Lessons in Digital Estimation Theory*, Lesson 11 (New York: Prentice-Hall, 1987).

Porat, B., and B. Friedlander [1986]. "Computation of the Exact Information Matrix of Gaussian Time Series with Stationary Random Components," *IEEE Trans ASSP* **ASSP-34**, pp. 118–130 (February 1986).

Sorenson, H. [1980]. *Parameter Estimation*, Chapter 5 (New York: Marcel-Dekker, 1980).

Steiglitz, K., and L. E. McBride [1965]. "A Technique for the Identification of Linear Systems," *IEEE Trans AC* **AC:10**, pp. 461–464 (October 1965).

Tourtier, P. J. P. [1986]. "Maximum Likelihood Estimation of Correlation Matrices for Spectrum Analysis," M.S. Thesis, University of Colorado, Boulder, 1986.

Tourtier, P. J. P., and L. L. Scharf [1987]. "Maximum Likelihood Identification of Correlation Matrices for Estimation of Power Spectra at Arbitrary Resolutions," *Proc Intl Conf on ASSP*, Dallas, pp. 2066–2069 (April 1987).

Van Trees, H. L. [1965]. *Detection, Estimation, and Modulation Theory*, Chapter 2 (New York: John Wiley & Sons, 1965).

Van Veen, B. D., and L. L. Scharf [1990]. "Estimation of Structured Covariance Matrices and Multiple Window Spectrum Analysis," *IEEE Trans ASSP* **ASSP:38** (August 1990).

Van Veen, B. D., and L. L. Scharf [1988]. "Spectrum Analysis Using Low-Rank Models and Structured Covariance Matrices," *Proc EUSIPCO-88*, Grenoble, pp. 171–174 (September 1988).

Zacks, S. [1971]. *The Theory of Statistical Inference*, Chapter 5 (New York: John Wiley & Sons, 1971).

PROBLEMS

6.1 Return to the maximum likelihood estimate $\hat{\theta} = \max(x_0, x_1)$ in Example 6.1. How is the estimator $\hat{\theta}$ distributed? Is $\hat{\theta}$ unbiased? What is its variance?

6.2 Let $X = (X_1, X_2, \ldots, X_M)$ denote a random sample of $U[0, \theta]$ random variables. Show that

a. $\hat{\theta}_{ML} = \max_m X_m$;

b. The density of $\hat{\theta}_{ML}$ is $f_{\hat{\theta}}(x) = \dfrac{M}{\theta^M} x^{M-1}$, $0 \le x \le \theta$;

c. $E\hat{\theta}_{ML} = \dfrac{M}{M+1} \theta$;

d. $\operatorname{var}(\hat{\theta}_{ML}) = \dfrac{M}{(M+2)(M+1)^2} \theta^2$.

(We say that $\hat{\theta}_{ML}$ is asymptotically unbiased and consistent.)

6.3 Return to the maximum likelihood estimates $\hat{\mu}$ and $\hat{\sigma}^2$ in Example 6.2. How are these estimators distributed? What are their means and variances?

6.4 Return to the maximum likelihood estimate $\hat{\lambda}$ in the Poisson problem of Example 6.3. How is the estimator $\hat{\lambda}$ distributed? What is its mean? What is its variance?

6.5 Let $\mathbf{X} = (\mathbf{X}_0 \quad \mathbf{X}_1 \quad \cdots \quad \mathbf{X}_{M-1})$ denote a random sample of normal random vectors $\mathbf{X}_i : N\left[\mathbf{0}, \sigma^2 \mathbf{R}(\boldsymbol{\theta})\right]$. Evaluate estimated entropy and compressed likelihood when σ^2 and $\boldsymbol{\theta}$ are unknown.

6.6 Let $\{\mathbf{X}_i\}_0^{M-1}$ denote a sequence of random vectors with mean $\mathbf{m}(\boldsymbol{\theta})$ and covariance $\mathbf{R}$. Define the matrix

$$\mathbf{S} = \frac{1}{M} \sum_{i=0}^{M-1} (\mathbf{X}_i - \mathbf{m})(\mathbf{X}_i - \mathbf{m})^{\mathrm{T}}.$$

Show

a. $E\mathbf{S} = \mathbf{R}; \ \mathbf{R} = E(\mathbf{X}_i - \mathbf{m})(\mathbf{X}_i - \mathbf{m})^{\mathrm{T}}; \ \mathbf{m} = \mathbf{m}(\boldsymbol{\theta})$

b. $E \dfrac{\partial \mathbf{S}}{\partial \theta_n} = \mathbf{0}$

c. $E \dfrac{\partial^2 \mathbf{S}}{\partial \theta_n \partial \theta_k} = \dfrac{\partial \mathbf{m}}{\partial \theta_k} \dfrac{\partial \mathbf{m}^{\mathrm{T}}}{\partial \theta_n} + \dfrac{\partial \mathbf{m}}{\partial \theta_n} \dfrac{\partial \mathbf{m}^{\mathrm{T}}}{\partial \theta_k}$

6.7   State and prove a theorem that captures the essence of the statement that "extra data can only improve the variance of an unbiased parameter estimator."

6.8   Derive the Fisher information matrix for the parameters $(A, \phi, \omega, \sigma^2)$ in the model

$$x_t = s_t + n_t; \qquad t = 0, 1, \ldots, N - 1$$

$$s_t = A\cos(\omega t - \phi); \qquad n_t : N[0, \sigma^2].$$

Plot your variance bounds versus SNR, parameterized by N and vice versa. This is the problem studied in Section 6.12.

6.9   Consider the zero-mean random vector

$$\mathbf{X} = \mathbf{U}\boldsymbol{\theta} + \mathbf{N}$$

with $\mathbf{U}^T\mathbf{U} = \mathbf{I}$, $E\boldsymbol{\theta} = \mathbf{0}$, $E\boldsymbol{\theta}\boldsymbol{\theta}^T = \boldsymbol{\Sigma}^2$, $E\mathbf{N} = \mathbf{0}$, $E\mathbf{N}\mathbf{N}^T = \sigma^2\mathbf{I}$, and $E\boldsymbol{\theta}\mathbf{N}^T = \mathbf{0}$. Find the covariance matrix for $\mathbf{X}$.

6.10 Begin with the random vector $\mathbf{x} : f_{\boldsymbol{\theta}}(\mathbf{x})$ and the statistic $\mathbf{t}(\mathbf{x})$. Prove

$$\mathbf{J}_{\mathbf{x}}(\boldsymbol{\theta}) \geq \mathbf{J}_{\mathbf{t}}(\boldsymbol{\theta})$$

with equality iff $\mathbf{t}(\mathbf{x})$ is sufficient for $\boldsymbol{\theta}$. ($\mathbf{J}_{\mathbf{x}}$ is the Fisher information matrix for $\mathbf{x}$ and $\mathbf{J}_{\mathbf{t}}$ is the Fisher information matrix for $\mathbf{t}$.)

6.11 Let $w = \boldsymbol{\theta}^T\boldsymbol{\theta}$ and assume $\mathbf{x} : N[\mathbf{H}\boldsymbol{\theta}, \mathbf{R}]$. Find the Fisher information matrix for w.

6.12 Let $\mathbf{R}$ denote a real circulant matrix of the form

$$\mathbf{R} = \begin{bmatrix} r_0 & r_1 & \cdots & r_{N-1} \\ r_{N-1} & r_0 & \ddots & \vdots \\ \vdots & \ddots & \ddots & r_1 \\ r_1 & \cdots & r_{N-1} & r_0 \end{bmatrix}.$$

a.   Show that the DFT eigenvectors diagonalize $\mathbf{R}$:

$$\mathbf{R}\mathbf{u}_k = \lambda_k\mathbf{u}_k; \qquad \sqrt{N}\,\mathbf{u}_k = (1\ e^{-j2\pi k/N} \cdots e^{-j2\pi(k/N)(N-1)})^T$$

$$\lambda_k = \sum_{n=0}^{N-1} r_n e^{-j2\pi(k/N)n}.$$

b.   Show $\lambda_k \geq 0$ if $\mathbf{R}$ is nonnegative definite.

c.   Show $\lambda_k = \lambda_{N-k}$ if $\mathbf{R}$ is symmetric.

6.13 Let $\mathbf{X} = (\mathbf{X}_1, \mathbf{X}_2, \ldots, \mathbf{X}_M)$ denote a random sample of $N[\mathbf{0}, \mathbf{R}]$ random vectors, with $\mathbf{R}$ a real, symmetric, nonnegative definite, circulant matrix (see Problem 6.12).

a.   Derive CR bounds on unbiased estimates of λ_k.

b.   Find ML estimates of λ_k.

c. Are your ML estimates unbiased, minimum variance, or efficient?

d. Find ML estimates of r_n.

e. Are your ML estimates of r_n unbiased, minimum variance, or efficient?

f. Can you connect your results for $\hat{\lambda}_k$ with the periodograms

$$P_i(\omega) = \frac{1}{N} \left| \sum_{n=0}^{N-1} (\mathbf{X}_i)_n e^{-jn\omega} \right|^2 ?$$

(Caution: The symmetry $\lambda_n = \lambda_{N-n}$ must be enforced in all of your calculations.)

6.14 Prove tr $[\mathbf{P}_A \mathbf{S}] \geq \sum_{i=p+1}^{N} \lambda_i^2$ whenever $\mathbf{S}$ is the positive-definite symmetric matrix $\mathbf{S} = \mathbf{U}\Lambda^2\mathbf{U}^T$ and $\mathbf{P}_A$ is an arbitrary rank $N - p$ projector. What can you say about the projector $\mathbf{P}_A$ that achieves the bound?

6.15 Let $\mathbf{X} = (\mathbf{X}_1, \mathbf{X}_2, \ldots, \mathbf{X}_M)$ denote a random sample of $N[\mathbf{m}, \mathbf{R}]$ random vectors, with $\mathbf{m}$ and $\mathbf{R}$ unknown but unstructured. Show that the ML estimates of $\mathbf{m}$ and $\mathbf{R}$ are

$$\hat{\mathbf{m}}_{\mathrm{ML}} = \hat{\mathbf{m}}$$

$$\hat{\mathbf{R}}_{\mathrm{ML}} = \mathbf{S}[\hat{\mathbf{m}}]$$

where $\hat{\mathbf{m}}$ and $\mathbf{S}[\hat{\mathbf{m}}]$ are the following sample mean and sample covariance:

$$\hat{\mathbf{m}} = \frac{1}{M} \sum_{i=1}^{M} \mathbf{x}_i$$

$$\mathbf{S}[\hat{\mathbf{m}}] = \frac{1}{M} \sum_{i=1}^{M} (\mathbf{x}_i - \hat{\mathbf{m}})(\mathbf{x}_i - \hat{\mathbf{m}})^T.$$

Solve for the Fisher information matrix. Are $\hat{\mathbf{m}}$ and $\mathbf{S}[\hat{\mathbf{m}}]$ unbiased, minimum variance, or efficient?

6.16 Prove that, if an efficient estimator for a parameter $\boldsymbol{\theta}$ exists, then it is an ML estimator. (The converse is not true; ML estimators are not necessarily efficient.)

6.17 Prove that an efficient estimator whose Fisher information matrix is independent of $\boldsymbol{\theta}$ is distributed as $\hat{\boldsymbol{\theta}} : N[\boldsymbol{\theta}, \mathbf{J}^{-1}]$.

6.18 Let $\{\mathbf{x}_i\}_0^{M-1}$ denote a random sample of $N[\mathbf{0}, \mathbf{R}]$ random vectors. Find the ML estimate of σ^2 in the covariance model

$$\mathbf{R} = \sigma^2 \mathbf{U}_p \boldsymbol{\Sigma}_p^2 \mathbf{U}_p^T; \qquad (\mathbf{U}_p, \boldsymbol{\Sigma}_p^2) \quad \text{known}$$

$$\mathbf{U}_p^T \mathbf{U}_p = \mathbf{I}.$$

Compute the Fisher information matrix. Is the ML estimate unbiased? minimum variance? efficient? Draw a block diagram for $\hat{\sigma}^2$ and interpret your result.

6.19 Let $\mathbf{X}$ denote a normal random vector with unknown mean $\boldsymbol{\theta}$ and known covariance $\mathbf{R}$. Find the maximum likelihood estimate of $(\boldsymbol{\theta}^{\mathrm{T}}\boldsymbol{\theta})^{1/2}$. What happens when $\mathbf{X}$ is replaced by the random sample $(\mathbf{X}_1, \mathbf{X}_2, \ldots, \mathbf{X}_M)$? Is your estimate of $(\boldsymbol{\theta}^{\mathrm{T}}\boldsymbol{\theta})^{1/2}$ unbiased, minimum variance, or efficient?

6.20 Let $X = (X_1, X_2, \ldots, X_N)$ denote a random sample of exponential random variables with unknown parameter θ:

$$f_\theta(x_n) = \frac{1}{\theta} e^{-x_n/\theta}; \qquad 0 \leq x_n < \infty, \quad 0 < \theta < \infty.$$

Compute the Fisher information matrix. Is the ML estimate unbiased? minimum variance? efficient?

6.21 Let $\mathbf{X}_n$ denote an $N \times 1$ random vector with distribution $f_\boldsymbol{\theta}(\mathbf{x}_n)$, score function $\mathbf{s}(\boldsymbol{\theta}, \mathbf{x}_n)$, and Fisher information matrix $\mathbf{J}_n$. Find the distribution, score, and Fisher information matrix for a random sample $\mathbf{X} = (\mathbf{X}_1, \mathbf{X}_2, \ldots, \mathbf{X}_M)$.

6.22 Write $\boldsymbol{\theta}^{\mathrm{T}} = (\boldsymbol{\theta}_r^{\mathrm{T}}, \theta_{r+1}, \ldots, \theta_p) = (\boldsymbol{\theta}_s^{\mathrm{T}}, \theta_{s+1}, \ldots, \theta_p)$, with $r \leq s$. Extend the argument advanced in the section on Nuisance Parameters to show

$$[\mathbf{J}_s^{-1}]_r - \mathbf{J}_r^{-1} \geq 0,$$

where $\mathbf{J}_r$ is the Fisher information matrix for $\boldsymbol{\theta}_r$ when θ_{r+1} through θ_p are known and $[\mathbf{J}_s^{-1}]_r$ is the upper $r \times r$ block of $\mathbf{J}_s$. Show

$$(\mathbf{J}_s^{-1})_{kk} \geq (\mathbf{J}_r^{-1})_{kk}$$

for $k \leq r$. Interpret your findings.

6.23 Is the estimator of Example 6.12 MVUB? Is it efficient?

6.24 The Fisher information matrix in the multivariate normal model is

$$\mathbf{J} = M \frac{\partial}{\partial \boldsymbol{\theta}} \mathbf{m}^{\mathrm{T}} \mathbf{R}^{-1} \left(\frac{\partial}{\partial \boldsymbol{\theta}} \mathbf{m}^{\mathrm{T}} \right)^{\mathrm{T}}$$

when $\mathbf{X} = (\mathbf{X}_1, \mathbf{X}_2, \ldots, \mathbf{X}_M)$, $\mathbf{X}_n : N[\mathbf{m}, \mathbf{R}]$, and $\mathbf{R}$ known. Evaluate this general result for

a. $\mathbf{m} = \boldsymbol{\theta}$;

b. $\mathbf{m} = \mathbf{H}\boldsymbol{\theta}$, $\mathbf{H}$ known;

c. $\mathbf{m} = \mathbf{H}(\omega)\boldsymbol{\theta}$, ω known;

d. $\mathbf{m} = \mathbf{H}(\omega)\boldsymbol{\theta}$, ω and $\boldsymbol{\theta}$ unknown. (In this case, $\boldsymbol{\theta}$ is replaced by $(\boldsymbol{\theta}, \omega)$.)

6.25 Let $\mathbf{X} = (\mathbf{X}_1, \mathbf{X}_2, \ldots, \mathbf{X}_M)$ denote a random sample of $N[\mathbf{m}, \mathbf{R}]$ random variables with $\mathbf{m}(\boldsymbol{\theta})$ and $\mathbf{R}$ unknown. Compute the compressed likelihood for $\boldsymbol{\theta}$ by plugging the maximum likelihood estimate of $\mathbf{R}$ into the likelihood function.

6.26 *Experimental Results.* In the sinusoidal model of Section 6.12, fix N at $N = 10$ and ω at $\pi/4$. This provides just over one cycle of s_t in the data window. Select a value of SNR and generate 100 experiments. Measure the sample standard deviation of $\hat{\omega}$ and plot it versus $10 \log_{10}$ SNR, the output signal-to-noise ratio in db.

6.27 Let $\mathbf{y} : N[\mathbf{x}, \sigma^2\mathbf{I}]$ denote a normal random vector with mean $\mathbf{x}$ and covariance $\sigma^2\mathbf{I}$. Write $\mathbf{x}$ as $\beta\mathbf{u}_x$, where β is the norm of $\mathbf{x}$ and $\mathbf{u}_x$ is a unit vector in the direction of $\mathbf{x}$. Find ML estimates of

  a. β when $\mathbf{u}_x$ is known (σ^2 can be known or unknown—it does not matter);

  b. $\mathbf{u}_x$ (β and σ^2 can be known or unknown);

  c. β and $\mathbf{u}_x$ (σ^2 can be known or unknown);

  d. σ^2 when β and $\mathbf{u}_x$ are known;

  e. σ^2 and β when $\mathbf{u}_x$ is known;

  f. σ^2 and $\mathbf{u}_x$ when β is known; and

  g. σ^2, β, and $\mathbf{u}_x$.

6.28 Compute the Fisher information matrix for each case of Problem 6.27.

6.29 Find distributions of $\hat{\sigma}^2$, $\hat{\beta}$, and $\hat{\mathbf{u}}_x$ from Problem 6.27.

6.30 Let $\{\mathbf{X}_i\}_0^{M-1}$ denote a sequence of i.i.d. $N[\mathbf{m}, \sigma^2\mathbf{I}]$ random vectors. Each $\mathbf{X}_i$ is a 2×1 vector whose mean is $\mathbf{m} = (m_1, m_2)$. Think of $\mathbf{X}_i$ as the ith estimate of the Cartesian coordinates (m_1, m_2) in a surveying experiment. Find the ML estimates of (m_1, m_2) and their CR bounds. Then compute the ML estimates of the range $r = (m_1^2 + m_2^2)^{1/2}$ and the angle $\theta = \tan^{-1}(m_2/m_1)$ to the coordinates (m_1, m_2). Compute the CR bound for estimating r and θ.

6.31 Let $\mathbf{x}$ be a measurement vector with density $f_\theta(\mathbf{x})$ parameterized by the vector $\boldsymbol{\theta}$. A biased estimator, $\hat{\boldsymbol{\theta}}(\mathbf{x})$, is used to estimate $\boldsymbol{\theta}$. Let $\mathbf{g}(\boldsymbol{\theta}) = E\{\hat{\boldsymbol{\theta}}(\mathbf{x})\}$ be the expected value of $\hat{\boldsymbol{\theta}}(\mathbf{x})$.

  a. Using the property $E[\hat{\boldsymbol{\theta}}(\mathbf{x}) - \mathbf{g}(\boldsymbol{\theta})]^{\mathrm{T}} = \mathbf{0}^{\mathrm{T}}$, show that

$$E[\mathbf{s}(\boldsymbol{\theta}, \mathbf{x})(\hat{\boldsymbol{\theta}}(\mathbf{x}) - \mathbf{g}(\boldsymbol{\theta}))^{\mathrm{T}}] = \int d\mathbf{x} f_\theta(\mathbf{x})\left(\frac{\partial}{\partial\boldsymbol{\theta}}\mathbf{g}^{\mathrm{T}}(\boldsymbol{\theta})\right)$$

$$= \frac{\partial}{\partial\boldsymbol{\theta}}\mathbf{g}^{\mathrm{T}}(\boldsymbol{\theta}).$$

  b. Let

$$\mathbf{z}(\boldsymbol{\theta}, \mathbf{x}) = \begin{bmatrix} \mathbf{I} & -[\frac{\partial}{\partial\boldsymbol{\theta}}\mathbf{g}^{\mathrm{T}}(\boldsymbol{\theta})]^{\mathrm{T}}\mathbf{J}^{-1}(\boldsymbol{\theta}) \\ \mathbf{0} & \mathbf{I} \end{bmatrix}\begin{bmatrix} \hat{\boldsymbol{\theta}}(\mathbf{x}) - \mathbf{g}(\boldsymbol{\theta}) \\ \mathbf{s}(\boldsymbol{\theta}, \mathbf{x}) \end{bmatrix}.$$

Using the property that $E[\mathbf{z}(\boldsymbol{\theta}, \mathbf{x})\mathbf{z}^{\mathrm{T}}(\boldsymbol{\theta}, \mathbf{x})]$ is nonnegative definite, show that the Cramer-Rao bound for biased estimators is given by

$$E\left[(\hat{\boldsymbol{\theta}}(\mathbf{x}) - \mathbf{g}(\boldsymbol{\theta}))(\hat{\boldsymbol{\theta}}(\mathbf{x}) - \mathbf{g}(\boldsymbol{\theta}))^{\mathrm{T}}\right] \geq \left[\frac{\partial}{\partial\boldsymbol{\theta}}\mathbf{g}^{\mathrm{T}}(\boldsymbol{\theta})\right]^{\mathrm{T}}\mathbf{J}^{-1}(\boldsymbol{\theta})\left[\frac{\partial}{\partial\boldsymbol{\theta}}\mathbf{g}^{\mathrm{T}}(\boldsymbol{\theta})\right]$$

where $\mathbf{J}(\boldsymbol{\theta}) = E\mathbf{s}(\boldsymbol{\theta}, \mathbf{x})\mathbf{s}^{\mathrm{T}}(\boldsymbol{\theta}, \mathbf{x})$ is the Fisher information matrix for $\boldsymbol{\theta}$. (Contributed by T. McWhorter.)

APPENDIX: VECTOR AND MATRIX GRADIENTS

Much of statistical signal processing comes down to the derivation of a nonnegative cost function followed by optimization of the cost function with respect to a vector or matrix of parameters. For this reason, vector and matrix gradients play an important role in statistical signal processing.

In this appendix we will review the definitions of vector and matrix gradients and state without proof several basic identities. Some are obvious and some are not.

Vector Gradients

Let $a(\boldsymbol{\theta})$ denote a scalar function of the $p \times 1$ vector $\boldsymbol{\theta}$. The gradient of a with respect to $\boldsymbol{\theta}$ is the $p \times 1$ vector $(\partial/\partial\boldsymbol{\theta})\, a$:

$$\frac{\partial}{\partial\boldsymbol{\theta}}\, a(\boldsymbol{\theta}) = \begin{bmatrix} \partial a(\boldsymbol{\theta})/\partial\theta_1 \\ \partial a(\boldsymbol{\theta})/\partial\theta_2 \\ \vdots \\ \partial a(\boldsymbol{\theta})/\partial\theta_p \end{bmatrix}.$$

This definition generalizes to the gradient of the $1 \times n$ row vector $\mathbf{a}^{\mathrm{T}}(\boldsymbol{\theta}) = [a_1(\boldsymbol{\theta}) \quad \cdots \quad a_n(\boldsymbol{\theta})]$ with respect to $\boldsymbol{\theta}$:

$$\frac{\partial}{\partial\boldsymbol{\theta}}\, \mathbf{a}^{\mathrm{T}} = \begin{bmatrix} \dfrac{\partial}{\partial\boldsymbol{\theta}}\, a_1(\boldsymbol{\theta}) & \cdots & \dfrac{\partial}{\partial\boldsymbol{\theta}}\, a_n(\boldsymbol{\theta}) \end{bmatrix}$$

$$= \begin{bmatrix} \partial a_1(\boldsymbol{\theta})/\partial\theta_1 & \cdots & \partial a_n(\boldsymbol{\theta})/\partial\theta_1 \\ \vdots & & \vdots \\ \partial a_1(\boldsymbol{\theta})/\partial\theta_p & \cdots & \partial a_n(\boldsymbol{\theta})/\partial\theta_p \end{bmatrix}.$$

Each column is a gradient of a scalar a_i with respect to the vector $\boldsymbol{\theta}$.

A number of special cases follow from these basic definitions:

1. $\dfrac{\partial}{\partial\boldsymbol{\theta}}\, \boldsymbol{\theta}^{\mathrm{T}} = \mathbf{I}$

2. $\dfrac{\partial}{\partial\boldsymbol{\theta}}\, \mathbf{b}^{\mathrm{T}}\boldsymbol{\theta} = \dfrac{\partial}{\partial\boldsymbol{\theta}}\, \boldsymbol{\theta}^{\mathrm{T}}\mathbf{b} = \mathbf{b}$

3. $\dfrac{\partial}{\partial\boldsymbol{\theta}}\, \mathbf{a}^{\mathrm{T}}(\boldsymbol{\theta})\mathbf{b}(\boldsymbol{\theta}) = \left(\dfrac{\partial}{\partial\boldsymbol{\theta}}\, \mathbf{a}^{\mathrm{T}}(\boldsymbol{\theta}) \right) \mathbf{b}(\boldsymbol{\theta}) + \left(\dfrac{\partial}{\partial\boldsymbol{\theta}}\, \mathbf{b}^{\mathrm{T}}(\boldsymbol{\theta}) \right) \mathbf{a}(\boldsymbol{\theta})$

4. $\dfrac{\partial}{\partial\boldsymbol{\theta}}\, \boldsymbol{\theta}^{\mathrm{T}}\mathbf{Q}\boldsymbol{\theta} = 2\mathbf{Q}\boldsymbol{\theta}, \qquad$ provided $\mathbf{Q}$ is independent of $\boldsymbol{\theta}$

5. $\dfrac{\partial}{\partial\boldsymbol{\theta}}\, \mathbf{m}^{\mathrm{T}}\mathbf{Q}\mathbf{m} = 2\left(\dfrac{\partial}{\partial\boldsymbol{\theta}}\, \mathbf{m}^{\mathrm{T}} \right) \mathbf{Q}\mathbf{m}, \qquad$ provided $\mathbf{Q}$ is independent of $\boldsymbol{\theta}$

6. $\dfrac{\partial}{\partial \boldsymbol{\theta}} \exp\left\{ -\dfrac{1}{2} \boldsymbol{\theta}^{\mathrm{T}} \mathbf{Q}^{-1} \boldsymbol{\theta} \right\} = -\exp\left\{ -\dfrac{1}{2} \boldsymbol{\theta}^{\mathrm{T}} \mathbf{Q}^{-1} \boldsymbol{\theta} \right\} \mathbf{Q}^{-1} \boldsymbol{\theta}$,   provided $\mathbf{Q}$ is independent of $\boldsymbol{\theta}$

7. $\dfrac{\partial}{\partial \boldsymbol{\theta}} \ln(\boldsymbol{\theta}^{\mathrm{T}} \mathbf{Q} \boldsymbol{\theta}) = 2(\boldsymbol{\theta}^{\mathrm{T}} \mathbf{Q} \boldsymbol{\theta})^{-1} \mathbf{Q} \boldsymbol{\theta}$,   provided $\mathbf{Q}$ is independent of $\boldsymbol{\theta}$.

Matrix Gradients

Let $a(\mathbf{R})$ denote a scalar function of the $p \times n$ matrix $\mathbf{R}$. The gradient of a with respect to the matrix $\mathbf{R}$ is the $p \times n$ matrix $(\partial/\partial\mathbf{R})a$:

$$\frac{\partial}{\partial \mathbf{R}} a = \begin{bmatrix} \dfrac{\partial a(\mathbf{R})}{\partial r_{11}} & \dfrac{\partial a(\mathbf{R})}{\partial r_{12}} & \cdots & \dfrac{\partial a(\mathbf{R})}{\partial r_{1n}} \\[2ex] \dfrac{\partial a(\mathbf{R})}{\partial r_{21}} & & & \vdots \\[2ex] \vdots & & & \\[2ex] \dfrac{\partial a(\mathbf{R})}{\partial r_{p1}} & \cdots & & \dfrac{\partial a(\mathbf{R})}{\partial r_{pn}} \end{bmatrix}.$$

Each column is the vector gradient of the scalar a with respect to a column of $\mathbf{R}$:

$$\frac{\partial}{\partial \mathbf{R}} a = \left(\frac{\partial}{\partial \mathbf{r}_1} a(\mathbf{R}) \quad \cdots \quad \frac{\partial}{\partial \mathbf{r}_n} a(\mathbf{R}) \right)$$

$$\mathbf{R} = (\mathbf{r}_1 \quad \cdots \quad \mathbf{r}_n)$$

$$\mathbf{r}_i = \begin{bmatrix} r_{1i} \\ r_{2i} \\ \vdots \\ r_{pi} \end{bmatrix}.$$

A number of basic identities follow from these definitions:

1. $\dfrac{\partial}{\partial \mathbf{R}} \mathrm{tr}\mathbf{R} = \mathbf{I}$

2. $\dfrac{\partial}{\partial \mathbf{R}} \mathrm{tr}[\mathbf{L}\mathbf{R}] = \dfrac{\partial}{\partial \mathbf{R}^{\mathrm{T}}} \mathrm{tr}[\mathbf{L}\mathbf{R}^{\mathrm{T}}] = \left(\dfrac{\partial}{\partial \mathbf{R}} \mathrm{tr}[\mathbf{L}\mathbf{R}^{\mathrm{T}}] \right)^{\mathrm{T}} = \mathbf{L}^{\mathrm{T}}$,   provided $\mathbf{L}$ is independent of $\mathbf{R}$

3. $\dfrac{\partial}{\partial \mathbf{R}} \mathrm{tr}[\mathbf{L}\mathbf{R}^{-1}] = \dfrac{\partial}{\partial \mathbf{R}} \mathrm{tr}[\mathbf{R}^{-1}\mathbf{L}] = (-\mathbf{R}^{-1}\mathbf{L}\mathbf{R}^{-1})^{\mathrm{T}}$,   provided $\mathbf{L}$ is independent of $\mathbf{R}$

4. $\dfrac{\partial}{\partial \mathbf{R}} \mathrm{tr}\mathbf{R}^n = n(\mathbf{R}^{n-1})^{\mathrm{T}}$

5. $\dfrac{\partial}{\partial \mathbf{R}} \, \text{tr}[\exp \mathbf{R}] \, = \, \exp \mathbf{R}$

6. $\dfrac{\partial}{\partial \mathbf{R}} \, \det \mathbf{R} \, = \, \det \mathbf{R}(\mathbf{R}^{-1})^{\text{T}}$

7. $\dfrac{\partial}{\partial \mathbf{R}} \, \ln \det \mathbf{R} \, = \, (\mathbf{R}^{-1})^{\text{T}}$

8. $\dfrac{\partial}{\partial \mathbf{R}} \, \det \mathbf{R}^{n} \, = \, n(\det \mathbf{R})^{n}(\mathbf{R}^{-1})^{\text{T}}.$

Bayes Estimators

In our study of maximum likelihood estimators, we have inferred the value of an unknown parameter θ by choosing $\hat{\theta}$ to be the parameter that maximizes the likelihood of the observed data $\mathbf{x}$. In the Bayesian theory of parameter estimation, the unknown parameter itself is treated as a realization of a random experiment and endowed with its own "prior distribution." The idea is to use this distribution for θ, together with the measurements $\mathbf{x}$ drawn from a measurement distribution, to turn the prior distribution for θ into a posterior distribution that may be used for inference. In broad outline, Bayes estimators are constrained by the prior probability distribution for θ to be a priori likely, whereas in maximum likelihood theory the estimator is unconstrained by any prior information.

7.1 BAYES RISK FOR PARAMETER ESTIMATION

Figure 7.1 illustrates the context for Bayesian parameter estimation. Mother Nature conducts a random experiment that generates a parameter θ from a probability distribution $P(\theta)$. This parameter then codes or parameterizes the conditional, or measure-

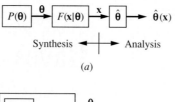

(a)

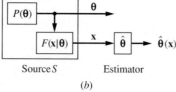

(b)

Figure 7.1 The context for Bayes
estimation. (*a*) Synthesis
and analysis. (*b*) Bayesian
source.

ment, distribution $F(\mathbf{x}|\boldsymbol{\theta})$. A random experiment generates a measurement $\mathbf{x}$ from this
distribution. The problem for the Bayesian experimentalist is to estimate $\boldsymbol{\theta}$ from $\mathbf{x}$
using the estimator $\hat{\boldsymbol{\theta}}(\mathbf{x})$. This description of Bayes estimation decomposes the problem
into a synthesis stage and an analysis stage, as illustrated in Figure 7.1(*a*). A slightly
more abstract description of Bayesian estimation is illustrated in Figure 7.1(*b*), where
a source of information S generates dependent random vectors $\boldsymbol{\theta}$ and $\mathbf{x}$. The problem
is to measure $\mathbf{x}$ and estimate $\boldsymbol{\theta}$.

Loss

The quality of the estimate $\hat{\boldsymbol{\theta}}(\mathbf{x})$ is measured by the real-valued loss function
$L[\boldsymbol{\theta}, \hat{\boldsymbol{\theta}}(\mathbf{x})]$. A typical example would be the quadratic function

$$L[\boldsymbol{\theta}, \hat{\boldsymbol{\theta}}(\mathbf{x})] = [\boldsymbol{\theta} - \hat{\boldsymbol{\theta}}(\mathbf{x})]^{\mathrm{T}}[\boldsymbol{\theta} - \hat{\boldsymbol{\theta}}(\mathbf{x})], \qquad\qquad 7.1$$

which assigns a loss equal to the Euclidean distance between the actual value of $\boldsymbol{\theta}$
and the estimated value $\hat{\boldsymbol{\theta}}(\mathbf{x})$.

Risk

Once the parameter $\boldsymbol{\theta}$ has been generated from the distribution $P(\boldsymbol{\theta})$, there remains
the randomness associated with the experiment that generates measurements from the
distribution $F(\mathbf{x}|\boldsymbol{\theta})$. It therefore makes sense to ask, "what is the average loss, or risk,
associated with the estimate $\hat{\boldsymbol{\theta}}(\mathbf{x})$?" The risk is the loss, averaged over the distribution

of the measurements, with $\boldsymbol{\theta}$ fixed:

$$R(\boldsymbol{\theta}, \hat{\boldsymbol{\theta}}) = E_{\boldsymbol{\theta}} L[\boldsymbol{\theta}, \hat{\boldsymbol{\theta}}(\mathbf{x})]$$

$$= \int L[\boldsymbol{\theta}, \hat{\boldsymbol{\theta}}(\mathbf{x})] \, dF(\mathbf{x}|\boldsymbol{\theta}). \qquad 7.2$$

The subscript on $E_{\boldsymbol{\theta}}$ indicates that the expectation is over the distribution of the random measurement $\mathbf{x}$, with $\boldsymbol{\theta}$ fixed. It is important to note that risk is a function only of $\boldsymbol{\theta}$ and the estimator $\hat{\boldsymbol{\theta}}$. The dependence on $\mathbf{x}$ that was evident in the loss function has been averaged out.

There is no estimator $\hat{\boldsymbol{\theta}}(\mathbf{x})$ that will minimize the risk uniformly in $\boldsymbol{\theta}$ for the simple reason that the estimator $\hat{\boldsymbol{\theta}}(\mathbf{x}) = \boldsymbol{\theta}_0$ is unbeatable at $\boldsymbol{\theta} = \boldsymbol{\theta}_0$ even though it ignores the data and performs badly for other values of the parameter $\boldsymbol{\theta}$.

Bayes Risk

Bayes risk is the risk, averaged over the prior distribution on $\boldsymbol{\theta}$:

$$R(\mathbf{p}, \hat{\boldsymbol{\theta}}) = ER(\boldsymbol{\theta}, \hat{\boldsymbol{\theta}})$$

$$= \int R(\boldsymbol{\theta}, \hat{\boldsymbol{\theta}}) \, dP(\boldsymbol{\theta}). \qquad 7.3$$

When the definition for risk is substituted into this equation, the result is

$$R(\mathbf{p}, \hat{\boldsymbol{\theta}}) = \int \left[\int L[\boldsymbol{\theta}, \hat{\boldsymbol{\theta}}(\mathbf{x})] \, dF(\mathbf{x}|\boldsymbol{\theta}) \right] dP(\boldsymbol{\theta})$$

$$= \int \int L[\boldsymbol{\theta}, \hat{\boldsymbol{\theta}}(\mathbf{x})] \, dP(\mathbf{x}, \boldsymbol{\theta})$$

$$dP(\mathbf{x}, \boldsymbol{\theta}) = dF(\mathbf{x}|\boldsymbol{\theta}) \, dP(\boldsymbol{\theta}).$$

This formula says that Bayes risk is the average of loss over the joint distribution for $\mathbf{x}$ and $\boldsymbol{\theta}$. When $F(\mathbf{x}|\boldsymbol{\theta})$ and $P(\boldsymbol{\theta})$ are continuous distributions, then the Bayes risk may be written

$$R(\mathbf{p}, \hat{\boldsymbol{\theta}}) = \int \int L[\boldsymbol{\theta}, \hat{\boldsymbol{\theta}}(\mathbf{x})] f(\mathbf{x}|\boldsymbol{\theta}) p(\boldsymbol{\theta}) \, d\mathbf{x} \, d\boldsymbol{\theta}$$

$$= \int \int L[\boldsymbol{\theta}, \hat{\boldsymbol{\theta}}(\mathbf{x})] f(\mathbf{x}, \boldsymbol{\theta}) \, d\mathbf{x} \, d\boldsymbol{\theta}. \qquad 7.4$$

The joint density function $f(\mathbf{x}, \boldsymbol{\theta})$ is the product of the conditional density $f(\mathbf{x}|\boldsymbol{\theta})$ and the prior density $p(\boldsymbol{\theta})$:

$$f(\mathbf{x}, \boldsymbol{\theta}) = f(\mathbf{x}|\boldsymbol{\theta}) p(\boldsymbol{\theta}). \qquad 7.5$$

The Bayes risk describes the average risk that the experimentalist runs when using the estimator $\hat{\boldsymbol{\theta}}(\mathbf{x})$ against Mother Nature, who is drawing her values of $\boldsymbol{\theta}$ from distribution $p(\boldsymbol{\theta})$.

Bayes Risk Estimator

The Bayes risk estimator minimizes Bayes risk:

$$\hat{\boldsymbol{\theta}}_B = \arg \min_{\hat{\boldsymbol{\theta}}} R(\mathbf{p}, \hat{\boldsymbol{\theta}}). \qquad 7.6$$

Read "arg min" as "the value of $\hat{\boldsymbol{\theta}}$ that minimizes," and call $\hat{\boldsymbol{\theta}}_B$ the Bayes estimator.

The Bayes risk estimator $\hat{\boldsymbol{\theta}}_B$ is a rule for mapping observations $\mathbf{x}$ into estimates $\hat{\boldsymbol{\theta}}_B(\mathbf{x})$. It depends on the conditional distribution of the measurements and on the prior distribution of the parameter. When this prior distribution is not known, then the minimax principle may be used.

Minimax Estimator

Suppose an experimentalist chooses an estimator and Mother Nature is allowed to choose her prior after the experimentalist has made his or her choice. If Mother Nature does not like the experimentalist, she will try to maximize the average risk for any choice of $\hat{\boldsymbol{\theta}}$:

$$\max_{\mathbf{p}} R(\mathbf{p}, \hat{\boldsymbol{\theta}}). \qquad 7.7$$

We can turn this into a game between Mother Nature and the experimentalist by allowing the experimentalist to observe the resulting average risk and permitting him or her to choose a decision rule to minimize this maximum average risk:

$$\min_{\hat{\boldsymbol{\theta}}} \max_{\mathbf{p}} R(\mathbf{p}, \hat{\boldsymbol{\theta}}). \qquad 7.8$$

The estimator that does this is called the minimax estimator $\tilde{\boldsymbol{\theta}}$:

$$\tilde{\boldsymbol{\theta}} = \arg \min_{\hat{\boldsymbol{\theta}}} \max_{\mathbf{p}} R(\mathbf{p}, \hat{\boldsymbol{\theta}}). \qquad 7.9$$

There is another way to play this game. Mother Nature goes first by selecting her prior. For each choice of a prior, the experimentalist is allowed to choose the Bayes risk estimator $\hat{\boldsymbol{\theta}}_B$ that minimizes Bayes risk. Mother Nature gets the last move: she is allowed to select the prior that maximizes the Bayes risk:

$$\max_{\mathbf{p}} \min_{\hat{\boldsymbol{\theta}}} R(\mathbf{p}, \hat{\boldsymbol{\theta}}) = \max_{\mathbf{p}} R(\mathbf{p}, \hat{\boldsymbol{\theta}}_B). \qquad 7.10$$

The resulting prior is called the least favorable prior:

$$\hat{\mathbf{p}} = \arg \max_{\mathbf{p}} R(\mathbf{p}, \hat{\boldsymbol{\theta}}_B). \qquad 7.11$$

Now here is the fundamental question: "When does $\min_{\hat{\boldsymbol{\theta}}} \max_{\mathbf{p}} R(\mathbf{p}, \hat{\boldsymbol{\theta}})$ equal

$\max_{\mathbf{p}} \min_{\hat{\boldsymbol{\theta}}} R(\mathbf{p}, \hat{\boldsymbol{\theta}})$?" This question is equivalent to the question, "when can the minimax estimator $\tilde{\boldsymbol{\theta}}$ be found by computing the Bayes estimator $\hat{\boldsymbol{\theta}}_B$ against the least favorable prior $\hat{\mathbf{p}}$?" These kinds of questions are fundamental to much of game theory. To answer them completely and unambiguously requires rather sophisticated mathematics. In the next section, we illustrate the considerations involved and consider cases where the minimax estimator is Bayes with respect to the least favorable prior. When this is true, then

$$\min_{\hat{\boldsymbol{\theta}}} \max_{\mathbf{p}} R(\mathbf{p}, \hat{\boldsymbol{\theta}}) = \max_{\mathbf{p}} \min_{\hat{\boldsymbol{\theta}}} R(\mathbf{p}, \hat{\boldsymbol{\theta}}), \qquad 7.12$$

and the risk for the estimator on the left (the minimax estimator) equals the risk for the estimator on the right (the Bayes risk estimator against the least favorable prior).

7.2   THE RISK SET

In our discussion of minimax and Bayes estimators, we need to admit the use of randomized estimators. There are many ways to construct such estimators, but we will be interested only in those that can be generated as random mixtures of deterministic estimators $\hat{\boldsymbol{\theta}}_1$ and $\hat{\boldsymbol{\theta}}_2$. These randomized estimators take the form

$$\hat{\boldsymbol{\theta}}(\mathbf{x}) = \begin{cases} \hat{\boldsymbol{\theta}}_1(\mathbf{x}), & \text{with probability } 0 \leq \gamma \leq 1 \\ \hat{\boldsymbol{\theta}}_2(\mathbf{x}), & \text{with probability } 0 \leq (1 - \gamma) \leq 1. \end{cases} \qquad 7.13$$

We can think of deterministic estimators as randomized estimators for which $\gamma = 1$. The risk associated with the randomized estimator $\hat{\boldsymbol{\theta}}$ is

$$R(\boldsymbol{\theta}, \hat{\boldsymbol{\theta}}) = \gamma R(\boldsymbol{\theta}, \hat{\boldsymbol{\theta}}_1) + (1 - \gamma)R(\boldsymbol{\theta}, \hat{\boldsymbol{\theta}}_2), \qquad 7.14$$

and the Bayes risk is

$$R(\mathbf{p}, \hat{\boldsymbol{\theta}}) = \gamma R(\mathbf{p}, \hat{\boldsymbol{\theta}}_1) + (1 - \gamma)R(\mathbf{p}, \hat{\boldsymbol{\theta}}_2). \qquad 7.15$$

Convexity

Suppose the parameter set is finite: $\boldsymbol{\theta} \in (\boldsymbol{\theta}_1, \ldots, \boldsymbol{\theta}_M)$ where each $\boldsymbol{\theta}_m$ is a $p \times 1$ vector. Assume the prior distribution $P(\boldsymbol{\theta})$ is characterized by

$$\mathbf{p} = (p_1, \ldots, p_M), \qquad 7.16$$

where p_m is the probability that Mother Nature selects parameter $\boldsymbol{\theta}_m$:

$$p_m = P[\boldsymbol{\theta} = \boldsymbol{\theta}_m]. \qquad 7.17$$

Corresponding to each estimator $\hat{\boldsymbol{\theta}}$ is the risk $R(\boldsymbol{\theta}, \hat{\boldsymbol{\theta}})$, which is completely specified by the risk vector $\mathbf{R}(\hat{\boldsymbol{\theta}}) = [R(\boldsymbol{\theta}_1, \hat{\boldsymbol{\theta}}), \ldots, R(\boldsymbol{\theta}_M, \hat{\boldsymbol{\theta}})]^T$. The mth coordinate of the risk vector is just the risk associated with parameter $\boldsymbol{\theta}_m$ and estimator $\hat{\boldsymbol{\theta}}$. The risk set is the set of all risk vectors that may be generated with randomized estimators:

$$\mathcal{R} = \{\mathbf{R}(\hat{\boldsymbol{\theta}}); \hat{\boldsymbol{\theta}} : \text{randomized estimator}\}. \qquad 7.18$$

The risk set $\mathcal{R}$ is convex. To see this, let $\hat{\boldsymbol{\theta}}_1$ and $\hat{\boldsymbol{\theta}}_2$ be two randomized estimators with respective risks $R(\boldsymbol{\theta}, \hat{\boldsymbol{\theta}}_1)$ and $R(\boldsymbol{\theta}, \hat{\boldsymbol{\theta}}_2)$. The estimator $\hat{\boldsymbol{\theta}} = \gamma\hat{\boldsymbol{\theta}}_1 + (1 - \gamma)\hat{\boldsymbol{\theta}}_2$ is another randomized estimator, so its risk

$$R(\boldsymbol{\theta}, \hat{\boldsymbol{\theta}}) = \gamma R(\boldsymbol{\theta}, \hat{\boldsymbol{\theta}}_1) + (1 - \gamma)R(\boldsymbol{\theta}, \hat{\boldsymbol{\theta}}_2) \qquad 7.19$$

is contained in the risk set $\mathcal{R}$. This is illustrated in Figure 7.2.

Bayes Rules

The Bayes risk associated with the prior distribution P and the estimator $\hat{\boldsymbol{\theta}}$ is

$$R(\mathbf{p}, \hat{\boldsymbol{\theta}}) = \mathbf{p}^\mathsf{T}\mathbf{R}(\hat{\boldsymbol{\theta}}). \qquad 7.20$$

The locus of risks for which Bayes risk is constant at the value r is illustrated in Figure 7.3. This locus intersects the $45°$ line of Figure 7.3 at the values $R(\boldsymbol{\theta}_1, \hat{\boldsymbol{\theta}}) = R(\boldsymbol{\theta}_2, \hat{\boldsymbol{\theta}}) = \cdots = R(\boldsymbol{\theta}_M, \hat{\boldsymbol{\theta}}) = r$, as illustrated. The Bayes risk estimator is found by moving the line $\mathbf{p}^\mathsf{T}\mathbf{R} = r$ perpendicular to $\mathbf{p}$ until the minimum intersection with the $45°$ line is found. This is illustrated in Figure 7.3. If Mother Nature is perverse, she adjusts her prior $\mathbf{p}$ so that the intersection with the $45°$ line is maximized, as illustrated in Figure 7.4. The corresponding prior $\mathbf{p}$ is least favorable, and the estimator $\hat{\boldsymbol{\theta}}$ is Bayes with respect to it. Note that the risk along each dimension in the risk set is equal:

$$R(\boldsymbol{\theta}_1, \hat{\boldsymbol{\theta}}) = \cdots = R(\boldsymbol{\theta}_M, \hat{\boldsymbol{\theta}}) = \max_{\mathbf{p}} \min_{\hat{\boldsymbol{\theta}}} R(\mathbf{p}, \hat{\boldsymbol{\theta}}). \qquad 7.21$$

The Bayes rule against the least favorable prior is said to be an equalizer rule because the risk is equal over $\boldsymbol{\theta}$.

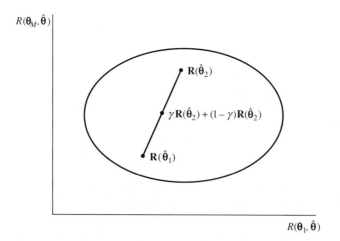

Figure 7.2 Convex risk set.

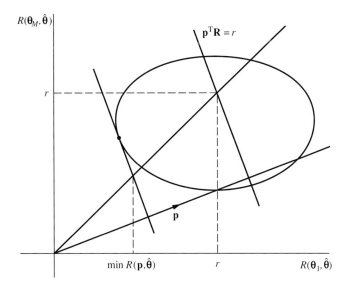

Figure 7.3 Bayes rules illustrated.

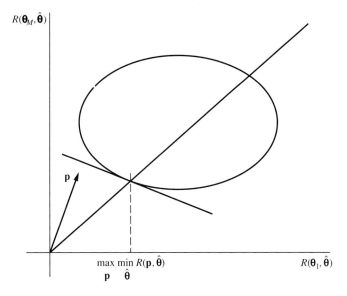

Figure 7.4 Least favorable prior illustrated.

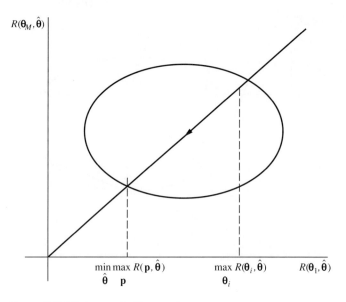

Figure 7.5 Minimax rule illustrated.

Minimax Rules

The worst-case risk associated with decision rule $\hat{\boldsymbol{\theta}}$ is

$$\max_{\boldsymbol{\theta}_m} R\left(\boldsymbol{\theta}_m, \hat{\boldsymbol{\theta}}\right). \qquad 7.22$$

Mother Nature can assign all of her probability mass to the maximizing value of $\boldsymbol{\theta}_m$ to achieve

$$\max_{\mathbf{p}} R\left(\mathbf{p}, \hat{\boldsymbol{\theta}}\right) = \max_{\boldsymbol{\theta}_m} R\left(\boldsymbol{\theta}_m, \hat{\boldsymbol{\theta}}\right). \qquad 7.23$$

Therefore, the minimax rule can be found as follows:

$$\tilde{\boldsymbol{\theta}} = \arg\min_{\hat{\boldsymbol{\theta}}} \max_{\mathbf{p}} R\left(\mathbf{p}, \hat{\boldsymbol{\theta}}\right) = \arg\min_{\hat{\boldsymbol{\theta}}} \max_{\boldsymbol{\theta}_m} R\left(\boldsymbol{\theta}_m, \hat{\boldsymbol{\theta}}\right). \qquad 7.24$$

Figure 7.5 shows that all risk vectors on a vertical or horizontal line have the same worst-case risk, namely $\max_{\boldsymbol{\theta}_m} R\left(\boldsymbol{\theta}_m, \hat{\boldsymbol{\theta}}\right)$. Geometrically, the minimax rule is found by crawling along the 45° line of Figure 7.5, as illustrated, to find the minimum value

$$\min_{\hat{\boldsymbol{\theta}}} \max_{\boldsymbol{\theta}_m} R\left(\boldsymbol{\theta}_m, \hat{\boldsymbol{\theta}}\right).$$

The solution point coincides with the solution point for the Bayes risk estimator against the least favorable prior, provided the risk set is bounded and closed. In this case, the minimax rule equals the Bayes rule against the least favorable prior.

7.3   COMPUTING BAYES RISK ESTIMATORS

The results in Sections 7.1 and 7.2 bring important information about the structure of Bayes and minimax estimators. However, they bring little insight about the actual computation of a Bayes estimator. In this section, we derive several important computation formulas for Bayes estimators and work out an example that illustrates how the formulas are used.

Continuous Case

Let $F(\mathbf{x}, \boldsymbol{\theta})$ denote the joint distribution of the measurement $\mathbf{x}$ and the parameter $\boldsymbol{\theta}$. Assume that the joint distribution has a probability density function, which we denote by $f(\mathbf{x}, \boldsymbol{\theta})$. The Bayes risk is

$$R(\mathbf{p}, \hat{\boldsymbol{\theta}}) = \int \int L[\boldsymbol{\theta}, \hat{\boldsymbol{\theta}}(\mathbf{x})] f(\mathbf{x}, \boldsymbol{\theta}) \, d\mathbf{x} \, d\boldsymbol{\theta}. \qquad 7.25$$

Sometimes the joint density function $f(\mathbf{x}, \boldsymbol{\theta})$ is specified a priori. However, more typically it is computed from a measurement density $f(\mathbf{x}|\boldsymbol{\theta})$ and a prior density $p(\boldsymbol{\theta})$:

$$f(\mathbf{x}, \boldsymbol{\theta}) = f(\mathbf{x}|\boldsymbol{\theta}) p(\boldsymbol{\theta}). \qquad 7.26$$

From Bayes rule for densities, we can also write the joint density as

$$f(\mathbf{x}, \boldsymbol{\theta}) = f(\boldsymbol{\theta}|\mathbf{x}) f(\mathbf{x}), \qquad 7.27$$

where $f(\boldsymbol{\theta}|\mathbf{x})$ is the posterior density for $\boldsymbol{\theta}$, given $\mathbf{x}$, and $f(\mathbf{x})$ is the marginal density for $\mathbf{x}$:

$$f(\boldsymbol{\theta}|\mathbf{x}) = \frac{f(\mathbf{x}, \boldsymbol{\theta})}{f(\mathbf{x})} = \frac{f(\mathbf{x}|\boldsymbol{\theta})}{f(\mathbf{x})} p(\boldsymbol{\theta}) \qquad 7.28$$

$$f(\mathbf{x}) = \int f(\mathbf{x}, \boldsymbol{\theta}) \, d\boldsymbol{\theta} = \int f(\mathbf{x}|\boldsymbol{\theta}) p(\boldsymbol{\theta}) \, d\boldsymbol{\theta}.$$

There is an important physical interpretation of this formula for the conditional density of $\boldsymbol{\theta}$, given $\mathbf{x}$. The prior density $p(\boldsymbol{\theta})$ is mapped into the posterior density $f(\boldsymbol{\theta}|\mathbf{x})$ by the ratio of the measurement density to the marginal density, as illustrated in Figure 7.6. These results are important because they allow us to recast Bayes risk in terms of the posterior density for $\boldsymbol{\theta}$, given $\mathbf{x}$:

$$R(\mathbf{p}, \hat{\boldsymbol{\theta}}) = \int \left[\int L[\boldsymbol{\theta}, \hat{\boldsymbol{\theta}}(\mathbf{x})] f(\boldsymbol{\theta}|\mathbf{x}) \, d\boldsymbol{\theta} \right] f(\mathbf{x}) \, d\mathbf{x}. \qquad 7.29$$

$$p(\boldsymbol{\theta}) \xrightarrow{\quad \mathbf{x} \quad} f(\boldsymbol{\theta}|\mathbf{x}) = \frac{f(\mathbf{x}|\boldsymbol{\theta})}{f(\mathbf{x})} p(\boldsymbol{\theta})$$

Figure 7.6 The data $\mathbf{x}$ used to map the prior density into the posterior density.

The marginal density $f(\mathbf{x})$ is nonnegative, so Bayes risk may be minimized by minimizing the inner integral for each value of $\mathbf{x}$:

$$\hat{\boldsymbol{\theta}}_B(\mathbf{x}) = \arg \min_{\hat{\boldsymbol{\theta}}} \int L[\boldsymbol{\theta}, \hat{\boldsymbol{\theta}}(\mathbf{x})] f(\boldsymbol{\theta}|\mathbf{x}) \, d\boldsymbol{\theta}. \qquad 7.30$$

This result says that the Bayes risk estimator $\hat{\boldsymbol{\theta}}_B$ is the estimator that minimizes the conditional risk. Conditional risk is the loss averaged over the conditional distribution of $\boldsymbol{\theta}$, given $\mathbf{x}$. To turn this general finding into a computing formula, we need to consider some typical loss functions.

Quadratic Loss and the Conditional Mean

When the loss function $L[\boldsymbol{\theta}, \hat{\boldsymbol{\theta}}(\mathbf{x})]$ is the quadratic form

$$L[\boldsymbol{\theta}, \hat{\boldsymbol{\theta}}(\mathbf{x})] = [\boldsymbol{\theta} - \hat{\boldsymbol{\theta}}(\mathbf{x})]^T [\boldsymbol{\theta} - \hat{\boldsymbol{\theta}}(\mathbf{x})], \qquad 7.31$$

we may write the conditional Bayes risk as follows:

$$\int [\boldsymbol{\theta} - \hat{\boldsymbol{\theta}}(\mathbf{x})]^T [\boldsymbol{\theta} - \hat{\boldsymbol{\theta}}(\mathbf{x})] f(\boldsymbol{\theta}|\mathbf{x}) \, d\boldsymbol{\theta}. \qquad 7.32$$

The gradients of this conditional risk, with respect to $\hat{\boldsymbol{\theta}}(\mathbf{x})$, are

$$\frac{\partial}{\partial \hat{\boldsymbol{\theta}}}(\ ) = -2 \int [\boldsymbol{\theta} - \hat{\boldsymbol{\theta}}(\mathbf{x})] f(\boldsymbol{\theta}|\mathbf{x}) \, d\boldsymbol{\theta}$$

$$\frac{\partial}{\partial \hat{\boldsymbol{\theta}}} \left[\frac{\partial}{\partial \hat{\boldsymbol{\theta}}}(\ ) \right]^T = \mathbf{I} > \mathbf{0}. \qquad 7.33$$

The first gradient may be equated to zero to give a computing formula for the Bayes risk estimator $\hat{\boldsymbol{\theta}}_B$, and the second may be used to establish that $\hat{\boldsymbol{\theta}}_B$ is a minimizing solution. The formula for $\hat{\boldsymbol{\theta}}_B$ is

$$\hat{\boldsymbol{\theta}}_B(\mathbf{x}) = \int \boldsymbol{\theta} f(\boldsymbol{\theta}|\mathbf{x}) \, d\boldsymbol{\theta} = E[\boldsymbol{\theta}|\mathbf{x}]. \qquad 7.34$$

We say, "the Bayes risk estimator under quadratic loss is the conditional mean of $\boldsymbol{\theta}$, given $\mathbf{x}$." In a nutshell, Bayes estimation under quadratic loss comes down to the computation of the mean of the conditional density $f(\boldsymbol{\theta}|\mathbf{x})$. Nonlinear filtering is a generic term for this calculation, because generally the answer is a nonlinear function of the measurement $\mathbf{x}$.

The Bayes risk under quadratic loss is, in fact, mean-squared error. So, the Bayes risk estimator, which is the conditional mean of $\boldsymbol{\theta}$, given $\mathbf{x}$, minimizes mean-squared error. Furthermore, the mean of $\hat{\boldsymbol{\theta}}_B(\mathbf{x})$ is

$$E \,\hat{\boldsymbol{\theta}}_B(\mathbf{x}) = \int \hat{\boldsymbol{\theta}}_B(\mathbf{x}) f(\mathbf{x}) \, d\mathbf{x} = \int \left[\int \boldsymbol{\theta} f(\boldsymbol{\theta}|\mathbf{x}) \, d\boldsymbol{\theta} \right] f(\mathbf{x}) \, d\mathbf{x}$$

$$= E\boldsymbol{\theta}. \qquad 7.35$$

So we say that the Bayes estimator under quadratic loss (i.e., the conditional mean estimator) is an unbiased estimator of $\boldsymbol{\theta}$. This means that it is also the minimum variance unbiased estimator of the random parameter $\boldsymbol{\theta}$.

Uniform Loss and the Maximum of the A Posteriori Density

Assume that the loss function is the following convex function of uniform errors:

$$L[\boldsymbol{\theta}, \hat{\boldsymbol{\theta}}(\mathbf{x})] = \boldsymbol{\pi}^{\mathsf{T}} \mathbf{w} \qquad 7.36$$

$$\mathbf{w}^{\mathsf{T}} = [w_1 \quad \cdots \quad w_p]; \; w_i = \begin{cases} 1, & |\theta_i - \phi_i(\mathbf{x})| > \varepsilon \\ 0, & \text{otherwise.} \end{cases}$$

The conditional risk is

$$\int d\boldsymbol{\theta} f(\boldsymbol{\theta}|\mathbf{x}) \boldsymbol{\pi}^{\mathsf{T}} \mathbf{w} = \sum_{i=1}^{p} \pi_i \left[1 - \int_{\phi_i(\mathbf{x})-\varepsilon}^{\phi_i(\mathbf{x})+\varepsilon} d\theta_i f(\theta_i|\mathbf{x}) \right]. \qquad 7.37$$

This is minimized by selecting

$$\phi_i(\mathbf{x}) = \arg\max \int_{\phi_i(\mathbf{x})-\varepsilon}^{\phi_i(\mathbf{x})+\varepsilon} d\theta_i f(\theta_i|\mathbf{x}). \qquad 7.38$$

For small ε, this produces the result

$$\phi_i(\mathbf{x}) = \arg\max_{\theta_i} f(\theta_i|\mathbf{x}), \qquad 7.39$$

which is the maximum of the a posteriori (or conditional) density $f(\theta_i|\mathbf{x})$. It is interesting to note that the computation of $\max f(\theta_i|\mathbf{x})$ may be performed by maximizing $\ln f(\theta_i|\mathbf{x})$. Using Bayes rule, this is

$$\begin{aligned} \phi_i(\mathbf{x}) &= \arg\max \left[\ln f(\mathbf{x}|\theta_i) + \ln p(\theta_i) - \ln f(\mathbf{x}) \right] \\ &= \arg\max_{\theta_i} \left[\ln f(\mathbf{x}|\theta_i) + \ln p(\theta_i) \right]. \end{aligned} \qquad 7.40$$

Both the density $f(\mathbf{x}|\theta_i)$ and the prior probability $p(\theta_i)$ play a role in the determination of the maximum a posteriori probability estimator.

Example 7.1 (Imperfect Geiger Counter)

The radioactive source of Figure 7.7 emits n radioactive particles, and an imperfect Geiger counter records $k \le n$ of them. Our problem is to estimate n from the measurement k. To proceed, we assume that n is drawn from a Poisson distribution with known parameter λ:

$$P[n] = e^{-\lambda} \frac{\lambda^n}{n!}; \qquad n \ge 0. \qquad 7.41$$

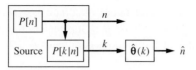

Figure 7.7 Imperfect Geiger counter.

This is Mother Nature's prior distribution for the parameter n. The number of recorded counts is binomial distributed:

$$P[k|n] = \binom{n}{k} p^k (1-p)^{n-k}; \qquad 0 \le k \le n. \qquad 7.42$$

This is the conditional, or measurement, density for the problem. The mean and variance of the binomial distribution are np and $np(1-p)$. In order to proceed with our Bayesian analysis, we need to compute the posterior distribution of n, given k:

$$P[n|k] = \frac{P[n,k]}{P[k]}. \qquad 7.43$$

This means that we need the joint distribution for n and k and the marginal distribution for k. Call $P[k,n]$ the joint distribution of k and n:

$$P[k,n] = P[k|n]P[n] = \binom{n}{k} p^k (1-p)^{n-k} e^{-\lambda} \frac{\lambda^n}{n!}; \quad 0 \le k \le n \dots, \quad n \ge 0. \qquad 7.44$$

The marginal distribution of k is therefore Poisson with rate parameter λp (see Problem 7.5):

$$P[k] = \sum_{n=k}^{\infty} P[k,n] = e^{-\lambda p} \frac{(\lambda p)^k}{k!}; \qquad k \ge 0. \qquad 7.45$$

From the joint and marginal distributions, we compute the posterior distribution of n, given k:

$$P[n|k] = \frac{\binom{n}{k} p^k (1-p)^{n-k} e^{-\lambda} \lambda^n / n!}{e^{-\lambda p} (\lambda p)^k / k!}$$

$$= \frac{1}{(n-k)!} [\lambda(1-p)]^{n-k} e^{-\lambda(1-p)}; \qquad n \ge k. \qquad 7.46$$

Jaynes [1958] calls this a Poisson distribution with parameter $\lambda(1-p)$ and displacement k. The mean and variance of this distribution are, respectively, $k + \lambda(1-p)$ and $\lambda(1-p)$. (See Problem 7.6.) Therefore the conditional mean and conditional variance of the random variable n, given the random variable k, may be written

$$E[n|k] = k + \lambda(1 - p) \qquad 7.47$$

$$E[(n - E[n|k])^2|k] = \lambda(1 - p).$$

Note that the conditional variance is independent of k. This often happens in Bayesian inference—but not always!

When the loss function is quadratic so that Bayes risk is mean-squared error, then the Bayes risk estimator is the conditional mean of n, given k:

$$\hat{n} = E[n|k] = k + \lambda(1 - p). \qquad 7.48$$

Can you see that the counts k are corrected by $\lambda(1 - p)$, the average number of missed counts? This estimator is unbiased, but it is *not* conditionally unbiased. What we mean is that $E\hat{n} = En$, but $E[\hat{n}|n] \neq n$:

$$E\hat{n} = Ek + \lambda(1 - p) = \lambda p + \lambda(1 - p) = \lambda = En \qquad 7.49$$

$$E[\hat{n}|n] = E[k|n] + \lambda(1 - p) = np + \lambda(1 - p) \neq n.$$

The mean-squared error of the estimator $\hat{n}$ is

$$E(n - \hat{n})^2 = EE[(n - E[n|k])^2|k]. \qquad 7.50$$

The inner expectation is just the conditional variance of n, given k, and the outer expectation is over the distribution of k, so we can write

$$E(n - \hat{n})^2 = E\lambda(1 - p) = \lambda(1 - p). \qquad 7.51$$

Note that the conditional mean-squared error is independent of k.

It is interesting to compare the conditional mean estimator $\hat{n}$ with the ad hoc estimator of n that corrects for the imperfect Geiger counter as follows:

$$\tilde{n} = \frac{k}{p}. \qquad 7.52$$

This estimator is unbiased *and* conditionally unbiased:

$$E\tilde{n} = \frac{1}{p}Ek = \frac{\lambda p}{p} = \lambda = En \qquad 7.53$$

$$E[\tilde{n}|n] = \frac{1}{p}E[k|n] = \frac{1}{p}np = n.$$

The mean-squared error of $\tilde{n}$ is

$$E[(n - \tilde{n})^2] = EE[(n - \tilde{n})^2|n] = E\frac{1}{p^2}E[(np - k)^2|n]. \qquad 7.54$$

The inner expectation on the far right is just the conditional variance of k, given n, so

$$E\left[(n - \tilde{n})^2\right] = E\,\frac{1}{p^2}\,np(1 - p) = \frac{\lambda}{p}(1 - p).$$ 7.55

This mean-squared error is greater than the mean-squared error of the conditional mean estimator $\hat{n}$. In fact,

$$E\left[(n - \tilde{n})^2\right] = \frac{1}{p}E\left[(n - \hat{n})^2\right].$$ 7.56

For bad geiger counters, $p \ll 1$, and this performance difference is substantial. ∎

7.4   BAYES SUFFICIENCY AND CONJUGATE PRIORS

The Bayes risk estimator minimizes the conditional risk:

$$\hat{\boldsymbol{\theta}}_B = \arg\min_{\hat{\boldsymbol{\theta}}} \int d\boldsymbol{\theta} f(\boldsymbol{\theta}|\mathbf{x})L[\boldsymbol{\theta}, \hat{\boldsymbol{\theta}}(\mathbf{x})].$$ 7.57

Clearly $\hat{\boldsymbol{\theta}}_B$ is a function of $\mathbf{x}$, but recall the way the conditional density $f(\boldsymbol{\theta}|\mathbf{x})$ is computed from the prior density $p(\boldsymbol{\theta})$:

$$f(\boldsymbol{\theta}|\mathbf{x}) = \frac{f(\mathbf{x}|\boldsymbol{\theta})p(\boldsymbol{\theta})}{\int f(\mathbf{x}|\boldsymbol{\theta})p(\boldsymbol{\theta})\,d\boldsymbol{\theta}}.$$ 7.58

If the measurement density can be factored as the product

$$f(\mathbf{x}|\boldsymbol{\theta}) = a(\mathbf{t})b(\mathbf{t}|\boldsymbol{\theta}),$$ 7.59

then according to our previous notion of sufficiency, we would say that $\mathbf{t} = \mathbf{T}(\mathbf{x})$ is sufficient for $\boldsymbol{\theta}$. But what is the implication in the context of Bayesian estimation? To answer this question, write the conditional density of $\boldsymbol{\theta}$, given $\mathbf{x}$, as follows:

$$\begin{aligned} f(\boldsymbol{\theta}|\mathbf{x}) &= \frac{a(\mathbf{t})b(\mathbf{t}|\boldsymbol{\theta})p(\boldsymbol{\theta})}{\int a(\mathbf{t})b(\mathbf{t}|\boldsymbol{\theta})p(\boldsymbol{\theta})\,d\boldsymbol{\theta}} \\ &= \frac{b(\mathbf{t}|\boldsymbol{\theta})p(\boldsymbol{\theta})}{\int b(\mathbf{t}|\boldsymbol{\theta})p(\boldsymbol{\theta})\,d\boldsymbol{\theta}} = f(\boldsymbol{\theta}|\mathbf{t}). \end{aligned}$$ 7.60

The sufficient statistic has the property that the conditional density for $\boldsymbol{\theta}$, given $\mathbf{t}$, is the same as the conditional density for $\boldsymbol{\theta}$, given $\mathbf{x}$. Therefore the conditioning on the sample $\mathbf{x}$ may be replaced by conditioning on the sufficient statistic $\mathbf{t}$. In fact, the Bayes risk estimator may be rewritten as

$$\hat{\boldsymbol{\theta}}_B = \arg\min_{\hat{\boldsymbol{\theta}}} \int d\boldsymbol{\theta} f[\boldsymbol{\theta}|\mathbf{t}]L[\boldsymbol{\theta}, \hat{\boldsymbol{\theta}}(\mathbf{t})].$$ 7.61

This makes $\hat{\boldsymbol{\theta}}_B$ a function of the sufficient statistic $\mathbf{t}$. We say that $\mathbf{t} = \mathbf{T}(\mathbf{x})$ is Bayes sufficient. A remarkable aspect of this is that sufficiency of $\mathbf{t}$ in the measurement density $f(\mathbf{x}|\boldsymbol{\theta})$ leads to sufficiency of $\mathbf{t}$ in the conditional, or a posteriori, density for $\boldsymbol{\theta}$.

There is one more wrinkle in this story. Assume that the prior density $p(\boldsymbol{\theta})$ is indexed or parameterized by the parameter $\boldsymbol{\theta}_0$. Denote this $p_{\boldsymbol{\theta}_0}(\boldsymbol{\theta})$. If the conditional density for $\boldsymbol{\theta}$, given $\mathbf{t}$, works out to be

$$f[\boldsymbol{\theta}|\mathbf{t}] = p_{\mathbf{t}}(\boldsymbol{\theta}), \qquad\qquad 7.62$$

then we say that the prior $p_{\boldsymbol{\theta}_0}(\boldsymbol{\theta})$ is a conjugate prior. The implication is that the prior density $p_{\boldsymbol{\theta}_0}(\boldsymbol{\theta})$ reproduces itself, with the data $\mathbf{x}$ updating the parameter $\boldsymbol{\theta}_0$ through the sufficient statistic $\mathbf{t} = \mathbf{T}(\mathbf{x})$.

Example 7.2

Let $\mathbf{X} = (X_0, X_1, \ldots, X_{N-1})$ denote a random sample of scalar exponential random variables:

$$f(x|\theta) = \theta e^{-\theta x}; \qquad x > 0, \quad \theta > 0. \qquad\qquad 7.63$$

The measurement density for the random sample is

$$f(\mathbf{x}|\theta) = \prod_{n=0}^{N-1} f(x_n|\theta) = \theta^N e^{-\theta t} \qquad\qquad 7.64$$

$$t = \sum_{n=0}^{N-1} x_n.$$

The statistic t is sufficient for θ.

Assume that the parameter θ is drawn from a gamma distribution with parameters α and β:

$$p(\theta) = \frac{\beta^\alpha}{\Gamma(\alpha)} \theta^{\alpha-1} e^{-\beta\theta}; \qquad (\theta > 0), \quad (\alpha, \beta > 0). \qquad\qquad 7.65$$

The marginal density for $\mathbf{x}$ is

$$f(\mathbf{x}) = \int_0^\infty \theta^N e^{-\theta t} \frac{\beta^\alpha}{\Gamma(\alpha)} \theta^{\alpha-1} e^{-\beta\theta} \, d\theta$$

$$= \frac{\beta^\alpha}{\Gamma(\alpha)} \int_0^\infty \theta^{N+\alpha-1} e^{-\theta(\beta+t)} \, d\theta \qquad\qquad 7.66$$

$$= \frac{\beta^\alpha}{\Gamma(\alpha)} \frac{\Gamma(N+\alpha)}{(\beta+t)^{N+\alpha}}.$$

The conditional density of θ, given $\mathbf{x}$, is therefore

$$
\begin{aligned}
f(\theta|\mathbf{x}) &= \frac{f(\mathbf{x}|\theta)p(\theta)}{f(\mathbf{x})} \\
&= \frac{[\beta + t]^{N+\alpha}}{\Gamma(N + \alpha)} \theta^{N+\alpha-1} e^{-\theta(\beta+t)} \qquad\qquad 7.67 \\
&= f(\theta|t).
\end{aligned}
$$

The posterior density for θ is also gamma but with parameters $N + \alpha$ and $\beta + t$, so the random sample of size N has updated the parameter α to $\alpha + N$, and the sufficient statistic t has updated the parameter β to $\beta + t$. The gamma prior is a conjugate prior.

The mean and variance of the posterior gamma density are

$$
\hat{\theta} = E[\theta|\mathbf{x}] = E[\theta|t] = \frac{\alpha + N}{\beta + t} \qquad\qquad 7.68
$$

$$
\sigma^2 = E\left[(\theta - \hat{\theta})^2|\mathbf{x}\right] = E\left[(\theta - \hat{\theta})^2|t\right] = \frac{\alpha + N}{[\beta + t]^2}. \qquad\qquad 7.69
$$

The Bayes risk estimator under squared error loss is $\hat{\theta}$. The Bayes risk, or variance, of the estimator is

$$
V = E\,\frac{\alpha + N}{[\beta + t]^2}. \qquad\qquad 7.70
$$

∎

7.5   THE MULTIVARIATE NORMAL MODEL

The theory of Bayesian inference produces its most elegant results when applied to the multivariate normal distribution. In order to cast these results in a form commonly found in the signal processing literature, we consider the experimental setup illustrated in Figure 7.8(a). Mother Nature and Father Nature produce the jointly distributed random vectors $\mathbf{x}$ and $\mathbf{y}$. Think of $\mathbf{x}$ as a random signal vector that is to be estimated from $\mathbf{y}$. It plays the same role that the parameter $\boldsymbol{\theta}$ played in earlier sections. Think of the vector $\mathbf{y}$ as the output of an imperfect channel or measuring device that produces a random measurement that is statistically dependent upon $\mathbf{x}$. The problem is to estimate $\mathbf{x}$ from $\mathbf{y}$. We shall assume that $\mathbf{x}$ is a $p \times 1$ vector and that $\mathbf{y}$ is an $N \times 1$ vector. When $p < N$ we have a filtering or enhancement problem, and when $p \geq N$ we have a reconstruction or interpolation problem.

Conditional Distribution of y, Given x

We assume that the random vectors $\mathbf{x}$ and $\mathbf{y}$ are jointly distributed according to the following normal distribution:

$$
\begin{bmatrix} \mathbf{x} \\ \mathbf{y} \end{bmatrix} = N\left[\begin{bmatrix} \mathbf{0} \\ \mathbf{0} \end{bmatrix}, \begin{bmatrix} \mathbf{R}_{xx} & \mathbf{R}_{xy} \\ \mathbf{R}_{yx} & \mathbf{R}_{yy} \end{bmatrix} \right]. \qquad\qquad 7.71
$$

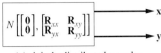

(a) jointly distributed **x** and **y**

(b) marginally distributed **x** and conditionally distributed **y**

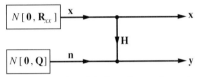

(c) channel model: linearly transformed **x** plus noise **n**

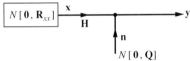

(d) signal plus noise model

Figure 7.8 Four equivalent channel representations for
the multivariate normal problem.

The auto- and cross-covariances are defined as follows:

$$\mathbf{R} = E \begin{bmatrix} \mathbf{x} \\ \mathbf{y} \end{bmatrix} [\, \mathbf{x}^T \quad \mathbf{y}^T \,]$$

$$= \begin{bmatrix} \mathbf{R}_{xx} & \mathbf{R}_{xy} \\ \mathbf{R}_{yx} & \mathbf{R}_{yy} \end{bmatrix}$$

$$\mathbf{R}_{xx} = E\mathbf{x}\mathbf{x}^T; \qquad \mathbf{R}_{yy} = E\mathbf{y}\mathbf{y}^T \qquad\qquad 7.72$$

$$\mathbf{R}_{xy} = E\mathbf{x}\mathbf{y}^T = \mathbf{R}_{yx}^T; \qquad \mathbf{R}_{yx} = E\mathbf{y}\mathbf{x}^T.$$

The random vector **x** is distributed as $N[\mathbf{0}, \mathbf{R}_{xx}]$, and the random vector **y** is distributed
as $N[\mathbf{0}, \mathbf{R}_{yy}]$. The conditional distribution of **y**, given **x**, may be computed from the
definition of conditional density:

$$f(\mathbf{y}|\mathbf{x}) = \frac{f(\mathbf{x}, \mathbf{y})}{f(\mathbf{x})}$$

$$= \frac{(2\pi)^{-N/2}(2\pi)^{-p/2}|\mathbf{R}|^{-1/2}\exp\{-(1/2)(\mathbf{x}^T \ \mathbf{y}^T)\mathbf{R}^{-1}\binom{\mathbf{x}}{\mathbf{y}}\}}{(2\pi)^{-p/2}|\mathbf{R}_{xx}|^{-1/2}\exp\{-(1/2)\mathbf{x}^T\mathbf{R}_{xx}^{-1}\mathbf{x}\}}. \qquad 7.73$$

In this formula, $\mathbf{R}$ is the structured covariance matrix

$$\mathbf{R} = \begin{bmatrix} \mathbf{R}_{xx} & \mathbf{R}_{xy} \\ \mathbf{R}_{yx} & \mathbf{R}_{yy} \end{bmatrix}. \qquad 7.74$$

The inverse of this structured matrix may be written in two different ways. The first is used to derive the conditional distribution of $\mathbf{y}$, given $\mathbf{x}$, and the second is used to derive the conditional distribution of $\mathbf{x}$, given $\mathbf{y}$. Recall from Chapter 2 the formulas for inverting a structured correlation matrix:

1. $\begin{bmatrix} \mathbf{R}_{xx} & \mathbf{R}_{xy} \\ \mathbf{R}_{yx} & \mathbf{R}_{yy} \end{bmatrix}^{-1} = \begin{bmatrix} \mathbf{R}_{xx}^{-1} & \mathbf{0} \\ \mathbf{0} & \mathbf{0} \end{bmatrix} + \begin{bmatrix} -\mathbf{R}_{xx}^{-1}\mathbf{R}_{xy} \\ \mathbf{I} \end{bmatrix} [\mathbf{Q}^{-1}] [-\mathbf{R}_{yx}\mathbf{R}_{xx}^{-1} \quad \mathbf{I}]$

$$\mathbf{Q} = \mathbf{R}_{yy} - \mathbf{R}_{yx}\mathbf{R}_{xx}^{-1}\mathbf{R}_{xy}, \qquad 7.75$$

2. $\begin{bmatrix} \mathbf{R}_{xx} & \mathbf{R}_{xy} \\ \mathbf{R}_{yx} & \mathbf{R}_{yy} \end{bmatrix}^{-1} = \begin{bmatrix} \mathbf{0} & \mathbf{0} \\ \mathbf{0} & \mathbf{R}_{yy}^{-1} \end{bmatrix} + \begin{bmatrix} \mathbf{I} \\ -\mathbf{R}_{yy}^{-1}\mathbf{R}_{yx} \end{bmatrix} [\mathbf{P}^{-1}] [\mathbf{I} \quad -\mathbf{R}_{xy}\mathbf{R}_{yy}^{-1}]$

$$\mathbf{P} = \mathbf{R}_{xx} - \mathbf{R}_{xy}\mathbf{R}_{yy}^{-1}\mathbf{R}_{yx}.$$

The corresponding determinant equations are

1. $\det \begin{bmatrix} \mathbf{R}_{xx} & \mathbf{R}_{xy} \\ \mathbf{R}_{yx} & \mathbf{R}_{yy} \end{bmatrix} = \det \mathbf{R}_{xx} \det \mathbf{Q},$

$$\qquad 7.76$$

2. $\det \begin{bmatrix} \mathbf{R}_{xx} & \mathbf{R}_{xy} \\ \mathbf{R}_{yx} & \mathbf{R}_{yy} \end{bmatrix} = \det \mathbf{R}_{yy} \det \mathbf{P}.$

The matrix $\mathbf{Q}$ is called the Schur complement of $\mathbf{R}_{yy}$, and the matrix $\mathbf{P}$ is called the Schur complement of $\mathbf{R}_{xx}$.

When the formula (1) for the inverse of $\mathbf{R}$ is substituted into the equation for the conditional density of $\mathbf{y}$, given $\mathbf{x}$, the following formula results:

$$f(\mathbf{y}|\mathbf{x}) = (2\pi)^{-N/2}|\mathbf{Q}|^{-1/2} \exp\left\{ -\frac{1}{2}(\mathbf{y} - \mathbf{R}_{yx}\mathbf{R}_{xx}^{-1}\mathbf{x})^{\mathsf{T}}\mathbf{Q}^{-1}(\mathbf{y} - \mathbf{R}_{yx}\mathbf{R}_{xx}^{-1}\mathbf{x}) \right\}. \qquad 7.77$$

We say that $\mathbf{y}$, given $\mathbf{x}$, is conditionally normal. Our notation is

$$\mathbf{y}|\mathbf{x} : N[\mathbf{R}_{yx}\mathbf{R}_{xx}^{-1}\mathbf{x}, \mathbf{Q}] \qquad 7.78$$

$$\mathbf{Q} = \mathbf{R}_{yy} - \mathbf{R}_{yx}\mathbf{R}_{xx}^{-1}\mathbf{R}_{xy}.$$

This means that the experimental setup illustrated in Figure 7.8(a) is entirely equivalent to the setup illustrated in Figure 7.8(b). In the latter setup, Mother Nature generates the random vector $\mathbf{x}$ from the distribution $N[\mathbf{0}, \mathbf{R}_{xx}]$ and Father Nature generates the measurement $\mathbf{y}$ from the conditional distribution $N[\mathbf{R}_{yx}\mathbf{R}_{xx}^{-1}\mathbf{x}, \mathbf{Q}]$. Figure 7.8($c$) is an even more illustrative diagram that shows $\mathbf{y}$ to be generated as the sum of the random vector $\mathbf{Hx} = \mathbf{R}_{yx}\mathbf{R}_{xx}^{-1}\mathbf{x}$ and an independent noise random variable $\mathbf{n}$ that is distributed as $N[\mathbf{0}, \mathbf{Q}]$. Figure 7.8($d$) is a redrawing of Figure 7.8(c) that emphasizes the "signal-plus-noise" nature of the decomposition. Every multivariate normal problem of the type illustrated in Figure 7.8(a) may be represented as the equivalent signal-plus-noise problems illustrated in Figures 7.8(c) and 7.8(d). We call Figures 7.8(c) and 7.8(d) channel models in which the observed signal $\mathbf{y}$ is related to the signal $\mathbf{x}$ through the linear statistical model

$$\mathbf{y} = \mathbf{Hx} + \mathbf{n} \, ; \qquad \mathbf{H} = \mathbf{R}_{yx}\mathbf{R}_{xx}^{-1} \qquad\qquad 7.79$$

$$\mathbf{x} : N[\mathbf{0}, \mathbf{R}_{xx}] \, ; \qquad \mathbf{n} : N[\mathbf{0}, \mathbf{Q}].$$

This linear statistical model may be organized into the matrix equations

$$\begin{bmatrix} \mathbf{x} \\ \mathbf{y} \end{bmatrix} = \begin{bmatrix} \mathbf{I} & \mathbf{0} \\ \mathbf{H} & \mathbf{I} \end{bmatrix} \begin{bmatrix} \mathbf{x} \\ \mathbf{n} \end{bmatrix} \qquad\qquad 7.80$$

with the corresponding inverse

$$\begin{bmatrix} \mathbf{x} \\ \mathbf{n} \end{bmatrix} = \begin{bmatrix} \mathbf{I} & \mathbf{0} \\ -\mathbf{H} & \mathbf{I} \end{bmatrix} \begin{bmatrix} \mathbf{x} \\ \mathbf{y} \end{bmatrix} . \qquad\qquad 7.81$$

From these matrix equations it is easy to check that our decomposition produces the correct covariance structures for $[\mathbf{x} \quad \mathbf{y}]$ *and* for $[\mathbf{x} \quad \mathbf{n}]$:

1. $$E \begin{bmatrix} \mathbf{x} \\ \mathbf{y} \end{bmatrix} \begin{bmatrix} \mathbf{x}^{\mathrm{T}} & \mathbf{y}^{\mathrm{T}} \end{bmatrix} = \begin{bmatrix} \mathbf{I} & \mathbf{0} \\ \mathbf{H} & \mathbf{I} \end{bmatrix} \begin{bmatrix} \mathbf{R}_{xx} & \mathbf{0} \\ \mathbf{0} & \mathbf{Q} \end{bmatrix} \begin{bmatrix} \mathbf{I} & \mathbf{H}^{\mathrm{T}} \\ \mathbf{0} & \mathbf{I} \end{bmatrix}$$

$$= \begin{bmatrix} \mathbf{R}_{xx} & \mathbf{R}_{xx}\mathbf{H}^{\mathrm{T}} \\ \mathbf{H}\mathbf{R}_{xx} & \mathbf{H}\mathbf{R}_{xx}\mathbf{H}^{\mathrm{T}} + \mathbf{Q} \end{bmatrix} = \begin{bmatrix} \mathbf{R}_{xx} & \mathbf{R}_{xy} \\ \mathbf{R}_{yx} & \mathbf{R}_{yy} \end{bmatrix} \qquad 7.82$$

2. $$E \begin{bmatrix} \mathbf{x} \\ \mathbf{n} \end{bmatrix} \begin{bmatrix} \mathbf{x}^{\mathrm{T}} & \mathbf{n}^{\mathrm{T}} \end{bmatrix} = \begin{bmatrix} \mathbf{I} & \mathbf{0} \\ -\mathbf{H} & \mathbf{I} \end{bmatrix} \begin{bmatrix} \mathbf{R}_{xx} & \mathbf{R}_{xy} \\ \mathbf{R}_{yx} & \mathbf{R}_{yy} \end{bmatrix} \begin{bmatrix} \mathbf{I} & -\mathbf{H}^{\mathrm{T}} \\ \mathbf{0} & \mathbf{I} \end{bmatrix}$$

$$= \begin{bmatrix} \mathbf{R}_{xx} & -\mathbf{R}_{xx}\mathbf{H}^{\mathrm{T}} + \mathbf{R}_{xy} \\ -\mathbf{H}\mathbf{R}_{xx} + \mathbf{R}_{yx} & (\mathbf{H}\mathbf{R}_{xx}\mathbf{H}^{\mathrm{T}} - \mathbf{H}\mathbf{R}_{xy} - \mathbf{R}_{yx}\mathbf{H}^{\mathrm{T}} + \mathbf{R}_{yy}) \end{bmatrix}$$

$$= \begin{bmatrix} \mathbf{R}_{xx} & \mathbf{0} \\ \mathbf{0} & \mathbf{Q} \end{bmatrix} .$$

The first set of these equations shows that the matrix $\mathbf{H}$ constructs correlated random vectors from uncorrelated ones. The second set shows the random vectors $\mathbf{x}$ and $\mathbf{n}$ are, indeed, independent by virtue of their zero cross-covariance in the multivariate normal model. In the derivation of these formulas we have used the identities

$$\mathbf{H}\mathbf{R}_{xx} = \mathbf{R}_{yx} ; \qquad \mathbf{R}_{xx}\mathbf{H}^{\mathrm{T}} = \mathbf{R}_{xy}$$

$$\mathbf{H}\mathbf{R}_{xx}\mathbf{H}^{\mathrm{T}} = \mathbf{R}_{yx}\mathbf{H}^{\mathrm{T}} = \mathbf{H}\mathbf{R}_{xy} ; \qquad \mathbf{Q} = \mathbf{R}_{yy} - \mathbf{H}\mathbf{R}_{xx}\mathbf{H}^{\mathrm{T}} = \mathbf{R}_{yy} - \mathbf{R}_{yx}\mathbf{R}_{xx}^{-1}\mathbf{R}_{xy}$$

Let's summarize. We have begun with the correlated multivariate normal random vectors $\mathbf{x}$ and $\mathbf{y}$, with no evident linear model connecting them. From the description of the correlation we have determined a linear transformation $\mathbf{H}$ and a noise vector

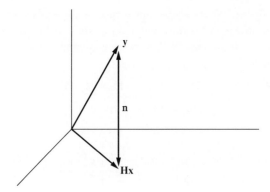

Figure 7.9 Orthogonal decomposition of $\mathbf{y}$ into $\mathbf{Hx} + \mathbf{n}$.

$\mathbf{n} : N[\mathbf{0}, \mathbf{Q}]$ with the property that $\mathbf{Hx}$ and $\mathbf{n}$ orthogonally decompose $\mathbf{y}$ into $\mathbf{Hx} + \mathbf{n}$. This decomposition also decomposes the covariance of $\mathbf{y}$:

$$\mathbf{R}_{yy} = \mathbf{H}\mathbf{R}_{xx}\mathbf{H}^\mathrm{T} + \mathbf{Q}. \qquad 7.83$$

The orthogonal decomposition of $\mathbf{y}$ into $\mathbf{Hx} + \mathbf{n}$ is illustrated geometrically in Figure 7.9. Algebraically, the matrices $\mathbf{H}$ and $\mathbf{Q}$ are the matrices required to take the covariance matrix $\mathbf{R}$ to block diagonal form, as indicated in Equation 7.82.

Conditional Distribution of x, Given y

In any Bayes strategy for estimating $\mathbf{x}$ from $\mathbf{y}$, it is the conditional distribution of $\mathbf{x}$, given $\mathbf{y}$, that is needed. This conditional density may be computed from the equation

$$f(\mathbf{x}|\mathbf{y}) = \frac{f(\mathbf{x}, \mathbf{y})}{f(\mathbf{y})}$$

$$= \frac{(2\pi)^{-N/2}(2\pi)^{-p/2}|\mathbf{R}|^{-1/2} \exp\left\{ -(1/2)(\mathbf{x}^\mathrm{T}\ \mathbf{y}^\mathrm{T})\mathbf{R}^{-1}\binom{\mathbf{x}}{\mathbf{y}}\right\}}{(2\pi)^{-N/2}|\mathbf{R}_{yy}|^{-1/2} \exp\left\{ -(1/2)\mathbf{y}^\mathrm{T}\mathbf{R}_{yy}^{-1}\mathbf{y}\right\}}. \qquad 7.84$$

As before, $\mathbf{R}$ is the structured covariance matrix

$$\mathbf{R} = \begin{bmatrix} \mathbf{R}_{xx} & \mathbf{R}_{xy} \\ \mathbf{R}_{yx} & \mathbf{R}_{yy} \end{bmatrix}. \qquad 7.85$$

We may use form (2) for the inverse of $\mathbf{R}$ in Equation 7.75 to obtain the result

$$f(\mathbf{x}|\mathbf{y}) = (2\pi)^{-p/2}|\mathbf{P}|^{-1/2} \exp\left\{ -\frac{1}{2}(\mathbf{x} - \mathbf{R}_{xy}\mathbf{R}_{yy}^{-1}\mathbf{y})^\mathrm{T}\mathbf{P}^{-1}(\mathbf{x} - \mathbf{R}_{xy}\mathbf{R}_{yy}^{-1}\mathbf{y})\right\}. \qquad 7.86$$

This means that the conditional distribution of $\mathbf{x}$, given $\mathbf{y}$, is normal:

$$\mathbf{x}|\mathbf{y} : N[\mathbf{R}_{xy}\mathbf{R}_{yy}^{-1}\mathbf{y}, \mathbf{P}] \qquad 7.87$$

$$\mathbf{P} = \mathbf{R}_{xx} - \mathbf{R}_{xy}\mathbf{R}_{yy}^{-1}\mathbf{R}_{yx}.$$

(a) jointly distributed **x** and **y**

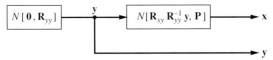

(b) marginally distributed **y** and conditionally distributed **x**

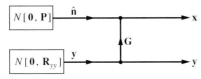

(c) estimated $\hat{\mathbf{x}}$ plus estimated noise $\hat{\mathbf{n}}$

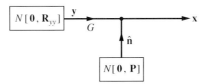

(d) estimator plus noise model

Figure 7.10 Four equivalent estimator, or posterior,
representations for the multivariate normal problem.

This equation is the dual of Equation 7.78. Therefore we may draw the dual diagrams of Figure 7.10. The experimental setup of Figure 7.10(a) is entirely equivalent to the setup illustrated in Figure 7.10(b). Figure 7.10(c) illustrates that the random vector **x** may be decomposed into the sum of the estimated random vector $\hat{\mathbf{x}} = \mathbf{G}\mathbf{y} = \mathbf{R}_{xy}\mathbf{R}_{yy}^{-1}\mathbf{y}$ and the independent error vector $\hat{\mathbf{n}} : N[\mathbf{0}, \mathbf{P}]$. The estimator-plus-noise model is illustrated in Figure 7.10(d). We call Figures 7.10(c) and (d) estimator, or posterior, representations and write their corresponding linear statistical model as

$$\mathbf{x} = \mathbf{G}\mathbf{y} + \hat{\mathbf{n}}; \qquad \mathbf{G} = \mathbf{R}_{xy}\mathbf{R}_{yy}^{-1} \qquad\qquad 7.88$$

$$\mathbf{y} : N[\mathbf{0}, \mathbf{R}_{yy}]; \qquad \hat{\mathbf{n}} : N[\mathbf{0}, \mathbf{P}].$$

This linear statistical model for the estimator representation may be organized into the matrix equation

$$\begin{bmatrix} \mathbf{x} \\ \mathbf{y} \end{bmatrix} = \begin{bmatrix} \mathbf{I} & \mathbf{G} \\ \mathbf{0} & \mathbf{I} \end{bmatrix} \begin{bmatrix} \hat{\mathbf{n}} \\ \mathbf{y} \end{bmatrix}, \qquad\qquad 7.89$$

with its corresponding inverse

$$\begin{bmatrix} \hat{\mathbf{n}} \\ \mathbf{y} \end{bmatrix} = \begin{bmatrix} \mathbf{I} & -\mathbf{G} \\ \mathbf{0} & \mathbf{I} \end{bmatrix} \begin{bmatrix} \mathbf{x} \\ \mathbf{y} \end{bmatrix}. \qquad\qquad 7.90$$

These equations produce the following results for the covariance structure of $[\mathbf{x} \quad \mathbf{y}]^T$ and $[\hat{\mathbf{n}} \quad \mathbf{y}]^T$:

$$E\begin{bmatrix} \mathbf{x} \\ \mathbf{y} \end{bmatrix}\begin{bmatrix} \mathbf{x}^T & \mathbf{y}^T \end{bmatrix} = \begin{bmatrix} \mathbf{I} & \mathbf{G} \\ \mathbf{0} & \mathbf{I} \end{bmatrix}\begin{bmatrix} \mathbf{P} & \mathbf{0} \\ \mathbf{0} & \mathbf{R}_{yy} \end{bmatrix}\begin{bmatrix} \mathbf{I} & \mathbf{0} \\ \mathbf{G}^T & \mathbf{I} \end{bmatrix}$$

$$\qquad\qquad 7.91$$

$$= \begin{bmatrix} \mathbf{R}_{xx} & \mathbf{R}_{xy} \\ \mathbf{R}_{yx} & \mathbf{R}_{yy} \end{bmatrix}$$

$$E\begin{bmatrix} \hat{\mathbf{n}} \\ \mathbf{y} \end{bmatrix}\begin{bmatrix} \hat{\mathbf{n}}^T & \mathbf{y}^T \end{bmatrix} = \begin{bmatrix} \mathbf{I} & -\mathbf{G} \\ \mathbf{0} & \mathbf{I} \end{bmatrix}\begin{bmatrix} \mathbf{R}_{xx} & \mathbf{R}_{xy} \\ \mathbf{R}_{yx} & \mathbf{R}_{yy} \end{bmatrix}\begin{bmatrix} \mathbf{I} & \mathbf{0} \\ -\mathbf{G}^T & \mathbf{I} \end{bmatrix}$$

$$= \begin{bmatrix} \mathbf{P} & \mathbf{0} \\ \mathbf{0} & \mathbf{R}_{yy} \end{bmatrix}.$$

These equations are the duals to Equation 7.82. Can you interpret the role of $\mathbf{G}$ and verify that $\hat{\mathbf{n}}$ and $\mathbf{y}$ are independent? In order to derive these results (you should carry out the details), we have used the identities

$$\mathbf{G}\mathbf{R}_{yy} = \mathbf{R}_{xy}; \qquad \mathbf{R}_{yy}\mathbf{G}^T = \mathbf{R}_{yx}$$

$$\mathbf{G}\mathbf{R}_{yy}\mathbf{G}^T = \mathbf{R}_{xy}\mathbf{G}^T = \mathbf{G}\mathbf{R}_{yx}; \qquad \mathbf{P} = \mathbf{R}_{xx} - \mathbf{G}\mathbf{R}_{yy}\mathbf{G}^T = \mathbf{R}_{xx} - \mathbf{R}_{xy}\mathbf{R}_{yy}^{-1}\mathbf{R}_{yx}.$$

Filtering Diagrams

The estimator diagram of Figure 7.10(c) may be concatenated with a filtering diagram to show how the measurement $\mathbf{y}$ and the vector $\mathbf{x}$ are mapped into the estimator $\hat{\mathbf{x}}$ and the estimator error $\hat{\mathbf{n}} = \mathbf{x} - \hat{\mathbf{x}}$. The result is Figure 7.11. The diagrams of Figure 7.11 show that the diagonal map from $[\hat{\mathbf{n}} \quad \mathbf{y}]$ to $[\hat{\mathbf{n}} \quad \hat{\mathbf{x}}]$ may actually be decomposed into the concatenation of two maps, one of which is the estimator map

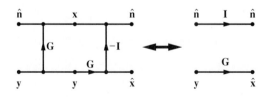

Figure 7.11 Filtering diagrams.

from $[\hat{\mathbf{n}} \quad \mathbf{y}]$ to $[\mathbf{x} \quad \mathbf{y}]$ and the other of which is the filtering map from $[\mathbf{x} \quad \mathbf{y}]$ to $[\hat{\mathbf{n}} \quad \hat{\mathbf{x}}]$:

$$
\begin{bmatrix} \hat{\mathbf{n}} \\ \hat{\mathbf{x}} \end{bmatrix} = \begin{bmatrix} \mathbf{I} & -\mathbf{G} \\ \mathbf{0} & \mathbf{G} \end{bmatrix} \begin{bmatrix} \mathbf{x} \\ \mathbf{y} \end{bmatrix} = \begin{bmatrix} \mathbf{I} & -\mathbf{G} \\ \mathbf{0} & \mathbf{G} \end{bmatrix} \begin{bmatrix} \mathbf{I} & \mathbf{G} \\ \mathbf{0} & \mathbf{I} \end{bmatrix} \begin{bmatrix} \hat{\mathbf{n}} \\ \mathbf{y} \end{bmatrix}
$$
$$
= \begin{bmatrix} \mathbf{I} & \mathbf{0} \\ \mathbf{0} & \mathbf{G} \end{bmatrix} \begin{bmatrix} \hat{\mathbf{n}} \\ \mathbf{y} \end{bmatrix}. \tag{7.92}
$$

This decomposition brings very important insights into the covariance structure of $[\hat{\mathbf{n}} \quad \hat{\mathbf{x}}]$ and the cross-covariance structure of $[\hat{\mathbf{n}} \quad \hat{\mathbf{x}}]$ and $[\mathbf{x} \quad \mathbf{y}]$:

1.
$$
E\begin{bmatrix} \hat{\mathbf{n}} \\ \hat{\mathbf{x}} \end{bmatrix} \begin{bmatrix} \hat{\mathbf{n}}^T & \hat{\mathbf{x}}^T \end{bmatrix} = \begin{bmatrix} \mathbf{I} & \mathbf{0} \\ \mathbf{0} & \mathbf{G} \end{bmatrix} \begin{bmatrix} \mathbf{P} & \mathbf{0} \\ \mathbf{0} & \mathbf{R}_{yy} \end{bmatrix} \begin{bmatrix} \mathbf{I} & \mathbf{0} \\ \mathbf{0} & \mathbf{G}^T \end{bmatrix} = \begin{bmatrix} \mathbf{P} & \mathbf{0} \\ \mathbf{0} & \mathbf{G}\mathbf{R}_{yy}\mathbf{G}^T \end{bmatrix}
$$
$$
= \begin{bmatrix} \mathbf{P} & \mathbf{0} \\ \mathbf{0} & \mathbf{R}_{xx} - \mathbf{P} \end{bmatrix} \tag{7.93}
$$

2.
$$
E\begin{bmatrix} \hat{\mathbf{n}} \\ \hat{\mathbf{x}} \end{bmatrix} \begin{bmatrix} \mathbf{x}^T & \mathbf{y}^T \end{bmatrix} = \begin{bmatrix} \mathbf{I} & -\mathbf{G} \\ \mathbf{0} & \mathbf{G} \end{bmatrix} \begin{bmatrix} \mathbf{R}_{xx} & \mathbf{R}_{xy} \\ \mathbf{R}_{yx} & \mathbf{R}_{yy} \end{bmatrix}
$$
$$
= \begin{bmatrix} \mathbf{R}_{xx} - \mathbf{G}\mathbf{R}_{yx} & \mathbf{R}_{xy} - \mathbf{G}\mathbf{R}_{yy} \\ \mathbf{G}\mathbf{R}_{yx} & \mathbf{G}\mathbf{R}_{yy} \end{bmatrix}
$$
$$
= \begin{bmatrix} \mathbf{P} & \mathbf{0} \\ \mathbf{R}_{xx} - \mathbf{P} & \mathbf{R}_{xy} \end{bmatrix}.
$$

These two results may be summarized with the following interpretations:

1. $\hat{\mathbf{x}}$ and $\hat{\mathbf{n}}$ are orthogonal decompositions of $\mathbf{x} = \hat{\mathbf{x}} + \hat{\mathbf{n}}$;
2. the covariance matrix for the error $\hat{\mathbf{n}}$ is $\mathbf{P}$;
3. the covariance matrix of $\hat{\mathbf{x}}$ is the same as the cross-covariance between $\hat{\mathbf{x}}$ and $\mathbf{x}$, namely $\mathbf{R}_{xx} - \mathbf{P}$;
4. the cross-covariance between $\hat{\mathbf{x}}$ and $\mathbf{y}$ is $\mathbf{R}_{xy}$, the same as the cross-covariance between $\mathbf{x}$ and $\mathbf{y}$;
5. the cross-covariance between $\hat{\mathbf{n}}$ and $\mathbf{y}$ is $\mathbf{0}$.

These results are illustrated in Figure 7.12. We call the decomposition of $\mathbf{x} : N[\mathbf{0}, \mathbf{R}_{xx}]$ into $\hat{\mathbf{x}} : N[\mathbf{0}, \mathbf{R}_{xx} - \mathbf{P}]$ plus $\hat{\mathbf{n}} : N[\mathbf{0}, \mathbf{P}]$ the Statisticians' Pythagorean Theorem. The orthogonality between $\hat{\mathbf{n}}$ and $\mathbf{y}$ is called the orthogonality condition. This orthogonality condition characterizes $\mathbf{G}$:

$$
E\hat{\mathbf{n}}\mathbf{y}^T = \mathbf{R}_{xy} - \mathbf{G}\mathbf{R}_{yy} = \mathbf{0} \tag{7.94}
$$
$$
\longrightarrow \quad \mathbf{G} = \mathbf{R}_{xy}\mathbf{R}_{yy}^{-1}.
$$

We shall have more to say about this condition in Chapter 9 when we study conditional mean estimators in more detail.

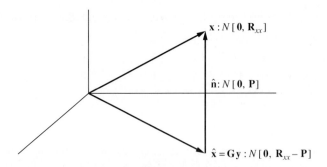

Figure 7.12 Statisticians' Pythagorean Theorem.

The results of this section may be summarized in the following Gauss-Markov theorem, which is a slight generalization of what we have done so far.

Gauss-Markov Theorem   Let $\mathbf{x}$ and $\mathbf{y}$ be random vectors that are distributed according to the multivariate distribution

$$\begin{bmatrix} \mathbf{x} \\ \mathbf{y} \end{bmatrix} = N\left[\begin{bmatrix} \mathbf{m}_x \\ \mathbf{m}_y \end{bmatrix}, \begin{bmatrix} \mathbf{R}_{xx} & \mathbf{R}_{xy} \\ \mathbf{R}_{yx} & \mathbf{R}_{yy} \end{bmatrix} \right]. \qquad 7.95$$

Then the conditional distribution of $\mathbf{x}$, given $\mathbf{y}$, is multivariate normal with conditional mean $\hat{\mathbf{x}} = \mathbf{m}_x + \mathbf{R}_{xy}\mathbf{R}_{yy}^{-1}(\mathbf{y} - \mathbf{m}_y)$ and conditional covariance $\mathbf{P} = \mathbf{R}_{xx} - \mathbf{R}_{xy}\mathbf{R}_{yy}^{-1}\mathbf{R}_{yx}$:

$$\mathbf{x}|\mathbf{y} : N[\hat{\mathbf{x}}, \mathbf{P}]$$

$$\hat{\mathbf{x}} = \mathbf{m}_x + \mathbf{R}_{xy}\mathbf{R}_{yy}^{-1}(\mathbf{y} - \mathbf{m}_y) \qquad 7.96$$

$$\mathbf{P} = \mathbf{R}_{xx} - \mathbf{R}_{xy}\mathbf{R}_{yy}^{-1}\mathbf{R}_{yx}.$$

Proof   The proof is a straightforward generalization of the techniques we have used in the previous two subsections. It is left as an exercise in the problems.   □

Interpretations

The estimator $\hat{\mathbf{x}}$ and the error $\hat{\mathbf{n}} = \mathbf{x} - \hat{\mathbf{x}}$ enjoy a number of remarkable properties. First, they are distributed as follows:

- $\hat{\mathbf{x}} : N[\mathbf{m}_x, \hat{\mathbf{P}}]$
  $$\hat{\mathbf{P}} = E[\hat{\mathbf{x}} - \mathbf{m}_x][\hat{\mathbf{x}} - \mathbf{m}_x]^T = \mathbf{R}_{xy}\mathbf{R}_{yy}^{-1}\mathbf{R}_{yx} = \mathbf{R}_{xx} - \mathbf{P};$$
- $\hat{\mathbf{n}} = \mathbf{x} - \hat{\mathbf{x}} : N[\mathbf{0}, \mathbf{P}]$ \qquad 7.97
  $$\mathbf{P} = \mathbf{R}_{xx} - \mathbf{R}_{xy}\mathbf{R}_{yy}^{-1}\mathbf{R}_{yx} = \mathbf{R}_{xx} - \hat{\mathbf{P}}.$$

We emphasize that these are marginal, or unconditional, distribution statements. The error $\hat{\mathbf{n}} = \mathbf{x} - \hat{\mathbf{x}}$ is uncorrelated with the estimate $\hat{\mathbf{x}}$ and with the measurement $\mathbf{y}$:

$$
\begin{aligned}
E[\mathbf{x} - \hat{\mathbf{x}}][\hat{\mathbf{x}} - \mathbf{m}_x]^{\mathrm{T}} &= E\left[(\mathbf{x} - \mathbf{m}_x) - (\hat{\mathbf{x}} - \mathbf{m}_x)\right][\hat{\mathbf{x}} - \mathbf{m}_x]^{\mathrm{T}} \\
&= E[\mathbf{x} - \mathbf{m}_x]\left[(\mathbf{y} - \mathbf{m}_y)^{\mathrm{T}}\mathbf{R}_{yy}^{-1}\mathbf{R}_{yx}\right] - \hat{\mathbf{P}} \\
&= \mathbf{R}_{xy}\mathbf{R}_{yy}^{-1}\mathbf{R}_{yx} - \hat{\mathbf{P}} \\
&= \mathbf{0}
\end{aligned}
$$

$$
\begin{aligned}
E[\mathbf{x} - \hat{\mathbf{x}}][\mathbf{y} - \mathbf{m}_y]^{\mathrm{T}} &= E\left[(\mathbf{x} - \mathbf{m}_x) - (\hat{\mathbf{x}} - \mathbf{m}_x)\right][\mathbf{y} - \mathbf{m}_y]^{\mathrm{T}} \\
&= \mathbf{R}_{xy} - E\mathbf{R}_{xy}\mathbf{R}_{yy}^{-1}[\mathbf{y} - \mathbf{m}_y][\mathbf{y} - \mathbf{m}_y]^{\mathrm{T}} \\
&= \mathbf{R}_{xy} - \mathbf{R}_{xy}\mathbf{R}_{yy}^{-1}\mathbf{R}_{yy} \\
&= \mathbf{0}.
\end{aligned}
$$

7.98

The estimator $\hat{\mathbf{x}}$ is correlated with $\mathbf{y}$ in just the same way that $\mathbf{x}$ and $\mathbf{y}$ are correlated:

$$
\begin{aligned}
E[\hat{\mathbf{x}} - \mathbf{m}_x][\mathbf{y} - \mathbf{m}_y]^{\mathrm{T}} &= E\mathbf{R}_{xy}\mathbf{R}_{yy}^{-1}[\mathbf{y} - \mathbf{m}_y][\mathbf{y} - \mathbf{m}_y]^{\mathrm{T}} \\
&= \mathbf{R}_{xy}.
\end{aligned}
$$

7.99

These properties are of inestimable value for solving complicated estimation problems. They tell us that the random vector $\mathbf{x}$ has an orthogonal (or uncorrelated) representation of the form

$$
\mathbf{x} = \hat{\mathbf{x}} + \hat{\mathbf{n}} : N[\mathbf{m}_x, \mathbf{R}_{xx}]
$$

$$
\hat{\mathbf{x}} : N[\mathbf{m}_x, \hat{\mathbf{P}}]
$$

$$
\hat{\mathbf{n}} = \mathbf{x} - \hat{\mathbf{x}} : N[\mathbf{0}, \mathbf{P}].
$$

7.100

This decomposition is a slight generalization of the Statisticians' Pythagorean Theorem illustrated in Figure 7.12. Note how means and covariances add. In this decomposition, also illustrated in Figure 7.10, $\hat{\mathbf{x}}$ is the estimable part of $\mathbf{x}$ and $\hat{\mathbf{n}}$ is the inestimable part. It is clear from Figure 7.10 that only $\hat{\mathbf{x}}$ carries information about $\mathbf{x}$.

Innovations

The estimator model for Figure 7.10(c) may be redrawn as in Figure 7.13 to produce the so-called innovations representation for $\mathbf{x}$. In this representation, $\mathbf{R}_{yy}^{1/2}$ is a Cholesky

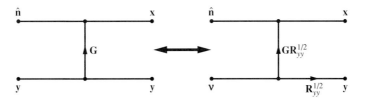

Figure 7.13 Innovations representation.

or a spectral square root of the covariance matrix $\mathbf{R}_{yy}$, and $\mathbf{R}_{yy}^{-1/2}$ is its inverse. That is, $\mathbf{R}_{yy} = \mathbf{R}_{yy}^{1/2}(\mathbf{R}_{yy}^{1/2})^T$ and $\mathbf{R}_{yy}^{-1} = (\mathbf{R}_{yy}^{-1/2})^T\mathbf{R}_{yy}^{-1/2}$. The innovations vector $\boldsymbol{v} = \mathbf{R}_{yy}^{-1/2}\mathbf{y}$ is a whitened version of $\mathbf{y}$ and $\mathbf{y} = \mathbf{R}_{yy}^{1/2}\boldsymbol{v}$ is a colored version of $\boldsymbol{v}$:

$$\boldsymbol{v} = \mathbf{R}_{yy}^{-1/2}\mathbf{y} \qquad\qquad 7.101$$

$$\mathbf{y} = \mathbf{R}_{yy}^{1/2}\mathbf{v}.$$

The Bayes estimate $\hat{\mathbf{x}}$ is then

$$\hat{\mathbf{x}} = \mathbf{G}\mathbf{R}_{yy}^{1/2}\boldsymbol{v}. \qquad\qquad 7.102$$

The innovations model for $\mathbf{x}$ and $\mathbf{y}$ is

$$\begin{bmatrix} \mathbf{x} \\ \mathbf{y} \end{bmatrix} = \begin{bmatrix} \mathbf{I} & \mathbf{G}\mathbf{R}_{yy}^{1/2} \\ \mathbf{0} & \mathbf{R}_{yy}^{1/2} \end{bmatrix} \begin{bmatrix} \hat{\mathbf{n}} \\ \boldsymbol{v} \end{bmatrix}. \qquad\qquad 7.103$$

From this representation, it is clear that the only information $\mathbf{y}$ carries about $\mathbf{x}$ is carried in $\mathbf{G}\mathbf{R}_{yy}^{1/2}\boldsymbol{v}$ with $\boldsymbol{v} = \mathbf{R}_{yy}^{-1/2}\mathbf{y}$. The inverse of the innovations representation is

$$\begin{bmatrix} \hat{\mathbf{n}} \\ \boldsymbol{v} \end{bmatrix} = \begin{bmatrix} \mathbf{I} & -\mathbf{G} \\ \mathbf{0} & \mathbf{R}_{yy}^{-1/2} \end{bmatrix} \begin{bmatrix} \mathbf{x} \\ \mathbf{y} \end{bmatrix}. \qquad\qquad 7.104$$

We shall use this representation in Section 7.9, where we derive the Wiener filter.

Example 7.3

Consider the signal-plus-noise model

$$\mathbf{y} = \mathbf{x} + \mathbf{n} \qquad\qquad 7.105$$

where $\mathbf{x} : N[\mathbf{0}, \mathbf{R}_{xx}]$ and $\mathbf{n} : N[\mathbf{0}, \mathbf{R}_{nn}]$. Assume that $\mathbf{x}$ and $\mathbf{n}$ are independent. The joint distribution of $\mathbf{x}$ and $\mathbf{n}$ may be written as

$$\begin{bmatrix} \mathbf{x} \\ \mathbf{n} \end{bmatrix} : N\left[\begin{bmatrix} \mathbf{0} \\ \mathbf{0} \end{bmatrix}, \begin{bmatrix} \mathbf{R}_{xx} & \mathbf{0} \\ \mathbf{0} & \mathbf{R}_{nn} \end{bmatrix} \right]. \qquad\qquad 7.106$$

The joint distribution of $\mathbf{x}$ and $\mathbf{y}$ is

$$\begin{bmatrix} \mathbf{x} \\ \mathbf{y} \end{bmatrix} = \begin{bmatrix} \mathbf{I} & \mathbf{0} \\ \mathbf{I} & \mathbf{I} \end{bmatrix} \begin{bmatrix} \mathbf{x} \\ \mathbf{n} \end{bmatrix} : N\left[\begin{bmatrix} \mathbf{0} \\ \mathbf{0} \end{bmatrix}, \begin{bmatrix} \mathbf{R}_{xx} & \mathbf{R}_{xx} \\ \mathbf{R}_{xx} & \mathbf{R}_{xx} + \mathbf{R}_{nn} \end{bmatrix} \right]. \qquad\qquad 7.107$$

The conditional distribution of $\mathbf{x}$, given $\mathbf{y}$, is

$$\mathbf{x}|\mathbf{y} : N[\hat{\mathbf{x}}, \mathbf{P}]; \qquad \hat{\mathbf{x}} = \mathbf{R}_{xx}(\mathbf{R}_{xx} + \mathbf{R}_{nn})^{-1}\mathbf{y} \qquad\qquad 7.108$$

$$\mathbf{P} = \mathbf{R}_{xx} - \mathbf{R}_{xx}(\mathbf{R}_{xx} + \mathbf{R}_{nn})^{-1}\mathbf{R}_{xx}.$$

The conditional mean $\hat{\mathbf{x}}$ minimizes mean-squared error and is, therefore, the Bayes estimator under quadratic loss. ∎

7.6   THE LINEAR STATISTICAL MODEL

We can modify the signal-plus-noise model of the previous example by linearly transforming the signal $\mathbf{x}$ and observing it in additive and independent noise:

$$\mathbf{y} = \mathbf{Hx} + \mathbf{n}. \qquad 7.109$$

We call this measurement model a linear statistical model. Of course, every multivariate normal problem may be brought to this form by deriving its appropriate channel model.

If the random vectors $\mathbf{x}$ and $\mathbf{n}$ are independent normal random vectors, then their joint distribution may be written

$$\begin{bmatrix} \mathbf{x} \\ \mathbf{n} \end{bmatrix} : N\left[\begin{bmatrix} \mathbf{0} \\ \mathbf{0} \end{bmatrix}, \begin{bmatrix} \mathbf{R}_{xx} & \mathbf{0} \\ \mathbf{0} & \mathbf{R}_{nn} \end{bmatrix}\right]. \qquad 7.110$$

The joint distribution of the signal $\mathbf{x}$ and the measurement $\mathbf{y}$ is then

$$\begin{bmatrix} \mathbf{x} \\ \mathbf{y} \end{bmatrix} = \begin{bmatrix} \mathbf{I} & \mathbf{0} \\ \mathbf{I} & \mathbf{H} \end{bmatrix}\begin{bmatrix} \mathbf{x} \\ \mathbf{n} \end{bmatrix} : N\left[\begin{bmatrix} \mathbf{0} \\ \mathbf{0} \end{bmatrix}, \begin{bmatrix} \mathbf{R}_{xx} & \mathbf{R}_{xx}\mathbf{H}^\mathrm{T} \\ \mathbf{HR}_{xx} & \mathbf{HR}_{xx}\mathbf{H}^\mathrm{T} + \mathbf{R}_{nn} \end{bmatrix}\right]. \qquad 7.111$$

The two conditional distributions of interest are

$$\mathbf{y}|\mathbf{x} : N[\mathbf{Hx}, \mathbf{R}_{nn}]$$

$$\mathbf{x}|\mathbf{y} : N[\hat{\mathbf{x}}, \mathbf{P}] \qquad 7.112$$

$$\hat{\mathbf{x}} = \mathbf{Gy}$$

$$\mathbf{G} = \mathbf{R}_{xy}\mathbf{R}_{yy}^{-1} = \mathbf{R}_{xx}\mathbf{H}^\mathrm{T}(\mathbf{HR}_{xx}\mathbf{H}^\mathrm{T} + \mathbf{R}_{nn})^{-1}$$

$$\mathbf{P} = \mathbf{R}_{xx} - \mathbf{GR}_{yx} = \mathbf{R}_{xx} - \mathbf{R}_{xx}\mathbf{H}^\mathrm{T}(\mathbf{HR}_{xx}\mathbf{H}^\mathrm{T} + \mathbf{R}_{nn})^{-1}\mathbf{HR}_{xx}.$$

The conditional distribution of $\mathbf{x}$, given $\mathbf{y}$, shows $\hat{\mathbf{x}}$ to be the conditional mean of $\mathbf{x}$, given $\mathbf{y}$, and $\mathbf{P}$ to be the conditional covariance of $\mathbf{x}$. In fact, $\mathbf{P}$ is also the unconditional mean-squared error of the estimator $\hat{\mathbf{x}}$.

This basic result for estimating $\mathbf{x}$ in the linear statistical model may be cast in a slightly different form by using the matrix inversion lemma of Chapter 2 to rewrite the error covariance matrix $\mathbf{P}$ as follows:

$$\mathbf{P} = (\mathbf{R}_{xx}^{-1} + \mathbf{H}^\mathrm{T}\mathbf{R}_{nn}^{-1}\mathbf{H})^{-1}. \qquad 7.113$$

Then, from the formula $\mathbf{P} = \mathbf{R}_{xx} - \mathbf{GR}_{yx} = \mathbf{R}_{xx} - \mathbf{GHR}_{xx}$, we may write $\mathbf{GHR}_{xx}$ as follows:

$$\mathbf{GHR}_{xx} = \mathbf{R}_{xx} - \mathbf{P} = \mathbf{P}(\mathbf{P}^{-1}\mathbf{R}_{xx} - \mathbf{I})$$

$$= \mathbf{P}\left[(\mathbf{R}_{xx}^{-1} + \mathbf{H}^\mathrm{T}\mathbf{R}_{nn}^{-1}\mathbf{H})\mathbf{R}_{xx} - \mathbf{I}\right] \qquad 7.114$$

$$= \mathbf{PH}^\mathrm{T}\mathbf{R}_{nn}^{-1}\mathbf{HR}_{xx}.$$

This result produces the identity

$$\mathbf{G} = \mathbf{PH}^\mathrm{T}\mathbf{R}_{nn}^{-1}. \qquad 7.115$$

With this identity, we can now write the conditional density of $\mathbf{x}$, given $\mathbf{y}$, as

$$\mathbf{x}|\mathbf{y} : N[\hat{\mathbf{x}}, \mathbf{P}] \qquad\qquad 7.116$$

where $\hat{\mathbf{x}}$ and $\mathbf{P}$ are given by the alternative formulas

$$\hat{\mathbf{x}} = \mathbf{P}\mathbf{H}^{\mathsf{T}}\mathbf{R}_{nn}^{-1}\mathbf{y} \qquad\qquad 7.117$$

$$\mathbf{P} = (\mathbf{R}_{xx}^{-1} + \mathbf{H}^{\mathsf{T}}\mathbf{R}_{nn}^{-1}\mathbf{H})^{-1}.$$

The so-called regression equations for $\hat{\mathbf{x}}$ and $\mathbf{G}$ are

$$\mathbf{P}^{-1}\hat{\mathbf{x}} = \mathbf{H}^{\mathsf{T}}\mathbf{R}_{nn}^{-1}\mathbf{y} \qquad\qquad 7.118$$

$$\mathbf{P}^{-1}\mathbf{G} = \mathbf{H}^{\mathsf{T}}\mathbf{R}_{nn}^{-1}\mathbf{I}.$$

These dual equations show that the ith column of the gain matrix $\mathbf{G}$ obeys the same equation as the estimator $\hat{\mathbf{x}}$, with $\mathbf{y}$ replaced by the ith coordinate vector $\mathbf{e}_i$:

$$\mathbf{P}^{-1}\hat{\mathbf{x}} = \mathbf{H}^{\mathsf{T}}\mathbf{R}_{nn}^{-1}\mathbf{y} \qquad\qquad 7.119$$

$$\mathbf{P}^{-1}\mathbf{g}_i = \mathbf{H}^{\mathsf{T}}\mathbf{R}_{nn}^{-1}\mathbf{e}_i; \qquad \mathbf{e}_i : i\text{th coordinate vector.}$$

Example 7.4 (Rank-One Signal Model)

Let $y_t = s + n_t$, $t = 0, 1, \ldots, N-1$, with $s : N[0, \sigma_s^2]$ and $\{n_t\}_0^{N-1}$ a sequence of independent, identically distributed $N[0, \sigma_n^2]$ random variables that are also independent of s. The measurement vector $\mathbf{y}$ may be written as

$$\mathbf{y} = s\mathbf{1} + \mathbf{n}$$

$$\mathbf{y} = [y_0 \quad \cdots \quad y_{N-1}]^{\mathsf{T}}; \qquad \mathbf{n} = [n_0 \quad \cdots \quad n_{N-1}]^{\mathsf{T}} \qquad 7.120$$

$$\mathbf{1} = [1 \quad \cdots \quad 1]^{\mathsf{T}}.$$

The joint distribution of s and $\mathbf{n}$ is

$$\begin{bmatrix} s \\ \mathbf{n} \end{bmatrix} : N\left[\begin{bmatrix} 0 \\ \mathbf{0} \end{bmatrix}, \begin{bmatrix} \sigma_s^2 & \mathbf{0}^{\mathsf{T}} \\ \mathbf{0} & \sigma^2\mathbf{I} \end{bmatrix}\right], \qquad 7.121$$

and the joint distribution of s and $\mathbf{y}$ is

$$\begin{bmatrix} s \\ \mathbf{y} \end{bmatrix} = \begin{bmatrix} 1 & \mathbf{0}^{\mathsf{T}} \\ \mathbf{1} & \mathbf{I} \end{bmatrix}\begin{bmatrix} s \\ \mathbf{n} \end{bmatrix} : N\left[\begin{bmatrix} 0 \\ \mathbf{0} \end{bmatrix}, \begin{bmatrix} \sigma_s^2 & \sigma_s^2\mathbf{1}^{\mathsf{T}} \\ \sigma_s^2\mathbf{1} & \sigma_n^2\mathbf{I} + \sigma_s^2\mathbf{1}\mathbf{1}^{\mathsf{T}} \end{bmatrix}\right]. \qquad 7.122$$

The conditional distribution of s, given $\mathbf{y}$, is

$$s|\mathbf{y} : N[\hat{s}, P]$$

$$\hat{s} = P\mathbf{1}^{\mathsf{T}}\frac{1}{\sigma_n^2}\mathbf{I}\mathbf{y} \qquad\qquad 7.123$$

$$P = \left(\frac{1}{\sigma_s^2} + \mathbf{1}^{\mathsf{T}}\frac{1}{\sigma_n^2}\mathbf{I}\mathbf{1}\right)^{-1}.$$

The expressions for $\hat{s}$ and P may be simplified as follows:

$$\hat{s} = \frac{\sigma_s^2/\sigma_n^2}{1 + N\sigma_s^2/\sigma_n^2} \mathbf{1}^T\mathbf{y} \qquad\qquad 7.124$$

$$P = \left(\frac{1}{\sigma_s^2} + \frac{N}{\sigma_n^2}\right)^{-1} = \frac{\sigma_s^2}{1 + N\sigma_s^2/\sigma_n^2}.$$

When the output signal-to-noise ratio $N\sigma_s^2/\sigma_n^2$ is much greater than 1, then these results become the maximum likelihood results

$$\hat{s} = \frac{1}{N}\sum_{n=0}^{N-1} y_n \qquad\qquad 7.125$$

$$P = \frac{\sigma_n^2}{N}. \qquad\qquad\blacksquare$$

7.7   SEQUENTIAL BAYES

The results of the previous section may be used to derive recursive estimates of the random vector $\mathbf{x}$ when the observation vector $\mathbf{y}_t = [y_0 \quad y_1 \quad \cdots \quad y_t]^T$ increases in dimension with increasing time. The basic idea is to write the measurement vector $\mathbf{y}_t$ as follows:

$$\mathbf{y}_t = \mathbf{H}_t\mathbf{x} + \mathbf{n}_t \qquad\qquad 7.126$$

$$\left[\begin{array}{c} \mathbf{y}_{t-1} \\ \hline y_t \end{array}\right] = \left[\begin{array}{c} \mathbf{H}_{t-1} \\ \hline \mathbf{c}_t^T \end{array}\right] \left[\mathbf{x}\right] + \left[\begin{array}{c} \mathbf{n}_{t-1} \\ n_t \end{array}\right].$$

This linear statistical model for $\mathbf{y}_t$ is simply a concise way of saying that the kth scalar measurement y_k may be written as

$$y_k = \mathbf{c}_k^T\mathbf{x} + n_k; \qquad k = 0, 1, \ldots. \qquad\qquad 7.127$$

In our treatment of sequential Bayes estimators, we shall assume that the signal vector $\mathbf{x}$ is distributed as $\mathbf{x} : N[\mathbf{0}, \mathbf{R}_{xx}]$ and that the noise vector $\mathbf{n}_t$ is distributed as $\mathbf{n}_t : N[\mathbf{0}, \mathbf{R}_t]$. We shall further assume that $\mathbf{x}$ and $\mathbf{n}_t$ are independent and that the noise covariance matrix $\mathbf{R}_t$ is diagonal:

$$\mathbf{R}_t = \left[\begin{array}{cc} \mathbf{R}_{t-1} & \mathbf{0} \\ \mathbf{0}^T & r_{tt} \end{array}\right]; \qquad R_0 = r_{00}. \qquad\qquad 7.128$$

This means that the inverse of $\mathbf{R}_t$ is

$$\mathbf{R}_t^{-1} = \left[\begin{array}{cc} \mathbf{R}_{t-1}^{-1} & \mathbf{0} \\ \mathbf{0}^T & r_{tt}^{-1} \end{array}\right]. \qquad\qquad 7.129$$

The joint distribution of $\mathbf{x}$ and $\mathbf{y}_t$ is

$$\begin{bmatrix} \mathbf{x} \\ \mathbf{y}_t \end{bmatrix} = N\left(\begin{bmatrix} \mathbf{0} \\ \mathbf{0} \end{bmatrix}, \begin{bmatrix} \mathbf{R}_{xx} & \mathbf{R}_{xx}\mathbf{H}_t^{\mathsf{T}} \\ \mathbf{H}_t\mathbf{R}_{xx} & \mathbf{H}_t\mathbf{R}_{xx}\mathbf{H}_t^{\mathsf{T}} + \mathbf{R}_t \end{bmatrix} \right). \qquad 7.130$$

The conditional distribution of $\mathbf{x}$, given $\mathbf{y}_t$, is

$$\mathbf{x}|\mathbf{y}_t : N[\hat{\mathbf{x}}_t, \mathbf{P}_t]$$

$$\hat{\mathbf{x}}_t = \mathbf{P}_t\mathbf{H}_t^{\mathsf{T}}\mathbf{R}_t^{-1}\mathbf{y}_t \qquad 7.131$$

$$\mathbf{P}_t^{-1} = \mathbf{R}_{xx}^{-1} + \mathbf{H}_t^{\mathsf{T}}\mathbf{R}_t^{-1}\mathbf{H}_t.$$

In these formulas, $\mathbf{H}_t$, $\mathbf{R}_t$, and $\mathbf{y}_t$ have dimensions that increase with increasing time. Conversely, the dimensions of $\hat{\mathbf{x}}_t$ and $\mathbf{P}_t$ are fixed at $p \times 1$ and $p \times p$, respectively, with the subscript indicating only that the values in $\hat{\mathbf{x}}_t$ and $\mathbf{P}_t$ are time varying.

In order to make these formulas recursive, we note that $\mathbf{P}_t^{-1}$ and $\mathbf{H}_t^{\mathsf{T}}\mathbf{R}_t^{-1}$ have the following simple recursions:

$$\mathbf{P}_t^{-1} = \mathbf{R}_{xx}^{-1} + \begin{bmatrix} \mathbf{H}_{t-1}^{\mathsf{T}} & \mathbf{c}_t \end{bmatrix} \begin{bmatrix} \mathbf{R}_{t-1}^{-1} & \mathbf{0} \\ \mathbf{0}^{\mathsf{T}} & r_{tt}^{-1} \end{bmatrix} \begin{bmatrix} \mathbf{H}_{t-1} \\ \mathbf{c}_t^{\mathsf{T}} \end{bmatrix}$$

$$= \mathbf{P}_{t-1}^{-1} + r_{tt}^{-1}\mathbf{c}_t\mathbf{c}_t^{\mathsf{T}} \qquad\qquad 7.132$$

$$\mathbf{H}_t^{\mathsf{T}}\mathbf{R}_t^{-1} = \begin{bmatrix} \mathbf{H}_{t-1}^{\mathsf{T}} & \mathbf{c}_t \end{bmatrix} \begin{bmatrix} \mathbf{R}_{t-1}^{-1} & \mathbf{0} \\ \mathbf{0}^{\mathsf{T}} & r_{tt}^{-1} \end{bmatrix}$$

$$= \begin{bmatrix} \mathbf{H}_{t-1}^{\mathsf{T}}\mathbf{R}_{t-1}^{-1} & r_{tt}^{-1}\mathbf{c}_t \end{bmatrix}.$$

With these recursions, we can write the estimate $\hat{\mathbf{x}}_t$ as

$$\mathbf{P}_t^{-1}\hat{\mathbf{x}}_t = \mathbf{H}_t^{\mathsf{T}}\mathbf{R}_t^{-1}\mathbf{y}_t$$

$$= \begin{bmatrix} \mathbf{H}_{t-1}^{\mathsf{T}}\mathbf{R}_{t-1} & r_{tt}^{-1}\mathbf{c}_t \end{bmatrix} \begin{bmatrix} \mathbf{y}_{t-1} \\ y_t \end{bmatrix}$$

$$= \mathbf{P}_{t-1}^{-1}\hat{\mathbf{x}}_{t-1} + r_{tt}^{-1}\mathbf{c}_t y_t \qquad\qquad 7.133$$

$$= \left(\mathbf{P}_t^{-1} - r_{tt}^{-1}\mathbf{c}_t\mathbf{c}_t^{\mathsf{T}} \right)\hat{\mathbf{x}}_{t-1} + r_{tt}^{-1}\mathbf{c}_t y_t.$$

It is a simple rearrangement to write these equations in a more familiar sequential Bayes recursion:

$$\hat{\mathbf{x}}_t = \hat{\mathbf{x}}_{t-1} + \mathbf{P}_t\mathbf{c}_t r_{tt}^{-1}(y_t - \mathbf{c}_t^{\mathsf{T}}\hat{\mathbf{x}}_{t-1}). \qquad 7.134$$

If we define the gain vector

$$\mathbf{k}_t = \mathbf{P}_t\mathbf{c}_t r_{tt}^{-1}, \qquad\qquad 7.135$$

then we get the following sequential Bayes solution, illustrated in Figure 7.14:

$$\hat{\mathbf{x}}_t = \hat{\mathbf{x}}_{t-1} + \mathbf{k}_t(y_t - \mathbf{c}_t^{\mathsf{T}}\hat{\mathbf{x}}_{t-1})$$

$$\mathbf{P}_t^{-1}\mathbf{k}_t = \mathbf{c}_t r_{tt}^{-1} \qquad\qquad 7.136$$

$$\mathbf{P}_t^{-1} = \mathbf{P}_{t-1}^{-1} + r_{tt}^{-1}\mathbf{c}_t\mathbf{c}_t^{\mathsf{T}}.$$

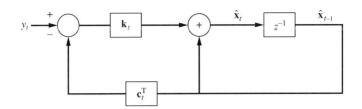

Figure 7.14 Network for implementing sequential Bayes estimate.

The regression equation for the gain bears a striking resemblance to the regression equation for the estimate $\hat{\mathbf{x}}_t$:

$$\mathbf{P}_t^{-1}\mathbf{k}_t = \mathbf{c}_t r_{tt}^{-1} \qquad\qquad 7.137$$

$$\mathbf{P}_t^{-1}\hat{\mathbf{x}}_t = \mathbf{H}_t^{\mathrm{T}}\mathbf{R}_t^{-1}\mathbf{y}_t.$$

In fact, the gain $\mathbf{k}_t$ is just the impulse response of the regression equation for $\hat{\mathbf{x}}_t$:

$$\mathbf{P}_t^{-1}\mathbf{k}_t = \mathbf{H}_t^{\mathrm{T}}\mathbf{R}_t^{-1}\mathbf{e}_t; \qquad \mathbf{e}_t = [0 \quad \cdots \quad 0 \quad 1]^{\mathrm{T}}$$
$$= \mathbf{c}_t r_{tt}^{-1}. \qquad\qquad 7.138$$

This property can also be illustrated with the network of Figure 7.14. If $\hat{\mathbf{x}}_{-1}$ is set to zero and $y_0 = 0$, $y_1 = 0, \ldots, y_{t-1} = 0$, $y_t = 1$, then $\hat{\mathbf{x}}_t = \mathbf{k}_t$.

7.8   THE KALMAN FILTER

The Kalman filter is an algorithm for recursively estimating dynamical state vectors $\mathbf{x}_t$ that evolve according to the difference equation

$$\mathbf{x}_{t+1} = \mathbf{A}\mathbf{x}_t + \mathbf{b}u_t. \qquad\qquad 7.139$$

In this difference equation, the initial state $\mathbf{x}_0$ is drawn from an $N[\mathbf{0}, \mathbf{R}_0]$ distribution. From there on, the difference equation is driven by a sequence of independent and identically distributed random variables $\{u_t\}$, each of which is independent of $\mathbf{x}_0$ and distributed as $u_t : N[0, \sigma_u^2]$. Because of the way $\mathbf{x}_t$ is constructed from a linear combination of normal random vectors, we know that it is a $p \times 1$ normal random vector with mean zero and covariance matrix $\mathbf{R}_t = E\mathbf{x}_t\mathbf{x}_t^{\mathrm{T}}$:

$$\mathbf{x}_t : N[\mathbf{0}, \mathbf{R}_t]. \qquad\qquad 7.140$$

The recursion for $\mathbf{R}_t$ is obtained from the difference equation 7.139:

$$\mathbf{R}_{t+1} = \mathbf{A}\mathbf{R}_t\mathbf{A}^{\mathrm{T}} + \sigma_u^2\mathbf{b}\mathbf{b}^{\mathrm{T}}. \qquad\qquad 7.141$$

The measurements available to us take the form

$$y_t = \mathbf{c}^{\mathrm{T}}\mathbf{x}_t + n_t, \qquad\qquad 7.142$$

where the noise sequence $\{n_t\}$ is a sequence of independent and identically distributed

$N[0, \sigma_n^2]$ random variables, each independent of the sequence $\{u_t\}$ and the state $\mathbf{x}_t$. The measurement y_t is normally distributed with mean zero and covariance σ_t^2:

$$y_t : N[0, \sigma_t^2] \qquad\qquad 7.143$$
$$\sigma_t^2 = \mathbf{c}^{\mathrm{T}}\mathbf{R}_t\mathbf{c} + \sigma_n^2.$$

A block diagram that illustrates how the variables $\mathbf{x}_t$ and y_t are generated is given in Figure 7.15.

Kalman posed for himself the problem of estimating the state $\mathbf{x}_t$ from the sequence of measurements $\mathbf{y}_t = [y_0 \quad y_1 \quad \cdots \quad y_t]^{\mathrm{T}}$. We shall derive the Kalman filter by using the Gauss-Markov theorem. We begin by noting that the conditional distribution of $\mathbf{x}_t$, given $\mathbf{y}_t$, will be normal:

$$\mathbf{x}_t|\mathbf{y}_t : N[\hat{\mathbf{x}}_{t|t}, \mathbf{P}_{t|t}]. \qquad\qquad 7.144$$

The problem, of course, is to find the a posteriori mean $\hat{\mathbf{x}}_{t|t}$ and the covariance $\mathbf{P}_{t|t}$.

Suppose we had measurements only up to time $t - 1$, namely $\mathbf{y}_{t-1} = [y_0 \quad y_1 \quad \cdots \quad y_{t-1}]^{\mathrm{T}}$. The conditional distribution of $\mathbf{x}_t$, given $\mathbf{y}_{t-1}$, would also be normal:

$$\mathbf{x}_t|\mathbf{y}_{t-1} : N[\hat{\mathbf{x}}_{t|t-1}, \mathbf{P}_{t|t-1}]. \qquad\qquad 7.145$$

While we do not know yet what form the "predictor" $\hat{\mathbf{x}}_{t|t-1}$ and prediction error covariance matrix $\mathbf{P}_{t|t-1}$ will take, we do know two important things about the predictor $\hat{\mathbf{x}}_{t|t-1}$ and prediction error $\hat{\mathbf{n}}_{t|t-1} = \mathbf{x}_t - \hat{\mathbf{x}}_{t|t-1}$:

1. The predictor $\hat{\mathbf{x}}_{t|t-1}$ is a sufficient statistic for $\mathbf{x}_t$. That is, $f(\mathbf{x}_t|\mathbf{y}_{t-1}) = f(\mathbf{x}_t|\hat{\mathbf{x}}_{t-1}) : N[\hat{\mathbf{x}}_{t|t-1}, \mathbf{P}_{t|t-1}]$, meaning the measurements $\mathbf{y}_{t-1}$ may be replaced by the predictor $\hat{\mathbf{x}}_{t|t-1}$.

2. The predictor $\hat{\mathbf{x}}_{t|t-1}$ and prediction error $\hat{\mathbf{n}}_{t|t-1} = \mathbf{x}_t - \hat{\mathbf{x}}_{t|t-1}$ orthogonally decompose $\mathbf{x}_t$. That is,

$$\mathbf{x}_t = \hat{\mathbf{x}}_{t|t-1} + \hat{\mathbf{n}}_{t|t-1}$$
$$\hat{\mathbf{x}}_{t|t-1} : N[\mathbf{0}, \mathbf{R}_t - \mathbf{P}_{t|t-1}] \qquad\qquad 7.146$$
$$\hat{\mathbf{n}}_{t|t-1} : N[\mathbf{0}, \mathbf{P}_{t|t-1}]$$
$$E\hat{\mathbf{x}}_{t|t-1}\hat{\mathbf{n}}_{t|t-1}^{\mathrm{T}} = \mathbf{0}.$$

These results are illustrated in the lower two branches of Figure 7.16.

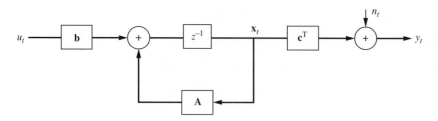

Figure 7.15 Block diagram for generating the state and the measurement.

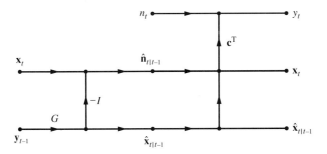

Figure 7.16 The state $\mathbf{x}_t$, predictor $\hat{\mathbf{x}}_{t|t-1}$, and measurement y_t.

Now let's suppose a new measurement y_t is obtained. This measurement is related to the state $\mathbf{x}_t$ by $y_t = \mathbf{c}^T\mathbf{x}_t + n_t$, but this may be written $y_t = \mathbf{c}^T(\hat{\mathbf{x}}_{t|t-1} + \hat{\mathbf{n}}_{t|t-1}) + n_t$ and illustrated with the top branch of Figure 7.16. With the aid of Figure 7.16 we may write down a linear model for the state $\mathbf{x}_t$, the predictor $\hat{\mathbf{x}}_{t|t-1}$, and the new measurement y_t, all in terms of the orthogonal variables $\hat{\mathbf{n}}_{t|t-1}$, $\hat{\mathbf{x}}_{t|t-1}$, and n_t:

$$
\begin{bmatrix} \mathbf{x}_t \\ \hat{\mathbf{x}}_{t|t-1} \\ y_t \end{bmatrix} = \begin{bmatrix} \mathbf{I} & \mathbf{I} & \mathbf{0} \\ \mathbf{0} & \mathbf{I} & \mathbf{0} \\ \mathbf{c}^T & \mathbf{c}^T & 1 \end{bmatrix} \begin{bmatrix} \hat{\mathbf{n}}_{t|t-1} \\ \hat{\mathbf{x}}_{t|t-1} \\ n_t \end{bmatrix}. \qquad 7.147
$$

The covariance matrix for the variables $\hat{\mathbf{n}}_{t|t-1}$, $\hat{\mathbf{x}}_{t|t-1}$, and n_t is

$$
E \begin{bmatrix} \hat{\mathbf{n}}_{t|t-1} \\ \hat{\mathbf{x}}_{t|t-1} \\ n_t \end{bmatrix} \begin{bmatrix} \hat{\mathbf{n}}_{t|t-1}^T & \hat{\mathbf{x}}_{t|t-1}^T & n_t^T \end{bmatrix} = \begin{bmatrix} \mathbf{P}_{t|t-1} & \mathbf{0} & \mathbf{0} \\ \mathbf{0} & \mathbf{R}_t - \mathbf{P}_{t|t-1} & \mathbf{0} \\ \mathbf{0} & \mathbf{0} & \sigma_n^2 \end{bmatrix}. \qquad 7.148
$$

When combined with the linear transformation of Equation 7.147, this diagonal covariance structure produces the following covariance matrix for $\mathbf{x}_t, \hat{\mathbf{x}}_{t|t-1}$, and y_t:

$$
E \begin{bmatrix} \mathbf{x}_t \\ \hat{\mathbf{x}}_{t|t-1} \\ y_t \end{bmatrix} \begin{bmatrix} \mathbf{x}_t^T & \hat{\mathbf{x}}_{t|t-1}^T & y_t \end{bmatrix} = \begin{bmatrix} \mathbf{R}_t & \hat{\mathbf{P}}_{t|t-1} & \mathbf{R}_t\mathbf{c} \\ \hat{\mathbf{P}}_{t|t-1} & \hat{\mathbf{P}}_{t|t-1} & \hat{\mathbf{P}}_{t|t-1}\mathbf{c} \\ \mathbf{c}^T\mathbf{R}_t & \mathbf{c}^T\hat{\mathbf{P}}_{t|t-1} & \sigma_t^2 \end{bmatrix}
$$

$$
= \left[\begin{array}{c|cc} \mathbf{R}_t & \hat{\mathbf{P}}_{t|t-1} & \mathbf{R}_t\mathbf{c} \\ \hline \hat{\mathbf{P}}_{t|t-1} & & \\ & \multicolumn{2}{c}{\mathbf{S}_t} \\ \mathbf{c}^T\mathbf{R}_t & & \end{array} \right] \qquad 7.149
$$

$$
\hat{\mathbf{P}}_{t|t-1} = \mathbf{R}_t - \mathbf{P}_{t|t-1}; \qquad \mathbf{S}_t = \begin{bmatrix} \hat{\mathbf{P}}_{t|t-1} & \hat{\mathbf{P}}_{t|t-1}\mathbf{c} \\ \mathbf{c}^T\hat{\mathbf{P}}_{t|t-1} & \sigma_t^2 \end{bmatrix}.
$$

This is a foreboding result, but it contains all of the information we need to compute the conditional distribution of $\mathbf{x}_t$, given $\hat{\mathbf{x}}_{t|t-1}$ and y_t:

$$\mathbf{x}_t \big| (\hat{\mathbf{x}}_{t|t-1}, y_t) : N[\hat{\mathbf{x}}_{t|t}, \mathbf{P}_{t|t}]$$

$$\hat{\mathbf{x}}_{t|t} = [\hat{\mathbf{P}}_{t|t-1} \quad \mathbf{R}_t \mathbf{c}] \mathbf{S}_t^{-1} \begin{bmatrix} \hat{\mathbf{x}}_{t|t-1} \\ y_t \end{bmatrix} \tag{7.150}$$

$$\mathbf{P}_{t|t} = \mathbf{R}_t - [\hat{\mathbf{P}}_{t|t-1} \quad \mathbf{R}_t \mathbf{c}] \mathbf{S}_t^{-1} \begin{bmatrix} \hat{\mathbf{P}}_{t|t-1} \\ \mathbf{c}^{\mathrm{T}} \mathbf{R}_t \end{bmatrix}.$$

The inverse of the patterned matrix $\mathbf{S}_t$ is

$$\mathbf{S}_t^{-1} = \begin{bmatrix} \hat{\mathbf{P}}_{t|t-1}^{-1} & \mathbf{0} \\ \mathbf{0}^{\mathrm{T}} & 0 \end{bmatrix} + \gamma_t^{-1} \begin{bmatrix} -\mathbf{c} \\ 1 \end{bmatrix} [-\mathbf{c}^{\mathrm{T}} \quad 1] \tag{7.151}$$

$$\gamma_t = \sigma_t^2 - \mathbf{c}^{\mathrm{T}} (\mathbf{R}_t - \mathbf{P}_{t|t-1}) \mathbf{c} = \mathbf{c}^{\mathrm{T}} \mathbf{P}_{t|t-1} \mathbf{c} + \sigma_n^2.$$

When this result is substituted into the formulas for $\hat{\mathbf{x}}_{t|t}$ and $\mathbf{P}_{t|t}$, the equations simplify as follows:

$$\hat{\mathbf{x}}_{t|t} = \hat{\mathbf{x}}_{t|t-1} + \mathbf{P}_{t|t-1} \mathbf{c} \gamma_t^{-1} \left(y_t - \mathbf{c}^{\mathrm{T}} \hat{\mathbf{x}}_{t|t-1} \right) \tag{7.152}$$

$$\mathbf{P}_{t|t} = \mathbf{P}_{t|t-1} - \frac{1}{\gamma_t} \mathbf{P}_{t|t-1} \mathbf{c} \mathbf{c}^{\mathrm{T}} \mathbf{P}_{t|t-1}.$$

These equations can be written a little more elegantly by defining the Kalman gain

$$\mathbf{k}_t = \mathbf{P}_{t|t-1} \mathbf{c} \gamma_t^{-1} \tag{7.153}$$

$$\gamma_t = \mathbf{c}^{\mathrm{T}} \mathbf{P}_{t|t-1} \mathbf{c} + \sigma_n^2.$$

Then the Kalman recursions are

$$\hat{\mathbf{x}}_{t|t} = \hat{\mathbf{x}}_{t|t-1} + \mathbf{k}_t \left(y_t - \mathbf{c}^{\mathrm{T}} \mathbf{x}_{t|t-1} \right)$$

$$\mathbf{P}_{t|t} = \mathbf{P}_{t|t-1} - \mathbf{k}_t \mathbf{c}^{\mathrm{T}} \mathbf{P}_{t|t-1} = \left(\mathbf{I} - \mathbf{k}_t \mathbf{c}^{\mathrm{T}} \right) \mathbf{P}_{t|t-1} \tag{7.154}$$

$$= \mathbf{P}_{t|t-1} - \gamma_t \mathbf{k}_t \mathbf{k}_t^{\mathrm{T}}.$$

These recursions are complete, except for the definitions of $\hat{\mathbf{x}}_{t|t-1}$ and $\mathbf{P}_{t|t-1}$. The predictor $\hat{\mathbf{x}}_{t|t-1}$ is the conditional mean of $\mathbf{x}_t$, given $\mathbf{y}_{t-1}$. It may be written as

$$\hat{\mathbf{x}}_{t|t-1} = E[\mathbf{x}_t|\mathbf{y}_{t-1}] = E[\mathbf{A}\mathbf{x}_{t-1} + \mathbf{b}u_{t-1}|\mathbf{y}_{t-1}] \tag{7.155}$$

$$= \mathbf{A}\hat{\mathbf{x}}_{t-1|t-1} + \mathbf{0}.$$

The error covariance $\mathbf{P}_{t|t-1}$ is defined as follows:

$$\mathbf{P}_{t|t-1} = E[\mathbf{x}_t - \hat{\mathbf{x}}_{t|t-1}][\mathbf{x}_t - \hat{\mathbf{x}}_{t|t-1}]^{\mathrm{T}}$$

$$= E[\mathbf{A}\mathbf{x}_{t-1} + \mathbf{b}u_{t-1} - \mathbf{A}\hat{\mathbf{x}}_{t-1|t-1}][\mathbf{A}\mathbf{x}_{t-1} + \mathbf{b}u_{t-1} - \mathbf{A}\hat{\mathbf{x}}_{t-1|t-1}]^{\mathrm{T}}$$

$$= E[\mathbf{A}(\mathbf{x}_{t-1} - \hat{\mathbf{x}}_{t-1|t-1}) + \mathbf{b}u_{t-1}][\mathbf{A}(\mathbf{x}_{t-1} - \hat{\mathbf{x}}_{t-1|t-1}) + \mathbf{b}u_{t-1}]^{\mathrm{T}} \tag{7.156}$$

$$= \mathbf{A}\mathbf{P}_{t-1|t-1}\mathbf{A}^{\mathrm{T}} + \sigma_u^2 \mathbf{b}\mathbf{b}^{\mathrm{T}}.$$

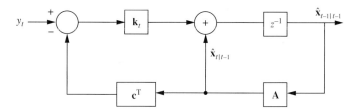

Figure 7.17 The Kalman filter.

So now we may summarize the celebrated equations of the Kalman filter:

$$\hat{\mathbf{x}}_{t|t} = \hat{\mathbf{x}}_{t|t-1} + \mathbf{k}_t\left(y_t - \mathbf{c}^{\mathrm{T}}\hat{\mathbf{x}}_{t|t-1}\right)$$

$$\hat{\mathbf{x}}_{t|t-1} = \mathbf{A}\hat{\mathbf{x}}_{t-1|t-1}$$

$$\mathbf{k}_t = \mathbf{P}_{t|t-1}\mathbf{c}\gamma_t^{-1} \qquad\qquad 7.157$$

$$\gamma_t = \mathbf{c}^{\mathrm{T}}\mathbf{P}_{t|t-1}\mathbf{c} + \sigma_n^2$$

$$\mathbf{P}_{t|t} = \mathbf{P}_{t|t-1} - \gamma_t\mathbf{k}_t\mathbf{k}_t^{\mathrm{T}}$$

$$\mathbf{P}_{t+1|t} = \mathbf{A}\mathbf{P}_{t|t}\mathbf{A}^{\mathrm{T}} + \sigma_u^2\mathbf{b}\mathbf{b}^{\mathrm{T}}.$$

From the prediction error covariance $\mathbf{P}_{t|t-1}$ we compute the Kalman gain $\mathbf{k}_t$ and the estimator error covariance $\mathbf{P}_{t|t}$. From $\mathbf{P}_{t|t}$, $\mathbf{P}_{t+1|t}$ is computed, and the recursion continues. The filter is depicted in Figure 7.17.

There is one final wrinkle to this story. From the Kalman recursion for the estimate $\hat{\mathbf{x}}_{t|t}$, we may construct the following error equation:

$$\mathbf{x}_t - \hat{\mathbf{x}}_{t|t} = \mathbf{x}_t - \hat{\mathbf{x}}_{t|t-1} - \mathbf{k}_t\left(\mathbf{c}^{\mathrm{T}}(\mathbf{x}_t - \hat{\mathbf{x}}_{t|t-1}) + n_t\right). \qquad 7.158$$

From this equation, we compute the covariance matrix $\mathbf{P}_{t|t}$:

$$\mathbf{P}_{t|t} = \mathbf{P}_{t|t-1} + \gamma_t\mathbf{k}_t\mathbf{k}_t^{\mathrm{T}} - \mathbf{P}_{t|t-1}\mathbf{c}\mathbf{k}_t^{\mathrm{T}} - \mathbf{k}_t\mathbf{c}^{\mathrm{T}}\mathbf{P}_{t|t-1}$$

$$= \mathbf{P}_{t|t-1} - \gamma_t\mathbf{k}_t\mathbf{k}_t^{\mathrm{T}}. \qquad\qquad 7.159$$

This reproduces our previously derived recursion and shows that the cross-covariance between the prediction error $\mathbf{x}_t - \hat{\mathbf{x}}_{t|t-1}$ and the residual term $y_t - \mathbf{c}^{\mathrm{T}}\hat{\mathbf{x}}_{t|t-1}$ is, in fact, the product of the variance γ_t and the Kalman gain $\mathbf{k}_t$:

$$E\left(\mathbf{x}_t - \hat{\mathbf{x}}_{t|t-1}\right)\left(y_t - \mathbf{c}^{\mathrm{T}}\hat{\mathbf{x}}_{t|t-1}\right)^{\mathrm{T}} = E[\mathbf{x}_t - \hat{\mathbf{x}}_{t|t-1}]\left[\mathbf{c}^{\mathrm{T}}(\mathbf{x}_t - \hat{\mathbf{x}}_{t|t-1}) + n_t\right]^{\mathrm{T}}$$

$$= \mathbf{P}_{t|t-1}\mathbf{c}$$

$$= \gamma_t\mathbf{k}_t. \qquad\qquad 7.160$$

7.9   THE WIENER FILTER

Recall from Section 7.5 that the inverse of the innovations representation for
$[\mathbf{x} \quad \mathbf{y}]$ is

$$\begin{bmatrix} \hat{\mathbf{n}} \\ \boldsymbol{\nu} \end{bmatrix} = \begin{bmatrix} \mathbf{I} & -\mathbf{G} \\ \mathbf{0} & \mathbf{R}_{yy}^{-1/2} \end{bmatrix} \begin{bmatrix} \mathbf{x} \\ \mathbf{y} \end{bmatrix}. \tag{7.161}$$

This representation produces the cross-covariance matrix

$$E \begin{bmatrix} \hat{\mathbf{n}} \\ \boldsymbol{\nu} \end{bmatrix} [\mathbf{x}^T \quad \mathbf{y}^T] = \begin{bmatrix} \mathbf{P} & \mathbf{0} \\ \mathbf{R}_{yy}^{-1/2}\mathbf{R}_{yx} & \left(\mathbf{R}_{yy}^{1/2}\right)^T \end{bmatrix}. \tag{7.162}$$

This result shows that the error $\hat{\mathbf{n}} = \mathbf{x} - \hat{\mathbf{x}}$ and the measurement $\mathbf{y}$ are orthogonal and
that $E\hat{\mathbf{n}}\mathbf{x}^T$ is the error covariance matrix $\mathbf{P}$.

Suppose we try to use a principle of orthogonality to determine the innovations
representation. That is, let's begin with the innovations representations

$$\begin{bmatrix} \mathbf{x} \\ \mathbf{y} \end{bmatrix} = \begin{bmatrix} \mathbf{I} & \mathbf{W} \\ \mathbf{0} & \mathbf{R}_{yy}^{1/2} \end{bmatrix} \begin{bmatrix} \hat{\mathbf{n}} \\ \boldsymbol{\nu} \end{bmatrix} \tag{7.163}$$

$$\begin{bmatrix} \hat{\mathbf{n}} \\ \boldsymbol{\nu} \end{bmatrix} = \begin{bmatrix} \mathbf{I} & -\mathbf{W}\mathbf{R}_{yy}^{-1/2} \\ \mathbf{0} & \mathbf{R}_{yy}^{-1/2} \end{bmatrix} \begin{bmatrix} \mathbf{x} \\ \mathbf{y} \end{bmatrix}$$

and try to solve for $\mathbf{W}$. The cross-covariance of this representation is

$$E \begin{bmatrix} \hat{\mathbf{n}} \\ \boldsymbol{\nu} \end{bmatrix} [\mathbf{x}^T \quad \mathbf{y}^T] = \begin{bmatrix} \mathbf{I} & -\mathbf{W}\mathbf{R}_{yy}^{-1/2} \\ \mathbf{0} & \mathbf{R}_{yy}^{-1/2} \end{bmatrix} \begin{bmatrix} \mathbf{R}_{xx} & \mathbf{R}_{xy} \\ \mathbf{R}_{yx} & \mathbf{R}_{yy} \end{bmatrix}$$

$$= \begin{bmatrix} \mathbf{R}_{xx} - \mathbf{W}\mathbf{R}_{yy}^{-1/2}\mathbf{R}_{yx} & \mathbf{R}_{xy} - \mathbf{W}\left(\mathbf{R}_{yy}^{1/2}\right)^T \\ \mathbf{R}_{yy}^{-1/2}\mathbf{R}_{yx} & \left(\mathbf{R}_{yy}^{1/2}\right)^T \end{bmatrix}. \tag{7.164}$$

The principle of orthogonality says that the NE term is zero:

$$\mathbf{W} = \mathbf{R}_{xy}\left(\mathbf{R}_{yy}^{-1/2}\right)^T, \tag{7.165}$$

and it produces this result for the error covariance matrix $\mathbf{P}$:

$$\mathbf{P} = \mathbf{R}_{xx} - \mathbf{R}_{xy}\left(\mathbf{R}_{yy}^{-1/2}\right)^T \mathbf{R}_{yy}^{-1/2}\mathbf{R}_{yx} \tag{7.166}$$

$$= \mathbf{R}_{xx} - \mathbf{R}_{xy}\mathbf{R}_{yy}^{-1}\mathbf{R}_{yx}.$$

The solution for the Bayes estimate is

$$\hat{\mathbf{x}} = \mathbf{W}\mathbf{R}_{yy}^{-1/2}\mathbf{y} \tag{7.167}$$

$$= \mathbf{R}_{xy}\mathbf{R}_{yy}^{-1}\mathbf{y}.$$

This basic idea may be used to derive the celebrated Wiener filter for a wide-sense stationary time series. We do not quite have all of the machinery to proceed rigorously, but we can proceed formally, even at this stage of our development.

Let $\mathbf{x}$ and $\mathbf{y}$ denote wide-sense stationary time series with power spectral densities $S_{xx}(z)$ and $S_{yy}(z)$ and with cross spectral matrices $S_{xy}(z)$ and $S_{yx}(z)$. Assume that $S_{yy}(z)$ has the spectral factorization

$$S_{yy}(z) = G(z)G(z^{-1}) \qquad\qquad 7.168$$

with $G(z)$ a causal and minimum phase transfer function. (This means that $1/G(z)$ is also causal.) Let's look for an innovations representation of the form

$$\begin{bmatrix} \mathbf{x} \\ \mathbf{y} \end{bmatrix} = \begin{bmatrix} 1 & W(z) \\ 0 & G(z) \end{bmatrix} \begin{bmatrix} \hat{\mathbf{n}} \\ \boldsymbol{\nu} \end{bmatrix} \qquad\qquad 7.169$$

$$\begin{bmatrix} \hat{\mathbf{n}} \\ \boldsymbol{\nu} \end{bmatrix} = \begin{bmatrix} 1 & -\dfrac{W(z)}{G(z)} \\ 0 & \dfrac{1}{G(z)} \end{bmatrix} \begin{bmatrix} \mathbf{x} \\ \mathbf{y} \end{bmatrix}.$$

The cross-covariance matrix of interest is characterized by the cross-spectrum matrix between the two representations:

$$\begin{bmatrix} S_{\hat{n}x}(z) & S_{\hat{n}y}(z) \\ S_{\nu x}(z) & S_{\nu y}(z) \end{bmatrix} = \begin{bmatrix} I & -\dfrac{W(z)}{G(z)} \\ 0 & \dfrac{1}{G(z)} \end{bmatrix} \begin{bmatrix} S_{xx}(z) & S_{xy}(z) \\ S_{yx}(z) & S_{yy}(z) \end{bmatrix}$$

$$\qquad\qquad 7.170$$

$$= \begin{bmatrix} S_{xx}(z) - \dfrac{W(z)}{G(z)} S_{yx}(z) & S_{xy}(z) - \dfrac{W(z)}{G(z)} S_{yy}(z) \\ \dfrac{1}{G(z)} S_{yx}(z) & \dfrac{S_{yy}(z)}{G(z)} \end{bmatrix}.$$

Now, what kind of orthogonality would be appropriate and what constraints on $W(z)$ are appropriate? We shall require $W(z)$ to be causal and the cross-spectrum $S_{\hat{n}y}(z)$ to be strictly anticausal, meaning that $\hat{\mathbf{n}}$ is orthogonal to the present and entire past of $\mathbf{y}$. Thus we require the NE term

$$S_{xy}(z) - \frac{W(z)}{G(z)} S_{yy}(z) \qquad\qquad 7.171$$

to be strictly anticausal, meaning that its inverse transform is zero for nonnegative indexes. Let's use the spectral factors of $S_{yy}(z)$ to write the NE term as

$$\left[\frac{S_{xy}(z)}{G(z^{-1})} - W(z) \right] G(z^{-1}). \qquad\qquad 7.172$$

The transform $G(z^{-1})$ is anticausal, meaning that the entire expression is strictly anticausal if

$$W(z) = \left[\frac{S_{xy}(z)}{G(z^{-1})} \right]_+$$ 7.173

where $[\ ]_+$ denotes the causal part of $[\ ]$. The Wiener filter for $\hat{\mathbf{x}}$ is now

$$\hat{\mathbf{x}} = \frac{W(z)}{G(z)} \mathbf{y}.$$ 7.174

More explicitly, the estimated sequence $\{\hat{x}_t\}$ is the following filtered version of $\{y_t\}$:

$$\{\hat{x}_t\} = \frac{1}{G(z)} \left[\frac{S_{xy}(z)}{G(z^{-1})} \right]_+ \{y_t\}.$$ 7.175

REFERENCES AND COMMENTS

Most of the results in this chapter are my adaptations and interpretations of standard results. My treatment of the risk set for Bayes estimators follows Ferguson pretty closely. The example of the leaky Geiger counter comes right out of E. T. Jaynes' delightful set of lectures. My treatment of the multivariate normal model is designed to illustrate the role played by linear models. The derivation of the Kalman filter is different from standard derivations. C. T. Mullis kindly helped me with the derivation of the Wiener filter.

Ferguson, T. J. [1967]. *Mathematical Statistics* (New York: Academic Press, 1967).

Jaynes, E. T. [1958]. "Probability Theory in Science and Engineering," *Colloquium Lectures in Pure and Applied Science*, No. 4 (Socony Mobil Oil Company: February 1958).

PROBLEMS

7.1   Let X denote a continuous random variable and N a discrete random variable. Prove that the conditional density for X, given $N = n$, is

$$f_{X|N}(x|n) = \frac{P[N = n|X = x]}{P[N = n]} f_X(x)$$

where $f_X(x)$ is the prior density for X. Interpret this result by explaining how the measurement $N = n$ transforms the prior density for X into a posterior density.

7.2   Let $X = (X_0, X_1, \ldots, X_{N-1})$ denote a random sample of scalar random variables X_n, each of which is Poisson distributed:

$$f(x_n|\lambda) = \frac{\lambda^{x_n}}{x_n!} e^{-\lambda}; \qquad x_n = 0, 1, 2, \ldots; \qquad \lambda \geq 0.$$

The parameter λ is exponentially distributed:

$$p(\lambda) = ae^{-a\lambda}; \qquad \lambda \ge 0; \quad a > 0.$$

a.   Show that $t = \sum_{n=0}^{N-1} x_n$ is sufficient for λ.

b.   Show that the marginal density for x is

$$f(x) = \frac{a}{\prod_{n=0}^{N-1} x_n!} \frac{\Gamma(t+1)}{(N+a)^{t-1}}.$$

c.   Show that the conditional density for λ, given x, is

$$f(\lambda|x) = e^{-(N+a)\lambda} \lambda^t \frac{(N+a)^{t+1}}{\Gamma(t+1)}.$$

This is a gamma density with parameters $t+1$ and $(N+a)^{-1}$; $\Gamma(\alpha)$ is $\int_0^\infty e^{-x} x^{\alpha-1}\, dx$.

d.   Show that the conditional mean of λ is $(t+1)/(N+a)$.

e.   Show that the conditional variance of λ is $(t+2)/(N+a)^2$.

f.   Find the mean-squared error between λ and its conditional mean.

7.3   Let $X = (X_0, X_1, \ldots, X_{N-1})$ denote a random sample of scalar random variables, each of which is uniformly distributed:

$$f(x_n|\theta) = \begin{cases} 1/\theta, & 0 \le x_n \le \theta, \ \theta > 0 \\ 0, & \text{otherwise.} \end{cases}$$

The parameter θ is exponentially distributed:

$$p(\theta) = ae^{-a\theta}; \qquad \theta > 0; \quad a > 0.$$

Find the Bayes estimator of θ that minimizes mean-squared error.

7.4   *Gauss-Gauss.* Let $X = (X_0, X_1, \ldots, X_{N-1})$ denote a random sample of scalar random variables, each of which is normally distributed:

$$f(x_n|\theta) = \frac{1}{(2\pi\sigma^2)^{1/2}} \exp\left\{ -\frac{1}{2\sigma^2} (x_n - \theta)^2 \right\}.$$

The parameter θ is also normally distributed:

$$p(\theta) = \frac{1}{(2\pi\sigma_\theta^2)^{1/2}} \exp\left\{ -\frac{1}{2\sigma_\theta^2} (\theta - m)^2 \right\}.$$

a.   Find the conditional density of θ, given x.

b.   Find the conditional mean and variance of θ, given x.

c.   Explain how this problem relates to the problem of estimating a Gaussian signal S when $X_n = S + N_n$ is measured and N_n is a Gaussian noise.

d.   Compare $E[\theta\,|\,x]$ to $\hat\theta_{\mathrm{ML}}$; compare the mean and variance of the two estimators.

7.5   *Imperfect Geiger Counter.* Consider the following Poisson source in cascade
with an imperfect counter that records each Poisson count with probability p:

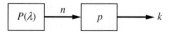

Show that k is Poisson $P[\lambda p]$ and $P[k|n]$ is $\binom{n}{k}p^k(1-p)^{n-k}$. This is the physical
model that underlies Example 7.1.

7.6   *Imperfect Geiger Counter.* For the leaky Geiger counter of Example 7.1,

a.   Prove that k is Poisson with parameter λp.

b.   Prove that the conditional mean and conditional variance of n, given k, are
$k + \lambda(1-p)$ and $\lambda(1-p)$.

c.   Lend intuitive insight to the conditional mean estimator $n = k + \lambda(1-p)$.
Why is $\lambda(1-p)$ a reasonable correction to k?

d.   Show that the mean-squared errors, conditioned on n, for the Bayes and ad
hoc estimators $\hat{n}$ and $\tilde{n}$ are

$$E\left[(n-\hat{n})^2|n\right] = (1-p)^2(n-\lambda)^2 + np(1-p)$$

$$E\left[(n-\tilde{n})^2|n\right] = \frac{n}{p}(1-p).$$

e.   Show that this conditional mean-squared error is smaller for $\tilde{n}$ than for $\hat{n}$
whenever

$$\frac{1+p}{\lambda p} < \left[\left(\frac{n}{\lambda}\right)^{1/2} - \left(\frac{n}{\lambda}\right)^{-1/2}\right]^2.$$

Plot the left- and right-hand sides above for insight and interpret your results.

f.   Show that the conditional mean-squared errors of (d) above produce the
correct answers for mean-squared error.

7.7   *Sum of Poissons.* Consider this source of information:

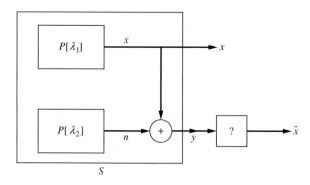

The random variables X and N are independent Poissons, and Y is their sum.

a.   Describe an experiment where this problem would apply.

b.   Find the distribution of Y.

c.   Find the conditional probability mass function for X, given Y.

d.   Find the minimum mean-squared error (MMSE) estimator of X.

e.   Compute the mean and mean-squared error of your MMSE estimator.

7.8   Consider this source of information:

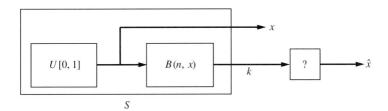

S

The uniformly distributed random variable X sets the success probability in a binomial random variable K; that is, $P[k \mid x] = \binom{n}{k} x^k (1 - x)^{n-k}$.

a.   Find $f_{X|K}(x|k)$, the (mixed) conditional density of continuous x, given discrete k.

b.   Find the MMSE estimator of X.

c.   Compute the mean and mean-squared error of the MMSE estimator.

7.9   *Bayesian Recursions.* Use the Bayesian recursions

$$f(\mathbf{x}_t|\mathbf{y}_t) = \frac{f(y_t|\mathbf{x}_t)f(\mathbf{x}_t|\mathbf{y}_{t-1})}{f(y_t|\mathbf{y}_{t-1})}$$

to derive the Kalman filtering equations. Hints:

a. $f(y_t|\mathbf{x}_t) : N[\mathbf{c}^\mathsf{T}\mathbf{x}_t, \sigma_n^2]$

b. $f(\mathbf{x}_t|\mathbf{y}_{t-1}) : N[\hat{\mathbf{x}}_{t/t-1}, \mathbf{P}_{t/t-1}]$

c. $f(\mathbf{x}_t|\mathbf{y}_t) : N[\hat{\mathbf{x}}_{t/t}, \mathbf{P}_{t/t}]$

d. $f(y_t|\mathbf{y}_{t-1}) : N[\mathbf{c}^\mathsf{T}\hat{\mathbf{x}}_{t/t-1}, \mathbf{c}^\mathsf{T}\mathbf{P}_{t/t-1}\mathbf{c} + \sigma_n^2]$.

(a) is obvious, (b) and (c) are definitions, and (d) must be derived.

7.10   *Innovations.* Show that the so-called innovations sequence of prediction errors in the Kalman filter,

$$\nu_t = y_t - \mathbf{c}^\mathsf{T}\hat{\mathbf{x}}_{t/t-1},$$

is a sequence of independent normal random variables, each distributed as ν_t : $N[0, \mathbf{c}^\mathsf{T}\mathbf{P}_{t/t-1}\mathbf{c} + \sigma_n^2]$.

7.11 *Linear Models.* Consider the following linear models:

$$\underset{(N\times1)}{\boxed{\mathbf{y}}} = \boxed{\mathbf{H}}\;\underset{(p\times1)}{\boxed{\mathbf{x}}} + \boxed{\mathbf{n}}$$

$$N>p$$

$$\underset{(N\times1)}{\boxed{\mathbf{y}}} = \boxed{\qquad\mathbf{H}\qquad}\;\underset{(p\times1)}{\boxed{\mathbf{x}}} + \boxed{\mathbf{n}}$$

$$N<p$$

Assume $\mathbf{n} : N[\mathbf{0}, \sigma^2\mathbf{I}]$ and $\mathbf{x} : N[\mathbf{0}, \mathbf{R}_{xx}]$. You have two sets of equations for estimating $\mathbf{x}$, given $\mathbf{y}$, and computing $\mathbf{P}$:

a. $\hat{\mathbf{x}} = \mathbf{R}_{xx}\mathbf{H}^{\mathrm{T}}(\mathbf{H}\mathbf{R}_{xx}\mathbf{H}^{\mathrm{T}} + \mathbf{R}_{nn})^{-1}\mathbf{y}$

   $\mathbf{P} = \mathbf{R}_{xx} - \mathbf{R}_{xx}\mathbf{H}^{\mathrm{T}}(\mathbf{H}\mathbf{R}_{xx}\mathbf{H}^{\mathrm{T}} + \mathbf{R}_{nn})^{-1}\mathbf{H}\mathbf{R}_{xx}$

b. $\hat{\mathbf{x}} = \mathbf{P}\mathbf{H}^{\mathrm{T}}\mathbf{R}_{nn}^{-1}\mathbf{y}$

   $\mathbf{P} = (\mathbf{R}_{xx}^{-1} + \mathbf{H}^{\mathrm{T}}\mathbf{R}_{nn}^{-1}\mathbf{H})^{-1}$.

Compare the computational complexity of these pairs of formulas. In which cases would you use set (a)? In which cases would you use set (b)?

7.12 *Prediction.* Let $\{x_t\}$ denote a wide-sense stationary (WSS), Gaussian time series with covariance sequence $\{r_t\} = \{\sigma^2 a^{|t|}\}$, $|a| < 1$. Find the minimum mean-squared error predictor of x_{t+s} from $\mathbf{x} = (x_{t-N}, \ldots, x_{t-1}, x_t)$. Does your result depend on N? What is the minimum mean-squared error predictor of x_{t+s} from the infinite past $(\ldots, x_{t-1}, x_t)$? What is the mean-squared prediction error?

7.13 From the time series of Problem 7.12, define the following measurements $\mathbf{y}$ and states $\mathbf{x}$:

a. $\mathbf{y} = x_1$ $\mathbf{x} = [x_3\quad x_4]^{\mathrm{T}}$

b. $\mathbf{y} = x_2$ $\mathbf{x} = [x_3\quad x_4]^{\mathrm{T}}$

c. $\mathbf{y} = [x_1\quad x_2]^{\mathrm{T}}$ $\mathbf{x} = [x_3\quad x_4]^{\mathrm{T}}$

d. $\mathbf{y} = [x_1\quad x_2]^{\mathrm{T}}$ $\mathbf{x} = x_3$

e. $\mathbf{y} = [x_1\quad x_2]^{\mathrm{T}}$ $\mathbf{x} = x_4$

For each case, find $\mathbf{x}|\mathbf{y} : N[\hat{\mathbf{x}}, \mathbf{P}]$. For each case, plot the contours of constant probability for $\mathbf{x} : N[\mathbf{0}, \mathbf{R}_{xx}]$ and for $\mathbf{x} - \hat{\mathbf{x}} : N[\mathbf{0}, \mathbf{P}]$.

7.14 You are told that the observed signal y_t obeys the equations

$$y_k = \sum_{i=1}^{p} h_{ki}x_i + n_k.$$

Write out a linear statistical model for $k = 0, 1, \ldots t - 1$:

$$\mathbf{y}_t = \mathbf{H}_t\mathbf{x} + \mathbf{n}_t$$

and show how the row vectors $\mathbf{c}_t$ used in the sequential Bayes estimator of $\mathbf{x} = [x_1 \ldots x_p]^T$ are defined.

7.15 *Rank One Signal Model.* Consider this measurement system:

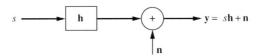

The random variable s is $N[0, \sigma_s^2]$ and the noise $\mathbf{n}$ is $N[\mathbf{0}, \sigma_n^2\mathbf{I}]$; s and $\mathbf{n}$ are independent. Assume $\mathbf{h}^T\mathbf{h} = 1$ and rework Example 7.4 (the Rank-One Signal Model) to find $\hat{s}$ and P.

7.16 In Example 7.4 and Problem 7.15, interpret σ_s^2/σ_n^2 as the input signal-to-noise ratio $(SNR)_{in}$ and $N\sigma_s^2/\sigma_n^2$ as the output SNR. Interpret the results for $\hat{s}$ and P in terms of these SNRs by showing what happens for low and high SNR.

7.17 *Gauss-Markov Theorem.* Prove the Gauss-Markov theorem:

$$\begin{bmatrix} \mathbf{x} \\ \mathbf{y} \end{bmatrix} : N\left[\begin{bmatrix} \mathbf{m}_x \\ \mathbf{m}_y \end{bmatrix}, \begin{bmatrix} \mathbf{R}_{xx} & \mathbf{R}_{xy} \\ \mathbf{R}_{yx} & \mathbf{R}_{yy} \end{bmatrix}\right]$$

$$\mathbf{x}|\mathbf{y} : N[\hat{\mathbf{x}}, \mathbf{P}]$$

$$\hat{\mathbf{x}} = \mathbf{m}_x + \mathbf{R}_{xy}\mathbf{R}_{yy}^{-1}(\mathbf{y} - \mathbf{m}_y)$$

$$\mathbf{P} = \mathbf{R}_{xx} - \mathbf{R}_{xy}\mathbf{R}_{yy}^{-1}\mathbf{R}_{yx}.$$

7.18 Use the formula

$$\mathbf{R}_{xx}^{-1}\hat{\mathbf{x}} = \mathbf{H}^T\mathbf{R}_{nn}^{-1}(\mathbf{y} - \mathbf{H}\hat{\mathbf{x}})$$

to compute the covariance of the innovations $\mathbf{y} - \mathbf{H}\hat{\mathbf{x}}$ and the cross-covariance between $\mathbf{y} - \mathbf{H}\hat{\mathbf{x}}$ and $\hat{\mathbf{x}}$.

7.19 Show that the innovations representation

$$\begin{bmatrix} \mathbf{x} \\ \mathbf{y} \end{bmatrix} = \begin{bmatrix} \mathbf{I} & \mathbf{R}_{xy}(\mathbf{R}_{yy}^{-1/2})^T \\ \mathbf{0} & \mathbf{R}_{yy}^{1/2} \end{bmatrix} \begin{bmatrix} \hat{\mathbf{n}} \\ \boldsymbol{\nu} \end{bmatrix}$$

$$\hat{\mathbf{n}} : N[\mathbf{0}, \mathbf{P}] ; \quad \boldsymbol{\nu} : N[\mathbf{0}, \mathbf{I}]$$

really does produce the covariance matrix

$$E\begin{bmatrix} \mathbf{x} \\ \mathbf{y} \end{bmatrix} [\mathbf{x}^T \ \mathbf{y}^T] = \begin{bmatrix} \mathbf{R}_{xx} & \mathbf{R}_{xy} \\ \mathbf{R}_{yx} & \mathbf{R}_{yy} \end{bmatrix}.$$

Why should we say that this representation is a "block Cholesky factorization" of

$$\begin{bmatrix} \mathbf{R}_{xx} & \mathbf{R}_{xy} \\ \mathbf{R}_{yx} & \mathbf{R}_{yy} \end{bmatrix}?$$

7.20 *Kalman Filters.* Specialize the Kalman filter for the following cases:

   a. $\mathbf{A} = \mathbf{I}$, $u_t = 0 \ \forall \ t \geq 0$;
   b. $\mathbf{A} = \mathbf{A}$, $u_t = 0 \ \forall \ t \geq 0$;
   c. $\mathbf{A} = \mathbf{I}$, $\mathbf{R}_0 = r_0\mathbf{I}$, $r_0 \to \infty$;
   d. $\mathbf{A} = \mathbf{A}$, $\mathbf{R}_0 = r_0\mathbf{I}$, $r_0 \to \infty$.

Draw a block diagram for each Kalman filter, and summarize the filter and covariance equations for each case.

7.21 Show how to initialize the Kalman filter. That is, compute $\mathbf{P}_{0|-1}$. Then show how to run the recursions for $t = 0, 1, 2, \ldots$.

7.22 We have two ways to phrase the Kalman filtering problem—as a batch process,

$$\hat{\mathbf{x}}_{t|t} = \mathbf{K}_t\mathbf{y}_t; \qquad \mathbf{P}_t^{-1}\mathbf{K}_t = \mathbf{H}_t^T; \qquad \mathbf{y}_t = [y_0 \quad y_1 \quad \cdots \quad y_t]^T$$

or as a sequential recursion,

$$\hat{\mathbf{x}}_{t|t} = \mathbf{A}\hat{\mathbf{x}}_{t|t-1} + \mathbf{k}_t\left(y_t - \mathbf{c}^T\mathbf{A}\hat{\mathbf{x}}_{t|t-1}\right).$$

Show that the Kalman gain is just the vector required to update the batch matrix $\mathbf{K}_t$ as follows:

$$\boxed{\quad\quad \mathbf{K}_t \quad\quad} \quad = \quad \boxed{(\mathbf{A} - \mathbf{k}_t\mathbf{c}^T)\mathbf{K}_{t-1} \quad | \quad \mathbf{k}_t}$$

7.23 *Prediction.* Let (x_n) denote a (wide-sense stationary) series of multivariate normal random variables with the following properties:

$$Ex_n = 0 \ \forall \ n$$

$$Ex_n x_{n+t} = r_t \ \forall \ (n, t).$$

Let $\mathbf{y} = [x_{t-p} \quad x_{t-p+1} \quad \cdots \quad x_{t-1}]$ represent the finite past of the sequence.

   a. Estimate x_t from $\mathbf{y}$: $\hat{x}_t = \mathbf{a}^T\mathbf{y}$; solve for $\mathbf{a}$.
   b. Compute the mean-squared error between x_t and $\hat{x}_t$.
   c. Replace $\mathbf{y}$ with $\mathbf{z} = \mathbf{y} + \mathbf{n}$, with $\mathbf{n} = [n_{t-p} \quad \cdots \quad n_{t-1}]^T$ and the n_t i.i.d. $N[0, \sigma_n^2]$ random variables. Write down the equation for the MMSE estimator of x_t for any t.
   d. Compute the mean-squared error between x_t and $\hat{x}_t$.

7.24 Define the loss function

$$L\left[\boldsymbol{\theta}, \hat{\boldsymbol{\theta}}(\mathbf{x})\right] = \boldsymbol{\pi}^{\mathrm{T}} \mathbf{a}\left[\boldsymbol{\theta}, \hat{\boldsymbol{\theta}}(\mathbf{x})\right]$$

where

$$\boldsymbol{\pi} = [\pi_1 \quad \pi_2 \quad \cdots \quad \pi_p]^{\mathrm{T}}; \qquad \pi_i > 0 \;\; \forall \; i$$

$$\mathbf{a}\left[\boldsymbol{\theta}, \hat{\boldsymbol{\theta}}(\mathbf{x})\right] = \left|\theta_i - \hat{\theta}_i(\mathbf{x})\right| \quad : \text{absolute error between } i\text{th component of } \boldsymbol{\theta}$$
$$\text{and } i\text{th component of } \hat{\boldsymbol{\theta}}(\mathbf{x}).$$

a.   Show that the conditional risk may be written as

$$\int d\boldsymbol{\theta} \, f(\boldsymbol{\theta}|\mathbf{x}) \boldsymbol{\pi}^{\mathrm{T}} \mathbf{a}\left[\boldsymbol{\theta}, \hat{\boldsymbol{\theta}}(\mathbf{x})\right]$$

$$= \sum_{i=1}^{p} \pi_i \left[-\int_{-\infty}^{\hat{\theta}_i(\mathbf{x})} d\theta_i \, f(\theta_i|\mathbf{x})[\theta_i - \hat{\theta}_i(\mathbf{x})] + \int_{\hat{\theta}_i(\mathbf{x})}^{\infty} d\theta_i \, f(\theta_i|\mathbf{x})[\theta_i - \hat{\theta}_i(\mathbf{x})] \right].$$

b.   Show that the minimizing estimator $\hat{\theta}_i(\mathbf{x})$ is the "median of the conditional density $f(\theta_i|\mathbf{x})$."

c.   How would you compute $f(\theta_i|\mathbf{x})$ from $f(\boldsymbol{\theta}|\mathbf{x})$?

7.25 *Channel Identification and Deconvolution.* Consider these two diagrams:

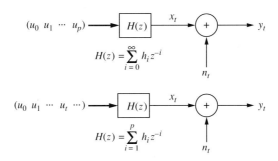

a.   In the top diagram, $H(z)$ is a known filter and $\mathbf{u} = [u_0 \quad u_1 \quad \cdots \quad u_p]^{\mathrm{T}}$ is an input to be estimated. Derive the sequential Bayes solution when $\mathbf{u} : N[\mathbf{0}, \sigma_u^2 \mathbf{I}]$ and the n_t are i.i.d. $N[0, \sigma_n^2]$ random variables.

b.   In the bottom diagram, $H(z)$ is an unknown finite impulse response filter and $\mathbf{u} = [u_0 \quad u_1 \quad \cdots]^{\mathrm{T}}$ is a known input. Derive the sequential Bayes solution for $\mathbf{h} = [h_0 \quad h_1 \quad \cdots \quad h_p]^{\mathrm{T}}$ when $\mathbf{h} : N[\mathbf{0}, \sigma_h^2 I]$.

7.26 (*Prediction in Poisson Process*) Let $N[0, t)$ denote the number of particles emit-
ted from a radioactive source on the time interval $[0, t)$. This is a Poisson random
variable:

$$P[N[0, t) = n] = e^{-\lambda t} \frac{(\lambda t)^n}{n!}$$

$$EN[0, t) = \lambda t; \qquad \text{var } N[0, t) = \lambda t.$$

Let $(t_1, t_2, \ldots, t_n)$ denote a sequence of times at which $N[0, t)$ is measured.

a. Find the MMSE estimator of $N(0, t_n)$, given $N[0, t_i)$, $i = 1, 2, \ldots, n-1$.
b. Find the mean of the MMSE estimator.
c. Find the mean-squared error of the MMSE estimator.

Minimum Mean-Squared Error Estimators

In Chapter 7 we learned that the *conditional mean estimator* minimizes *mean-squared error*, regardless of the joint distribution for a vector $\mathbf{x}$ and a measurement $\mathbf{y}$. We learned that the *conditional mean* of $\mathbf{x}$ is a *linear function* of the measurement $\mathbf{y}$ when $\mathbf{x}$ and $\mathbf{y}$ are *jointly normal*. This is the Gauss-Markov theorem. These results are so fundamental that we pursue their consequences further in this chapter. In particular, we explore the theory of *linear minimum mean-squared error* (LMMSE) estimators, wherein the estimator of $\mathbf{x}$ is forced to be a linear function of the measurement $\mathbf{y}$, whether or not $\mathbf{x}$ and $\mathbf{y}$ are jointly normal. This theory produces the *Wiener-Hopf equations* as the fundamental design equations. We find that these equations may be generated by a fundamental orthogonality condition that simplifies the study of LMMSE estimators. We apply the theory of LMMSE estimators to low-rank Wiener filtering, linear prediction, source modeling, and Kalman filtering. We apply the more general theory of MMSE estimators to scalar and block quantizers.

8.1   CONDITIONAL EXPECTATION AND ORTHOGONALITY

We found in Chapter 7 that the conditional expection $E[\mathbf{x}|\mathbf{y}]$ plays a fundamental role in minimum mean-squared error estimation. In this section, we review a number of its important properties.

Let $\mathbf{x}$ and $\mathbf{y}$ be jointly distributed random vectors. The suggestion is that $\mathbf{y}$ brings information about $\mathbf{x}$ and that the observation $\mathbf{y}$ may be used to estimate unobserved $\mathbf{x}$. The dimensions of $\mathbf{x}$ and $\mathbf{y}$ need not be identical. For example, $\mathbf{x}$ may be a $p \times 1$ random parameter, and $\mathbf{y}$ may be an $N \times 1$ random measurement. Depending on the application, it will be appropriate to call $\mathbf{x}$ a signal, a parameter, or a state.

Define the statistic $\mathbf{g}(\mathbf{y})$. Its dimension may differ from that of $\mathbf{x}$ and $\mathbf{y}$. The expectation of the outer product of $\mathbf{x}$ and $\mathbf{g}(\mathbf{y})$ may be written

$$
\begin{aligned}
E\mathbf{x}\mathbf{g}^{\mathrm{T}}(\mathbf{y}) &= EE\big[\mathbf{x}\,\mathbf{g}^{\mathrm{T}}(\mathbf{y})|\mathbf{y}\big] \\
&= EE[\mathbf{x}|\mathbf{y}]\mathbf{g}^{\mathrm{T}}(\mathbf{y}) \\
&= E\hat{\mathbf{x}}\,\mathbf{g}^{\mathrm{T}}(\mathbf{y})
\end{aligned}
\tag{8.1}
$$

where $\hat{\mathbf{x}}$ is the conditional expectation

$$
\hat{\mathbf{x}} = E[\mathbf{x}|\mathbf{y}].
\tag{8.2}
$$

We call $\hat{\mathbf{x}}$ the conditional mean, conditional mean estimator, or conditional expectation. We emphasize that it is a function only of $\mathbf{y}$.

This basic result for conditional expectation may be rewritten as

$$
E(\mathbf{x} - \hat{\mathbf{x}})\mathbf{g}^{\mathrm{T}}(\mathbf{y}) = \mathbf{0},
\tag{8.3}
$$

meaning that the error between $\mathbf{x}$ and its conditional mean estimator $\hat{\mathbf{x}}$ is orthogonal to every measurable function of $\mathbf{y}$. There are several striking consequences of this basic result. We list and annotate them here.

1. **Expectation.** Let $\mathbf{g}(\mathbf{y}) = \mathbf{I}$. Then

$$
\begin{aligned}
E\mathbf{x} &= E\hat{\mathbf{x}} \\
&= EE(\mathbf{x}|\mathbf{y}).
\end{aligned}
\tag{8.4}
$$

This relation says that $\hat{\mathbf{x}}$ is an unbiased estimate of $\mathbf{x}$. It says, further, that the expectation of $\mathbf{x}$ may be computed by averaging over the marginal distribution of $\mathbf{x}$ (the left-hand side) or by averaging over the conditional distribution of $\mathbf{x}$, given $\mathbf{y}$, and then averaging over the marginal distribution of $\mathbf{y}$.

2. **Orthogonality.** Let $\mathbf{g}(\mathbf{y}) = \mathbf{y}$. Then

$$
E(\mathbf{x} - \hat{\mathbf{x}})\mathbf{y}^{\mathrm{T}} = \mathbf{0}.
\tag{8.5}
$$

This relation says that the error between $\mathbf{x}$ and its conditional mean estimator $\hat{\mathbf{x}}$ is orthogonal to the random vector $\mathbf{y}$ from which the conditional mean was computed. The result also says that the correlation between $\mathbf{x}$ and $\mathbf{y}$ equals the correlation between $\hat{\mathbf{x}}$ and $\mathbf{y}$. That is,

$$
\mathbf{R}_{xy} = \mathbf{R}_{\hat{x}y}
\tag{8.6}
$$

where $\mathbf{R}_{xy}$ and $\mathbf{R}_{\hat{x}y}$ are the following correlation matrices:

$$\mathbf{R}_{xy} = E\mathbf{x}\mathbf{y}^\mathrm{T} \tag{8.7}$$

$$\mathbf{R}_{\hat{x}y} = E\hat{\mathbf{x}}\mathbf{y}^\mathrm{T}.$$

3. **Orthogonality between Estimator and Estimator Error.** Let $\mathbf{g}(\mathbf{y}) = \hat{\mathbf{x}}$. Then

$$E(\mathbf{x} - \hat{\mathbf{x}})\hat{\mathbf{x}}^\mathrm{T} = \mathbf{0}, \tag{8.8}$$

meaning that the estimator error is orthogonal to the estimator. Furthermore, the following correlations are equal:

$$\mathbf{R}_{x\hat{x}} = \mathbf{R}_{\hat{x}\hat{x}}. \tag{8.9}$$

The orthogonality condition (3) is illustrated in Figure 8.1.

We often say that the conditional mean estimator $\hat{\mathbf{x}}$ brings an orthogonal representation of $\mathbf{x}$, which we write as follows:

$$\mathbf{x} = \hat{\mathbf{x}} + \hat{\mathbf{n}} \tag{8.10}$$

$$\hat{\mathbf{n}} = \mathbf{x} - \hat{\mathbf{x}}.$$

This orthogonal representation allows us to decompose $\mathbf{R}_{xx}$ as follows:

$$\mathbf{R}_{xx} = \mathbf{R}_{\hat{x}\hat{x}} + \mathbf{R}_{\hat{n}\hat{n}}. \tag{8.11}$$

This result may be used to show that the "energy" in $\hat{\mathbf{x}}$ is always less than or equal to the energy in $\mathbf{x}$:

$$E\mathbf{x}^\mathrm{T}\mathbf{x} = \mathrm{tr}\,\mathbf{R}_{xx}$$
$$= E\hat{\mathbf{x}}^\mathrm{T}\hat{\mathbf{x}} + E\hat{\mathbf{n}}^\mathrm{T}\hat{\mathbf{n}}. \tag{8.12}$$

4. **Projection.** The conditional expectation is a (generally nonlinear) projection. The argument goes as follows. Write conditional expectation as the operator $\mathbf{P}(\mathbf{y})$:

$$\mathbf{P}(\mathbf{y}) = E[\mathbf{x}|\mathbf{y}]. \tag{8.13}$$

Then note that

$$\mathbf{P}\big(\mathbf{P}(\mathbf{y})\big) = E\big[E[\mathbf{x}|\mathbf{y}]\big|\mathbf{y}\big] = E[\mathbf{x}|\mathbf{y}] = \mathbf{P}(\mathbf{y}). \tag{8.14}$$

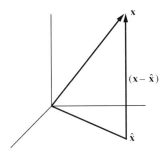

Figure 8.1 Orthogonality.

We say that the operator $\mathbf{P}(\mathbf{y})$ is idempotent, meaning that $\mathbf{P}$ composed with itself is just $\mathbf{P}$. So, once $\mathbf{x}$ is projected, it remains in its position $\hat{\mathbf{x}}$ under continued projection.

8.2   MINIMUM MEAN-SQUARED ERROR ESTIMATORS

The conditional mean estimator $\hat{\mathbf{x}}$ is also the minimum mean-squared error estimator of $\mathbf{x}$ that can be constructed from $\mathbf{y}$. To prove this claim, we proceed as follows. Let $\mathbf{g}(\mathbf{y})$ be an arbitrary estimator of $\mathbf{x}$ and let $\hat{\mathbf{x}}$ be the conditional mean estimator of $\mathbf{x}$. Both estimators are orthogonal to the error $\mathbf{x} - \hat{\mathbf{x}}$, as shown in the previous section. This means that we can decompose the error between $\mathbf{x}$ and $\mathbf{g}(\mathbf{y})$ into two orthogonal components:

$$\mathbf{x} - \mathbf{g}(\mathbf{y}) = (\mathbf{x} - \hat{\mathbf{x}}) + \left(\hat{\mathbf{x}} - \mathbf{g}(\mathbf{y})\right). \qquad 8.15$$

This decomposition is illustrated in Figure 8.2.

The mean-squared error between $\mathbf{x}$ and $\mathbf{g}(\mathbf{y})$ is

$$\begin{aligned} E\big[\mathbf{x} - \mathbf{g}(\mathbf{y})\big]^{\mathrm{T}}\big[\mathbf{x} - \mathbf{g}(\mathbf{y})\big] &= E(\mathbf{x} - \hat{\mathbf{x}})^{\mathrm{T}}(\mathbf{x} - \hat{\mathbf{x}}) + E\big[\hat{\mathbf{x}} - \mathbf{g}(\mathbf{y})\big]^{\mathrm{T}}\big[\hat{\mathbf{x}} - \mathbf{g}(\mathbf{y})\big] \\ &\geq E(\mathbf{x} - \hat{\mathbf{x}})^{\mathrm{T}}(\mathbf{x} - \hat{\mathbf{x}}) \end{aligned} \qquad 8.16$$

with equality iff $\mathbf{g}(\mathbf{y}) = \hat{\mathbf{x}}$. This result says that the conditional mean estimator $\hat{\mathbf{x}}$ minimizes mean-squared error. The conditional mean estimator is generally a nonlinear function of $\mathbf{y}$, and the computation of $\hat{\mathbf{x}}$ is often called nonlinear filtering. The conditional mean estimator produces errors that are orthogonal to $\mathbf{y}$.

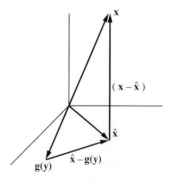

Figure 8.2 Decomposition of the
error between $\mathbf{x}$ and
$\mathbf{g}(\mathbf{y})$.

8.3   LINEAR MINIMUM MEAN-SQUARED ERROR (LMMSE) ESTIMATORS

Let us now assume that our estimator of $\mathbf{x}$ is constrained to be a *linear function* of $\mathbf{y}$. The problem is to find the matrix $\mathbf{K}$ that minimizes the mean-squared error between $\mathbf{x}$ and the *linear* estimator $\mathbf{Ky}$. The mean-squared error of any linear estimator may be compared with the mean-squared error of the conditional mean estimator by writing

$$
\begin{aligned}
E(\mathbf{x} - \mathbf{Ky})^{\mathsf{T}}(\mathbf{x} - \mathbf{Ky}) &= E(\mathbf{x} - \hat{\mathbf{x}} + \hat{\mathbf{x}} - \mathbf{Ky})^{\mathsf{T}}(\mathbf{x} - \hat{\mathbf{x}} + \hat{\mathbf{x}} - \mathbf{Ky}) \\
&= E(\mathbf{x} - \hat{\mathbf{x}})^{\mathsf{T}}(\mathbf{x} - \hat{\mathbf{x}}) + E(\hat{\mathbf{x}} - \mathbf{Ky})^{\mathsf{T}}(\hat{\mathbf{x}} - \mathbf{Ky}).
\end{aligned}
\tag{8.17}
$$

The cross-terms on the right-hand side have vanished because the error $\mathbf{x} - \hat{\mathbf{x}}$ is orthogonal to every measurable function of $\mathbf{y}$. Two points are evident:

1.   The mean-squared error of a linear estimator is never smaller than the mean-squared error of the conditional mean estimator; and

2.   The linear minimum mean-squared error estimator of $\mathbf{x}$ is also the linear minimum mean-squared error estimator of the conditional mean $\hat{\mathbf{x}}$.

Wiener-Hopf Equation

By analogy with our results for the minimum mean-squared error estimator, let us hypothesize that the linear minimum mean-squared error estimator of $\mathbf{x}$ satisfies the orthogonality condition

$$
E(\mathbf{x} - \mathbf{Ky})\mathbf{y}^{\mathsf{T}} = \mathbf{0}.
\tag{8.18}
$$

This condition may be rewritten as

$$
\mathbf{R}_{xy} - \mathbf{K}\mathbf{R}_{yy} = \mathbf{0},
\tag{8.19}
$$

meaning that the matrix $\mathbf{K}$ satisfies a Wiener-Hopf equation and has the solution

$$
\mathbf{K} = \mathbf{R}_{xy}\mathbf{R}_{yy}^{-1}.
\tag{8.20}
$$

The error associated with the linear estimator $\mathbf{Ky}$ has covariance matrix $\mathbf{P}$:

$$
\begin{aligned}
\mathbf{P} = E(\mathbf{x} - \mathbf{Ky})(\mathbf{x} - \mathbf{Ky})^{\mathsf{T}} &= \mathbf{R}_{xx} - \mathbf{K}\mathbf{R}_{yx} \\
&= \mathbf{R}_{xx} - \mathbf{R}_{xy}\mathbf{R}_{yy}^{-1}\mathbf{R}_{yx}.
\end{aligned}
\tag{8.21}
$$

Does this choice of $\mathbf{Ky}$ really minimize the mean-squared error among all linear functions of $\mathbf{y}$? Try the competing linear estimator $\mathbf{Ly}$. Write its mean-squared error as

$$
\begin{aligned}
E(\mathbf{x} - \mathbf{Ly})^{\mathsf{T}}(\mathbf{x} - \mathbf{Ly}) &= E(\mathbf{x} - \mathbf{Ky} + \mathbf{Ky} - \mathbf{Ly})^{\mathsf{T}}(\mathbf{x} - \mathbf{Ky} + \mathbf{Ky} - \mathbf{Ly}) \\
&= E(\mathbf{x} - \mathbf{Ky})^{\mathsf{T}}(\mathbf{x} - \mathbf{Ky}) + E(\mathbf{Ky} - \mathbf{Ly})^{\mathsf{T}}(\mathbf{Ky} - \mathbf{Ly}) \\
&\geq E(\mathbf{x} - \mathbf{Ky})^{\mathsf{T}}(\mathbf{x} - \mathbf{Ky}).
\end{aligned}
\tag{8.22}
$$

Again, the cross-terms vanish because $\mathbf{x} - \mathbf{Ky}$ is orthogonal to $\mathbf{y}$.

Summary and Interpretations

Begin with a source of information that generates correlated random vectors $\mathbf{x}$ and $\mathbf{y}$ with correlation matrix

$$E \begin{bmatrix} \mathbf{x} \\ \mathbf{y} \end{bmatrix} [\mathbf{x}^T \quad \mathbf{y}^T] = \begin{bmatrix} \mathbf{R}_{xx} & \mathbf{R}_{xy} \\ \mathbf{R}_{yx} & \mathbf{R}_{yy} \end{bmatrix}. \tag{8.23}$$

This source is illustrated in Figure 8.3(a). One of the Schur complements of this correlation matrix is

$$\mathbf{Q} = \mathbf{R}_{yy} - \mathbf{R}_{yx}\mathbf{R}_{xx}^{-1}\mathbf{R}_{xy}. \tag{8.24}$$

This, in fact, is the correlation matrix of a noise vector $\mathbf{n}$ that may be added to $\mathbf{R}_{yx}\mathbf{R}_{xx}^{-1}\mathbf{x}$ to produce $\mathbf{y}$. That is, the linear transformation

$$\begin{bmatrix} \mathbf{x} \\ \mathbf{y} \end{bmatrix} = \begin{bmatrix} \mathbf{I} & \mathbf{0} \\ \mathbf{R}_{yx}\mathbf{R}_{xx}^{-1} & \mathbf{I} \end{bmatrix} \begin{bmatrix} \mathbf{x} \\ \mathbf{n} \end{bmatrix} \tag{8.25}$$

(a) correlated source

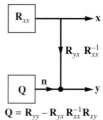

$\mathbf{Q} = \mathbf{R}_{yy} - \mathbf{R}_{yx}\mathbf{R}_{xx}^{-1}\mathbf{R}_{xy}$

(b) prior signal-plus-noise model

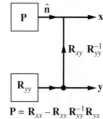

$\mathbf{P} = \mathbf{R}_{xx} - \mathbf{R}_{xy}\mathbf{R}_{yy}^{-1}\mathbf{R}_{yx}$

(c) posterior signal-plus-noise model

Figure 8.3 Correlated sources of
information.

with

$$E\begin{bmatrix} \mathbf{x} \\ \mathbf{n} \end{bmatrix} [\mathbf{x}^T \quad \mathbf{n}^T] = \begin{bmatrix} \mathbf{R}_{xx} & \mathbf{0} \\ \mathbf{0} & \mathbf{Q} \end{bmatrix} \qquad 8.26$$

reproduces the correct correlation matrix for $(\mathbf{x}, \mathbf{y})$:

$$E\begin{bmatrix} \mathbf{x} \\ \mathbf{y} \end{bmatrix} [\mathbf{x}^T \quad \mathbf{y}^T] = \begin{bmatrix} \mathbf{I} & \mathbf{0} \\ \mathbf{R}_{yx}\mathbf{R}_{xx}^{-1} & \mathbf{I} \end{bmatrix} \begin{bmatrix} \mathbf{R}_{xx} & \mathbf{0} \\ \mathbf{0} & \mathbf{Q} \end{bmatrix} \begin{bmatrix} \mathbf{I} & \mathbf{R}_{xx}^{-1}\mathbf{R}_{xy} \\ \mathbf{0} & \mathbf{I} \end{bmatrix}$$

$$= \begin{bmatrix} \mathbf{R}_{xx} & \mathbf{R}_{xy} \\ \mathbf{R}_{yx} & \mathbf{R}_{yy} \end{bmatrix}. \qquad 8.27$$

This "prior" signal-plus-noise model is illustrated in Figure 8.3(*b*).

The linear estimator that minimizes mean-squared error is

$$\hat{\mathbf{x}} = \mathbf{R}_{xy}\mathbf{R}_{yy}^{-1}\mathbf{y}. \qquad 8.28$$

This estimator produces the "posterior" signal-plus-noise model

$$\mathbf{x} = \hat{\mathbf{x}} + \hat{\mathbf{n}} \qquad 8.29$$

where $\hat{\mathbf{x}}$ and $\hat{\mathbf{n}}$ are uncorrelated and where the estimation error $\hat{\mathbf{n}}$ has covariance matrix $\mathbf{P}$:

$$\mathbf{P} = E\hat{\mathbf{n}}\hat{\mathbf{n}}^T = \mathbf{R}_{xx} - \mathbf{R}_{xy}\mathbf{R}_{yy}^{-1}\mathbf{R}_{yx}. \qquad 8.30$$

These results, illustrated in Figure 8.3(*c*), may be written as

$$\begin{bmatrix} \mathbf{x} \\ \mathbf{y} \end{bmatrix} = \begin{bmatrix} \mathbf{I} & \mathbf{R}_{xy}\mathbf{R}_{yy}^{-1} \\ \mathbf{0} & \mathbf{I} \end{bmatrix} \begin{bmatrix} \hat{\mathbf{n}} \\ \mathbf{y} \end{bmatrix} \qquad 8.31$$

with

$$E\begin{bmatrix} \hat{\mathbf{n}} \\ \mathbf{y} \end{bmatrix} [\hat{\mathbf{n}}^T \quad \mathbf{y}] = \begin{bmatrix} \mathbf{P} & \mathbf{0} \\ \mathbf{0} & \mathbf{R}_{yy} \end{bmatrix}. \qquad 8.32$$

The correlation matrix of $(\mathbf{x}, \mathbf{y})$ is

$$E\begin{bmatrix} \mathbf{x} \\ \mathbf{y} \end{bmatrix} [\mathbf{x}^T \quad \mathbf{y}^T] = \begin{bmatrix} \mathbf{I} & \mathbf{R}_{xy}\mathbf{R}_{yy}^{-1} \\ \mathbf{0} & \mathbf{I} \end{bmatrix} \begin{bmatrix} \mathbf{P} & \mathbf{0} \\ \mathbf{0} & \mathbf{R}_{yy} \end{bmatrix} \begin{bmatrix} \mathbf{I} & \mathbf{0} \\ \mathbf{R}_{yy}^{-1}\mathbf{R}_{yx} & \mathbf{I} \end{bmatrix}$$

$$= \begin{bmatrix} \mathbf{R}_{xx} & \mathbf{R}_{xy} \\ \mathbf{R}_{yy} & \mathbf{R}_{yy} \end{bmatrix}. \qquad 8.33$$

In summary, any correlated source of variables $(\mathbf{x}, \mathbf{y})$ has two signal-plus-noise representations, illustrated in Figures 8.3(*b*) and (*c*). In the posterior diagram, the variable $\hat{\mathbf{x}} = \mathbf{R}_{xy}\mathbf{R}_{yy}^{-1}\mathbf{y}$ is the linear estimator that minimizes mean-squared error and $\hat{\mathbf{n}}$ is its error.

8.4   LOW-RANK WIENER FILTER

Let's summarize our findings from the previous section with the diagram of Figure
8.4. The left-hand side of the diagram is the posterior representation of the pair $(\mathbf{x}, \mathbf{y})$
and the right-hand side is the filtering diagram for producing the Wiener estimate
$\hat{\mathbf{x}} = \mathbf{Gy}$, and its error $\hat{\mathbf{n}} = \mathbf{x} - \hat{\mathbf{x}}$, from the measurement $\mathbf{y}$. The Wiener-Hopf
equation $\mathbf{GR}_{yy} = \mathbf{R}_{xy}$ produces the filter $\mathbf{G} = \mathbf{R}_{xy}\mathbf{R}_{yy}^{-1}$, which is $\mathbf{K}$ in Section 8.3.

Suppose we wish to approximate the filter $\mathbf{G}$ with a low-rank approximation,
called $\mathbf{G}_r$. How should we choose it? To answer this question, we replace $\mathbf{G}$ on the
right-hand side of Figure 8.4 by $\mathbf{G}_r$ and write out expressions for the resulting filter
$\hat{\mathbf{x}}_r$ and error $\hat{\mathbf{n}}_r$:

$$
\begin{bmatrix} \hat{\mathbf{n}}_r \\ \hat{\mathbf{x}}_r \end{bmatrix} = \begin{bmatrix} \mathbf{I} & -\mathbf{G}_r \\ \mathbf{0} & \mathbf{G}_r \end{bmatrix} \begin{bmatrix} \mathbf{x} \\ \mathbf{y} \end{bmatrix} = \begin{bmatrix} \mathbf{I} & -\mathbf{G}_r \\ \mathbf{0} & \mathbf{G}_r \end{bmatrix} \begin{bmatrix} \mathbf{I} & \mathbf{G} \\ \mathbf{0} & \mathbf{I} \end{bmatrix} \begin{bmatrix} \hat{\mathbf{n}} \\ \mathbf{y} \end{bmatrix}
$$
$$
= \begin{bmatrix} \mathbf{I} & \mathbf{G} - \mathbf{G}_r \\ \mathbf{0} & \mathbf{G}_r \end{bmatrix} \begin{bmatrix} \hat{\mathbf{n}} \\ \mathbf{y} \end{bmatrix}.
$$

8.34

From this equation it is a simple matter to compute the covariance matrix for the error
$\hat{\mathbf{n}}_r$ and the low-rank estimator $\hat{\mathbf{x}}_r$:

$$
E\begin{bmatrix} \hat{\mathbf{n}}_r \\ \hat{\mathbf{x}}_r \end{bmatrix} [\hat{\mathbf{n}}_r^{\mathrm{T}} \quad \hat{\mathbf{x}}_r^{\mathrm{T}}] = \begin{bmatrix} \mathbf{I} & \mathbf{G} - \mathbf{G}_r \\ \mathbf{0} & \mathbf{G}_r \end{bmatrix} \begin{bmatrix} \mathbf{P} & \mathbf{0} \\ \mathbf{0} & \mathbf{R}_{yy} \end{bmatrix} \begin{bmatrix} \mathbf{I} & \mathbf{0} \\ \mathbf{G}^{\mathrm{T}} - \mathbf{G}_r^{\mathrm{T}} & \mathbf{G}_r^{\mathrm{T}} \end{bmatrix}
$$
$$
= \begin{bmatrix} \mathbf{P} + (\mathbf{G} - \mathbf{G}_r)\mathbf{R}_{yy}(\mathbf{G} - \mathbf{G}_r)^{\mathrm{T}} & \mathbf{0} \\ \mathbf{0} & \mathbf{G}_r\mathbf{R}_{yy}\mathbf{G}_r^{\mathrm{T}} \end{bmatrix}.
$$

8.35

A reasonable strategy for reducing the rank of $\mathbf{G}$ is to choose the rank r approximation
to $\mathbf{G}$ that minimizes the trace of the extra covariance:

$$
\min_{\mathbf{G}_r} \mathrm{tr}[(\mathbf{G} - \mathbf{G}_r)\mathbf{R}_{yy}(\mathbf{G} - \mathbf{G}_r)^{\mathrm{T}}] = \min_{\mathbf{G}_r} \mathrm{tr}[(\mathbf{GR}_{yy}^{1/2} - \mathbf{G}_r\mathbf{R}_{yy}^{1/2})(\mathbf{GR}_{yy}^{1/2} - \mathbf{G}_r\mathbf{R}_{yy}^{1/2})^{\mathrm{T}}].
$$

8.36

The solution is to make $\mathbf{G}_r\mathbf{R}_{yy}^{1/2}$ the best low-rank approximation of $\mathbf{GR}_{yy}^{1/2}$. To this
end we give $\mathbf{GR}_{yy}^{1/2}$ an SVD representation:

$$
\mathbf{GR}_{yy}^{1/2} = \mathbf{U\Sigma V}^{\mathrm{T}}; \qquad \mathbf{\Sigma} = \left[\begin{array}{c|c} \mathbf{\Sigma}_r & \mathbf{0} \\ \hline \mathbf{0} & \star \end{array} \right].
$$

8.37

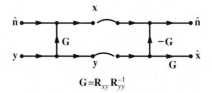

Figure 8.4 Linear minimum mean-squared
error (or Wiener) filtering of $\mathbf{y}$ for $\mathbf{x}$.

Then $\mathbf{G}_r\mathbf{R}_{yy}^{1/2}$ and $\mathbf{G}_r$ are

$$
\mathbf{G}_r\mathbf{R}_{yy}^{1/2} = \mathbf{U} \left[\begin{array}{c|c} \Sigma_r & \mathbf{0} \\ \hline \mathbf{0} & \mathbf{0} \end{array} \right] \mathbf{V}^\mathrm{T}
$$

$$
\mathbf{G}_r = \mathbf{U} \left[\begin{array}{c|c} \Sigma_r & \mathbf{0} \\ \hline \mathbf{0} & \mathbf{0} \end{array} \right] \mathbf{V}^\mathrm{T}\mathbf{R}_{yy}^{-1/2}.
$$

8.38

This completes the specification of the low-rank Wiener filter. Can you compute the resulting error covariance matrix for $\hat{\mathbf{x}}_r$? See Problem 8.26.

8.5   LINEAR PREDICTION

Linear prediction is a standard term applied to the problem of predicting a random variable x from a finite linear combination of the correlated random variables contained in the random vector $\mathbf{y} = [y_1 \quad \cdots \quad y_p]^\mathrm{T}$. The random variables are typically drawn from a time series $\{x_t\}$, but the theory of linear prediction really applies to a wide variety of problems where one set of observed variables brings information about an unobserved variable that is to be predicted.

In Chapter 10 we take up the details of linear prediction in earnest when we study the many fast algorithms that apply when the time series $\{x_t\}$ is stationary. In this section, we simply advance the rudiments of linear prediction and show how the theory of LMMSE estimation applies.

Consider the correlation between the random variable x and the measurements $\mathbf{y} = [y_1 \quad \cdots \quad y_p]^\mathrm{T}$:

$$
E \begin{bmatrix} x \\ \mathbf{y} \end{bmatrix} [x \quad \mathbf{y}^\mathrm{T}] = \left[\begin{array}{c|c} \sigma^2 & \mathbf{r}^\mathrm{T} \\ \hline \mathbf{r} & \mathbf{R} \end{array} \right]
$$

8.39

$$
\sigma^2 = Ex^2 \qquad \mathbf{r} = E\mathbf{y}x
$$

$$
\mathbf{R} = E\mathbf{y}\mathbf{y}^\mathrm{T}.
$$

The linear estimator that minimizes mean-squared error, and its error covariance matrix, are

$$
\hat{x} = \mathbf{a}^\mathrm{T}\mathbf{y}
$$

$$
\mathbf{a}^\mathrm{T} = \mathbf{r}^\mathrm{T}\mathbf{R}^{-1}; \quad (\mathbf{R}\mathbf{a} = \mathbf{r})
$$

8.40

$$
P = \sigma^2 - \mathbf{r}^\mathrm{T}\mathbf{R}^{-1}\mathbf{r}
$$

$$
= \sigma^2 - \mathbf{a}^\mathrm{T}\mathbf{r}.
$$

If we denote the vector **a** by the elements $\mathbf{a} = [-a_1 \quad -a_2 \quad \cdots \quad -a_p]^{\mathrm{T}}$, then the estimator $\hat{x}$ is written as

$$\hat{x} = -\sum_{i=1}^{p} a_i y_i. \tag{8.41}$$

This is standard notation in the theory of linear prediction.

Now suppose x is the time series value x_t and $\mathbf{y}$ is the vector of preceding values $[x_{t-1} \quad \cdots \quad x_{t-p}]^{\mathrm{T}}$. Assume that the correlations are independent of time:

$$E\begin{bmatrix} x_t \\ x_{t-1} \\ \vdots \\ x_{t-p} \end{bmatrix} \begin{bmatrix} x_t & x_{t-1} & \cdots & x_{t-p} \end{bmatrix} = \left[\begin{array}{c|ccccc} r_0 & r_1 & \cdots & & r_p \\ \hline r_1 & r_0 & r_1 & \cdots & r_{p-1} \\ \vdots & r_1 & \ddots & \ddots & \vdots \\ & \vdots & \ddots & & r_1 \\ r_p & r_{p-1} & \cdots & r_1 & r_0 \end{array}\right] = \left[\begin{array}{c|c} r_0 & \mathbf{r}^{\mathrm{T}} \\ \hline \mathbf{r} & \mathbf{R} \end{array}\right]. \tag{8.42}$$

The LMMSE estimator of x_t from the preceding time series values $(x_{t-1}, \ldots, x_{t-p})$ is

$$\hat{x}_t = -\sum_{i=1}^{p} a_i x_{t-i}$$

$$\mathbf{Ra} = \mathbf{r} \tag{8.43}$$

$$\mathbf{a} = [-a_1 \quad \cdots \quad -a_p)^{\mathrm{T}}.$$

The mean-squared error is

$$\begin{aligned} P &= r_0 - \mathbf{r}^{\mathrm{T}}\mathbf{R}^{-1}\mathbf{r} \\ &= r_0 - \mathbf{a}^{\mathrm{T}}\mathbf{r} \\ &= \sum_{i=0}^{p} a_i r_i; \qquad a_0 = 1. \end{aligned} \tag{8.44}$$

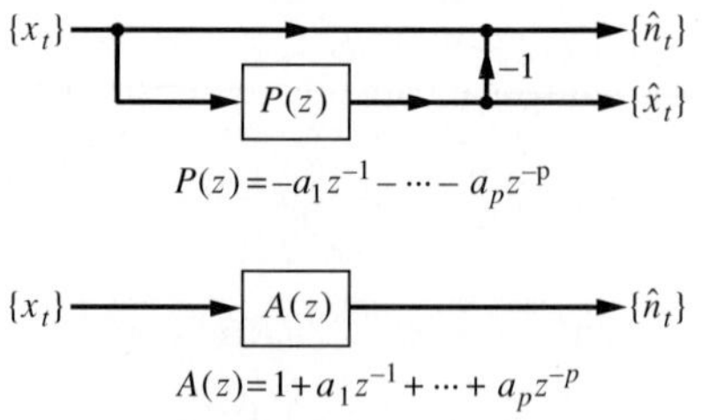

$\{x_t\}$
$\{\hat{n}_t\}$
-1
$P(z)$
$\{\hat{x}_t\}$
$P(z) = -a_1 z^{-1} - \cdots - a_p z^{-p}$
$\{x_t\}$
$A(z)$
$\{\hat{n}_t\}$
$A(z) = 1 + a_1 z^{-1} + \cdots + a_p z^{-p}$

Figure 8.5 Linear prediction of a time series.

This result is illustrated in Figure 8.5. In the figure, $P(z)$ is often called the prediction filter, and $A(z) = 1 - P(z)$ is the prediction error filter:

$$P(z) = -a_1 z^{-1} \quad \cdots \quad - a_p z^{-p} \qquad 8.45$$

$$A(z) = 1 - P(z) = 1 + a_1 z^{-1} + \cdots + a_p z^{-p}.$$

Then we say that the predicted time series $\{\hat{x}_t\}$ and the prediction error sequence $\{\hat{n}_t\}$ are described by the filtering relations

$$\{\hat{x}_t\} = P(z)\{x_t\} \qquad 8.46$$

$$\{\hat{n}_t\} = \{x_t\} - P(z)\{x_t\}$$

$$= A(z)\{x_t\},$$

meaning that the sequence $\{x_t\}$ is filtered by the moving average filter $P(z)$ to produce the predicted sequence $\{\hat{x}_t\}$ or by the moving average prediction error filter $A(z)$ to produce the prediction errors $\{\hat{n}_t\}$.

8.6   THE KALMAN FILTER

In Section 7.8 we derived the Kalman filter by assuming that the state $\mathbf{x}_t$ and the measurement sequence $\mathbf{y}_t^T = [\mathbf{y}_{t-1}^T, y_t]$ were multivariate normal. In this section we return to the Kalman filter. However, this time we make no assumptions about the joint or marginal distributions of $\mathbf{x}_t$ and $\mathbf{y}_t$. Rather, we assume only that their means are zero and their covariances are known. Then we require that the estimator of $\mathbf{x}_t$ from $\mathbf{y}_t$ be linear, and we derive the LMMSE estimator of $\mathbf{x}_t$. Our procedure will be to apply the orthogonality condition over and over again.

Let's review our modeling assumptions. The state $\mathbf{x}_t$ is a dynamical state that evolves according to the first-order recursion

$$\mathbf{x}_t = \mathbf{A}\mathbf{x}_{t-1} + \mathbf{b}u_{t-1}. \qquad 8.47$$

The initial state $\mathbf{x}_0$ has mean zero and covariance $E\mathbf{x}_0\mathbf{x}_0^T = \mathbf{R}_0$; the sequence $\{u_t\}$ that drives the recursion is a white sequence with mean zero and covariance $E u_t u_s = \delta_{t-s}$. The sequence $\{u_t\}$ is uncorrelated with $\mathbf{x}_0$: $E\mathbf{x}_0 u_t = \mathbf{0}$. The measurements are linear combinations of the state plus uncorrelated noise:

$$y_t = \mathbf{c}^T\mathbf{x}_t + n_t. \qquad 8.48$$

The noise sequence $\{n_t\}$ is a white sequence with mean zero and covariance $E n_t n_s = \sigma_n^2 \delta_{t-s}$. The noise sequence $\{n_t\}$ is uncorrelated with the noise sequence $\{u_t\}$ and with the initial state $\mathbf{x}_0$: $E u_s n_t = 0$ and $E\mathbf{x}_0 n_t = 0$.

Prediction

Let's begin by assuming that the LMMSE estimator of $\mathbf{x}_{t-1}$ from $\mathbf{y}_{t-1}$ is known. Then, by orthogonality,

$$E(\mathbf{x}_{t-1} - \hat{\mathbf{x}}_{t-1|t-1})\mathbf{y}_{t-1}^T = \mathbf{0}. \qquad 8.49$$

From the structure of the dynamical equation for $\mathbf{x}_t$ and the measurement equation for y_t, we hypothesize that the LMMSE predictor of $\mathbf{x}_t$ from $\mathbf{y}_{t-1}$ and the LMMSE predictor of y_t from $\mathbf{y}_{t-1}$ are

$$\hat{\mathbf{x}}_{t|t-1} = \mathbf{A}\hat{\mathbf{x}}_{t-1|t-1} \tag{8.50}$$

$$\hat{y}_{t|t-1} = \mathbf{c}^T\hat{\mathbf{x}}_{t|t-1}.$$

In order to check these hypotheses, we need only check for orthogonality:

$$E(\mathbf{x}_t - \hat{\mathbf{x}}_{t|t-1})\mathbf{y}_{t-1}^T = E(\mathbf{A}\mathbf{x}_{t-1} + \mathbf{b}u_{t-1} - \mathbf{A}\hat{\mathbf{x}}_{t-1|t-1})\mathbf{y}_{t-1}^T$$

$$= \mathbf{A}E(\mathbf{x}_{t-1} - \hat{\mathbf{x}}_{t-1|t-1})\mathbf{y}_{t-1}^T + \mathbf{b}E u_{t-1}\mathbf{y}_{t-1}^T \tag{8.51}$$

$$= \mathbf{0} + \mathbf{0}.$$

(Do you see why $\mathbf{b}E u_{t-1}\mathbf{y}_{t-1}^T = \mathbf{0}$?) Similarly, the orthogonality between $y_t - \hat{y}_{t|t-1}$ and $\mathbf{y}_{t-1}$ is checked as follows:

$$E(y_t - \hat{y}_{t|t-1})\mathbf{y}_{t-1}^T = E(\mathbf{c}^T\mathbf{A}\mathbf{x}_{t-1} + \mathbf{c}^T\mathbf{b}u_{t-1} + n_t - \mathbf{c}^T\mathbf{A}\hat{\mathbf{x}}_{t-1|t-1})\mathbf{y}_{t-1}^T$$

$$= \mathbf{c}^T\mathbf{A}E(\mathbf{x}_{t-1} - \hat{\mathbf{x}}_{t-1|t-1})\mathbf{y}_{t-1}^T + \mathbf{c}^T\mathbf{b}E u_{t-1}\mathbf{y}_{t-1}^T + E n_t\mathbf{y}_{t-1}^T$$

$$= \mathbf{0} + \mathbf{0} + \mathbf{0}. \tag{8.52}$$

(Do you see why $E n_t\mathbf{y}_{t-1}^T = \mathbf{0}$?)

At this point in our development we have the following orthogonal (or a posteriori) models for $\mathbf{x}_t$ and y_t:

$$\mathbf{x}_t = \hat{\mathbf{x}}_{t|t-1} + \hat{\mathbf{n}}_{t|t-1}; \qquad E\hat{\mathbf{n}}_{t|t-1}\hat{\mathbf{x}}_{t|t-1}^T = \mathbf{0} \tag{8.53}$$

$$y_t = \hat{y}_{t|t-1} + \hat{n}_{t|t-1}; \qquad E\hat{n}_{t|t-1}\hat{y}_{t|t-1} = 0.$$

This representation is illustrated in Figure 8.6. In the figure, the linear map $\mathbf{G}\mathbf{y}_{t-1}$ (with $\mathbf{G}$ currently unknown to us) produces $\hat{\mathbf{x}}_{t|t-1} = \mathbf{G}\mathbf{y}_{t-1}$ and $\hat{y}_{t|t-1} = \mathbf{c}^T\mathbf{G}\mathbf{y}_{t-1}$. These predictions, together with the orthogonal errors $\hat{\mathbf{n}}_{t|t-1}$ and $\hat{n}_{t|t-1}$, synthesize $\mathbf{x}_t$ and the new measurement y_t. In this new model for $\mathbf{x}_t$ and y_t, the predictions $\hat{\mathbf{x}}_{t|t-1}$ and $\hat{y}_{t|t-1}$ subsume the measurements $\mathbf{y}_{t-1}$. That is, they replace $\mathbf{y}_{t-1}$ as "sufficient statistics."

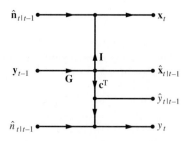

Figure 8.6 Orthogonal, or posterior, models for $\mathbf{x}_t$ and y_t.

Estimation

The most general estimator of $\mathbf{x}_t$ from $\mathbf{y}_t$ that we can construct is the estimator

$$\hat{\mathbf{x}}_{t|t} = \hat{\mathbf{x}}_{t|t-1} + \mathbf{k}_t(y_t - \hat{y}_{t|t-1}). \qquad 8.54$$

(If you are skeptical about this claim, you will try $\mathbf{K}\hat{\mathbf{x}}_{t|t-1}$ and add $\mathbf{L}y_{t-1} + \mathbf{l}y_t$ to this linear combination and find that $\mathbf{K} = \mathbf{I}$, $\mathbf{L} = \mathbf{0}$, and $\mathbf{l} = \mathbf{0}$. See Problem 8.11.) This solution will work if orthogonality holds. Therefore we require

1. $E(\mathbf{x}_t - \hat{\mathbf{x}}_{t|t})\mathbf{y}_{t-1}^T = \mathbf{0}$;
2. $E(\mathbf{x}_t - \hat{\mathbf{x}}_{t|t})y_t = \mathbf{0}$.

Condition (1) holds by virtue of the orthogonality between $\mathbf{x}_t - \hat{\mathbf{x}}_{t|t-1}$ and $\mathbf{y}_{t-1}$ and the orthogonality between $y_t - \hat{y}_{t|t-1}$ and $\mathbf{y}_{t-1}$. Condition (2) requires some work, and in fact it holds only for a very special choice of the Kalman gain $\mathbf{k}_t$. In order to proceed with the evaluation of condition (2), let's write y_t two ways:

$$\begin{aligned} y_t &= \hat{y}_{t|t-1} + (y_t - \hat{y}_{t|t-1}) \\ &= \mathbf{c}^T\hat{\mathbf{x}}_{t|t-1} + \left[\mathbf{c}^T(\mathbf{x}_t - \hat{\mathbf{x}}_{t|t-1}) + n_t\right]. \end{aligned} \qquad 8.55$$

The expectations in condition (2) are now

$$\begin{aligned} E(\mathbf{x}_t - \hat{\mathbf{x}}_{t|t})y_t &= E\left[\mathbf{x}_t - \hat{\mathbf{x}}_{t|t-1} - \mathbf{k}_t(y_t - \hat{y}_{t|t-1})\right]y_t \\ &= E(\mathbf{x}_t - \hat{\mathbf{x}}_{t|t-1})\left[\hat{\mathbf{x}}_{t|t-1}^T\mathbf{c} + (\mathbf{x}_t - \hat{\mathbf{x}}_{t|t-1})^T\mathbf{c} + n_t\right] \\ &\quad - \mathbf{k}_t E(y_t - \hat{y}_{t|t-1})\left[\hat{y}_{t|t-1} + (y_t - \hat{y}_{t|t-1})\right] \\ &= E(\mathbf{x}_t - \hat{\mathbf{x}}_{t|t-1})(\mathbf{x}_t - \hat{\mathbf{x}}_{t|t-1})^T\mathbf{c} - \mathbf{k}_t E(y_t - \hat{y}_{t|t-1})(y_t - \hat{y}_{t|t-1}). \end{aligned} \qquad 8.56$$

In order for this to equal zero, we require the gain $\mathbf{k}_t$ to satisfy the Wiener-Hopf equation

$$\sigma_{t|t-1}^2 \mathbf{k}_t = \mathbf{P}_{t|t-1}\mathbf{c}, \qquad 8.57$$

where $\sigma_{t|t-1}^2$ is the prediction error variance for y_t and $\mathbf{P}_{t|t-1}$ is the prediction error covariance for $\mathbf{x}_t$:

$$\sigma_{t|t-1}^2 = E(y_t - \hat{y}_{t|t-1})^2 \qquad 8.58$$

$$\mathbf{P}_{t|t-1} = E(\mathbf{x}_t - \hat{\mathbf{x}}_{t|t-1})(\mathbf{x}_t - \hat{\mathbf{x}}_{t|t-1})^T.$$

Let's summarize what we know so far. We know that the estimator $\hat{\mathbf{x}}_{t|t}$ is

$$\hat{\mathbf{x}}_{t|t} = \hat{\mathbf{x}}_{t|t-1} + \mathbf{k}_t(y_t - \hat{y}_{t|t-1})$$

$$\hat{\mathbf{x}}_{t|t-1} = \mathbf{A}\hat{\mathbf{x}}_{t-1|t-1} \qquad 8.59$$

$$\hat{y}_{t|t-1} = \mathbf{c}^T\hat{\mathbf{x}}_{t|t-1}$$

$$\sigma_{t|t-1}^2 \mathbf{k}_t = \mathbf{P}_{t|t-1}\mathbf{c}.$$

This represents real progress—but we do not have $\mathbf{P}_{t|t-1}$ or $\sigma^2_{t|t-1}$! Let's try to find them.

Covariance Recursions

The prediction error $\mathbf{x}_t - \hat{\mathbf{x}}_{t|t-1}$ may be written

$$\begin{aligned}
\mathbf{x}_t - \hat{\mathbf{x}}_{t|t-1} &= \mathbf{A}\mathbf{x}_{t-1} + \mathbf{b}u_{t-1} - \mathbf{A}\hat{\mathbf{x}}_{t-1|t-1} \\
&= \mathbf{A}(\mathbf{x}_{t-1} - \hat{\mathbf{x}}_{t-1|t-1}) + \mathbf{b}u_{t-1}.
\end{aligned} \tag{8.60}$$

Therefore the error covariance $\mathbf{P}_{t|t-1}$ may be written as

$$\mathbf{P}_{t|t-1} = \mathbf{A}\mathbf{P}_{t-1|t-1}\mathbf{A}^{\mathrm{T}} + \mathbf{b}\mathbf{b}^{\mathrm{T}}, \tag{8.61}$$

where $\mathbf{P}_{t-1|t-1}$ is the error covariance of $\mathbf{x}_{t-1} - \hat{\mathbf{x}}_{t-1|t-1}$:

$$\mathbf{P}_{t-1|t-1} = E(\mathbf{x}_{t-1} - \hat{\mathbf{x}}_{t-1|t-1})(\mathbf{x}_{t-1} - \hat{\mathbf{x}}_{t-1|t-1})^{\mathrm{T}}. \tag{8.62}$$

This allows us to move from the estimator covariance $\mathbf{P}_{t-1|t-1}$ to the predictor covariance $\mathbf{P}_{t|t-1}$. But how about the move from the predictor covariance $\mathbf{P}_{t|t-1}$ to the estimator covariance $\mathbf{P}_{t|t}$ so that we may keep the recursion going? Well, note

$$\mathbf{x}_t - \hat{\mathbf{x}}_{t|t} = \mathbf{x}_t - \hat{\mathbf{x}}_{t|t-1} - \mathbf{k}_t(y_t - \hat{y}_{t|t-1}). \tag{8.63}$$

Therefore, the error covariance $\mathbf{P}_{t|t}$ is

$$\begin{aligned}
\mathbf{P}_{t|t} &= E(\mathbf{x}_t - \hat{\mathbf{x}}_{t|t})(\mathbf{x}_t - \hat{\mathbf{x}}_{t|t})^{\mathrm{T}} \\
&= E(\mathbf{x}_t - \hat{\mathbf{x}}_{t|t-1})(\mathbf{x}_t - \hat{\mathbf{x}}_{t|t-1})^{\mathrm{T}} - 2\mathbf{k}_t E(y_t - \hat{y}_{t|t-1})(\mathbf{x}_t - \hat{\mathbf{x}}_{t|t-1})^{\mathrm{T}} \\
&\quad + \mathbf{k}_t E(y_t - \hat{y}_{t|t-1})^2 \mathbf{k}_t^{\mathrm{T}} \\
&= \mathbf{P}_{t|t-1} - 2\mathbf{k}_t E\big[\mathbf{c}^{\mathrm{T}}(\mathbf{x}_t - \hat{\mathbf{x}}_{t|t-1}) + n_t\big](\mathbf{x}_t - \hat{\mathbf{x}}_{t|t-1}^{\mathrm{T}}) \\
&\quad + \mathbf{k}_t \sigma^2_{t|t-1} \mathbf{k}_t^{\mathrm{T}} \\
&= \mathbf{P}_{t|t-1} - 2\mathbf{k}_t \mathbf{c}^{\mathrm{T}} \mathbf{P}_{t|t-1} + \mathbf{k}_t \sigma^2_{t|t-1} \mathbf{k}_t^{\mathrm{T}} \\
&= \mathbf{P}_{t|t-1} - \sigma^2_{t|t-1} \mathbf{k}_t \mathbf{k}_t^{\mathrm{T}} \\
&= (\mathbf{I} - \mathbf{k}_t \mathbf{c}^{\mathrm{T}}) \mathbf{P}_{t|t-1}.
\end{aligned} \tag{8.64}$$

Finally, the recursion for $\sigma^2_{t|t-1}$ is

$$\begin{aligned}
\sigma^2_{t|t-1} &= E(y_t - \hat{y}_{t|t-1})^2 \\
&= E\big[\mathbf{c}^{\mathrm{T}}(\mathbf{x}_t - \hat{\mathbf{x}}_{t|t-1}) + n_t\big]\big[(\mathbf{x}_t - \hat{\mathbf{x}}_{t|t-1})^{\mathrm{T}}\mathbf{c} + n_t\big] \\
&= \mathbf{c}^{\mathrm{T}}\mathbf{P}_{t|t-1}\mathbf{c} + \sigma^2_n.
\end{aligned} \tag{8.65}$$

The Kalman Recursions

In summary, the Kalman recursions are

$$\hat{\mathbf{x}}_{t|t} = \hat{\mathbf{x}}_{t|t-1} + \mathbf{k}_t(y_t - \hat{y}_{t|t-1})$$

$$\hat{\mathbf{x}}_{t|t-1} = \mathbf{A}\mathbf{x}_{t-1|t-1}$$

$$\hat{y}_{t|t-1} = \mathbf{c}^{\mathrm{T}}\hat{\mathbf{x}}_{t|t-1}$$

$$\sigma^2_{t|t-1}\mathbf{k}_t = \mathbf{P}_{t|t-1}\mathbf{c}$$

$$\mathbf{P}_{t|t-1} = \mathbf{A}\mathbf{P}_{t-1|t-1}\mathbf{A}^{\mathrm{T}} + \mathbf{b}\mathbf{b}^{\mathrm{T}}$$

$$\sigma^2_{t|t-1} = \mathbf{c}^{\mathrm{T}}\mathbf{P}_{t|t-1}\mathbf{c} + \sigma^2_n \qquad \text{8.66}$$

$$\mathbf{P}_{t|t} = \mathbf{P}_{t|t-1} - \sigma^2_{t|t-1}\mathbf{k}_t\mathbf{k}_t^{\mathrm{T}}.$$

In Problem 8.12 you are asked to determine the initial conditions for these recursions and to iterate them.

8.7 LOW-RANK APPROXIMATION OF RANDOM VECTORS

In this section we take up the question of approximating a random vector $\mathbf{x}$ with a low-rank random vector $\mathbf{x}_r$, as illustrated in Figure 8.7(a). We represent the low-rank approximant as

$$\mathbf{x}_r = \mathbf{K}\mathbf{x}, \qquad \text{8.67}$$

where $\mathbf{K}$ is an $N \times N$ rank-r matrix. The mean-squared error between $\mathbf{x}$ and $\mathbf{x}_r$ is

$$e^2(\mathbf{K}) = \operatorname{tr} E(\mathbf{x} - \mathbf{x}_r)(\mathbf{x} - \mathbf{x}_r)^{\mathrm{T}}$$
$$= \operatorname{tr}(\mathbf{I} - \mathbf{K})\mathbf{R}_{xx}(\mathbf{I} - \mathbf{K})^{\mathrm{T}}. \qquad \text{8.68}$$

(a) approximating $\mathbf{x}$

(b) coder, editor, decoder

(c) noisy problem

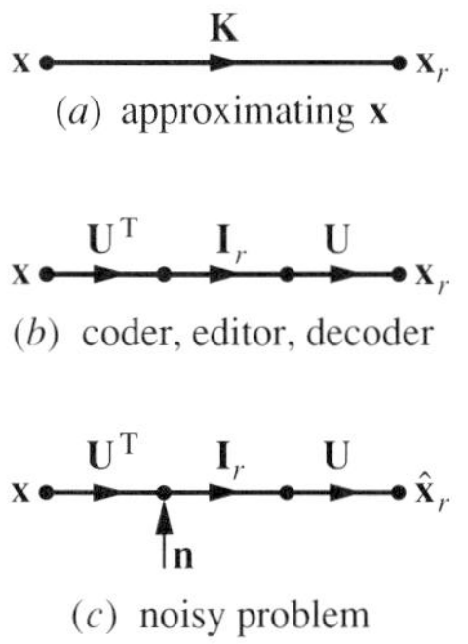

Figure 8.7 Approximating a random vector by a low-rank random vector.

Every rank-r symmetric matrix has the orthogonal decomposition

$$\mathbf{K} = \mathbf{U}\mathbf{\Lambda}_r\mathbf{U}^\mathrm{T} = \sum_{n=1}^{r} \lambda_n \mathbf{u}_n \mathbf{u}_n^\mathrm{T}.$$

$$\mathbf{U}^\mathrm{T}\mathbf{U} = \mathbf{U}\mathbf{U}^\mathrm{T} = \mathbf{I}; \qquad \mathbf{\Lambda}_r = \mathrm{diag}[\lambda_1 \quad \lambda_2 \quad \cdots \quad \lambda_r \quad 0 \quad \cdots \quad 0].$$

8.69

When the representation is substituted into the equation for mean-squared error, then we obtain this expansion for $e^2(\mathbf{K})$:

$$\begin{aligned} e^2(\mathbf{K}) &= \mathrm{tr}\,\mathbf{U}(\mathbf{I} - \mathbf{\Lambda}_r)\mathbf{U}^\mathrm{T}\mathbf{R}_{xx}\mathbf{U}(\mathbf{I} - \mathbf{\Lambda}_r)\mathbf{U}^\mathrm{T} \\ &= \mathrm{tr}\,\mathbf{U}^\mathrm{T}\mathbf{R}_{xx}\mathbf{U}(\mathbf{I} - \mathbf{\Lambda}_r)^2 \\ &= \sum_{n=1}^{r} \mathbf{u}_n^\mathrm{T}\mathbf{R}_{xx}\mathbf{u}_n(1 - \lambda_n)^2 + \sum_{n=r+1}^{N} \mathbf{u}_n^\mathrm{T}\mathbf{R}_{xx}\mathbf{u}_n. \end{aligned}$$

8.70

This mean-squared error may be minimized (see Problem 8.2) by selecting the $\mathbf{u}_n$ to be eigenvectors of $\mathbf{R}_{xx}$ and the λ_n to be unity. Then the rank-r matrix $\mathbf{K}$ and the corresponding mean-squared error are

$$\mathbf{K} = \mathbf{P}_r = \mathbf{U}\mathbf{I}_r\mathbf{U}^\mathrm{T} = \sum_{n=1}^{r} \mathbf{u}_n \mathbf{u}_n^\mathrm{T}$$

8.71

$$e^2(\mathbf{K}) = \sum_{n=r+1}^{N} \sigma_n^2.$$

The matrix $\mathbf{K}$ is a rank-r projection, and the σ_n^2, $n \geq r+1$, are the trailing eigenvalues of $\mathbf{R}_{xx}$. With this solution we redraw Figure 8.7(a) as 8.7(b), where the rank-r matrix $\mathbf{K}$ has been decomposed into a coder $\mathbf{U}^\mathrm{T}$, an editor $\mathbf{I}_r$, and a decoder $\mathbf{U}$. Can you justify this terminology?

Interpretation

In summary, here is what we have done. The vector $\mathbf{x}$ with covariance matrix $\mathbf{R}_{xx} = E\mathbf{x}\mathbf{x}^\mathrm{T}$ may be synthesized in terms of the eigenvectors of $\mathbf{R}_{xx}$ and a white vector $\boldsymbol{\theta}$ whose covariance $\mathbf{R}_{\theta\theta} = E\boldsymbol{\theta}\boldsymbol{\theta}^\mathrm{T}$ is determined from the eigenvalues of $\mathbf{R}_{xx}$:

$$\mathbf{x} = \mathbf{U}\boldsymbol{\theta} \iff \boldsymbol{\theta} = \mathbf{U}^\mathrm{T}\mathbf{x}$$

8.72

$$\mathbf{R}_{xx} = \mathbf{U}\mathbf{\Sigma}^2\mathbf{U}^\mathrm{T} \iff \mathbf{\Sigma}^2 = \mathbf{U}^\mathrm{T}\mathbf{R}_{xx}\mathbf{U} = \mathrm{diag}[\sigma_1^2 \quad \cdots \quad \sigma_r^2 \quad \cdots \quad \sigma_N^2].$$

When expanded, the equation for the random vector $\mathbf{x}$ is

$$\mathbf{x} = \sum_{n=1}^{r} \mathbf{u}_n \theta_n + \sum_{n=r+1}^{N} \mathbf{u}_n \theta_n$$

8.73

$$E\theta_n^2 = \sigma_n^2.$$

If the trailing $N - r$ variances $\{\sigma_n^2\}_{r+1}^N$ are small compared with the leading r variances $\{\sigma_n^2\}_1^r$, then the second sum in the expansion for $\mathbf{x}$ may be dropped to produce the approximation

$$\mathbf{x}_r = \sum_{n=1}^{r} \mathbf{u}_n \theta_n = \sum_{n=1}^{r} \mathbf{u}_n \mathbf{u}_n^T \mathbf{x}$$

$$= \mathbf{K}\mathbf{x}.$$

8.74

The process of rank reduction may be interpreted as coding $\mathbf{x}$ with $\mathbf{U}^T\mathbf{x}$, editing the elements of $\mathbf{U}^T\mathbf{x}$ for the dominant modes $\mathbf{I}_r\mathbf{U}^T\mathbf{x}$, followed by decoding with $\mathbf{U}$:

$$\mathbf{x}_r = \mathbf{U}\big[\mathbf{I}_r(\mathbf{U}^T\mathbf{x})\big].$$

8.75

Order Selection

Typically the pattern of eigenvalues for $\mathbf{R}_{xx}$ splits the eigenvectors of $\mathbf{U}$ into dominant and subdominant sets. Then the choice of r is more or less obvious. But there is another way to proceed that brings additional insight. Assume, as illustrated in Figure 8.7(c), that white noise $\mathbf{n}$, with covariance $\sigma^2\mathbf{I}$, is introduced into a channel between the two halves of the projector $\mathbf{K}$. The resulting approximant to $\mathbf{x}$ is

$$\hat{\mathbf{x}}_r = \mathbf{U}\mathbf{I}_r(\mathbf{U}^T\mathbf{x} + \mathbf{n}) = \mathbf{U}\mathbf{I}_r\mathbf{U}^T\mathbf{x} + \mathbf{U}\mathbf{I}_r\mathbf{n}.$$

8.76

The error between $\mathbf{x}$ and $\hat{\mathbf{x}}_r$ is now

$$\mathbf{x} - \hat{\mathbf{x}}_r = \mathbf{U}(\mathbf{I} - \mathbf{I}_r)\mathbf{U}^T\mathbf{x} - \mathbf{U}\mathbf{I}_r\mathbf{n},$$

8.77

and the mean-squared error is

$$e^2(r) = \operatorname{tr}\mathbf{U}(\mathbf{I} - \mathbf{I}_r)\mathbf{U}^T\mathbf{R}_{xx}\mathbf{U}(\mathbf{I} - \mathbf{I}_r)\mathbf{U}^T + \operatorname{tr}\mathbf{U}\mathbf{I}_r\sigma^2\mathbf{I}\mathbf{I}_r\mathbf{U}^T$$

$$= \sum_{i=r+1}^{N} \sigma_i^2 + r\sigma^2.$$

8.78

The best choice of rank minimizes $e^2(r)$.

8.8 OPTIMUM SCALAR QUANTIZERS

The main idea in quantization theory is to replace a continuous variable with a discrete approximation. When the variable is scalar, then the quantizer that produces the discrete approximation is called a scalar quantizer. When the variable is a vector, then the quantizer is called a vector quantizer. A special form of the vector quantizer is called a block quantizer. In this section, we will derive the scalar quantizer that minimizes mean-squared error. In Section 8.9, we will derive the block quantizer that minimizes mean-squared error, and in Section 8.10, we will imbed the block quantizer into a more general theory of reduced-rank block quantizers.

Scalar Quantizers

Typically, the real variable x is approximated by the discrete variable x_k whenever x is contained in the interval I_k:

$$Q(x) = \hat{x}_k \qquad \text{whenever } x \in I_k; \qquad k = 1, 2, \ldots, K. \qquad 8.79$$

This is illustrated in Figure 8.8. The interval $I_k = [x_k, x_{k+1})$ is called a cell of the quantizer, and the discrete approximant $\hat{x}_k$ is called a value of the quantizer. Sometimes $\hat{x}_k$ itself has a binary representation, but more generally each $\hat{x}_k$, $k = 1, 2, \ldots, K$, is represented by one of K addresses, each of which has a binary representation consisting of $\log_2 K$ bits.

When x is drawn from a probability distribution $F(x)$, then the values $\hat{x}_k$ are drawn from a discrete distribution P that has probability mass function

$$p_k = P[Q(X) = \hat{x}_k] = P[X \in I_k]$$

$$= \int_{I_k} dF(x). \qquad 8.80$$

The mean-squared error between the real random variable X and the discrete random variable $\hat{X} = Q(X)$ is

$$\sigma^2 = E[X - \hat{X}]^2$$

$$= \int [x - Q(x)]^2 \, dF(x)$$

$$= \sum_k \int_{I_k} (x - \hat{x}_k)^2 \, dF(x). \qquad 8.81$$

Think of $Q(X)$ as an estimator of the random variable X and think of $\hat{x}_k$ as the value assumed by the estimator when X lies in the cell I_k. We suspect from our study of minimum mean-squared error estimators that $Q(X)$ will be a conditional mean estimator. However, let's proceed pedantically to prove this result constructively.

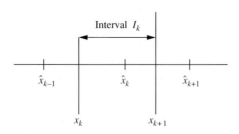

Figure 8.8 Intervals and values in a quantizer.

The mean-squared error is minimized with respect to $\hat{x}_k$ as follows:

$$\frac{\partial \sigma^2}{\partial \hat{x}_k} = \int_{I_k} (x - \hat{x}_k)\, dF(x) = 0 \qquad\qquad 8.82$$

$$\hat{x}_k = \frac{1}{p_k} \int_{I_k} x\, dF(x).$$

We claim that this result is the conditional mean of X, given that X lies in I_k:

$$\hat{x}_k = E[X|X \in I_k]. \qquad\qquad 8.83$$

To establish this interpretation, we note that the conditional probability that $X \le x$, given $X \in I_k$, may be written as

$$G_k(x) = P[X \le x | X \in I_k]$$

$$= P[X \le x, X \in I_k]/P[X \in I_k]$$

$$= \begin{cases} 0, & x < x_k \\ (1/p_k)[F(x) - F(x_k)], & x_k \le x < x_{k+1} \\ 1, & x \ge x_{k+1}. \end{cases} \qquad\qquad 8.84$$

The differential $dG_k(x)$ is

$$dG_k(x) = \begin{cases} 0, & x < x_k \\ (1/p_k)\, dF(x), & x_k \le x < x_{k+1} \\ 0, & x \ge x_{k+1}. \end{cases} \qquad\qquad 8.85$$

Therefore, the quantized value $\hat{x}_k$ may be written as

$$\hat{x}_k = \frac{1}{p_k} \int_{I_k} x\, dF(x)$$

$$\qquad\qquad 8.86$$

$$= \int x\, dG_k(x),$$

which is just the conditional mean of X, given $X \in I_k$. In summary, the scalar quantizer that minimizes mean-squared error is the conditional mean quantizer

$$Q(X) = \hat{x}_k \qquad \text{when } X \in I_k \qquad\qquad 8.87$$

$$\hat{x}_k = E[X|X \in I_k].$$

We know a lot about the conditional mean estimator. Recall that $X - Q(X)$ is orthogonal to $Q(X)$:

$$X = Q(X) + [X - Q(X)] \qquad\qquad 8.88$$

$$EQ(X)[X - Q(X)] = 0.$$

It follows that the variance of the random variable X decomposes as

$$\sigma_X^2 = \sigma_Q^2 + \sigma^2 \qquad\qquad 8.89$$

where the σ^2's are the following variances:

$$\sigma_X^2 = EX^2$$
$$\sigma_Q^2 = EQ^2(X) \qquad\qquad 8.90$$
$$\sigma^2 = E[X - Q(X)]^2.$$

This means that the quantizer is lossy in the sense that the variance, or power, of $Q(X)$ is less than that of X.

There is one more observation we can now make about our solution. The slicing levels x_k must be midway between $\hat{x}_k$ and $\hat{x}_{k-1}$, as illustrated in Figure 8.8. Why? Because displacement of x_k to the right or left of the midpoint would increase the mean-squared error by increasing the distance between any point x and its nearest quantized value. In summary,

$$x_k = \frac{1}{2}(\hat{x}_k + \hat{x}_{k-1}) \qquad\qquad 8.91$$

$$\hat{x}_k = E[X | X \in I_k].$$

These results tie the values of the quantizer to the slicing levels and vice versa. The representation value $\hat{x}_k$ is the conditional mean of X, given that X lies between the slicing levels of x_k and x_{k+1}, and the slicing level x_k lies midway between representation points $\hat{x}_k$ and $\hat{x}_{k+1}$. But these connections are circular in the sense that $\hat{x}_k$ is designed in terms of the x_k and vice versa. Neither one nor the other is pinned down.

Designing the Optimum Quantizer

In the paragraphs to follow, we show how to select the slicing levels x_k and the representation points $\hat{x}_k$ when the quantizer is arbitrarily fine with respect to the distribution $F(x)$. Figure 8.9 establishes our conventions. The new variables introduced in Figure 8.9 are $\tilde{x}_k$, the midpoint of interval I_k, and h_k, the length of interval I_k:

$$\tilde{x}_k = (x_{k+1} + x_k)/2 \qquad\qquad 8.92$$

$$h_k = x_{k+1} - x_k.$$

These variables are important because, in an arbitrarily fine quantizer, the midpoint is the conditional mean estimator and the resulting mean-squared error is just a function of the interval lengths. These findings are established as follows. Write the conditional

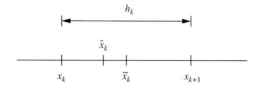

Figure 8.9 Notational conventions for the asymptotic quantizer.

mean $\hat{x}_k$ as

$$
\hat{x}_k = \frac{\displaystyle\int_{x_k}^{x_{k+1}} x \, dF(x)}{\displaystyle\int_{x_k}^{x_{k+1}} dF(x)}
$$

$$
\cong \frac{\displaystyle f(\tilde{x}_k) \int_{x_k}^{x_{k+1}} x \, dx}{\displaystyle f(\tilde{x}_k) \int_{x_k}^{x_{k+1}} dx} \tag{8.93}
$$

$$
= \frac{(1/2)(x_{k+1}^2 - x_k^2)}{x_{k+1} - x_k} = \tilde{x}_k.
$$

The same kind of computation shows that the mean-squared error may be approximated with a simple function of the interval lengths h_k:

$$
\sigma^2 = \sum_k \int_{x_k}^{x_{k+1}} (x - \hat{x}_k)^2 \, dF(x)
$$

$$
\cong \sum_k f(\tilde{x}_k) \int_{x_k}^{x_{k+1}} (x - \tilde{x}_k)^2 \, dx
$$

$$
= \sum_k f(\tilde{x}_k) \int_{-(1/2)h_k}^{(1/2)h_k} u^2 \, du
$$

$$
= \frac{1}{12} \sum_k f(\tilde{x}_k) h_k^3.
$$

$$\tag{8.94}$$

With these results, we may minimize the mean-squared error with respect to the interval lengths h_k.

We note from Figure 8.9 that the slicing levels are causally dependent on the interval lengths h_n:

$$
x_{k+1} = x_k + h_k = \cdots = x_1 + \sum_{n=1}^{k} h_n. \tag{8.95}
$$

Similarly, the midpoints depend on the interval lengths:

$$
\tilde{x}_k = x_k + \frac{1}{2} h_k. \tag{8.96}
$$

This formula shows that the derivative of $\tilde{x}_k$ with respect to h_n is

$$
\frac{\partial}{\partial h_n} \tilde{x}_k = \begin{cases} 0, & n > k \\ 1/2, & n = k \\ 1, & n < k. \end{cases} \tag{8.97}
$$

Let's now minimize the mean-squared error under the constraint that the interval lengths h_k span the range of the quantizer:

$$\sum_n h_n = x_{K+1} - x_1.$$

8.98

Form the Lagrangian

$$J = \sigma^2 - \lambda\left[\sum_n h_n - (x_{K+1} - x_1)\right].$$

8.99

Except for irrelevant scale constants, we may write the derivative of J with respect to h_n as

$$\frac{\partial J}{\partial h_n} = \frac{\partial \sigma^2}{\partial h_n} - \lambda$$

$$= 3f(\tilde{x}_n)h_n^2 + \sum_k \frac{\partial f(\tilde{x}_k)}{\partial h_n} h_k^3 - \lambda$$

$$= 3f(\tilde{x}_n)h_n^2 + \sum_k \frac{\partial f(\tilde{x}_k)}{\partial x} \frac{\partial \tilde{x}_k}{\partial h_n} h_k^3 - \lambda$$

8.100

$$= 3f(\tilde{x}_n)h_n^2 + 1/2\frac{\partial f(\tilde{x}_n)}{\partial x} h_n^3 + \sum_{k=n+1}^{K} \frac{\partial f(\tilde{x}_k)}{\partial x} h_k^3 - \lambda.$$

Approximate the partial derivatives of $f(\ )$ to write

$$\frac{\partial J}{\partial h_n} \cong 3f(\tilde{x}_n)h_n^2 + \frac{1}{2}\left[f(x_{n+1}) - f(x_n)\right]h_n^2 + \sum_{k=n+1}^{K} \left[f(x_{n+1}) - f(x_n)\right]h_n^2 - \lambda$$

$$\cong 3f(\tilde{x}_n)h_n^2 + \int_{\tilde{x}_n}^{x_{K+1}} df(x)h^2(x) - \lambda.$$

8.101

This result says that the gradient of the Lagrangian equals zero when

$$3f(\tilde{x}_n)h_n^2 + \int_{\tilde{x}_n}^{x_{K+1}} df(x)h^2(x) = \lambda$$

8.102

for each n. For this to happen, the integral must be a perfect differential that, when integrated, produces a term to annihilate $3f(\tilde{x}_n)h_n^2$; that is,

$$df(x)h^2(x) = 3 d\left[f(x)h^2(x)\right]$$
$$= 3 df(x)h^2(x) + 3f(x)2h(x)\,dh(x).$$

8.103

Then

$$3f(\tilde{x}_n)h_n^2 + \int_{\tilde{x}_n}^{x_{K+1}} 3 d\left[f(x)h^2(x)\right] = 3f(x_{K+1})h^2(x_{K+1}) = \lambda.$$

More importantly, we find from Equation 8.103 for the perfect differential that

$$\frac{d\,h(x)}{h(x)} = -\frac{1}{3}\frac{df(x)}{f(x)}. \tag{8.104}$$

The connection between the density $f(x)$ and the interval length $h(x)$ is therefore

$$h(x) = \frac{c}{f^{1/3}(x)} \tag{8.105}$$

$$c = h(x)f^{1/3}(x).$$

The constant c is determined by noting that

$$c = \frac{1}{K}\sum_{k=1}^{K} c$$

$$= \frac{1}{K}\sum_{k=1}^{K} h_k f^{1/3}(x_k)$$

$$= \frac{1}{K}\sum_{k=1}^{K} f^{1/3}(x_k)(x_{k+1} - x_k) \tag{8.106}$$

$$= \frac{1}{K}\sum_{k=1}^{K} f^{1/3}(x_k)\int_{x_k}^{x_{k+1}} dx$$

$$\cong \frac{1}{K}\int_{x_1}^{x_{K+1}} f^{1/3}(x)\,dx.$$

In summary, the asymptotic answer for the optimum interval width is

$$h(x) = \frac{\int_{x_1}^{x_{K+1}} f^{1/3}(x)\,dx}{Kf^{1/3}(x)}. \tag{8.107}$$

The corresponding mean-squared error is

$$\sigma^2 = \frac{1}{12}\sum_{k=1}^{K} f(x_k)h_k^3$$

$$= \sum_{k=1}^{K} \frac{c^3}{12} \tag{8.108}$$

$$= \frac{1}{12}Kc^3 = \frac{1}{12K^2}\left[\int_{x_1}^{x_{K+1}} f^{1/3}(x)\,dx\right]^3.$$

Note that asymptotically the optimum quantizer assigns equal mean-squared error to each interval I_k:

$$\sigma^2 = \sum_{k=1}^{K} \sigma_k^2 \qquad\qquad 8.109$$

$$\sigma_k^2 = \frac{1}{12} f(\tilde{x}_k) h_k^3 = \frac{c^3}{12}.$$

The results were first derived in quite a different way by Lloyd and Max.

8.9   OPTIMUM BLOCK QUANTIZERS

Let's now assume that we have not just a single random variable x, but a sequence of zero-mean random variables $x_0, x_1, \ldots, x_{N-1}$ that are to be quantized. "Block" them into a random vector $\mathbf{x} = [x_0 \quad x_1 \quad \cdots \quad x_{N-1}]^T$, which is to be quantized. The vector $\mathbf{x}$ has covariance matrix $\mathbf{R}_{xx}$:

$$E\mathbf{x}\mathbf{x}^T = \mathbf{R}_{xx} \qquad\qquad 8.110$$

$$\mathbf{R}_{xx} = (r_{ij}); \quad r_{nn} = Ex_n^2.$$

As a first step in our discussion of block quantizers, we propose a scalar quantizer that quantizes each random variable x_n independently of any other and that uses a uniform rounding quantizer with B_n bits for variable x_n. The quantizer for x_n is then

$$Q(x_n) = k\beta 2^{-B_n} \quad \text{when} \quad \left(\frac{2k-1}{2}\right)\beta 2^{-B_n} < x_n \le \left(\frac{2k+1}{2}\right)\beta 2^{-B_n}. \qquad 8.111$$

The mean-squared quantization error for variable x_n is then (see Problem 8.18)

$$\sigma_n^2 = \beta^2 r_{nn} 2^{-2B_n}. \qquad\qquad 8.112$$

This quantizer is illustrated in Figure 8.10(a). The question we pose is, "What is the optimum allocation of bits B_n to variable x_n when there is a constraint that a total of B bits be used to code all N random variables?"

To answer this question, we begin with the following measure of mean-squared quantization error:

$$\sigma^2 = E(\mathbf{x} - \hat{\mathbf{x}})^T(\mathbf{x} - \hat{\mathbf{x}}) = \sum_{n=0}^{N-1}(x_n - \hat{x}_n)^2$$

$$\qquad\qquad\qquad\qquad\qquad\qquad 8.113$$

$$= \beta^2 \sum_{n=0}^{N-1} r_{nn} 2^{-2B_n}.$$

The mean-squared quantization error for the quantized vector in this element-by-element scalar quantizing scheme is just the sum of the mean-squared errors for each variable.

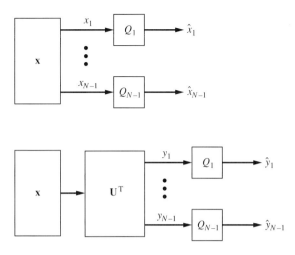

Figure 8.10 Block quantizers: (*a*) quantizing **x** and
(*b*) quantizing **y**.

We now minimize σ^2 under the constraint that $\sum_{n=0}^{N-1} B_n = B$. To do so, we
form the Lagrangian

$$J = \beta^2 \sum_{n=0}^{N-1} r_{nn} 2^{-2B_n} - \lambda^{1/N} \left(\sum_{n=0}^{N-1} (B_n - B) \right) \qquad 8.114$$

$$\frac{\partial J}{\partial B_n} = r_{nn} 2^{-2B_n} - \lambda^{1/N} = 0.$$

The solution for B_n in terms of λ is

$$r_{nn} 2^{-2B_n} = \lambda^{1/N}. \qquad 8.115$$

Note that the mean-squared errors $\beta^2 r_{nn} 2^{-2B_n}$ are equalized by this choice of B_n. To
invoke the constraint, we note that λ may be written

$$\lambda = \prod_{n=0}^{N-1} r_{nn} 2^{-2B_n} = 2^{-2B} \prod_{n=0}^{N-1} r_{nn}. \qquad 8.116$$

This result determines $\lambda^{1/N}$ and produces these solutions for $r_{nn} 2^{-2B_n}$ and B_n:

$$r_{nn} 2^{-2B_n} = 2^{-2B/N} \left(\prod_{n=0}^{N-1} r_{nn} \right)^{1/N} \qquad 8.117$$

or

$$B_n = \frac{1}{2} \log_2 \frac{r_{nn}}{\left(\prod_{n=0}^{N-1} r_{nn} \right)^{1/N}} + \frac{B}{N}. \qquad 8.118$$

The bit allocation is B/N (the average number of bits per variable), plus or minus a correction that depends on the log ratio between the variance r_{nn} for variable x_n and the geometric mean of variances for the variables $(x_n)_0^{N-1}$. When $r_{nn} = r_{00}$ for all n, then

$$B_n = \frac{B}{N} \ \forall \ n.$$ 8.119

The accumulated squared error using this optimum bit-allocation scheme is

$$\sigma^2 = \beta^2 \sum_{n=0}^{N-1} r_{nn} 2^{-2B_n} = N\beta^2 \lambda^{1/N}$$

$$= N\beta^2 2^{-2B/N} \left(\prod_{n=0}^{N-1} r_{nn} \right)^{1/N}.$$ 8.120

We summarize our results as follows:

$$B_n = \frac{B}{N} + \frac{1}{2} \log_2 \frac{r_{nn}}{\left(\prod_{n=0}^{N-1} r_{nn} \right)^{1/N}}$$ 8.121

$$\sigma^2 = N\beta^2 2^{-2B/N} \left(\prod_{n=0}^{N-1} r_{nn} \right)^{1/N}.$$

It is natural to ask whether this result can be improved by *dependently* quantizing the variables. One way to quantize dependently is to linearly transform variables x_n into intermediate variables y_n that are then independently quantized. The idea is to preprocess the variables x_n, quantize intermediate variables y_n, and then reconstruct $\hat{x}_n$. To this end, define the vector

$$\mathbf{y} = \mathbf{U}^\mathsf{T}\mathbf{x}$$ 8.122

$$\mathbf{x} = \mathbf{U}\mathbf{y}$$

where $\mathbf{U}$ is an orthogonal matrix:

$$\mathbf{U}^\mathsf{T}\mathbf{U} = \mathbf{U}\mathbf{U}^\mathsf{T} = \mathbf{I}.$$ 8.123

The correlation matrix of $\mathbf{y}$ of

$$E\mathbf{y}\mathbf{y}^\mathsf{T} = \mathbf{U}^\mathsf{T}\mathbf{R}\mathbf{U} = \mathbf{Q}$$ 8.124

$$\mathbf{Q} = (q_{ij}); \quad q_{nn} = E y_n^2.$$

Let's suppose that the variables y_n are quantized as illustrated in Figure 8.10(*b*) and used to construct the $\hat{x}_n$ as illustrated:

$$\hat{\mathbf{x}} = \mathbf{U}\hat{\mathbf{y}}.$$ 8.125

The mean-squared error between $\mathbf{x}$ and $\hat{\mathbf{x}}$ is just the mean-squared error between $\mathbf{y}$ and $\hat{\mathbf{y}}$:

$$\sigma^2 = E(\mathbf{x} - \hat{\mathbf{x}})^{\mathrm{T}}(\mathbf{x} - \hat{\mathbf{x}}) = E(\mathbf{y} - \hat{\mathbf{y}})^{\mathrm{T}}\mathbf{U}^{\mathrm{T}}\mathbf{U}(\mathbf{y} - \hat{\mathbf{y}})$$
$$= E(\mathbf{y} - \hat{\mathbf{y}})^{\mathrm{T}}(\mathbf{y} - \hat{\mathbf{y}}).$$

8.126

If the y_n are independently quantized with a uniform rounding quantizer in which bits are allocated optimally, then the mean-squared quantization error for $\hat{\mathbf{y}}$ is

$$\sigma^2 = N\beta^2 2^{-2B/N}\left(\prod_{n=0}^{N-1} q_{nn}\right)^{1/N}.$$

8.127

Hadamard's inequality says

$$\prod_{n=0}^{N-1} q_{nn} \geq \det \mathbf{Q} = \det \mathbf{U}^{\mathrm{T}}\mathbf{R}\mathbf{U}$$
$$= \det \mathbf{R} = \prod_{n=0}^{N-1} \lambda_n^2$$

8.128

with equality if and only if $\mathbf{Q}$ is diagonal. (The last equality in Equation 8.128 follows from the fact that orthogonal transformations of the form $\mathbf{U}^{\mathrm{T}}\mathbf{R}\mathbf{U}$ leave eigenvalues unchanged.) This means that the mean-squared error can be made no smaller than

$$\sigma^2 = N\beta^2 2^{-2B/N}\left(\prod_{n=0}^{N-1} \lambda_n^2\right)^{1/N},$$

8.129

and this is the value achieved when the orthogonal matrix $\mathbf{U}$ diagonalizes $\mathbf{R}$:

$$\mathbf{Q} = \mathbf{U}^{\mathrm{T}}\mathbf{R}\mathbf{U} = \mathrm{diag}[\lambda_0^2 \quad \lambda_1^2 \quad \cdots \quad \lambda_{N-1}^2].$$

8.130

This means that the variables y_n are uncorrelated and $Ey_n^2 = q_{nn} = \lambda_n^2$.

In conclusion, the optimum uniform rounding block quantizer consists of an orthogonalizing transformation $\mathbf{U}^{\mathrm{T}}$ of the vector $\mathbf{x}$ that presents uncorrelated random variables y_n to uniform rounding quantizers. Each quantizer allocates bits independently to the y_n according to the rule

$$B_n = \frac{B}{N} + \frac{1}{2} \log_2 \frac{\lambda_{nn}^2}{\left(\prod_{n=0}^{N-1} \lambda_{nn}^2\right)^{1/N}}.$$

8.131

The quantized vector $\hat{\mathbf{y}}$ is then used to represent the quantized vector $\hat{\mathbf{x}} = \mathbf{U}\hat{\mathbf{y}}$.

8.10   RANK REDUCTION AND RATE DISTORTION

We may couple our results for low-rank approximation of random vectors with our results for the MMSE, uniform rounding, block quantizer to derive reduced-rank MMSE quantizers. The problem is illustrated in Figure 8.11(*a*), where the source vector $\mathbf{x}$ is to be coded with $\mathbf{U}_r^T$, quantized with the uniform rounding quantizer $\mathbf{Q}_r$, and decoded with $\mathbf{U}_r$. We may think of this as a variation on the problem of representing the random vector $\mathbf{x}$ by $\mathbf{x}_r = \mathbf{U}\mathbf{I}_r\mathbf{U}^T\mathbf{x}$, as in Section 8.7. In this variation, the editor $\mathbf{I}_r$, which *selects* the components of $\mathbf{U}^T\mathbf{x}$ that have dominant eigenvalues, is replaced by a quantizer $\mathbf{Q}_r$ that *quantizes* the components of $\mathbf{U}^T\mathbf{x}$ that have dominant eigenvalues. We call this quantizer a *reduced-rank block quantizer* and illustrate it as in Figure 8.11(a). The quantizer $\mathbf{Q}_r$ is a noise source that introduces independent white noise $\mathbf{n}$, whose mean is zero, and whose covariance is

$$\mathbf{R}_{nn} = E\mathbf{n}\mathbf{n}^T = \mathrm{diag}[\sigma_1^2 \quad \sigma_2^2 \quad \cdots \quad \sigma_r^2] \qquad 8.132$$

$$\sigma_n^2 = \lambda_n^2 2^{-2B_n}.$$

See Figure 8.11(b). The mean-squared error between $\mathbf{x}$ and $\hat{\mathbf{x}}_r$ is

$$\sigma^2 = \mathrm{tr}\, E(\mathbf{x} - \hat{\mathbf{x}}_r)(\mathbf{x} - \hat{\mathbf{x}}_r)^T$$

$$= \mathrm{tr}\, E\big[(\mathbf{I} - \mathbf{U}_r\mathbf{U}_r^T)\mathbf{x} - \mathbf{U}_r\mathbf{n}\big]^T\big[(\mathbf{I} - \mathbf{U}_r\mathbf{U}_r^T)\mathbf{x} - \mathbf{U}_r\mathbf{n}\big]^T$$

$$= \mathrm{tr}\,(\mathbf{I} - \mathbf{U}_r\mathbf{U}_r^T)\mathbf{R}_{xx}(\mathbf{I} - \mathbf{U}_r\mathbf{U}_r^T) + \mathrm{tr}\,\mathbf{U}_r\mathbf{R}_{nn}\mathbf{U}_r^T \qquad 8.133$$

$$= \sum_{n=r+1}^{N} \lambda_n^2 + \sum_{n=1}^{r} \lambda_n^2 2^{-2B_n}.$$

There are two problems of interest:

1. Fix the number of bits $B = \sum_{n=1}^{r} B_n$ and then find B_n to minimize mean-squared error σ^2; this is the rate-distortion, or (B, σ^2), problem;

2. Fix the distortion σ^2 and find B_n to minimize $B = \sum_{n=1}^{r} B_n$; this is the distortion-rate, or (σ^2, B), problem.

The rate-distortion and distortion-rate problems are solved as follows. Minimize σ^2 with respect to B_n under the constraint that $\sum_{n=1}^{r} B_n = B$:

$$\frac{\partial}{\partial B_n}\left[\sum_{n=r+1}^{N}\lambda_n^2 + \sum_{n=1}^{r}\lambda_n^2 2^{-2B_n} + 2\theta\left(\sum_{n=1}^{r}B_n - B\right)\right] = -2\lambda_n^2 2^{-2B_n} + 2\theta = 0.$$

$$8.134$$

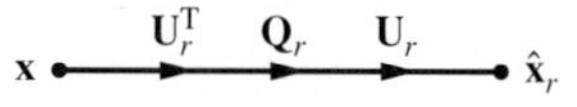

(a)

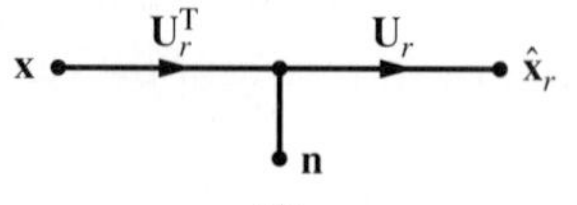

(b)

Figure 8.11 Reduced-rank block quantizer:
(a) block quantizer and
(b) equivalent noise model.

The result for B_n is

$$B_n = \frac{1}{2} \log_2 \frac{\lambda_n^2}{\theta}.$$
8.135

This is nonnegative for $\lambda_n^2/\theta \geq 1$. So, evidently the rank r is $r = \max n[\lambda_n^2/\theta > 1]$. The resulting solutions for rank, bit allocation, and mean-squared error are

$$r = \max n\left(\frac{\lambda_n^2}{\theta} > 1\right)$$

$$B_n = \frac{1}{2} \log_2 \frac{\lambda_n^2}{\theta}$$
8.136

$$B = \sum_{n=1}^{r} B_n = \sum_{n=1}^{N} \max\left(0, \frac{1}{2} \log_2 \frac{\lambda_n^2}{\theta}\right)$$

$$\sigma^2 = \sum_{n=r+1}^{N} \lambda_n^2 + \sum_{n=1}^{r} \theta = \sum_{n=1}^{N} \min(\theta, \lambda_n^2).$$

The pair (B, σ^2) is the rate-distortion pair and the pair (σ^2, B) is the distortion-rate pair. If B is fixed, then the formula for B is used to determine θ, which in turn determines r, B_n, and σ^2. If σ^2 is fixed, then the formula for σ^2 is used to determine θ, which in turn determines r, B_n, and B. You can see that the solutions to these two problems are determined by the Lagrange multiplier θ, which slices eigenvalues into dominant ones and subdominant ones according to the formula $\lambda_N^2 < \cdots < \lambda_{r+1}^2 < \theta < \lambda_r^2 < \cdots < \lambda_1^2$. Of course, the slicing level θ depends on B in the rate-distortion problem and on σ^2 in the distortion-rate problem.

REFERENCES AND COMMENTS

In this chapter we have tried to reinforce the role played by orthogonality in the theory of MMSE and LMMSE estimators. We have used orthogonality to derive the Wiener filter, the linear predictor, and the Kalman filter. We shall have much more to say about the linear predictor in Chapter 10.

Our derivation of the Kalman filter is a little different than Kalman's original derivation, but we were strongly influenced by his profound 1960 paper. The development of the theory of Kalman filtering was one of the most elegant successes of control and estimation theory during the 1960s and 70s. For excellent reviews of Kalman filtering and related work in prediction and filtering, I recommend Anderson and Moore [1979]; Kailath [1974]; and Sorenson [1970].

Our perspective on low-rank and low-rank quantized approximations of vectors was reported in Scharf and Tufts [1987]. In this paper we followed the lead of Sullivan and Liu [1984] in their paper on related problems. We drew upon Segall's paper [1976] to develop our results for reduced-rank quantized approximations. Much more on scalar and vector quantizers may be found in standard texts and reprint series such as Jayant [1976] and Jayant and Noll [1984].

Anderson, B. D. O., and J. B. Moore [1979]. *Optimal Filtering* (Englewood Cliffs, NJ: Prentice-Hall, 1979).

Jayant, N. S., ed. [1976]. *Waveform Quantization* (New York: IEEE Press, 1976).

Jayant, N. S., and P. Noll [1984]. *Digital Coding of Waveforms* (Englewood Cliffs, NJ: Prentice-Hall, 1984).

Kailath, T. [1974]."A View of Three Decades of Linear Filtering Theory," *IEEE Trans Info Th* **IT-20**, pp. 145–181 (March 1974).

Kalman, R. E. [1960]."A New Approach to Linear Filtering and Prediction Problems," *J Basic Engr, Trans ASME*, pp. 35–45 (March 1960).

Scharf, L. L., and D. W. Tufts [1987]. "Rank Reduction for Modeling Stationary Signals," *IEEE Trans ASSP* **ASSP-35**, pp. 350–355 (March 1987).

Segall, A. [1976]. "Bit Allocation and Encoding for Vector Sources," *IEEE Trans Infor Th* **IT-22**, pp. 167–169 (March 1976).

Sorenson, H. E. [1970]."Least Square Estimation: From Gauss to Kalman," *IEEE Spectrum*, pp. 63–68 (July 1970).

Sullivan, B. J., and B. Liu [1984]. "On the Use of Singular Value Decomposition and Decimation in Discrete-Time Band-Limited Signal Extrapolation," *IEEE Trans ASSP* **ASSP-32**, pp. 1201–1212 (December 1984).

PROBLEMS

8.1   Let $\mathbf{R}$ denote an $N \times N$ positive definite covariance matrix, and let $\mathbf{P}_A$ denote an $N \times N$ rank $N - p$ projection matrix. Prove

$$\text{tr}[\mathbf{P}_A \mathbf{R} \mathbf{P}_A] = \text{tr}[\mathbf{P}_A \mathbf{R}] \geq \sum_{n=p+1}^{N} \lambda_n^2$$

where $\lambda_1^2 \geq \lambda_2^2 \geq \cdots \geq \lambda_N^2$ are the ordered eigenvalues of $\mathbf{R}$. Find a representation for a $\mathbf{P}_A$ that achieves the bound.

8.2   *Low-Rank Approximation.* Let $\mathbf{x}$ denote a random vector with covariance $\mathbf{R}_{xx} = E\mathbf{x}\mathbf{x}^T$, and let $\mathbf{x}_r = \mathbf{K}\mathbf{x}$ denote a low-rank approximation of $\mathbf{x}$.

  a.   Show $e^2(\mathbf{K}) = E(\mathbf{x} - \mathbf{x}_r)^T(\mathbf{x} - \mathbf{x}_r) = \text{tr}\,(\mathbf{I} - \mathbf{K})^T(\mathbf{I} - \mathbf{K})\mathbf{R}_{xx}$.

  b.   Show $e^2(\mathbf{K}) = e^2(\mathbf{K}^T)$.

  c.   Prove that the rank-r symmetric matrix $\mathbf{K}$ that minimizes $e^2(\mathbf{K})$ is $\mathbf{K} = \mathbf{U}\mathbf{I}_r\mathbf{U}^T$ where $\mathbf{R}_{xx} = \mathbf{U}\mathbf{\Sigma}^2\mathbf{U}^T$.

  d.   Prove $E(\mathbf{x} - \mathbf{x}_r)\mathbf{x}_r^T = \mathbf{0}$.

  e.   Examine $E(\mathbf{x} - \mathbf{x}_r)\mathbf{x}^T$ and explain what you observe.

8.3   Extend the results of Problem 8.2 by not requiring apriori that the rank r matrix $\mathbf{K}$ be symmetric.

8.4 Prove that the noise introduced in Figure 8.7(c) could also be introduced at the input to $\mathbf{U}^{\mathrm{T}}$, provided its covariance is $\sigma^2 \mathbf{I}$, without changing the conclusion regarding order selection in Section 8.7. Why is this?

8.5 Hypothesize a set of eigenvalues $\{\sigma_i^2\}_1^N$ and plot

$$e^2(r) = \sum_{i=r+1}^{N} \sigma_i^2 + r\sigma^2$$

to show how order selection for r proceeds when approximating a random vector $\mathbf{x}$ by the low-rank vector $\mathbf{x}_r = \mathbf{K}\mathbf{x}$. When is $r = 0$ better than $r = N$?

8.6 *Asymptotics.* When $\mathbf{R}_{xx}$ is Toeplitz and $N \to \infty$, then the eigenvalues of $\mathbf{R}_{xx}$ are

$$\sigma_n^2 \longrightarrow \mathbf{S}_{xx}(e^{j\omega_n})$$

$$\omega_n = \frac{2\pi n}{N},$$

where $\mathbf{S}_{xx}(e^{j\theta})$ is the power spectral density of the wide-sense stationary time series $\{x_t\}$ from which the snapshot $\mathbf{x} = (x_0 \quad x_1 \quad \cdots \quad x_{N-1})^{\mathrm{T}}$ is drawn. Show that the minimum achievable mean-squared error $e^2(r) = \sum_{i=r+1}^{N} \sigma_i^2 + r\sigma^2$ may be written as

$$\lim_{N\to\infty} \frac{1}{N} e^2(r) = \lim_{N\to\infty} \left(\frac{1}{N} \sum_{r+1}^{N} S_{xx}(e^{j\omega_n}) + \frac{1}{N} \sigma^2 \sum_{n=1}^{r} 1 \right)$$

$$= \int_{\omega:S(e^{j\omega})<\sigma^2} S_{xx}(e^{j\omega}) \frac{d\omega}{2\pi} + \int_{\omega:S(e^{j\omega})>\sigma^2} \sigma^2 \frac{d\omega}{2\pi}.$$

Interpret this finding by plotting the terms of $(1/N)\, e^2(r)$.

8.7 *Asymptotics.* Repeat Problem 8.6, but this time compute $\lim_{N\to\infty} \sigma^2$ where σ^2 is the mean-squared error of the reduced-rank, uniformly quantized, block quantizer studied in Section 8.10. Illustrate your findings by plotting and interpreting terms.

8.8 *Bivariate Quantizer.* To illustrate the importance of the block quantizer results, we consider the following problem. Let x_0 and x_1 denote two random variables whose correlation is $0 \leq \rho < 1$ and whose respective variances are 1:

$$\mathbf{R}_{xx} = E \begin{bmatrix} x_0 \\ x_1 \end{bmatrix} [x_0 \quad x_1] = \begin{bmatrix} 1 & \rho \\ \rho & 1 \end{bmatrix}.$$

a. Show that the eigenvalues of R are $1 + \rho$ and $1 - \rho$, and the determinant of $\mathbf{R}_{xx}$ is $(1 - \rho)(1 + \rho) = 1 - \rho^2$, the product of eigenvalues. Note the Hadamard inequality

$$r_{00}r_{11} = 1 \geq 1 - \rho^2.$$

b. Show that the bit-allocation formula when each x_n is quantized independently is

$$B_0 = \frac{B}{2} + \frac{1}{2} \log_2 1 = \frac{B}{2}$$

$$B_1 = \frac{B}{2}.$$

Show that the mean-squared error is

$$\sigma^2 = 2\beta^2 2^{-B}.$$

c. Find the bit-allocation formula for $\mathbf{y}$ when $[y_1 \quad y_2]^T = \mathbf{U}^T [x_1 \quad x_2]^T$ is constructed from the orthogonal matrix that diagonalizes $\mathbf{R}_{xx}$: $\mathbf{U}^T \mathbf{R} \mathbf{U} = \text{diag}[\lambda_0^2 \quad \lambda_1^2]$. Prove $\Sigma B_n = B$.

d. Show that the mean-squared quantization error when quantizing $\mathbf{y}$ and constructing $\mathbf{x}$ as $\mathbf{U}^T \mathbf{y}$ is

$$\sigma^2 = 2\beta^2 2^{-B}[(1 + \rho)(1 - \rho)]^{1/2}$$

and the gain of the block quantizer over the independent quantizer is

$$G = \frac{1}{[(1 + \rho)(1 - \rho)]^{1/2}}.$$

Evaluate these formulas for $\rho = 1 - 1/16$. How much gain do you get?

8.9   For the scalar quantizer $Q(X) = E(X|X \in I_k)$, evaluate $EX[X - Q(X)]$ and interpret your findings.

8.10 Prove Hadamard's inequality:

$$\prod_{n=0}^{N-1} r_{nn} \geq \det \mathbf{R} = \prod_{n=0}^{N-1} \lambda_{nn}^2.$$

8.11 Hypothesize that the most general LMMSE estimator of $\mathbf{x}_t$ from $\mathbf{y}_t$ in the Kalman filter is

$$\hat{\mathbf{x}}_t t = \mathbf{K}\hat{\mathbf{x}}_{t|t-1} + \mathbf{L}\mathbf{y}_{t-1} + \mathbf{k}_t(\mathbf{y}_t - \hat{\mathbf{y}}_{t|t-1}) + \mathbf{l}\mathbf{y}_t.$$

Then use orthogonality to show $\mathbf{K} = \mathbf{I}$, $\mathbf{L} = 0$, and $\mathbf{l} = \mathbf{0}$.

8.12 *Initializing the Kalman Filter.* Show how to initialize the Kalman recursions of Equation 8.66. Step through the recursions to show how they go.

8.13 Let x_t be the first-order dynamical system

$$x_t = ax_{t-1} + bu_{t-1}; \qquad t = 1, 2, \ldots$$

and y_t the "signal-plus-noise" measurement

$$y_t = x_t + n_t.$$

Assume $E x_0 = 0$, $E x_0^2 = 0$, $E u_t = 0$, $E u_t^2 = 1$, $E u_t u_s = \delta_{t-s}$, $E n_t = 0$, $E n_t^2 - \sigma_n^2$, $E n_t n_s - 0$, and $E u_t n_s - 0$.

a. Derive the equations of the Kalman filter.

b. Show what happens when $b^2/\sigma_n^2 \ll 1$ and when $b^2/\sigma_n^2 \gg 1$.

c. Let $t \to \infty$ to obtain a stationary Wiener filter; determine k_t, $\sigma_{t|t-1}^2$, $P_{t|t-1}$, and $P_{t|t}$ as $t \to \infty$.

8.14 Repeat Problem 8.13 when $E x_0^2 = r_0 > 0$. Explain how your Kalman filter changes.

8.15 *Kalman.* A particle leaves the origin at $t = 0$ with a random velocity. The position of the particle is measured (imperfectly) in additive autoregressive noise. Find the LMMSE estimate of the position and velocity of the particle at time t. *Hint*: Let $\mathbf{x}(t) = \begin{bmatrix} x_t(1) & x_t(2) & x_t(3) \end{bmatrix}^\mathrm{T}$, with $x_t(1) = x_{t-1}(1) + x_{t-1}(2)$ the position at time t and $x_t(2) = x_{t-1}(2)$ the constant velocity. Let $x_t(3) = a x_{t-1}(3) + b u_t$ be the correlated noise, and let the measurement be $y_t = x_t(1) + x_t(3)$. The initial conditions are $E x_0^2(1) = E x_0(2) = 0$; $E x_0^2(2) = \sigma_v^2 > 0$; $E u_t = 0$; $E u_t^2 = \sigma_u^2$.

8.16 In the previous problem, show $\hat{x}_{t|t}(3) \cong y_t - \hat{x}_{t|t}(1)$ for $t \gg 1$; explain Kalman's comment, "One would, of course, expect something like this since the problem is analogous to fitting a straight line to an increasing number of points."

8.17 Specialize the equations of the Kalman filter for

a. $\mathbf{A} = \mathbf{I}$, $\mathbf{b} = \mathbf{0}$;

b. $\mathbf{A} = \mathbf{I}$, $\mathbf{b} = \mathbf{0}$, $\sigma_n^2 = 0$.

Interpret your findings and connect your results with sequential Bayes.

8.18 *Uniform Rounding Quantizer.* Begin with a scalar random variable X that is uniformly distributed on $[0, L]$.

a. Show that the MMSE quantizer $Q(\ )$ is the *uniform rounding quantizer*

$$Q(x) = \hat{x}_k = k\Delta \quad \text{when} \quad \left(\frac{2k-1}{2} \right)\Delta < x \le \left(\frac{2k+1}{2} \right)\Delta$$

for $k = 1, 2, \ldots, K$. How is the step size Δ chosen?

b. Derive an equation for the mean-squared quantization error.

c. What is the mean-squared error when $K = 2^B$, where B is the number of bits used to code the 2^B representation points $\hat{x}_1, \hat{x}_2, \ldots, \hat{x}_{2^B}$?

8.19 Let's suppose the uniform rounding quantizer is used to quantize an arbitrary random variable. Show that, for arbitrarily large K, the mean-squared error is

$$\sigma^2 = \frac{\Delta^2}{12}.$$

If $\Delta = 2^{-B} L$, then $\sigma^2 = (L^2/12) 2^{-2B}$.

8.20 *Quantizer Errors.* Apply the uniform rounding quantizer to an arbitrary scalar random variable. Define the quantization error to be $e = x - \hat{x}$. Show that the density function for e is the aliased density

$$f(e) = \sum_n f_x(e + n\Delta); \qquad \frac{-\Delta}{2} < e \le \frac{\Delta}{2}.$$

Are quantization errors generally uniformly distributed? If not, when are they approximately uniformly distributed?

8.21 Derive a condition on the characteristic function of a random variable that ensures that its uniform rounding quantization errors are uniformly distributed.

8.22 *Lloyd-Max Quantizer.* Derive an improvement algorithm for iteratively finding the representation points $\hat{x}_k$ and slicing levels x_k that determine the MMSE quantizer of a random variable whose density is $f(x)$.

8.23 *Wiener Sequence.* Let $\{x_t\}$ denote a time series with correlation sequence $r_{t,s} = Ex_t x_s = \min(t, s)$. Let $\mathbf{x}_t^{\mathrm{T}} = (x_1 \quad \cdots \quad x_t)$ and define $\mathbf{R}_t = E\mathbf{x}_t \mathbf{x}_t^{\mathrm{T}}$.

a. Show

$$\mathbf{R}_t = E\mathbf{x}_t\mathbf{x}_t^{\mathrm{T}} = \begin{bmatrix} 1 & 1 & \cdots & 1 \\ 1 & 2 & \cdots & 2 \\ \vdots & \vdots & 3 & \vdots \\ & & & \ddots & \\ 1 & 2 & \cdots & t \end{bmatrix} = \left[\begin{array}{ccc|c} & & & 1 \\ & \mathbf{R}_{t-1} & & \vdots \\ & & & t-1 \\ \hline 1 & 2 & \cdots\ t-1 & t \end{array} \right].$$

This is called a Wiener matrix.

b. Prove that $\mathbf{x}_t$ may be synthesized as

$$\mathbf{x}_t = \begin{bmatrix} 1 & & & & & \\ 1 & 1 & & & \mathbf{0} & \\ 1 & 1 & 1 & & & \\ & & & \ddots & & \\ \vdots & & & & 1 & \\ 1 & & \cdots & & 1 & 1 \end{bmatrix} \mathbf{u}_t,$$

where $E\mathbf{u}_t\mathbf{u}_t^{\mathrm{T}} = \mathbf{I}$.

c. Prove

$$\mathbf{R}_t \begin{bmatrix} 1 & -1 & & & \\ & 1 & -1 & & \mathbf{0} \\ & & 1 & \ddots & \\ & \mathbf{0} & & \ddots & -1 \\ & & & & 1 \end{bmatrix} = \begin{bmatrix} 1 & & & & & \\ 1 & 1 & & & \mathbf{0} & \\ 1 & 1 & 1 & & & \\ & & & \ddots & & \\ \vdots & & & & 1 & \\ 1 & & \cdots & & 1 & 1 \end{bmatrix}.$$

d.   Prove

$$\begin{bmatrix} 1 & 0 & & & \\ -1 & 1 & & \mathbf{0} & \\ & -1 & \ddots & & \\ & \mathbf{0} & \ddots & & \\ & & & -1 & 1 \end{bmatrix} \mathbf{R}_t \begin{bmatrix} 1 & -1 & & & \\ & 1 & -1 & \mathbf{0} & \\ & & 1 & \ddots & \\ & \mathbf{0} & & \ddots & -1 \\ & & & & 1 \end{bmatrix} = \mathbf{I}.$$

e.   Find $\mathbf{R}_t^{-1}$.

8.24 *Prediction in Wiener Sequence.* Return to Problem 8.23. Find the LMMSE predictor of x_t from $\mathbf{x}_{t-1}$; find σ_t^2.

8.25 *Companding.* The asymptotic Lloyd-Max quantizer may be constructed by companding x, uniformly quantizing, and expanding, as illustrated below:

Show that $C(x)$ is the function

$$C(x) = \frac{x_{K+1} - x_1}{\int_{x_1}^{x_{K+1}} f^{1/3}(x)\,dx} \int_{x_1}^{x} f^{1/3}(\xi)\,d\xi.$$

Compare the performance of $C(x)$ with the compander

$$C'(x) = \int_{-\infty}^{x} f(\xi)\,d\xi$$

by computing mean-squared error for each compander.

8.26 *Low-Rank Wiener Filter.* Compute $E(\mathbf{x} - \hat{\mathbf{x}}_r)(\mathbf{x} - \hat{\mathbf{x}}_r)^T$ for $\hat{\mathbf{x}}_r$, the low-rank Wiener estimate of Section 8.4. That is, evaluate $\mathbf{P} + (\mathbf{G} - \mathbf{G}_r)\mathbf{R}_{yy}(\mathbf{G} - \mathbf{G}_r)^T$ for $\mathbf{G}_r$, the low-rank Wiener filter of Section 8.4.

Least Squares

In the previous eight chapters, we have used statistical reasoning to guide our choice of good detectors and estimators. In Chapter 8 we relaxed our adherence to statistical reasoning somewhat by deriving *linear* estimators that were designed to minimize mean-squared error. For the design of such estimators, we required only the covariance (or second-order) properties of the data and the parameter to be estimated, not the detailed distributions.

In this chapter we deviate even further from statistical reasoning and use a simple least squares criterion to guide our choice of good estimators. The idea behind least squares is to fit a model to measurements in such a way that weighted errors between the measurements and the model are minimized. Least squares may be used in linear or nonlinear modeling. In this chapter we treat both types of problems. The linear problems illustrate the basic principles of least squares and bring a wealth of physical and geometrical insight.

9.1   THE LINEAR MODEL

The linear model says that observations $\mathbf{y} = [y_1 \quad y_2 \quad \cdots \quad y_N]^{\mathrm{T}}$ consist of a model, or signal, component $\mathbf{x} = [x_1 \quad x_2 \quad \cdots \quad x_N]^{\mathrm{T}}$, plus an error component $\mathbf{n} = [n_1 \quad n_2 \quad \cdots \quad n_N]^{\mathrm{T}}$:

$$\mathbf{y} = \mathbf{x} + \mathbf{n}. \qquad\qquad 9.1$$

The model says, further, that the signal component obeys the linear equation

$$\mathbf{x} = \mathbf{H\theta} \qquad\qquad 9.2$$

where $\mathbf{H}$ is an $N \times p$ matrix and $\mathbf{\theta}$ is a $p \times 1$ vector:

$$\mathbf{H} = \begin{bmatrix} h_{11} & h_{12} & \cdots & h_{1p} \\ h_{21} & & & \\ \vdots & & & \vdots \\ h_{N1} & \cdots & & h_{Np} \end{bmatrix} \qquad\qquad 9.3$$

$$\mathbf{\theta} = [\theta_1 \quad \theta_2 \quad \cdots \quad \theta_p]^{\mathrm{T}}.$$

The vector $\mathbf{\theta}$ is unknown. The matrix $\mathbf{H}$ is known when the mode structure, or dynamics, of a system are known, but unknown otherwise.

Interpretations

Sometimes we interpret the vector of errors $\mathbf{n}$ as errors or residuals in fitting the model $\mathbf{H\theta}$ to the data $\mathbf{y}$. Then the equation $\mathbf{y} = \mathbf{H\theta} + \mathbf{n}$ is called an equation error model for the ideal model $\mathbf{x} = \mathbf{H\theta}$. Sometimes $\mathbf{x} = \mathbf{H\theta}$ is the signal component in a signal-plus-noise model $\mathbf{y} = \mathbf{x} + \mathbf{n}$, where $\mathbf{n}$ is additive noise. In either of these interpretations, we may think of the matrix $\mathbf{H}$ as composed of columns $\mathbf{h}_n$

or rows $\mathbf{c}_n^T$:

$$\mathbf{H} = [\mathbf{h}_1, \mathbf{h}_2, \ldots, \mathbf{h}_p]$$

$$= \begin{bmatrix} \mathbf{c}_1^T \\ \vdots \\ \mathbf{c}_N^T \end{bmatrix} \tag{9.4}$$

$$\mathbf{h}_n = [h_{1n} \quad h_{2n} \quad \cdots \quad h_{Nn}]^T$$

$$\mathbf{c}_n^T = [c_{n1} \quad c_{n2} \quad \cdots \quad c_{np}].$$

Call each column a mode of the signal $\mathbf{x}$. Then the signal $\mathbf{x}$ consists of a linear combination of these modes:

$$\mathbf{x} = \sum_{n=1}^{p} \theta_n \mathbf{h}_n. \tag{9.5}$$

It is precisely the weights in this linear combination that we wish to determine. Alternatively, think of x_n, the nth term in $\mathbf{x}$, as the scalar variable

$$x_n = \mathbf{c}_n^T \boldsymbol{\theta} \tag{9.6}$$

where $\mathbf{c}_n^T$ is row n of the model matrix $\mathbf{H}$. In this interpretation, x_n is a linear combination of the parameters.

Example 9.1 (Polynomial Regression)

Let $\mathbf{y}$ denote readings of a continuous curve $y(t)$ at points $t = T, 2T, \ldots, NT$:

$$\mathbf{y} = [y_1 \quad y_2 \quad \cdots \quad y_N]^T \tag{9.7}$$

$$y_k = y(kT).$$

This is illustrated in Figure 9.1. We would like to fit a polynomial model of the following form to the curve $y(t)$:

$$x(t) = \sum_{n=1}^{p} \theta_n t^{n-1}. \tag{9.8}$$

The samples of this polynomial are

$$x_k = x(kT) = \sum_{n=1}^{p} \theta_n (kT)^{n-1}. \tag{9.9}$$

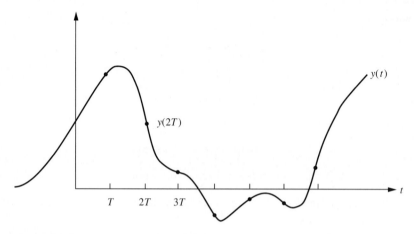

Figure 9.1 Readings of a continuous curve.

At the sample points $t = kT$, with $T = 1$, this model produces the following equation for $\mathbf{x} = [x_1 \quad x_2 \quad \cdots \quad x_N]^T$:

$$\mathbf{x} = \mathbf{H}\boldsymbol{\theta}$$

$$\mathbf{H} = \begin{bmatrix} 1 & 1 & 1 \\ 1 & 2 & 2^{p-1} \\ \vdots & 3 & \vdots \\ & \vdots & \\ 1 & N & N^{p-1} \end{bmatrix} \qquad\qquad 9.10$$

$$\mathbf{h}_n = \begin{bmatrix} 1^{n-1} & 2^{n-1} & \cdots & N^{n-1} \end{bmatrix}^T$$

$$\mathbf{c}_n^T = \begin{bmatrix} 1 & n & n^2 & \cdots & n^{p-1} \end{bmatrix}.$$

The least squares problem is one of finding the parameters $\boldsymbol{\theta}$ that least squares fit this model to the measurements $\mathbf{y}$. ∎

Example 9.2 (Complex Exponential Model)

Consider a resonant structure that supports complex exponential modes $h_n(t)$:

$$h_n(t) = \exp(j\omega_n t); \qquad \omega_n: n\text{th angular frequency (radians/sec).} \qquad 9.11$$

A typical mode is illustrated in Figure 9.2. At sample points $t = kT$, these modes take the form

$$h_{kn} = h_n(kT) = z_n^k \qquad\qquad 9.12$$

$$z_n = \exp(j\omega_n T).$$

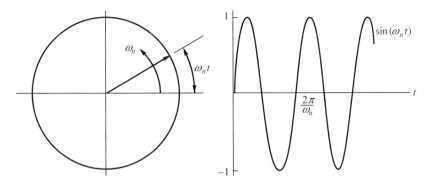

Figure 9.2 A typical mode.

A model for the measurements $\mathbf{y} = [y_0 \quad y_1 \quad \cdots \quad y_{N-1}]^T$ is obtained by using a linear combination of the modes to build a model of the following form:

$$\mathbf{x} = \mathbf{H\theta}$$

$$\mathbf{x} = [x_0 \quad x_1 \quad \cdots \quad x_{N-1}]^T \qquad\qquad 9.13$$

$$H = \begin{bmatrix} 1 & 1 & \cdots & 1 \\ z_1 & z_2 & \cdots & z_p \\ z_1^2 & \vdots & & \vdots \\ \vdots & & & \\ z_1^{N-1} & z_2^{N-1} & \cdots & z_p^{N-1} \end{bmatrix}$$

$$\mathbf{h}_n = \begin{bmatrix} z_n^0 & \cdots & z_n^{N-1} \end{bmatrix}^T.$$

The vector $\mathbf{x}$ is a linear combination of modes $\mathbf{h}_n$. The least squares problem is one of finding the parameters $\mathbf{\theta}$ that least squares fit this model to the measurements $\mathbf{y}$. ■

In the study of least squares problems, it is necessary to distinguish among three cases: (1) overdetermined, (2) determined, and (3) underdetermined. Let's define these cases and discuss them.

Overdetermined Case (p < N)

When the number of measurements exceeds the number of free parameters, an exact fit of the model to the data is not possible. This is the most interesting case because it leads naturally to least squares fitting. The model is $\mathbf{x} = \mathbf{H\theta}$, with $\mathbf{H}$ the $N \times p$ matrix $\mathbf{H} = [\mathbf{h}_1 \quad \mathbf{h}_2 \quad \cdots \quad \mathbf{h}_p]$.

Determined Case (p = N)

When the number of free parameters equals the number of measurements, then the measurements can be exactly fit provided that there is no linear dependence among the columns of **H**. The model is $\mathbf{x} = \mathbf{H}\boldsymbol{\theta}$, with **H** the $N \times N$ matrix $\mathbf{H} = [\mathbf{h}_1 \quad \mathbf{h}_2 \quad \cdots \quad \mathbf{h}_N]$.

Underdetermined Case (p > N)

When the number of free parameters exceeds the number of measurements, then there is not enough data to uniquely determine $\boldsymbol{\theta}$. In fact, there are generally an infinite number of solutions that reproduce the measurements, and only side conditions can select one or more desirable solutions. The model is $\mathbf{x} = \mathbf{H}\boldsymbol{\theta}$, with **H** the $N \times p$ matrix $\mathbf{H} = [\mathbf{h}_1 \quad \mathbf{h}_2 \quad \cdots \quad \mathbf{h}_p]$.

In this preliminary discussion of cases, we have temporarily ignored the rank of the system matrix **H**. As subsequent developments will show, this is of paramount importance because it determines the structure of the system model $\mathbf{x} = \mathbf{H}\boldsymbol{\theta}$.

The Normal Error Model

We require no statistical description of the parameter vector $\boldsymbol{\theta}$ or the noise vector **n** to proceed with the theory of least squares. However, when the error vector $\mathbf{n} : N[\mathbf{0}, \mathbf{R}]$ *is* a realization of a normal random vector, we say that the linear model is a linear statistical model with normal errors. It follows that the measurements **y** are realizations in the normal model $N[\mathbf{H}\boldsymbol{\theta}, \mathbf{R}]$. The density function for **y** is then

$$f_{\boldsymbol{\theta}}(\mathbf{y}) = (2\pi)^{-N/2}|\mathbf{R}|^{-1/2}\exp\left\{-\frac{1}{2}(\mathbf{y} - \mathbf{H}\boldsymbol{\theta})^{\mathrm{T}}\mathbf{R}^{-1}(\mathbf{y} - \mathbf{H}\boldsymbol{\theta})\right\}. \qquad 9.14$$

The theory of maximum likelihood would have us minimize the weighted squared errors $(\mathbf{y} - \mathbf{H}\boldsymbol{\theta})^{\mathrm{T}}\mathbf{R}^{-1}(\mathbf{y} - \mathbf{H}\boldsymbol{\theta})$. We recall from the factorization theorem of Chapter 3 that the statistic

$$\mathbf{H}^{\mathrm{T}}\mathbf{R}^{-1}\mathbf{y} : N[\mathbf{H}^{\mathrm{T}}\mathbf{R}^{-1}\mathbf{H}\boldsymbol{\theta}, \mathbf{H}^{\mathrm{T}}\mathbf{R}^{-1}\mathbf{H}] \qquad 9.15$$

is a sufficient statistic for $\boldsymbol{\theta}$. The maximum likelihood and minimum variance unbiased estimators are identical:

$$\hat{\boldsymbol{\theta}} = (\mathbf{H}^{\mathrm{T}}\mathbf{R}^{-1}\mathbf{H})^{-1}\mathbf{H}^{\mathrm{T}}\mathbf{R}^{-1}\mathbf{y} : N[\boldsymbol{\theta}, (\mathbf{H}^{\mathrm{T}}\mathbf{R}^{-1}\mathbf{H})^{-1}]. \qquad 9.16$$

We will find this solution to be identical with a weighted least squares solution, indicating that least squares is no philosophical departure from statistical methodology when dealing with normal errors in the model $\mathbf{y} = \mathbf{H}\boldsymbol{\theta} + \mathbf{n}$. After we have computed a least squares solution, we often use a normal error model to find the distribution for $\hat{\boldsymbol{\theta}}$ and determine the performance of least squares.

9.2   LEAST SQUARES SOLUTIONS

Consider the linear model

$$\mathbf{y} = \mathbf{H}\boldsymbol{\theta} + \mathbf{n}. \qquad 9.17$$

For a given estimate of $\boldsymbol{\theta}$, the squared error between $\mathbf{y}$ and the model $\mathbf{H}\boldsymbol{\theta}$ is defined to be

$$e^2 = tr\left[(\mathbf{y} - \mathbf{H}\boldsymbol{\theta})(\mathbf{y} - \mathbf{H}\boldsymbol{\theta})^{\mathrm{T}}\right]$$
$$= (\mathbf{y} - \mathbf{H}\boldsymbol{\theta})^{\mathrm{T}}(\mathbf{y} - \mathbf{H}\boldsymbol{\theta}) = \mathbf{n}^{\mathrm{T}}\mathbf{n}. \qquad 9.18$$

This is minimized to obtain the least squares estimate. The appropriate equations for the gradient and second gradient of e^2 with respect to $\boldsymbol{\theta}$ are

$$\frac{\partial}{\partial\boldsymbol{\theta}} e^2 = 2\mathbf{H}^{\mathrm{T}}(\mathbf{y} - \mathbf{H}\boldsymbol{\theta}) \qquad 9.19$$

$$\frac{\partial}{\partial\boldsymbol{\theta}}\left(\frac{\partial}{\partial\boldsymbol{\theta}} e^2\right)^{\mathrm{T}} = 2\mathbf{H}^{\mathrm{T}}\mathbf{H} \geq \mathbf{0}.$$

The least squares estimate equates the gradient to zero to produce the solution

$$\hat{\boldsymbol{\theta}} = (\mathbf{H}^{\mathrm{T}}\mathbf{H})^{-1}\mathbf{H}^{\mathrm{T}}\mathbf{y}. \qquad 9.20$$

This solution exists and is unique, provided that the inverse of $\mathbf{H}^{\mathrm{T}}\mathbf{H}$ exists.

We say that the least squares solution for $\hat{\boldsymbol{\theta}}$ obeys the *normal equations*

$$\mathbf{H}^{\mathrm{T}}\mathbf{H}\hat{\boldsymbol{\theta}} = \mathbf{H}^{\mathrm{T}}\mathbf{y}, \qquad 9.21$$

where the matrix $\mathbf{G} = \mathbf{H}^{\mathrm{T}}\mathbf{H}$ is called the Grammian or Gram matrix. If the Gram matrix is nonsingular, the solution for $\hat{\boldsymbol{\theta}}$ is unique. If it is singular, we have many solutions. The following theorem establishes conditions for the Gram matrix to be nonsingular.

Theorem (Nonsingular Gram Matrix)   The Gram matrix is singular if and only if the columns of $\mathbf{H}$ are linearly dependent. Equivalently, the Gram matrix is nonsingular if and only if the columns of $\mathbf{H}$ are linearly independent.

Proof   See the proof of Theorem 2.1 in Chapter 2.   □

Projections

Assume that the Gram matrix is nonsingular. Then the least squares estimate of the parameter $\boldsymbol{\theta}$ is

$$\hat{\boldsymbol{\theta}} = (\mathbf{H}^{\mathrm{T}}\mathbf{H})^{-1}\mathbf{H}^{\mathrm{T}}\mathbf{y}. \qquad 9.22$$

We may use $\hat{\boldsymbol{\theta}}$ to construct an estimate of the "signal" $\mathbf{x} = \mathbf{H}\boldsymbol{\theta}$:

$$\hat{\mathbf{x}} = \mathbf{H}\hat{\boldsymbol{\theta}}. \tag{9.23}$$

In Problem 9.1 you are asked to show that $\hat{\mathbf{x}}$ minimizes the squared error $e^2 = (\mathbf{y} - \mathbf{x})^{\mathrm{T}}(\mathbf{y} - \mathbf{x})$ under the constraint that $\mathbf{x} = \mathbf{H}\boldsymbol{\theta}$. The solution for $\hat{\mathbf{x}}$ may be written as

$$\hat{\mathbf{x}} = \mathbf{P}_H \mathbf{y} \tag{9.24}$$

where P_H is the following orthogonal projection:

$$\mathbf{P}_H = \mathbf{H}(\mathbf{H}^{\mathrm{T}}\mathbf{H})^{-1}\mathbf{H}^{\mathrm{T}}.$$

The difference between the measurements $\mathbf{y}$ and the estimated model $\hat{\mathbf{x}}$ is the fitting error, or estimated noise, $\hat{\mathbf{n}}$:

$$\hat{\mathbf{n}} = \mathbf{y} - \hat{\mathbf{x}} = (\mathbf{I} - \mathbf{P}_H)\mathbf{y} = \mathbf{P}_A \mathbf{y} \tag{9.25}$$

$$\mathbf{P}_A = \mathbf{I} - \mathbf{P}_H.$$

The orthogonal projections $\mathbf{P}_H$ and $\mathbf{P}_A$ share some remarkable properties:

1. $\mathbf{P}_H = \mathbf{P}_H^{\mathrm{T}}$ (symmetric); $\mathbf{P}_A = \mathbf{P}_A^{\mathrm{T}}$ (symmetric); $\qquad\qquad\qquad$ 9.26
2. $\mathbf{P}_H\mathbf{P}_H = \mathbf{P}_H$ (idempotent); $\mathbf{P}_A\mathbf{P}_A = \mathbf{P}_A$ (idempotent).

Furthermore, $\mathbf{P}_H$ and $\mathbf{P}_A$ are orthogonal matrices that decompose identity:

$$\mathbf{P}_H\mathbf{P}_A = \mathbf{P}_A\mathbf{P}_H = \mathbf{0} \quad \text{(orthogonal)} \tag{9.27}$$

$$\mathbf{P}_H + \mathbf{P}_A = \mathbf{I}. \quad \text{(decomposition of identity)}$$

The orthogonal projections $\mathbf{P}_H$ and $\mathbf{P}_A$ trace their heritage to the model matrix $\mathbf{H}$. It is easy to check that $\mathbf{P}_H$ is transparent to $\mathbf{H}$ and $\mathbf{P}_A$ is opaque to $\mathbf{H}$:

1. $\mathbf{P}_H\mathbf{H} = \mathbf{H}$; $\qquad\qquad\qquad\qquad\qquad\qquad\qquad\qquad\qquad\qquad$ 9.28
2. $\mathbf{P}_A\mathbf{H} = \mathbf{0}$.

These two results say that the columns of $\mathbf{H}$ remain unchanged as they are projected through $\mathbf{P}_H$ but are annihilated when they are projected through $\mathbf{P}_A$. The results say also that $\mathbf{P}_H$ perfectly images $\mathbf{x}$ and $\mathbf{P}_A = \mathbf{I} - \mathbf{P}_H$ annihilates $\mathbf{x}$, provided $\mathbf{x} = \mathbf{H}\boldsymbol{\theta}$:

1. $\mathbf{P}_H\mathbf{x} = \mathbf{x}$; $\qquad\qquad\qquad\qquad\qquad\qquad\qquad\qquad\qquad\qquad$ 9.29
2. $\mathbf{P}_A\mathbf{x} = \mathbf{0}$.

Signal and Orthogonal Subspaces

The projectors $\mathbf{P}_H$ and $\mathbf{P}_A$ are illustrated in Figure 9.3. The plane of the figure represents all vectors $\mathbf{x} = \mathbf{H}\boldsymbol{\theta}$ that may be represented as linear combinations of the columns of $\mathbf{H}$. Call this the *signal subspace* spanned by $\mathbf{H}$, and denote it by $\langle \mathbf{H} \rangle$. The vertical axis of the figure represents all vectors $\mathbf{u}$ that are orthogonal to the columns of $\mathbf{H}$. We call this the *orthogonal subspace* $\langle \mathbf{A} \rangle$. When the p columns of $\mathbf{H}$ are lin-

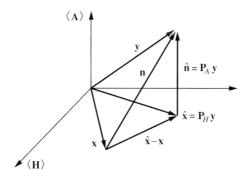

Figure 9.3 Illustrating projections $\mathbf{P}_H$ and $\mathbf{P}_A$, and
the orthogonal decompositions of $\mathbf{y}$ and $\mathbf{n}$.

early independent, then there are just $N - p$ linearly independent vectors that can be orthogonal to the p columns of $\mathbf{H}$. Organize them into the $N \times (N - p)$ matrix $\mathbf{A}$:

$$\mathbf{A} = \begin{bmatrix} a_{11} & a_{12} & \cdots & a_{1(N-p)} \\ a_{21} & & & \vdots \\ \vdots & & & \\ a_{N1} & \cdots & & a_{N(N-p)} \end{bmatrix} = [\mathbf{a}_1 \quad \mathbf{a}_2 \quad \cdots \quad \mathbf{a}_{N-p}]. \qquad 9.30$$

Each of the column vectors $\mathbf{a}_i$ that comprise $\mathbf{A}$ must be orthogonal to each of the column vectors $\mathbf{h}_j$ that comprise $\mathbf{H}$. We summarize these orthogonality conditions as follows:

$$\mathbf{a}_i^{\mathsf{T}} \mathbf{h}_j = 0 \qquad 9.31$$

$$\mathbf{A}^{\mathsf{T}} \mathbf{H} = \mathbf{0}.$$

Each vector $\mathbf{u}$ in the orthogonal subspace is a linear combination of the vectors $\mathbf{a}_n$:

$$\mathbf{u} = \mathbf{A}\boldsymbol{\phi} \qquad 9.32$$

$$\boldsymbol{\phi} = [\phi_1 \quad \phi_2 \quad \cdots \quad \phi_{N-p}]^{\mathsf{T}}.$$

This suggests that a projection onto the orthogonal subspace $\langle \mathbf{A} \rangle$ should look like this:

$$\mathbf{P}_A = \mathbf{A}(\mathbf{A}^{\mathsf{T}}\mathbf{A})^{-1}\mathbf{A}^{\mathsf{T}}. \qquad 9.33$$

Note that $\mathbf{P}_A$ is, indeed, an orthogonal projection. We say that the representation

$$\mathbf{I} = \mathbf{P}_H + \mathbf{P}_A = \mathbf{H}(\mathbf{H}^{\mathsf{T}}\mathbf{H})^{-1}\mathbf{H}^{\mathsf{T}} + \mathbf{A}(\mathbf{A}^{\mathsf{T}}\mathbf{A})^{-1}\mathbf{A}^{\mathsf{T}} \qquad 9.34$$

splits Euclidean space into orthogonal subspaces, the first spanned by the columns of $\mathbf{H}$ and the second spanned by the columns of $\mathbf{A}$. The projection $\mathbf{P}_H = \mathbf{H}(\mathbf{H}^{\mathsf{T}}\mathbf{H})^{-1}\mathbf{H}^{\mathsf{T}}$ projects onto the first subspace, and the projection $\mathbf{P}_A = \mathbf{A}(\mathbf{A}^{\mathsf{T}}\mathbf{A})^{-1}\mathbf{A}^{\mathsf{T}}$ projects onto the second. (See Problem 9.2.)

An arbitrary vector $\mathbf{y}$ may be decomposed into the components

$$\mathbf{y} = \mathbf{Iy} = \mathbf{P}_H \mathbf{y} + \mathbf{P}_A \mathbf{y}$$
$$= \hat{\mathbf{x}} + \hat{\mathbf{n}}$$

9.35

where $\hat{\mathbf{x}}$ and $\hat{\mathbf{n}}$ are the estimated signal and estimated noise:

$$\hat{\mathbf{x}} = \mathbf{P}_H \mathbf{y}; \qquad \hat{\mathbf{n}} = \mathbf{P}_A \mathbf{y}.$$

9.36

The estimated signal $\hat{\mathbf{x}}$ lies in the span of $\mathbf{H}$, and the estimated noise $\hat{\mathbf{n}}$ lies in the span of $\mathbf{A}$:

$$\hat{\mathbf{x}} = \mathbf{H}\hat{\boldsymbol{\theta}}; \qquad \hat{\boldsymbol{\theta}} = (\mathbf{H}^T\mathbf{H})^{-1}\mathbf{H}^T\mathbf{y}$$
$$\hat{\mathbf{n}} = \mathbf{A}\hat{\boldsymbol{\phi}}; \qquad \hat{\boldsymbol{\phi}} = (\mathbf{A}^T\mathbf{A})^{-1}\mathbf{A}^T\mathbf{y}.$$

9.37

Another way of saying that the estimates $\hat{\mathbf{x}}$ and $\hat{\mathbf{n}}$ lie in subspaces spanned by $\mathbf{H}$ and $\mathbf{A}$, respectively, is to say that projections leave them unchanged:

$$\mathbf{P}_H \hat{\mathbf{x}} = \hat{\mathbf{x}}$$

9.38

$$\mathbf{P}_A \hat{\mathbf{n}} = \hat{\mathbf{n}}.$$

Orthogonality

In the decomposition of $\mathbf{y}$ into $\mathbf{y} = \hat{\mathbf{x}} + \hat{\mathbf{n}}$, the estimated noise $\hat{\mathbf{n}}$ is orthogonal to every vector $\mathbf{x}$ in the signal subspace, including $\hat{\mathbf{x}}$:

$$\hat{\mathbf{n}}^T \hat{\mathbf{x}} = \mathbf{y}^T \mathbf{P}_A \mathbf{P}_H \mathbf{y} = 0.$$

9.39

We say that $\mathbf{y} = \hat{\mathbf{x}} + \hat{\mathbf{n}}$ is an orthogonal decomposition of $\mathbf{y}$. The norm of the fitting error is therefore

$$\hat{\mathbf{n}}^T \hat{\mathbf{n}} = \mathbf{y}^T \mathbf{P}_A \mathbf{y} = \mathbf{y}^T (\mathbf{I} - \mathbf{P}_H)\mathbf{y} = \mathbf{y}^T \mathbf{y} - \hat{\mathbf{x}}^T \hat{\mathbf{x}}.$$

9.40

So, we have a Pythagorean theorem at work:

$$\mathbf{y}^T \mathbf{y} = \hat{\mathbf{x}}^T \hat{\mathbf{x}} + \hat{\mathbf{n}}^T \hat{\mathbf{n}}.$$

9.41

There is no other value of $\hat{\mathbf{x}}$ in the signal subspace that provides a smaller norm for $\hat{\mathbf{n}}^T \hat{\mathbf{n}}$. To see this, give and take $\hat{\mathbf{x}}$ in the equation for $\mathbf{n}^T\mathbf{n}$ to obtain

$$\mathbf{n}^T\mathbf{n} = (\mathbf{y} - \mathbf{x} + \hat{\mathbf{x}} - \hat{\mathbf{x}})^T(\mathbf{y} - \mathbf{x} + \hat{\mathbf{x}} - \hat{\mathbf{x}})$$
$$= \left[\hat{\mathbf{n}} - (\mathbf{x} - \hat{\mathbf{x}})\right]^T\left[\hat{\mathbf{n}} - (\mathbf{x} - \hat{\mathbf{x}})\right].$$

9.42

Use the orthogonality between $\hat{\mathbf{n}}$ and $(\mathbf{x} - \hat{\mathbf{x}})$ to write

$$\mathbf{n}^T\mathbf{n} = \hat{\mathbf{n}}^T\hat{\mathbf{n}} + (\mathbf{x} - \hat{\mathbf{x}})^T(\mathbf{x} - \hat{\mathbf{x}}) \geq \hat{\mathbf{n}}^T\hat{\mathbf{n}}$$

9.43

with equality if and only if $\mathbf{x} = \hat{\mathbf{x}}$. This result also states that $\mathbf{n} = \hat{\mathbf{n}} + (\hat{\mathbf{x}} - \mathbf{x})$ is an orthogonal decomposition of $\mathbf{n}$.

We may refer back to Figure 9.3 and think about these orthogonal projections in the following way: The prior model $\mathbf{y} - \mathbf{H\theta} + \mathbf{n}$ has been replaced by a posterior model $\mathbf{y} = \mathbf{P}_H\mathbf{y} + \mathbf{P}_A\mathbf{y}$ in which the projection $\hat{\mathbf{x}} = \mathbf{Py}$ onto the signal subspace has produced the fitting error $\hat{\mathbf{n}} = \mathbf{P}_A\mathbf{y} = (\mathbf{y} - \hat{\mathbf{x}})$ with minimum norm. In the posterior model, $\mathbf{P}_H\mathbf{y}$ and $\mathbf{P}_A\mathbf{y}$ are orthogonal. These ideas are illustrated in Figures 9.3 and 9.4. Figure 9.3 illustrates the orthogonal decomposition of $\mathbf{y}$ into $\hat{\mathbf{x}} + \hat{\mathbf{n}}$ and the decomposition of $\mathbf{n}$ into the orthogonal components $\hat{\mathbf{n}}$ and $(\mathbf{x} - \hat{\mathbf{x}})$. Figure 9.4 illustrates the same thing algebraically. We say that the posterior model $\mathbf{y} = \hat{\mathbf{x}} + \hat{\mathbf{n}}$ splits the data into orthogonal subspaces.

Example 9.3 (Complex Exponential Model Again)

In the complex exponential model, the model matrix $\mathbf{H}$ is

$$\mathbf{H} = \begin{bmatrix} 1 & 1 & \cdots & 1 \\ z_1 & z_2 & & z_p \\ \vdots & \vdots & & \vdots \\ z_1^{N-1} & z_2^{N-1} & & z_p^{N-1} \end{bmatrix},$$

　9.44

and the projector $\mathbf{P}_H$ is

$$\mathbf{P}_H = \mathbf{H}(\mathbf{H}^T\mathbf{H})^{-1}\mathbf{H}^T.$$

　9.45

From the roots z_n, $n = 1, 2, \ldots, p$, form the polynomial

$$A(z) = \prod_{n=1}^{p}\left(1 - z_n z^{-1}\right) = \sum_{n=0}^{p} a_n z^{-n}; \qquad a_0 = 1.$$

　9.46

Note the following properties:

$$A(z_n) = 0$$

　9.47

$$z_n^t A(z_n) = \sum_{n=0}^{p} a_n z^{(t-n)} = 0.$$

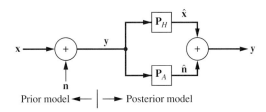

Figure 9.4 Orthogonal projections split the data into orthogonal subspaces.

For $t = p, p + 1, \ldots, N - 1$, Equation 9.47 may be written out as follows:

$$
\begin{array}{l}
(t = p) \\
(t = p + 1) \\
\vdots \\
(t = N - 1)
\end{array}
\begin{bmatrix}
a_p & a_{p-1} & \cdots & a_1 & 1 & 0 & \cdots & 0 \\
0 & a_p & a_{p-1} & \cdots & a_1 & 1 & & \vdots \\
\vdots & \vdots & & & & & & 0 \\
0 & 0 & \cdots & a_p & a_{p-1} & \cdots & a_1 & 1
\end{bmatrix}
$$

$$
\begin{bmatrix}
1 & 1 & \cdots & 1 \\
z_1 & z_2 & \cdots & z_p \\
\vdots & \vdots & & \vdots \\
z_1^{N-1} & z_2^{N-1} & \cdots & z_p^{N-1}
\end{bmatrix} = \mathbf{0}. \qquad 9.48
$$

The corresponding matrix equation is

$$
\mathbf{A}^T\mathbf{H} = \mathbf{0} \qquad 9.49
$$

where $\mathbf{A}$ is the following $N \times (N - p)$ matrix:

$$
A = \begin{bmatrix}
a_p & 0 & \cdots & 0 \\
a_{p-1} & a_p & & \vdots \\
\vdots & a_{p-1} & \ddots & \\
a_1 & \vdots & \ddots & a_p \\
1 & a_1 & & a_{p-1} \\
& 1 & \ddots & \vdots \\
\vdots & \vdots & \ddots & a_1 \\
0 & 0 & \cdots & 1
\end{bmatrix} = [\mathbf{a}_1 \quad \mathbf{a}_2 \quad \cdots \quad \mathbf{a}_{N-p}] \qquad 9.50
$$

$$
\mathbf{a}_n = [0 \quad \cdots \quad 0 \quad a_p \quad \cdots \quad a_1 \quad 1 \quad 0 \quad \cdots \quad 0]^T.
$$

We have constructed a matrix $\mathbf{A}$ whose columns are orthogonal to $\mathbf{H}$ and that spans the orthogonal subspace. The matrix

$$
\mathbf{P}_A = \mathbf{A}(\mathbf{A}^T\mathbf{A})^{-1}\mathbf{A}^T \qquad 9.51
$$

is a projection onto this subspace.   ■

We now know a lot about the subspaces $\mathbf{H}$ and $\mathbf{A}$ that orthogonally decompose Euclidean space and about their corresponding projections that orthogonally decompose identity. But there is more yet to be learned, and to learn it we must investigate to one more level of detail the structures of $\mathbf{H}$, $\mathbf{A}$, $\mathbf{P}_H$, and $\mathbf{P}_A$. Our program in the next section is to explore orthogonal decompositions of each space $\mathbf{H}$ and $\mathbf{A}$ while preserving the orthogonality between them.

9.3   STRUCTURE OF SUBSPACES IN LEAST SQUARES PROBLEMS

Let's summarize our findings thus far. The least squares estimates of the parameter $\boldsymbol{\theta}$, the signal $\mathbf{x}$, and the noise $\mathbf{n}$ are

$$\hat{\boldsymbol{\theta}} = (\mathbf{H}^{\mathsf{T}}\mathbf{H})^{-1}\mathbf{H}^{\mathsf{T}}\mathbf{y}$$

$$\hat{\mathbf{x}} = \mathbf{H}(\mathbf{H}^{\mathsf{T}}\mathbf{H})^{-1}\mathbf{H}^{\mathsf{T}}\mathbf{y} = \mathbf{P}_H\mathbf{y} \qquad\qquad 9.52$$

$$\hat{\mathbf{n}} = \mathbf{y} - \hat{\mathbf{x}} = \mathbf{A}(\mathbf{A}^{\mathsf{T}}\mathbf{A})^{-1}\mathbf{A}^{\mathsf{T}}\mathbf{y} = \mathbf{P}_A\mathbf{y}.$$

Each of these estimates shows explicit dependence on the system matrix $\mathbf{H}$ and its Grammian $\mathbf{H}^{\mathsf{T}}\mathbf{H}$. In fact, $\hat{\mathbf{x}}$ is just a linear combination of the columns of $\mathbf{H}$ with each column $\mathbf{h}_i$ weighted by the least squares estimate of θ_i. In general, the columns of $\mathbf{H}$ are far from orthogonal, and consequently the Grammian of $\mathbf{H}$ can have very small eigenvalues. The same can be said of $\mathbf{A}$. It is natural to ask whether there is another representation for $\mathbf{H}$ and $\mathbf{A}$ that would generate orthogonal representations for $\hat{\boldsymbol{\theta}}$, $\hat{\mathbf{x}}$, and $\hat{\mathbf{n}}$. So the idea is to replace the basis $(\mathbf{h}_1, \mathbf{h}_2, \dots, \mathbf{h}_p)$ for $\mathbf{H}$ and the basis $(\mathbf{a}_1, \dots, \mathbf{a}_{N-p})$ for $\mathbf{A}$ with orthogonal bases.

Let's assemble the columns of $\mathbf{H}$ and the columns of $\mathbf{A}$ into a system matrix $\mathbf{S}$ whose N linearly independent columns span Euclidean N-space:

$$\mathbf{S} = \begin{bmatrix} \mathbf{H} \,|\, \mathbf{A} \end{bmatrix}. \qquad\qquad 9.53$$

The matrix $\mathbf{S}$ is $N \times N$; the matrices $\mathbf{H}$ and $\mathbf{A}$ are, respectively, $N \times p$ and $N \times (N-p)$. The Grammian of $\mathbf{S}$ is

$$\mathbf{G} = \mathbf{S}^{\mathsf{T}}\mathbf{S} = \begin{bmatrix} \mathbf{H}^{\mathsf{T}}\mathbf{H} & \mathbf{0} \\ \mathbf{0} & \mathbf{A}^{\mathsf{T}}\mathbf{A} \end{bmatrix}. \qquad\qquad 9.54$$

Now consider the following, seemingly clumsy, way to write identity:

$$\mathbf{I} = \mathbf{S}\mathbf{S}^{-1}\mathbf{S}^{-\mathsf{T}}\mathbf{S}^{\mathsf{T}} = \mathbf{S}(\mathbf{S}^{\mathsf{T}}\mathbf{S})^{-1}\mathbf{S}^{\mathsf{T}}. \qquad\qquad 9.55$$

($\mathbf{S}^{-1}$ exists because $\mathbf{S}$ constructed from $\mathbf{H}$ and $\mathbf{A}$ is full-rank.) Think of $\mathbf{I}$ as the full-rank projector onto the Euclidean space spanned by $\mathbf{S}$ and think of $\mathbf{S}^{\mathsf{T}}\mathbf{S}$ as the Grammian of the span $\mathbf{S}$. From our representation of $\mathbf{S}$, we have the following representation of identity:

$$\mathbf{I} = \begin{bmatrix} \mathbf{H} \,|\, \mathbf{A} \end{bmatrix} \begin{bmatrix} (\mathbf{H}^{\mathsf{T}}\mathbf{H})^{-1} & \mathbf{0} \\ \mathbf{0} & (\mathbf{A}^{\mathsf{T}}\mathbf{A})^{-1} \end{bmatrix} \begin{bmatrix} \mathbf{H}^{\mathsf{T}} \\ \mathbf{A}^{\mathsf{T}} \end{bmatrix}$$

$$= \mathbf{H}(\mathbf{H}^{\mathsf{T}}\mathbf{H})^{-1}\mathbf{H}^{\mathsf{T}} + \mathbf{A}(\mathbf{A}^{\mathsf{T}}\mathbf{A})^{-1}\mathbf{A}^{\mathsf{T}} \qquad\qquad 9.56$$

$$= \mathbf{P}_H + \mathbf{P}_A.$$

The question that now arises is, "Can we find a different representation for $\mathbf{S}$—that is, for $\mathbf{H}$ and for $\mathbf{A}$—that produces orthogonal representations for the Grammian $\mathbf{S}^{\mathsf{T}}\mathbf{S}$ *and* for the projection $\mathbf{I}$?" If we had such a representation, we would have a representation in which each component of **H** and each component of **A** would be orthogonal to every other. We can then study the relative importance of orthogonal modes without other modes interfering.

9.4   SINGULAR VALUE DECOMPOSITION

In our search for a doubly orthogonal representation for $\mathbf{S}^T\mathbf{S}$ and for $\mathbf{I} = \mathbf{S}(\mathbf{S}^T\mathbf{S})^{-1}\mathbf{S}$, we encounter the *singular value decomposition* (SVD) studied in Section 2.6 of Chapter 2. We propose to write **S** as

$$\mathbf{S} = [\mathbf{H}\,|\,\mathbf{A}] = \left[\mathbf{U}\boldsymbol{\Sigma}_H\mathbf{V}_H^T\,|\,\mathbf{U}\boldsymbol{\Sigma}_A\mathbf{V}_A^T\right]$$

$$= \mathbf{U}\begin{bmatrix} \boldsymbol{\Sigma}_H & 0 \\ \hline 0 & \boldsymbol{\Sigma}_A \end{bmatrix}\begin{bmatrix} \mathbf{V}_H^T & 0 \\ \hline 0 & \mathbf{V}_A^T \end{bmatrix} = \begin{bmatrix} \mathbf{U}_H & \mathbf{U}_A \end{bmatrix}\begin{bmatrix} \boldsymbol{\Sigma}_H & 0 \\ \hline 0 & \boldsymbol{\Sigma}_A \end{bmatrix}\begin{bmatrix} \mathbf{V}_H^T & 0 \\ \hline 0 & \mathbf{V}_A^T \end{bmatrix},$$

$$\text{9.57}$$

where $\mathbf{U}_H$, $\mathbf{U}_A$, $\mathbf{V}_H$, and $\mathbf{V}_A$ are orthogonal matrices, and $\boldsymbol{\Sigma}_H$ and $\boldsymbol{\Sigma}_A$ are diagonal matrices:

$$\mathbf{U}_H = \begin{bmatrix} \mathbf{u}_1 & \cdots & \mathbf{u}_p \end{bmatrix} \in R^{N\times p}; \quad \mathbf{U}_A = \begin{bmatrix} \mathbf{u}_{p+1} & \cdots & \mathbf{u}_N \end{bmatrix} \in R^{N\times(N-p)}$$

$$\mathbf{V}_H = \begin{bmatrix} \mathbf{v}_1 & \cdots & \mathbf{v}_p \end{bmatrix} \in R^{p\times p}; \quad \mathbf{V}_A = \begin{bmatrix} \mathbf{w}_{p+1} & \cdots & \mathbf{w}_N \end{bmatrix} \in R^{(N-p)\times(N-p)}$$

$$\boldsymbol{\Sigma}_H = \mathrm{diag}\begin{bmatrix} \sigma_1 & \cdots & \sigma_p \end{bmatrix} \in R^{p\times p}; \quad \boldsymbol{\Sigma}_A = \mathrm{diag}\begin{bmatrix} \sigma_{p+1} & \cdots & \sigma_N \end{bmatrix} \in R^{(N-p)\times(N-p)}.$$

$$\text{9.58}$$

This decomposition produces the following doubly orthogonal representations for $\mathbf{S}^T\mathbf{S}$ and $\mathbf{I}$:

$$\mathbf{S}^T\mathbf{S} = \begin{bmatrix} \mathbf{G}_H & 0 \\ \hline 0 & \mathbf{G}_A \end{bmatrix} = \begin{bmatrix} \mathbf{V}_H\boldsymbol{\Sigma}_H^2\mathbf{V}_H^T & 0 \\ \hline 0 & \mathbf{V}_A\boldsymbol{\Sigma}_A^2\mathbf{V}_A^T \end{bmatrix}. \qquad \text{9.59}$$

The Grammians $\mathbf{G}_H$ and $\mathbf{G}_A$ and the projectors $\mathbf{P}_H$ and $\mathbf{P}_A$ may be written out in terms of the vectors $\mathbf{u}_i$, $\mathbf{v}_i$, and $\mathbf{w}_i$ defined in Equation 9.58:

$$\mathbf{G}_H = \mathbf{V}_H\boldsymbol{\Sigma}_H^2\mathbf{V}_H^T = \sum_{i=1}^{p}\sigma_i^2\mathbf{v}_i\mathbf{v}_i^T \qquad \mathbf{G}_A = \mathbf{V}_A\boldsymbol{\Sigma}_A^2\mathbf{V}_A^T = \sum_{i=p+1}^{N}\sigma_i^2\mathbf{w}_i\mathbf{w}_i^T$$

$$\mathbf{I} = \mathbf{P}_H + \mathbf{P}_A = \mathbf{U}_H\mathbf{U}_H^T + \mathbf{U}_A\mathbf{U}_A^T \qquad \text{9.60}$$

$$\mathbf{P}_H = \mathbf{U}_H\mathbf{U}_H^T = \sum_{i=1}^{p}\mathbf{u}_i\mathbf{u}_i^T \qquad \mathbf{P}_A = \mathbf{U}_A\mathbf{U}_A^T = \sum_{i=p+1}^{N}\mathbf{u}_i\mathbf{u}_i^T.$$

Synthesis Representation

This representation for $\mathbf{S}$ is called the *SVD representation*. (See Section 2.6 in Chapter 2 for more discussion on its properties.) The original matrices $\mathbf{H}$ and $\mathbf{A}$ may be read out of the representation for $\mathbf{S}$ in the following way:

$$\mathbf{H} = \mathbf{S}\begin{bmatrix} \mathbf{I} \\ \mathbf{0} \end{bmatrix} = \mathbf{U}_H \mathbf{\Sigma}_H \mathbf{V}_H^T = \sum_{i=1}^{p} \mathbf{u}_i \sigma_i \mathbf{v}_i^T \qquad 9.61$$

$$\mathbf{A} = \mathbf{S}\begin{bmatrix} \mathbf{0} \\ \mathbf{I} \end{bmatrix} = \mathbf{U}_A \mathbf{\Sigma}_A \mathbf{V}_A^T = \sum_{i=p+1}^{N} \mathbf{u}_i \sigma_i \mathbf{w}_i^T.$$

These are *synthesis representations* for $\mathbf{H}$ and $\mathbf{A}$. The corresponding Grammians and projections are given in Equation 9.60.

Analysis Representations

The analysis representations are obtained by inverting the singular value decompositions of $\mathbf{H}$ and $\mathbf{A}$. Pre and postmultiply $\mathbf{H} = \mathbf{U}_H \mathbf{\Sigma}_H \mathbf{V}_H^T$ and $\mathbf{A} = \mathbf{U}_A \mathbf{\Sigma}_A \mathbf{V}_A^T$ to obtain the results

$$\mathbf{U}_H^T \mathbf{H} \mathbf{V}_H = \mathbf{\Sigma}_H \qquad 9.62$$

$$\mathbf{U}_A^T \mathbf{A} \mathbf{V}_A = \mathbf{\Sigma}_A.$$

This same trick may be applied to the Grammians and projections:

$$\mathbf{V}_H^T \mathbf{G}_H \mathbf{V}_H = \mathbf{\Sigma}_H^2$$

$$\mathbf{V}_A^T \mathbf{G}_A \mathbf{V}_A = \mathbf{\Sigma}_A^2 \qquad 9.63$$

$$\mathbf{U}^T \mathbf{P}_H \mathbf{U} = \left[\begin{array}{c|c} \mathbf{I} & \mathbf{0} \\ \hline \mathbf{0} & \mathbf{0} \end{array}\right]$$

$$\mathbf{U}^T \mathbf{P}_A \mathbf{U} = \left[\begin{array}{c|c} \mathbf{0} & \mathbf{0} \\ \hline \mathbf{0} & \mathbf{I} \end{array}\right].$$

In summary, these singular value decompositions simultaneously diagonalize $\mathbf{S}$ and $\mathbf{I}$. They produce orthogonal decompositions for the Grammians $\mathbf{G}_H$ and $\mathbf{G}_A$ and for the projections $\mathbf{P}_H$ and $\mathbf{P}_A$. We often say that the Grammians $\mathbf{G}_H$ and $\mathbf{G}_A$ are, respectively, orthogonally similar to the diagonal matrices $\mathbf{\Sigma}_H^2$ and $\mathbf{\Sigma}_A^2$ and that the projectors $\mathbf{P}_H$ and $\mathbf{P}_A$ are, respectively, orthogonally similar to the low-rank identities

$$\begin{bmatrix} \mathbf{I} & \mathbf{0} \\ \mathbf{0} & \mathbf{0} \end{bmatrix} \quad \text{and} \quad \begin{bmatrix} \mathbf{0} & \mathbf{0} \\ \mathbf{0} & \mathbf{I} \end{bmatrix}.$$

9.5   SOLVING LEAST SQUARES PROBLEMS

We wish to study solution techniques for the normal equations

$$\mathbf{H}^T\mathbf{H}\boldsymbol{\theta} = \mathbf{H}^T\mathbf{y} \qquad\qquad 9.64$$

$$\det(\mathbf{H}^T\mathbf{H}) \neq 0.$$

We have found in Section 9.1 that the Gram matrix $\mathbf{H}^T\mathbf{H}$ is nonsingular if and only if the columns of $\mathbf{H}$ are linearly independent. In the paragraphs that follow, we will review three approaches to solving least squares problems.

Cholesky-Factoring the Gram Matrix

Write the Gram matrix $\mathbf{H}^T\mathbf{H}$ as the product of lower and upper triangular factors

$$\mathbf{H}^T\mathbf{H} = \mathbf{L}_H\mathbf{L}_H^T$$

$$\mathbf{L}_H = \begin{bmatrix} l_{11} & & & \\ l_{21} & l_{22} & & \mathbf{0} \\ \vdots & & & \\ l_{p1} & l_{p2} & \cdots & l_{pp} \end{bmatrix} = [\mathbf{l}_1\mathbf{l}_2\cdots\mathbf{l}_p]. \qquad 9.65$$

Then the normal equations may be written as

$$\mathbf{L}_H\mathbf{L}_H^T\hat{\boldsymbol{\theta}} = \mathbf{H}^T\mathbf{y}. \qquad\qquad 9.66$$

These equations may be solved in two steps:

$$1. \quad \begin{bmatrix} l_{11} & & & \\ l_{21} & l_{22} & & \mathbf{0} \\ \vdots & \vdots & & \\ l_{p1} & l_{p2} & \cdots & l_{pp} \end{bmatrix} \begin{bmatrix} w_1 \\ w_2 \\ \vdots \\ w_p \end{bmatrix} = \mathbf{H}^T\mathbf{y} \qquad 9.67$$

$$2. \quad \begin{bmatrix} l_{11} & l_{21} & \cdots & l_{p1} \\ & l_{22} & \cdots & l_{p2} \\ & \mathbf{0} & & \vdots \\ & & & l_{pp} \end{bmatrix} \begin{bmatrix} \hat{\theta}_1 \\ \hat{\theta}_2 \\ \vdots \\ \hat{\theta}_p \end{bmatrix} = \begin{bmatrix} w_1 \\ w_2 \\ \vdots \\ w_p \end{bmatrix}.$$

Equation (1) is solved from top to bottom for $\mathbf{w}$, and equation (2) is solved from bottom to top for $\hat{\boldsymbol{\theta}}$. So how do we factor $\mathbf{H}^T\mathbf{H}$ into its Cholesky factors $L_H L_H^T$?

Cholesky's Algorithm   Write the Gram matrix as

$$\mathbf{G} = \begin{bmatrix} \mathbf{g}_1 & \cdots & \mathbf{g}_p \end{bmatrix} \qquad\qquad 9.68$$

where the nth column is

$$\mathbf{g}_n = \begin{bmatrix} g_{1n} \\ g_{2n} \\ \vdots \\ g_{pn} \end{bmatrix}.$$ 9.69

The Cholesky factorization is obtained by writing out $\mathbf{G} = \mathbf{L}_H \mathbf{L}_H^{\mathrm{T}}$:

$$\mathbf{g}_1 = \mathbf{l}_1 l_{11}$$
$$\mathbf{g}_2 = \mathbf{l}_1 l_{21} + \mathbf{l}_2 l_{22}$$
                                                                                         9.70
$$\vdots$$
$$\mathbf{g}_p = \mathbf{l}_1 l_{p1} + \mathbf{l}_2 l_{p2} + \cdots + \mathbf{l}_p l_{pp}.$$

These are solved for the columns $\mathbf{l}_n$ and the diagonal terms l_{nn} as follows:

$$l_{11}^2 = g_{11}$$
$$l_{11}\mathbf{l}_1 = \mathbf{g}_1$$

$$\vdots$$

$$l_{nn}^2 = g_{nn} - \sum_{i=1}^{n-1} l_{ni}^2$$ 9.71

$$l_{nn}\mathbf{l}_n = \mathbf{g}_n - \sum_{i=1}^{n-1} \mathbf{l}_i l_{ni}$$

$$\vdots$$

These recursions constitute the Cholesky factorization of the Gram matrix $\mathbf{H}^{\mathrm{T}}\mathbf{H}$. The next solution technique proceeds directly from the model matrix $\mathbf{H}$.   □

QR-Factoring the Model Matrix

The matrix $\mathbf{H}$ may be written as the product of an orthogonal matrix $\mathbf{U}_H$ and an upper triangular matrix $\mathbf{L}_H^{\mathrm{T}}$:

$$\mathbf{H} = \mathbf{U}_H \mathbf{L}_H^{\mathrm{T}}$$ 9.72

$$\mathbf{U}_H = [\mathbf{u}_1 \mathbf{u}_2 \cdots \mathbf{u}_p]; \qquad \mathbf{L}_H = \begin{bmatrix} l_{11} & & & \\ l_{21} & l_{22} & & \mathbf{0} \\ \vdots & & \ddots & \\ l_{p1} & & \cdots & l_{pp} \end{bmatrix}.$$

This factorization simply represents the nth column of $\mathbf{H}$ by a linear combination of orthogonal columns:

$$\mathbf{h}_n = \sum_{i=1}^{n} \mathbf{u}_i l_{ni}. \qquad 9.73$$

With this factorization, the normal equations $(\mathbf{H}^T\mathbf{H})\boldsymbol{\theta} = \mathbf{H}^T\mathbf{y}$ may be written

$$\mathbf{L}_H\mathbf{L}_H^T\hat{\boldsymbol{\theta}} = \mathbf{L}_H\mathbf{U}_H^T\mathbf{y}, \qquad 9.74$$

with solution

$$\mathbf{L}_H^T\hat{\boldsymbol{\theta}} = \mathbf{U}_H^T\mathbf{y}. \qquad 9.75$$

This equation is simply solved from bottom to top for the vector $\hat{\boldsymbol{\theta}}$. The solution for $\hat{\mathbf{x}}$ is then

$$\hat{\mathbf{x}} = \mathbf{H}\hat{\boldsymbol{\theta}} = \mathbf{U}_H\mathbf{L}_H^T\hat{\boldsymbol{\theta}} = \mathbf{U}_H\mathbf{U}_H^T\mathbf{y}. \qquad 9.76$$

In Section 2.3 of Chapter 2 we discussed Gram-Schmidt, Householder, and Givens transformations for QR-factoring the matrix $\mathbf{H}$.

Singular Value Decomposition

When the matrix $\mathbf{H}$ is written as the singular value decomposition $\mathbf{H} = \mathbf{U}_H\boldsymbol{\Sigma}_H\mathbf{V}_H^T$, then the normal equations $\mathbf{H}^T\mathbf{H}\boldsymbol{\theta} = \mathbf{H}^T\mathbf{y}$ may be written as

$$\mathbf{V}_H\boldsymbol{\Sigma}_H^T\boldsymbol{\Sigma}_H\mathbf{V}_H^T\hat{\boldsymbol{\theta}} = \mathbf{V}_H\boldsymbol{\Sigma}_H^T\mathbf{U}_H^T\mathbf{y}. \qquad 9.77$$

These equations are solved as follows:

$$\hat{\boldsymbol{\theta}} = \mathbf{V}_H\boldsymbol{\Sigma}_H^{-1}\mathbf{U}_H^T\mathbf{y} = \sum_{i=1}^{p} \mathbf{v}_i\sigma_i^{-1}\mathbf{u}_i^T\mathbf{y} \qquad 9.78$$

where $\boldsymbol{\Sigma}_H^{-1}$ is the inverse $\boldsymbol{\Sigma}_H^{-1} = \mathrm{diag}[\sigma_1^{-1} \quad \cdots \quad \sigma_p^{-1}]$. The least squares solution for $\hat{\mathbf{x}}$ is

$$\hat{\mathbf{x}} = \mathbf{U}_H\mathbf{U}_H^T\mathbf{y} = \sum_{i=1}^{p} \mathbf{u}_i\mathbf{u}_i^T\mathbf{y}. \qquad 9.79$$

Can you see that the columns of $\mathbf{V}_H$ span the solution space for $\hat{\boldsymbol{\theta}}$ and that the first p columns of $\mathbf{U}$, namely $\mathbf{U}_H$, span the solution space for $\hat{\mathbf{x}}$? Recall from Chapter 2 that we call $\mathbf{H}^\# = \mathbf{V}_H\boldsymbol{\Sigma}_H^{-1}\mathbf{U}_H^T$ the pseudoinverse of $\mathbf{H}$, because $\mathbf{H}\mathbf{H}^\#\mathbf{H} = \mathbf{H}$ and $\mathbf{H}^\#\mathbf{H}\mathbf{H}^\# = \mathbf{H}^\#$. The SVD representation for $\hat{\boldsymbol{\theta}}$ in Equation 9.78 shows that the measurement $\mathbf{y}$ is correlated with the spanning vectors of $\langle\mathbf{H}\rangle$, namely the $\mathbf{u}_i$, these correlations are then weighted by the inverse eigenvalues σ_i^{-1}, and finally the weighted correlations $\sigma_i^{-1}\mathbf{u}_i^T\mathbf{y}$ are used to build $\hat{\boldsymbol{\theta}}$ from the spanning vectors for the parameter space, namely the $\mathbf{v}_i$.

9.6   PERFORMANCE OF LEAST SQUARES

For our discussion of performance, we return to the linear statistical model

$$\mathbf{y} = \mathbf{H\theta} + \mathbf{n}. \tag{9.80}$$

We further assume that $\mathbf{n}$ is a realization of the normal random vector $\mathbf{n} : N[\mathbf{0}, \sigma^2\mathbf{I}]$. Then $\mathbf{y}$ is a realization of the normal random vector $\mathbf{y} : N[\mathbf{H\theta}, \sigma^2\mathbf{I}]$. The prior model for $\mathbf{y}$ is illustrated in Figure 9.5(a).

Posterior Model

The posterior model for $\mathbf{y}$ is

$$\mathbf{y} = \hat{\mathbf{x}} + \hat{\mathbf{n}}$$

$$\hat{\mathbf{x}} = \mathbf{P}_H\mathbf{y} \tag{9.81}$$

$$\hat{\mathbf{n}} = (\mathbf{I} - \mathbf{P}_H)\mathbf{y} = \mathbf{P}_A\mathbf{y}.$$

The estimator $\hat{\mathbf{x}}$ and the error $\hat{\mathbf{n}}$ are linear transformations on the multivariate normal random vector $\mathbf{y}$. Therefore they are distributed as follows:

$$\hat{\mathbf{x}} : N\left[\mathbf{P}_H\mathbf{H\theta}, \sigma^2\mathbf{P}_H\right] : N\left[\mathbf{H\theta}, \sigma^2\mathbf{P}_H\right] \tag{9.82}$$

$$\hat{\mathbf{n}} : N\left[\mathbf{P}_A\mathbf{H\theta}, \sigma^2\mathbf{P}_A\right] : N\left[\mathbf{0}, \sigma^2\mathbf{P}_A\right].$$

These normal random variables are uncorrelated and therefore independent:

$$E\hat{\mathbf{x}}\hat{\mathbf{n}}^T = E\mathbf{P}_H\mathbf{y}\mathbf{y}^T\mathbf{P}_A$$

$$= \mathbf{P}_H(\mathbf{H\theta\theta}^T\mathbf{H}^T + \sigma^2\mathbf{I})\mathbf{P}_A \tag{9.83}$$

$$= \mathbf{0}.$$

This means that in the posterior model $\mathbf{y}$ is the sum of independent normal random variables $\hat{\mathbf{x}}$ and $\hat{\mathbf{n}}$. The posterior model for $\mathbf{y}$ is illustrated in Figure 9.5(b). Note that the means and covariances of $\hat{\mathbf{x}}$ and $\hat{\mathbf{n}}$ add to produce the appropriate distribution for $\mathbf{y}$.

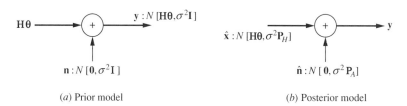

(a) Prior model   (b) Posterior model

Figure 9.5 Prior and posterior models for $\mathbf{y} = \mathbf{x} + \mathbf{n}$.

Performance

The estimator $\hat{\mathbf{x}}$ is distributed as $N[\mathbf{H\theta}, \sigma^2\mathbf{P}_H]$. Therefore it is unbiased with error covariance matrix $\sigma^2\mathbf{P}_H$:

$$E\hat{\mathbf{x}} = \mathbf{H\theta} = \mathbf{x} \qquad\qquad 9.84$$

$$E(\hat{\mathbf{x}} - \mathbf{H\theta})(\hat{\mathbf{x}} - \mathbf{H\theta})^{\mathrm{T}} = \sigma^2\mathbf{P}_H = \sigma^2\mathbf{U}_H\mathbf{U}_H^{\mathrm{T}}.$$

The estimator variance is

$$\xi^2 = E(\hat{\mathbf{x}} - \mathbf{H\theta})^{\mathrm{T}}(\hat{\mathbf{x}} - \mathbf{H\theta}) = \mathrm{tr}\left[E(\hat{\mathbf{x}} - \mathbf{H\theta})(\hat{\mathbf{x}} - \mathbf{H\theta})^{\mathrm{T}}\right]$$

$$= \sigma^2\,\mathrm{tr}\,[\mathbf{P}_H] = \sigma^2\,\mathrm{tr}\left[\mathbf{U}_H\mathbf{U}_H^{\mathrm{T}}\right] = \sigma^2\,\mathrm{tr}\left[U_H^{\mathrm{T}}U_H\right] \qquad 9.85$$

$$= p\sigma^2$$

where $\mathrm{tr}\,[\mathbf{P}_H] = p$ is the dimension of the signal subspace.

The parameter estimator $\hat{\mathbf{\theta}}$ is also normally distributed:

$$\hat{\mathbf{\theta}} = (\mathbf{H}^{\mathrm{T}}\mathbf{H})^{-1}\mathbf{H}^{\mathrm{T}}\mathbf{y} : N[\mathbf{0}, \sigma^2(\mathbf{H}^{\mathrm{T}}\mathbf{H})^{-1}]. \qquad 9.86$$

It is unbiased with covariance matrix $\sigma^2(\mathbf{H}^{\mathrm{T}}\mathbf{H})^{-1}$:

$$E\hat{\mathbf{\theta}} = \mathbf{0} \qquad\qquad 9.87$$

$$E(\hat{\mathbf{\theta}} - \mathbf{\theta})(\hat{\mathbf{\theta}} - \mathbf{\theta})^{\mathrm{T}} = \sigma^2(\mathbf{H}^{\mathrm{T}}\mathbf{H})^{-1} = \sigma^2\left(\mathbf{V}_H\mathbf{\Sigma}_H^2\mathbf{V}_H^{\mathrm{T}}\right)^{-1}$$

$$= \sigma^2\mathbf{V}_H\mathbf{\Sigma}_H^{-2}\mathbf{V}_H^{\mathrm{T}}.$$

Note that the error covariance matrix is the inverse of the Grammian, and this inverse depends on the inverse eigenvalues of the Grammian.

9.7   GOODNESS OF FIT

The error vector $\hat{\mathbf{n}} = \mathbf{y} - \hat{\mathbf{x}}$ is distributed as $\hat{\mathbf{n}} : N[\mathbf{0}, \sigma^2\mathbf{P}_A]$. Therefore the scaled error $\sigma^{-1}\hat{\mathbf{n}}$ is distributed as $\sigma^{-1}\hat{\mathbf{n}} : N[\mathbf{0}, \mathbf{P}_A]$. From our discussion of quadratic forms in normal random vectors in Section 2.11 of Chapter 2, we know that the quadratic form $\sigma^{-2}\hat{\mathbf{n}}^{\mathrm{T}}\hat{\mathbf{n}} = \sigma^{-2}\mathbf{n}^{\mathrm{T}}\mathbf{P}_A\mathbf{n}^{\mathrm{T}}$ is distributed as a chi-squared random variable with $N - p$ degrees of freedom. We shall denote this as

$$\hat{\mathbf{n}}^{\mathrm{T}}\hat{\mathbf{n}} : \sigma^2\chi_{N-p}^2, \qquad\qquad 9.88$$

meaning $\sigma^{-2}\hat{\mathbf{n}}^{\mathrm{T}}\hat{\mathbf{n}} : \chi_{N-p}^2$. So, when fitting the model $\hat{\mathbf{x}} = \mathbf{P}_H\mathbf{y}$ to the data $\mathbf{y}$, the normalized error vector should assume norm-squared values consistent with the $\sigma^2\chi_{N-p}^2$ distribution with $N - p$ degrees of freedom. If the actual value of $\hat{\mathbf{n}}^{\mathrm{T}}\hat{\mathbf{n}}$ is very improbably observed in a $\sigma^2\chi_{N-p}^2$ distribution, then this is a good tip that the model $\mathbf{x} = \mathbf{H\theta}$ is inaccurate. So, by testing errors against the $\sigma^2\chi_{N-p}^2$ distribution, we are testing the goodness of fit for our model.

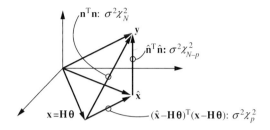

Figure 9.6 Orthogonal decomposition of the noise **n**.

Statistician's Pythagorean Theorem

The same line of argument may be used to show that the norm squared of the noise **n** in the prior model $\mathbf{y} = \mathbf{H\theta} + \mathbf{n}$ is $\chi^2(N)$:

$$\mathbf{n} : N[\mathbf{0}, \sigma^2\mathbf{I}] \tag{9.89}$$

$$\mathbf{n}^T\mathbf{n} : \sigma^2\chi_N^2 .$$

The same argument holds for the estimator error $\hat{\mathbf{x}} - \mathbf{H\theta}$:

$$\hat{\mathbf{x}} - \mathbf{H\theta} : N[\mathbf{0}, \sigma^2\mathbf{P}_H] \tag{9.90}$$

$$(\hat{\mathbf{x}} - \mathbf{H\theta})^T(\hat{\mathbf{x}} - \mathbf{H\theta}) : \sigma^2\chi_p^2 .$$

Combining these two results with the result that the norm squared of $\hat{\mathbf{n}}$ is $\sigma^2\chi_{N-p}^2$ gives us the following Pythagorean relation:

$$\hat{\mathbf{x}} - \mathbf{H\theta} + \hat{\mathbf{n}} = \mathbf{n} \tag{9.91}$$

$$\sigma^2\chi_p^2 + \sigma^2\chi_{N-p}^2 = \sigma^2\chi_N^2 .$$

The latter expression is just a shorthand way of saying that χ_N^2 is the distribution of the sum of independent χ_p^2 and χ_{N-p}^2 random variables. These relations are illustrated in Figure 9.6.

9.8   IMPROVEMENT IN SNR

In the prior linear statistical model, we observe $\mathbf{y} : N[\mathbf{x}, \sigma^2\mathbf{I}]$. Let's agree to call the ratio of the mean-squared of **y** to the variance, the input signal-to-noise ratio $(\text{SNR})_{\text{in}}$:

$$(\text{SNR})_{\text{in}} = \frac{\mathbf{x}^T\mathbf{x}}{\text{tr}\,[\sigma^2 I]} = \frac{\mathbf{x}^T\mathbf{x}}{(N\sigma^2)} . \tag{9.92}$$

(If we interpret $\mathbf{x}^T\mathbf{x}/N$ to be the per sample energy (or power) of **x**, then we can think of $(\text{SNR})_{\text{in}}$ as the signal-to-noise ratio, per sample.) In the posterior model, the estimator

$\hat{\mathbf{x}}$ is distributed as $\hat{\mathbf{x}} : N[\mathbf{x}, \sigma^2 \mathbf{P}_H]$. The signal-to-noise ratio in this model is called the output signal-to-noise ratio (SNR):

$$\text{SNR} = \frac{\mathbf{x}^T\mathbf{x}}{\text{tr}\,[\sigma^2 \mathbf{P}_H]} = \frac{\mathbf{x}^T\mathbf{x}}{p\sigma^2}. \qquad 9.93$$

The gain in signal-to-noise ratio is the ratio of SNR to $(\text{SNR})_{\text{in}}$:

$$G = \frac{\text{SNR}}{(\text{SNR})_{\text{in}}} = \frac{N}{p}. \qquad 9.94$$

This result is fundamental. It is only when N, the dimension of the data vector, substantially exceeds p, the dimension of the parameter vector, that least squares estimation brings a significant improvement in signal-to-noise ratio. For this reason, we value simple models for which p is small and rich measurement plans for which N is large.

9.9   RANK REDUCTION

In the normal error model the error $\hat{\mathbf{n}}$ is distributed as $\hat{\mathbf{n}} : N[\mathbf{0}, \sigma^2 \mathbf{P}_A]$ and the squared error is distributed as $\hat{\mathbf{n}}^T\hat{\mathbf{n}} : \sigma^2 \chi^2_{N-p}$. The mean-squared error is $\sigma^2(N - p)$. The idea behind rank reduction is to replace the full-rank estimator $\hat{\mathbf{x}}$ by a low-rank estimator $\hat{\mathbf{x}}_r$. The resulting error $\hat{\mathbf{n}}_r$ will no longer have zero mean, but its covariance matrix will be smaller. This suggests that there will be a bias/variance trade that can produce smaller mean-squared error than $\sigma^2(N - p)$. Here's the idea.

Replace the least squares estimator $\hat{\mathbf{x}} = \mathbf{P}_H \mathbf{x}$ by the reduced-rank estimator

$$\hat{\mathbf{x}}_r = \mathbf{P}_r \mathbf{y}, \qquad 9.95$$

where $\mathbf{P}_r$ is a reduced-rank version of $\mathbf{P}_H$ that is obtained by discarding $p - r$ of the orthogonal vectors that comprise $\mathbf{U}_H$ in the decomposition $\mathbf{P}_H = \mathbf{U}_H \mathbf{U}_H^T$:

$$\mathbf{P}_H = \mathbf{U}_H \mathbf{U}_H^T; \qquad \mathbf{U}_H = \begin{bmatrix} \mathbf{u}_1 & \mathbf{u}_2 & \cdots & \mathbf{u}_p \end{bmatrix} \qquad 9.96$$

$$\mathbf{P}_r = \mathbf{U}_r \mathbf{U}_r^T; \qquad \mathbf{U}_r = \begin{bmatrix} \mathbf{u}_{(1)} & \mathbf{u}_{(2)} & \cdots & \mathbf{u}_{(r)} \end{bmatrix}.$$

In this characterization of $\mathbf{P}_r$, the vector $\mathbf{u}_{(i)}$ is the ith *ordered* orthogonal vector, which is not necessarily the ith vector. We shall discuss the ordering rule in the next section. The estimator $\hat{\mathbf{x}}_r$ is distributed as $\hat{\mathbf{x}}_r : N[\mathbf{x}_r, \sigma^2 \mathbf{P}_r]$ where $\mathbf{x}_r$ is the projection of $\mathbf{x}$ onto the span of $\mathbf{U}_r$:

$$\mathbf{x}_r = \mathbf{P}_r \mathbf{x}. \qquad 9.97$$

The estimator error $\mathbf{x} - \hat{\mathbf{x}}_r$ is distributed as $\mathbf{x} - \hat{\mathbf{x}}_r : N[\mathbf{b}_r, \sigma^2 \mathbf{P}_r]$ where $\mathbf{b}_r$ is the *bias* of $\mathbf{x}_r$ (the mean of $\mathbf{x} - \mathbf{x}_r$) and $\sigma^2 \mathbf{P}_r$ is the covariance of $\mathbf{x} - \hat{\mathbf{x}}_r$. That is,

$$\mathbf{x} - \hat{\mathbf{x}}_r : N[\mathbf{b}_r, \sigma^2 \mathbf{P}_r] \qquad 9.98$$

$$\mathbf{b}_r = \mathbf{x} - \mathbf{x}_r = (\mathbf{P}_H - \mathbf{P}_r)\mathbf{x}.$$

The mean-squared error between $\mathbf{x}$ and $\hat{\mathbf{x}}_r$ is

$$\text{mse}(r) = E[\mathbf{x} - \hat{\mathbf{x}}_r]^T[\mathbf{x} - \hat{\mathbf{x}}_r] = \mathbf{b}_r^T\mathbf{b}_r + \sigma^2 \text{ tr } [\mathbf{P}_r]$$

$$= \mathbf{x}^T(\mathbf{P}_H - \mathbf{P}_r)\mathbf{x} + r\sigma^2. \qquad 9.99$$

In this formula for the mean-squared error, $\mathbf{x}^T(\mathbf{P}_H - \mathbf{P}_r)\mathbf{x}$ is the bias-squared and $r\sigma^2$ is the variance. Think of bias-squared as energy discarded in the low-rank approximation of $\mathbf{x}$, and think of variance as variability that remains in the approximation.

We may write out the mean-squared error, $\text{mse}(r)$, as

$$\text{mse}(r) = \sum_{i=r+1}^{p} |\mathbf{u}_{(i)}^T\mathbf{x}|^2 + r\sigma^2, \qquad 9.100$$

where the $\mathbf{u}_{(i)}$, $i = r + 1, \ldots, p$, are the eigenvectors of the rank-p span $\langle \mathbf{U}_H \rangle$ that are *discarded* to form the span $\langle \mathbf{U}_r \rangle$. Rank reduction will reduce mean-squared error whenever $\text{mse}(r) \leq p\sigma^2$, where $p\sigma^2$ is the mean-squared error of the full-rank estimator. This condition may be rewritten as

$$\sum_{i=r+1}^{p} |\mathbf{u}_{(i)}^T\mathbf{x}|^2 \leq (p - r)\sigma^2. \qquad 9.101$$

That is, we should reduce rank whenever the energy discarded in the low-rank approximation (the left-hand side) is less than the variance saved (the right-hand side). It is interesting to note that for $r = 0$ the condition for reducing rank is

$$\sum_{i=0}^{p} |\mathbf{u}_{(i)}^T\mathbf{x}|^2 \leq p\sigma^2 \qquad 9.102$$

or

$$\text{SNR} \leq 1. \qquad 9.103$$

This result says that the rank-0 estimator, namely $\mathbf{x}_r = \mathbf{0}$, improves on the full-rank estimator whenever the output SNR $= \mathbf{x}^T\mathbf{x}/p\sigma^2$ is less than 1! Can you interpret this finding?

The optimum choice of rank is

$$r^* = \arg\min_r \left[\mathbf{x}^T(\mathbf{P}_H - \mathbf{P}_r)\mathbf{x} + r\sigma^2\right]. \qquad 9.104$$

In the next section we study practical ways of finding the optimum rank when the signal $\mathbf{x}$ is unknown (as it always is).

9.10   ORDER SELECTION IN THE LINEAR STATISTICAL MODEL

The basic problem in rank reduction is to determine the rank r that minimizes $\text{mse}(r)$. This determination can only be approximate because $\mathbf{x}$, and therefore the bias $\mathbf{b}_r =$

$(\mathbf{P}_H - \mathbf{P}_r)\mathbf{x}$, are unknown. There are two statistics that come to mind as estimators of $\mathbf{b}_r = (\mathbf{P}_H - \mathbf{P}_r)\mathbf{y}$. They are

1. $\hat{\mathbf{n}}_r = (\mathbf{I} - \mathbf{P}_r)\mathbf{y} : N[\mathbf{b}_r, \sigma^2(\mathbf{I} - \mathbf{P}_r)]$ 9.105

2. $\hat{\mathbf{b}}_r = \hat{\mathbf{x}} - \hat{\mathbf{x}}_r = (\mathbf{P}_H - \mathbf{P}_r)\mathbf{y} : N[\mathbf{b}_r, \sigma^2(\mathbf{P}_H - \mathbf{P}_r)]$.

The estimator $\hat{\mathbf{n}}_r$ is the fitting error between $\mathbf{y}$ and the low-rank estimator $\hat{\mathbf{x}}_r$, and the estimator $\hat{\mathbf{b}}_r$ is the difference between the full-rank estimator $\hat{\mathbf{x}}$ and the low-rank estimator $\hat{\mathbf{x}}_r$. These two estimators are illustrated in Figure 9.7. The estimator $\hat{\mathbf{b}}_r$ is clearly superior, as it operates entirely in the subspace $\langle \mathbf{U}_H \rangle$ where its covariance is $\sigma^2(\mathbf{P}_H - \mathbf{P}_r)$. Conversely, the estimator $\hat{\mathbf{n}}_r$ operates in and out of the subspace, where its covariance is $\sigma^2[(\mathbf{I} - \mathbf{P}_H) + (\mathbf{P}_H - \mathbf{P}_r)]$. The component of $\hat{\mathbf{n}}_r$ outside the subspace is useless because it contains no information about $\mathbf{x}$ and it contributes extra covariance $\sigma^2(\mathbf{I} - \mathbf{P})$. In fact, the statistic $\hat{\mathbf{b}}_r$ is just $\hat{\mathbf{n}}_r$ projected onto $\langle \mathbf{U}_H \rangle$:

$$\hat{\mathbf{b}}_r = \mathbf{P}_H \hat{\mathbf{n}}_r = (\mathbf{P}_H - \mathbf{P}_r)\mathbf{y}.   \qquad 9.106$$

The statistic $\hat{\mathbf{b}}_r$ is an *unbiased minimum variance* estimator of the *bias* $\mathbf{b}_r$. The mean-squared error of the estimator is

$$E(\hat{\mathbf{b}}_r - \mathbf{b}_r)^{\mathrm{T}}(\hat{\mathbf{b}}_r - \mathbf{b}_r) = \mathrm{tr}\left[\sigma^2(\mathbf{P}_H - \mathbf{P}_r)\right]$$
$$= \sigma^2(p - r).   \qquad 9.107$$

This result shows that the estimator $\hat{\mathbf{b}}_r$ is a low-variance estimator of $\mathbf{b}_r$ when r is near p. The result also shows that the statistic $\hat{\mathbf{b}}_r^{\mathrm{T}} \hat{\mathbf{b}}_r$ is a *biased* estimate of the bias-squared $\mathbf{b}_r^{\mathrm{T}} \mathbf{b}_r$. That is,

$$E(\hat{\mathbf{b}}_r - \mathbf{b}_r)^{\mathrm{T}}(\hat{\mathbf{b}}_r - \mathbf{b}_r) = E\hat{\mathbf{b}}_r^{\mathrm{T}}\hat{\mathbf{b}}_r - \mathbf{b}_r^{\mathrm{T}}\mathbf{b}_r - \mathbf{b}_r^{\mathrm{T}}\mathbf{b}_r + \mathbf{b}_r^{\mathrm{T}}\mathbf{b}_r$$
$$= E\hat{\mathbf{b}}_r^{\mathrm{T}}\hat{\mathbf{b}}_r - \mathbf{b}_r^{\mathrm{T}}\mathbf{b}_r = \sigma^2(p - r)   \qquad 9.108$$

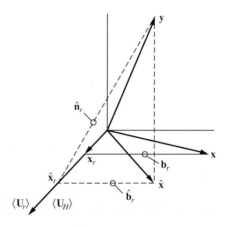

Figure 9.7 Contrasting the two estimators of
bias, $\hat{\mathbf{b}}_r$ and $\hat{\mathbf{n}}_r$.

or

$$E\hat{\mathbf{b}}_r^T\hat{\mathbf{b}}_r \;=\; \mathbf{b}_r^T\mathbf{b}_r + \sigma^2(p-r). \tag{9.109}$$

This is a very significant result because it shows that the estimator $\hat{\mathbf{b}}_r^T\hat{\mathbf{b}}_r$ must be corrected by $-\sigma^2(p-r)$ if it is to be an unbiased estimator of the bias-squared $\mathbf{b}_r^T\mathbf{b}_r$.

For our estimator of the mse(r) we propose the unbiased estimator

$$\begin{aligned}
\hat{\mathrm{mse}}(r) &= \hat{\mathbf{b}}_r^T\hat{\mathbf{b}}_r - \sigma^2(p-r) + r\sigma^2 \\
&= \hat{\mathbf{b}}_r^T\hat{\mathbf{b}}_r + (2r-p)\sigma^2.
\end{aligned} \tag{9.110}$$

From our study of the statistic $\hat{\mathbf{b}}_r^T\hat{\mathbf{b}}_r$ we know that $\hat{\mathrm{mse}}(r)$ is unbiased:

$$E\,\hat{\mathrm{mse}}(r) \;=\; \mathbf{b}_r^T\mathbf{b}_r + r\sigma^2 \;=\; \mathrm{mse}(r). \tag{9.111}$$

The order-selection rule for rank reduction is

$$r^* \;=\; \arg\min_r \hat{\mathrm{mse}}(r), \tag{9.112}$$

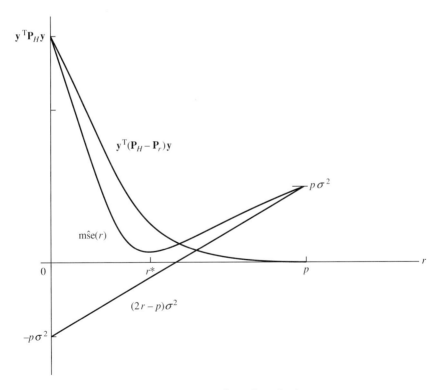

Figure 9.8 Typical plot of $\hat{\mathrm{mse}}(r)$ versus r for order selection.

where the estimated mean-squared error $\hat{\text{mse}}(r)$ is actually computed as follows:

$$\hat{\text{mse}}(r) = \hat{\mathbf{b}}_r^T\hat{\mathbf{b}}_r + (2r - p)\sigma^2$$

$$= \mathbf{y}^T(\mathbf{P}_H - \mathbf{P}_r)\mathbf{y} + (2r - p)\sigma^2 \qquad \text{9.113}$$

$$= \sum_{i=r+1}^{p} |\mathbf{u}_{(i)}^T\mathbf{y}|^2 + (2r - p)\sigma^2.$$

This form for the estimated mean-squared error shows that the eigenvectors of $\mathbf{U}_H$ should be ordered according to the rule

$$|\mathbf{u}_{(1)}^T\mathbf{y}|^2 \geq \cdots \geq |\mathbf{u}_{(r)}^T\mathbf{y}|^2 \geq \cdots \geq |\mathbf{u}_{(p)}^T\mathbf{y}|^2 \qquad \text{9.114}$$

and the dominant $\mathbf{u}_{(i)}$, $i = 1, 2, \ldots, r$, should be used to construct the rank-r projector $\mathbf{P}_r$. Beginning with $r = 0$, the estimated mean-squared error $\hat{\text{mse}}(r)$ is computed and plotted versus r, as in Figure 9.8. The selected rank r^* for the low-rank estimator $\hat{\mathbf{x}}_r$ is the value of r that minimizes $\hat{\text{mse}}(r)$.

9.11   SEQUENTIAL LEAST SQUARES

The idea behind sequential least squares is to sequentially, or recursively, update the least squares estimate as new measurements are acquired. Let's begin with the linear statistical model

$$\mathbf{y} = \mathbf{H}\boldsymbol{\theta} + \mathbf{n}. \qquad \text{9.115}$$

Our purpose now is to fix the dimension of $\boldsymbol{\theta}$ at $p \times 1$ but to allow $\mathbf{y}$ to increase in dimension. In this way, we allow measurements to be taken sequentially (in time or space). To lend some physical reality to the problem, let's write the measurement vector $\mathbf{y}_t$ as follows:

$$\begin{bmatrix} y_1 \\ y_2 \\ \vdots \\ y_{t-1} \\ \hline y_t \end{bmatrix} = \begin{bmatrix} \mathbf{H}_{t-1} \\ \hline \mathbf{c}_t^T \end{bmatrix} \begin{bmatrix} \theta_1 \\ \vdots \\ \theta_p \end{bmatrix} + \begin{bmatrix} n_1 \\ n_2 \\ \vdots \\ n_{t-1} \\ n_t \end{bmatrix} \qquad \text{9.116}$$

$$\mathbf{y}_t = \mathbf{H}_t\boldsymbol{\theta} + \mathbf{n}_t.$$

We gain considerable insight by writing these equations in a form that shows how past measurements $\mathbf{y}_{t-1}$ and the current measurement y_t depend on $\boldsymbol{\theta}$:

$$\mathbf{y}_t = \begin{bmatrix} \mathbf{y}_{t-1} \\ \hline y_t \end{bmatrix}$$

$$\mathbf{y}_{t-1} = \mathbf{H}_{t-1}\boldsymbol{\theta} + \mathbf{n}_{t-1} \qquad \text{9.117}$$

$$y_t = \mathbf{c}_t^T\boldsymbol{\theta} + n_t.$$

So really the linear statistical model just summarizes the scalar equations

$$y_t = \mathbf{c}_t^T \boldsymbol{\theta} + n_t; \qquad t = 1, 2, \ldots \qquad\qquad 9.118$$

where the time varying vector $\mathbf{c}_t^T$ is the tth row of $\mathbf{H}_t$. The least squares solution for $\boldsymbol{\theta}$, given the observation vector $\mathbf{y}_t$, is denoted $\hat{\boldsymbol{\theta}}_t$. It satisfies the regression equation

$$\mathbf{P}_t^{-1}\hat{\boldsymbol{\theta}}_t = \mathbf{H}_t^T \mathbf{y}_t \qquad\qquad 9.119$$

where $\mathbf{P}_t^{-1}$ and $\mathbf{H}_t^T\mathbf{y}_t$ may be written as follows:

$$\begin{aligned}
\mathbf{P}_t^{-1} &= \mathbf{H}_t^T\mathbf{H}_t = \left(\mathbf{H}_{t-1}^T\mathbf{H}_{t-1} + \mathbf{c}_t\mathbf{c}_t^T\right) \\
&= \mathbf{P}_{t-1}^{-1} + \mathbf{c}_t\mathbf{c}_t^T
\end{aligned} \qquad\qquad 9.120$$

$$\mathbf{H}_t^T\mathbf{y}_t = \mathbf{H}_{t-1}^T\mathbf{y}_{t-1} + \mathbf{c}_t y_t.$$

Note that $\mathbf{P}_t^{-1}$ is just the Grammian of $\mathbf{H}_t$.

The least squares solution may now be written

$$\hat{\boldsymbol{\theta}}_t = \left(\mathbf{P}_{t-1}^{-1} + \mathbf{c}_t\mathbf{c}_t^T\right)^{-1}\mathbf{H}_{t-1}^T\mathbf{y}_{t-1} + \left(\mathbf{P}_{t-1}^{-1} + \mathbf{c}_t\mathbf{c}_t^T\right)^{-1}\mathbf{c}_t y_t. \qquad 9.121$$

Use the matrix-inversion lemma of Section 2.9 in Chapter 2 to write the inverse of $\mathbf{P}_{t-1}^{-1} + \mathbf{c}_t\mathbf{c}_t^T$ as

$$\begin{aligned}
\left(\mathbf{P}_{t-1}^{-1} + \mathbf{c}_t\mathbf{c}_t^T\right)^{-1} &= \mathbf{P}_{t-1} - \gamma_t\mathbf{P}_{t-1}\mathbf{c}_t\mathbf{c}_t^T\mathbf{P}_{t-1} \\
\gamma_t^{-1} &= 1 + \mathbf{c}_t^T\mathbf{P}_{t-1}\mathbf{c}_t \qquad\qquad 9.122 \\
\left(\gamma_t\mathbf{c}_t^T\mathbf{P}_{t-1}\mathbf{c}_t \right. &= \left. 1 - \gamma_t\right).
\end{aligned}$$

Substitute this recursion into the least squares solution for $\hat{\boldsymbol{\theta}}_t$ to obtain

$$\begin{aligned}
\hat{\boldsymbol{\theta}}_t &= \left(\mathbf{P}_{t-1} - \gamma_t\mathbf{P}_{t-1}\mathbf{c}_t\mathbf{c}_t^T\mathbf{P}_{t-1}\right)\mathbf{H}_{t-1}^T\mathbf{y}_{t-1} + (\mathbf{P}_{t-1} - \gamma_t\mathbf{P}_{t-1}\mathbf{c}_t\mathbf{c}_t^T\mathbf{P}_{t-1})\mathbf{c}_t y_t \\
&= \mathbf{P}_{t-1}\mathbf{H}_{t-1}^T\mathbf{y}_{t-1} - \gamma_t\mathbf{P}_{t-1}\mathbf{c}_t\mathbf{c}_t^T\mathbf{P}_{t-1}\mathbf{H}_{t-1}^T\mathbf{y}_{t-1} + \mathbf{P}_{t-1}\mathbf{c}_t y_t - \mathbf{P}_{t-1}\mathbf{c}_t(1 - \gamma_t)y_t \\
&= \hat{\boldsymbol{\theta}}_{t-1} + \gamma_t\mathbf{P}_{t-1}\mathbf{c}_t\left[y_t - \mathbf{c}_t^T\hat{\boldsymbol{\theta}}_{t-1}\right]. \qquad\qquad 9.123
\end{aligned}$$

We may define the gain vector $\mathbf{k}_t = \gamma_t\mathbf{P}_{t-1}\mathbf{c}_t$ and summarize the recursive equations of sequential least squares for estimating $\boldsymbol{\theta}$:

$$\begin{aligned}
\hat{\boldsymbol{\theta}}_t &= \hat{\boldsymbol{\theta}}_{t-1} + \mathbf{k}_t\left[y_t - \mathbf{c}_t^T\hat{\boldsymbol{\theta}}_{t-1}\right] \\
\mathbf{P}_{t-1}^{-1}\mathbf{k}_t &= \gamma_t\mathbf{c}_t \\
\mathbf{P}_t^{-1} &= \mathbf{P}_{t-1}^{-1} + \mathbf{c}_t\mathbf{c}_t^T \qquad\qquad 9.124 \\
\mathbf{P}_t &= \mathbf{P}_{t-1} - \gamma_t\mathbf{P}_{t-1}\mathbf{c}_t\mathbf{c}_t^T\mathbf{P}_{t-1} \\
\gamma_t^{-1} &= 1 + \mathbf{c}_t^T\mathbf{P}_{t-1}\mathbf{c}_t.
\end{aligned}$$

A block diagram for these recursions is illustrated in Figure 9.9.

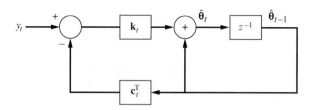

Figure 9.9 Block diagram for sequential least squares.

Recall that, if $\mathbf{n}_t : N[\mathbf{0}, \mathbf{I}]$, then $\hat{\boldsymbol{\theta}}_t : N[\boldsymbol{\theta}, \mathbf{P}_t]$. So $\mathbf{P}_t$ is just the covariance matrix for the estimator $\hat{\boldsymbol{\theta}}_t$! Not only do we have a recursion for our estimator, but we also have a recursion for the estimator covariance.

9.12   WEIGHTED LEAST SQUARES

Weighted least squares is a generalization of least squares. The problem is to

$$\min_{\boldsymbol{\theta}}(\mathbf{y} - \mathbf{H}\boldsymbol{\theta})^{\mathrm{T}}\mathbf{W}(\mathbf{y} - \mathbf{H}\boldsymbol{\theta}) \qquad 9.125$$

when $\mathbf{W}$ is a nonsingular symmetric matrix. There is no loss in generality by assuming $\mathbf{W}$ is symmetric. Can you see why? The weighted least squares solution for $\boldsymbol{\theta}$ satisfies the regression equation

$$\mathbf{H}^{\mathrm{T}}\mathbf{W}(\mathbf{y} - \mathbf{H}\boldsymbol{\theta}) = \mathbf{0} \qquad 9.126$$

with solution

$$\hat{\boldsymbol{\theta}}_{\mathrm{WLS}} = (\mathbf{H}^{\mathrm{T}}\mathbf{W}\mathbf{H})^{-1}\mathbf{H}^{\mathrm{T}}\mathbf{W}\mathbf{y}. \qquad 9.127$$

The matrix $\mathbf{H}^{\mathrm{T}}\mathbf{W}\mathbf{H}$ is invertible provided $\mathbf{H}$ has full-rank p and $\mathbf{W}$ has full-rank N.

If the observations $\mathbf{y}$ are drawn from the normal distribution $\mathbf{y} : N[\mathbf{H}\boldsymbol{\theta}, \mathbf{R}]$, then the weighted least squares solution is distributed as follows:

$$\hat{\boldsymbol{\theta}}_{\mathrm{WLS}} : N\left[\boldsymbol{\theta}, (\mathbf{H}^{\mathrm{T}}\mathbf{W}\mathbf{H})^{-1}\mathbf{H}^{\mathrm{T}}\mathbf{W}\mathbf{R}\mathbf{W}^{\mathrm{T}}\mathbf{H}(\mathbf{H}^{\mathrm{T}}\mathbf{W}\mathbf{H})^{-1}\right]. \qquad 9.128$$

If the symmetric weight matrix $\mathbf{W}$ is chosen to be the inverse of the covariance matrix $\mathbf{R}$, $\mathbf{W} = \mathbf{R}^{-1}$, then the weighted least squares solution and its distribution are

$$\hat{\boldsymbol{\theta}}_{\mathrm{WLS}} = (\mathbf{H}^{\mathrm{T}}\mathbf{R}^{-1}\mathbf{H})^{-1}\mathbf{H}^{\mathrm{T}}\mathbf{R}^{-1}\mathbf{y} \qquad 9.129$$

$$\hat{\boldsymbol{\theta}}_{\mathrm{WLS}} : N[\boldsymbol{\theta}, (\mathbf{H}^{\mathrm{T}}\mathbf{R}^{-1}\mathbf{H})^{-1}].$$

This weighted least squares solution is identical to the maximum likelihood solution in the observation model $\mathbf{y} : N[\mathbf{H}\boldsymbol{\theta}, \mathbf{R}]$.

9.13   CONSTRAINED LEAST SQUARES

What if we do not have complete freedom in our choice of $\boldsymbol{\theta}$ to minimize the quadratic form $(\mathbf{y} - \mathbf{H}\boldsymbol{\theta})^T(\mathbf{y} - \mathbf{H}\boldsymbol{\theta})$? One way to phrase our lack of freedom is to say that every candidate $\boldsymbol{\theta}$ must satisfy the linear constraints

$$\mathbf{C}^T\boldsymbol{\theta} = \mathbf{c} \qquad\qquad 9.130$$

where $\mathbf{C}^T$ is a known $r \times p$ matrix $(r \leq p)$ and $\mathbf{c}$ is a known $r \times 1$ vector. In this way, we are imposing r linear constraints on the elements of $\boldsymbol{\theta}$. The problem now is to

$$\min_{\boldsymbol{\theta}}(\mathbf{y} - \mathbf{H}\boldsymbol{\theta})^T(\mathbf{y} - \mathbf{H}\boldsymbol{\theta}) \qquad\qquad 9.131$$

under the constraint $\mathbf{C}^T\boldsymbol{\theta} = \mathbf{c}$. Form the following Lagrangian J:

$$J = \frac{1}{2}(\mathbf{y} - \mathbf{H}\boldsymbol{\theta})^T(\mathbf{y} - \mathbf{H}\boldsymbol{\theta}) - (\mathbf{C}^T\boldsymbol{\theta} - \mathbf{c})^T\boldsymbol{\lambda}. \qquad\qquad 9.132$$

Equate the gradient of J with respect to $\boldsymbol{\theta}$ to zero in order to find the solution for $\boldsymbol{\theta}$:

$$\frac{\partial}{\partial\boldsymbol{\theta}} J = \mathbf{H}^T\mathbf{H}\boldsymbol{\theta} - \mathbf{H}^T\mathbf{y} - \mathbf{C}\boldsymbol{\lambda} = \mathbf{0}. \qquad\qquad 9.133$$

$$\hat{\boldsymbol{\theta}}_{CLS} = (\mathbf{H}^T\mathbf{H})^{-1}(\mathbf{H}^T\mathbf{y} + \mathbf{C}\boldsymbol{\lambda})$$

$$= \hat{\boldsymbol{\theta}}_{LS} + (\mathbf{H}^T\mathbf{H})^{-1}\mathbf{C}\boldsymbol{\lambda}.$$

The conclusion is that the constrained solution $\hat{\boldsymbol{\theta}}_{CLS}$ is a modification of the unconstrained least squares solution $\hat{\boldsymbol{\theta}}_{LS} = (\mathbf{H}^T\mathbf{H})^{-1}\mathbf{H}^T\mathbf{y}$. The trick is to solve for the value of $\boldsymbol{\lambda}$ that makes $\hat{\boldsymbol{\theta}}_{CLS}$ satisfy the constraints. Invoke the constraint and solve for $\boldsymbol{\lambda}$:

$$\mathbf{C}^T\hat{\boldsymbol{\theta}}_{CLS} = \mathbf{C}^T\hat{\boldsymbol{\theta}}_{LS} + \mathbf{C}^T(\mathbf{H}^T\mathbf{H})^{-1}\mathbf{C}\boldsymbol{\lambda} = \mathbf{c} \qquad\qquad 9.134$$

$$\boldsymbol{\lambda} = \left[\mathbf{C}^T(\mathbf{H}^T\mathbf{H})^{-1}\mathbf{C}\right]^{-1}(\mathbf{c} - \mathbf{C}^T\hat{\boldsymbol{\theta}}_{LS}).$$

The corresponding solution for the constrained least squares solution is

$$\hat{\boldsymbol{\theta}}_{CLS} = \hat{\boldsymbol{\theta}}_{LS} + (\mathbf{H}^T\mathbf{H})^{-1}\mathbf{C}\left[\mathbf{C}^T(\mathbf{H}^T\mathbf{H})^{-1}\mathbf{C}\right]^{-1}(\mathbf{c} - \mathbf{C}^T\hat{\boldsymbol{\theta}}_{LS}). \qquad 9.135$$

When $\mathbf{c} = \mathbf{0}$, the constraints are said to be homogeneous. Then the solution $\hat{\boldsymbol{\theta}}_{CLS}$ is just a projection of $\hat{\boldsymbol{\theta}}_{LS}$:

$$\hat{\boldsymbol{\theta}}_{CLS} = \mathbf{P}\hat{\boldsymbol{\theta}}_{LS}$$

$$\mathbf{P} = \mathbf{I} - (\mathbf{H}^T\mathbf{H})^{-1}\mathbf{C}\left[\mathbf{C}^T(\mathbf{H}^T\mathbf{H})^{-1}\mathbf{C}\right]^{-1}\mathbf{C}^T. \qquad\qquad 9.136$$

This solution is actually general enough for every constrained problem because non-homogenous constraints can always be written as homogenous constraints by writing $[\mathbf{C}^T| - \mathbf{c}][\boldsymbol{\theta}^T \quad 1]^T = \mathbf{0}$.

Interpretations

It is natural to ask whether there is a projection at work in the measurement space. To answer this question, we assume that the constraints are homogeneous and consider the estimate of $\hat{\mathbf{x}} = \mathbf{H}\hat{\boldsymbol{\theta}}_{\text{CLS}}$ obtained from the constrained least squares solution for $\boldsymbol{\theta}$:

$$\hat{\mathbf{x}}_{\text{CLS}} = \mathbf{H}\hat{\boldsymbol{\theta}}_{\text{CLS}} = \mathbf{H}\mathbf{P}\hat{\boldsymbol{\theta}}_{\text{LS}}. \qquad 9.137$$

This estimate may be written as

$$\hat{\mathbf{x}}_{\text{CLS}} = (\mathbf{P}_H - \mathbf{P}_C)\mathbf{y}$$

$$\mathbf{P}_H = \mathbf{H}(\mathbf{H}^T\mathbf{H})^{-1}\mathbf{H}^T \qquad 9.138$$

$$\mathbf{P}_C = \mathbf{H}(\mathbf{H}^T\mathbf{H})^{-1}\mathbf{C}[\mathbf{C}^T(\mathbf{H}^T\mathbf{H})^{-1}\mathbf{C}]^{-1}\mathbf{C}^T(\mathbf{H}^T\mathbf{H})^{-1}\mathbf{H}^T.$$

The projector $\mathbf{P}_H$ is the usual rank-p projector constructed from $\mathbf{H}$. The projector $\mathbf{P}_C$ is a rank-r projector constructed from $\mathbf{H}(\mathbf{H}^T\mathbf{H})^{-1}\mathbf{C}$. Therefore,

$$\mathbf{P}_H\mathbf{H} = \mathbf{H} \qquad 9.139$$

$$\mathbf{P}_C\mathbf{H}(\mathbf{H}^T\mathbf{H})^{-1}\mathbf{C} = \mathbf{H}(\mathbf{H}^T\mathbf{H})^{-1}\mathbf{C}.$$

The projector $\mathbf{P}_C$ is, in fact, a subprojector of $\mathbf{P}_H$:

$$\mathbf{P}_H\mathbf{P}_C\mathbf{P}_H = \mathbf{P}_H\mathbf{P}_C = \mathbf{P}_C\mathbf{P}_H = \mathbf{P}_C. \qquad 9.140$$

This means that the constrained least squares solution may be written in a number of equivalent ways:

$$\hat{\mathbf{x}}_{\text{CLS}} = \mathbf{P}_H(\mathbf{I} - \mathbf{P}_C)\mathbf{P}_H\mathbf{y}$$

$$= \mathbf{P}_H(\mathbf{y} - \mathbf{P}_C\hat{\mathbf{x}}_{\text{LS}}) \qquad 9.141$$

$$= (\mathbf{I} - \mathbf{P}_C)\hat{\mathbf{x}}_{\text{LS}}.$$

These forms of the estimator are illustrated in Figure 9.10. You can see that constraining is another way of projecting down into a corner of the subspace spanned by $\mathbf{H}$.

Condition Adjustment

Condition adjustment is a simple variation on constrained least squares. The matrix $\mathbf{H}$ is identity, and the problem is to

$$\min_{\boldsymbol{\theta}}(\mathbf{y} - \boldsymbol{\theta})^T(\mathbf{y} - \boldsymbol{\theta}) \qquad 9.142$$

under the constraint that $\boldsymbol{\theta}$ satisfy the linear equations

$$\mathbf{C}^T\boldsymbol{\theta} = \mathbf{c}. \qquad 9.143$$

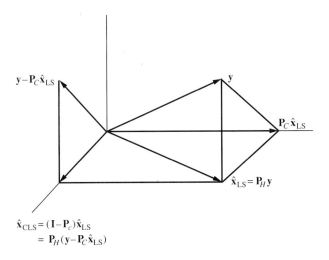

$$\hat{\mathbf{x}}_{CLS} = (\mathbf{I} - \mathbf{P}_c)\hat{\mathbf{x}}_{LS}$$
$$= \mathbf{P}_H(\mathbf{y} - \mathbf{P}_C\hat{\mathbf{x}}_{LS})$$

Figure 9.10 Interpretations of constrained least squares estimates.

The solution is

$$\hat{\boldsymbol{\theta}}_{CLS} = \left[\mathbf{I} - \mathbf{C}(\mathbf{C}^T\mathbf{C})^{-1}\mathbf{C}^T\right]\mathbf{y} + \mathbf{C}(\mathbf{C}^T\mathbf{C})^{-1}\mathbf{c}$$

$$= \hat{\boldsymbol{\theta}}_{LS} + \mathbf{C}(\mathbf{C}^T\mathbf{C})^{-1}\!\left(\mathbf{c} - \mathbf{C}^T\hat{\boldsymbol{\theta}}_{LS}\right) \qquad\qquad 9.144$$

$$\hat{\boldsymbol{\theta}}_{LS} = \mathbf{y}.$$

The solution for condition adjustment is just the unconstrained least squares solution $\hat{\boldsymbol{\theta}}_{LS} = \mathbf{y}$, corrected to satisfy the constraints. Why is this called *condition adjustment*? Because we are adjusting the measurements $\mathbf{y}$ to the adjusted measurements $\hat{\boldsymbol{\theta}}_{CLS}$ that satisfy the conditions $\mathbf{C}^T\hat{\boldsymbol{\theta}}_{CLS} = \mathbf{c}$. The fitting error $\mathbf{y} - \hat{\boldsymbol{\theta}}_{CLS}$ is

$$\hat{\mathbf{n}} = \mathbf{y} - \hat{\boldsymbol{\theta}}_{CLS} = -\mathbf{C}(\mathbf{C}^T\mathbf{C})^{-1}(\mathbf{c} - \mathbf{C}^T\mathbf{y}). \qquad\qquad 9.145$$

This is as close to the data as we can get (in squared error) under the constraint that $\mathbf{C}^T\boldsymbol{\theta} = \mathbf{c}$. Condition adjustment is regularly applied to digitized maps in order to smooth them.

9.14   QUADRATIC MINIMIZATION WITH LINEAR CONSTRAINTS

There is a variation on constrained least squares that arises over and over again in array processing and time-series analysis. This variation may be obtained, formally, from the general problem by setting $\mathbf{y} = \mathbf{0}$, in which case the problem $\min_{\boldsymbol{\theta}}(\mathbf{y} - \mathbf{H}\boldsymbol{\theta})^T(\mathbf{y} - \mathbf{H}\boldsymbol{\theta})$ under the constraint $\mathbf{C}^T\boldsymbol{\theta} = \mathbf{c}$ becomes

$$\min_{\boldsymbol{\theta}} \boldsymbol{\theta}^T\mathbf{H}^T\mathbf{H}\boldsymbol{\theta} \quad \text{subject to } \mathbf{C}^T\boldsymbol{\theta} = \mathbf{c}. \qquad\qquad 9.146$$

In order to bring our treatment of this problem in line with the engineering literature, we shall replace $\boldsymbol{\theta}$ with $\mathbf{w}$ and $\mathbf{H}^T\mathbf{H}$ with $\mathbf{R}$. We shall further assume that $\mathbf{R}$ is nonnegative definite. The problem we shall study is therefore

$$\min_{\mathbf{w}} \mathbf{w}^T\mathbf{R}\mathbf{w} \quad \text{subject to } \mathbf{C}^T\mathbf{w} = \mathbf{c}. \qquad 9.147$$

The solution is found by constructing the Lagrangian $\mathbf{w}^T\mathbf{R}\mathbf{w}-(\mathbf{C}^T\mathbf{w}-\mathbf{c})^T\boldsymbol{\lambda}$ and solving for $\mathbf{w}$ to minimize the Lagrangian and $\boldsymbol{\lambda}$ to satisfy the constraints. The solution is

$$\mathbf{w}_0 = \mathbf{R}^{-1}\mathbf{C}(\mathbf{C}^T\mathbf{R}^{-1}\mathbf{C})^{-1}\mathbf{c}, \qquad 9.148$$

and the resulting minimum value of $\mathbf{w}^T\mathbf{R}\mathbf{w}$ is

$$\mathbf{w}_0^T\mathbf{R}\mathbf{w}_0 = \mathbf{c}^T(\mathbf{C}^T\mathbf{R}^{-1}\mathbf{C})^{-1}\mathbf{c}. \qquad 9.149$$

It is easy to check that this solution obeys the constraints:

$$\mathbf{C}^T\mathbf{w}_0 = \mathbf{c}. \qquad 9.150$$

In order to interpret the nature of this solution for $\mathbf{w}_0$, we note that it may be decomposed into two orthogonal components, one of which lies in the subspace spanned by the columns of $\mathbf{C}$ and the other of which lies in the orthogonal subspace:

$$\mathbf{w}_0 = \mathbf{P}_C\mathbf{w}_0 + (\mathbf{I} - \mathbf{P}_C)\mathbf{w}_0; \qquad \mathbf{P}_C = \mathbf{C}(\mathbf{C}^T\mathbf{C})^{-1}\mathbf{C}^T. \qquad 9.151$$

But $\mathbf{P}_C\mathbf{w}_0 = \mathbf{C}(\mathbf{C}^T\mathbf{C})^{-1}\mathbf{c} = \mathbf{w}_C$, where $\mathbf{w}_C$ is the minimum-norm solution for $\mathbf{w}$ that will satisfy the constraints $\mathbf{C}^T\mathbf{w} = \mathbf{c}$. It follows that $\mathbf{w}_0$ is

$$\mathbf{w}_0 = \mathbf{w}_C + (\mathbf{I} - \mathbf{P}_C)\mathbf{w}_0 \qquad 9.152$$

where $\mathbf{w}_C$ is the minimum-norm solution

$$\mathbf{w}_C = \mathbf{P}_C\mathbf{w}_0 = \mathbf{C}(\mathbf{C}^T\mathbf{C})^{-1}\mathbf{c}. \qquad 9.153$$

This finding, illustrated in Figure 9.11, means that we could have set up the equivalent,

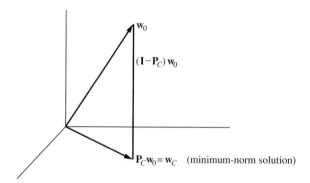

Figure 9.11 Decomposing the linearly constrained solution
$\mathbf{w}_0$ into the minimum-norm solution $\mathbf{w}_C$ plus
the orthogonal vector $\mathbf{w}_0 - \mathbf{w}_C$.

unconstrained, minimization problem

$$\min_{\mathbf{w}}\left[\mathbf{w}_C + (\mathbf{I} - \mathbf{P}_C)\mathbf{w}\right]^{\mathrm{T}}\mathbf{R}\left[\mathbf{w}_C + (\mathbf{I} - \mathbf{P}_C)\mathbf{w}\right] \qquad 9.154$$

and solved for **w**. In order to further illustrate the interpretation of this finding we consider the problem, illustrated in Figure 9.12(a), of minimizing the mean-square value of the scalar $\mathbf{w}^{\mathrm{T}}\mathbf{x}$ when the random vector **x** has correlation matrix **R** and the "weight vector" **w** is constrained to satisfy the "multiple linear constraints" $\mathbf{C}^{\mathrm{T}}\mathbf{w} = \mathbf{c}$ or $\mathbf{w}^{\mathrm{T}}\mathbf{C} = \mathbf{c}^{\mathrm{T}}$. The resulting vector **w** is called a "minimum mean-squared error, linearly constrained beamformer" in the array-processing literature. We know that **w** may be written as

$$\mathbf{w} = \mathbf{w}_C + (\mathbf{I} - \mathbf{P}_C)\mathbf{w}, \qquad 9.155$$

meaning that the diagram of Figure 9.12(b) is equivalent to 9.12(a). Diagram 9.12(b) is called a "generalized sidelobe canceler (GSC)." The mean-squared value of $\mathbf{w}^{\mathrm{T}}\mathbf{x}$ may now be written as

$$\begin{aligned}
e^2 &= E\mathbf{w}^{\mathrm{T}}\mathbf{x}\mathbf{x}^{\mathrm{T}}\mathbf{w} = \mathbf{w}^{\mathrm{T}}\mathbf{R}\mathbf{w} \\
&= E\left[\mathbf{w}_C + (\mathbf{I} - \mathbf{P}_C)\mathbf{w}\right]^{\mathrm{T}}\mathbf{R}\left[\mathbf{w}_C + (\mathbf{I} - \mathbf{P}_C)\mathbf{w}\right] \qquad 9.156 \\
&= \mathbf{w}_C^{\mathrm{T}}\mathbf{R}\mathbf{w}_C + 2\mathbf{w}^{\mathrm{T}}(\mathbf{I} - \mathbf{P}_C)\mathbf{R}\mathbf{w}_C + \mathbf{w}^{\mathrm{T}}(\mathbf{I} - \mathbf{P}_C)\mathbf{R}(\mathbf{I} - \mathbf{P}_C)\mathbf{w}.
\end{aligned}$$

This is precisely the mean-squared error we would get if we tried to estimate the scalar $\mathbf{w}_C^{\mathrm{T}}\mathbf{P}_C\mathbf{x}$ using the scalar $-\mathbf{w}^{\mathrm{T}}(\mathbf{I} - \mathbf{P}_C)\mathbf{x}$. So, as illustrated in Figure 9.12(c),

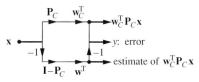

(a) linearly constrained beamformer

(b) equivalent generalized sidelobe canceler (GSC)

(c) equivalent beamformer

Figure 9.12 Interpretations of the minimum mean-squared error, linearly constrained beamformer.

the minimum mean-squared error, linearly constrained beamformer is equivalent to a "beamformer" that estimates the output of the "upper beam" $\mathbf{w}_C^T \mathbf{P}_C \mathbf{x}$ with the "lower beam" $-\mathbf{w}^T(\mathbf{I} - \mathbf{P}_C)\mathbf{x}$. The practical importance of this result is that $\mathbf{w}$ may be adaptively computed without constraints, using only data in the lower branch of Figure 9.12(c).

9.15   TOTAL LEAST SQUARES

In the theory of least squares (LS) the prior model $\mathbf{y} = \mathbf{x} + \mathbf{n}$, with $\mathbf{x} = \mathbf{H\theta}$, is replaced by the posterior model

$$\begin{aligned} \mathbf{y} &= \mathbf{P}_H \mathbf{y} + (\mathbf{I} - \mathbf{P}_H)\mathbf{y} \\ &= \hat{\mathbf{x}} + \hat{\mathbf{n}}. \end{aligned} \qquad 9.157$$

The projector $\mathbf{P}_H = \mathbf{H}(\mathbf{H}^T\mathbf{H})^{-1}\mathbf{H}^T$ is chosen because it minimizes the sum of the squares of the elements of $\hat{\mathbf{n}} = (\mathbf{I}-\mathbf{P}_H)\mathbf{y}$ (thus the term *least squares*). In the posterior model, $\hat{\mathbf{x}}$ is the estimated signal component and $\hat{\mathbf{n}}$ is the estimated noise component:

$$\hat{\mathbf{x}} = \mathbf{P}_H \mathbf{y} = \mathbf{x} + \mathbf{P}_H \mathbf{n} \qquad 9.158$$

$$\hat{\mathbf{n}} = (\mathbf{I} - \mathbf{P}_H)\mathbf{y} = (\mathbf{I} - \mathbf{P}_H)\mathbf{n}.$$

The rightmost equalities follow from the important fact that the projector $\mathbf{P}_H$ is perfectly matched to the subspace $\langle \mathbf{H} \rangle$, meaning that $\mathbf{P}_H \mathbf{H} = \mathbf{H}$ (or $\mathbf{P}_H \mathbf{x} = \mathbf{x}$). From $\mathbf{x} = \mathbf{H\theta}$, we deduce that the LS estimate of $\mathbf{\theta}$ is

$$\hat{\mathbf{\theta}} = (\mathbf{H}^T\mathbf{H})^{-1}\mathbf{H}^T\hat{\mathbf{x}} = (\mathbf{H}^T\mathbf{H})^{-1}\mathbf{H}^T\mathbf{y}. \qquad 9.159$$

The least squares solution is rather restrictive because the projector $\mathbf{P}_H$ forces $\hat{\mathbf{x}}$ to lie exactly in the subspace $\langle \mathbf{H} \rangle$. This implies that we have a lot of confidence in the signal model $\mathbf{x} = \mathbf{H\theta}$. But suppose there are errors in the model matrix $\mathbf{H}$. We may be led to this conclusion, for example, by the failure of a goodness-of-fit test on the modeling error $\hat{\mathbf{n}}$. Is there a way to move away from the model $\mathbf{H}$ and, if so, is there a theory for guiding the move? There is a way, and the theory of total least squares (TLS), discussed by Golub and Van Loan, is the appropriate theory.

In the theory of total least squares, the prior model $\mathbf{y} = \mathbf{x} + \mathbf{n}$, $\mathbf{x} = \mathbf{H\theta}$ is replaced by the posterior model

$$\begin{aligned} \mathbf{y} &= \mathbf{P}_s \mathbf{y} + (\mathbf{I} - \mathbf{P}_s)\mathbf{y} \\ &= \hat{\mathbf{x}}_{\text{TLS}} + \hat{\mathbf{n}}_{\text{TLS}}. \end{aligned} \qquad 9.160$$

The projector $\mathbf{P}_s$ is chosen to minimize the sum of squares of the elements in $\hat{\mathbf{n}}_{\text{TLS}} = (\mathbf{I} - \mathbf{P}_s)\mathbf{y}$ *plus* the sum of squares of the elements in $\mathbf{\Delta}_H = (\mathbf{I} - \mathbf{P}_s)\mathbf{H}$ (thus the term *total least squares*). In the posterior model, $\mathbf{P}_s\mathbf{H}$ is the corrected model $\hat{\mathbf{H}}$, $\hat{\mathbf{x}}_{\text{TLS}}$ is the estimated signal component, and $\hat{\mathbf{n}}_{\text{TLS}}$ is the estimated noise component:

$$\hat{\mathbf{x}}_{\text{TLS}} = \mathbf{P}_s\mathbf{y} = \mathbf{P}_s\mathbf{x} + \mathbf{P}_s\mathbf{n} = \mathbf{P}_s\mathbf{H\theta} + \mathbf{P}_s\mathbf{n} \qquad 9.161$$

$$\hat{\mathbf{n}}_{\text{TLS}} = (\mathbf{I} - \mathbf{P}_s)\mathbf{y} = (\mathbf{I} - \mathbf{P}_s)\mathbf{x} + (\mathbf{I} - \mathbf{P}_s)\mathbf{n}.$$

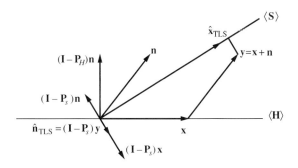

Figure 9.13 The two components of fitting error in total least squares.

The rightmost equalities differ from the LS case because the projector $\mathbf{P}_s$ is not matched to the subspace $\langle\mathbf{H}\rangle$ but is instead matched to $\langle\hat{\mathbf{H}}\rangle$, a kind of compromise between $\mathbf{H}$ and the observation $\mathbf{y}$. The fitting error now has two components, as illustrated in Figure 9.13. The method of TLS tries to minimize the norm-squared of $(\mathbf{I} - \mathbf{P}_s)\mathbf{y}$ plus the Frobenius norm of $(\mathbf{I} - \mathbf{P}_s)\mathbf{H}$. This sum is written as

$$\xi^2 = \text{tr}\begin{bmatrix}\mathbf{H}^T \\ \mathbf{y}^T\end{bmatrix}[\mathbf{I} - \mathbf{P}_s]\begin{bmatrix}\mathbf{H} & \mathbf{y}\end{bmatrix}. \qquad 9.162$$

The squared error ξ^2 is minimized by writing the SVD of $[\mathbf{H}\,\mathbf{y}]$ as $\mathbf{U\Sigma V}^T$ and constructing $\mathbf{P}_s$ from all but the singular vector that corresponds to the smallest singular value of $[\mathbf{H}\,\mathbf{y}]$:

$$[\mathbf{H}\,\mathbf{y}] = \mathbf{U\Sigma V}^T$$
$$\mathbf{U} = [\mathbf{U}_s\,\mathbf{u}] \qquad 9.163$$
$$\mathbf{P}_s = \mathbf{U}_s\mathbf{U}_s^T.$$

From Equation 9.161 we deduce that the TLS estimate of $\boldsymbol{\theta}$ is

$$\hat{\boldsymbol{\theta}}_{\text{TLS}} = \left[(\mathbf{P}_s\mathbf{H})^T(\mathbf{P}_s\mathbf{H})\right]^{-1}(\mathbf{P}_s\mathbf{H})^T\hat{\mathbf{x}}_{\text{TLS}}$$
$$= (\mathbf{H}^T\mathbf{P}_s\mathbf{H})^{-1}\mathbf{H}^T\mathbf{P}_s\mathbf{y}. \qquad 9.164$$

This is exactly the solution that would be obtained for the weighted least squares problem with $\mathbf{W} = \mathbf{P}_s = \mathbf{U}_s\mathbf{U}_s^T$ constructed from the dominant eigenvectors of $[\mathbf{H}\,\mathbf{y}]$.

9.16   INVERSE PROBLEMS AND UNDERDETERMINED LEAST SQUARES

Underdetermined least squares problems arise whenever one tries to estimate a high-dimensional vector $\boldsymbol{\theta}$ from a low-dimensional measurement $\mathbf{y}$. Such problems are also called "inverse problems," because the problem is to invert $\mathbf{y}$ to obtain $\boldsymbol{\theta}$. There is no unique inverse, and this is what makes the problem so fascinating. In a very real sense, the "solution" to the inverse problem is determined by the prior information, or the constraints, that one imposes on $\boldsymbol{\theta}$. This information selects out an "optimum" solution for $\boldsymbol{\theta}$ from the infinite class of vectors that satisfy the equation $\mathbf{y} = \mathbf{H}\boldsymbol{\theta}$.

Characterizing the Class of Solutions

We wish to characterize solutions to the equation

$$\mathbf{y} = \mathbf{H}\boldsymbol{\theta} \qquad\qquad 9.165$$

where $\mathbf{y} \in R^N$, $\boldsymbol{\theta} \in R^p$, $\mathbf{H} \in R^{N \times p}$, and $p \geq N$. In order to do so, let's write $\mathbf{H}$ in terms of its SVD:

$$\mathbf{H} = \mathbf{U}\boldsymbol{\Sigma}\mathbf{V}^T \qquad\qquad 9.166$$

$$\mathbf{U} \in R^{N \times N}; \qquad \boldsymbol{\Sigma} \in R^{N \times N}; \qquad \mathbf{V} \in R^{p \times N}.$$

The matrix $\mathbf{H}^{\#} = \mathbf{V}\boldsymbol{\Sigma}^{-1}\mathbf{U}^T$ is called a *pseudoinverse* of $\mathbf{H}$ because it satisfies the equations $\mathbf{H}\mathbf{H}^{\#} = \mathbf{I}$ and $\mathbf{H}^{\#}\mathbf{H} = \mathbf{V}\mathbf{V}^T$ or, equivalently,

$$\mathbf{H}\mathbf{H}^{\#}\mathbf{H} = \mathbf{H}; \qquad \mathbf{H}^{\#}\mathbf{H}\mathbf{H}^{\#} = \mathbf{H}^{\#}. \qquad\qquad 9.167$$

A solution to the equation $\mathbf{y} = \mathbf{H}\boldsymbol{\theta}$ is

$$\boldsymbol{\theta}_0 = \mathbf{H}^{\#}\mathbf{y} = \mathbf{V}\boldsymbol{\Sigma}^{-1}\mathbf{U}^T\mathbf{y}$$

$$= \sum_{i=1}^{N} \mathbf{v}_i \frac{1}{\sigma_i} \mathbf{u}_i^T \mathbf{y}. \qquad\qquad 9.168$$

This is a rank N solution for the p-dimensional vector $\boldsymbol{\theta}$. It is easy to see that $\mathbf{H}\boldsymbol{\theta}_0 = \mathbf{y}$:

$$\mathbf{H}\boldsymbol{\theta}_0 = \mathbf{H}\mathbf{H}^{\#}\mathbf{y} = \mathbf{U}\boldsymbol{\Sigma}\mathbf{V}^T\mathbf{V}\boldsymbol{\Sigma}^{-1}\mathbf{U}^T\mathbf{y}$$

$$= \mathbf{y}. \qquad\qquad 9.169$$

As we shall see shortly, this solution is also the minimum-norm solution for $\boldsymbol{\theta}$. However, for now we note simply that any correction to $\boldsymbol{\theta}_0$ that lies in the rank $p - N$ sub-

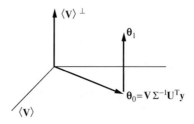

Figure 9.14 Characterizing solutions
to $\mathbf{y} = \mathbf{H}\boldsymbol{\theta}$.

space that is orthogonal to the span of $\mathbf{V}$ is also a solution. That is, choose any $\boldsymbol{\theta}_1$ such that $\mathbf{V}^T\boldsymbol{\theta}_1 = \mathbf{0}$. Then $\mathbf{H}(\boldsymbol{\theta}_0 + \boldsymbol{\theta}_1) = \mathbf{y} + \mathbf{0}$. We say that any $\boldsymbol{\theta}$ composed as $\boldsymbol{\theta} = \boldsymbol{\theta}_0 + \boldsymbol{\theta}_1$, with $\mathbf{H}\boldsymbol{\theta}_0 = \mathbf{y}$ and $\mathbf{H}\boldsymbol{\theta}_1 = \mathbf{0}$, is a solution to $\mathbf{y} = \mathbf{H}\boldsymbol{\theta}$. These solutions are illustrated in Figure 9.14.

The natural question now is, "How do we determine a unique solution?" There are many ways to proceed. The efficacy of each procedure comes down to deciding which procedure is best matched to the physical situation under study.

Minimum-Norm Solution

The *minimum-norm solution* is found by minimizing $(1/2)\,\boldsymbol{\theta}^T\boldsymbol{\theta}$ under the constraint that $\mathbf{H}\boldsymbol{\theta} = \mathbf{y}$:

$$\min \frac{1}{2}\boldsymbol{\theta}^T\boldsymbol{\theta} \qquad \text{subject to} \quad \mathbf{H}\boldsymbol{\theta} - \mathbf{y} = \mathbf{0}. \qquad 9.170$$

We may equate the gradient of a Lagrangian to zero and solve for $\boldsymbol{\theta}$:

$$\frac{\partial}{\partial\boldsymbol{\theta}}\left\{\frac{1}{2}\boldsymbol{\theta}^T\boldsymbol{\theta} - (\mathbf{H}\boldsymbol{\theta} - \mathbf{y})^T\boldsymbol{\lambda}\right\} = \mathbf{0} \qquad 9.171$$

$$\boldsymbol{\theta} = \mathbf{H}^T\boldsymbol{\lambda}.$$

The constraint is satisfied for $\mathbf{H}\boldsymbol{\theta} = \mathbf{H}\mathbf{H}^T\boldsymbol{\lambda} = \mathbf{y}$. This produces the value of $\boldsymbol{\lambda} = (\mathbf{H}\mathbf{H}^T)^{-1}\mathbf{y}$, and the corresponding solution for $\boldsymbol{\theta}$ is

$$\boldsymbol{\theta} = \mathbf{H}^T(\mathbf{H}\mathbf{H}^T)^{-1}\mathbf{y}. \qquad 9.172$$

In terms of the SVD for $\mathbf{H}$, this result may be written as

$$\begin{aligned}\boldsymbol{\theta} &= \mathbf{V}\boldsymbol{\Sigma}\mathbf{U}^T(\mathbf{U}\boldsymbol{\Sigma}\mathbf{V}^T\mathbf{V}\boldsymbol{\Sigma}\mathbf{U}^T)^{-1}\mathbf{y} \\ &= \mathbf{V}\boldsymbol{\Sigma}^{-1}\mathbf{U}^T\mathbf{y}.\end{aligned} \qquad 9.173$$

This is just the pseudoinverse solution that we found previously. We say that the pseudoinverse solution is the minimum-norm solution, a property that is more or less obvious from Figure 9.14.

Reducing Rank

The estimator $\hat{\boldsymbol{\theta}}$ is a rank-N estimator $\boldsymbol{\theta}$ that reproduces the measurements $\mathbf{y}$. However, the solution can be very sensitive to slight changes in the measurements, as the following argument shows. Replace $\mathbf{y}$ by $\mathbf{y} + \mathbf{n}$. The change in the minimum-norm solution for $\boldsymbol{\theta}$ is

$$\Delta\boldsymbol{\theta} = \mathbf{V}\boldsymbol{\Sigma}^{-1}\mathbf{U}^T\mathbf{n} = \sum_{i=1}^{N}\mathbf{v}_i\frac{1}{\sigma_i}\mathbf{u}_i^T\mathbf{n}. \qquad 9.174$$

If $\mathbf{u}_i^T\mathbf{n}$ is not zero and σ_i is very small, then $\Delta\boldsymbol{\theta}$ will have a large component $\mathbf{v}_i\,(1/\sigma_i)\,\mathbf{u}_i^T\mathbf{n}$. This suggests that small singular values give rise to large sensitivities in the solution for $\boldsymbol{\theta}$. Perhaps there is a systematic way to exclude them.

Let's suppose that the disturbance $\mathbf{n}$ is $N[\mathbf{0}, \sigma^2\mathbf{I}]$. Then $\Delta\boldsymbol{\theta}$ is $N[\mathbf{0}, \sigma^2\mathbf{V}\boldsymbol{\Sigma}^{-2}\mathbf{V}^T]$. The mean-squared value of $\Delta\boldsymbol{\theta}$ is

$$E(\Delta\boldsymbol{\theta})^T(\Delta\boldsymbol{\theta}) = \sigma^2 \sum_{i=1}^{N} \frac{1}{\sigma_i^2}, \qquad 9.175$$

again showing sensitivity to small singular values. Suppose we replace the rank-N estimator $\hat{\boldsymbol{\theta}}$ by the rank $r < N$ estimator $\hat{\boldsymbol{\theta}}_r$:

$$\hat{\boldsymbol{\theta}}_r = \mathbf{V}\boldsymbol{\Sigma}_r^{-1}\mathbf{U}^T\mathbf{y}; \qquad \boldsymbol{\Sigma}_r^{-1}: \text{ a rank-}r \text{ approximant to } \boldsymbol{\Sigma}^{-1} \text{ wherein} \qquad 9.176$$
$$N - r \text{ of the } \sigma_i^{-1} \text{ are set to zero.}$$

The measurements will not quite be reproduced:

$$\mathbf{y}_r = \mathbf{H}\hat{\boldsymbol{\theta}}_r = \mathbf{U}\boldsymbol{\Sigma}\mathbf{V}^T\mathbf{V}\boldsymbol{\Sigma}_r^{-1}\mathbf{U}^T\mathbf{y}$$
$$\qquad\qquad\qquad\qquad\qquad\qquad 9.177$$
$$= \mathbf{U}\mathbf{I}_r\mathbf{U}^T\mathbf{y} = \mathbf{P}_r\mathbf{y}$$

$$\mathbf{I}_r: \quad \text{rank-}r \text{ identity}$$

The error in the reconstruction of measurements is $\mathbf{b}_r = \mathbf{y} - \mathbf{y}_r$, and the squared reconstruction error is

$$b_r^2 = (\mathbf{y} - \mathbf{y}_r)^T(\mathbf{y} - \mathbf{y}_r)$$
$$\qquad\qquad\qquad\qquad\qquad 9.178$$
$$= \mathbf{y}^T(\mathbf{I} - \mathbf{P}_r)\mathbf{y}.$$

The sensitivity to measurement noise is now characterized by the error

$$\Delta\boldsymbol{\theta}_r = \mathbf{V}\boldsymbol{\Sigma}_r^{-1}\mathbf{U}^T\mathbf{n}$$
$$\qquad\qquad\qquad\qquad\qquad 9.179$$
$$: N[\mathbf{0}, \sigma^2\mathbf{V}\boldsymbol{\Sigma}_r^{-2}\mathbf{V}^T],$$

whose mean-squared value is

$$E(\Delta\boldsymbol{\theta}_r)^T(\Delta\boldsymbol{\theta}_r) = \sigma^2 \sum_{i=1}^{r} \frac{1}{\sigma_{(i)}^2}. \qquad 9.180$$

Here the $\sigma_{(i)}^{-1}$ are the singular values selected by $\boldsymbol{\Sigma}_r^{-1}$.

A reasonable "bias-variance trade" would trade bias (fitting error) for variance (measurement sensitivity). Thus we reduce rank in order to minimize

$$V = \mathbf{y}^T(\mathbf{I} - \mathbf{P}_r)\mathbf{y} + \sigma^2 \sum_{i=1}^{r} \frac{1}{\sigma_{(i)}^2}. \qquad 9.181$$

The resulting low-rank solution for $\boldsymbol{\theta}$ is

$$\hat{\boldsymbol{\theta}}_r = \mathbf{V}\boldsymbol{\Sigma}_r^{-1}\mathbf{U}^T\mathbf{y}. \qquad 9.182$$

Bayes

The Bayes theory developed in Chapter 7 forms another basis for solving inverse problems. Recall the structure of the Bayes problem when the measurement obeys a linear statistical model:

$$\mathbf{y} = \mathbf{H}\boldsymbol{\theta} + \mathbf{n} \qquad\qquad 9.183$$

$$\boldsymbol{\theta} : N[\mathbf{0}, \mathbf{R}_{\theta\theta}]; \qquad \mathbf{n} : N[\mathbf{0}, \mathbf{R}_{nn}].$$

The prior model for $\boldsymbol{\theta}$ is normal, with mean $\mathbf{0}$ and covariance $\mathbf{R}_{\theta\theta}$. Typically such a model is based on a detailed understanding of the physical system that generates $\boldsymbol{\theta}$.

The Bayes solution for $\hat{\boldsymbol{\theta}}$ and the corresponding error covariance matrix $\mathbf{P} = E[\boldsymbol{\theta} - \hat{\boldsymbol{\theta}}][\boldsymbol{\theta} - \hat{\boldsymbol{\theta}}]^{\mathrm{T}}$ are most efficiently written as follows for the case $p > N$ (see problem 7.11):

$$\hat{\boldsymbol{\theta}} = \mathbf{R}_{\theta\theta}\mathbf{H}^{\mathrm{T}}\left(\mathbf{H}\mathbf{R}_{\theta\theta}\mathbf{H}^{\mathrm{T}} + \mathbf{R}_{nn}\right)^{-1}\mathbf{y} \qquad\qquad 9.184$$

$$\mathbf{P} = \mathbf{R}_{\theta\theta} - \mathbf{R}_{\theta\theta}\mathbf{H}^{\mathrm{T}}\left(\mathbf{H}\mathbf{R}_{\theta\theta}\mathbf{H}^{\mathrm{T}} + \mathbf{R}_{nn}\right)^{-1}\mathbf{H}\mathbf{R}_{\theta\theta}.$$

This is the general Bayes solution. Recall from our discussion of orthogonality that $\hat{\boldsymbol{\theta}}$ represents $\boldsymbol{\theta}$ as

$$\boldsymbol{\theta} = \hat{\boldsymbol{\theta}} + (\boldsymbol{\theta} - \hat{\boldsymbol{\theta}}), \qquad\qquad 9.185$$

where the estimator $\hat{\boldsymbol{\theta}}$ and the estimator error $\boldsymbol{\theta} - \hat{\boldsymbol{\theta}}$ are orthogonal, normal random vectors:

$$\boldsymbol{\theta} : N[\mathbf{0}, \mathbf{R}_{\theta\theta}] \qquad\qquad 9.186$$

$$\hat{\boldsymbol{\theta}} : N[\mathbf{0}, \mathbf{R}_{\theta\theta} - \mathbf{P}]; \qquad \boldsymbol{\theta} - \hat{\boldsymbol{\theta}} : N[\mathbf{0}, \mathbf{P}].$$

In the special case where $\mathbf{R}_{\theta\theta} = \mathbf{I}$ and $\mathbf{R}_{nn} = \mathbf{0}$, the Bayes solution reduces to the minimum-norm solution

$$\hat{\boldsymbol{\theta}} = \mathbf{H}^{\mathrm{T}}(\mathbf{H}\mathbf{H}^{\mathrm{T}})^{-1}\mathbf{y} = \mathbf{V}\boldsymbol{\Sigma}^{-1}\mathbf{U}^{\mathrm{T}}\mathbf{y} \qquad\qquad 9.187$$

$$\mathbf{P} = \mathbf{I} - \mathbf{H}^{\mathrm{T}}(\mathbf{H}\mathbf{H}^{\mathrm{T}})^{-1}\mathbf{H} = \mathbf{I} - \mathbf{V}\mathbf{V}^{\mathrm{T}}.$$

In this case, the orthogonal decomposition of $\boldsymbol{\theta}$ is

$$\boldsymbol{\theta} = \hat{\boldsymbol{\theta}} + (\boldsymbol{\theta} - \hat{\boldsymbol{\theta}}), \qquad\qquad 9.188$$

where $\hat{\boldsymbol{\theta}}$ and $(\boldsymbol{\theta} - \hat{\boldsymbol{\theta}})$ are orthogonal, normal random vectors:

$$\boldsymbol{\theta} : N[\mathbf{0}, \mathbf{I}] \qquad\qquad 9.189$$

$$\hat{\boldsymbol{\theta}} : N[\mathbf{0}, \mathbf{V}\mathbf{V}^{\mathrm{T}}]; \qquad \boldsymbol{\theta} - \hat{\boldsymbol{\theta}} : N[\mathbf{0}, \mathbf{I} - \mathbf{V}\mathbf{V}^{\mathrm{T}}].$$

The matrix $\mathbf{V}\mathbf{V}^{\mathrm{T}}$ is a rank-N ($N \leq p$) projection. Therefore we have the decomposition of identity into two orthogonal projections

$$\mathbf{I} = \mathbf{V}\mathbf{V}^{\mathrm{T}} + (\mathbf{I} - \mathbf{V}\mathbf{V}^{\mathrm{T}}), \qquad\qquad 9.190$$

where $\mathbf{I} - \mathbf{V}\mathbf{V}^\mathrm{T}$ is the error covariance matrix and $\mathbf{V}\mathbf{V}^\mathrm{T}$ is the covariance matrix of the rank-N estimator $\hat{\boldsymbol{\theta}} = \mathbf{V}\boldsymbol{\Sigma}^{-1}\mathbf{U}^\mathrm{T}\mathbf{y}$. The trace of the error covariance matrix is the mean-squared error $E[\hat{\boldsymbol{\theta}} - \boldsymbol{\theta}]^\mathrm{T}[\hat{\boldsymbol{\theta}} - \boldsymbol{\theta}]$:

$$E[\hat{\boldsymbol{\theta}} - \boldsymbol{\theta}]^\mathrm{T}[\hat{\boldsymbol{\theta}} - \boldsymbol{\theta}] = \mathrm{tr}\left[\mathbf{I} - \mathbf{V}\mathbf{V}^\mathrm{T}\right] = p - N \geq 0. \qquad 9.191$$

Any reduced-rank approximation to $\hat{\boldsymbol{\theta}}$ would use

$$\hat{\boldsymbol{\theta}}_r = \mathbf{V}\boldsymbol{\Sigma}_r^{-1}\mathbf{U}^\mathrm{T}\mathbf{y}. \qquad 9.192$$

The error covariance would then be $\mathbf{I} - \mathbf{P}_r$, with $\mathbf{P}_r$ a rank-r $(r < N)$ projection. The trace of the error covariance would be $p - r \geq p - N$.

Maximum Entropy

The minimum-norm and Bayes solutions find solutions $\boldsymbol{\theta}$ that are bipolar and more or less concentrated around $\boldsymbol{\theta} = \mathbf{0}$. In many applications, such as image or tomographic reconstruction, this solution is not acceptable. The principle of maximum entropy provides a way of constraining the $\boldsymbol{\theta}_i$ to be nonnegative and maximizing a quantity of great utility for many branches of applied science. The idea behind maximum entropy is to

$$\max_{\boldsymbol{\theta}} \left(-\sum_{i=1}^{p} \theta_i \log_2 \theta_i \right) \qquad \text{subject to} \quad \mathbf{y} = \mathbf{H}\boldsymbol{\theta}. \qquad 9.193$$

Before solving this problem, we offer Roy Frieden's justification for its use in the case where $\boldsymbol{\theta}$ is an image (arranged in lexicographic row ordering) and $\mathbf{H}$ is the linear transformation induced by an imaging system.

Image Formation

Let's visualize an experiment in which photons strike a detecting surface to form an image $\boldsymbol{\theta} = [\theta_1 \quad \theta_2 \quad \cdots \quad \theta_p]$. The number $\theta_i \geq 0$ records the intensity associated with θ_i detections in pixel i. Assume that M photons strike the pixel array to form an image. Then

$$\sum_{i=1}^{p} \theta_i = M. \qquad 9.194$$

There are p^M possible outcomes that may be observed in this experiment. Of these outcomes, there are

$$N(\boldsymbol{\theta}) = \frac{M!}{\theta_1!\theta_2!\cdots\theta_p!} \qquad 9.195$$

that produce the same image $\boldsymbol{\theta} = [\theta_1 \quad \theta_2 \quad \cdots \quad \theta_p]^\mathrm{T}$. For large M, $N(\boldsymbol{\theta})$ approaches the asymptotic result

$$N(\boldsymbol{\theta}) \cong 2^{M\hat{\varepsilon}(\boldsymbol{\theta})} \qquad 9.196$$

$$\hat{\varepsilon}(\boldsymbol{\theta}) = -\sum_{i=1}^{p} \frac{\theta_i}{M} \log_2 \frac{\theta_i}{M}.$$

Assuming that the photons randomly strike the pixels, the *probability* of seeing image $\boldsymbol{\theta}$ is described by the multinomial law,

$$P[\boldsymbol{\theta}] = N(\boldsymbol{\theta}) \prod_{i=1}^{p} p_i^{\theta_i}, \qquad 9.197$$

where p_i is the probability that a photon strikes pixel i. This multinomial distribution may be approximated as

$$P[\boldsymbol{\theta}] \cong 2^{M\hat{\varepsilon}(\boldsymbol{\theta})} 2^{\left\{ \log_2 \prod_{i=1}^{p} p_i^{\frac{M\theta_i}{M}} \right\}} \qquad 9.198$$

$$= 2^{\left\{ -M \sum_{i=1}^{p} \frac{\theta_i}{M} \log_2 \frac{\theta_i}{M} \right\}} 2^{\left\{ M \sum_{i=1}^{p} \frac{\theta_i}{M} \log_2 p_i \right\}}.$$

This probability is 1 when $\theta_i / M = p_i$, meaning that all of the probability mass is consumed by the small fraction of experimental outcomes that produce the typical image $\boldsymbol{\theta} = [Mp_1, \quad Mp_2, \quad \cdots \quad Mp_p]$. The fraction of experimental outcomes that produce this typical image is

$$\beta = \frac{2^{M\varepsilon(\mathbf{p})}}{p^M} \qquad 9.199$$

$$\varepsilon(\mathbf{p}) = -\sum_{i=1}^{p} p_i \log_2 p_i.$$

The principle of maximum entropy argues that what we observe must be typical and, furthermore, that this typical outcome will be very difficult for Mother Nature to generate unless the entropy $\hat{\varepsilon}(\boldsymbol{\theta})$ is large enough for the typical outcome to be generated in a large number of ways. In short, maximum entropy gives Mother Nature the maximum number of ways to generate a typical image.

The Maximum Entropy Solution

Our problem is to find the saddle points of the Lagrangian:

$$\mathcal{L} = -\sum_{i=1}^{p} \frac{\theta_i}{M} \log_2 \frac{\theta_i}{M} - (\mathbf{y} - \mathbf{H}\boldsymbol{\theta})^{\mathrm{T}} \boldsymbol{\mu}$$

$$= -\frac{1}{M} \sum_{i=1}^{p} \theta_i \log_2 \theta_i + \sum_{i=1}^{N} \frac{\theta_i}{M} \log_2 M - \sum_{n=1}^{N} \mu_n (y_n - \mathbf{c}_n^{\mathrm{T}} \boldsymbol{\theta}). \qquad 9.200$$

We may replace this Lagrangian with

$$\mathcal{L}' = -\sum_{i=1}^{p} \theta_i \ln \theta_i - \sum_{n=1}^{N} \lambda_n \left(y_n - \sum_{i=1}^{p} c_{ni} \theta_i \right). \tag{9.201}$$

The gradients with respect to θ_i are

$$\frac{\partial}{\partial \theta_i} : -\ln \theta_i - 1 + \sum_{n=1}^{N} \lambda_n c_{ni}. \tag{9.202}$$

When equated to zero, the solution is

$$\hat{\theta}_i = \exp\left\{ -1 + \sum_{n=1}^{N} \lambda_n c_{ni} \right\}. \tag{9.203}$$

In this maximum entropy solution for θ_i, θ_i is a nonnegative number that is parameterized by $N < p$ parameters $\{\lambda_n\}_1^N$. By choosing the λ_n to satisfy the constraints, we invert for $\boldsymbol{\theta}$ and interpolate $\mathbf{y} = \mathbf{H}\boldsymbol{\theta}$.

To solve for the constraints, we define the so-called partition function

$$Z = \sum_{i=1}^{p} \theta_i = \sum_{i=1}^{p} \exp\left\{ -1 + \sum_{n=1}^{N} \lambda_n c_{ni} \right\} = M. \tag{9.204}$$

The gradient with respect to λ_n produces the constraint equation

$$\frac{\partial Z}{\partial \lambda_n} = \sum_{i=1}^{p} c_{ni} \theta_i = y_n; \qquad n = 1, 2, \dots, N. \tag{9.205}$$

We think of these equations as N linear constraints. Our problem is to find the value of $\boldsymbol{\lambda} = [\lambda_1 \quad \lambda_2 \quad \cdots \quad \lambda_N]^T$ for which the solution

$$\theta_i = \exp\{-1 + \boldsymbol{\lambda}^T \mathbf{h}_i\} \tag{9.206}$$

satisfies the measurement constraints

$$\mathbf{H}\boldsymbol{\theta} = \mathbf{y}. \tag{9.207}$$

Newton-Raphson

Let's think of $\boldsymbol{\theta}$ as $\boldsymbol{\theta}(\boldsymbol{\lambda})$ and define $\mathbf{n}(\boldsymbol{\lambda})$ to be the corresponding measurement error:

$$\mathbf{n}(\boldsymbol{\lambda}) = \mathbf{y} - \mathbf{H}\boldsymbol{\theta}(\boldsymbol{\lambda}) \tag{9.208}$$

Our problem is to drive $\mathbf{n}(\boldsymbol{\lambda})$ to zero by finding the right $\boldsymbol{\lambda}$. Call $\boldsymbol{\lambda}_t$ the tth guess at $\boldsymbol{\lambda}$, denote $\mathbf{n}(\boldsymbol{\lambda}_t)$ by $\mathbf{n}_t$ and $\boldsymbol{\theta}(\boldsymbol{\lambda}_t)$ by $\boldsymbol{\theta}_t$; then approximate $\mathbf{n}_t$ as follows:

$$\mathbf{n}_{t+1} = \mathbf{n}_t + \left[\frac{\partial}{\partial \boldsymbol{\lambda}} \mathbf{n}_t^T \right] [\boldsymbol{\lambda}_{t+1} - \boldsymbol{\lambda}_t]$$

$$= \mathbf{n}_t - \left[\frac{\partial}{\partial \boldsymbol{\lambda}} \boldsymbol{\theta}_t^T \mathbf{H}^T \right] [\boldsymbol{\lambda}_{t+1} - \boldsymbol{\lambda}_t]. \tag{9.209}$$

The gradient $(\partial/\partial\boldsymbol{\lambda})\,\boldsymbol{\theta}_t^{\mathrm{T}}$ is

$$\frac{\partial}{\partial\boldsymbol{\lambda}}\,\boldsymbol{\theta}_t^{\mathrm{T}} = [\mathbf{h}_1\theta_1 \cdots \mathbf{h}_p\theta_p] = \mathbf{H}\,\mathrm{diag}[\theta_1 \quad \theta_2 \quad \cdots \quad \theta_p] \qquad 9.210$$

$$\theta_i = \exp\left\{-1 + \sum_{n=1}^{N} \lambda_n(t)c_{ni}\right\}.$$

(Note the dependence of θ_i on $\boldsymbol{\lambda}_t$.) We may insert this into the expression for $\mathbf{n}_{t+1}$ and equate $\mathbf{n}_{t+1}$ to zero to obtain the Newton-Raphson map:

$$\boldsymbol{\lambda}_{t+1} = \boldsymbol{\lambda}_t + \left[\mathbf{H}\,\mathrm{diag}[\theta_1 \quad \theta_2 \quad \cdots \quad \theta_p]\mathbf{H}^{\mathrm{T}}\right]^{-1}\mathbf{n}_t. \qquad 9.211$$

When iterated, this map converges to a numerical estimate of $\boldsymbol{\lambda}$, and the corresponding maximum-entropy solution for $\boldsymbol{\theta} = [\theta_1 \quad \theta_2 \quad \cdots \quad \theta_p]^{\mathrm{T}}$ is

$$\theta_i = \exp\{-1 + \boldsymbol{\lambda}^{\mathrm{T}}\mathbf{h}_i\}. \qquad 9.212$$

9.17   MODE IDENTIFICATION IN THE LINEAR STATISTICAL MODEL

Every problem we have considered so far concerns the estimation of $\boldsymbol{\theta}$ in the linear model $\mathbf{x} = \mathbf{H}\boldsymbol{\theta}$ when $\mathbf{H}$ is known. More generally, $\mathbf{H}$ is also unknown and the problem is to simultaneously estimate $\boldsymbol{\theta}$ and $\mathbf{H}$. We proceed by replacing $\boldsymbol{\theta}$ in the linear statistical model by its least squares estimate:

$$\mathbf{y} = \mathbf{H}\hat{\boldsymbol{\theta}} + \hat{\mathbf{n}}$$
$$= \mathbf{P}_H\mathbf{y} + \mathbf{P}_A\mathbf{y}. \qquad 9.213$$

The norm squared of the fitting error $\hat{\mathbf{n}}$ is then

$$e^2 = \hat{\mathbf{n}}^{\mathrm{T}}\hat{\mathbf{n}} = \mathbf{y}^{\mathrm{T}}\mathbf{P}_A\mathbf{y} = \mathbf{y}^{\mathrm{T}}(\mathbf{I} - \mathbf{P}_H)\mathbf{y}. \qquad 9.214$$

If the matrix $\mathbf{H}$ is unknown, then it too must be estimated to minimize e^2:

$$\hat{\mathbf{H}} = \arg \max_{\mathbf{H}} \mathbf{y}^{\mathrm{T}}\mathbf{P}_H\mathbf{y}. \qquad 9.215$$

Equivalently, the problem may be stated as a minimization problem with respect to $\mathbf{A}$:

$$\hat{\mathbf{A}} = \arg \min_{\mathbf{A}} \mathbf{y}^{\mathrm{T}}\mathbf{P}_A\mathbf{y}; \qquad \mathbf{P}_A = \mathbf{I} - \mathbf{P}_H \qquad 9.216$$

Actually, as it stands, this problem is ill-posed because $\hat{\mathbf{H}}$ can be selected to be the matrix $\hat{\mathbf{H}} = [\mathbf{y}, \hat{\mathbf{h}}_2, \ldots, \hat{\mathbf{h}}_p]$ where $\hat{\mathbf{h}}_2 \quad \cdots \quad \hat{\mathbf{h}}_p$ are any $p - 1$ vectors that are orthogonal to $\mathbf{y}$. Then $\mathbf{P}_{\hat{H}}\mathbf{y} = \mathbf{y}$ and the fitting error is zero. The problem is that we have given ourselves too much freedom in the selection of $\mathbf{H}$. The much more typical case arises when $\mathbf{H}$ or $\mathbf{A}$ must be optimized within a parametric class.

Let's denote the parametric class for $\mathbf{H}$ by $\mathbf{H}(\mathbf{z})$ and the parametric class for $\mathbf{A}$

by $\mathbf{A}(\mathbf{a})$. Then the problem of identifying $\mathbf{H}$ or $\mathbf{A}$ becomes

$$\hat{\mathbf{H}} = \mathbf{H}(\hat{\mathbf{z}}) \tag{9.217}$$

$$\hat{\mathbf{z}} = \arg \max_{\mathbf{z}} \mathbf{y}^{\mathsf{T}} \mathbf{H}(\mathbf{z}) \big[\mathbf{H}^{\mathsf{T}}(\mathbf{z}) \mathbf{H}(\mathbf{z}) \big]^{-1} \mathbf{H}^{\mathsf{T}}(\mathbf{z}) \mathbf{y}$$

or

$$\hat{\mathbf{A}} = \mathbf{A}(\hat{\mathbf{a}}) \tag{9.218}$$

$$\hat{\mathbf{a}} = \arg \min_{\mathbf{a}} \big[\mathbf{y}^{\mathsf{T}} \mathbf{A}(\mathbf{a}) \big[\mathbf{A}^{\mathsf{T}}(\mathbf{a}) \mathbf{A}(\mathbf{a}) \big]^{-1} \mathbf{A}^{\mathsf{T}}(\mathbf{a}) \mathbf{y} \big].$$

These equations summarize the problem of complete least squares parameter estimation in the linear statistical model when $\mathbf{H}$ or $\mathbf{A}$ lies in a parametric class. The example of the following section illustrates a procedure for a concrete problem.

9.18   IDENTIFICATION OF AUTOREGRESSIVE MOVING AVERAGE SIGNALS AND SYSTEMS

An autoregressive moving average (ARMA) system is illustrated in Figure 9.15. If the system is excited with the unit pulse sequence $\{\delta_t\}$, then the response is the unit pulse response sequence $\{h_t\}$:

$$\{h_t\} = H(z)\{\delta_t\}. \tag{9.219}$$

Here $H(z)$ is the following ARMA transfer function:

$$H(z) = \frac{B(z)}{A(z)}$$

$$B(z) = b_0 + b_1 z^{-1} + \cdots + b_{p-1} z^{-(p-1)} \tag{9.220}$$

$$A(z) = 1 + a_1 z^{-1} + \cdots + a_p z^{-p}.$$

The equation for the unit pulse response may also be written $A(z)\{h_t\} = B(z)\{u_t\}$, which means that $\{h_t\}$ obeys the recursions

$$h_t = \begin{cases} 0, & t < 0 \\[2mm] -\displaystyle\sum_{n=1}^{p} a_n h_{t-n} + b_t, & 0 \le t < p \\[4mm] -\displaystyle\sum_{n=1}^{p} a_n h_{t-n}, & t \ge p. \end{cases} \tag{9.221}$$

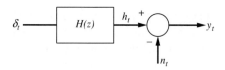

Figure 9.15 Impulse response of ARMA system.

We may think of this as a model for an ARMA signal. These equations, written out in the matrix form

$$
\begin{bmatrix}
1 & & & & & \\
a_1 & 1 & & & \mathbf{0} & \\
\vdots & \ddots & \ddots & & & \\
a_{p-1} & & & 1 & & \\
a_p & \cdots & & a_1 & 1 & \\
& \ddots & & & \ddots & \ddots \\
\mathbf{0} & & a_p & \cdots & & a_1 & 1
\end{bmatrix}
\begin{bmatrix}
h_0 \\
h_1 \\
\vdots \\
h_{p-1} \\
h_p \\
\vdots \\
h_{N-1}
\end{bmatrix}
=
\begin{bmatrix}
b_0 \\
b_1 \\
\vdots \\
b_{p-1} \\
0 \\
\vdots \\
0
\end{bmatrix},
\qquad 9.222
$$

produce the analysis model

$$
\mathbf{K}\mathbf{h}^{-1} = \begin{bmatrix} \mathbf{b} \\ \mathbf{0} \end{bmatrix}.
\qquad 9.223
$$

The definitions of $\mathbf{K}^{-1}$, $\mathbf{h}$, and $\mathbf{b}$ are obvious:

$$
\mathbf{K}^{-1} =
\left[
\begin{array}{ccccccc}
1 & & & & & & \\
a_1 & 1 & & & & & \\
\vdots & \ddots & \ddots & & & \mathbf{0} & \\
a_{p-1} & \cdots & a_1 & 1 & & & \\
a_p & a_{p-1} & \cdots & a_1 & 1 & & \\
& a_p & a_{p-1} & \cdots & a_1 & 1 & \\
\mathbf{0} & & \ddots & \ddots & \vdots & \ddots & \ddots \\
& & & a_p & a_{p-1} & \cdots & a_1 & 1
\end{array}
\right]
=
\begin{bmatrix} \star \\ \mathbf{A}^{\mathrm{T}} \end{bmatrix}
\qquad 9.224
$$

$$
\mathbf{h} = [h_0 h_1 \cdots h_{N-1}]^{\mathrm{T}}
$$

$$
\mathbf{b} = [b_0 \cdots b_{p-1}]^{\mathrm{T}}.
$$

These equations define the "prediction error" matrix $\mathbf{A}^{\mathrm{T}}$.

The matrix $\mathbf{K}^{-1}$ is invertible, so $\mathbf{h}$ also has the synthesis representation

$$
\mathbf{h} = \mathbf{K} \begin{bmatrix} \mathbf{b} \\ \mathbf{0} \end{bmatrix}
\qquad 9.225
$$

$$
= \mathbf{H}\mathbf{b}
$$

where $\mathbf{H}$ consists of the first p columns of $\mathbf{K}$:

$$
\mathbf{K} = \begin{bmatrix} \mathbf{H} & \star \end{bmatrix}.
\qquad 9.226
$$

From the analysis model of Equation 9.223 and the synthesis model of Equation 9.225 we deduce

$$
\mathbf{K}^{-1}\mathbf{K} = \begin{bmatrix} \star \\ \mathbf{A} \end{bmatrix} \begin{bmatrix} \mathbf{H} & \star \end{bmatrix} = \mathbf{I},
\qquad 9.227
$$

This result shows that the matrix $\mathbf{H}$ in the synthesis model and the matrix $\mathbf{A}$ in the analysis model are orthogonal to each other:

$$\mathbf{A}^T\mathbf{H} = \mathbf{0}. \qquad 9.228$$

As $\mathbf{H}$ is rank p and $\mathbf{A}$ is rank $N - p$, these matrices span the respective rank p and rank $N - p$ subspaces $\langle \mathbf{H} \rangle$ and $\langle \mathbf{A} \rangle$. Together, $\langle \mathbf{H} \rangle$ and $\langle \mathbf{A} \rangle$ span R^N. We call $\langle \mathbf{H} \rangle$ the signal subspace and $\langle \mathbf{A} \rangle$ the orthogonal subspace. The "prediction error matrix" $\mathbf{A}$ that characterizes the orthogonal subspace $\langle \mathbf{A} \rangle$ is defined in Equation 9.224. The "impulse response matrix" $\mathbf{H}$ that characterizes $\langle \mathbf{H} \rangle$ is obtained by noting that $\mathbf{K}$ is the matrix of impulse responses for the autoregressive filter $1/A(z)$:

$$\mathbf{K} = \begin{bmatrix} g_0 & & & \mathbf{0} \\ g_1 & g_0 & & \\ \vdots & & \ddots & \\ g_{N-1} & & g_1 & g_0 \end{bmatrix}; \quad \mathbf{H} = \begin{bmatrix} g_0 & & \\ g_1 & g_0 & \mathbf{0} \\ \vdots & \vdots & g_0 \\ & & \vdots \\ g_{N-1} & & g_{N-p} \end{bmatrix}; \qquad 9.229$$

$$A(z)\{g_t\} = \{\delta_t\}; \qquad \sum_{n=0}^{p} a_n g_{t-n} = \delta_t.$$

If $\mathbf{h}$ is observed in noise, then we have the linear statistical model

$$\mathbf{y} = \mathbf{h} + \mathbf{n}. \qquad 9.230$$

The least squares estimates of $\mathbf{b}$ and $\mathbf{h}$ are

$$\hat{\mathbf{b}} = (\mathbf{H}^T\mathbf{H})^{-1}\mathbf{H}^T\mathbf{y} \qquad 9.231$$

$$\hat{\mathbf{h}} = \mathbf{P}_H\mathbf{y}; \qquad \mathbf{P}_H = \mathbf{H}(\mathbf{H}^T\mathbf{H})^{-1}\mathbf{H}^T. \qquad 9.232$$

The norm squared of the fitting error is then

$$\hat{\mathbf{n}}^T\hat{\mathbf{n}} = \mathbf{y}^T(\mathbf{I} - \mathbf{P}_H)\mathbf{y}$$
$$= \mathbf{y}^T\mathbf{P}_A\mathbf{y}, \qquad 9.233$$

where $\mathbf{P}_A = \mathbf{A}(\mathbf{A}^T\mathbf{A})^{-1}\mathbf{A}^T$. The least squares estimate of $\mathbf{A}$ is found from the nonlinear optimization:

$$\hat{\mathbf{A}} = \arg\min \mathbf{y}^T\mathbf{A}(\mathbf{A}^T\mathbf{A})^{-1}\mathbf{A}^T\mathbf{y} \qquad 9.234$$

where $\mathbf{A}^T$ is the prediction error matrix

$$\mathbf{A}^T = \begin{bmatrix} a_p & \cdots & 1 & & \mathbf{0} \\ & \ddots & & \ddots & \\ \mathbf{0} & & a_p & \cdots & 1 \end{bmatrix}. \qquad 9.235$$

From here on, the story is identical to that told in Section 6.10, Maximum-Likelihood Estimation of ARMA Parameters.

9.19   LINEAR PREDICTION AND PRONY'S METHOD

Let's return to the least squares solution for the matrix $\mathbf{A}$ in the previous section:

$$\hat{\mathbf{A}} = \arg \min \mathbf{y}^T \mathbf{P}_A \mathbf{y} \qquad\qquad 9.236$$

$$\mathbf{P}_A = \mathbf{A}(\mathbf{A}^T \mathbf{A})^{-1} \mathbf{A}^T.$$

The matrix $\mathbf{P}_A$ is the familiar projection onto the orthogonal subspace spanned by $\mathbf{A}$. Recall that $\mathbf{P}_A \mathbf{y}$ is the estimated noise $\hat{\mathbf{n}}$ in the least squares problem and the least squares solution for $\mathbf{A}$ minimizes the norm of the estimated noise $\hat{\mathbf{n}}$.

Modified Least Squares

The inverse $(\mathbf{A}^T \mathbf{A})^{-1}$ is the offending term in the minimization problem because it makes the quadratic form $\mathbf{y}^T \mathbf{P}_A \mathbf{y}$ non-quadratic in the variables $(a_1, \ldots, a_p)$. If the inverse is ignored, then the least squares problem becomes the modified least squares problem

$$\hat{\mathbf{A}} = \arg \min \mathbf{y}^T \mathbf{A} \mathbf{A}^T \mathbf{y} = \arg \min \mathbf{e}^T \mathbf{e} \qquad\qquad 9.237$$

$$\mathbf{e} = \mathbf{A}^T \mathbf{y}$$

where $\mathbf{e}$ is called a modified error.

Equation Error

The modified least squares problem is also called an equation error problem. To see why, recall the analysis model

$$\begin{bmatrix} \star \\ \hline \mathbf{A}^T \end{bmatrix} \mathbf{h} = \begin{bmatrix} \mathbf{b} \\ \mathbf{0} \end{bmatrix}. \qquad\qquad 9.238$$

The trailing zeros on the right-hand side show that the vector $\mathbf{h}$ satisfies the homogeneous equation

$$\mathbf{A}^T \mathbf{h} = \mathbf{0}. \qquad\qquad 9.239$$

If $\mathbf{h}$ is replaced by $\mathbf{y} = \mathbf{h} + \mathbf{n}$, then this equation no longer holds. The idea behind equation error is to replace the signal $\mathbf{h}$ by the measurement $\mathbf{y}$ and represent the resulting error in the equation as $\mathbf{e}$:

$$\mathbf{A}^T \mathbf{y} = \mathbf{e}. \qquad\qquad 9.240$$

Connection Between Errors and Modified Errors

What is the connection between the norm squared error $\hat{\mathbf{n}}^T \hat{\mathbf{n}}$ that is minimized in least squares and the norm squared error $\mathbf{e}^T \mathbf{e}$ that is minimized in modified least squares or equation error? To answer this question, write out $\hat{\mathbf{n}}$ and $\hat{\mathbf{n}}^T\hat{\mathbf{n}}$ as follows:

$$\hat{\mathbf{n}} = \mathbf{A}(\mathbf{A}^T\mathbf{A})^{-1}\mathbf{A}^T\mathbf{y} = \mathbf{A}(\mathbf{A}^T\mathbf{A})^{-1}\mathbf{e} \qquad 9.241$$

$$\hat{\mathbf{n}}^T\hat{\mathbf{n}} = \mathbf{e}^T(\mathbf{A}^T\mathbf{A})^{-1}\mathbf{A}^T\mathbf{A}(\mathbf{A}^T\mathbf{A})^{-1}\mathbf{e} = \mathbf{e}^T(\mathbf{A}^T\mathbf{A})^{-1}\mathbf{e}.$$

We say that the norm squared of $\hat{\mathbf{n}}$ is a weighted norm squared of $\mathbf{e}$. The other connection is obtained by writing out $\mathbf{e}$ and $\mathbf{e}^T\mathbf{e}$:

$$\mathbf{e} = \mathbf{A}^T\mathbf{y} = \mathbf{A}^T\mathbf{A}(\mathbf{A}^T\mathbf{A})^{-1}\mathbf{A}^T\mathbf{y} = \mathbf{A}^T\mathbf{P}_A\mathbf{y} = \mathbf{A}^T\hat{\mathbf{n}} \qquad 9.242$$

$$\mathbf{e}^T\mathbf{e} = \hat{\mathbf{n}}^T(\mathbf{A}\mathbf{A}^T)\mathbf{n}.$$

Again, the norm squared of the modified error $\mathbf{e}$ is a weighted norm squared of the error $\hat{\mathbf{n}}$.

Linear Prediction

The modified errors $\mathbf{e} = \mathbf{A}^T\mathbf{y}$ have two different representations:

1.
$$\mathbf{e} = \mathbf{A}^T\mathbf{y} = \begin{bmatrix} a_p & \cdots & a_1 & 1 & & \mathbf{0} \\ & \ddots & & \ddots & \ddots & \\ \mathbf{0} & & a_p & \cdots & a_1 & 1 \end{bmatrix} \begin{bmatrix} y_1 \\ y_2 \\ \vdots \\ \\ \\ y_N \end{bmatrix}$$

$$9.243$$

2.
$$\mathbf{e} = \begin{bmatrix} y_1 & \cdots & y_p \\ y_2 & \cdots & y_{p+1} \\ \vdots & & \vdots \\ y_{N-p} & \cdots & y_{N-1} \end{bmatrix} \begin{bmatrix} a_p \\ \vdots \\ a_2 \\ a_1 \end{bmatrix} + \begin{bmatrix} y_{p+1} \\ \vdots \\ \\ y_N \end{bmatrix}.$$

The first representation (1) says that each element of $\mathbf{e}$ is a prediction error of the form

$$\mathbf{e} = \begin{bmatrix} e_{p+1} & e_{p+2} & \cdots & e_N \end{bmatrix}^T \qquad 9.244$$

$$e_t = y_t - \left(-\sum_{n=1}^{p} a_n y_{t-n} \right),$$

where $-\sum_{n=1}^{p} a_n y_{t-n}$ is a linear prediction of y_t based on the previous values $y_{t-1}, \ldots, y_{t-p}$. The modified squared error is the sum of squared prediction errors:

$$\mathbf{e}^T\mathbf{e} = \sum_{t=p+1}^{N} e_t^2. \qquad 9.245$$

The second representation (2) says that **e** is the error vector in the following linear statistical model:

$$\mathbf{z} = -\mathbf{Ya} + \mathbf{e}$$

$$\mathbf{z} = \begin{bmatrix} y_{p+1} & \cdots & y_N \end{bmatrix}^{\mathrm{T}} \qquad\qquad 9.246$$

$$\mathbf{Y} = \begin{bmatrix} y_1 & y_2 & \cdots & y_p \\ y_2 & y_3 & \cdots & y_{p+1} \\ \vdots & & & \vdots \\ y_p & & & y_{2p} \\ \vdots & & & \vdots \\ y_{N-p} & & \cdots & y_{N-1} \end{bmatrix}$$

$$\mathbf{a} = \begin{bmatrix} a_p & \cdots & a_2 & a_1 \end{bmatrix}^{\mathrm{T}}.$$

These equations are often called the covariance equations of linear prediction.

The modified least squares estimate of **a** minimizes $\mathbf{e}^{\mathrm{T}}\mathbf{e}$, and the modified least squares estimate of **A** is $\mathbf{A}(\hat{\mathbf{a}})$:

$$\hat{\mathbf{a}} = -(\mathbf{Y}^{\mathrm{T}}\mathbf{Y})^{-1}\mathbf{Y}^{\mathrm{T}}\mathbf{z}$$

$$\mathbf{A}(\hat{\mathbf{a}}) = \begin{bmatrix} \hat{a}_p & \cdots & \hat{a}_1 & 1 & & \mathbf{0} \\ \mathbf{0} & \ddots & & \ddots & \ddots & \\ & & \hat{a}_p & \cdots & \hat{a}_1 & 1 \end{bmatrix}. \qquad 9.247$$

Another way to interpret our solution is to say that we have used the equation error $\mathbf{A}^{\mathrm{T}}\mathbf{y} = \mathbf{e}$ to estimate the subspace **A**, which is orthogonal to the original subspace **H**. This estimated subspace may be used in a number of ways to estimate the signal subspace **H**. This brings us to Prony's Method.

Prony's Method

In 1795 at Ecolé Polytechnique, while studying the effects of alcohol vapor pressures, le Baron de Prony observed a number of things. Among them was the property that linear combinations of complex exponentials obey homogeneous linear recursions. These recursions may be used to interpolate or to least squares-fit data, although Prony was apparently only aware of the interpolating value of his results. We shall now show that Prony's method was the antecedent of the modern-day variations that we have outlined in the previous section.

Prony began with $2p$ measurements $\mathbf{y} = [y_0 \ y_2 \ \cdots \ y_p \ y_{p+1} \ \cdots \ y_{2p-1}]^{\mathrm{T}}$. He wished to fit a curve to these measurements that would pass through them and that could be used to interpolate between and extrapolate beyond them. Prony proposed the complex exponential model

$$y_t = \sum_{i=1}^{p} c_i z_i^t. \tag{9.248}$$

When written out for $t = 0, 1, \ldots, 2p-1$, this model produces the synthesis equations

$$\mathbf{y} = \mathbf{Vc}$$

$$\mathbf{V} = \begin{bmatrix} 1 & 1 & \cdots & 1 \\ z_1 & z_2 & \cdots & z_p \\ \vdots & \vdots & & \vdots \\ z_1^{2p-1} & z_2^{2p-1} & \cdots & z_p^{2p-1} \end{bmatrix}; \qquad \mathbf{c} = [c_1 \quad c_2 \quad \cdots \quad c_p]^{\mathrm{T}}. \tag{9.249}$$

Prony recognized that these synthesis equations could be annihilated with the predictor matrix $\mathbf{A}^{\mathrm{T}}$, provided $\mathbf{A}^{\mathrm{T}}$ was properly selected. That is,

$$\mathbf{A}^{\mathrm{T}}\mathbf{y} = \mathbf{0} \tag{9.250}$$

$$\mathbf{A}^{\mathrm{T}} = \begin{bmatrix} a_p & \cdots & a_1 & 1 & & \mathbf{0} \\ & \ddots & & & \ddots & \\ \mathbf{0} & & a_p & \cdots & a_1 & 1 \end{bmatrix},$$

Prony solved these linear equations for the a_i and used them as coefficients in the following predictor polynomial:

$$A(z) = \sum_{n=0}^{p} a_n z^{-n} = \prod_{n=1}^{p} (1 - z_n z^{-1}) = 0; \qquad a_0 = 1 \tag{9.251}$$

$$A(z_n) = 0.$$

The zeros z_n are precisely the roots of the Vandermonde matrix $\mathbf{V}$. That is, $\mathbf{A}^{\mathrm{T}}\mathbf{V} = \mathbf{0}$ when z_n is a zero of $A(z)$. In this way Prony used the null space $\mathbf{A}$ to build the orthogonal space $\mathbf{V}$.

In summary, Prony's method writes the equation $\mathbf{A}^{\mathrm{T}}\mathbf{y} = \mathbf{0}$ as

$$\mathbf{z} = -\mathbf{Ya} \tag{9.252}$$

$$\mathbf{z} = \begin{bmatrix} y_p & \cdots & y_{2p-1} \end{bmatrix}; \qquad \mathbf{Y} = \begin{bmatrix} y_0 & y_1 & \cdots & y_{p-1} \\ y_1 & y_2 & \cdots & y_p \\ \vdots & & & \vdots \\ y_{p-1} & & \cdots & y_{2p-2} \end{bmatrix}; \qquad \mathbf{a} = \begin{bmatrix} a_p & \cdots & a_1 \end{bmatrix}^{\mathrm{T}}$$

and solves for $\mathbf{a}$. The polynomial $A(z)$ is rooted for the z_i, and the Vandermonde matrix $\mathbf{V}$ is constructed. The synthesis equation $\mathbf{y} = \mathbf{V}\mathbf{c}$ is then used to solve for the parameters $\mathbf{c}$:

$$\mathbf{c} = (\mathbf{V}^H\mathbf{V})^{-1}\mathbf{V}^H\mathbf{y}. \qquad 9.253$$

Can you see that $\mathbf{A}^T\mathbf{y} = \mathbf{0}$ forces a homogeneous recursion on the tail of $\{y_t\}$? In Chapter 11, Modal Analysis, we treat least squares versions of Prony's method.

9.20   LEAST SQUARES ESTIMATION OF STRUCTURED CORRELATION MATRICES

It is commonplace in signal processing to estimate a correlation matrix from data. The estimated correlation matrix, call it $\mathbf{S}$, is typically nonnegative definite, but beyond that, it often has no special structure. So the question is, "How can a structured correlation matrix, parameterized by a small number of parameters, be fitted to the estimated correlation matrix?" In this section, we give one answer based on least squares. The results are applied to the problem of least squares estimating Toeplitz and rank-deficient correlation matrices.

The problem of least squares-fitting a structured correlation matrix $\mathbf{R}$ to an estimated correlation matrix $\mathbf{S} = (s_{ij})$ is simply one of minimizing the squared error

$$e^2 = \operatorname{tr}\left[(\mathbf{S} - \mathbf{R})^T(\mathbf{S} - \mathbf{R})\right]. \qquad 9.254$$

This is the Frobenius norm of $\mathbf{S} - \mathbf{R}$, namely the sum of squares of all differences $(s_{ij} - r_{ij})$.

Let's denote the structured matrix $\mathbf{R}$ by $\mathbf{R}(\boldsymbol{\theta})$ and minimize e^2 with respect to θ_n:

$$\frac{\partial e^2}{\partial \theta_n} = -2\operatorname{tr}\left[\mathbf{S}^T\frac{\partial \mathbf{R}}{\partial \theta_n}\right] + 2\operatorname{tr}\left[\mathbf{R}^T\frac{\partial \mathbf{R}}{\partial \theta_n}\right]. \qquad 9.255$$

The regression equation to be solved for the θ_n is

$$\operatorname{tr}\left[(\mathbf{S} - \mathbf{R})^T\frac{\partial \mathbf{R}}{\partial \theta_n}\right] = 0. \qquad 9.256$$

The second derivative of e^2 with respect to θ_n is

$$\operatorname{tr}\left[\left(\frac{\partial \mathbf{R}}{\partial \theta_n}\right)^T\frac{\partial \mathbf{R}}{\partial \theta_n}\right] + \operatorname{tr}\left[(\mathbf{S} - \mathbf{R})^T\frac{\partial^2 \mathbf{R}}{\partial \theta_n^2}\right]. \qquad 9.257$$

Linear Structure

When the parameterized correlation matrix $\mathbf{R}(\boldsymbol{\theta})$ has the linear structure

$$\mathbf{R} = \sum_{i=1}^{p} \theta_i\mathbf{Q}_i, \qquad 9.258$$

then we can write the regression equation and the second derivative as follows:

$$\text{tr}\left[(\mathbf{S} - \mathbf{R})^{\text{T}}\mathbf{Q}_n\right] = 0 \tag{9.259}$$

$$\text{tr}\left[\mathbf{Q}_n^{\text{T}}\mathbf{Q}_n\right] \geq 0.$$

This means that the solution for θ_n is a minimizing solution.

When written out, the regression equation is

$$\text{tr}\left[\mathbf{S}^{\text{T}}\mathbf{Q}_n\right] = \sum_{i=1}^{p} \theta_i \, \text{tr}\left[\mathbf{Q}_i^{\text{T}}\mathbf{Q}_n\right] = \sum_{i=1}^{p} \text{tr}\left[\mathbf{Q}_n^{\text{T}}\mathbf{Q}_i\right]\theta_i. \tag{9.260}$$

For $n = 1, 2, \ldots, p$, these equations may be organized into the system of linear equations

$$\mathbf{G}\boldsymbol{\theta} = \mathbf{g}, \tag{9.261}$$

where

$$\mathbf{G} = (g_{ni}); \qquad g_{ni} = \text{tr}\,\mathbf{Q}_n^{\text{T}}\mathbf{Q}_i \tag{9.262}$$

$$\mathbf{g} = \left[g_1 \cdots g_p\right]^{\text{T}}; \qquad g_i = \text{tr}\left[\mathbf{S}^{\text{T}}\mathbf{Q}_i\right].$$

These equations characterize the unique least squares solution when $\mathbf{G}$ is nonsingular:

$$\hat{\boldsymbol{\theta}} = \mathbf{G}^{-1}\mathbf{g}. \tag{9.263}$$

Toeplitz Matrix

In the Toeplitz case, the structured correlation matrix $\mathbf{R}$ is represented by

$$\mathbf{R} = \begin{bmatrix} r_0 & r_1 & \cdots & r_{N-1} \\ r_1 & & & \vdots \\ \vdots & & & r_1 \\ r_{N-1} & \cdots & r_1 & r_0 \end{bmatrix} = \sum_{n=0}^{N-1} r_n \mathbf{Q}_n. \tag{9.264}$$

$$\mathbf{Q}_n = \begin{bmatrix} & & 1 & & \mathbf{0} \\ & & & \ddots & \\ 1 & & \mathbf{0} & & 1 \\ & \ddots & & & \\ \mathbf{0} & & 1 & & \end{bmatrix} \quad \text{(a matrix with ones on its } \pm n\text{th diagonals and zeros elsewhere)}$$

The unknown parameters are the correlations $\{r_n\}_0^{N-1}$. The derivative of $\mathbf{R}$ with respect to r_n is $\mathbf{Q}_n$. Therefore, the regression equation is

$$\text{tr}\left[(\mathbf{S} - \mathbf{R})^{\text{T}}\frac{\partial \mathbf{R}}{\partial r_n}\right] = \text{tr}\left[(\mathbf{S} - \mathbf{R})^{\text{T}}\mathbf{Q}_n\right] = 0. \tag{9.265}$$

and the solution is

$$\mathrm{tr}\left[\mathbf{R}^{\mathrm{T}}\mathbf{Q}_n\right] = \mathrm{tr}\left[\mathbf{S}^{\mathrm{T}}\mathbf{Q}_n\right].$$ 9.266

Using the definitions of $\mathbf{R}$, $\mathbf{S}$, and $\mathbf{Q}_n$, we obtain the solutions

$$N\hat{r}_0 = \sum_{i=1}^{N} s_{ii}$$ 9.267

$$2(N-n)\hat{r}_n = \sum_{j=n+1}^{N} s_{(j-n)j} + \sum_{i=n+1}^{N} s_{i(i-n)}; \qquad (n \neq 0).$$

This solution builds an estimated Toeplitz matrix $\hat{\mathbf{R}} = \sum_{n=0}^{N-1} \hat{r}_n \mathbf{Q}_n$ from a non-Toeplitz matrix $\mathbf{S}$. The solution simply averages $\mathbf{S}$ along the $\pm n$th diagonals to get $\hat{r}_n$.

Low-Rank Matrix

In the so-called low-rank case, the structured correlation matrix is represented by

$$\mathbf{R} = \sum_{n=1}^{p} \sigma_n^2 \mathbf{u}_n \mathbf{u}_n^{\mathrm{T}}.$$ 9.268

The rank-p matrix $\sum_{n=1}^{p} \sigma_n^2 \mathbf{u}_n \mathbf{u}_n^{\mathrm{T}}$ is usually associated with a low-rank signal model. The model for $\mathbf{R}$ may also be written as

$$\mathbf{R} = \mathbf{U}_p \mathbf{\Sigma}_p^2 \mathbf{U}_p^{\mathrm{T}},$$ 9.269

where the matrix $\mathbf{U}_p$ contains the p-linearly independent column vectors $(\mathbf{u}_n)_1^p$ and $\mathbf{\Sigma}_p^2$ is a diagonal matrix:

$$\mathbf{U}_p = [\mathbf{u}_1 \quad \cdots \quad \mathbf{u}_p]$$ 9.270

$$\mathbf{\Sigma}_p^2 = \mathrm{diag}[\sigma_1^2 \quad \cdots \quad \sigma_p^2].$$

We shall assume that the vectors $(\mathbf{u}_n)_1^p$ are known but not necessarily orthogonal, and the parameters $(\sigma_n{}^2)_1^p$ are unknown. The least squares estimate of $\mathbf{R}$ is

$$\hat{\mathbf{R}} = \mathbf{U}_p \hat{\mathbf{\Sigma}}_p^2 \mathbf{U}_p^{\mathrm{T}},$$ 9.271

with $\hat{\mathbf{\Sigma}}_p^2$ chosen to minimize e^2.

The regression equations for the σ_k^2, $k = 1, 2, \ldots, p$ are

$$\mathrm{tr}\left[(\mathbf{S} - \mathbf{R})^{\mathrm{T}} \frac{\partial \mathbf{R}}{\partial \sigma_k^2}\right] = 0$$ 9.272

$$\frac{\partial \mathbf{R}}{\partial \sigma_k^2} = \mathbf{u}_k \mathbf{u}_k^{\mathrm{T}}.$$

The solutions for the σ_k^2 are therefore

$$\mathbf{u}_k^T \left(\sum_{n=1}^{p} \sigma_n^2 \mathbf{u}_n \mathbf{u}_n^T \right) \mathbf{u}_k = \mathbf{u}_k^T \mathbf{S} \mathbf{u}_k \qquad 9.273$$

or

$$\sum_{n=1}^{p} \sigma_n^2 \left(\mathbf{u}_k^T \mathbf{u}_n \right)^2 = \mathbf{u}_k^T \mathbf{S} \mathbf{u}_k . \qquad 9.274$$

These solutions may be combined into the following system of equations, which may be solved for the $(\sigma_n^2)_1^p$:

$$\begin{bmatrix} \left(\mathbf{u}_1^T \mathbf{u}_1 \right)^2 & \cdots & \left(\mathbf{u}_1^T \mathbf{u}_p \right)^2 \\ \vdots & & \vdots \\ \left(\mathbf{u}_p^T \mathbf{u}_1 \right)^2 & \cdots & \left(\mathbf{u}_p^T \mathbf{u}_p \right)^2 \end{bmatrix} \begin{bmatrix} \sigma_1^2 \\ \vdots \\ \sigma_p^2 \end{bmatrix} = \begin{bmatrix} \mathbf{u}_1^T \mathbf{S} \mathbf{u}_1 \\ \vdots \\ \mathbf{u}_p^T \mathbf{S} \mathbf{u}_p \end{bmatrix} . \qquad 9.275$$

With the solution to these equations, the least squares estimate of $\mathbf{R}$ is

$$\hat{\mathbf{R}} = \sum_{n=1}^{p} \hat{\sigma}_n^2 \mathbf{u}_n \mathbf{u}_n^T .$$

Example 9.4 (Sparse DFT Representation)
Consider the following representation for data $\mathbf{x} = [x_0 x_1 \ldots x_{N-1}]^T$:

$$x_k = \sum_{n=n_1}^{n_p} X_n e^{j(2\pi/M)nk} . \qquad 9.276$$

Assume that the random variables X_n are i.i.d. random variables with zero means and variances σ_n^2. The correlation matrix for $\mathbf{x}$ is

$$\mathbf{R} = E\mathbf{x}\mathbf{x}^T = \sum_{n=n_1}^{n_p} \sigma_n{}^2 \mathbf{u}_n \mathbf{u}_n{}^T \qquad 9.277$$

$$\mathbf{u}_n = \begin{bmatrix} 1 \\ e^{j(2\pi/M)n} \\ \vdots \\ e^{j(2\pi/M)n(N-1)} \end{bmatrix} .$$

This is a sparse DFT representation in which a small number of frequency components, on a narrow band from $(2\pi/M)n_1$ to $(2\pi/M)n_p$, are used to represent the data. If the data record is short compared to the resolution of the representation ($N < M$), the representation vectors are not orthogonal. Nonetheless, the solution derived earlier applies, and the result is a high-resolution analyzer of the σ_n^2. ∎

Orthonormal Case

When the $(\mathbf{u}_n)_1^p$ are orthonormal, the correlation model is

$$\mathbf{R} = \mathbf{U}_p \mathbf{\Sigma}_p^2 \mathbf{U}_p^T \qquad\qquad 9.278$$

$$\mathbf{U}_p^T \mathbf{U}_p = \mathbf{I}.$$

The least squares solutions for the σ_n^2 are

$$\sigma_n^2 = \mathbf{u}_n^T \mathbf{S} \mathbf{u}_n \qquad n = 1, 2, \ldots, p. \qquad\qquad 9.279$$

If the $\mathbf{u}_n$ are chosen to be the first p eigenvectors in an orthogonal decomposition of $\mathbf{S} = \mathbf{U}\mathbf{D}^2\mathbf{U}^T$, then the solution specializes even further to $\hat{\sigma}_n^2 = d_n^2$, where $\mathbf{D}^2 = \text{diag}[d_1^2 \quad \cdots \quad d_p^2]$.

More on the Orthonormal Case

Let's generalize the orthonormal case to the model

$$\mathbf{R} = \mathbf{U}(\mathbf{\Sigma}^2 + \sigma^2\mathbf{I})\mathbf{U}^T; \qquad \mathbf{U} = [\mathbf{U}_p \,|\, \mathbf{U}_{N-p}]$$

$$\mathbf{\Sigma}^2 = \text{diag}[\sigma_1^2 \quad \cdots \quad \sigma_p^2 \quad \mathbf{0}^T] \qquad\qquad 9.280$$

$$\mathbf{U}^T\mathbf{U} = \mathbf{U}\mathbf{U}^T.$$

The covariance matrix $\mathbf{U}\mathbf{\Sigma}^2\mathbf{U}^T$ is usually associated with a low-rank signal component, and the covariance matrix $\mathbf{U}\sigma^2\mathbf{I}\mathbf{U}^T = \sigma^2\mathbf{I}$ is the covariance matrix for additive white noise. Oftentimes, it is known that signal variances σ_1^2 through σ_p^2 are equal. The model for the correlation matrix is then

$$\mathbf{R} = \mathbf{U}(\mathbf{\Sigma}^2 + \sigma^2\mathbf{I})\mathbf{U}^T \qquad\qquad 9.281$$

$$\mathbf{\Sigma}^2 = \text{diag}[\sigma_s^2 \quad \cdots \quad \sigma_s^2 \quad \mathbf{0}^T].$$

It is actually more illuminating to write the representation for $\mathbf{R}$ as follows:

$$\mathbf{R} = \sigma_s^2 \mathbf{U}_p \mathbf{U}_p^T + \sigma_n^2 \mathbf{U}\mathbf{U}^T$$

$$= (\sigma_s^2 + \sigma_n^2)\mathbf{U}_p\mathbf{U}_p^T + \sigma_n^2(\mathbf{I} - \mathbf{U}_p\mathbf{U}_p^T). \qquad\qquad 9.282$$

The matrix $\mathbf{U}_p$ contains the first p columns of the orthogonal matrix $\mathbf{U}$. This makes $\mathbf{U}_p\mathbf{U}_p^T$ a projection operator, which we denote by $\mathbf{P}$:

$$\mathbf{R} = (\sigma_s^2 + \sigma_n^2)\mathbf{P} + \sigma_n^2(\mathbf{I} - \mathbf{P}). \qquad\qquad 9.283$$

With this representation for $\mathbf{R}$, the regression equation for σ_n^2 may be written

$$\text{tr}\left[(\mathbf{S} - \mathbf{R})^T \frac{\partial \mathbf{R}}{\partial \sigma_n^2}\right] = \text{tr}\left[(\mathbf{S} - \mathbf{R})^T\mathbf{I}\right] = 0, \qquad\qquad 9.284$$

with the solution

$$\text{tr}[\mathbf{S}] = (\sigma_s^2 + \sigma_n^2)\,\text{tr}[\mathbf{P}] + \sigma_n^2\,\text{tr}[\mathbf{I} - \mathbf{P}]$$

$$= p\sigma_s^2 + N\sigma_n^2. \tag{9.285}$$

The regression equation for σ_s^2 is

$$\text{tr}\left[(\mathbf{S} - \mathbf{R})^\text{T}\,\frac{\partial \mathbf{R}}{\partial \sigma_s^2}\right] = \text{tr}\left[(\mathbf{S} - \mathbf{R})^\text{T}\mathbf{P}\right] = 0, \tag{9.286}$$

with the solution

$$\text{tr}[\mathbf{PS}] = \text{tr}\left(\sigma_s^2 + \sigma_n^2\right)\mathbf{P}^2$$

$$= p\sigma_s^2 + p\sigma_n^2. \tag{9.287}$$

These equations produce the following solutions for $\hat{\sigma}_n^2$ and $\hat{\sigma}_s^2$:

$$\hat{\sigma}_n^2 = \frac{1}{N - p}\,\text{tr}(\mathbf{I} - \mathbf{P})\mathbf{S}$$

$$= \frac{1}{N - p}\left[\text{tr}[\mathbf{S}] - \sum_{i=1}^p \mathbf{u}_i^\text{T}\mathbf{S}\mathbf{u}_i\right] \tag{9.288}$$

$$\hat{\sigma}_s^2 = \frac{1}{p}\,\text{tr}\,\mathbf{PS} - \hat{\sigma}_n^2$$

$$= \frac{1}{p}\sum_{i=1}^p \mathbf{u}_i^\text{T}\mathbf{S}\mathbf{u}_i - \hat{\sigma}_n^2.$$

The quantity $p\hat{\sigma}_s^2$ is the estimated *signal* power in the signal subspace. The quantity $p(\hat{\sigma}_s^2 + \hat{\sigma}_n^2) = \text{tr}\,\mathbf{PS} = \sum_{i=1}^p \mathbf{u}_i^\text{T}\mathbf{S}\mathbf{u}_i$ is the estimated *total* power in the signal subspace.

Example 9.5

Consider a rank-one signal correlation matrix plus a full-rank noise correlation matrix:

$$\mathbf{R} = \sigma_s^2 \mathbf{e}\mathbf{e}^\text{H} + \sigma_n^2 \mathbf{I} \tag{9.289}$$

$$\mathbf{e}^\text{H} = \begin{bmatrix} 1 & e^{-j\mathbf{k}\cdot\mathbf{r}_1} & \cdots & e^{-j\mathbf{k}\cdot\mathbf{r}_N} \end{bmatrix}.$$

The signal correlation matrix is that of a random amplitude, random phase, plane wave observed at sensors located in positions $(\mathbf{r}_i)_1^N$ when the wavenumber for the propagating field is $\mathbf{k}$. Write $\mathbf{R}$ as

$$\mathbf{R} = \sigma_s^2 \mathbf{e}^\text{H}\mathbf{e}\,\frac{\mathbf{e}\mathbf{e}^\text{H}}{\mathbf{e}^\text{H}\mathbf{e}} + \sigma_n^2 \mathbf{I} \tag{9.290}$$

$$= (\sigma_s^2 \mathbf{e}^\text{H}\mathbf{e} + \sigma_n^2)\mathbf{u}\mathbf{u}^\text{H} + \sigma_n^2(\mathbf{I} - \mathbf{u}\mathbf{u}^\text{H})$$

$$\mathbf{u}\mathbf{u}^\text{H} = \mathbf{e}\mathbf{e}^\text{H}/\mathbf{e}^\text{H}\mathbf{e} \quad \text{(rank one projection)}.$$

The estimate of the power in the subspace $\mathbf{u}$ is obtained directly from our previous results:

$$(\mathbf{e}^H\mathbf{e}\,\hat{\sigma}_s^2 + \hat{\sigma}_n^2) = \frac{\mathbf{e}^H\mathbf{S}\mathbf{e}}{\mathbf{e}^H\mathbf{e}}.$$
9.291

■

REFERENCES AND COMMENTS

Much of the material in this chapter may be found in a classic such as Lawson and Hanson [1974]. However, my treatment of subspaces is personal, if not original. The sections on improving SNR, reducing rank, and selecting order first appeared in Scharf [1985]. The interpretation of maximum entropy is drawn from the delightful writings of Roy Frieden [1972] and Edwin Jaynes [1982]. A more complete treatment of rank reduction in the linear statistical model may be found in Thorpe and Scharf [1989].

Freiden, R. [1972]. "Restoring with Maximum Likelihood and Maximum Entropy," *J Opt Soc Amer* **62**, pp. 511–518 (April 1972).

Jaynes, E. T. [1982]. "On the Rationale of Maximum Entropy Methods," *Proc IEEE* **70**, pp. 939–952 (September 1982).

Lawson, C. L. and R. J. Hanson [1974]. *Solving Least Squares Problems* (Englewood Cliffs, NJ: Prentice Hall, 1974).

Scharf, L. L. [1985]. "Topics in Statistical Signal Processing," Chapter 2 in the *Les Houches Lectures*, Grenoble, J. L. Lacoume, T. Durrani, and R. Stora, eds. (1985).

Thorpe, A. and L. L. Scharf [1989]. "Reduced Rank Methods for Solving Least Squares Problems," pp. 609–613, *Proceedings of the 23rd Asilomar Conference on Signals and Systems*, Pacific Grove, CA (November 1989).

PROBLEMS

9.1   Consider the quadratic form

$$(\mathbf{y} - \mathbf{x})^T(\mathbf{y} - \mathbf{x}).$$

Minimize it under the constraint that $\mathbf{x} = \mathbf{H}\boldsymbol{\theta}$ to obtain the solution

$$\hat{\mathbf{x}} = \mathbf{P}_H\mathbf{y}$$

$$\mathbf{P}_H = \mathbf{H}(\mathbf{H}^T\mathbf{H})^{-1}\mathbf{H}^T.$$

Generalize your result to the quadratic form $(\mathbf{y} - \mathbf{x})^T\mathbf{W}(\mathbf{y} - \mathbf{x})$ with $\mathbf{W}$ a full-rank symmetric matrix.

9.2   Begin with $\mathbf{H} \in R^{N \times p}$ and $\mathbf{A} \in R^{N \times (N-p)}$. Assume that the columns of $\mathbf{H}$ are linearly independent, the columns of $\mathbf{A}$ are linearly independent, and the columns of $\mathbf{H}$ are orthogonal to the columns of $\mathbf{A}$. Find projections $\mathbf{P}_H$ and $\mathbf{P}_A$ that produce $\mathbf{P}_H\mathbf{P}_A = \mathbf{0}$ and $\mathbf{P}_H + \mathbf{P}_A = \mathbf{I}$. You must *prove* that $\mathbf{P}_H + \mathbf{P}_A = \mathbf{I}$.

9.3   Show that in the Gram-Schmidt representations of $\mathbf{H}$ and $\mathbf{A}$, the projections $\mathbf{P}_H$ and $\mathbf{P}_A$ have orthogonal representations

$$\mathbf{P}_H = \mathbf{U}_H\mathbf{U}_H^T$$

$$\mathbf{P}_A = \mathbf{U}_A\mathbf{U}_A^T.$$

9.4   Try to diagonalize $\mathbf{S}^T\mathbf{S}$ and $\mathbf{S}(\mathbf{S}^T\mathbf{S})^{-1}\mathbf{S}^T$, when $\mathbf{S} = [\mathbf{H}|\mathbf{A}]$, by using the QR factors $\mathbf{H} = \mathbf{U}_H\mathbf{R}_H$ and $\mathbf{A} = \mathbf{U}_A\mathbf{R}_A$. Show

a. $\mathbf{S}^T\mathbf{S} = \begin{bmatrix} \mathbf{R}_H^T\mathbf{R}_H & \mathbf{0} \\ \mathbf{0} & \mathbf{R}_A^T\mathbf{R}_A \end{bmatrix};$

b. $\mathbf{S}(\mathbf{S}^T\mathbf{S})^{-1}\mathbf{S}^T = \mathbf{U}_H\mathbf{U}_H^T + \mathbf{U}_A\mathbf{U}_A^T.$

This means that the QR factors provide a singly orthogonal representation of $\mathbf{H}$ and $\mathbf{A}$ that orthogonalizes $\mathbf{I}$ but not the Grammian $\mathbf{S}^T\mathbf{S}$.

9.5   Consider the quadratic minimization

$$\min_{\mathbf{x}} \mathbf{x}^T\mathbf{Q}\mathbf{x} \quad \text{subject to} \quad \mathbf{C}^T\mathbf{x} = \mathbf{c}.$$

Derive the solution for $\mathbf{x}$. What conditions do you require on $\mathbf{Q}$ and $\mathbf{C}$?

9.6   Consider the weighted squares

$$(\mathbf{y} - \mathbf{H}\boldsymbol{\theta})^T\mathbf{Q}(\mathbf{y} - \mathbf{H}\boldsymbol{\theta}).$$

The matrix $\mathbf{Q}$ is generally nonsymmetric. Show that any such weighted squares may be replaced by

$$(\mathbf{y} - \mathbf{H}\boldsymbol{\theta})^T\mathbf{W}(\mathbf{y} - \mathbf{H}\boldsymbol{\theta})$$

where $\mathbf{W}$ *is* symmetric.

9.7   We have claimed that Prony's method is the antecedent to the linear prediction techniques of Section 9.19. Tell precisely how the least squares solutions of linear prediction are variations on Prony's method.

9.8   Suppose $x(t)$ is a waveform that is known to consist of a linear combination of p complex exponentials:

$$x(t) = \sum_{n=1}^{p} A_n \exp(\beta_n t + j\omega_n t)$$

β_n (damping coefficient if negative, and expanding coefficient if positive)
ω_n (radian frequency of nth mode).

Sample the waveform at the uniform times $t = kT$, $k = 0, 1, \ldots, N - 1$. Use these samples to solve for the coefficients $(A_1 \cdots A_p)$ and the modes $(z_1 \cdots z_p)$

where $z_n = \exp(\beta_n T + j\omega_n T)$. Discuss Prony's method and least squares linear prediction. How would you determine β_n and ω_n from z_n?

9.9 For each of the following correlation models, give an interpretation of the signal and noise components and find least squares estimates of the parameters to fit $\mathbf{R}$ to an estimated correlation matrix $\mathbf{S}$:

a. $\mathbf{R} = \mathbf{U} \begin{bmatrix} \mathbf{D} & \mathbf{0} \\ \mathbf{0} & \mathbf{N} \end{bmatrix} \mathbf{U}^T; \qquad \mathbf{D} = \mathrm{diag}[\sigma_1^2 \quad \cdots \quad \sigma_p^2],$
$\mathbf{N} = \mathrm{diag}[0 \quad \cdots \quad 0]$

b. $\mathbf{R} = \mathbf{U} \begin{bmatrix} \mathbf{D} & \mathbf{0} \\ \mathbf{0} & \mathbf{N} \end{bmatrix} \mathbf{U}^T; \qquad \mathbf{D} = \mathrm{diag}[\sigma_1^2 \quad \cdots \quad \sigma_p^2],$
$\mathbf{N} = \mathrm{diag}[\sigma_{p+1}^2 \quad \cdots \quad \sigma_N^2]$

c. $\mathbf{R} = \mathbf{U} \begin{bmatrix} \mathbf{D} & \mathbf{0} \\ \mathbf{0} & \mathbf{N} \end{bmatrix} \mathbf{U}^T; \qquad \mathbf{D} = \mathrm{diag}[\sigma_1^2 \quad \cdots \quad \sigma_p^2],$
$\mathbf{N} = \mathrm{diag}[\sigma^2 \quad \cdots \quad \sigma^2]$

d. $\mathbf{R} = \mathbf{U} \begin{bmatrix} \mathbf{D} & \mathbf{0} \\ \mathbf{0} & \mathbf{N} \end{bmatrix} \mathbf{U}^T; \qquad \mathbf{D} = \mathrm{diag}[(\sigma_1^2 + \sigma^2) \quad \cdots \quad (\sigma_p^2 + \sigma^2)],$
$\mathbf{N} = \mathrm{diag}[\sigma^2 \quad \cdots \quad \sigma^2]$

e. $\mathbf{R} = \mathbf{U} \begin{bmatrix} \mathbf{D} & \mathbf{0} \\ \mathbf{0} & \mathbf{N} \end{bmatrix} \mathbf{U}^T; \qquad \mathbf{D} = \mathrm{diag}[\sigma_s^2 \quad \cdots \quad \sigma_s^2],$
$\mathbf{N} = \mathrm{diag}[\sigma^2 \quad \cdots \quad \sigma^2]$

f. $\mathbf{R} = \mathbf{U} \begin{bmatrix} \mathbf{D} & \mathbf{0} \\ \mathbf{0} & \mathbf{N} \end{bmatrix} \mathbf{U}^T; \qquad \mathbf{D} = \mathrm{diag}[(\sigma_s^2 + \sigma^2) \quad \cdots \quad (\sigma_s^2 + \sigma^2)],$
$\mathbf{N} = \mathrm{diag}[\sigma^2 \quad \cdots \quad \sigma^2]$

9.10 Show that the constrained least squares solution $\hat{\boldsymbol{\theta}}_{\mathrm{CLS}}$ may be written as

$$\hat{\boldsymbol{\theta}}_{\mathrm{CLS}} = \mathbf{P}\hat{\boldsymbol{\theta}}_{\mathrm{LS}} + \boldsymbol{\theta}_0,$$

where $\mathbf{P}$ is an oblique (nonsymmetric) projection and $\boldsymbol{\theta}_0$ is orthogonal to $\mathbf{P}$:

$$\mathbf{P} = \mathbf{I} - (\mathbf{H}^T\mathbf{H})^{-1}\mathbf{C}[\mathbf{C}^T(\mathbf{H}^T\mathbf{H})^{-1}\mathbf{C}]^{-1}\mathbf{C}^T = \mathbf{P}^2$$
$$\boldsymbol{\theta}_0 = (\mathbf{H}^T\mathbf{H})^{-1}\mathbf{C}[\mathbf{C}^T(\mathbf{H}^T\mathbf{H})^{-1}\mathbf{C}]^{-1}\mathbf{c}.$$

9.11 Prove that $\mathbf{P}_s = \mathbf{U}_s\mathbf{U}_s^T$ is the projector that minimizes the "total sum of squares"

$$\mathrm{tr}\begin{bmatrix} \mathbf{H}^T \\ \mathbf{y}^T \end{bmatrix}[\mathbf{I} - \mathbf{P}_s][\mathbf{H} \; \mathbf{y}]$$

when

$$\begin{bmatrix} \mathbf{H} \\ \mathbf{y} \end{bmatrix} = \begin{bmatrix} \mathbf{U}_s & \mathbf{u}_N \end{bmatrix} \begin{bmatrix} \mathbf{\Sigma} & \mathbf{0} \\ \mathbf{0} & \sigma_N \end{bmatrix} \begin{bmatrix} \mathbf{V}_s^\mathsf{T} \\ \mathbf{v}_N^\mathsf{T} \end{bmatrix} \qquad \text{and } \sigma_N < \sigma_{N-1} < \cdots < \sigma_1.$$

9.12 Let $\mathbf{y} = [y_1 y_2 \ldots y_N]^\mathsf{T}$ denote N measurements that must be modeled by $\mathbf{x} = \mathbf{H\theta}$. Call $(y_{n_1}, y_{n_2}, \ldots, y_{n_r})$ a subset of the measurements that must be *interpolated* by the model (that is, $(\mathbf{H\theta})_{n_i} = y_{n_i}$):

   a. Set up a constrained minimization problem of the form $\min_\theta (\mathbf{y} - \mathbf{H\theta})^\mathsf{T}(\mathbf{y} - \mathbf{H\theta})$ subject to $\mathbf{C}^\mathsf{T}\mathbf{\theta} = \mathbf{c}$; identify the appropriate $r \times p$ matrix $\mathbf{C}^\mathsf{T}$ and $r \times 1$ vector $\mathbf{c}$;

   b. solve for $\hat{\mathbf{\theta}}_{\mathrm{CLS}}$ and $\hat{\mathbf{x}}_{\mathrm{CLS}}$;

   c. verify that $(\hat{\mathbf{x}}_{\mathrm{CLS}})_{n_i} = y_{n_i}$;

   d. write $\hat{\mathbf{x}}_{\mathrm{CLS}}$ as $(\mathbf{I} - \mathbf{P}_C)\hat{\mathbf{x}}_{\mathrm{LS}}$; write out $\mathbf{P}_C$ and $\mathbf{I} - \mathbf{P}_C$.

9.13 Let $\mathbf{x} = [x_1 \cdots x_N]^\mathsf{T}$ denote an experimental data set and let $[x_{(1)} x_{(2)} \cdots x_{(r)}]^\mathsf{T}$ denote a sparse set of these data that must be interpolated. The problem is to create a modified data set that has small variation. Our definition of variation is

$$e^2 = \mathbf{x}^\mathsf{T}\mathbf{A}^\mathsf{T}\mathbf{A}\mathbf{x}$$

where $\mathbf{A}$ is the matrix

$$A = \begin{bmatrix} -2 & 1 & & & \\ 1 & -2 & 1 & & \mathbf{0} \\ & \ddots & \ddots & \ddots & \\ \mathbf{0} & & 1 & -2 & 1 \\ & & & 1 & -2 \end{bmatrix}.$$

   a. Show that $e^2 = \sum_{n=1}^N (x_{n+1} - 2x_n + x_{n-1})^2$ with $x_0 = 0 = x_{N+1}$; interpret e^2.

   b. Minimize e^2 under the constraint that the modified data set interpolates $(x_{(1)}, x_{(2)}, \ldots, x_{(r)})$.

9.14 In Equation 9.182 we propose the low-rank estimator $\hat{\mathbf{\theta}}_r = \mathbf{V}\mathbf{\Sigma}_r^{-1}\mathbf{U}^\mathsf{T}\mathbf{y}$, with $\mathbf{\Sigma}_r^{-1}$ selected to minimize

$$V = \mathbf{y}^\mathsf{T}(\mathbf{I} - \mathbf{P}_r)\mathbf{y} + \sigma^2 \sum_{i=1}^r \frac{1}{\sigma_{(i)}^2}$$

$$\mathbf{P}_r = \mathbf{U}\mathbf{I}_r\mathbf{U}^\mathsf{T} \qquad \text{(rank-}r \text{ projector corresponding to } \mathbf{\Sigma}_r\text{)}.$$

Show that the best rank-r estimator constructed by this procedure may be obtained by ordering singular values according to the rule

$$(\mathbf{u}_{(1)}^\mathsf{T}\mathbf{y})^2 - \frac{\sigma^2}{\sigma_{(1)}^2} \geq \cdots \geq (\mathbf{u}_{(r)}^\mathsf{T}\mathbf{y})^2 - \frac{\sigma^2}{\sigma_{(r)}^2} \geq \cdots \geq (\mathbf{u}_{(N)}^\mathsf{T}\mathbf{y})^2 - \frac{\sigma^2}{\sigma_{(N)}^2}$$

and then selecting the r dominant triples ($\mathbf{u}_{(i)}$, $\sigma_{(i)}$, and $\mathbf{v}_{(i)}$) to build $\hat{\mathbf{\theta}}_r$.

9.15 The real variables $a_1, a_2, \ldots, a_{p/2}, b_1, b_2, \ldots, b_{p/2}$ are used to modulate a set of parallel tones:

$$x_t = a_1 \cos \omega t + a_2 \cos 2\omega t + \cdots + a_{p/2} \cos(p/2)\omega t + b_1 \sin \omega t + b_2 \sin 2\omega t$$
$$+ \cdots + b_{p/2} \sin(p/2)\omega t.$$

The received signal is noisy:

$$y_t = x_t + n_t.$$

The problem is to observe $y_0, y_1, \ldots, y_{N-1}$, $N > p$, and estimate the symbols $a_1, \ldots, a_{p/2}, b_1, \ldots, b_{p/2}$.

a. Set up the linear model for this problem.

b. Find the least squares estimate of $\boldsymbol{\theta} = [a_1 \quad \cdots \quad a_{p/2} \quad \cdots \quad b_{p/2}]^{\mathrm{T}}$.

c. Find the least squares estimate of $\mathbf{x} = [x_0 \quad x_1 \quad \ldots \quad x_{N-1}]^{\mathrm{T}}$.

d. Compute the (i,j)th element of the Grammian for the linear model. When is it diagonal?

Now assume the n_t are i.i.d. $N[0, 1]$.

e. What is the distribution of $\hat{\boldsymbol{\theta}}$?

f. What is the distribution of $\hat{\mathbf{x}}$?

g. What is the distribution of $\mathbf{y} - \hat{\mathbf{x}}$?

h. What is the distribution of $N^{-1}(\mathbf{y} - \hat{\mathbf{x}})^{\mathrm{T}}(\mathbf{y} - \hat{\mathbf{x}})$?

i. Explain how the least squares estimate is related to the ML and MVUB estimates for this problem.

9.16 *Channel Identification.* A system consisting of a known source, unknown channel, and additive noise is illustrated in Figure P9.1. The problem is to identify the channel unit-pulse response $(h_n)_0^p$ from a noisy record of measurements of the form

$$y_t = \sum_{n=0}^{p} h_n u_{t-n} + n_t.$$

We assume that the input sequence $\{u_n\}$ is zero for $n < 0$ and for $n > N - 1$.

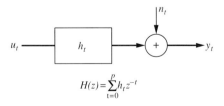

$$H(z) = \sum_{t=0}^{p} h_t z^{-t}$$

Figure P9.1 Channel identification.

We further assume that measurements of $\{y_n\}$ are taken for $0 \le n \le N - 1 + p$. The measurements may be written out as

$$
\begin{bmatrix} y_0 \\ \vdots \\ y_{p-1} \\ \hline y_p \\ \vdots \\ y_{N-1} \\ \hline y_N \\ \vdots \\ y_{N-1+p} \end{bmatrix}
=
\begin{bmatrix}
u_0 & & & \mathbf{0} \\
\vdots & \ddots & & \\
u_{p-1} & \cdots & u_0 & \\
\hline
u_p & \cdots & u_1 & u_0 \\
\vdots & & \ddots & \ddots \\
u_{N-1} & \cdots & & u_{N-1-p} \\
\hline
& u_{N-1} & \cdots & u_{N-p} \\
& & \ddots & \vdots \\
\mathbf{0} & & & u_{N-1}
\end{bmatrix}
\begin{bmatrix} h_0 \\ h_1 \\ \vdots \\ h_p \end{bmatrix} + \mathbf{n}
$$

$$\mathbf{y} = \mathbf{Uh} + \mathbf{n}.$$

This is just one of several ways to define measurements. Sometimes the first and/or last p equations are discarded, leading to the so-called prewindowed, postwindowed, or covariance methods of least squares.

a. What do we mean when we interpret the first p equations as transient, the next $N - p$ as steady-state, and the last p as transient?

b. Solve for the least squares estimate of $\mathbf{h}$.

c. How is $\hat{\mathbf{h}}$ distributed when $\mathbf{n} : N[\mathbf{0}, \mathbf{I}]$?

d. How does the least squares estimate compare with the maximum-likelihood estimate of $\mathbf{h}$ when $\mathbf{n} : N[\mathbf{0}, \mathbf{I}]$?

9.17 The least squares estimate of $\mathbf{h}$ in Problem 9.16 has covariance matrix $(\mathbf{U}^T\mathbf{U})^{-1}$ when $\mathbf{n} : N[\mathbf{0}, \mathbf{I}]$. Show

a. $E(\hat{\mathbf{h}} - \mathbf{h})^T(\hat{\mathbf{h}} - \mathbf{h}) = \operatorname{tr}[(\mathbf{U}^T\mathbf{U})^{-1}]$;

b. $\mathbf{U}^T\mathbf{U} :$ Toeplitz;

c. $\operatorname{tr}[\mathbf{U}^T\mathbf{U}] = (p + 1)\mathbf{u}^T\mathbf{u} = (p + 1)$, when $\mathbf{u}^T\mathbf{u} = 1$;

d. $\operatorname{tr}[(\mathbf{U}^T\mathbf{U})^{-1}] \ge (p + 1)$
    (*hint:* $\operatorname{tr}[\mathbf{U}^T\mathbf{U}] \operatorname{tr}[(\mathbf{U}^T\mathbf{U})^{-1}] = \sum_{ij} \lambda_i/\lambda_j = p + 1 + \sum_{i>j}(\lambda_j/\lambda_i + \lambda_i/\lambda_j)$
    and $\lambda_j/\lambda_i + \lambda_i/\lambda_j \ge 2$ with equality iff $\lambda_i = \lambda_j$);

e. Show $\mathbf{u} = [u_0 \ u_1 \ \cdots \ u_{N-1}]^T = [1 \ 0 \ \cdots \ 0]$ produces $\operatorname{tr}[(\mathbf{U}^T\mathbf{U})^{-1}] = p + 1$. Does it bother you that $\mathbf{u} = [1 \ 0 \ \cdots \ 0]$ is a good sequence? Why?

9.18 *Deconvolution.* The dual of the channel identification problem is the *deconvolution* problem. The unit pulse response of the channel $H(z)$ is known to be $\{h_t\}$, and the input is to be determined. We assume that the input sequence $\{u_t\}$

is zero for $t < 0$ and $t > p$. Then the noisy observations generated from the experimental setup of Figure P9.2 are

$$y_t = \sum_{n=0}^{p} h_{t-n} u_n + n_t; \qquad 0 \le t \le N-1.$$

a. Write out the measurements as $\mathbf{y} = \mathbf{H}\mathbf{u} + \mathbf{n}$; determine $\mathbf{H}$ and $\mathbf{u}$.
b. Find the least squares estimate of $\mathbf{u}$.
c. Find the distribution of $\hat{\mathbf{u}}$ when $\mathbf{n} : N[\mathbf{0}, \sigma^2 \mathbf{I}]$.
d. Discuss reduced-rank procedures for approximating $\hat{\mathbf{u}}$.

9.19 *Data Communication.* There is a variation on Problem 9.18 that is very important for data communication. Suppose, as illustrated in Figure P9.3, that a source sends a sequence of information-bearing symbols $(u_0 u_1 \cdots u_p)$. The symbol u_n is actually "carried" by the causal signal $\{h_t\}_0^\infty$ so that the transmitted sequence corresponding to symbol u_n is

$$\{u_n h_{t-nT}\}$$

The spacing between consecutive symbols is called the baud interval, in this case T. The composite source output may be written

$$s_t = \sum_{n=0}^{p} u_n h_{t-nT}; \qquad t = 0, 1, \ldots, \qquad (T > 1).$$

This kind of signal model also arises when symbols are transmitted over a channel whose unit pulse response is $\{(h_t\}_0^\infty$. Generally the response to u_n overlays the response to u_{n+1} so that we have intersymbol interference.

The noisy measurements corresponding to the p transmitted symbols are

$$y_t = \sum_{n=0}^{p} u_n h_{t-nT} + n_t; \qquad t = 0, 1, \ldots.$$

a. Sketch a typical signal $\{s_t\}$.
b. Organize measurements $\{y_t\}$ into a vector $\mathbf{y}$ and write it as $\mathbf{y} = \mathbf{H}\mathbf{u} + \mathbf{n}$. Show carefully how the matrix $\mathbf{H}$ is constructed.

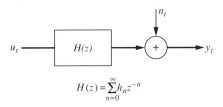

$$H(z) = \sum_{n=0}^{\infty} h_n z^{-n}$$

Figure P9.2 Deconvolution.

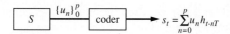

Figure P9.3 Communication source.

c. Find the least squares solution for **u**.

d. Show that the nth element of the matrix $\mathbf{H}^T\mathbf{y}$ in the least squares solution for **u** may be written

$$(\mathbf{H}^T\mathbf{y})_n = [\mathbf{0}^T \cdots \mathbf{0}^T\mathbf{h}^T]\mathbf{y}.$$

This is an abstract way of saying that the vector **h** is correlated with delayed elements of **y**.

e. Show how the least squares estimate of **u** may be implemented with a digital filter.

9.20 Prove that the unconstrained minimization of $\left(\mathbf{w}_C + (\mathbf{I}-\mathbf{P}_C)\mathbf{w}\right)^T\mathbf{R}\left(\mathbf{w}_C + (\mathbf{I}-\mathbf{P}_C)\mathbf{w}\right)$ really does produce $(\mathbf{I}-\mathbf{P}_C)\mathbf{w} = (\mathbf{I}-\mathbf{P}_C)\mathbf{w}_0$ where $\mathbf{w}_0$ is the constrained solution $\mathbf{w}_0 = \mathbf{R}^{-1}\mathbf{C}(\mathbf{C}^T\mathbf{R}^{-1}\mathbf{C})^{-1}\mathbf{c}$.

9.21 Prove that the mean-squared error between $\mathbf{w}_C^T\mathbf{P}_C\mathbf{x}$ and $-\mathbf{w}_C^T(\mathbf{I} - \mathbf{P}_C)\mathbf{x}$ is the mean-squared error of Equation 9.156 when $E\mathbf{x}\mathbf{x}^T = \mathbf{R}$.

9.22 Rewrite the constraints for condition adjustment (Section 9.13) so that they are homogeneous. Then rewrite the solution for $\hat{\boldsymbol{\theta}}_{\text{CLS}}$ as a projection of the measurement **y**. Interpret your results.

Linear Prediction

with Cédric Demeure

In this chapter we study the classical and modern theories of linear prediction for stationary time series. The classical problem is to predict a value of the time series x_t from the infinite past $\{x_s, \ s < t\}$. The Wold and Kolmogorov representations, which we call *stationary* innovations representations, may be used to derive elegant formulas for prediction error polynomials and for prediction error variances.

The modern problem is to predict x_t from the finite past $\{x_s, \ 0 \le s < t\}$. The Wold and Kolmogorov representations are replaced by *nonstationary* innovations representations that Cholesky-factor the covariance matrix, and its inverse, for the vector $\mathbf{x}_t = [x_0 \quad x_1 \quad \cdots \quad x_t]^T$. These nonstationary innovations representations are computed with the Schur and Levinson algorithms for factoring a Toeplitz correlation matrix and its inverse, respectively.

In the least squares theory of linear prediction, the problem is to estimate x_t from the finite past $\{x_s,\ 0 \le s < t\}$ when *only* experimental data, and *not* mathematical correlations, are available. This problem is closely related to the nonstationary prediction problem. Either the Levinson or Schur recursions may be used to solve for the least squares predictor. However, there is a third algorithm, called a lattice algorithm, that is a square-root version of these. Its numerical properties are superior to the Schur and Levinson algorithms.

When the underlying time series is autoregressive-moving-average (ARMA), then the stationary and nonstationary prediction formulas take a very special and illuminating form. In particular, the stationary innovations representation produces a Markovian representation, and the nonstationary innovations representation produces a nonstationary Markovian representation. The Schur recursions produce the Morf-Sidhu-Kailath (MSK) recursions for computing Kalman gains in the nonstationary innovations and predictor representations of an ARMA time series. The MSK recursions produce fast Kalman predictor formulas.

Finally, we show how the representation formulas of linear prediction theory may be used to derive likelihood formulas for snapshots, $\mathbf{x}^T = [x_0 \quad x_1 \quad \cdots \quad x_t]^T$, of stationary Gaussian time series.

10.1   STATIONARY TIME SERIES

In this section, we review a few fundamental notions from the theory of stationary time series. Let $\{x_t\}$ denote a real, wide-sense stationary time series with zero mean and symmetric correlation sequence $\{r_t\}$:

$$Ex_t = 0 \qquad\qquad 10.1$$

$$Ex_n x_{n+t} = r_t = r_{-t}.$$

We shall assume that the correlation sequence is absolutely summable: $\sum_{t=-\infty}^{\infty} |r_t| < \infty$. The power spectral density for the time series is the Fourier transform of the correlation sequence:

$$S(e^{j\theta}) = \sum_{t=-\infty}^{\infty} r_t e^{-jt\theta} \qquad\qquad 10.2$$

$$r_t = \int_0^{2\pi} S(e^{j\theta}) e^{jt\theta} \frac{d\theta}{2\pi}.$$

These transform relations go by the names of Bochner [1933], Herglotz [1960], Yaglom [1952], Khinchin [1934], and Wiener [1947], all of whom showed that every nonnegative definite correlation sequence is the inverse discrete-time Fourier transform of a nonnegative power spectral density.

Rather surprisingly, the log spectrum plays a fundamental role in the classical theory of linear prediction. The reason is that the log spectrum may be used to derive an infinite moving average representation for a stationary time series. We shall assume that the log spectrum has a uniformly convergent Fourier series expansion in which its coefficients are absolutely summable:

$$\log S(e^{j\theta}) = \sum_{t=-\infty}^{\infty} c_t e^{-jt\theta}$$

$$c_t = \int_0^{2\pi} \log S(e^{j\theta}) e^{jt\theta} \frac{d\theta}{2\pi} \tag{10.3}$$

$$\sum_{t=-\infty}^{\infty} |c_t| < \infty.$$

The Fourier series coefficients c_t are called cepstral coefficients, and the sequence $\{c_t\}$ is called the cepstrum (pronounced *kepstrum*). The condition that the cepstrum be absolutely summable ensures two other conditions that we shall need shortly:

$$1. \; -\infty < c_0 = \int_0^{2\pi} \log S(e^{j\theta}) \frac{d\theta}{2\pi} < \infty; \tag{10.4}$$

$$2. \; -\infty < \sum_{t=0}^{\infty} c_t z^{-t} < \infty \qquad \text{for all } |z| > 1.$$

The power spectral density is nonnegative and real. Therefore it may be written as the magnitude squared of a complex function $H(e^{j\theta})$:

$$S(e^{j\theta}) = H(e^{j\theta})H^{\star}(e^{j\theta}). \tag{10.5}$$

The log spectrum is then

$$\log S(e^{j\theta}) = \log H(e^{j\theta}) + \log H^{\star}(e^{j\theta})$$

$$= \sum_{t=-\infty}^{\infty} c_t e^{-jt\theta}. \tag{10.6}$$

We may arbitrarily define $\log H(e^{j\theta})$ to be the causal Fourier series

$$\log H(e^{j\theta}) = \frac{c_0}{2} + \sum_{t=1}^{\infty} c_t e^{-jt\theta} \tag{10.7}$$

$$H(e^{j\theta}) = \exp\left\{ \frac{c_0}{2} + \sum_{t=1}^{\infty} c_t e^{-jt\theta} \right\}.$$

This Fourier series expansion for $\log H(e^{j\theta})$ is uniformly convergent, as is the expansion for $H(e^{j\theta})$.

This procedure for characterizing the function $H(e^{j\theta})$ is usually credited to Kolmogorov [1939]. Its significance is that it produces an infinite moving average representation for the time series $\{x_t\}$. To illustrate, let's write $H(e^{j\theta})$ as follows:

$$H(e^{j\theta}) = \exp\left\{ \frac{c_0}{2} + \sum_{t=1}^{\infty} c_t e^{-jt\theta} \right\}$$

$$= \sum_{t=0}^{\infty} h_t e^{-jt\theta}.$$
$$(10.8)$$

But this is just the complex frequency response of a digital filter whose transfer function is $H(z)$ and whose unit pulse response sequence is $\{h_t\}$. If this filter is excited by a sequence of zero-mean uncorrelated random variables $\{u_t\}$, then the output of the filter may be written as

$$\{x_t\} = H(z)\{u_t\} \tag{10.9}$$

$$x_t = \sum_{n=0}^{\infty} h_n u_{t-n}$$

$$H(z) = \sum_{n=0}^{\infty} h_n z^{-n}$$

We say that the time series $\{x_t\}$ may be synthesized by passing white noise $\{u_t\}$ through the infinite moving average filter $H(z)$. The correlation sequence and the power spectral density for the time series $\{x_t\}$ are

$$r_t = E x_n x_{n+t}$$

$$= \sum_{n=0}^{\infty} h_n h_{n+t}$$
$$(10.10)$$

$$S(e^{j\theta}) = H(e^{j\theta})H^{\star}(e^{j\theta}).$$

So, the filter $H(z) = \sum_{t=0}^{\infty} h_t z^{-t}$, with the coefficients chosen according to the Kolmogorov procedure, may be used to synthesize a sequence $\{x_t\}$ with the desired power spectral density $S(e^{j\theta})$. The power spectrum $S(e^{j\theta})$ is said to be an infinite moving average spectrum.

Wold Representation

The representation that we have developed in the previous paragraphs, namely

$$\{x_t\} = H(z)\{u_t\}$$

$$x_t = \sum_{n=0}^{\infty} h_n u_{t-n}$$
$$10.11$$

$$E u_n u_{n+t} = \delta_t$$

is called the Wold representation for a stationary time series [Wold, 1954]. It is also called a synthesis or coloring representation because it synthesizes a time series $\{x_t\}$, with colored spectrum, from a time series $\{u_t\}$ that has the white spectrum

$$S_{uu}(e^{j\theta}) = \sum_{t=-\infty}^{\infty} \delta_t e^{-jt\theta} = 1. \tag{10.12}$$

The input-output representation of the power spectrum $S(e^{j\theta})$ is

$$S(e^{j\theta}) = \left|H(e^{j\theta})\right|^2 S_{uu}(e^{j\theta}). \tag{10.13}$$

This result shows that every uniformly convergent power spectrum is an infinite moving average spectrum.

The filter $H(z)$ in the Wold representation is stable *and* minimum phase. That is,

1. $\displaystyle\sum_{t=0}^{\infty} |h_t| < \infty;$ \hfill (10.14)

2. $\displaystyle\sum_{t=0}^{\infty} h_t z^{-t} \neq 0, \qquad |z| > 1.$

Condition (1) is the stability condition and condition (2) is the minimum phase condition that says all zeros of $H(z)$ lie inside the unit circle $|z| = 1$. Stability follows immediately from the definition for h_t. The minimum phase condition follows from the fact that, for $|z| > 1$, $\left|\log H(z)\right|$ may be written as

$$\left|\log \sum_{t=0}^{\infty} h_t z^{-t}\right| = \left|\frac{c_0}{2} + \sum_{t=1}^{\infty} c_t z^{-1}\right| \leq \frac{c_0}{2} + \sum_{t=1}^{\infty} |c_t| < \infty. \tag{10.15}$$

Kolmogorov Representation

The inverse of the stable minimum phase filter $H(z)$ is also stable and minimum phase. That is, $H(z)$ has the inverse representation

$$H(z) = \frac{h_0}{A(z)} \tag{10.16}$$

$$A(z) = \frac{h_0}{H(z)} = \sum_{n=0}^{\infty} a_n z^{-n}; \qquad a_0 = 1.$$

From the condition $H(z)A(z) = h_0$, we see that the zeros of $H(z)$ are the poles of $A(z)$, making $A(z)$ stable and minimum phase. This means that the Wold representation

$$\{x_t\} = H(z)\{u_t\} \tag{10.17}$$

may be rewritten as the Kolmogorov representation

$$A(z)\{x_t\} = h_0\{u_t\} \tag{10.18}$$

$$\sum_{n=0}^{\infty} a_n x_{t-n} = h_0 u_t.$$

We say that the time series $\{x_t\}$ may be synthesized by passing white noise through the infinite autoregressive filter $h_0/A(z)$. The Kolmogorov representation is called an analysis or whitening representation because the sequence $\{x_t\}$, with colored spectrum $S(e^{j\theta})$, is analyzed to produce the sequence $\{u_t\}$ that has a white spectrum. The input-output representation of the power spectrum $S(e^{j\theta})$ is

$$\left|A(e^{j\theta})\right|^2 S(e^{j\theta}) = h_0^2 \tag{10.19}$$

or

$$S(e^{j\theta}) = \frac{h_0^2}{\left|A(e^{j\theta})\right|^2}. \tag{10.20}$$

This result shows that every uniformly convergent power spectrum is also an infinite autoregressive spectrum.

Filtering Interpretations

Let's summarize. Every wide-sense stationary (WSS) time series $\{x_t\}$ with uniformly convergent power spectral density $S(e^{j\theta})$ has two causal, stable, and minimum phase representations

1. $x_t = \sum_{n=0}^{\infty} h_n u_{t-n}$

    $\{x_t\} = H(z)\{u_t\}; \tag{10.21}$

2. $\sum_{n=0}^{\infty} a_n x_{t-n} = h_0 u_t; \qquad a_0 = 1$

    $A(z)\{x_t\} = h_0\{u_t\}.$

The filter $H(z)$, illustrated in Figure 10.1(a), is an infinite moving average (MA) filter that synthesizes or colors $\{x_t\}$. The filter $A(z)$, illustrated in Figure 10.1(b), is an infinite MA filter that analyzes or whitens $\{x_t\}$. Of course, we may also say that $h_0/A(z)$ is an infinite autoregressive (AR) filter that colors $\{x_t\}$ and that $1/H(z)$ is an infinite AR filter that whitens $\{x_t\}$:

- $\{x_t\} = \dfrac{h_0}{A(z)}\{u_t\}; \tag{10.22}$

- $\dfrac{1}{H(z)}\{x_t\} = \{u_t\}.$

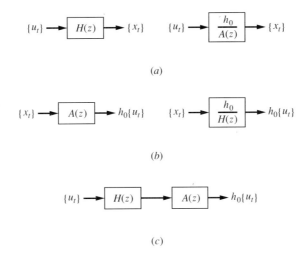

Figure 10.1 Synthesis (a), analysis (b), and inverse (c) filters.

These filters are also illustrated in Figure 10.1(a) and (b). The filters $H(z)$ and $A(z)$ are related in the following way:

$$A(z)H(z) = h_0. \qquad\qquad 10.23$$

We say that $A(z)$ and $H(z)$ are inverse filters, as illustrated in Figure 10.1(c).

10.2   STATIONARY PREDICTION THEORY

What is the linear, minimum mean-squared error predictor of x_t, given the infinite past of the time series $\{x_s,\ s < t\}$? The answer to this question generates a wealth of classical results from the theory of stationary linear prediction.

Recall the Kolmogorov representation for the stationary time series $\{x_t\}$:

$$\sum_{n=0}^{\infty} a_n x_{t-n} = h_0 u_t; \qquad a_0 = 1. \qquad\qquad 10.24$$

This representation may be rewritten in the "predictor" form

$$x_t = \hat{x}_{t|t-1} + h_0 u_t \qquad\qquad 10.25$$

$$\hat{x}_{t|t-1} = -\sum_{n=1}^{\infty} a_n x_{t-n}$$

The variable $\hat{x}_{t|t-1}$ is a linear function of the past of the time series $\{x_s,\ s < t\}$. The error between x_t and $\hat{x}_{t|t-1}$ is the random variable $h_0 u_t$:

$$h_0 u_t = x_t - \hat{x}_{t|t-1}. \qquad\qquad 10.26$$

Now consider the correlation between this error and the past of the time series $\{x_s, s < t\}$:

$$E(x_t - \hat{x}_{t|t-1})x_s = Eh_0 u_t x_s, \qquad s < t. \tag{10.27}$$

From the Wold decomposition, we know that x_s is a linear combination of the past of $\{u_s, s < t\}$. Consequently,

$$E(x_t - \hat{x}_{t|t-1})x_s = Eh_0 u_t \sum_{n=0}^{\infty} h_n u_{s-n} \tag{10.28}$$

$$= 0, \qquad t > s.$$

The linear estimator $\hat{x}_{t|t-1}$ generates errors that are orthogonal to the past of the time series. Consequently, from the principle of orthogonality studied in Chapter 8 the linear estimator

$$\hat{x}_{t|t-1} = -\sum_{n=1}^{\infty} a_n x_{t-n} \tag{10.29}$$

is the *linear minimum mean-squared error* predictor of x_t from its past $\{x_s, s < t\}$.

Prediction Error Variance

The variance of the predictor $\hat{x}_{t|t-1}$ is

$$\sigma^2 = E(x_t - \hat{x}_{t|t-1})^2 = E(h_0 u_t)^2 \tag{10.30}$$

$$= h_0^2.$$

This result makes the zeroth coefficient in the Wold decomposition a quantity of fundamental importance. However, we can do even more in the way of interpretation. From the Kolmogorov procedure for determining the synthesis filter $H(z)$ from the log spectrum, we observe that h_0 is related to the zeroth cepstral coefficient as follows:

$$h_0 = \exp\left\{\frac{c_0}{2}\right\}. \tag{10.31}$$

But the zeroth cepstral coefficient is the integral of the log spectrum, so we may write the prediction error as

$$\sigma^2 = h_0^2 = \exp(c_0) \tag{10.32}$$

$$= \exp\left\{\int_0^{2\pi} \log S(e^{j\theta}) \frac{d\theta}{2\pi}\right\}.$$

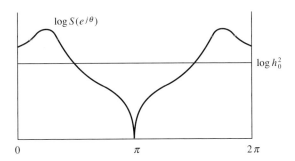

Figure 10.2 Averaged log spectrum equals log of prediction error variance.

This is the celebrated Szegö-Kolmogorov-Krein theorem for prediction error [Whittle, 1954]. Note that it may be written in the following two ways:

1. $\displaystyle \int_0^{2\pi} \log S(e^{j\theta}) \, \frac{d\theta}{2\pi} = \log h_0^2;$ 10.33

2. $\displaystyle \int_0^{2\pi} \log \frac{S(e^{j\theta})}{h_0^2} \, \frac{d\theta}{2\pi} = 0.$

The first of these equations says that the average of the log spectrum equals the log of the prediction error. This is illustrated in Figure 10.2. The second expression says that the normalized log spectrum integrates to zero.

Prediction Error Variance and Poles and Zeros

When the filter $H(z)$ is autoregressive-moving-average with p poles and q zeros, then the prediction error may be written in terms of the poles and zeros of the spectrum $S(z)$. The idea is to write $S(z) = H(z)H(z^{-1})$ in the form

$$S(z) = z^{-q+p}G(z), \qquad\qquad 10.34$$

where $G(z)$ has no poles or zeros at $z = 0$, and then to use Jensen's formula [Lang, 1977] in the equation for prediction error:

$$\sigma^2 = \exp\left\{ \int_0^{2\pi} \log S(e^{j\theta}) \, \frac{d\theta}{2\pi} \right\} = \exp\left\{ \int_0^{2\pi} \log \left| S(e^{j\theta}) \right| \, \frac{d\theta}{2\pi} \right\}$$

$$ = \exp\left\{ \int_0^{2\pi} \log \left| G(e^{j\theta}) \right| \, \frac{d\theta}{2\pi} \right\}$$

10.35

Jensen's formula shows this to equal

$$\sigma^2 = \left| G(0) \frac{\prod\limits_{i=1}^{p} p_i}{\prod\limits_{i=1}^{q} z_i} \right|.$$ 10.36

In this formula, the z_i and p_i are zeros and poles of $G(z)$ that lie *inside* the unit circle; these are the zeros and poles of $H(z)$. This result may be illustrated with two examples.

Example 10.1

Let $\{r_t\} = \{\rho^{|t|}\}$, with $|\rho| < 1$, denote the covariance sequence for a wide-sense stationary time series. The corresponding spectrum is

$$S(z) = \sum_{t=-\infty}^{\infty} \rho^{|t|} z^{-t} = \frac{1 - \rho^2}{(1 - \rho z^{-1})(1 - \rho z)}$$ 10.37

$$= zG(z); \qquad G(z) = \frac{1 - \rho^2}{(z - \rho)(1 - \rho z)}.$$

The prediction error is then

$$\sigma^2 = |G(0)\rho| = 1 - \rho^2.$$ 10.38

You can see that, as $\rho \to 1$, the process is highly correlated over large lags, and the prediction error variance is zero; conversely, when $\rho \to 0$, the process becomes uncorrelated and the prediction error variance is 1. ∎

Example 10.2

When the rational factors of $S(z)$ are known, we may write $S(z) = H(z)H(z^{-1})$. Then $H(z)$ and $H(z^{-1})$ may be written as

$$H(z) = \frac{\sum\limits_{n=0}^{q} b_n z^{-n}}{\sum\limits_{n=0}^{p} a_n z^{-n}}; \qquad a_0 = 1$$

10.39

$$= \frac{b_0 \prod\limits_{n=1}^{q} (1 - z_n z^{-1})}{\prod\limits_{n=1}^{p} (1 - p_n z^{-1})} = b_0 z^{-q+p} \frac{\prod\limits_{n=1}^{q} (z - z_n)}{\prod\limits_{n=1}^{p} (z - p_n)}$$

$$H(z^{-1}) = \frac{b_0 \prod\limits_{n=1}^{q} (1 - z_n z)}{\prod\limits_{n=1}^{p} (1 - p_n z)}.$$

The spectrum $S(z)$ then has the form

$$S(z) = H(z)H(z^{-1}) = b_0^2 z^{-q+p} \frac{\prod\limits_{n=1}^{q} (z - z_n)(1 - z_n z)}{\prod\limits_{n=1}^{p} (z - p_n)(1 - p_n z)} \qquad 10.40$$

$$= z^{-q+p} G(z)$$

where $G(z)$ is

$$G(z) = b_0^2 \frac{\prod\limits_{n=1}^{q} (z - z_n)(1 - z_n z)}{\prod\limits_{n=1}^{p} (z - p_n)(1 - p_n z)}. \qquad 10.41$$

The function $G(z)$ has no poles or zeros at $z = 0$, and $G(0)$ is

$$|G(0)| = b_0^2 \frac{\prod\limits_{n=1}^{q} |z_n|}{\prod\limits_{n=1}^{p} |p_n|}. \qquad 10.42$$

The prediction error is therefore

$$\sigma^2 = \frac{\left| G(0) \prod\limits_{i=1}^{p} p_i \right|}{\prod\limits_{i=1}^{q} z_i} = b_0^2. \qquad 10.43$$

This is a perfectly general result for ARMA time series. Of course, $b_0^2 = h_0^2$, so this result is consistent with the more general result of Equation 10.30. ∎

Spectral Flatness

The prediction error h_0^2 is bounded from above by r_0, the variance of the time series. That is,

$$r_0 = \sum_{t=0}^{\infty} h_t^2 = h_0^2 + \sum_{t=1}^{\infty} h_t^2 \geq h_0^2 \qquad 10.44$$

$$r_0 = \int_0^{2\pi} S(e^{j\theta}) \frac{d\theta}{2\pi}.$$

The bound is achieved when $S(e^{j\theta})$ is the flat, or white, spectrum $S(e^{j\theta}) = r_0$, in which case $h_0^2 = r_0$:

$$h_0^2 = \exp\left\{\int_0^{2\pi} \log r_0 \, \frac{d\theta}{2\pi}\right\} \qquad 10.45$$

$$= r_0.$$

In this case, we say that the past brings no information about the future. This suggests that we use the ratio h_0^2 to r_0 to measure the flatness of a spectrum:

$$\eta = h_0^2/r_0$$

$$= \frac{\exp\left\{\int_0^{2\pi} \log S(e^{j\theta}) \, \frac{d\theta}{2\pi}\right\}}{\int_0^{2\pi} S(e^{j\theta}) \, \frac{d\theta}{2\pi}} \leq 1. \qquad 10.46$$

This measure of spectral flatness has found widespread application in speech processing.

Filtering Interpretations

The linear prediction formulas we have derived may be given a number of interesting filtering interpretations. The Wold representation may be written in the predictor form

$$\{x_t\} = \{\hat{x}_{t|t-1}\} + h_0\{u_t\} \qquad 10.47$$

$$\{\hat{x}_{t|-1}\} = P_1(z)\{u_t\}$$

where $P_1(z)$ is the following predictor polynomial:

$$P_1(z) = H(z) - h_0 = \sum_{n=1}^{\infty} h_n z^{-n}. \qquad 10.48$$

The filtering diagrams of these equations are illustrated in Figure 10.3(a) and (b). In Figure 10.3(b), the input $\{u_t\}$ has been synthesized as a prediction error between $\{x_t\}$ and $\{\hat{x}_{t|t-1}\}$.

The Kolmogorov representation may be written in the predictor form

$$\{x_t\} - \{\hat{x}_{t|t-1}\} = h_0\{u_t\} \qquad 10.49$$

$$\{\hat{x}_{t|t-1}\} = P_2(z)\{x_t\}$$

where $P_2(z)$ is the predictor polynomial

$$P_2(z) = 1 - A(z) = -\sum_{n=1}^{\infty} a_n z^{-n}. \qquad 10.50$$

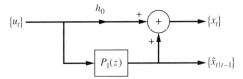

(*a*) synthesis and predictor filters associated with the Wold representation

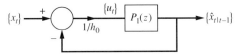

(*b*) feedback-predictor diagram associated with the Wold representation

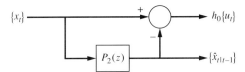

(*c*) analysis and predictor filters associated with the Kolmogorov representation

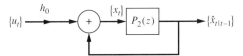

(*d*) feedback diagram associated with the Kolmogorov representation

Figure 10.3 Prediction filters.

These equations are illustrated in Figure 10.3(*c*) and (*d*). In Figure 10.3(*d*), the input $\{x_t\}$ has been synthesized from the sum of $h_0\{u_t\}$ and $\{\hat{x}_{t|t-1}\}$.

The predictor diagrams of Figure 10.3(*b*) and (*c*) are equivalent. That is,

$$P_2(z) = \frac{P_1(z)}{h_0 + P_1(z)} \tag{10.51}$$

$$H(z)P_2(z) = P_1(z).$$

Similarly, the prediction diagrams of Figure 10.3(*a*) and (*d*) are equivalent; that is,

$$P_1(z) = \frac{h_0 P_2(z)}{1 - P_2(z)} \tag{10.52}$$

$$A(z)P_1(z) = h_0 P_2(z).$$

Figure 10.3(*b*) and (*c*) are practical filtering diagrams that show two equivalent ways to predict $\{x_t\}$ from its past. The first is a feedback diagram that uses the coefficients of a synthesis filter in its feedforward loop, and the second is a feedforward diagram that uses the coefficients of an analysis filter.

10.3   MAXIMUM ENTROPY AND LINEAR PREDICTION

Suppose you are given the first $p+1$ values of an autocorrelation sequence $\{r_t\}_{-\infty}^{\infty}$ and are asked to say something about the rest of the sequence or, equivalently, about the corresponding power spectral density. There are an infinity of answers you could give, because there are an infinity of ways to extend the given sequence while ensuring that the result is nonnegative definite. Among the possible extensions perhaps there is a favored one, and perhaps the favored extension produces an interesting power spectral density.

We have seen in our discussion of stationary linear prediction that a WSS time series with a white power spectrum is the least predictable time series with a specified variance of r_0. In this case, the specification of r_0 delivers an unshaped spectrum for the least predictable time series. Perhaps the specification of r_0 and subsequent correlations $r_1, r_2, \ldots, r_p$ will deliver a sequence of "conservatively" shaped spectra for the corresponding least predictable sequence. This mode of thought motivates us to pose the question, "What is the power spectral density of the least predictable time series $\{x_t\}$ whose first $p+1$ autocorrelations are $\{r_0, r_1, \ldots, r_p\}$?" This question is equivalent to the question, "What extension of $\{r_0, \ldots, r_p\}$ produces the correlation sequence for the least predictable time series $\{x_t\}$?"

In order to answer these questions, we define the Lagrangian

$$\mathcal{L} = \int_0^{2\pi} \log S(e^{j\theta}) \frac{d\theta}{2\pi} - \sum_{n=0}^{p} \lambda_n(\hat{r}_n - r_n) \qquad 10.53$$

where $S(e^{j\theta})$ is the power spectral density

$$S(e^{j\theta}) = \sum_{n=-\infty}^{\infty} \hat{r}_n e^{-jn\theta} = \hat{r}_0 + \sum_{n=1}^{\infty} \hat{r}_n(e^{-jn\theta} + e^{jn\theta}) \qquad 10.54$$

$$\hat{r}_n = \int_0^{2\pi} S(e^{j\theta})e^{jn\theta} \frac{d\theta}{2\pi} = \frac{1}{2} \int_0^{2\pi} S(e^{j\theta})(e^{jn\theta} + e^{-jn\theta}) \frac{d\theta}{2\pi}.$$

The first term in the Lagrangian $\mathcal{L}$ is the log of the prediction error variance for a time series whose power spectral density is $S(e^{j\theta})$. The second term is a constraint term that will force $\hat{r}_n$ to equal the given autocorrelation r_n for $n = 0, \ldots, p$, provided λ_n is chosen properly. In our notation, $\{\hat{r}_n\}$ is the autocorrelation sequence for the time series $\{x_t\}$, and $\{r_n\}_0^p$ are specified values for the first $p+1$ values of the sequence.

The prediction error variance is maximized with respect to the autocorrelation sequence by taking partials of $\mathcal{L}$ with respect to $\hat{r}_n$ and equating these partials to zero:

1. $\dfrac{\partial \mathcal{L}}{\partial \hat{r}_0} = \displaystyle\int_0^{2\pi} \frac{1}{S(e^{j\theta})} \frac{d\theta}{2\pi} - \lambda_0 = 0$

2. $$\frac{\partial \mathscr{L}}{\partial \hat{r}_n} = \frac{1}{2} \int_0^{2\pi} \left(\frac{e^{-jn\theta} + e^{jn\theta}}{S(e^{j\theta})} \right) \frac{d\theta}{2\pi} - \lambda_n = 0, \ 1 \le n \le p \qquad 10.55$$

3. $$\frac{\partial \mathscr{L}}{\partial \hat{r}_n} = \frac{1}{2} \int_0^{2\pi} \left(\frac{e^{-jn\theta} + e^{jn\theta}}{S(e^{j\theta})} \right) \frac{d\theta}{2\pi} = 0, \ n > p.$$

The function $1/S(e^{j\theta})$ is an even function with its own Fourier series:

$$\frac{1}{S(e^{j\theta})} = \sum_{n=-\infty}^{\infty} s_n e^{-jn\theta} = s_0 + \sum_{n=1}^{\infty} s_n(e^{-jn\theta} + e^{jn\theta}) \qquad 10.56$$

$$s_n = \int_0^{2\pi} \left(\frac{e^{jn\theta}}{S(e^{j\theta})} \right) \frac{d\theta}{2\pi} = \frac{1}{2} \int_0^{2\pi} \left(\frac{e^{-jn\theta} + e^{jn\theta}}{S(e^{j\theta})} \right) \frac{d\theta}{2\pi} = s_{-n}$$

The equation for $\partial \mathscr{L}/\partial \hat{r}_n$, $n > p$, says that $s_n = 0$, and the equation for $\partial \mathscr{L}/\partial \hat{r}_n$, $1 \le n \le p$, says that $s_n = \lambda_n$. Therefore, the inverse spectrum has the representation

$$\frac{1}{S(e^{j\theta})} = \sum_{n=-p}^{p} \lambda_n e^{-jn\theta}. \qquad 10.57$$

This makes $1/S(e^{j\theta})$ a moving average spectrum of order p. Such a spectrum can be factored as the product $1/S(e^{j\theta}) = A(e^{j\theta})A(e^{-j\theta})/\sigma^2$. The spectrum itself is the autoregressive spectrum

$$S(e^{j\theta}) = \frac{\sigma^2}{A(e^{j\theta})A(e^{-j\theta})} \qquad 10.58$$

with $A(z) = 1 + a_1 z^{-1} + \cdots + a_p z^{-p}$. The polynomial $A(z)$ is unique when forced to have all its roots inside the unit circle. As there is a one-to-one correspondence between the set of Lagrangian coefficients $\{\lambda_n\}_{n=0}^{p}$ and the set of filter coefficients $\{\sigma^2, (a_i)_1^p\}$, this latter set may be used to enforce the constraints. Write the spectrum $S(e^{j\theta})$ as

$$A(e^{j\theta})S(e^{j\theta}) = \sigma^2 H(e^{-j\theta}) \qquad 10.59$$

$$H(e^{-j\theta}) = \frac{1}{A(e^{-j\theta})}.$$

The inverse transform of this equation is

$$\sum_{n=0}^{p} a_n \hat{r}_{t-n} = \sigma^2 h_{-t}; \qquad a_0 = 1$$

$$\qquad\qquad\qquad\qquad = \sigma^2 \delta_t, \qquad \text{if } t \ge 0 \qquad\qquad\qquad 10.60$$

where h_t is the causal inverse of $H(e^{j\theta}) = 1/A(e^{j\theta})$ and h_{-t} is its anticausal time reversal.

This system of equations may be written out for $t = 0, 1, 2, \ldots, p$ to produce the Toeplitz normal equations (note $\hat{r}_t = r_t$ for $t \leq p$)

$$
\begin{bmatrix}
r_0 & r_1 & \cdots & r_p \\
r_1 & r_0 & \ddots & \vdots \\
\vdots & \ddots & \ddots & r_1 \\
r_p & \cdots & r_1 & r_0
\end{bmatrix}
\begin{bmatrix}
a_p \\
\vdots \\
a_1 \\
1
\end{bmatrix}
=
\begin{bmatrix}
0 \\
\vdots \\
0 \\
\sigma^2
\end{bmatrix}.
\qquad 10.61
$$

When solved for σ^2 and $(a_i)_1^p$, these equations produce the solution to our constrained minimization. If the Toeplitz matrix $\mathbf{R}$ is positive definite, the solution is unique. All of this says that the maximum entropy spectrum is the autoregressive spectrum

$$
S(e^{j\theta}) = \frac{\sigma^2}{A(e^{j\theta})A(e^{-j\theta})}
\qquad 10.62
$$

where σ^2 and $A(e^{j\theta}) = \sum_{n=0}^{p} a_n e^{-jn\theta}$ are determined from the Toeplitz normal equations. This spectrum matches the covariances $\{r_0, r_1, \ldots, r_p\}$ and extends the covariance sequence autoregressively according to the equation $\sum_{n=0}^{p} a_n r_{t-n} = 0$, $t > p$. There is no difference between maximum entropy extension of a covariance sequence and autoregressive extension.

10.4   NONSTATIONARY LINEAR PREDICTION

The problem we address here is the linear, minimum mean-squared error prediction of x_t from its *finite* past $\{x_s,\ 0 \leq s < t\}$. The results that we obtain apply to very general correlations between the variables. However, when the vector $\mathbf{x}_t = [x_0 \ x_1 \ \cdots \ x_t]^{\mathrm{T}}$ is a snapshot from a stationary time series, the prediction error filters and prediction error variances take on special recursive forms.

For each value of time t, the linear predictor of x_t is a linear combination of variables $\{x_s,\ 0 \leq s < t\}$, using predictor coefficients a_n^t that depend on t:

$$
\hat{x}_{t|t-1} = -\sum_{n=1}^{t} a_n^t x_{t-n}.
\qquad 10.63
$$

As t increases, the coefficients a_n^t of the predictor change and the predictor order increases. We call $P_t(z)$ the order t predictor and $A_t(z)$ the order t whitener:

$$
P_t(z) = -\sum_{n=1}^{t} a_n^t z^{-n}
\qquad 10.64
$$

$$
A_t(z) = \sum_{n=0}^{t} a_n^t z^{-n} = 1 - P_t(z).
$$

The simplest statement of the linear prediction problem is to minimize the mean-squared error between x_t and $\hat{x}_{t|t-1}$ with respect to the predictor coefficients:

$$\min_{P_t(z)} \sigma_t^2 \qquad\qquad 10.65$$

$$\sigma_t^2 = E\left(x_t + \sum_{n=1}^{t} a_n^t x_{t-n}\right)^2.$$

The prediction error variance may be written in the compact notation

$$\sigma_t^2 = E\mathbf{a}_t^T\mathbf{x}_t\mathbf{x}_t^T\mathbf{a}_t$$
$$= \mathbf{a}_t^T\mathbf{R}_t\mathbf{a}_t \qquad\qquad 10.66$$

where $\mathbf{a}_t$ is the prediction error vector and $\mathbf{R}_t$ is the covariance matrix for $\mathbf{x}_t = [x_0 \quad x_1 \quad \cdots \quad x_t]^T$:

$$\mathbf{a}_t = [a_t^t \quad \cdots \quad a_1^t \quad 1]^T$$

$$\mathbf{R_t} = E\mathbf{x}_t\mathbf{x}_t^T = E\begin{bmatrix} \mathbf{x}_{t-1} \\ x_t \end{bmatrix}[\mathbf{x}_{t-1}^T \quad x_t]$$

$$= \begin{bmatrix} \mathbf{R}_{t-1} & \mathbf{r}_t \\ \mathbf{r}_t^T & r_0 \end{bmatrix} \qquad\qquad 10.67$$

$$\mathbf{r}_t = E\mathbf{x}_{t-1}x_t = \begin{bmatrix} r_t \\ \vdots \\ r_1 \end{bmatrix}.$$

The prediction error variance σ_t^2 is a quadratic form in the correlation matrix $\mathbf{R}_t$. This quadratic form may be minimized with respect to the predictor coefficients by minimizing with respect to $\mathbf{a}_t$ under the constraint that $a_0^t = 1$. In order to do so, we form the Lagrangian

$$\mathcal{L} = \mathbf{a}_t^T\mathbf{R}_t\mathbf{a}_t - \lambda(\mathbf{a}_t^T\boldsymbol{\delta}_t - 1) \qquad\qquad 10.68$$

$$\boldsymbol{\delta}_t = [0 \quad \cdots \quad 0 \quad 1]^T$$

and equate the gradient with respect to $\mathbf{a}_t$ to zero. This produces the normal equations

$$\mathbf{R}_t\mathbf{a}_t = \lambda\boldsymbol{\delta}_t. \qquad\qquad 10.69$$

But $\mathbf{a}_t^T\mathbf{R}_t\mathbf{a}_t = \lambda$ is the prediction error, so the normal equations may be rewritten as

$$\mathbf{R}_t\mathbf{a}_t = \sigma_t^2\boldsymbol{\delta}_t \qquad\qquad 10.70$$

with solution

$$\mathbf{a}_t = \sigma_t^2 \mathbf{R}_t^{-1} \boldsymbol{\delta}_t.$$

10.71

The explicit formula for prediction error is then

$$\sigma_t^2 = \mathbf{a}_t^{\mathrm{T}} \mathbf{R}_t \mathbf{a}_t = \sigma_t^2 \boldsymbol{\delta}_t^{\mathrm{T}} \mathbf{R}_t^{-1} \mathbf{R}_t \sigma_t^2 \mathbf{R}_t^{-1} \boldsymbol{\delta}_t$$
$$= \sigma_t^4 \boldsymbol{\delta}_t^{\mathrm{T}} \mathbf{R}_t^{-1} \boldsymbol{\delta}_t$$

10.72

or

$$\sigma_t^2 = \frac{1}{\boldsymbol{\delta}_t^{\mathrm{T}} \mathbf{R}_t^{-1} \boldsymbol{\delta}_t} = \frac{\det \mathbf{R}_t}{\det \mathbf{R}_{t-1}}.$$

10.73

It is now natural to ask what relationship this solution for the prediction error vector $\mathbf{a}_t$ and prediction error variance σ_t^2 bears to solutions for lower and higher orders of t. In order to explore this relationship, we first proceed along different lines to show that the prediction error vectors $\mathbf{a}_t$, and their corresponding prediction error variances σ_t^2, may be associated with nonstationary synthesis and analysis equations in much the same way as the prediction error filter is associated with the Wold and Kolmogorov representations of Sections 10.1 and 10.2.

Synthesis

Recall that $\mathbf{R}_t$ is the correlation matrix for the vector $\mathbf{x}_t = [x_0 \quad x_1 \quad \cdots \quad x_t]^{\mathrm{T}}$. Let's denote its lower-diagonal-upper (LDU) or Cholesky factor as follows:

$$\mathbf{R}_t = \mathbf{H} \mathbf{D}^2 \mathbf{H}^{\mathrm{T}}$$

$$\mathbf{H} = \begin{bmatrix} 1 & & & & \\ h_1^0 & 1 & & \mathbf{0} & \\ h_2^0 & h_1^1 & 1 & & \\ \vdots & & & \ddots & \\ h_t^0 & h_{t-1}^1 & & & 1 \end{bmatrix}$$

10.74

$$\mathbf{D} = \mathrm{diag}[\sigma_0^2 \quad \sigma_1^2 \quad \cdots \quad \sigma_t^2].$$

This Cholesky factor is unique when the diagonal elements of $\mathbf{H}$ are set to unity. The Cholesky factor gives us an equation for synthesizing the vector $\mathbf{x}_t$ from a vector of uncorrelated random variables $\mathbf{u}_t = [u_0 \quad u_1 \quad \cdots \quad u_t]^{\mathrm{T}}$:

$$\mathbf{x}_t = \mathbf{H} \mathbf{D} \mathbf{u}_t; \quad E\mathbf{u}_t \mathbf{u}_t^{\mathrm{T}} = \mathbf{I}$$

10.75

$$E\mathbf{x}_t \mathbf{x}_t^{\mathrm{T}} = \mathbf{H} \mathbf{D}^2 \mathbf{H}^{\mathrm{T}} = \mathbf{R}_t.$$

In this synthesis equation, the cross-covariance between the output $\mathbf{x}_t$ and the input $\mathbf{u}_t$ is

$$E\mathbf{x}_t\mathbf{u}_t^\mathrm{T} = \mathbf{HD}. \qquad 10.76$$

Nonstationary Innovations Representation

When the synthesis equation $\mathbf{x}_t = \mathbf{HDu}_t$ is written out explicitly, we obtain the following equation for x_t:

$$x_t = \sum_{n=0}^{t} h_n^{t-n}\sigma_{t-n}u_{t-n}. \qquad 10.77$$

This equation is a nonstationary innovations representation of $\{x_t\}$ that uses a nonstationary filter h_n^t to synthesize the data x_t from nonstationary innovations $\sigma_t u_t$. The nonstationary filter may be interpreted as a sequence of order-increasing MA filters. An alternative representation is

$$\mathbf{x}_t = \sum_{n=0}^{t} \begin{bmatrix} \mathbf{0} \\ \mathbf{h}_n \end{bmatrix} \sigma_n u_n \qquad 10.78$$

where $[\,\mathbf{0}^\mathrm{T} \quad \mathbf{h}_n^\mathrm{T}\,]^\mathrm{T}$ is the $(n+1)$st column of $\mathbf{H}$.

In summary, the synthesis or "Wold" representation for a finite-dimensional snapshot of a stationary time series is nonstationary. This nonstationary representation produces a stationary innovations representation for $\{x_t\}_0^\infty$ when the length of the snapshot increases without bound.

Analysis

Let's denote the upper-diagonal-lower (UDL) or Cholesky factor of $\mathbf{R}^{-1}$ as follows:

$$\mathbf{R}_t^{-1} = \mathbf{AD}^{-2}\mathbf{A}^\mathrm{T} \qquad 10.79$$

$$\mathbf{A} = \begin{bmatrix} 1 & a_1^1 & a_2^2 & \cdots & a_t^t \\ & 1 & a_1^2 & & \\ & & 1 & \ddots & \vdots \\ & \mathbf{0} & & \ddots & a_1^t \\ & & & & 1 \end{bmatrix}.$$

This UDL factorization of $\mathbf{R}_t^{-1}$ may be written as $\mathbf{A}^\mathrm{T}\mathbf{R}_t\mathbf{A} = \mathbf{D}^2$. It is related to the previous LDU factorization of $\mathbf{R}_t$ by the equation $\mathbf{R}_t = \mathbf{A}^{-T}\mathbf{D}^2\mathbf{A}^{-1}$, which means that $\mathbf{H}^{-T} = \mathbf{A}$. This factorization produces an analysis representation for $\mathbf{x}_t$:

$$\mathbf{A}^T\mathbf{x}_t = \mathbf{D}\mathbf{u}_t; \qquad E\mathbf{u}_t\mathbf{u}_t^T = \mathbf{I} \qquad\qquad 10.80$$

$$E(\mathbf{A}^T\mathbf{x}_t)(\mathbf{A}^T\mathbf{x}_t)^T = \mathbf{A}^T\mathbf{R}_t\mathbf{A} = \mathbf{D}^2.$$

In this analysis equation, the matrix $\mathbf{A}$ contains a family of nonstationary, order-increasing analysis filters.

Nonstationary Predictor Representation

From the equation $\mathbf{A}^T\mathbf{x}_t = \mathbf{D}\mathbf{u}_t$, the value u_t may be written as

$$\sigma_t u_t = \sum_{n=0}^{t} a_n^t x_{t-n}. \qquad\qquad 10.81$$

As $a_0^t = 1$, this equation can be rewritten as the synthesis equation

$$x_t = \hat{x}_{t|t-1} + \sigma_t u_t \qquad\qquad 10.82$$

$$\hat{x}_{t|t-1} = -\sum_{n=1}^{t} a_n^t x_{t-n}.$$

The equation for $\hat{x}_{t|t-1}$ is a nonstationary, minimum mean-squared error, linear predictor equation, so we call this representation the *nonstationary predictor representation* of the time series $\{x_t\}$. Note that $\sigma_0^2 \geq \sigma_1^2 \geq \cdots \geq \sigma_t^2$, as the predictor error variance cannot increase as samples are added: "More data cannot hurt."

10.5   FAST ALGORITHMS OF THE LEVINSON TYPE

The normal equations that describe the linear prediction problem for any order i are

$$\mathbf{R}_i\mathbf{a}_i = \sigma_i^2\boldsymbol{\delta}_i; \qquad \mathbf{a}_i = \left[a_i^i, a_{i-1}^i, \ldots, a_1^i, 1\right]^T \qquad\qquad 10.83$$

$$\boldsymbol{\delta}_i = \begin{bmatrix} 0 & \cdots & 0 & 1 \end{bmatrix}^T$$

where $\mathbf{a}_i$ is the prediction error vector for predicting x_{i+1} from the past $(x_0, x_1, \ldots, x_i)$. For $i = 0, 1, \ldots, t$, these normal equations may be organized into the single equation

$$\mathbf{R}_t\mathbf{A} = \mathbf{H}\mathbf{D}^2 \qquad\qquad 10.84$$

where the matrices $\mathbf{R}_t$, $\mathbf{A}$, $\mathbf{H}$, and $\mathbf{D}^2$ are defined as follows:

$$\mathbf{R}_t = \begin{bmatrix} r_0 & r_1 & \cdots & r_t \\ r_1 & r_0 & & \vdots \\ \vdots & & \ddots & r_1 \\ r_t & \cdots & r_1 & r_0 \end{bmatrix} = \left[\begin{array}{c|c} \mathbf{R}_i & \star \\ \hline \star & \star \end{array} \right]$$

$$\mathbf{A} = \begin{bmatrix} \mathbf{a}_0 & & & \\ & \mathbf{a}_1 & & \\ \mathbf{0} & & \ddots & \\ & & & \mathbf{a}_t \end{bmatrix} = \begin{bmatrix} 1 & & & \\ & 1 & & \star \\ \mathbf{0} & & \ddots & \\ & & & 1 \end{bmatrix} \qquad 10.85$$

$$\mathbf{H} = \begin{bmatrix} \mathbf{h}_0 & & & \\ & \mathbf{h}_1 & & \mathbf{0} \\ & & \ddots & \\ & & & \mathbf{h}_t \end{bmatrix} = \begin{bmatrix} 1 & & & \\ & 1 & & \mathbf{0} \\ \star & & \ddots & \\ & & & 1 \end{bmatrix}$$

$$\mathbf{D}^2 = \text{diag}[\sigma_0^2 \quad \sigma_1^2 \quad \cdots \quad \sigma_t^2].$$

The matrix $\mathbf{A}$ is an upper triangular matrix of prediction error vectors $\mathbf{a}_i$. The matrix $\mathbf{H}$ is a lower triangular matrix of synthesis vectors $\mathbf{h}_i$. If we multiply both sides of Equation 10.84 by $\mathbf{A}^T$, we obtain the identity

$$\mathbf{A}^T \mathbf{R}_t \mathbf{A} = \mathbf{A}^T \mathbf{H} \mathbf{D}^2. \qquad 10.86$$

The left-hand side is symmetric and the right-hand side is lower-triangular. This means that $\mathbf{A}^T \mathbf{H} = \mathbf{I}$ or $\mathbf{A}^{-1} = \mathbf{H}^T$. Therefore Equation 10.84 may be rewritten as

$$\mathbf{R}_t = \mathbf{H} \mathbf{D}^2 \mathbf{H}^T \qquad 10.87$$

$$\mathbf{R}_t^{-1} = \mathbf{A} \mathbf{D}^{-2} \mathbf{A}^T.$$

From this fundamental result, we conclude that the order-increasing predictors $\mathbf{a}_i$ that solve the linear prediction problem for $i = 0, 1, \ldots, t$ actually Cholesky-factor $\mathbf{R}_t^{-1}$ into upper-diagonal-lower (UDL) factors. The Levinson recursions are a fast way to do this.

The ith column of the normal equation $\mathbf{R}_t \mathbf{A} = \mathbf{H} \mathbf{D}^2$ may be read out as follows:

$$\mathbf{R}_t \begin{bmatrix} \mathbf{a}_i \\ \mathbf{0} \end{bmatrix} = \sigma_i^2 \begin{bmatrix} \mathbf{0} \\ \mathbf{h}_i \end{bmatrix}. \qquad 10.88$$

These normal equations define the linear minimum mean-squared error predictor coefficients. The leading submatrix of $\mathbf{R}_t$, namely $\mathbf{R}_i$, may be written out in terms of $\mathbf{R}_{i-1}$:

$$
\mathbf{R}_i = \begin{bmatrix} r_0 & r_1 & \cdots & r_i \\ r_1 & r_0 & & \\ \vdots & & \ddots & \vdots \\ r_i & & \cdots & r_0 \end{bmatrix} = \left[\begin{array}{c|c} & r_i \\ \mathbf{R}_{i-1} & \vdots \\ & r_1 \\ \hline r_i & \cdots \;\; r_1 \;\; r_0 \end{array} \right] = \left[\begin{array}{c|c} r_0 & r_1 \;\; \cdots \;\; r_i \\ \hline r_1 & \\ \vdots & \mathbf{R}_{i-1} \\ r_i & \end{array} \right] . \qquad 10.89
$$

Due to the special structure of $\mathbf{R}_i$ and its simple dependence on $\mathbf{R}_{i-1}$, one can expect a simple relationship to exist between $\mathbf{a}_i$ and $\mathbf{a}_{i-1}$.

Now let's reverse the direction of time by operating on $\mathbf{x}_t$ with the exchange matrix $\mathbf{J}$:

$$
\mathbf{J} \begin{bmatrix} x_0 \\ \vdots \\ x_t \end{bmatrix} = \begin{bmatrix} x_t \\ \vdots \\ x_0 \end{bmatrix} . \qquad 10.90
$$

$$
\mathbf{J} = \begin{bmatrix} \mathbf{0} & & 1 \\ & \cdot^{\cdot^{\cdot}} & \\ 1 & & \mathbf{0} \end{bmatrix} , \qquad \mathbf{J} = \mathbf{J}^{\mathrm{T}} = \mathbf{J}^{-1} .
$$

As the statistical properties of a stationary time series are not modified by reversing time, we have the following identity for Toeplitz matrices:

$$
\mathbf{J}\mathbf{R}_t\mathbf{J} = \mathbf{R}_t . \qquad 10.91
$$

This means that $\mathbf{R}_t$ is J-symmetric and the normal equations of Equation 10.83 may be rewritten as

$$
\mathbf{J}\mathbf{R}_i\mathbf{J}\mathbf{J}\mathbf{a}_i = \mathbf{R}_i\mathbf{J}\mathbf{a}_i = \sigma_i^2 \mathbf{J}\boldsymbol{\delta}_i . \qquad 10.92
$$

We shall use these two forms of the normal equation, namely Equations 10.83 and 10.92, to derive the Levinson recursions.

If the solution $\mathbf{a}_{i-1}$ is used in the normal equations of Equation 10.83 in place of $\mathbf{a}_i$, then the normal equations are nearly satisfied:

$$
\mathbf{R}_i \begin{bmatrix} 0 \\ \\ \mathbf{a}_{i-1} \end{bmatrix} = \left[\begin{array}{c|c} r_0 & r_1 \;\; \cdots \;\; r_i \\ \hline r_1 & \\ \vdots & \mathbf{R}_{i-1} \\ r_i & \end{array} \right] \begin{bmatrix} 0 \\ a_{i-1}^{i-1} \\ \vdots \\ 1 \end{bmatrix} = \begin{bmatrix} \gamma_i \\ 0 \\ \vdots \\ 0 \\ \sigma_{i-1}^2 \end{bmatrix} . \qquad 10.93
$$

This shows that $\mathbf{a}_{i-1}$ is almost the desired solution. Similarly, using the alternate normal equations of Equation 10.92 and replacing $J\mathbf{a}_i$ by $J\mathbf{a}_{i-1}$, we have

$$
\mathbf{R}_i \begin{bmatrix} J\mathbf{a}_{i-1} \\ \\ 0 \end{bmatrix} = \left[\begin{array}{c|c} & r_i \\ & \vdots \\ \mathbf{R}_{i-1} & \\ & r_1 \\ \hline r_i \cdots r_1 & r_0 \end{array} \right] \begin{bmatrix} 1 \\ a_1^{i-1} \\ \vdots \\ a_{i-1}^{i-1} \\ 0 \end{bmatrix} = \begin{bmatrix} \sigma_{i-1}^2 \\ 0 \\ \vdots \\ 0 \\ \gamma_i \end{bmatrix} \qquad 10.94
$$

$$
\gamma_i = \sum_{j=0}^{i-1} a_j^{i-1} r_{i-j}.
$$

A linear combination of $\mathbf{a}_{i-1}$ and $J\mathbf{a}_{i-1}$ produces a near solution:

$$
\mathbf{a}_i = \begin{bmatrix} 0 \\ \mathbf{a}_{i-1} \end{bmatrix} + k_i \begin{bmatrix} J\mathbf{a}_{i-1} \\ 0 \end{bmatrix}; \qquad J\mathbf{a}_i = \begin{bmatrix} J\mathbf{a}_{i-1} \\ 0 \end{bmatrix} + k_i \begin{bmatrix} 0 \\ \mathbf{a}_{i-1} \end{bmatrix} \qquad 10.95
$$

$$
\mathbf{R}_i \mathbf{a}_i = \begin{bmatrix} k_i \sigma_{i-1}^2 + \gamma_i \\ 0 \\ \vdots \\ 0 \\ \sigma_{i-1}^2 + k_i \gamma_i \end{bmatrix}.
$$

The multipler k_i, called the *reflection coefficient*, may be chosen to make the term $k_i \sigma_{i-1}^2 + \gamma_i = 0$:

$$
k_i \sigma_{i-1}^2 + \gamma_i = 0 \implies k_i = \frac{-\gamma_i}{\sigma_{i-1}^2}. \qquad 10.96
$$

In this solution, the first element of $\mathbf{a}_i$, namely a_i^i, equals the reflection coefficient k_i. The prediction error variance for the ith order predictor is read out from the last line of Equation 10.95:

$$
\sigma_i^2 = \sigma_{i-1}^2 + k_i \gamma_i = \sigma_{i-1}^2 - \sigma_{i-1}^2 k_i^2
$$
$$
= \sigma_{i-1}^2 (1 - k_i^2). \qquad 10.97
$$

This procedure is repeated from $i = 1$ to $i = t$ to get the Levinson [1947] recursions. The initialization is obtained with $J\mathbf{a}_0 = \mathbf{a}_0 = 1$ and $\sigma_0^2 = r_0$. Each step of the recursion requires roughly i operations to compute k_i and i operations to update $\mathbf{a}_i$, so the total complexity, or operations count, is

$$
\sum_{i=0}^{p} 2i = 2 \frac{p(p+1)}{2} = p^2 + 0(p). \qquad 10.98
$$

Example 10.3

Let $t = 2$. Then we have the system

$$
\begin{bmatrix} r_0 & r_1 & r_2 \\ r_1 & r_0 & r_1 \\ r_2 & r_1 & r_0 \end{bmatrix} \begin{bmatrix} a_2^2 \\ a_1^2 \\ 1 \end{bmatrix} = \begin{bmatrix} 0 \\ 0 \\ \sigma_2^2 \end{bmatrix}.
\tag{10.99}
$$

We start with $\sigma_0^2 = r_0$ and $a_0^0 = 1$, so the next step of the recursion is

$$
\begin{bmatrix} a_1^1 \\ 1 \end{bmatrix} = \begin{bmatrix} 0 \\ 1 \end{bmatrix} + k_1 \begin{bmatrix} 1 \\ 0 \end{bmatrix} = \begin{bmatrix} k_1 \\ 1 \end{bmatrix},
\tag{10.100}
$$

with $k_1 = -r_1/r_0$ and $\sigma_1^2 = \sigma_0^2(1 - k_1^2)$. The next step is

$$
\begin{bmatrix} a_2^2 \\ a_1^2 \\ 1 \end{bmatrix} = \begin{bmatrix} 0 \\ a_1^1 \\ 1 \end{bmatrix} + k_2 \begin{bmatrix} 1 \\ a_1^1 \\ 0 \end{bmatrix} = \begin{bmatrix} k_2 \\ (1 + k_2)k_1 \\ 1 \end{bmatrix},
\tag{10.101}
$$

$$
k_2 = \frac{-\gamma_2}{\sigma_1^2} = \frac{-(r_2 + a_1^1 r_1)}{\sigma_1^2}
\tag{10.102}
$$

$$
\sigma_2^2 = \sigma_1^2(1 - k_2^2).
\tag{10.103}
$$

■

Interpretation

At each order, the Levinson algorithm gives the predictor $\mathbf{a}_i$ and the corresponding predictor variance σ_i^2. But the normal equations $\mathbf{R}_i \mathbf{J} \mathbf{a}_i = \sigma_i^2 \mathbf{J} \boldsymbol{\delta}$ are identical to the normal equations of Equation 10.61 that describe the maximum entropy spectrum. Therefore, the Levinson algorithm also identifies the autoregressive spectrum of order i that fits the covariances $\{r_0, \ldots, r_i\}$. The associated covariance sequence and power spectrum are

$$
\sum_{n=0}^{i} a_n^i r_{t-n}^i = \sigma_i^2 \delta_t
\tag{10.104}
$$

$$
S_i(e^{j\theta}) = \frac{\sigma_i^2}{\left| A_i(e^{j\theta}) \right|^2},
$$

where $A_i(z)$ is the prediction error filter

$$
A_i(z) = \sum_{n=0}^{i} a_n^i z^{-n}.
\tag{10.105}
$$

The covariance sequence r_t^i matches the original covariance sequence for $|t| \le i$ and extends it autoregressively for $|t| > i$. It is evident that the linear prediction problem produces the same solution as the maximum entropy problem.

Backward Form

The Levinson recursions of Equation 10.95 may be organized into the equation

$$\mathbf{a}_i = (\mathbf{I} + k_i \mathbf{J}) \begin{bmatrix} 0 \\ \mathbf{a}_{i-1} \end{bmatrix}.$$ 10.106

The inverse of this equation is

$$\begin{bmatrix} 0 \\ \mathbf{a}_{i-1} \end{bmatrix} = (\mathbf{I} + k_i \mathbf{J})^{-1} \mathbf{a}_i$$ 10.107

$$= \frac{1}{1 - k_i^2} (\mathbf{I} - k_i \mathbf{J}) \mathbf{a}_i.$$

The Levinson recursions may then be run backward from any prediction error vector of order p, namely $\mathbf{a}_p$, with the reflection coefficients obtained as $k_i = a_i^i$. The forward and backward forms of the Levinson algorithm allow one to go from the set $\{1, a_1^p, \ldots, a_p^p\}$ to the set $\{k_1, \ldots, k_p\}$ and back. The Levinson algorithm is usually run backward to test the stability of the given filter $A_p(z)$. This is equivalent to running the Jury stability test. If $|k_i| < 1$ for $i = p, \ldots, 1$, then the roots of $A_p(z)$ are guaranteed to lie inside the unit circle.

Filtering Interpretations

The ith order predictor coefficient $\mathbf{a}_i$ characterizes a moving average whitening filter whose transfer function is

$$A_i(z) = \sum_{n=0}^{i} a_n^i z^{-n}$$ 10.108

Similarly, the reversed vector $\mathbf{J}\mathbf{a}_i$ has transfer function

$$B_i(z) = \sum_{n=0}^{i} a_{i-n}^i z^{-n} = z^{-i} A_i(z^{-1}).$$ 10.109

The Levinson recursions connect these transfer functions through the coupled recursions

$$A_i(z) = A_{i-1}(z) + k_i z^{-1} B_{i-1}(z)$$ 10.110

$$B_i(z) = z^{-1} B_{i-1}(z) + k_i A_{i-1}(z).$$

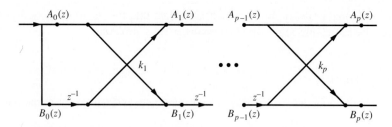

Figure 10.4 Moving average lattice filter.

These equations may be organized in the matrix form,

$$\begin{bmatrix} B_i(z) \\ A_i(z) \end{bmatrix} = \begin{bmatrix} z^{-1} & k_i \\ k_i z^{-1} & 1 \end{bmatrix} \begin{bmatrix} B_{i-1}(z) \\ A_{i-1}(z) \end{bmatrix} \qquad 10.111$$

Therefore beginning with $A_0(z) = B_0(z) = 1$, we may implement the filters $A_p(z)$ and $B_p(z)$ as shown in Figure 10.4.

Such a filter is called a *lattice filter*. It is formed from p cells, each of which takes the form shown in Figure 10.5. Each basic cell contains two equal reflection coefficients. A lattice implementation of the filter $A_p(z)$ contains twice the number of multipliers contained in a traditional tap-delay line structure. The real advantage of the structure lies in its fixed-point implementation of the reflection coefficients (recall $|k_i| < 1$) and in its robustness to coefficient quantization.

AR Synthesis Lattice

When the time series $\{x_t\}$ is passed through the MA lattice filter $A_p(z)$, the output is the time series $\sigma_p\{u_t\}$:

$$A_p(z)\{x_t\} = \sigma_p\{u_t\}. \qquad 10.112$$

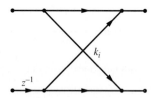

Figure 10.5 Moving average lattice cell.

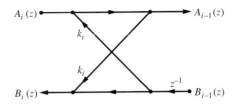

Figure 10.6 AR lattice cell.

The time series $\{u_t\}$ is not white unless $p \to \infty$ or unless the time series $\{x_t\}$ is an order-p autoregression. Let's assume the latter and ask how the MA lattice can be inverted to produce the autoregressive synthesis equation $\{x_t\} = 1/A_p(z)\sigma_p\{u_t\}$. What we are after is a lattice implementation of the inverse filter $1/A_p(z)$ for synthesizing the time series $\{x_t\}$.

Recall the lattice recursions of Equation 10.110:

$$A_i(z) = A_{i-1}(z) + k_i z^{-1} B_{i-1}(z) \qquad\qquad 10.113$$

$$B_i(z) = z^{-1} B_{i-1}(z) + k_i A_{i-1}(z).$$

Invert the top equation to obtain the following equations, which have an inverse lattice representation:

$$A_{i-1}(z) = A_i(z) - k_i z^{-1} B_{i-1}(z) \qquad\qquad 10.114$$

$$B_i(z) = z^{-1} B_{i-1}(z) + k_i A_{i-1}(z).$$

The new lattice cell is represented in Figure 10.6. These cells may be concatenated to obtain the inverse filter $1/A_p(z)$ shown in Figure 10.7.

The two lattice filters of Figures 10.4 and 10.7 may be concatenated to get the filter illustrated in Figure 10.8.

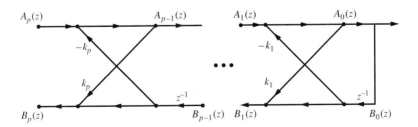

Figure 10.7 AR lattice filter.

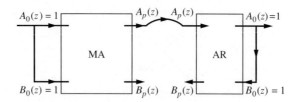

Figure 10.8 Combined AR and MA lattice filters.

The overall transfer function is 1, so the second transfer function must be equal to $1/A_p(z)$. The final synthesis lattice filter is illustrated in Figure 10.9. Can you explain why the transfer functions at the bottom of the lattice are $\sigma_p/A_p(z)$, $\sigma_p B_1(z)/A_p(z)$, and so on?

ARMA Lattice

These results may be extended to the implementation of the ARMA filter $C(z)/A(z)$ when the degree of $C(z)$ is less than or equal to the degree of $A(z)$. Form the AR lattice filter that implements $1/A(z)$ by running the Levinson recursions backward. Then decompose the numerator polynomial onto the basis formed by the set of increasing degree polynomials $\{B_0(z), B_1(z), \ldots, B_p(z)\}$, where $B_i(z) = z^{-i}A_i(z^{-1})$:

$$C(z) = \sum_{i=0}^{p} \alpha_i B_i(z). \qquad 10.115$$

As shown in Figure 10.9, the internal variables on the lower line of the AR synthesis lattice implement $B_i(z)/A(z)$. Therefore the ARMA lattice for the ARMA filter $C(z)/A(z) = \sum_{i=0}^{p} \alpha_i B_i(z)/A(z)$ takes the form shown in Figure 10.10. In Problems 10.10 and 10.11 you are asked to design ARMA lattices.

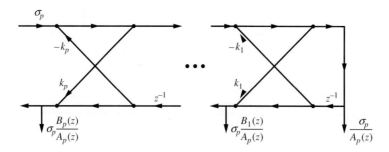

Figure 10.9 AR synthesis filter in lattice form.

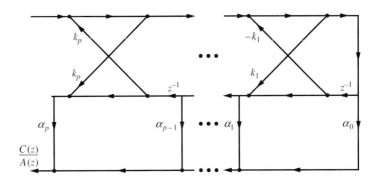

Figure 10.10 ARMA lattice structure.

10.6   FAST ALGORITHMS OF THE SCHUR TYPE

Equation 10.87 shows that the correlation matrix $\mathbf{R}_t$ has Cholesky factors $\mathbf{HD}^2\mathbf{H}^T$ and that $\mathbf{R}_t^{-1}$ has the Cholesky factors $\mathbf{AD}^{-2}\mathbf{A}^T$. The Levinson algorithm finds the latter factors. Perhaps there is an algorithm to find the former. The desired algorithm is the Schur algorithm, which computes the lower triangular matrix $\mathbf{H}$ column by column in the Cholesky factorization $\mathbf{R}_t = \mathbf{HD}^2\mathbf{H}^T$.

If $\mathbf{R}_t$ is Toeplitz, we have the equation $\mathbf{R}_t\mathbf{A} = \mathbf{HD}^2$, where the Levinson recursions compute the columns of $\mathbf{A}$. Write out the ith column of the matrix equation as

$$\mathbf{R}_t \begin{bmatrix} \mathbf{a}_i \\ \mathbf{0} \end{bmatrix} = \begin{bmatrix} \mathbf{0} \\ \mathbf{h}_i \end{bmatrix} \sigma_i^2 = \mathbf{H}_i \qquad 10.116$$

and reverse $\mathbf{a}_i$ to write

$$\mathbf{R}_t \begin{bmatrix} \mathbf{Ja}_i \\ \mathbf{0} \end{bmatrix} = \begin{bmatrix} 1 \\ \mathbf{0} \\ \mathbf{g}_i \end{bmatrix} \sigma_i^2 = \begin{bmatrix} \sigma_i^2 \\ \mathbf{0} \end{bmatrix} + \mathbf{G}_i. \qquad 10.117$$

Recursions for $\hat{\mathbf{h}}_i$ and $\hat{\mathbf{g}}_i$ may be deduced by substituting the Levinson recursions for $\mathbf{a}_i$ and $\mathbf{Ja}_i$. The result is (see Problem 10.7)

$$\mathbf{H}_{i+1} = \mathbf{ZH}_i + k_i\mathbf{G}_i; \qquad \mathbf{H}_0 = \begin{bmatrix} r_0 \\ r_1 \\ \vdots \\ r_t \end{bmatrix} \qquad 10.118$$

$$\mathbf{G}_{i+1} = \mathbf{G}_i + k_i\mathbf{ZH}_i; \qquad \mathbf{G}_0 = \begin{bmatrix} 0 \\ r_1 \\ \vdots \\ r_{t-1} \end{bmatrix},$$

where $\mathbf{Z}$ is the delay matrix

$$
\mathbf{Z} = \left[\begin{array}{c|c} \mathbf{0}^{\mathrm{T}} & 0 \\ \hline \mathbf{I} & \mathbf{0} \end{array} \right].
$$

10.119

There is no inner product necessary in the algorithm, as

$$
k_{i+1} = -\frac{\mathbf{G}_i(i+1)}{\mathbf{H}_i(i)},
$$

10.120

where $\mathbf{G}_i(i+1)$ is the $(i+1)$st element of $\mathbf{G}_i$ and $\mathbf{H}_i(i)$ is the ith element of $\mathbf{H}_i$. Once more, the structure of the algorithm is exactly the same as the Levinson recursions, which means that a lattice structure implementation is possible. Just as in the lattice algorithm, variables in the Schur algorithm may be scaled to allow a fixed-point implementation of the algorithm [LeRoux and Gueguen, 1977].

10.7 LEAST SQUARES THEORY OF LINEAR PREDICTION

In the least squares theory of linear prediction, no assumptions are made about the correlation sequence for the time series $\{x_t\}$. In order to distinguish the least squares theory from the minimum mean-squared error theory of the previous sections, we replace x_t by y_t and give ourselves only the snapshot $\{y_t\}_0^{N-1}$. From this snapshot, we would like to derive a pth-order prediction error filter

$$
A_p(z) = \sum_{n=0}^{p} a_n^p z^{-n}; \qquad a_0^p = 1
$$

10.121

that produces small errors when applied to the snapshot. The prediction errors are

$$
\{u_t^p\} = A_p(z)\{y_t\}
$$

$$
u_t^p = y_t - \hat{y}_t^p
$$

10.122

$$
\hat{y}_t^p = -\sum_{i=1}^{p} a_i^p y_{t-i}.
$$

In these formulas, $y_t = 0$ for $t < 0$ and $t > N - 1$.

The prediction errors of Equation 10.122 may be written out in matrix form as follows:

$$
\begin{array}{c}
(t = N - 1 + p) \\
\\
\\
(t = N - 1) \\
\\
\\
\\
(t = p) \\
\\
\\
\\
(t = 0)
\end{array}
\begin{bmatrix}
y_{N-1} & & & & \\
y_{N-2} & y_{N-1} & & \mathbf{0} & \\
\vdots & & \ddots & & \\
y_{N-1-p} & & & y_{N-1} & \\
\vdots & & & \vdots & \\
y_{N-1-i} & & & & \\
\vdots & & & & \\
y_0 & & & y_p & \\
& & & \vdots & \\
& & & y_i & \\
& \mathbf{0} & \ddots & \vdots & \\
& & & y_0
\end{bmatrix}
\begin{bmatrix}
a_p^p \\
a_{p-1}^p \\
\vdots \\
a_1^p \\
1
\end{bmatrix}
=
\begin{bmatrix}
u_{N-1+p}^p \\
\vdots \\
u_{N-1}^p \\
\vdots \\
u_p^p \\
\vdots \\
u_i^p \\
\vdots \\
u_0^p
\end{bmatrix}
$$

$$10.123$$

The superscripts on the a_i^p make it clear that we are talking about a predictor of order p. This distinction is important to make because shortly we will study predictors of increasing order i as a way of developing fast algorithms. The prediction error u_t^p is the error in predicting y_t with a pth order predictor.

We may measure the quality of the predictor polynomial $A_p(z)$ by measuring accumulated squared errors over some range of time indexes:

$$\sigma_p^2 = \sum_{t \in T} (u_t^p)^2. \qquad 10.124$$

Each choice of index set T determines a subset of the error variables u_t^p on the right-hand side of Equation 10.123 that are to be squared and summed. Alternatively, we can say that each choice of T determines a horizontal cut through the matrix equation of Equation 10.123. Four such cuts are illustrated in Figure 10.11, corresponding to the following methods of linear prediction:

1. correlation method: $T = \{t : 0 \le t \le N - 1 + p\}$
2. prewindowed method: $T = \{t : 0 \le t \le N - 1\}$ 10.125
3. post-windowed method: $T = \{t : p \le t \le N - 1 + p\}$
4. covariance method: $T = \{t : p \le t \le N - 1\}$.

In the pre- and postwindowed methods, the cardinality of the index set T is N, independent of the predictor order p. In the correlation method, the cardinality is $N + p$, and in the covariance method it is $N - p$.

Regardless of how the index set T is chosen, the resulting set of error equations

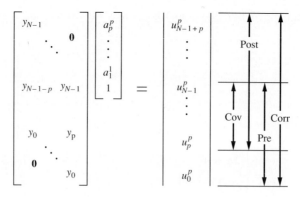

Figure 10.11 Slicing the Toeplitz data matrix to obtain different methods of linear prediction.

defined in Equation 10.123 and in Figure 10.11 may be written as

$$\mathbf{Y}\mathbf{a}_p = \mathbf{u}_p; \qquad \mathbf{a}_p = \begin{bmatrix} a^p_p \\ \vdots \\ a^p_1 \\ 1 \end{bmatrix}. \qquad\qquad 10.126$$

In the *correlation* method of linear prediction, $\mathbf{Y}$ and $\mathbf{u}_p$ are defined as follows:

$$\mathbf{Y} = \begin{bmatrix} y_{N-1} & & \\ \vdots & \ddots & \mathbf{0} \\ y_{N-1-p} & \cdots & y_{N-1} \\ \vdots & & \vdots \\ y_0 & \cdots & y_p \\ \mathbf{0} & \ddots & \vdots \\ & & y_0 \end{bmatrix}; \qquad \mathbf{u}_p = \begin{bmatrix} u^p_{N-1+p} \\ \vdots \\ \vdots \\ u^p_0 \end{bmatrix}. \qquad 10.127$$

In the *covariance* method, $\mathbf{Y}$ and $\mathbf{u}_p$ are

$$\mathbf{Y} = \begin{bmatrix} y_{N-1-p} & \cdots & y_{N-1} \\ \vdots & & \vdots \\ y_0 & \cdots & y_p \end{bmatrix}; \qquad \mathbf{u}_p = \begin{bmatrix} u^p_{N-1} \\ \vdots \\ u^p_p \end{bmatrix}. \qquad 10.128$$

The matrix equation 10.126 for the errors $\mathbf{u}_p$ may also be cast in the form

$$\mathbf{y}_p = \mathbf{X}\boldsymbol{\theta}_p + \mathbf{u}_p, \qquad\qquad 10.129$$

where $\mathbf{X}$ consists of the first p columns of $\mathbf{Y}$, $\mathbf{y}_p$ is the last column of $\mathbf{Y}$, and $-\boldsymbol{\theta}_p$ consists of the first p elements of $\mathbf{a}_p$:

$$\mathbf{Y} = [\mathbf{X} \ \mathbf{y}_p]; \qquad \mathbf{a}_p = \begin{bmatrix} -\boldsymbol{\theta}_p \\ 1 \end{bmatrix}. \qquad\qquad 10.130$$

In the correlation method of linear prediction, $\mathbf{y}_p$ is

$$\mathbf{y}_p = \begin{bmatrix} 0 \\ \vdots \\ 0 \\ y_{N-1} \\ \vdots \\ y_0 \end{bmatrix} \qquad (p \text{ leading zeros}), \qquad\qquad 10.131$$

and in the covariance method, $\mathbf{y}_p$ is

$$\mathbf{y}_p = \begin{bmatrix} y_{N-1} \\ \vdots \\ y_p \end{bmatrix}. \qquad\qquad 10.132$$

Equation 10.129 shows that the predictor $\boldsymbol{\theta}_p = [-a_p^p \quad \cdots \quad -a_1^p]^T$ is really trying to represent the vector $\mathbf{y}_p$ in the subspace $\langle \mathbf{X} \rangle$, and $\mathbf{u}_p$ is the resulting error. The accumulated squared error may be written in two ways:

$$\begin{aligned} \sigma_p^2 &= \mathbf{a}_p^T \mathbf{Y}^T \mathbf{Y} \mathbf{a}_p \\ &= (\mathbf{y}_p - \mathbf{X}\boldsymbol{\theta}_p)^T (\mathbf{y}_p - \mathbf{X}\boldsymbol{\theta}_p). \end{aligned} \qquad\qquad 10.133$$

The minimization of σ_p^2 with respect to the predictor coefficients $(a_i^p)_1^p$ may be written in two ways as well:

1. $\min_{\mathbf{a}_p} \mathbf{a}_p^T \mathbf{Y}^T \mathbf{Y} \mathbf{a}_p$ under constraint $a_0^p = 1$; $\qquad\qquad$ 10.134
2. $\min_{\boldsymbol{\theta}_p} (\mathbf{y}_p - \mathbf{X}\boldsymbol{\theta}_p)^T (\mathbf{y}_p - \mathbf{X}\boldsymbol{\theta}_p).$

From our study of least squares problems, we know that the solutions for $\hat{\boldsymbol{\theta}}_p$ and $\hat{\mathbf{y}}_p = \mathbf{X}\hat{\boldsymbol{\theta}}_p$ are

$$\hat{\boldsymbol{\theta}}_p = \mathbf{X}^{\#} \mathbf{y}_p; \qquad \mathbf{X}^{\#} = (\mathbf{X}^T \mathbf{X})^{-1} \mathbf{X}^T \qquad\qquad 10.135$$

$$\hat{\mathbf{y}}_p = \mathbf{P}_X \mathbf{y}_p; \qquad \mathbf{P}_X = \mathbf{X}\mathbf{X}^{\#} = \mathbf{X}(\mathbf{X}^T \mathbf{X})^{-1} \mathbf{X}^T.$$

The resulting error is $\hat{\mathbf{u}}_p$, and the resulting value of $\hat{\sigma}_p^2$ is $\hat{\mathbf{u}}_p^T \hat{\mathbf{u}}_p$:

$$\hat{\mathbf{u}}_p = \mathbf{y}_p - \hat{\mathbf{y}}_p = (\mathbf{I} - \mathbf{P}_X)\mathbf{y}_p \qquad\qquad 10.136$$

$$\hat{\sigma}_p^2 = \mathbf{y}_p^T (\mathbf{I} - \mathbf{P}_X)\mathbf{y}_p.$$

These results are illustrated in Figure 10.12. The error $\hat{\mathbf{u}}_p$ is orthogonal to $\langle \mathbf{X} \rangle$, and the inner product between $\mathbf{y}_p$ and $\hat{\mathbf{u}}_p$ is $\hat{\sigma}_p^2$:

$$\mathbf{X}^T \hat{\mathbf{u}}_p = \mathbf{X}^T (\mathbf{I} - \mathbf{P}_X)\mathbf{y}_p = \mathbf{0} \qquad\qquad 10.137$$

$$\mathbf{y}_p^T \hat{\mathbf{u}}_p = \mathbf{y}_p^T (\mathbf{I} - \mathbf{P}_X)\mathbf{y}_p = \hat{\sigma}_p^2.$$

These two results are organized into the matrix equation

$$\mathbf{Y}^T \hat{\mathbf{u}}_p = \begin{bmatrix} \mathbf{0} \\ \hat{\sigma}_p^2 \end{bmatrix}; \qquad \mathbf{Y} = [\mathbf{X} \quad \mathbf{y}_p]. \qquad\qquad 10.138$$

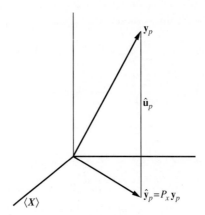

Figure 10.12 Orthogonality in linear prediction.

In summary, the prediction error equation

$$\mathbf{Y}\hat{\mathbf{a}}_p = \hat{\mathbf{u}}_p \qquad\qquad 10.139$$

produces orthogonality conditions and the corresponding normal equations:

$$\mathbf{Y}^\mathrm{T}\hat{\mathbf{u}}_p = \begin{bmatrix} \mathbf{0} \\ \hat{\sigma}_p^2 \end{bmatrix} \qquad \text{(orthogonality)} \qquad\qquad 10.140$$

$$\mathbf{Y}^\mathrm{T}\mathbf{Y}\hat{\mathbf{a}}_p = \mathbf{Y}^\mathrm{T}\hat{\mathbf{u}}_p = \begin{bmatrix} \mathbf{0} \\ \hat{\sigma}_p^2 \end{bmatrix} \qquad \text{(normal equation).} \qquad\qquad 10.141$$

These normal equations may be solved for $\hat{\mathbf{a}}_p$ and $\hat{\sigma}_p^2$.

QR Factors and Sliding Windows

Let's now try to adapt our discussion of least squares to a sequence of prediction error polynomials whose orders increase from $i = 0$ to $i = p$. We intend to retain the error equations of Equation 10.123 but to replace the prediction error vector $\mathbf{a}_p$ by the following *sequence* of prediction error vectors:

$$\mathbf{A}_0 = \begin{bmatrix} 1 \\ \mathbf{0} \end{bmatrix}, \dots, \mathbf{A}_i = \begin{bmatrix} \mathbf{a}_i \\ \mathbf{0} \end{bmatrix}, \dots, \mathbf{A}_p = \begin{bmatrix} \mathbf{a}_p \end{bmatrix}. \qquad\qquad 10.142$$

The ith prediction error vector $\mathbf{A}_i$ has the order i prediction error vector $\mathbf{a}_i$ in its first $i + 1$ elements and zeros from then on. When these prediction error vectors are substituted into Equation 10.123, the equations become

$$
\begin{bmatrix}
y_{N-1} & y_N & \cdots & y_{N-1+p} \\
\vdots & & & \vdots \\
y_{N-1+p} & & \cdots & y_{N-1} \\
\vdots & & & \vdots \\
y_0 & & \cdots & y_p \\
\vdots & & & \vdots \\
y_{-p} & & \cdots \; y_{-1} & y_0
\end{bmatrix}
\begin{bmatrix} \mathbf{A}_0 & \cdots & \mathbf{A}_i & \cdots & \mathbf{A}_p \end{bmatrix}
$$

$$
=
\begin{bmatrix}
u^0_{N-1} & \cdots & u^i_{N-1+i} & \cdots & u^p_{N-1+p} \\
\vdots & & \vdots & & \\
& & & & \\
u^0_{-p} & \cdots & u^i_{-p+i} & \cdots & u^p_0
\end{bmatrix} .
$$

$$10.143$$

In these equations we have shown the measurements y_t for $t < 0$ and $t > N - 1$, even though they are zero, because the equations produce nonzero prediction errors for these quantities that are nonzero. For example, u^i_{-p+i} and u^i_{N-1+i} are nonzero prediction errors for y_{-p+i} and y_{N-1+i}. These equations show that a *constant* length window of *equation errors* is being swept through the data as the order i increases from 0 to p. This is illustrated in Figure 10.13. In the correlation method of linear prediction, a window of length $N - 1 + p$ is swept through the data, beginning at $-p$ for order 0 and continuing to 0 for order p. In the covariance method of linear prediction, a window of length $N - 1$ is swept through the data, beginning at 0 for order 0 and continuing to p for order p.

Let's isolate the ith equation in 10.143, involving the ith order prediction error equation:

$$
\mathbf{Y} \begin{bmatrix} \mathbf{a}_i \\ \mathbf{0} \end{bmatrix} = \mathbf{u}_i .
$$

$$10.144$$

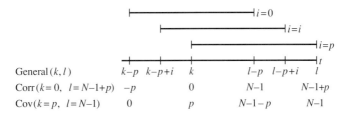

Figure 10.13 Sliding windows.

This may, as well, be written as the linear model

$$\mathbf{y}_i = \mathbf{X}_i \boldsymbol{\theta}_i + \mathbf{u}_i \qquad\qquad 10.145$$

where $\mathbf{Y}$ is parsed as follows:

$$\mathbf{Y} = [\mathbf{X}_i \quad \mathbf{y}_i \quad \star \ ], \qquad \mathbf{a}_i = \begin{bmatrix} -\boldsymbol{\theta}_i \\ 1 \end{bmatrix}. \qquad\qquad 10.146$$

The least squares estimates of $\boldsymbol{\theta}_i$ and $\mathbf{u}_i$ are, as before,

$$\boldsymbol{\theta}_i = \mathbf{X}_i^{\#} \mathbf{y}_i \qquad\qquad 10.147$$

$$\mathbf{u}_i = (\mathbf{I} - \mathbf{P}_{X_i})\mathbf{y}_i.$$

The orthogonality conditions of Equation 10.137 are retained:

$$\mathbf{X}_i^{\mathrm{T}} \mathbf{u}_i = \mathbf{0} \qquad\qquad 10.148$$

$$\mathbf{y}_i^{\mathrm{T}} \mathbf{u}_i = \hat{\sigma}_i^2.$$

This means that we may multiply $\mathbf{u}_i$ by $\mathbf{Y}^{\mathrm{T}}$ to obtain the equation

$$\mathbf{Y}^{\mathrm{T}} \mathbf{u}_i = \begin{bmatrix} \mathbf{0} \\ \sigma_i^2 \\ \star \end{bmatrix}. \qquad\qquad 10.149$$

We may proceed in this way to generate solutions for $\boldsymbol{\theta}_i$ and $\mathbf{u}_i$ and organize our orthogonality conditions as follows:

$$\mathbf{Y}^{\mathrm{T}}[\mathbf{u}_0 \quad \mathbf{u}_1 \quad \cdots \quad \mathbf{u}_p] = \begin{bmatrix} \sigma_0^2 & & & & & \\ & \sigma_1^2 & & & \mathbf{0} & \\ & & \ddots & & & \\ & & & \sigma_i^2 & & \\ & \star & & & \ddots & \\ & & & & & \sigma_p^2 \end{bmatrix} \qquad\qquad 10.150$$

$$\mathbf{Y}^{\mathrm{T}} \mathbf{U} = \mathbf{H} \mathbf{D}^2; \qquad \mathbf{D}^2 = \mathrm{diag}[\sigma_0^2 \quad \cdots \quad \sigma_p^2].$$

Let's compare this equation with Equation 10.143, written out as follows:

$$\mathbf{Y} \begin{bmatrix} \mathbf{a}_0 & & & \\ & \mathbf{a}_1 & & \\ & & \mathbf{0} & \ddots \\ & & & \mathbf{a}_p \end{bmatrix} = [\mathbf{u}_0 \quad \mathbf{u}_1 \quad \cdots \quad \mathbf{u}_p] \qquad\qquad 10.151$$

$$\mathbf{Y} \mathbf{A} = \mathbf{U}.$$

If we premultiply Equation 10.151 by $\mathbf{Y}^T$ and use the orthogonality of Equation 10.150, we get the result

$$\mathbf{Y}^T\mathbf{Y}\begin{bmatrix} \mathbf{a}_0 & & & \\ & \mathbf{a}_1 & & \\ & 0 & \ddots & \\ & & & \mathbf{a}_p \end{bmatrix} = \begin{bmatrix} \sigma_0^2 & & & \\ & \sigma_1^2 & 0 & \\ & \star & \ddots & \\ & & & \sigma_p^2 \end{bmatrix} \qquad 10.152$$

$$\mathbf{Y}^T\mathbf{Y}\mathbf{A} = \mathbf{H}\mathbf{D}^2; \qquad \mathbf{H} = \begin{bmatrix} \mathbf{h}_0 & & & \\ & \mathbf{h}_1 & 0 & \\ & & \ddots & \\ & & & \mathbf{h}_p \end{bmatrix}.$$

Premultiply this equation by the lower triangular matrix $\mathbf{A}^T$. The left-hand side is symmetric and the right-hand side is lower triangular. Therefore, $\mathbf{A}^T\mathbf{H} = \mathbf{I}$ and we have the results

1. $\mathbf{A}^T\mathbf{Y}^T\mathbf{Y}\mathbf{A} = \mathbf{D}^2 \;\Rightarrow\; (\mathbf{Y}^T\mathbf{Y})^{-1} = \mathbf{A}\mathbf{D}^{-2}\mathbf{A}^T$ 10.153

2. $\mathbf{Y}^T\mathbf{Y} = \mathbf{H}\mathbf{D}^2\mathbf{H}^T$.

But we know that $\mathbf{Y}\mathbf{A} = \mathbf{U}$, so evidently $\mathbf{U}^T\mathbf{U} = \mathbf{D}^2$, meaning that $\mathbf{U}$ is an orthogonal matrix. This tells us that our sequence of least squares solutions has produced the QR factorizations

$$\mathbf{Y}\mathbf{A} = \mathbf{U}; \qquad \mathbf{U}^T\mathbf{U} = \mathbf{D}^2 \qquad 10.154$$

$$\mathbf{Y} = \mathbf{U}\mathbf{H}^T; \qquad \mathbf{H}^T = \mathbf{A}^{-1}.$$

Summary and Interpretations

Let's summarize our findings by restating the important equations. In our restatement of them, we define the orthonormal matrix $\mathbf{Q}$:

$$\mathbf{U} = \mathbf{Q}\mathbf{D} \;\Rightarrow\; \mathbf{U}^T\mathbf{U} = \mathbf{D}^2; \qquad \mathbf{Q}^T\mathbf{Q} = \mathbf{I}, \qquad 10.155$$

Then the fundamental equations for least squares linear prediction, and their interpretations, are

1. $\mathbf{A}^T\mathbf{Y}^T\mathbf{Y}\mathbf{A} = \mathbf{D}^2 \;\Rightarrow\; (\mathbf{Y}^T\mathbf{Y})^{-1} = \mathbf{A}\mathbf{D}^{-2}\mathbf{A}^T$
    The order-increasing whiteners (or prediction error filters) in $\mathbf{A}$ provide a Cholesky factorization of the inverse of the experimental correlation matrix $\mathbf{Y}^T\mathbf{Y}$;

2. $\mathbf{Y}^T\mathbf{Y} = \mathbf{H}\mathbf{D}^2\mathbf{H}^T$
    The order-increasing synthesizers in $\mathbf{H}$ provide a Cholesky factorization of $\mathbf{Y}^T\mathbf{Y}$;

3. $\mathbf{Y}^T\mathbf{Y}\mathbf{A} = \mathbf{H}\mathbf{D}^2$
    The order-increasing whiteners in $\mathbf{A}$ solve the normal equations;

4. $\mathbf{Y}\mathbf{A} = \mathbf{Q}\mathbf{D}$
    The order-increasing whiteners in $\mathbf{A}$ provide an inverse "$\mathbf{QDR}$" factorization of

the data matrix $\mathbf{Y}$, and the orthogonal columns of $\mathbf{QD}$ are forward prediction errors;

5. $\mathbf{Y}^T\mathbf{QD} = \mathbf{HD}^2$

    The forward prediction errors are causally orthogonal to the data matrix $\mathbf{Y}$;

6. $\mathbf{Y} = \mathbf{QDH}^T$

    The order-increasing synthesizers in $\mathbf{H}$ provide a direct "$\mathbf{QDR}$" factorization of $\mathbf{Y}$.

It is instructive to write out equations 3, 4, and 5 column-by-column:

3. $$\mathbf{Y}^T\mathbf{Y}\begin{bmatrix} \mathbf{a}_i \\ \mathbf{0} \end{bmatrix} = \sigma_i^2 \begin{bmatrix} \mathbf{0} \\ \mathbf{h}_i \end{bmatrix}$$

4. $$\mathbf{Y}\begin{bmatrix} \mathbf{a}_i \\ \mathbf{0} \end{bmatrix} = \sigma_i \mathbf{q}_i \qquad\qquad\qquad 10.156$$

5. $$\mathbf{Y}^T\sigma_i\mathbf{q}_i = \sigma_i^2 \begin{bmatrix} \mathbf{0} \\ \mathbf{h}_i \end{bmatrix}.$$

These results are fundamentally important because, reading from top to bottom, they suggest that the whiteners $\mathbf{a}_i$ may be determined from the normal equations, the orthogonal prediction errors $\mathbf{q}_i$ may be determined from the whiteners $\mathbf{a}_i$, and the synthesizers $\mathbf{h}_i$ may be determined from the prediction errors $\mathbf{q}_i$:

$$\mathbf{a}_i \longrightarrow \mathbf{q}_i \longrightarrow \mathbf{h}_i. \qquad\qquad 10.157$$

This brings us to the lattice algorithms of the next section.

10.8   LATTICE ALGORITHMS

In this section we exploit the connections between the whiteners $\mathbf{a}_i$, the orthogonal error vectors $\mathbf{q}_i$, and the synthesizers $\mathbf{h}_i$ in order to obtain fast algorithms for $\mathbf{q}_i$ and $\mathbf{h}_i$. The fast algorithm for the prediction errors $\sigma_i\mathbf{q}_i$ is called a lattice algorithm for reasons that will become obvious.

We begin our discussion by assuming that the data matrix $\mathbf{Y}$ in the error equations of Equation 10.123 is the data matrix for the *correlation method* of linear prediction. (This assumption may be relaxed to cover all of the cases of linear prediction. See [Demeure and Scharf, 1990].) Then the normal equations in the first equation of Equation 10.156 may be solved by using the Levinson recursions of Equation 10.95:

$$\mathbf{Ja}_{i+1} = \begin{bmatrix} \mathbf{Ja}_i \\ 0 \end{bmatrix} + k_{i+1}\begin{bmatrix} 0 \\ \mathbf{a}_i \end{bmatrix} \qquad\qquad 10.158$$

$$\mathbf{a}_{i+1} = \begin{bmatrix} 0 \\ \mathbf{a}_i \end{bmatrix} + k_{i+1}\begin{bmatrix} \mathbf{Ja}_i \\ 0 \end{bmatrix}.$$

The second equation of Equation 10.156 suggests that we premultiply these recursions by $\mathbf{Y}$ in an attempt to generate recursions for $\mathbf{q}_i$:

$$
\mathbf{Y}
\begin{bmatrix} \mathbf{Ja}_{i+1} \\ \mathbf{0} \end{bmatrix}
= \mathbf{Y}
\begin{bmatrix} \mathbf{Ja}_i \\ \mathbf{0} \end{bmatrix}
+ k_{i+1}\mathbf{Y}
\begin{bmatrix} 0 \\ \mathbf{a}_i \\ 0 \end{bmatrix}
\qquad\qquad 10.159
$$

$$
\mathbf{Y}
\begin{bmatrix} \mathbf{a}_{i+1} \\ \mathbf{0} \end{bmatrix}
= \mathbf{Y}
\begin{bmatrix} 0 \\ \mathbf{a}_i \\ 0 \end{bmatrix}
+ k_{i+1}\mathbf{Y}
\begin{bmatrix} \mathbf{Ja}_i \\ \mathbf{0} \end{bmatrix} .
$$

It is clear from Equation 10.123 that the second of these equations characterizes the *forward* prediction errors $\mathbf{u}_{i+1}$ and the first characterizes *backward* prediction errors $\mathbf{v}_{i+1}$. (Check this by replacing $\mathbf{a}_i$ with $\mathbf{Ja}_i$ in Equation 10.123.) For $i < p$, the following identity holds:

$$
\mathbf{Y}
\begin{bmatrix} 0 \\ \mathbf{a}_i \\ 0 \end{bmatrix}
= \mathbf{YZ}
\begin{bmatrix} \mathbf{a}_i \\ 0 \end{bmatrix}
= \mathbf{ZY}
\begin{bmatrix} \mathbf{a}_i \\ 0 \end{bmatrix}
\qquad\qquad 10.160
$$

$$
\mathbf{Z} =
\begin{bmatrix}
0 \cdots 0 & 0 \\
\hline
 & 0 \\
\mathbf{I} & \vdots \\
 & 0
\end{bmatrix}
\quad \text{(delay matrix)}.
$$

This identity produces the following lattice recursions for the backward prediction errors $\mathbf{v}_i = \mathbf{Y}[(\mathbf{Ja}_i)^{\mathrm{T}} \quad \mathbf{0}^{\mathrm{T}}]^{\mathrm{T}}$ and the forward prediction errors $\mathbf{u}_i = \mathbf{Y}[\mathbf{a}_i^{\mathrm{T}} \quad \mathbf{0}^{\mathrm{T}}]^{\mathrm{T}}$:

$$
\mathbf{v}_{i+1} = \mathbf{v}_i + k_{i+1}\mathbf{Zu}_i \qquad\qquad 10.161
$$

$$
\mathbf{u}_{i+1} = \mathbf{Zu}_i + k_{i+1}\mathbf{v}_i .
$$

In these recursions, the error vectors $\mathbf{u}_i$ are orthogonal with norm-squared σ_i^2:

$$
\mathbf{u}_i^{\mathrm{T}}\mathbf{u}_j = \sigma_i^2\delta_{ij} . \qquad\qquad 10.162
$$

The backward prediction errors are not orthogonal, but their norm-squared is σ_i^2 as well:

$$
\mathbf{v}_i^{\mathrm{T}}\mathbf{v}_i =
\left(\mathbf{Y}\begin{bmatrix} \mathbf{Ja}_i \\ 0 \end{bmatrix} \right)^{\mathrm{T}}
\left(\mathbf{Y}\begin{bmatrix} \mathbf{Ja}_i \\ 0 \end{bmatrix} \right)
$$

$$
= [\,\mathbf{a}_i^{\mathrm{T}} \quad \mathbf{0}^{\mathrm{T}}\,]\mathbf{JY}^{\mathrm{T}}\mathbf{YJ}
\begin{bmatrix} \mathbf{a}_i \\ 0 \end{bmatrix}
\qquad\qquad 10.163
$$

$$
= [\,\mathbf{a}_i^{\mathrm{T}} \quad \mathbf{0}^{\mathrm{T}}\,]\mathbf{Y}^{\mathrm{T}}\mathbf{Y}
\begin{bmatrix} \mathbf{a}_i \\ 0 \end{bmatrix}
= \sigma_i^2 .
$$

Initialization

The lattice recursions are initialized as follows:

$$
\mathbf{u}_0 = \mathbf{Y} \begin{bmatrix} \mathbf{a}_0 \\ \mathbf{0} \end{bmatrix} = \begin{bmatrix} y_{N-1} \\ \vdots \\ y_0 \\ 0 \\ \vdots \\ 0 \end{bmatrix} = \mathbf{Y} \begin{bmatrix} \mathbf{Ja}_0 \\ \mathbf{0} \end{bmatrix} = \mathbf{v}_0. \qquad 10.164
$$

Recursions for k_i

There are three different ways to solve for the reflection coefficients. The first comes from the formula for k_i in the Levinson recursions, and the second and third come directly from the lattice recursions:

1. $\sigma_i^2 k_{i+1} = -\displaystyle\sum_{n=0}^{i} a_n^i r_{i+1-n}$

$$
= -[r_0 \quad \cdots \quad r_{i+1} \quad \cdots \quad r_p] \begin{bmatrix} 0 \\ \mathbf{a}_i \\ \mathbf{0} \end{bmatrix} \qquad 10.165
$$

$$
= -\mathbf{u}_0^T \mathbf{Y} \begin{bmatrix} 0 \\ \mathbf{a}_i \\ \mathbf{0} \end{bmatrix}
$$

$\Rightarrow \sigma_i^2 k_{i+1} = -\mathbf{u}_0^T \mathbf{Z} \mathbf{u}_i;$

2. $\mathbf{u}_i^T \mathbf{u}_{i+1} = 0 = \mathbf{u}_i^T \mathbf{Z} \mathbf{u}_i + k_{i+1} \mathbf{u}_i^T \mathbf{v}_i$

$$
\Rightarrow k_{i+1} = -\frac{\mathbf{u}_i^T \mathbf{Z} \mathbf{u}_i}{\mathbf{u}_i^T \mathbf{v}_i}; \qquad 10.166
$$

3. $\mathbf{v}_i^T \mathbf{u}_{i+1} = 0 = \mathbf{v}_i^T \mathbf{Z} \mathbf{u}_i + k_{i+1}$

$$
\Rightarrow k_{i+1} = -\mathbf{v}_i^T \mathbf{Z} \mathbf{u}_i. \qquad 10.167
$$

Can you see why $\mathbf{v}_i$ and $\mathbf{u}_{i+1}$ are orthogonal?

Solving for σ_i^2

The obvious way to solve for σ_i^2 is to measure the norm-squared of $\mathbf{u}_i$ or $\mathbf{v}_i$:

$$
\sigma_i^2 = \mathbf{u}_i^T \mathbf{u}_i = \mathbf{v}_i^T \mathbf{v}_i. \qquad 10.168
$$

But the last equation of Equation 10.156 provides another method:

$$\sigma_i = \mathbf{y}_i^T \mathbf{q}_i. \qquad 10.169$$

Algebraic Interpretations

Substitute the formula for the reflection coefficient

$$k_{i+1} = -\mathbf{v}_i^T \mathbf{Z} \mathbf{u}_i = -(\mathbf{Z} \mathbf{u}_i)^T \mathbf{v}_i \qquad 10.170$$

into the lattice recursions to obtain the coupled recursions

$$\mathbf{v}_{i+1} = \left[\mathbf{I} - (\mathbf{Z} \mathbf{u}_i)(\mathbf{Z} \mathbf{u}_i)^T \right] \mathbf{v}_i \qquad 10.171$$

$$\mathbf{u}_{i+1} = \left[\mathbf{I} - \mathbf{v}_i \mathbf{v}_i^T \right] \mathbf{Z} \mathbf{u}_i.$$

We may think of the lattice recursions as a sequence of rank-one modifications of identity, successively applied.

Lattice Interpretations

Let's define the prediction error vectors $\mathbf{u}_i$ and $\mathbf{v}_i$ to be $\mathbf{u}_i = [u_i(0) \cdots u_i(N-1+p)]^T$ and $\mathbf{v}_i = [v_i(0) \cdots v_i(N-1+p)]^T$. Then the lattice recursions may be written

$$v_{i+1}(t) = v_i(t) + k_{i+1} u_i(t-1) \qquad 10.172$$

$$u_{i+1}(t) = u_i(t-1) + k_{i+1} v_i(t).$$

These errors may be organized in the lattice structure of Figure 10.14. At any time t, the variables in the lattice of Figure 10.14 are forward and backward prediction errors. The errors observed at the input to the lattice are $v_0(t) = u_0(t) = y_{N-1-t}$. Therefore, the inputs arrive in reverse order.

In summary, the Lattice algorithm computes the forward and backward errors $u_i(t)$ and $v_i(t)$ and uses them to compute reflection coefficients k_{i+1} and variances σ_i^2. The reflection coefficients keep the recursions going, and they are computed from prediction errors according to the formula

$$k_{i+1} = -\mathbf{v}_i^T \mathbf{Z} \mathbf{u}_i = -\sum_{t=1}^{N-1} v_i(t) u_i(t-1). \qquad 10.177$$

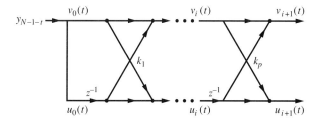

Figure 10.14 Lattice interpretation of the lattice recursions.

In this formula, the errors $u_i(t)$ and $v_i(t)$ are data-dependent errors that appear at cell i of the lattice at time t. In Problem 10.19 you are asked to relate them to the prediction errors of Equation 10.143.

10.9 PREDICTION IN AUTOREGRESSIVE MOVING AVERAGE TIME SERIES

When the time series $\{x_t\}$ is autoregressive moving average (ARMA), then it satisfies the difference equation

$$\sum_{n=0}^{p} a_n x_{t-n} = \sum_{n=0}^{p} b_n u_{t-n}; \qquad a_0 = 1. \tag{10.178}$$

The input time series $\{u_t\}$ is assumed to be a zero-mean WSS sequence with correlation $E[u_t u_s] = \delta_{t-s}$. The operator notation

$$A(z)\{x_t\} = B(z)\{u_t\} \tag{10.179}$$

produces the analysis and synthesis representations

$$\{x_t\} = \frac{B(z)}{A(z)}\{u_t\}; \qquad \{u_t\} = \frac{A(z)}{B(z)}\{x_t\}. \tag{10.180}$$

These representations are illustrated as digital filters in Figure 10.15. They are the Wold and Kolmogorov representations for an ARMA time series.

The predictors $P_1(z)$ and $P_2(z)$ defined in Section 10.2 take the form

$$\{\hat{x}_{t|t-1}\} = P_1(z)\{u_t\}$$

$$P_1(z) = \frac{B(z)}{A(z)} - h_0 = \frac{B(z) - b_0 A(z)}{A(z)} \tag{10.181}$$

$$\{\hat{x}_{t|t-1}\} = P_2(z)\{x_t\}$$

$$P_2(z) = 1 - h_0 \frac{A(z)}{B(z)} = \frac{B(z) - b_0 A(z)}{B(z)}.$$

These predictors are illustrated in Figure 10.16.

Stationary State-Space Representations

State-space representations may be derived for the ARMA time series by defining a state variable at the output of each delay element z^{-1} in Figure 10.16. The synthesis

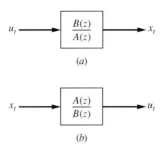

Figure 10.15 Synthesis (a) and analysis (b)
filters for an ARMA
time series.

equations are obtained from the predictor diagram of Figure 10.16(a):

$$\mathbf{x}_{t+1} = \mathbf{A}\mathbf{x}_t + \mathbf{b}u_t$$

$$\hat{x}_{t|t-1} = \mathbf{c}^{\mathrm{T}}\mathbf{x}_t \qquad\qquad 10.182$$

$$x_t = \hat{x}_{t|t-1} + b_0 u_t$$

$$\mathbf{A} = \begin{bmatrix} 0 & 1 & & \\ & \ddots & \ddots & \\ & & 0 & 1 \\ -a_p & & \cdots & -a_1 \end{bmatrix}; \qquad \mathbf{b} = \begin{bmatrix} 0 \\ \vdots \\ 0 \\ 1 \end{bmatrix}$$

$$\mathbf{c}^{\mathrm{T}} = \begin{bmatrix} (b_1 - b_0 a_1) & \cdots & (b_p - b_0 a_p) \end{bmatrix}.$$

(a) predictor $\{\hat{x}_{t|t-1}\} = P_1(z)\{u_t\}$

(b) predictor $\{\hat{x}_{t|t-1}\} = P_2(z)\{x_t\}$

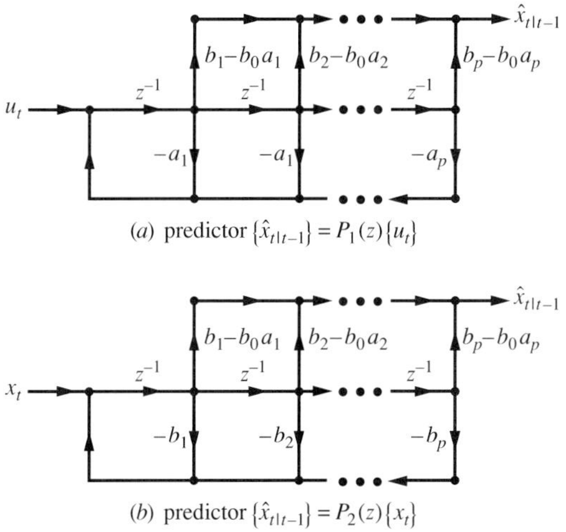

Figure 10.16 Predictor filters for an ARMA time series.

Similarly, the analysis state equations are obtained from the predictor diagram of Figure 10.16(*b*):

$$\mathbf{x}_{t+1} = \mathbf{A}\mathbf{x}_t + \mathbf{b}x_t$$

$$\hat{x}_{t|t-1} = \mathbf{c}^{\mathrm{T}}\mathbf{x}_t \qquad\qquad 10.183$$

$$b_0 u_t = -\hat{x}_{t/t-1} + x_t$$

$$\mathbf{A} = \begin{bmatrix} 0 & 1 & & & \\ & \ddots & & \ddots & \\ & & 0 & 1 \\ -b_p & & \cdots & & -b_1 \end{bmatrix}; \qquad \mathbf{b} = \begin{bmatrix} 0 \\ \vdots \\ 0 \\ 1 \end{bmatrix}$$

$$\mathbf{c}^{\mathrm{T}} = \begin{bmatrix} (b_1 - b_0 a_1) & \cdots & (b_p - b_0 a_p) \end{bmatrix}.$$

These representations are called *canonical representations*. The state vector $\mathbf{x}_t$ summarizes all of the information about the infinite past needed to perform the prediction.

Markovian State-Space Model

An alternate form of the synthesis state-space representation is obtained by generalizing the one-step predictor to the *s*-step predictor. Use the Wold representation for x_{t+s}:

$$x_{t+s} = \sum_{n=0}^{\infty} h_n u_{t+s-n} = \hat{x}_{t+s|t-1} + h_s u_t + \sum_{n=0}^{s-1} h_n u_{t+s-n}$$

$$\qquad\qquad 10.184$$

$$= \hat{x}_{t+s|t} + \sum_{n=0}^{s-1} h_n u_{t+s-n}.$$

$\hat{x}_{t+s|t-1}$ is the linear minimum mean-squared error (LMMSE) estimator of x_{t+s} given the infinite past up to time $t-1$, and $\hat{x}_{t+s|t}$ is the LMMSE estimator given the infinite past up to time t. The connection between $\hat{x}_{t+s|t}$ and $\hat{x}_{t+s|t-1}$ is

$$\hat{x}_{t+s|t} = \hat{x}_{t+s|t-1} + h_s u_t. \qquad\qquad 10.185$$

When $s = p$, the autoregressive moving average model for $\{x_t\}$ may be used to write

$$x_{t+p} = -\sum_{i=1}^{p} a_i x_{t+p-i} + \sum_{j=0}^{p} b_j u_{t+p-j}. \qquad\qquad 10.186$$

Take conditional expectation on both sides, given the infinite past up to time t, to produce the prediction equation:

$$\hat{x}_{t+p|t} = -\sum_{i=1}^{p} a_i x_{t+p-i|t} + b_p u_t. \qquad\qquad 10.187$$

Use the connection between $\hat{x}_{t+s|t}$ and $\hat{x}_{t+s|t-1}$ and the result that $h_p + \sum_{i=1}^{p} a_i h_{p-i} = b_p$ to write

$$\hat{x}_{t+p|p} = -\sum_{i=1}^{p} a_i \hat{x}_{t+p-i|t-1} - \sum_{i=1}^{p} a_i h_{p-i} u_t + b_p u_t$$

$$= -\sum_{i=1}^{p} a_i \hat{x}_{t+p-i|t-1} + h_p u_t.$$

$\qquad$ 10.188

These predictions may be organized into a p-dimensional state vector:

$$\begin{bmatrix} \hat{x}_{t+1|t} \\ \vdots \\ \vdots \\ \hat{x}_{t+p|t} \end{bmatrix} = \begin{bmatrix} 0 & 1 & & \\ \vdots & & \mathbf{0} & \\ 0 & & \ddots & \\ \vdots & & & 1 \\ -a_p & \cdots & & -a_1 \end{bmatrix} \begin{bmatrix} \hat{x}_{t|t-1} \\ \hat{x}_{t+1|t-1} \\ \vdots \\ \hat{x}_{t+p-1|t-1} \end{bmatrix} + \begin{bmatrix} h_1 \\ \vdots \\ h_p \end{bmatrix} u_t.$$

$\qquad$ 10.189

The Markovian state-space representation is obtained by defining the state $\mathbf{x}_t$ to be $[\,\hat{x}_{t+1|t} \quad \cdots \quad \hat{x}_{t+p|t}\,]^{\mathrm{T}}$:

$$\mathbf{x}_{t+1} = \mathbf{A}\mathbf{x}_t + \mathbf{b}u_t$$

$$x_t = \mathbf{c}^{\mathrm{T}}\mathbf{x}_t + b_0 u_t$$

$\qquad$ 10.190

$$\mathbf{A} = \begin{bmatrix} 0 & 1 & & \\ \vdots & \ddots & \ddots & \\ & & 0 & 1 \\ -a_p & \cdots & -a_2 & -a_1 \end{bmatrix}; \quad \mathbf{c} = \begin{bmatrix} 1 \\ 0 \\ \vdots \\ 0 \end{bmatrix}; \quad \mathbf{b} = \begin{bmatrix} h_1 \\ h_2 \\ \vdots \\ h_p \end{bmatrix}.$$

The impulse response values in $\mathbf{b}$ may be computed using the linear equations

$$\begin{bmatrix} 1 & & \mathbf{0} \\ a_1 & 1 & \\ \vdots & & \ddots \\ a_p & \cdots & a_1 & 1 \end{bmatrix} \begin{bmatrix} h_0 \\ h_1 \\ \vdots \\ h_p \end{bmatrix} = \begin{bmatrix} b_0 \\ b_1 \\ \vdots \\ b_p \end{bmatrix}.$$

$\qquad$ 10.191

The Markovian state-space model is illustrated in Figure 10.17, where double lines represent vector connections. The impulse response of this model is

$$h_t = \begin{cases} 0, & t < 0 \\ h_0, & t = 0 \\ \mathbf{c}^{\mathrm{T}}\mathbf{A}^{t-1}\mathbf{b}, & t > 0. \end{cases}$$

$\qquad$ 10.192

The state covariance $\mathbf{Q}_t$ at time t satisfies the recursion

$$\mathbf{Q}_t = E[\mathbf{x}_t \mathbf{x}_t^{\mathrm{T}}] = \mathbf{A}\mathbf{Q}_{t-1}\mathbf{A}^{\mathrm{T}} + \sigma^2 \mathbf{b}\mathbf{b}^{\mathrm{T}}$$

$\qquad$ 10.193

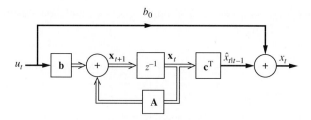

Figure 10.17 Markovian state-space synthesis model.

where σ^2 is the input-noise variance. The state is stationary, so $\mathbf{Q}_t = \mathbf{Q}_{t-1}$. This means that the state-covariance matrix satisfies the Lyapunov equation

$$\mathbf{Q} = \mathbf{AQA}^\mathrm{T} + \sigma^2 \mathbf{bb}^\mathrm{T}. \qquad 10.194$$

From this result, the output covariance is

$$r_t = E[x_s x_{t+s}] = \mathbf{c}^\mathrm{T} \mathbf{A}^t \mathbf{Qc} + \sigma^2 h_0 h_t. \qquad 10.195$$

This synthesis representation may be inverted to get the analysis Markovian representation given in Figure 10.18. In both the analysis and synthesis Markovian representations, the state vector $\mathbf{x}_t$ contains all of the information about the infinite past. The state at time $t = 0$ has stationary covariance $\mathbf{Q}$. If we wish to synthesize a wide-sense stationary time series $\{x_t\}$, whose covariance is $\{r_t\}$, then $\mathbf{x}_0$ must be drawn from a distribution whose covariance is $\mathbf{Q}$.

Nonstationary (or Innovations) State-Space Representations

Perhaps we can replace the stationary Markovian model with a nonstationary model whose initial state is zero. The new representation is a time varying representation called the *innovations representation*. The initial state is identically set to zero, and the equations are modified to replace $\mathbf{b}$ by a time varying vector $\mathbf{k}_t$ and the unit-variance input u_t by an input $\sigma_t u_t$ with variance σ_t^2. The resulting state equations are then

$$\mathbf{x}_{t+1} = \mathbf{Ax}_t + \mathbf{k}_t \sigma_t u_t \qquad 10.196$$

$$x_t = \mathbf{c}^\mathrm{T} \mathbf{x}_t + h_0 \sigma_t u_t$$

where $\mathbf{k}_t$ is called the Kalman gain vector, and $\sigma_t u_t$ is the input-noise sequence of variance σ_t^2.

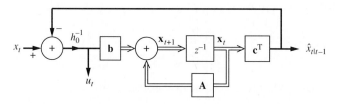

Figure 10.18 Markovian nonstationary state-space analysis model.

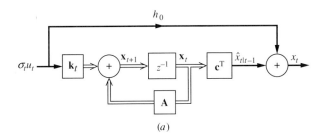

(a)

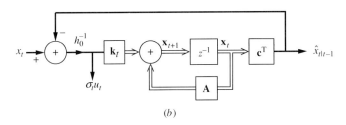

(b)

Figure 10.19 Synthesis (a) and analysis (b) innovations representations.

The correlation sequence for $\{x_t\}$ is then

$$r_n(t) = E[x_t x_{t+n}] = \mathbf{c}^T \mathbf{A}^n \mathbf{Q}_t \mathbf{c} + h_0 \mathbf{c}^T \mathbf{A}^{n-1} \mathbf{k}_t \sigma_t^2 \qquad 10.197$$

where $\mathbf{Q}_t = E\mathbf{x}_t\mathbf{x}_t^T$ is the zero-lag covariance matrix of the state at time t. If this correlation is to equal the stationary correlation r_n, then we require $\mathbf{k}_t$ and σ_t^2 to satisfy the equations

$$\mathbf{k}_t \sigma_t^2 = -\mathbf{A}(\mathbf{Q}_t - \mathbf{Q})\mathbf{c} + \mathbf{b} \qquad 10.198$$

$$\sigma_t^2 = \mathbf{c}^T(\mathbf{Q} - \mathbf{Q}_t)\mathbf{c} + h_0^2 = r_0 - \mathbf{c}^T\mathbf{Q}_t\mathbf{c}.$$

The state covariance is updated from the state equations in order to keep the recursion going:

$$\mathbf{Q}_{t+1} = E\mathbf{x}_{t+1}\mathbf{x}_{t+1}^T = \mathbf{A}\mathbf{Q}_t\mathbf{A}^T + \sigma_t^2\mathbf{k}_t\mathbf{k}_t^T. \qquad 10.199$$

The innovations representation permits us to replace, without altering the second-order properties of the stationary random process, the time-invariant Markovian representation by a time-varying innovations representation that produces the same mean value and correlation sequence and has the enormous advantage that the initial state is deterministically set to zero.

The synthesis innovations representation may be inverted to obtain an analysis (or predictor) innovations representation. Both representations are shown in Figure 10.19.

10.10   FAST ALGORITHMS OF THE
MORF-SIDHU-KAILATH TYPE

Computation of the Kalman gain vector is quite tedious. At each time t a Ricatti equation must be solved. In this section we show how the ARMA recursions allow for a large reduction in the computation count by reducing the Schur algorithm to the Morf-Sidhu-Kailath (MSK) algorithm.

The time varying impulse response of the nonstationary Markovian state-space model is given by

$$h_i^t = \begin{cases} 0, & i < 0 \\ h_0, & i = 0 \\ \mathbf{c}^T \mathbf{A}^{i-1} \mathbf{k}_t, & i > 0. \end{cases} \qquad 10.200$$

Because of the companion form for $\mathbf{A}$, the Kalman gain vector may be read off from the h_i^t:

$$\mathbf{k}_t = \begin{bmatrix} h_1^t & h_2^t & \cdots & h_p^t \end{bmatrix}^T. \qquad 10.201$$

The output of the innovations representation for the time series $\{x_t\}$ may be written

$$x_t = \sum_{i=0}^{t} h_i^{t-i} \sigma_{t-i} u_{t-i}. \qquad 10.202$$

This equation may be written in matrix form as

$$\mathbf{x} = \begin{bmatrix} x_0 \\ x_1 \\ \vdots \\ x_t \end{bmatrix} = \begin{bmatrix} h_0^0 & 0 & \cdots & 0 \\ h_1^0 & h_0^1 & \ddots & \vdots \\ \vdots & \vdots & \ddots & 0 \\ h_t^0 & h_{t-1}^1 & \cdots & h_0^t \end{bmatrix} \mathbf{D} \begin{bmatrix} u_0 \\ u_1 \\ \vdots \\ u_t \end{bmatrix} \qquad 10.203$$

with $\mathbf{D} = \text{diag}[\sigma_0 \quad \sigma_1 \quad \cdots \quad \sigma_t]$. Thus, $E[\mathbf{x}\mathbf{x}^T] = \mathbf{R} = \mathbf{H}\mathbf{D}^2\mathbf{H}^T$ and $E[\mathbf{u}\mathbf{u}^T] = \mathbf{I}$. This means that the Kalman gain vector may be read out of the Cholesky factor of $\mathbf{R}$. Recall the coupled Schur recursions for the columns of $\mathbf{HD}$:

$$\mathbf{H}_{n+1} = \mathbf{Z}\mathbf{H}_n + k_n \mathbf{G}_n \qquad 10.204$$

$$\mathbf{G}_{n+1} = \mathbf{G}_n + k_n \mathbf{Z}\mathbf{H}_n.$$

Within $\mathbf{H}_{n+1}$ is embedded the Kalman gain vector $\mathbf{k}_{n+1}$, and within $\mathbf{G}_{n+1}$ is embedded the auxiliary Namlak gain vector $\hat{\mathbf{k}}_{n+1}$:

$$\hat{\mathbf{k}}_t = [g_1^t g_2^t \dots g_p^t]^T. \qquad 10.205$$

The Schur recursions are

$$
\sigma^2_{n+1}
\begin{bmatrix} 0 \\ \vdots \\ 0 \\ 1 \\ h^{n+1}_1 \\ \vdots \\ h^{n+1}_p \\ h^{n+1}_{p+1} \\ \vdots \\ h^{n+1}_{t-n-1} \end{bmatrix}
= \sigma^2_n
\begin{bmatrix} 0 \\ \vdots \\ 0 \\ 1 \\ h^n_1 \\ \vdots \\ h^n_p \\ h^n_{p+1} \\ \vdots \\ h^n_{t-n-1} \end{bmatrix}
+ k_n \sigma^2_n
\begin{bmatrix} 0 \\ \vdots \\ 0 \\ g^n_1 \\ g^n_2 \\ \vdots \\ g^n_{p+1} \\ g^n_{p+2} \\ \vdots \\ g^n_{t-n} \end{bmatrix}
\qquad 10.206
$$

and

$$
\sigma^2_{n+1}
\begin{bmatrix} 0 \\ \vdots \\ 0 \\ 0 \\ g^{n+1}_1 \\ \vdots \\ g^{n+1}_{p+1} \\ g^{n+1}_{p+2} \\ \vdots \\ g^{n+1}_{t-n-1} \end{bmatrix}
= \sigma^2_n
\begin{bmatrix} 0 \\ \vdots \\ 0 \\ g^n_1 \\ g^n_2 \\ \vdots \\ g^n_{p+1} \\ g^n_{p+2} \\ \vdots \\ g^n_{t-n} \end{bmatrix}
+ k_n \sigma^2_n
\begin{bmatrix} 0 \\ \vdots \\ 0 \\ 1 \\ h^n_1 \\ \vdots \\ h^n_p \\ h^n_{p+1} \\ \vdots \\ h^n_{t-n-1} \end{bmatrix} .
$$

Thus we have the following recursions for $\mathbf{k}_n$ and $\hat{\mathbf{k}}_n$:

$$
\sigma^2_{n+1} \hat{\mathbf{k}}_{n+1} = \sigma^2_n \mathbf{Z}^{\mathrm{T}} \hat{\mathbf{k}}_n + \sigma^2_n k_{n+1} \mathbf{k}_n + \sigma^2_n k_{n+1} \begin{bmatrix} \mathbf{0} \\ g^n_{p+1} \end{bmatrix}
\qquad 10.207
$$

$$
\sigma^2_{n+1} \mathbf{k}_{n+1} = \sigma^2_n \mathbf{k}_n + \sigma^2_n k_{n+1} \mathbf{Z}^{\mathrm{T}} \hat{\mathbf{k}}_n + \sigma^2_n \begin{bmatrix} \mathbf{0} \\ g^n_{p+1} \end{bmatrix} .
$$

Recall, however, the definition of g^n_i:

$$
\sigma^2_n
\begin{bmatrix} g^n_1 \\ \vdots \\ g^n_{p+1} \end{bmatrix}
=
\begin{bmatrix} r_{n+1} & \cdots & r_1 \\ \vdots & & \vdots \\ r_{n+p+1} & \cdots & r_{p+1} \end{bmatrix}
\begin{bmatrix} 1 \\ a^n_1 \\ \vdots \\ a^n_n \end{bmatrix} .
\qquad 10.208
$$

If $\{r_n\}$ is ARMA (p, p), then

$$[a_p \quad \cdots \quad a_1 1] \begin{bmatrix} r_{n+1} & \cdots & r_1 \\ \vdots & & \vdots \\ r_{n+p+1} & \cdots & r_{p+1} \end{bmatrix} = [0 \quad \cdots \quad 0], \qquad 10.209$$

so we have this recursion on g_{p+1}^n:

$$g_{p+1}^n = -\sum_{i=1}^{p} a_i g_i^n. \qquad 10.210$$

This recursion can be incorporated into the recursion for $\mathbf{k}_n$ and $\hat{\mathbf{k}}_n$ by replacing $\mathbf{Z}^T$ by $\mathbf{A}$ to get

$$\sigma_{n+1}^2 \begin{bmatrix} \hat{\mathbf{k}}_{n+1} \\ \mathbf{k}_{n+1} \end{bmatrix} = \sigma_n^2 \begin{bmatrix} \mathbf{A} & k_{n+1}\mathbf{I} \\ k_{n+1}\mathbf{A} & \mathbf{I} \end{bmatrix} \begin{bmatrix} \hat{\mathbf{k}}_n \\ \mathbf{k}_n \end{bmatrix} \qquad 10.211$$

where $k_{n+1} = -\mathbf{k}_n(1)$ and $\sigma_{n+1}^2 = \sigma_n^2(1 - k_{n+1}^2)$. The initialization is

$$\sigma_0^2 \mathbf{k}_0 = \sigma_0^2 \hat{\mathbf{k}}_0 = \begin{bmatrix} r_1 \\ \vdots \\ r_p \end{bmatrix}, \qquad \sigma_0^2 = r_0. \qquad 10.212$$

This leads to the minimum computation to get the Kalman gain vector in the ARMA (p, p) case. These recursions are the same as Pearlman's scalar version [1980] of the Morf-Sidhu-Kailath algorithm [1974]. The nonstationary Markovian representation, together with the fast algorithms for the Kalman gain, constitute a fast Kalman predictor because the innovations representation may be inverted to obtain the nonstationary predictor representation.

10.11   LINEAR PREDICTION AND LIKELIHOOD FORMULAS

Let $\mathbf{x} = [x_0 \quad \cdots \quad x_{N-1}]^T$ denote a finite snapshot of a zero-mean, wide-sense stationary Gaussian sequence $\{x_t\}$. The log-likelihood function for a realization of $\mathbf{x}$ is

$$L(\mathbf{x}) = -\frac{N}{2} \log(2\pi) - \frac{1}{2} \log \det(\mathbf{R}) - \frac{1}{2} \mathbf{x}^T \mathbf{R}^{-1} \mathbf{x} \qquad 10.213$$

where $\mathbf{R}$ is the $N \times N$ Toeplitz correlation matrix and $\det(\mathbf{R})$ is its determinant.

Let's consider the Markovian representation for the time series $\{x_t\}$. If we knew the initial state $\mathbf{x}_0$, we could run the Markovian recursions to produce the variables $\hat{x}_{0|-1}, \ldots, \hat{x}_{N|N-1}$ from the formulas

$$\hat{x}_{t|t-1} = \mathbf{c}^T \mathbf{x}_t \qquad 10.214$$

$$\mathbf{x}_{t+1} = \mathbf{A}\mathbf{x}_t + \mathbf{b}[x_t - \hat{x}_{t|t-1}].$$

Then, given $\mathbf{x}_0$, the distribution of x_t is $N[\hat{x}_{t|t-1}, \sigma^2]$, and the distribution of u_t is $N[0, \sigma^2]$. The $\{u_t\}_0^{N-1}$ are a sequence of independent random variables, so we can write the conditional likelihood of $\{x_t\}_0^{N-1}$ as

$$l(\mathbf{x}|\mathbf{x}_0) = \prod_{t=0}^{N-1} N_{x_t}[\hat{x}_{t|t-1}, \sigma^2] \qquad 10.215$$

$$N_{x_t}[\hat{x}_{t|t-1}, \sigma^2] = (2\pi\sigma^2)^{-1/2} \exp\left\{-\frac{1}{2\sigma^2}(x_t - \hat{x}_{t|t-1})^2\right\}.$$

The state $\mathbf{x}_0$ itself is distributed as $N[\mathbf{0}, \mathbf{Q}]$:

$$N_{\mathbf{x}_0}[\mathbf{0}, \mathbf{Q}] = (2\pi)^{-p/2}|\mathbf{Q}|^{-1/2} \exp\left\{-\frac{1}{2}\mathbf{x}_0\mathbf{Q}^{-1}\mathbf{x}_0\right\}. \qquad 10.216$$

The unconditional likelihood of $\mathbf{x}$ is obtained by integrating over all the realizations of the random initial state $\mathbf{x}_0$:

$$l(\mathbf{x}) = \int N_{\mathbf{x}_0}[\mathbf{0}, \mathbf{Q}] \prod_{t=0}^{N-1} N_{x_t}[\hat{x}_{t|t-1}, \sigma^2] \, d\mathbf{x}_0. \qquad 10.217$$

This is only a representation formula. It is of no computational value because of the integral over the distribution of the initial state $\mathbf{x}_0$. This is where the innovations representation, whose initial state is identically zero, comes into play.

In the innovations representation, we may set $\mathbf{x}_0 = \mathbf{0}$ and run the recursions for $\hat{x}_{0|-1}, \ldots, \hat{x}_{N|N-1}$ as follows:

$$\hat{x}_{t|t-1} = \mathbf{c}^{\mathsf{T}}\mathbf{x}_t \qquad 10.218$$

$$\mathbf{x}_{t+1} = A\mathbf{x}_t + \mathbf{k}_t\sigma_t[x_t - \hat{x}_{t|t-1}].$$

The distribution of x_t is $N[\hat{x}_{t|t-1}, \sigma_t^2]$, and the distribution of $u_t = x_t - \hat{x}_{t|t-1}$ is $N[0, \sigma_t^2]$. These prediction errors are independent, so the likelihood function for $\mathbf{x}$ is

$$l(\mathbf{x}) = \prod_{t=0}^{N-1} (2\pi\sigma_t^2)^{-1} \exp\left\{-\frac{1}{2\sigma_t^2}(x_t - \hat{x}_{t|t-1})^2\right\}. \qquad 10.219$$

This is a recursive way of computing the exact likelihood function of an ARMA process.

10.12   DIFFERENTIAL PCM

One of the most important applications of linear prediction is the design of differential pulse code modulation (DPCM) systems. The basic idea is illustrated in Figure 10.20.

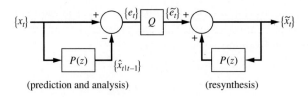

(prediction and analysis)   (resynthesis)

Figure 10.20 Differential PCM.

The time series $\{x_t\}$ is predicted with the predictor $P(z)$ to produce the predicted series $\{\hat{x}_{t|t-1}\}$. The error sequence is generated as

$$\{e_t\} = \left(1 - P(z)\right)\{x_t\}$$
$$= A(z)\,\{x_t\}.$$

                                                                                                   10.220

The polynomial $A(z)$ is, of course, the prediction error polynomial. From the error sequence $\{e_t\}$, the time series may be resynthesized, as illustrated in the figure. The resynthesized sequence depends on the error sequence as follows:

$$\{x_t\} = \frac{1}{1 - P(z)}\,\{e_t\}.$$

                                                                                                   10.221

If the error sequence is quantized, then e_t is replaced by $\tilde{e}_t = Q(e_t)$, and the resynthesized sequence is replaced by

$$\{\tilde{x}_t\} = \frac{1}{1 - P(z)}\,\{\tilde{e}_t\}.$$

                                                                                                   10.222

The error between $\{x_t\}$ and $\{\tilde{x}_t\}$ is

$$\{n_t\} = \{x_t\} - \{\tilde{x}_t\} = \left(\frac{1}{1 - P(z)}\right)\left(\{e_t\} - \{\tilde{e}_t\}\right)$$

                                                                                                   10.223

$$= \frac{1}{A(z)}\,\{u_t\},$$

where $u_t = e_t - \tilde{e}_t$ is the quantizing error. The error spectrum for $\{n_t\}$ is the white error spectrum of u_t shaped by $1/|A(e^{j\theta})|^2$, the spectrum of $1/A(z)$.

Let's redraw Figure 10.20, as illustrated in Figure 10.21, and quantize the error internally. The internal signal $\tilde{x}_t$ in the prediction and analysis diagram is related to the quantized error as follows:

$$\left(1 - P(z)\right)\{\tilde{x}_t\} = \{\tilde{e}_t\}.$$

                                                                                                   10.224

But this is exactly how $\{\tilde{x}_t\}$ depends on the quantized error $\{\tilde{e}_t\}$ in the resynthesis diagram, meaning that $\tilde{x}_t$ is the same variable in the two block diagrams of Figure 10.21.

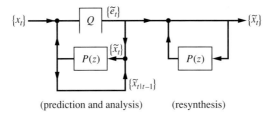

(prediction and analysis)   (resynthesis)

Figure 10.21 Differential PCM.

Furthermore, the internal variable $\tilde{x}_t$ may be written as

$$\{\tilde{x}_t\} = P(z)\{\tilde{x}_t\} + Q\big[\{x_t\} - P(z)\{\tilde{x}_t\}\big]$$
$$= P(z)\{\tilde{x}_t\} + \{x_t\} - P(z)\{\tilde{x}_t\} + \{u_t\} \qquad\qquad 10.225$$
$$= \{x_t\} + \{u_t\}.$$

This result says that the "quantization error for x_t is just the quantization error of the quantizer Q":

$$\tilde{x}_t - x_t = u_t. \qquad\qquad 10.226$$

REFERENCES AND COMMENTS

Even in a chapter of this length, one can only piece together the most prominent features of a topic as broad as linear prediction. In our writing we have been strongly influenced by Claude Gueguen [1979] and his beautiful set of unpublished notes.

The first fast algorithm for computing whiteners and reflection coefficients for the correlation method of linear prediction was discovered by Levinson [1947], who was studying practical ways to approximate Wiener filters. The first fixed-point algorithm for computing reflection coefficients (in the correlation method of linear prediction) was discovered by LeRoux and Gueguen [1977].

The first practical algorithm for estimating reflection coefficients from data was published by Burg [1975]. Morf et al. [1977] were the first to derive fast order- and time-update equations for computing whiteners and reflection coefficients in the covariance method of linear prediction. This work was generalized by Friedlander et al. [1979] who introduced the idea of displacement ranks for matrices that are close to Toeplitz. An excellent review of displacement tank is Gueguen's [1987] chapter.

There is a wealth of literature on ways to recursively update fixed-order whiteners (or predictors) in time. The recursive updates are usually called fast recursive least squares algorithms. The topic does not quite fit in the story we have told in this chapter, so we have omitted it. The interested reader may read [Ljung, Morf, and Falconer, 1978; Carayannis, Manolakis, and Kalouptsidis 1983; and Bellanger, 1987].

Recently, Rialan and I found a way to derive Schur and QR recursions from Levinson recursions using order updates *only* in the correlation method of linear prediction [Rialan and Scharf, 1988]. The QR algorithm is a lattice algorithm that had been discovered earlier by Sweet [1984] and by Cybenko [1983]. Demeure and I extended the technique to produce QR factors and corresponding lattice algorithms for all of the cases of linear prediction [Demeure and Scharf, 1990].

The original work on fast algorithms for Kalman gains was done by Morf, Sidhu, and Kailath [1974].

The original work on likelihood ratios was done by Schweppe [1965] and by Kailath et al. [1978]. For our perspective on these topics, see Dugré, Scharf, and Gueguen [1986]; and Gueguen and Scharf [1980].

Bellanger, M. [1987]. *Adaptive Digital Filters and Signal Analysis* (New York and Basel: Marcel Dekker, Inc., 1987).

Bochner, S. [1933]."Monotone Funktionen, Stieltjes Integrale und Harmonische Analyse," *Math Annal* **108**, p. 378 (1933). English translation as supplement to S. Bochner: *Lectures on Fourier Integrals,* translated by M. Tenenbaum and H. Pollard, Princeton University Press, Princeton, New Jersey, 1959.

Burg, J. P. [1975]. "Maximum Entropy Spectral Analysis," Ph.D. Thesis, Stanford University, Stanford, CA (1975).

Carayannis, G., D. G. Manolakis, and N. Kalouptsidis [1983]. "A Fast Sequential Algorithm for Least Squares Filtering and Prediction," *IEEE Trans. Acoust., Speech, and Sign. Proc.,* **ASSP-31**, pp. 1394–1402 (December 1983).

Cybenko, G. [1983]. "A General Orthogonalization Technique with Applications to Time Series Analysis and Signal Processing," *Math Comput* **40**:161, pp. 323–336 (January 1983).

Demeure C. J., and L. L. Scharf [1990]. "Sliding Windows and Lattice Algorithms for Computing QR Factors in the Least Squares Theory of Linear Prediction" *IEEE Trans ASSP* **ASSP-38**:4, pp. 721–725 (April 1990).

Dugré, J.-P., L. L. Scharf, and C. J. Gueguen [1986]. "Exact Likelihood for Stationary Vector Autoregressive Moving Average Processes," *Signal Proc* **11**, pp. 105-118 (1986).

Friedlander, B., M. Morf, T. Kailath, and L. Ljung [1979]."New Inversion Formulas for Matrices Classified in Terms of their Distance from Toeplitz Matrices," *Lin Alg Appl* **27**, pp. 31–60 (1979).

Gueguen, C. J. [1987]. "An Introduction to Displacement Ranks and Related Fast Algorithms," Course 12 in Signal Processing, Volume II, J.-L. Lacoume, T. Durrani, and R. Stora, eds. (Amsterdam: North Holland, 1987).

Gueguen, C. J. [1979]. "Advanced Signal Processing," Course Notes, Spring 1979, Colorado State University, Fort Collins, Colorado 80526.

Gueguen, C. J. and L. L. Scharf [1980]. "Exact Maximum Likelihood Identification of ARMA Models: A Signal Processing Perspective," *ONR Technical Report #36* (September 1980).

Itakura, F., and S. Saito [1971]. "Digital Filtering Techniques for Speech Analysis and Synthesis," *Proc 7th Intl Congr Acoust*, Budapest, pp. 262–264 (1971).

Kailath, T., B. Levy, L. Ljung, and M. Morf [1978]. "Fast Time-Invariant Implementation of Gaussian Signal Detectors" *IEEE Trans Infor Th* **IT-24**:4 (July 1978).

Khinchin, A. Y. [1934]. "Korrelationstheorie der Stationären Stochastichen Prozesse," *Math Annal* **109**, pp. 604–615 (1934).

Kolmogorov, A. N. [1939]. "Interpolation and Extrapolation of Stationary Random Sequences," *Bull de l'Académie des Sciences de l'URSS*, Ser. Math 5, p. 3–14 (1941). Also in *C R Acad Sciences* **208**, p. 2043, 1939.

Lang, S. [1977]. *Complex Analysis* (Reading, MA: Addison-Wesley, 1977), p. 223.

LeRoux, J., and C. J. Gueguen [1977]. "A Fixed Point Computation of Partial Correlation Coefficients," *IEEE Trans ASSP* **ASSP-25**, pp. 257–259 (June 1977).

Levinson, N. [1947]. "The Wiener RMS Criterion in Filter Design and Prediction," *J Math Phys* **25**, pp. 261–278 (1947).

Ljung, L., M. Morf, and J. J. Falconer [1978]. "Fast Calculation of Gain Matrices for Recursive Estimation Schemes," *Fut J Contr*, pp. 1–19 (January 1978).

Loéve, M. [1960]. *Probability Theory*, Second Ed. (Princeton, NJ: D. Van Nostrand Company, 1960).

Makhoul, J. [1977]. "Stable and Efficient Lattice Methods for Linear Prediction," *IEEE Trans ASSP* **25**, pp. 423–428 (October 1977).

Morf, M., G. S. Sidhu, and T. Kailath [1974]. "Some New Algorithms for Recursive Estimation on Constant Linear Discrete Time Systems," *IEEE Trans Autom Contr* **AC-19**, pp. 315–323 (August 1974).

Morf, M., T. Kailath, B. Dickinson, and A. Vieira [1977]. "Efficient Solution of Covariance Equations for Linear Prediction," *IEEE Trans ASSP* **ASSP-25**:2, pp. 224–235 (April 1977).

Pearlman, J. G. [1980]. "An Algorithm for the Exact Likelihood of a High Order ARMA Process," *Biometrika* **67**:1, pp. 232–233 (1980).

Rialan C. P., and L. L. Scharf [1988]. "Fast Algorithms for Computing QR and Cholesky Factors of Toeplitz Operators," *IEEE Trans ASSP* **ASSP-36**:11, pp. 1740–1748 (November 1988).

Schweppe, F. C. [1965]. "Evaluation of Likelihood Functions of Gaussian Signals," *IEEE Trans Infor Th* **IT-11**, pp. 61–70 (1965).

Sweet, D. R. [1984]. "Fast Toeplitz Orthogonalization," *Numerische Mathematik* **43**, pp. 1–4 (1984).

Whittle, P. [1954]. "Some Recent Contributions to the Theory of Stationary Processes," Appendix 2 in H. Wold, *A Study in the Analysis of Stationary Time
Series* (Stockholm: Almquist and Wiksell, 1954).

Wiener, N. [1949]. *Extrapolation, Interpolation and Smoothing of Stationary Time
Series with Engineering Applications* (New York: Technology Press and Wiley,
1949).

Yaglom, A. M. [1952]. *An Introduction to the Theory of Stationary Random Functions*, in Russian (1952). Translated by R. A. Silverman, Prentice Hall, Englewood Cliffs, New Jersey, 1962.

PROBLEMS

10.1 Let $\{r_t\} = \{\rho^t \cos t\theta\}$ denote the covariance sequence for a WSS time series.
Use Jensen's formula to find the prediction error σ^2 in terms of $z_1 = \rho e^{j\theta}$ and
$z_1^{\star} = \rho e^{-j\theta}$.

10.2 For the covariance sequence of Problem 10.1, run the Levinson and Schur
recursions for $i = 0, 1, \ldots, 4$. Explain your findings.

10.3 Use the lattice recursions in the equation

$$\mathbf{Y}^{\mathrm{T}}\mathbf{u}_i = \sigma_i^2 \begin{bmatrix} \mathbf{0} \\ \mathbf{h}_i \end{bmatrix}$$

to derive Schur recursions for the vector $[\mathbf{0} \quad \mathbf{h}_i]^{\mathrm{T}}$. Show how reflection coefficients and variances are computed.

10.4 *Burg's Algorithm.* The first algorithm for estimating reflection coefficients directly from the data is known as *Burg's Algorithm* [Burg, 1975]. This algorithm
computes the reflection coefficients by minimizing, for each order, a sum of
squared forward and backward linear prediction errors in the lattice of Figure
10.14. These errors are defined in Equation 10.172. Burg's idea was to choose
the reflection coefficient k_{i+1} in order to minimize the accumulated squared
errors at the output of cell $i + 1$ for $t \geq i + 1$:

$$\sigma_{i+1}^2 = \frac{1}{2(N - 1 - i)} \sum_{t=i+1}^{N-1} \left[u_{i+1}^2(t) + v_{i+1}^2(t) \right].$$

   a.   Use the lattice equations of Equation 10.172 to show that σ_{i+1}^2 may be
written as

$$\sigma_{i+1}^2 = \frac{(1 + k_{i+1}^2)}{2(N - 1 - i)} \sum_{t=i+1}^{N-1} \left[v_i^2(t) + u_i^2(t - 1) \right]$$

$$+ \frac{2k_{i+1}}{N - 1 - i} \sum_{t=i+1}^{N-1} v_i(t) u_i(t - 1).$$

b. Show that the solution for k_{i+1} to minimize σ_{i+1}^2 is

$$k_{i+1} = \frac{-2 \sum\limits_{t=i+1}^{N-1} v_i(t)u_i(t-1)}{\sum\limits_{t=i+1}^{N-1} \left[v_i^2(t) + u_i^2(t-1)\right]}.$$

The algorithm is initialized with $u_0(t) = v_0(t) = y_{N-1-t}$.

c. Prove $|k_{i+1}| < 1$.

d. Interpret the formula for k_{i+1}.

10.5 Show that the denominator of Burg's solution for the reflection coefficient may be written as the recursion

$$\text{den}(i+1) = (1 - k_i^2)\,\text{den}(i) - v_i^2(i) - u_i^2(N-1).$$

Several modified versions of Burg's algorithm are also available, such as a windowed version using

$$k_{i+1} = \frac{-2 \sum\limits_{t=i+1}^{N-1} w_i(t)v_i(t)u_i(t-1)}{\sum\limits_{t=i+1}^{N-1} w_{i+1}(t)\left[v_i^2(t) + u_i^2(t-1)\right]},$$

where $w_{i+1}(t)$ is a window with nonnegative coefficients chosen to ensure stability. Itakura and Saito [1971] proposed the alternative

$$k_{i+1} = \frac{-\sum\limits_{t=i+1}^{N-1} v_i(t)u_i(t-1)}{\left(\sum\limits_{t=i+1}^{N-1} v_i^2(t)\right)^{1/2}\left(\sum\limits_{t=i+1}^{N-1} u_i^2(t-1)\right)^{1/2}}.$$

Other formulas can be found in Makhoul [1977].

10.6 Let $x_1, x_2, \ldots, x_m, \ldots, x_N$ be zero-mean, unit variance random variables:

$$= E\mathbf{x}\mathbf{x}^T = \{r_{ij}\}; \qquad r_{ii} = 1$$

$$\mathbf{x} = (x_1 \quad x_2 \quad \cdots \quad x_N)^T.$$

Assume $\mathbf{R} = \{r_{ij}\}$ is nonsingular. We want to estimate x_m from all of the other x_i:

$$\hat{x}_m = -\sum_{i \neq m} a_i x_i.$$

The error is

$$e_m = x_m - \hat{x}_m = \mathbf{a}^T\mathbf{x}$$

$$\mathbf{a}^T = \begin{bmatrix} a_1 & \cdots & 1 & \cdots & a_N \end{bmatrix}.$$

Show

a.   the minimum value of $\sigma_m^2 = E e_m^2$ is ≤ 1;

b. σ_m^2 is minimized by $\mathbf{a} = [1/\delta_m^T \mathbf{R}^{-1} \delta_m] \mathbf{R}^{-1} \delta_m$;

c.   the minimum σ_m^2 is $1/\delta_m^T \mathbf{R}^{-1} \delta_m \leq 1$;

d.   the diagonal elements of $\mathbf{R}^{-1}$ are ≥ 1.

10.7 Use the Levinson recursions to derive the Schur recursions of Equation 10.118 from Equations 10.116 and 10.117. What identity do you have to use for $\mathbf{R}_t \mathbf{Z} - \mathbf{Z} \mathbf{R}_t$ to get the Schur recursions?

10.8 Run the Schur recursions of Equation 10.118 for $i = 0, 1, 2$. Show

a. $\mathbf{H}_0 = \begin{bmatrix} r_0 \\ r_1 \\ r_2 \end{bmatrix}$; $\mathbf{G}_0 = \begin{bmatrix} 0 \\ r_1 \\ r_2 \end{bmatrix}$;

b. $k_1 = -r_1/r_0$;

c. $\mathbf{H}_1 = \begin{bmatrix} 0 \\ \sigma_1^2 \\ r_1 + k_1 r_2 \end{bmatrix}$; $\mathbf{G}_1 = \begin{bmatrix} 0 \\ 0 \\ r_2 + k_1 r_1 \end{bmatrix}$;

d. $k_2 = -(r_2 + k_1 r_1)/\sigma_1^2$;

e. $\mathbf{H}_2 = \begin{bmatrix} 0 \\ 0 \\ \sigma_3^2 \end{bmatrix}$; $\mathbf{G}_2 = \begin{bmatrix} 0 \\ 0 \\ 0 \end{bmatrix}$.

10.9 Begin with the ARMA transfer function

$$H(z) = \frac{C(z)}{A(z)}$$

$$C(z) = c_0 + c_1 z^{-1} + c_2 z^{-2};$$

$$A(z) = 1 + a_1 z^{-1} + a_2 z^{-2}.$$

a.   Show that the reflection coefficients for the inverse AR lattice that implements $1/A(z)$ are $k_2 = a_2$ and $k_1 = a_1/(1 + a_2)$; draw this inverse AR lattice.

b.   Show that the coefficients in the expansion $C(z) = \sum_{i=0}^{p} \alpha_i B_i(z)$, with $B_i(z) = z^{-i} A_i(z^{-1})$, may be obtained by solving the equation

$$\begin{bmatrix} 1 & k_1 & a_2 \\ 0 & 1 & a_1 \\ 0 & 0 & 1 \end{bmatrix} \begin{bmatrix} \alpha_0 \\ \alpha_1 \\ \alpha_2 \end{bmatrix} = \begin{bmatrix} c_0 \\ c_1 \\ c_2 \end{bmatrix};$$

find the solutions for α_0, α_1, and α_2.

c.   Draw the ARMA lattice structure for $H(z)$.

10.10 Determine the ARMA lattice filter for

$$H(z) = \frac{0.3572(1 - 0.4662z^{-1})}{(1 - 0.9512z^{-1})(1 - 0.6703z^{-1})}.$$

10.11 Determine the ARMA lattice filter for

$$H(z) = \frac{(1 - z_1 z^{-1})(1 - z_2 z^{-1})(1 - z_3 z^{-1})}{(1 - p_1 z^{-1})(1 - p_1^* z^{-1})(1 - p_2 z^{-1})(1 - p_2^* z^{-1})},$$

where $z_1 = -0.1226$, $z_2 = -0.5353$, $z_3 = -0.0091$, $p_1 = 0.9583 + j0.0888$, and $p_2 = 0.9111 + j0.0349$.

10.12 *Wiener Sequence.* Consider the Wiener sequence

$$x_t = \sum_{n=1}^{t} u_n$$

where $\{u_n\}$ is a sequence of zero-mean, unit variance, uncorrelated random variables.

a.   Find $E x_t x_s$.

b.   Write $\mathbf{x}_t = (x_1 \quad x_2 \quad \cdots \quad x_t)^T$ as $\mathbf{x}_t = \mathbf{H}\mathbf{u}_t$ with $\mathbf{u}_t = (u_1 \quad u_2 \quad \cdots \quad u_t)^T$. What is $\mathbf{H}$?

c.   Compute $\mathbf{R}_t = E\mathbf{x}_t \mathbf{x}_t^T$. (This is called a Wiener matrix.)

d.   Construct the normal equations to compute the MMSE predictor of x_i from $\mathbf{x}_{i-1}$. Compute σ_i^2.

e.   Repeat (iv) for $i = 1, 2, \ldots, t$ and organize your prediction error vectors into the matrix $\mathbf{A}$. Show

$$\mathbf{R}_t \mathbf{A} = \mathbf{H}\mathbf{D}.$$

f.   Prove $\mathbf{A}^T \mathbf{R}_t \mathbf{A} = \mathbf{D}$.

g.   Compute $\mathbf{R}_t^{-1}$ and use it to verify that $\sigma_i^2 = 1$ for $i \geq 1$.

10.13 Write out pseudo-code for the Levinson recursions, in their forward and backward forms.

10.14 Write out pseudo-code for the Schur recursions.

10.15 Write out pseudo-code for the lattice recursions of Section 10.8.

10.16 From the equation for the impulse response in Equation 10.200, prove that the Kalman gain $\mathbf{k}_t$ is $\mathbf{k}_t = [h_1^t \quad \cdots \quad h_p^t]^T$.

10.17 Draw the predictor filters of Figure 10.16 for the $H(z)$ given in Problem 10.10.

10.18 Draw the stationary and nonstationary Markovian representations for the $H(z)$ given in Problem 10.10. Show both the synthesis and analysis representations.

10.19 In the lattice diagram of Figure 10.14, replace t by $N - 1 - t$. Then the input is y_t and the lattice variables are $v_i(N - 1 - t)$ and $u_i(N - 1 - t)$. Interpret these lattice variables in terms of the forward and backward prediction errors $u_{N-1+i-t}^i$ and v_{N-1-t}^i defined in Equation 10.143.

Modal Analysis

with Cédric Demeure

This chapter is dedicated to modal analysis. By and large the chapter is an application of the theory we have developed in previous chapters. But there are a few new wrinkles. In particular, we mix linear prediction theory and modal representations with the singular value decomposition in order to study the principal components—or minimum norm—theory of Tufts and Kumaresan [1983] and the MUSIC theory of Schmidt [1981] and others. The style of presentation is that of a review paper.

The data to be analyzed in this chapter is assumed to consist of a sum of exponentials in additive white noise. When the exponentials have complex arguments, then the data contains "complex modes," or damped sines and cosines. Such models arise in all domains of applied science and engineering, including radar, sonar, geophysics, and speech processing. The model is also well suited to the study of transient signals, as a transient is often modeled as the complex exponential impulse response of a finite dimensional linear system.

11.1   SIGNAL MODEL

Let's model the signal component of a received, or measured, data record as a finite sum of q damped cosines:

$$y_t = \sum_{i=1}^{q} A_i \rho_i^t \cos(2\pi f_i t + \phi_i); \qquad t = 0, 1, \ldots, N-1. \qquad 11.1$$

The parameter A_i is the amplitude of the ith cosine (in the signal units), ρ_i is its damping factor (dimensionless), f_i is the frequency (in hertz), and ϕ_i is the phase (in radians). A cosinusoidal component of the data is then defined by the quadruplet $(A_i, \rho_i, f_i, \phi_i)$. One way to obtain such a signal is to sample a continuous time signal $A_i e^{-\alpha_i t} \cos(\omega_i t + \phi_i)$ at $t = nT$ to produce $A_i (e^{-\alpha_i T})^n \cos(\omega_i T n + \phi_i)$. In this case, $\rho_i = e^{-\alpha_i T}$ and $2\pi f_i T = \omega_i T$. When the frequency of a cosine is an integer multiple of half the sampling frequency ($f_i = k/2T$), then the phase is taken to be equal to zero (or incorporated in the amplitude).

The signal y_t may be represented as a finite sum of p complex modes

$$y_t = \sum_{i=1}^{p} B_i z_i^t, \qquad 11.2$$

where any complex mode pair (B_i, z_i) is matched by a complex pair (B_i^*, z_i^*). The number of complex mode pairs p is equal to twice the number of cosines for which $f_i \neq k/2T$ plus the number of real mode pairs for which z_i is real. The amplitude B_i is complex (real) if the pole z_i is complex (real) with the relations

$$z_i = \rho_i \quad \text{with } B_i = A_i, \quad \text{when } f_i = kT/2$$

$$z_i = \rho_i e^{j2\pi f_i T} \quad \text{with } B_i = \frac{A_i}{2} e^{j\phi_i}, \quad \text{otherwise.} \qquad 11.3$$

In most applications, the problem is to estimate the $4p$ real parameters (ρ_i, f_i, A_i, ϕ_i, with $i = 1, \ldots, p$) from N samples of the time series $\{y_t\}_{t=0}^{N-1}$. In some instances, such as interferometry, the data are covariances $\{r_t\}_{t=0}^{k}$. We shall have more to say about this in Section 11.7. A closely related problem is to estimate the $2p$ filter parameters in an ARMA representation of y_t. This is a Padé approximation problem [Baker and Graves-Morris, 1981].

There is an important difference between the parameters B_i and z_i; the signal model is linear in the first and nonlinear in the second. This observation allows us to anticipate that the estimation of the parameters z_i is going to be much more difficult than the estimation of the parameters B_i. In the following several paragraphs we establish three equivalent models for the signal y_t: (1) a modal decomposition, (2) an ARMA model, and (3) a linear prediction model.

Modal Decomposition

Equation 11.2 is a modal representation for the signal $\{y_t\}$. The snapshot $\mathbf{y} = [y_0 \ \ y_1 \ \ \cdots \ \ y_{N-1}]^T$ has the corresponding modal decomposition

$$
\mathbf{y} = \begin{bmatrix} y_0 \\ y_1 \\ \vdots \\ y_{N-1} \end{bmatrix} = \mathbf{VB}, \tag{11.4}
$$

where $\mathbf{V}$ is a complex Vandermonde matrix of dimension $N \times p$ and $\mathbf{B}$ is a $p \times 1$ vector of mode weights:

$$
\mathbf{V} = \begin{bmatrix} 1 & 1 & \cdots & 1 \\ z_1 & z_2 & \cdots & z_p \\ \vdots & \vdots & & \vdots \\ z_1^{N-1} & z_2^{N-1} & \cdots & z_p^{N-1} \end{bmatrix} ; \quad \mathbf{B} = \begin{bmatrix} B_1 \\ B_2 \\ \vdots \\ B_p \end{bmatrix} . \tag{11.5}
$$

We call the ith column of $\mathbf{V}$ the ith mode and denote it by $\boldsymbol{\psi}(z_i)$:

$$
\boldsymbol{\psi}(z_i) = \begin{bmatrix} 1 \\ z_i \\ \vdots \\ z_i^{N-1} \end{bmatrix} . \tag{11.6}
$$

In general, columns containing complex conjugate modes are adjacent in $\mathbf{V}$.

From the complex conjugate poles $(z_1, z_2, \ldots, z_p)$, we may form the real polynomial $A(z)$:

$$
A(z) = \prod_{i=1}^{p} (1 - z_i z^{-1}) = \sum_{i=0}^{p} a_i z^{-i}; \qquad a_0 = 1 \tag{11.7}
$$

and the corresponding "whitening" or "prediction error" matrix $\mathbf{A}^T$ of dimension $(N - p) \times N$:

$$
\mathbf{A}^T = \begin{bmatrix} a_p & a_{p-1} & \cdots & a_1 & 1 & 0 & \cdots & 0 \\ 0 & a_p & & & \ddots & \ddots & \ddots & \vdots \\ \vdots & \ddots & \ddots & & & a_1 & 1 & 0 \\ 0 & \cdots & 0 & a_p & \cdots & a_2 & a_1 & 1 \end{bmatrix} . \tag{11.8}
$$

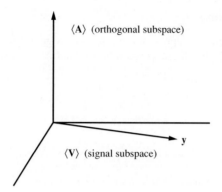

Figure 11.1 Signal and orthogonal subspaces.

The complex poles z_i are the zeros of the polynomial $A(z)$. Therefore we have the property $\mathbf{A}^T\mathbf{V} = \mathbf{0}$. The polynomial coefficients in the rows of $\mathbf{A}^T$ annihilate the columns of $\mathbf{V}$. Furthermore, the $N - p$ rows of $\mathbf{A}^T$ are linearly independent, as are the p columns of $\mathbf{V}$, provided $z_i \neq z_j$. This means that the matrices $\mathbf{V}$ and $\mathbf{A}$ span orthogonal subspaces as illustrated in Figure 11.1. We call $\langle\mathbf{V}\rangle$ the signal subspace spanned by the modes of $\mathbf{V}$ and $\langle\mathbf{A}\rangle$ the orthogonal subspace spanned by the whiteners of $\mathbf{A}$. The projector onto the subspace spanned by the columns of $\mathbf{V}$ is given by the matrix $\mathbf{P}_V = \mathbf{V}(\mathbf{V}^H\mathbf{V})^{-1}\mathbf{V}^H$, and the projector onto the subspace spanned by the columns of $\mathbf{A}$ is given by the matrix $\mathbf{P}_A = \mathbf{A}(\mathbf{A}^T\mathbf{A})^{-1}\mathbf{A}^T$. Furthermore, $\mathbf{P}_V + \mathbf{P}_A = \mathbf{I}$. These two matrices are useful in a geometrical understanding of the problem.

ARMA Impulse Response

The measurement record $\mathbf{y} = [y_0 \quad y_1 \quad \cdots \quad y_{N-1}]^T$ may also be interpreted as a snapshot of the impulse response for an ARMA $(p, p - 1)$ discrete-time system. That is, the sequence $\{y_t\}$ may be generated as

$$\{y_t\} = H(z)\{\delta_t\}, \tag{11.9}$$

where $H(z)$ is the ARMA transfer function illustrated in Figure 11.2:

$$H(z) = \frac{B(z)}{A(z)}$$

$$= \sum_{i=0}^{p-1} b_i z^{-i} \Big/ \sum_{i=0}^{p} a_i z^{-i}; \qquad a_0 = 1 \tag{11.10}$$

$$= \sum_{i=1}^{p} \frac{B_i}{1 - z_i z^{-1}}.$$

The modal form of the transfer function $H(z)$ produces the modal decomposition of the impulse response:

$$\{y_t\} = \sum_{i=1}^{p} \frac{B_i}{1 - z_i z^{-1}} \{\delta_t\}$$

$$= \sum_{i=1}^{p} B_i z_i^t .$$

11.11

The rational form of the transfer function produces the difference equation for the impulse response:

$$\{y_t\} = \frac{B(z)}{A(z)} \{\delta_t\}$$

$$A(z)\{y_t\} = B(z)\{\delta_t\}$$

$$\sum_{i=0}^{p} a_i y_{t-i} = \begin{cases} 0, & t < 0 \\ b_t, & 0 \le t < p \\ 0, & t \ge p. \end{cases}$$

11.12

If we think of these filtering equations as

$$\{y_t\} = B(z)H_{AR}(z)\{\delta_t\}$$

11.13

where $H_{AR}(z)$ is the AR transfer function $1/A(z)$, then we may also write y_t as

$$y_t = \sum_{i=0}^{p} b_i g_{t-i},$$

11.14

where $\{g_t\}$ is the impulse response of $1/A(z)$:

$$\{g_t\} = \frac{1}{A(z)} \{\delta_t\}.$$

11.15

In these filtering equations, the polynomial $A(z)$ is called a prediction error polynomial because it predicts and annihilates $\{y_t\}$ for $t \ge p$.

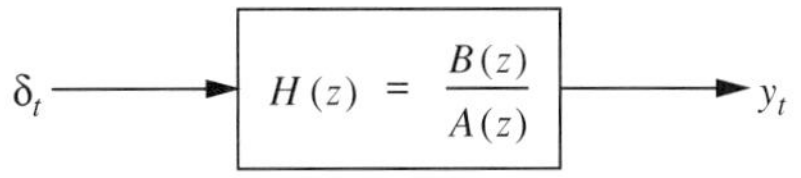

Figure 11.2 Signal model as the impulse response of an ARMA filter.

Linear Prediction

The ARMA difference equation of Equation 11.12 may be written out for $0 \le t \le N - 1$ to produce the linear equation

$$
\begin{bmatrix}
1 & & & & \\
a_1 & 1 & & \mathbf{0} & \\
\vdots & & \ddots & & \\
a_p & \cdots & a_1 & 1 & \\
& \ddots & & & 1 \\
\mathbf{0} & & a_p & \cdots & a_1 & 1
\end{bmatrix}
\begin{bmatrix}
y_0 \\
y_1 \\
\vdots \\
y_p \\
\vdots \\
y_{N-1}
\end{bmatrix}
=
\begin{bmatrix}
b_0 \\
b_1 \\
\vdots \\
b_{p-1} \\
0 \\
\vdots \\
0
\end{bmatrix} .
\qquad 11.16
$$

We define the matrix on the left as $\mathbf{G}^{-1}$ and rewrite these equations as

$$
\mathbf{G}^{-1}\mathbf{y} = \begin{bmatrix} \mathbf{b} \\ \mathbf{0} \end{bmatrix} ; \qquad
\mathbf{G}^{-1} =
\begin{bmatrix}
1 & & & & \\
a_1 & 1 & & \mathbf{0} & \\
& & \ddots & & \\
a_{p-1} & \cdots & a_1 & 1 & \\
& & & \mathbf{A}^{\mathsf{T}}
\end{bmatrix} .
\qquad 11.17
$$

There are analysis equations that show how $\mathbf{y}$ may be analyzed to produce the MA coefficients and, ultimately, zeros. The synthesis form of the equations is

$$
\mathbf{y} = \mathbf{G} \begin{bmatrix} \mathbf{b} \\ \mathbf{0} \end{bmatrix} = \mathbf{H}\mathbf{b}
$$

$$
=
\begin{bmatrix}
g_0 & & & & \\
g_1 & g_0 & & \mathbf{0} & \\
\vdots & \vdots & \ddots & & \\
& & & g_0 & \\
& & & \vdots & \\
g_{N-1} & \cdots & & g_{N-p}
\end{bmatrix}
\begin{bmatrix}
b_0 \\
\vdots \\
b_{p-1}
\end{bmatrix} .
\qquad 11.18
$$

The matrix $\mathbf{H}$ is comprised of the first p columns in $\mathbf{G}$. The entries in H are the impulse responses of $1/A(z)$. In fact, Equation 11.18 is just a fancy way to write out the ARMA equations $\{y_t\} = B(z)[1/A(z)]\{\delta_t\}$.

The analysis equations of Equation 11.16 may be written as linear prediction equations by organizing the coefficients a_i into a vector and the data into a matrix:

$$
\begin{bmatrix}
& & & & y_0 \\
& \mathbf{0} & & y_0 & y_1 \\
& & \ddots & & \vdots \\
y_0 & y_1 & \cdots & y_p \\
y_1 & & & \vdots \\
\vdots & & & \\
y_{N-1-p} & \cdots & & y_{N-1}
\end{bmatrix}
\begin{bmatrix}
a_p \\
\vdots \\
a_1 \\
1
\end{bmatrix}
=
\begin{bmatrix}
b_0 \\
b_1 \\
b_{p-1} \\
\vdots \\
0 \\
\vdots \\
0
\end{bmatrix} .
\qquad 11.19
$$

The homogeneous tail of these equations produces the so-called covariance equations of linear prediction:

$$
\begin{bmatrix}
y_0 & y_1 & \cdots & y_p \\
 & & & \\
y_1 & & & y_{p+1} \\
\vdots & & & \vdots \\
 & & & \\
y_{N-1-p} & & \cdots & y_{N-1}
\end{bmatrix}
\begin{bmatrix}
a_p \\
\vdots \\
a_1 \\
1
\end{bmatrix}
=
\begin{bmatrix}
0 \\
\vdots \\
\vdots \\
0
\end{bmatrix}.
\qquad 11.20
$$

These equations follow directly from the orthogonality of the linear prediction matrix $\mathbf{A}^T$ and the data $\mathbf{y}$ or, equivalently, from the linear prediction matrix $\mathbf{A}^T$ and the modal matrix $\mathbf{V}$. The covariance equations show that the prediction error vector $\mathbf{a}^T = [a_p \quad \cdots \quad a_1 \quad 1]$ may be found from linear prediction equations or from the mode matrix $\mathbf{V}$, as outlined in Equation 11.7.

11.2   THE ORIGINAL PRONY METHOD

In the absence of additive noise, the polynomial $A(z)$ is an annihilator of the time series $\{y_t\}$. That is, the synthesis model of Equation 11.12 may be multiplied on both sides by $A(z)$ to annihilate the right-hand side:

$$
\{y_t\} = H(z)\{\delta_t\} \quad \Rightarrow \quad A(z)\{y_t\} = \begin{cases} B(z)\{\delta_t\} \\ 0, & \text{for } t \geq p. \end{cases}
\qquad 11.21
$$

The equation $A(z)\{y_t\} = B(z)\{\delta_t\}$ may be written out for $t = 0, 1, \ldots, n$ to produce the linear prediction equations of Equation 11.19 and the following covariance equations of linear prediction:

$$
\begin{bmatrix}
y_0 & y_1 & \cdots & y_p \\
y_1 & \cdots & y_p & y_{p+1} \\
\vdots & & & \vdots \\
y_{N-1-p} & & \cdots & y_{N-1}
\end{bmatrix}
\begin{bmatrix}
a_p \\
\vdots \\
a_1 \\
1
\end{bmatrix}
=
\begin{bmatrix}
0 \\
0 \\
\vdots \\
0
\end{bmatrix}
$$

$$
\mathbf{Ya} = \mathbf{0}. \qquad 11.22
$$

The column rank of $\mathbf{Y}$ is evidently p and not $p + 1$. That is, the last column of $\mathbf{Y}$ lies in the range of the first p columns of $\mathbf{Y}$, meaning N may be chosen to equal $2p + 1$ and the resulting system of equations solved for $\mathbf{a}$. If there is no noise added, this system gives the desired coefficients $\{a_i\}$. This is the original Prony method [1795]. The determination of the coefficients is reduced to solving a nonsymmetric Hankel system of equations using a fast algorithm [Trench, 1964; and Zohar, 1974 and 1969].

Filter Coefficients

If only the filter representation is desired, then the numerator coefficients may be obtained without computing the roots of $A(z)$ by using the head of the linear prediction equations of Equation 11.19 to solve for the $\{b_i\}$:

$$\begin{bmatrix} & & & y_0 \\ \mathbf{0} & & y_0 & y_1 \\ & \cdot^{\cdot^{\cdot}} & & \vdots \\ y_0 & & \cdots & y_p \end{bmatrix} \begin{bmatrix} a_p \\ \vdots \\ a_1 \\ 1 \end{bmatrix} = \begin{bmatrix} b_0 \\ b_1 \\ \vdots \\ b_{p-1} \end{bmatrix}. \qquad 11.23$$

This method is equivalent to computing the Padé approximation of the polynomial associated with the data values $Y(z)$. That is, the polynomial

$$Y(z) = y_0 + y_1 z^{-1} + \cdots + y_n z^{-(2p-1)} \qquad 11.24$$

is modeled as

$$Y(z) = \frac{B(z)}{A(z)} \iff A(z)Y(z) = B(z) \iff \sum_{n=0}^{p} a_n y_{t-n} = b_t. \qquad 11.25$$

For more details on rational fraction approximation, see [Padé, 1892; or Baker and Morris-Graves, 1981], and for an implementation using the Euclid algorithm, see [McEliece and Shearer, 1978].

Modes

The roots of the polynomial $A(z)$ determine the frequencies and associated damping factors of the poles $z_i = \rho_i e^{j2\pi f_i T}$. If the modes themselves are desired, then a root-finding routine (see [Conte and de Boor, 1972; Jenkins and Traub, 1972; or Jenkins, 1975]) must be used to compute the roots of the polynomial $A(z)$. The magnitude and phase of the B_i parameters may then be computed by writing out the modal equation $\mathbf{y} = \mathbf{VB}$ and solving the following Vandermonde system of equations:

$$\begin{bmatrix} 1 & 1 & \cdots & 1 \\ z_1 & z_2 & \cdots & z_p \\ z_1^2 & z_2^2 & \cdots & z_p^2 \\ \vdots & \vdots & & \vdots \\ z_1^{p-1} & z_2^{p-1} & \cdots & z_p^{p-1} \end{bmatrix} \begin{bmatrix} B_1 \\ B_2 \\ B_3 \\ \vdots \\ B_p \end{bmatrix} = \begin{bmatrix} y_0 \\ y_1 \\ y_2 \\ \vdots \\ y_{p-1} \end{bmatrix} \qquad 11.26$$

This system has p equations and p unknowns. The algorithm of Björk and Pereyra [1970] may be used to solve the system in $O(p^2)$ operations. This completes the identification of (z_i, B_i) or (a_i, b_i) in the complex exponential model, using $2p + 1$ measurements.

Example 11.1

Suppose we have the signal

$$y_t = A\rho^t \cos(2\pi f t) = \frac{A}{2} z_0^t + \frac{A}{2} (z_0^*)^t, \qquad \text{with } z_0 = \rho e^{j2\pi f}. \qquad 11.27$$

Then if a second-order prediction error polynomial is used, we have the equation

$$
\begin{bmatrix} 0 & 0 & y_0 \\ 0 & y_0 & y_1 \\ y_0 & y_1 & y_2 \\ y_1 & y_2 & y_3 \end{bmatrix} \begin{bmatrix} a_2 \\ a_1 \\ 1 \end{bmatrix} = \begin{bmatrix} b_0 \\ b_1 \\ 0 \\ 0 \end{bmatrix}.
\tag{11.28}
$$

The coefficients of the prediction error polynomial $A(z)$ are obtained by solving for a_1 and a_2 from the homogeneous part of this equation:

$$
\begin{bmatrix} y_0 & y_1 \\ y_1 & y_2 \end{bmatrix} \begin{bmatrix} a_2 \\ a_1 \end{bmatrix} = - \begin{bmatrix} y_2 \\ y_3 \end{bmatrix}
\tag{11.29}
$$

$$
\begin{bmatrix} A & A\rho \cos 2\pi f \\ A\rho \cos 2\pi f & A\rho^2 \cos 4\pi f \end{bmatrix} \begin{bmatrix} a_2 \\ a_1 \end{bmatrix} = - \begin{bmatrix} A\rho^2 \cos 4\pi f \\ A\rho^3 \cos 6\pi f \end{bmatrix}.
$$

The solution for a_1 and a_2 is

$$
a_1 = -2\rho \cos(2\pi f); \qquad a_2 = \rho^2,
\tag{11.30}
$$

and the corresponding prediction error polynomial is

$$
\begin{aligned}
A(z) &= 1 - 2\rho \cos(2\pi f)z^{-1} + \rho^2 z^{-2} \\
&= (1 - \rho e^{j2\pi f} z^{-1})(1 - \rho e^{-j2\pi f} z^{-1}) \\
&= (1 - z_0 z^{-1})(1 - z_0^* z^{-1}).
\end{aligned}
\tag{11.31}
$$

The amplitude and phase are obtained using the Vandermonde equation

$$
\begin{bmatrix} 1 & 1 \\ z_0 & z_0^* \end{bmatrix} \begin{bmatrix} B_0 \\ B_1 \end{bmatrix} = \begin{bmatrix} y_0 \\ y_1 \end{bmatrix}
\tag{11.32}
$$

$$
\begin{bmatrix} 1 & 1 \\ \rho e^{j2\pi f} & \rho e^{-j2\pi f} \end{bmatrix} \begin{bmatrix} B_0 \\ B_1 \end{bmatrix} = \begin{bmatrix} A \\ A\rho \cos 2\pi \end{bmatrix}.
$$

The solution is $B_0 = B_1 = A/2$, and the identified model is

$$
\begin{aligned}
y_t &= B_0 z_0^t + B_1 (z_0^*)^t = \frac{A}{2} \rho^t (e^{j2\pi f t} + e^{-j2\pi f t}) \\
&= A\rho^t \cos(2\pi f t). \qquad \blacksquare
\end{aligned}
\tag{11.33}
$$

11.3   LEAST SQUARES PRONY METHOD

The Prony procedure works perfectly well when no noise is present in the data, but when noise (or measurement imprecision) is introduced, then this method performs very poorly, largely due to the extreme sensitivity of root locations z_i to the coefficients of the polynomial $A(z)$. It seems natural to use more measurements of $\{y_t\}$ to determine the polynomial coefficients, just as in linear prediction more than n measurements are used to determine a predictor of order n. This is the foundation of the extended Prony method or least squares Prony method [Hildebrand, 1956].

If $N > 2p + 1$, then the system of equations used to compute the coefficients of the polynomial $A(z)$ is overdetermined, and a least squares solution to the system $\mathbf{Ya} = \mathbf{0}$ is used, with the constraint that the zeroth coefficient of the vector $\mathbf{a}$ is equal to one. The least squares equations and their solution are $\mathbf{Ya} = \mathbf{0}$ and $\mathbf{a} = (\mathbf{Y}^T\mathbf{Y})^{-1}\sigma^2\boldsymbol{\delta}$, where σ^2 is the minimized sum of squared errors. This is just the least squares solution of linear prediction.

Choice of the Data Matrix

By analogy with Chapter 10, various data matrices may be used, with different initial and final conditions. In Chapter 10, the linear prediction model assumed that the data was the output of a linear system driven by white noise. In modal analysis, the data is modeled as the noisy output of a linear system driven by a pulse train. This distinction is illustrated in Figure 11.3. If the system $1/A(z)$ is stable, then the prediction error polynomial has zeros inside the unit circle. The backward prediction error filter $z^{-p}A(z^{-1})$ has zeros outside the unit circle and noise zeros inside the unit circle [Kumaresan, 1984]. If the system is quasi-stable, meaning that its poles are on the unit circle, then forward, backward, or forward-backward linear prediction may be used, and the latter is usually preferred as it produces more equations. In the case of poles on the unit circle, the polynomial must be symmetrical. The model is then forced to have symmetrical coefficients ($\mathbf{a} = \mathbf{Ja}$). This condition is necessary to have unit magnitude roots, but it is not sufficient.

Fast Algorithms

The fast algorithms of Chapter 10, and the references therein, may be used to compute the filter coefficients $\mathbf{a}$ for each choice of the data matrix $\mathbf{Y}$. The minimization of a quadratic error is not the only possible choice for a minimization criterion. Other

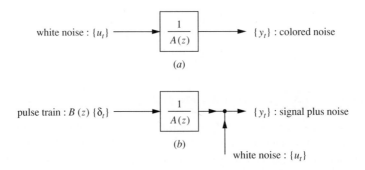

Figure 11.3 Distinction between linear prediction (a) and modal analysis (b).

norms can be used. They force the use of an iterative search technique to implement the minimization step, as no closed form solution is available. See [Barrodale and Phillips, 1975] for an algorithm to solve this minimization problem when a Chebyschev norm (L_∞ norm) is used.

Solving for the Mode Weights

Once the poles $\{z_i\}_{i=1}^p$ have been computed, the system $\mathbf{y} = \mathbf{VB}$, is solved for the complex weights $\{B_i\}_{i=1}^p$, in the least squares sense. The algorithm developed by Demeure [1988] uses a fast QR factorization of the Vandermonde matrix $\mathbf{V}$. Alternative versions of this algorithm for complex conjugate poles and for poles with multiplicity greater than or equal to one may also be found in [Demeure, 1988].

Solving for the Filter Coefficients

If only the filter coefficients are desired, then the roots of $A(z)$ are not computed and the numerator polynomial $B(z)$ may be estimated from the synthesis equations given in Equation 11.18:

$$
\begin{bmatrix} y_0 \\ \\ y_1 \\ \\ = \\ \cdots \\ \\ \\ y_{N-1} \end{bmatrix}
=
\begin{bmatrix} g_0 & & & & \\ \vdots & g_0 & & \mathbf{0} & \\ \vdots & \vdots & \ddots & & \\ & & & g_0 & \\ & & & \vdots & \\ g_{N-1} & & \cdots & & g_{N-p} \end{bmatrix}
\begin{bmatrix} b_0 \\ b_1 \\ \vdots \\ b_p \end{bmatrix}
\qquad 11.34
$$

$$\mathbf{y} = \mathbf{Hb}$$

Here $\{g_t\}$ is the impulse response of the autoregressive transfer function $1/A(z)$. The matrix $\mathbf{H}$ is Hankel, and the solution for the $\{b_i\}$ is computed in the least squares sense. This procedure for finding $B(z)$ is known as Shank's method [Shank, 1967; Kopec, Oppenheim, and Tribolet, 1977; and Kumaresan and Tufts, 1982]. It is really just a least squares version of the Padé approximation problem. We emphasize that this solution is not the exact least squares solution for the coefficients $\{a_i, b_i\}$. Rather it is a modified least squares estimate of the coefficients $\{a_i\}$ using the tail of the data, followed by a least squares estimate of the coefficients $\{b_i\}$.

11.4   MAXIMUM LIKELIHOOD AND EXACT LEAST SQUARES

The maximum likelihood estimator (MLE) chooses the values of $\mathbf{B}$ and $\mathbf{V}$ that maximize $L(\mathbf{B}, \mathbf{V}; \mathbf{y})$, the likelihood function for the data. In modal analysis, the model is

$$\mathbf{y} = \mathbf{VB} + \mathbf{n}, \qquad 11.35$$

where the noise vector $\mathbf{n}$ is $N[\mathbf{0}, \sigma^2\mathbf{I}]$ and $\mathbf{V}$ is a complex Vandermonde matrix. The likelihood function is

$$L(\mathbf{B}, \mathbf{V}; \mathbf{y}) = \frac{1}{\pi^N \sigma^{2N}} \exp\left(-\frac{1}{\sigma^2}(\mathbf{y} - \mathbf{VB})^H(\mathbf{y} - \mathbf{VB})\right). \qquad 11.36$$

The MLEs of $\mathbf{B}$ and $\mathbf{V}$ are found by minimizing the quadratic form in the exponent:

$$\min_{\mathbf{V}, \mathbf{B}} \left((\mathbf{y} - \mathbf{VB})^H(\mathbf{y} - \mathbf{VB})\right). \qquad 11.37$$

The minimization is equivalent in this case to the nonlinear least squares problem

$$\min_{B_k, z_k} \sum_{t=0}^{N-1} \left| y_t - \sum_{k=1}^{p} B_k z_k^t \right|^2. \qquad 11.38$$

This problem was first studied by Aigrain and Williams [1949]. Successive approximation techniques must be used to solve this problem. For a detailed discussion of optimization techniques, see the book by Dennis and Schnabel [1983].

Compressed Likelihood

The minimization problem is linear in the parameter $\mathbf{B}$, so that given the solution for the Vandermonde matrix $\mathbf{V}$, the solution for the linear parameter is $\hat{\mathbf{B}} = (\mathbf{V}^H\mathbf{V})^{-1}\mathbf{V}^H\mathbf{y}$. The problem is then reduced to the following equivalent formulations:

$$\min_{\mathbf{V}} \mathbf{y}^T(\mathbf{I} - \mathbf{P}_V)\mathbf{y} = \min_{A} \mathbf{y}^T\mathbf{P}_A\mathbf{y}$$

$$= \max_{z_i} \mathbf{y}^T\mathbf{P}_V\mathbf{y} \qquad 11.39$$

$$= \min_{\mathbf{a}_p} \mathbf{y}^T\mathbf{P}_A\mathbf{y},$$

where $\mathbf{P}_V = \mathbf{V}(\mathbf{V}^H\mathbf{V})^{-1}\mathbf{V}^H$ is the projection onto the space spanned by the columns of the matrix $\mathbf{V}$, and $\mathbf{P}_A = \mathbf{I} - \mathbf{P}_V$ is the projection onto its orthogonal complement. Equation 11.39 is a compressed maximum likelihood formula. Using the prediction error matrix $\mathbf{A}$ defined in Equation 11.8, the maximization of $\mathbf{y}^T\mathbf{P}_A\mathbf{y}$ is restated as

$$\min_{\mathbf{a}} \mathbf{y}^T\mathbf{A}(\mathbf{A}^T\mathbf{A})^{-1}\mathbf{A}^T\mathbf{y}. \qquad 11.40$$

Once $\mathbf{a}$ is estimated, its roots are computed to obtain the modes z_i, and the parameters $\{b_i\}_0^{p-1}$, or the amplitudes $\{B_i\}_1^p$, are obtained using linear least squares estimates. Still another equivalent and more useful formulation is obtained by rewriting $\mathbf{Ya}$ as $\mathbf{A}^T\mathbf{y}$:

$$\begin{bmatrix} y_0 & \cdots & y_{p-1} & y_p \\ y_1 & \cdots & y_p & y_{p+1} \\ \vdots & & \vdots & \vdots \\ y_{N-1-p} & \cdots & y_{N-2} & y_{N-1} \end{bmatrix} \begin{bmatrix} a_p \\ \vdots \\ a_1 \\ 1 \end{bmatrix} = \begin{bmatrix} a_p & \cdots & 1 & 0 & \cdots & 0 \\ 0 & a_p & 1 & \ddots & \vdots \\ \vdots & \ddots & \ddots & & \ddots & 0 \\ 0 & \cdots & 0 & a_p & \cdots & 1 \end{bmatrix} \begin{bmatrix} y_0 \\ y_1 \\ \vdots \\ y_{N-1} \end{bmatrix}.$$

$$11.41$$

Then the minimization with respect to $\mathbf{a}$ is

$$\min_{\mathbf{a}} \mathbf{a}^T \mathbf{Y}^T (\mathbf{A}^T \mathbf{A})^{-1} \mathbf{Y} \mathbf{a}. \qquad 11.42$$

This problem was first studied by Evans and Fischl [1973] in a filter design context.

An iterative solution of the minimization problem is presented in [Kumaresan, Scharf, and Shaw, 1986]. The idea is to iteratively minimize the quadratic form

$$(\mathbf{a}_k)^T \mathbf{Y}^T (\mathbf{A}_{k-1}^T \mathbf{A}_{k-1})^{-1} \mathbf{Y} \mathbf{a}_k, \qquad 11.43$$

where $\mathbf{a}_k$ is the approximation of $\mathbf{a}$ at the kth iteration and $\mathbf{A}_{k-1}$ is built using the coefficients in $\mathbf{a}_{k-1}$. At each iteration, the new approximation is obtained with a simple quadratic minimization. An equivalent algorithm, called IQML for Iterative Quadratic Maximum Likelihood, is published in [Bresler and Macovski, 1986]. McClellan and Lee [1990] have recently shown that the KiSS algorithm of Kumaresan, Scharf, and Shaw [1986], and the IQML algorithm are equivalent to the original Steiglitz-McBride [1974] algorithm. Still another approach was introduced by Kay [1984] and called the Iterative Filtering Algorithm (IFA) in the case of pure sinusoids. The IFA is a method for computing an approximate MLE of $\mathbf{a}$. It replaces the criterion above by estimating $(\mathbf{A}^T \mathbf{A})^{-1}$ as the correlation matrix of an MA process. See [Kay, 1988 and 1984] for more details on the filtering interpretation and the implementation of the IFA.

11.5 TOTAL LEAST SQUARES

In the traditional least squares solution of linear prediction equations, the whitener $\mathbf{a}$ is chosen to minimize the L_2 norm of $\mathbf{e}$ in the linear prediction equation $\mathbf{Y}\mathbf{a} = \mathbf{e}$.

Equivalently, $\mathbf{Y}$ is written as $[\mathbf{X} \quad \mathbf{y}]$ and $\mathbf{a}$ is written as $\mathbf{a} = [-\boldsymbol{\theta}^T \quad 1]^T$, and the equations $\mathbf{Y}\mathbf{a} = \mathbf{y} - \mathbf{X}\boldsymbol{\theta} = -\mathbf{e}$ are solved. The total least squares method of solving an overdetermined system of linear equations is a refinement over the least squares method that allows for the introduction of errors in the data matrix $\mathbf{X}$ as well as errors in $\mathbf{y}$.

The conventional least squares method solves the system $\mathbf{y} = \mathbf{X}\boldsymbol{\theta}$ by minimizing the L_2 norm of the error vector $\mathbf{X}\boldsymbol{\theta} - \mathbf{y} = \mathbf{e}$ while forcing $\mathbf{y} + \mathbf{e}$ to lie in the range space of the columns of the data matrix $\mathbf{X}$. The total least squares method solves the equations by supposing that there is some error in both $\mathbf{X}$ and $\mathbf{y}$ so that the model is

$$(\mathbf{X} + \mathbf{dX})\boldsymbol{\theta} = \mathbf{y} + \mathbf{e}. \qquad 11.44$$

This equation may be written as

$$([\mathbf{X}|\mathbf{y}] + [\mathbf{dX}|\mathbf{e}]) \begin{bmatrix} -\boldsymbol{\theta} \\ 1 \end{bmatrix} = \mathbf{0}. \qquad 11.45$$

This equation says that the vector $[-\boldsymbol{\theta}^T \quad 1]^T$ lies in the null space of the augmented matrix $([\mathbf{X}|\mathbf{y}] + [\mathbf{dX}|\mathbf{e}])$. This, in turn, says that the augmented matrix must be rank-deficient, meaning $[\mathbf{dX}|\mathbf{e}]$ must be chosen to reduce the rank of $[\mathbf{X}|\mathbf{y}]$. The method of total least squares consists of finding the matrix $[\mathbf{dX}|\mathbf{e}]$ of minimum Frobenius norm that reduces the rank of $[\mathbf{X}|\mathbf{y}]$. This, of course, forces $\mathbf{y} + \mathbf{e}$ to lie in the range of the columns of the matrix $\mathbf{X} + \mathbf{dX}$. Once more the SVD is useful. Take the SVD of the augmented matrix $[\mathbf{X}|\mathbf{y}]$:

$$[\mathbf{X}|\mathbf{y}] = \mathbf{U}\boldsymbol{\Sigma}\mathbf{V}^{\mathrm{T}} = \sum_{i=1}^{L+1} \mathbf{u}_i\sigma_i\mathbf{v}_i^{\mathrm{T}}. \qquad 11.46$$

Call $[\mathbf{dX}|\mathbf{e}] = \mathbf{u}_{L+1}\sigma_{L+1}\mathbf{v}_{L+1}^{\mathrm{T}}$. This is the minimum norm correction to $[\mathbf{X}|\mathbf{y}]$. Then $[-\boldsymbol{\theta} \quad 1]^{\mathrm{T}}$ may be chosen to be $\mathbf{v}_{L+1}$ to within a constant:

$$\begin{bmatrix} -\boldsymbol{\theta} \\ 1 \end{bmatrix} = \lambda\mathbf{v}_{L+1}. \qquad 11.47$$

The last element of $\mathbf{v}_{L+1}$ is assumed unequal to zero, as it equals the inverse of λ. We also assume that the smallest singular value σ_{L+1} has multiplicity one. If σ_{L+1} is a repeated singular value, then the solution is not unique and may be obtained from any vector in the space spanned by the eigenvectors associated with the smallest singular value. In this case, a minimum norm solution can be singled out.

This method is discussed by Golub and Van Loan [1980] (see also Chapter 12 in [Golub and Van Loan, 1983]). Many applications of this method in signal processing are described in [Zoltowsky, 1987]. The total least squares method is used for the problem of frequency estimation by Rahman and Yu [1987] and in system identification by Van Huffel and Vandewalle [1985]. A partial SVD algorithm to efficiently compute the total least squares solution even in the nongeneric (repeated smallest eigenvalue) case is presented by Van Huffel and Vanderwalle [1988].

11.6   PRINCIPAL COMPONENT METHOD (TUFTS AND KUMARESAN)

The Principal Component method is an extension of the least squares Prony technique. The model order for the prediction error filter $A(z)$ is increased in order to introduce extra roots z_i and corresponding modes z_i^l to account for the noise in the measurements. A rank reduction procedure is used to force the data (or equivalently, the sample correlation) matrix to have a rank corresponding to the number of exponentials in the signal. This technique originated with Tufts and Kumaresan [1983]. A review of the technique is available in [Vaccaro, Tufts, and Boudreaux-Bartels, 1988].

Without any knowledge of the exact number of modes in the signal, it is preferable to overfit the model (in other words, to suppose that there are many modes) and then to keep only the useful ones. The subset selection procedure presented in [Kumaresan, Tufts, and Scharf, 1984] allows for the separation of signal modes from noise modes.

Information Criteria

When the number of modes is unknown and the data matrix $\mathbf{X}$ is noise-free, then the number of nonzero singular values of $\mathbf{X}$ determines the number of modes. When the matrix $\mathbf{X}$ is noisy, then the number of modes may be estimated by detecting a gap in the singular values. A statistical approach to this problem is based on the use of hypothesis tests on the eigenvalues of the sample covariance matrix [Bartlett, 1954; Lawley, 1956]. For each hypothesis, a likelihood ratio statistic is computed and compared to a threshold; the hypothesis accepted is the first one for which the threshold is crossed. The problem consists then in the choice of thresholds.

Other methods are based on the information theoretic criteria (AIC) for model order selection introduced in the context of linear prediction by Akaike [1974 and 1973] and the minimum description length (MDL) criterion introduced by Schwartz [1978] and Rissanen [1978]. The number of signals is then determined as the value for which the criterion is minimized. Both criteria were adapted to the exponential model-fitting problem, and to the direct use of the data matrix (instead of the sample covariance matrix) by Wax and Kailath [1985]. The function to be minimized is then

$$\text{AIC}(k) = -2 \, \log\left(\left(\prod_{i=k+1}^{L} \lambda_i\right)^N \Big/ \left(\frac{1}{(L-k)} \sum_{i=k+1}^{L} \lambda_i\right)^{(L-k)N}\right) + 2k(2L - k)$$

$$11.48$$

for the AIC criterion and

$$\text{MDL}(k) = -\log\left(\left(\prod_{i=k+1}^{L} \lambda_i\right)^N \Big/ \left(\frac{1}{(L-k)} \sum_{i=k+1}^{L} \lambda_i\right)^{(L-k)N}\right) + \frac{1}{2}k(2L - k)\log(N)$$

$$11.49$$

for the MDL criterion. The $\{\lambda_i\}_{i=1}^{L}$ are the singular values of the Toeplitz data matrix $\mathbf{X}$, and N is the number of data points. The variable p is taken to be the value of k that minimizes either $\text{AIC}(k)$ or $\text{MDL}(k)$. For more on the performance of these criteria, see [Wax and Kailath, 1985].

Overfitting

Linear prediction with overfitting leads to the overdetermined system of equations $\mathbf{Ya}_L = \mathbf{0}$, where $\mathbf{a}_L$ contains the whitening coefficients $\mathbf{a}_L = [a_L \quad \cdots \quad a_1 \quad 1]^T$ or, equivalently, $\mathbf{X\theta}_L = \mathbf{y}$ with $\mathbf{Y} = [\mathbf{X}, \mathbf{y}]$ and $\mathbf{a}_L = [-\mathbf{\theta}_L^T \quad 1]^T$. Using a singular value decomposition (SVD) of the data matrix $\mathbf{X} = \mathbf{U\Sigma V}^T$, where the matrices $\mathbf{U}$ and $\mathbf{V}$ are orthonormal and $\mathbf{\Sigma}$ is a diagonal matrix containing the singular values, the least squares solution $\mathbf{\theta}_L$ may then be computed using the equation [Golub and Reinsch, 1970]

$$\mathbf{\theta}_L = -\mathbf{V\Sigma}^{-1}\mathbf{U}^T\mathbf{y}. \qquad 11.50$$

The method of Tufts and Kumaresan [1983] allows for a reduction in the rank of the inverse of $\mathbf{X}^T\mathbf{X}$ by replacing $\mathbf{\Sigma}^{-1}$ with $\mathbf{\Sigma}_p^{-1}$, another diagonal matrix with p nonzero terms instead of L, where only the p singular values with largest magnitude are kept. The matrix $\hat{\mathbf{X}} = \mathbf{U\Sigma}_p\mathbf{V}^T$ is the rank-p least squares approximant of $\mathbf{X}$ [Eckart and Young, 1936]. This reduced-rank method works well as long as the SNR does not fall below threshold. As the noise level increases, so do the noise singular values, in which case a wrong (signal) singular value may be thrown out instead of a noise singular value. When this switching occurs, a catastrophic decrease in performance is observed. This threshold behavior is analyzed in [Tufts, Kot, and Vacarro, 1987].

Order Selection

Once the linear prediction model of order L is estimated and its L roots have been computed, then we have L complex roots or q "root pairs." (By a root pair, we mean a pair of complex conjugate roots or a single real root, so that q will not be equal to $L/2$ when there are real roots.) We have q candidate root pairs from which we want to select p root pairs in order to represent the signal. Once more the least squares technique is used to discriminate between all the possible subsets of p modes out of the q candidates and also to compute the associated amplitudes and initial phases associated with these modes [Kumaresan, Tufts, and Scharf, 1984]. To illustrate this technique, we form the Vandermonde matrix $\mathbf{V}_p$ that contains some set of p modes and their complex conjugates. The reconstructed signal is then given by

$$\hat{\mathbf{y}} = \mathbf{V}_p \hat{\mathbf{B}}, \qquad\qquad 11.51$$

where $\hat{\mathbf{B}} = (\mathbf{V}_p^H \mathbf{V}_p)^{-1} \mathbf{V}_p^H \mathbf{y}$ is the parameter vector that produces the minimum squared error

$$\hat{e}^2 = (\mathbf{y} - \hat{\mathbf{y}})^T (\mathbf{y} - \hat{\mathbf{y}}) = \mathbf{y}^T (\mathbf{I} - \mathbf{P}_p) \mathbf{y}. \qquad\qquad 11.52$$

Here $\mathbf{P}_p$ is the projection onto the subspace spanned by the columns of $\mathbf{V}_p$:

$$\mathbf{P}_p = \mathbf{V}_p (\mathbf{V}_p^H \mathbf{V}_p)^{-1} \mathbf{V}_p^H. \qquad\qquad 11.53$$

The subset of p modes that leads to the minimum value of the error e^2 over all the possible subsets of p modes is then kept. A Grey code recursion may be used to generate all the possible subsets of size p out of q. See [Liu and Tang, 1973] for an algorithm that generates all these subsets using the theory of Grey codes. A simpler approach is to use the procedure of Hocking and Leslie [1967], as suggested in [Kumaresan, Tufts, and Scharf, 1984]. In this procedure, the best subset is usually found without searching through all the possible subsets.

Real Arithmetic

To avoid complex arithmetic, it is possible to replace any two complex conjugate columns in the matrix $\mathbf{V}_p$ by two real columns containing the associated real and imaginary parts of one of the complex columns. In other words, if a complex root is equal to

$$z_i = \rho_i e^{j 2\pi f_i T}, \qquad \text{with } f_i \neq k/2\mathrm{T}, \qquad\qquad 11.54$$

then replace the column of $\mathbf{V}_p$ containing the mode z_i^k and the column containing its conjugate mode $(z_i^k)^*$ by a column containing the mode $\rho_i^k \cos(2\pi k f_i T)$ and another containing the mode $\rho_i^k \sin(2\pi k f_i T)$. This simplification allows for a savings by a factor of four in complexity. Instead of containing pairs of complex conjugate weights B_i and B_i^*, the parameter vector contains weights $A_i \cos(\phi_i)$ and $-A_i \sin(\phi_i)$. A QR factorization of the real version of the matrix $\mathbf{V}_p$ is then used to solve a least squares system associated with any set of p modes [Demeure, 1988]. The QR factorization of $\mathbf{V}_p$ is $\mathbf{V}_p = \mathbf{QR}$, where $\mathbf{Q}$ is an orthonormal matrix ($\mathbf{Q}^T\mathbf{Q} = \mathbf{I}$) and $\mathbf{R}$ is an upper triangular matrix. The least squares solution for $\mathbf{B}$ is

$$\mathbf{RB} = \mathbf{Q}^T\mathbf{y}, \qquad\qquad\qquad 11.55$$

and the fitting error is

$$\hat{e}^2 = \mathbf{y}^T(I - \mathbf{P}_p)\mathbf{y}, \qquad \text{with } \mathbf{P}_p = \mathbf{QQ}^T. \qquad\qquad 11.56$$

Example 11.2

Suppose we have the signal

$$y_t = A\rho^t \cos(2\pi f t + \phi). \qquad\qquad\qquad 11.57$$

The modes are $z_0 = \rho e^{j2\pi f}$ and $z_1 = z_0^* = \rho e^{-j2\pi f}$. The complex amplitudes are obtained from the Vandermonde system

$$\begin{bmatrix} 1 & 1 \\ z_0 & z_1 \end{bmatrix} \begin{bmatrix} B_0 \\ B_1 \end{bmatrix} = \begin{bmatrix} y_0 \\ y_1 \end{bmatrix} \qquad\qquad 11.58$$

$$\begin{bmatrix} 1 & 1 \\ \rho e^{j2\pi f} & \rho e^{-j2\pi f} \end{bmatrix} \begin{bmatrix} B_0 \\ B_1 \end{bmatrix} = \begin{bmatrix} A\,\cos\,\phi \\ A\rho\,\cos(2\pi f t + \phi) \end{bmatrix}.$$

The solution is $B_0 = A/2\, e^{j\phi}$ and $B_1 = A/2\, e^{-j\phi}$. The alternate real system of equations is

$$\begin{bmatrix} 1 & 0 \\ \rho\cos(2\pi f) & \rho\sin(2\pi f) \end{bmatrix} \begin{bmatrix} \alpha \\ \beta \end{bmatrix} = \begin{bmatrix} A\cos\phi \\ A\rho\cos(2\pi f + \phi) \end{bmatrix}, \qquad 11.59$$

which has the solution $\alpha = A\cos\phi$ and $\beta = -A\sin\phi$. The relationship between the two solutions is $B_0 = (1/2)(\alpha - j\beta)$ and $B_1 = (1/2)(\alpha + j\beta)$. The amplitude A and the phase ϕ are computed using $A = (\alpha^2 + \beta^2)^{1/2}$ and $\tan \phi = -\beta/\alpha$. ■

11.7   PISARENKO AND MUSIC METHODS

The Pisarenko and MUSIC methods predate the principal component methods for mode indentification. Both methods may be applied when the data is assumed to be complex and undamped:

$$y_t = \sum_{i=1}^{p} A_i e^{j(2\pi f_i t + \phi_i)} + n_t. \qquad\qquad 11.60$$

The phases ϕ_i are assumed to be independent $U[0, 2\pi)$ random variables, and the added noise is assumed to be independent of the signal. The autocorrelation sequence of the data sequence is

$$r_k = E[y_t y_{t+k}] = \sum_{i=1}^{p} A_i^2 e^{j2\pi f_i k} + \sigma^2 \delta_k, \qquad 11.61$$

where σ^2 is the noise variance. Let's denote the power of the ith exponential by $\sigma_i^2 = A_i^2$ and call $\mathbf{e}_i$ the Fourier vector

$$\mathbf{e}_i = [1 \ e^{j2\pi f_i} \ e^{j4\pi f_i} \ \cdots \ e^{j2\pi(N-1)f_i}]^{\mathrm{T}}. \qquad 11.62$$

The $N \times N$ autocorrelation matrix $\mathbf{R} = E[\mathbf{yy}^{\mathrm{T}}]$ for the data vector $\mathbf{y} = [y_0 \ y_1 \ \cdots \ y_{N-1}]^{\mathrm{T}}$ is the Toeplitz matrix

$$\mathbf{R} = E[\mathbf{yy}^{\mathrm{T}}] = \{r_{|i-j|}\} = \sum_{i=1}^{p} \sigma_i^2 \mathbf{e}_i \mathbf{e}_i^{\mathrm{H}} + \sigma^2 \mathbf{I}. \qquad 11.63$$

The matrix $\mathbf{R} - \sigma^2 \mathbf{I}$ is rank-deficient for $N > p$. The eigendecomposition of the Hermitian matrix $\mathbf{R}$ shows the same rank deficiency:

$$\mathbf{R} = \sum_{i=1}^{p} (\lambda_i + \sigma^2) \mathbf{v}_i \mathbf{v}_i^{\mathrm{H}} + \sum_{p+1}^{N} \sigma^2 \mathbf{v}_i \mathbf{v}_i^{\mathrm{H}} \qquad 11.64$$

$$\mathbf{R} - \sigma^2 I = \sum_{i=1}^{p} \lambda_i \mathbf{v}_i \mathbf{v}_i^{\mathrm{H}}$$

This result shows that the first p Fourier vectors $\{\mathbf{e}_i\}_{i=1}^{p}$ and the first p eigenvectors $\{\mathbf{v}_i\}_{i=1}^{p}$ span the same "signal subspace."

The Pisarenko Method

The Pisarenko method uses the rank deficiency of $\mathbf{R}$, with $N = p + 1$. The orthogonal subspace then has dimension one. As the vector $\mathbf{v}_{p+1}$ is orthonormal to all the other eigenvectors and to the Fourier vectors $\mathbf{e}_i$, we may write

$$\mathbf{v}_{p+1}^{\mathrm{T}} \mathbf{e}_i = 0; \qquad i = 1, \ldots, p. \qquad 11.65$$

This means that, given $\mathbf{v}_{p+1}$, we can find the frequencies f_i by computing the p roots of the polynomial

$$V_{p+1}(z) = \mathbf{v}_{p+1}^{\mathrm{T}} \boldsymbol{\psi}(z) = \sum_{n=0}^{p} [\mathbf{v}_{p+1}]_{n+1} z^{-n}, \qquad 11.66$$

with $\boldsymbol{\psi}(z) = [1 \ z^{-1} \ \cdots \ z^{-n}]^{\mathrm{T}}$. The roots all lie on the unit circle, and the angles are equal to $2\pi f_i$. This is essentially a restatement of the Caratheodory theorem

[Grenander and Szegö, 1958], which says that, given any N complex numbers r_k, there exists a set of exponentials such that

$$r_k = \sum_{i=1}^{p} \rho_i e^{j\omega_i k}; \qquad k = 0, 1, \ldots, N - 1, \qquad 11.67$$

where the real parameters ρ_i and ω_i are unique if p is minimized. The Toeplitz property of **R** is fundamental in the derivation of this result, as it ensures that the roots of the polynomial $V_{p+1}(z)$ lie on the unit circle. That is, every rank-deficient Toeplitz correlation matrix has the representation $\sum_{i=1}^{p} \sigma_i^2 \mathbf{e}_i \mathbf{e}_i^H$.

The Pisarenko [1973] method uses an estimated autocorrelation sequence in place of **R**. The biased Toeplitz estimate (corresponding to the correlation method of linear prediction) is preferred because it maintains the property that the zeros of the last eigenvector lie on the unit circle. If the data is real, then the size of the matrix **R** is taken as $2p + 1$ where $2p$ is the number of complex conjugate exponentials to be identified. In the undamped real data case (pure sinusoids in noise), some simplification arises from the fact that the prediction polynomial is symmetric; that is, $\mathbf{a} = \mathbf{Ja}$ [Chan, Lavoie, and Plant, 1981].

Several practical remarks should be made about this technique. The Pisarenko technique is limited to the estimation of purely undamped modes. It is computationally demanding, as it uses estimated correlations, an eigenvalue computation, and also a root-finding routine. More importantly, the technique does not work very well in practice. In fact, its performance is worse than the extended Prony method of the previous section, which is much less demanding. For a statistical analysis, see [Sakai, 1984], and for an experimental comparison and some extensions, see [Kay, 1988].

Example 11.3

Suppose that the data contains a single exponential in noise:

$$y_t = A \exp j(2\pi f t + \phi) + n_t. \qquad 11.68$$

The correlation sequence is

$$r_k = A^2 \exp j(2\pi f k) + \sigma^2 \delta_k. \qquad 11.69$$

Form the correlation matrix **R**

$$R = \begin{bmatrix} r_0 & r_1^* \\ r_1 & r_0 \end{bmatrix}$$

$$= \begin{bmatrix} A^2 + \sigma^2 & A^2 e^{-j2\pi f} \\ A^2 e^{j2\pi f} & A^2 + \sigma^2 \end{bmatrix} \qquad 11.70$$

$$= A^2 \begin{bmatrix} 1 \\ e^{j2\pi f} \end{bmatrix} \begin{bmatrix} 1 & e^{-j2\pi f} \end{bmatrix} + \sigma^2 I.$$

The eigenvalues of $\mathbf{R}$ are $r_0 \pm |r_1|$ so that $\lambda = 2|r_1|$ and $\sigma^2 = r_0 - |r_1|$. The associated eigenvectors are

$$\mathbf{v}_1 = \frac{1}{\sqrt{2}} \begin{bmatrix} 1 \\ r_1/|r_1| \end{bmatrix}; \qquad \mathbf{v}_2 = \frac{1}{\sqrt{2}} \begin{bmatrix} 1 \\ -r_1/|r_1| \end{bmatrix}. \qquad \text{11.71}$$

Once the eigendecomposition of $\mathbf{R}$ is computed, the Pisarenko method builds the polynomial:

$$V_2(z) = \mathbf{v}_2^T \begin{bmatrix} 1 \\ z^{-1} \end{bmatrix}$$

$$= \frac{1}{\sqrt{2}} \left(1 - \frac{r_1 z^{-1}}{|r_1|} \right). \qquad \text{11.72}$$

The root of the polynomial is $z = e^{j2\pi f}$. ∎

MUSIC

The multiple signal classification (MUSIC) method [Schmidt, 1981] is based on the same equations as the Pisarenko method. That is, p Fourier vectors $\mathbf{e}_i$ are sought that are orthogonal to $N - p$ vectors $\mathbf{v}_j$ that span an orthogonal subspace:

$$\mathbf{v}_j^T \mathbf{e}_i = 0 \quad \text{for} \quad \begin{cases} i = 1, \ldots, p \\ j = p + 1, \ldots, N. \end{cases} \qquad \text{11.73}$$

Here N is not generally equal to $p + 1$. These orthogonal relations may be organized into the vector

$$\mathbf{f}(e^{j\omega}) = \begin{bmatrix} \mathbf{v}_{p+1}^T \\ \vdots \\ \mathbf{v}_N^T \end{bmatrix} \mathbf{e} = \mathbf{0}, \qquad \text{where } \mathbf{e} = [1 \; e^{j\omega} \; e^{j2\omega} \; \cdots \; e^{j(L-1)\omega}]^T \qquad \text{11.74}$$
$$\text{and } \omega = 2\pi f.$$

The equality will not hold if there is noise in the data. The MUSIC method is then based on a least squares version of the Pisarenko method:

$$\min_{\omega} \|\mathbf{f}(e^{j\omega})\|_2 = \min_{\omega} \sum_{i=p+1}^{N} |\mathbf{v}_i^T \mathbf{e}|^2. \qquad \text{11.75}$$

This minimization problem may be replaced by

$$\min_{\omega} \sum_{i=p+1}^{N} \mathbf{v}_i^T \mathbf{e} \mathbf{e}^H \mathbf{v}_i = \min_{\omega} \mathbf{e}^H \left(\mathbf{I} - \sum_{i=1}^{p} \mathbf{v}_i \mathbf{v}_i^H \right) \mathbf{e}, \qquad \text{11.76}$$

which in turn leads to

$$\max_{\omega} \sum_{i=1}^{p} |\mathbf{v}_i^{\mathrm{T}}\mathbf{e}|^2.$$ 11.77

The MUSIC method produces frequency estimates as the peaks of the "spectral" estimator

$$P(e^{j\omega}) = \left(\frac{1}{\displaystyle\sum_{i=p+1}^{N} |\mathbf{e}^{\mathrm{H}}\mathbf{v}_i|^2} \right).$$ 11.78

When ω is one of the sinusoidal frequencies, the "spectral" estimator will have a very sharp peak. But, as pointed out in [Kay and Demeure, 1984], the peak sharpness has nothing to do with resolution. The optimization procedure to estimate ω is computationally demanding and is replaced in practice by a DFT estimation technique.

11.8   PENCIL METHODS

ESPRIT (Estimation of Signal Parameters via Rotation Invariance Techniques) is a technique that may be used to estimate frequencies or directions of arrival [Paulraj, Roy, Kailath, 1985; Roy, Paulraj, and Kailath, 1986; and Hua and Sarkar, 1988]. In the single channel case, ESPRIT is based on the following observation: there is a constant delay (and attenuation) between any two adjacent samples in a uniformly sampled time or space series. This observation leads to the following idea. Let $\mathbf{Y}_0$ and $\mathbf{Y}_1$ be two data matrices defined by

$$\mathbf{Y}_0 = \begin{bmatrix} y_{p-1} & y_{p-2} & \cdots & y_0 \\ y_p & y_{p-1} & \cdots & y_1 \\ \vdots & \vdots & & \vdots \\ y_{N-2} & y_{N-3} & \cdots & y_{N-p-1} \end{bmatrix}; \quad \mathbf{Y}_1 = \begin{bmatrix} y_p & y_{p-1} & \cdots & y_1 \\ y_{p+1} & y_p & \cdots & y_2 \\ \vdots & \vdots & & \vdots \\ y_{N-1} & y_{N-2} & \cdots & y_{N-p} \end{bmatrix},$$

11.79

where p is the overfitting parameter. If the data consists entirely of complex exponentials in the case of a time series or entirely of planes waves in the case of a space series, then these matrices may be written as the product of Vandermonde matrices:

$$\mathbf{Y}_0 = \mathbf{V}_1\mathbf{B}\mathbf{V}_2^{\mathrm{T}}; \qquad \mathbf{Y}_1 = \mathbf{V}_1\mathbf{B}\mathbf{D}\mathbf{V}_2^{\mathrm{T}},$$ 11.80

where $\mathbf{B}$ is a diagonal matrix containing the complex amplitudes, $\mathbf{D}$ is a diagonal matrix containing the modes, and $\mathbf{V}_1$ and $\mathbf{V}_2$ are two Vandermonde matrices:

$$\mathbf{B} = \mathrm{diag}\begin{bmatrix} B_1 & B_2 & \cdots & B_p \end{bmatrix}; \qquad \mathbf{D} = \mathrm{diag}\begin{bmatrix} z_1 & z_2 & \cdots & z_p \end{bmatrix}$$

$$\mathbf{V}_1 = \begin{bmatrix} 1 & 1 & \ldots & 1 \\ z_1 & z_2 & \ldots & z_p \\ \vdots & \vdots & & \vdots \\ z_1^{N-p-1} & z_2^{N-p-1} & \ldots & z_p^{N-p-1} \end{bmatrix}; \qquad \mathbf{V}_2 = \begin{bmatrix} z_1^{p-1} & z_2^{p-1} & \ldots & z_p^{p-1} \\ \vdots & \vdots & & \vdots \\ z_1 & z_2 & \ldots & z_p \\ 1 & 1 & \ldots & 1 \end{bmatrix}.$$

$$11.81$$

The roots of the pencil of matrices $\mathbf{Y}_0 - z\mathbf{Y}_1 = \mathbf{V}_1\mathbf{B}(\mathbf{D} - z\mathbf{I})\mathbf{V}_2^{\mathsf{T}}$ produce the roots z_i. In other words, the nonzero generalized eigenvalues of $\mathbf{Y}_0 - z\mathbf{Y}_1$ are the complex numbers $\{z_i\}_{i=1}^{p}$. This means that a generalized eigenvalue algorithm may be used to compute the values of z, and the associated generalized eigenvectors $\mathbf{v}$, for which

$$\mathbf{Y}_0\mathbf{v} = z\mathbf{Y}_1\mathbf{v}. \qquad\qquad 11.82$$

This characterization of ESPRIT is attributed to Hua and Sarkar [1988]. The original second-order characterization is given by Roy, Paulraj, and Kailath [1986], using exact correlation and cross-correlation matrices. A total least squares variation can be found in [Roy and Kailath, 1987]. For more details on the computation of a generalized eigenvalue decomposition, see [Golub and Van Loan, 1983; and Moler and Stewart, 1973]. Several algorithms can be found in EISPACK [Garbow et al., 1972; Schmidt et al., 1976].

In the case of a pure sinusoidal signal, it is common to use forward and backward data matrices, so that the original pencil $\mathbf{Y}_0 - z\mathbf{Y}_1$ is replaced by $\mathbf{Y}_0^{\mathrm{FB}} - z\mathbf{Y}_1^{\mathrm{FB}}$, with

$$\mathbf{Y}_0^{\mathrm{FB}} = \begin{bmatrix} y_{p-1} & y_{p-2} & \cdots & y_0 \\ y_p & y_{p-1} & \cdots & y_1 \\ \vdots & \vdots & & \vdots \\ y_{N-2} & y_{N-3} & \cdots & y_{N-p-1} \\ y_1 & y_2 & \cdots & y_p \\ y_2 & y_3 & \cdots & y_{p+1} \\ \vdots & \vdots & & \vdots \\ y_{N-p} & y_{N-p+1} & \cdots & y_{N-1} \end{bmatrix} \qquad 11.83$$

$$\mathbf{Y}_1^{\mathrm{FB}} = \begin{bmatrix} y_p & y_{p-1} & \cdots & y_1 \\ y_{p+1} & y_p & \cdots & y_2 \\ \vdots & \vdots & & \vdots \\ y_{N-1} & y_{N-2} & \cdots & y_{N-p} \\ y_0 & y_1 & \cdots & y_{p-1} \\ y_1 & y_2 & \cdots & y_p \\ \vdots & \vdots & & \vdots \\ y_{N-p-1} & y_{N-p} & \cdots & y_{N-2} \end{bmatrix}.$$

This method does not ensure that the generalized eigenvalues will lie exactly on the unit circle, but they are usually very close. Only the p generalized eigenvalues closest to the unit circle are actually used; the others are set to zero.

Example 11.4

Suppose the data contains two modes

$$y_t = B_1 z_1^t + B_2 z_2^t. \qquad\qquad 11.84$$

The matrices

$$\mathbf{Y}_0 = \begin{bmatrix} y_1 & y_0 \\ y_2 & y_1 \end{bmatrix}; \qquad \mathbf{Y}_1 = \begin{bmatrix} y_2 & y_1 \\ y_3 & y_2 \end{bmatrix} \qquad\qquad 11.85$$

may be factored as

$$\mathbf{Y}_0 = \begin{bmatrix} 1 & 1 \\ z_1 & z_2 \end{bmatrix} \begin{bmatrix} B_1 & 0 \\ 0 & B_2 \end{bmatrix} \begin{bmatrix} z_1 & 1 \\ z_2 & 1 \end{bmatrix} = \mathbf{V}_1 \mathbf{B} \mathbf{V}_2^{\mathsf{T}} \qquad\qquad 11.86$$

$$\mathbf{Y}_1 = \begin{bmatrix} 1 & 1 \\ z_1 & z_2 \end{bmatrix} \begin{bmatrix} B_1 & 0 \\ 0 & B_2 \end{bmatrix} \begin{bmatrix} z_1 & 0 \\ 0 & z_2 \end{bmatrix} \begin{bmatrix} z_1 & 1 \\ z_2 & 1 \end{bmatrix} = \mathbf{V}_1 \mathbf{B} \mathbf{D} \mathbf{V}_2^{\mathsf{T}}.$$

The pencil is then

$$\mathbf{Y}_0 - z \mathbf{Y}_1 = \mathbf{V}_1 \begin{bmatrix} z_1 - z & 0 \\ 0 & z_2 - z \end{bmatrix}. \qquad\qquad 11.87$$

The roots z_1 and z_2 are the generalized eigenvalues of the pair $(\mathbf{Y}_0, \mathbf{Y}_1)$ or, equivalently, the eigenvalues of $\mathbf{Y}_1^{-1}\mathbf{Y}_0$. When more measurements are available, the number of rows of the matrices $\mathbf{Y}_0$ and $\mathbf{Y}_1$ increase. In this case, a pseudo-inverse of $\mathbf{Y}_1$ may be used, so that the roots are the eigenvalues of the matrix $(\mathbf{Y}_1^{\mathsf{H}}\mathbf{Y}_1)^{-1}\mathbf{Y}_1^{\mathsf{H}}\mathbf{Y}_0$. When overfitting is used, the number of columns of $\mathbf{Y}_0$ and $\mathbf{Y}_1$ increases and a rank reduction technique is used to get a low-rank estimate of $\mathbf{Y}_1$. ∎

11.9   A FREQUENCY-DOMAIN VERSION OF PRONY'S METHOD (KUMARESAN)

In this section we outline Kumaresan's recent discovery of a *frequency-domain version of Prony's method*. This version provides a method to estimate the coefficients of a rational transfer function from its complex frequency response samples. The problem may be summarized as follows. Given N samples of the complex frequency response $H(z_n)$, $z_n = e^{j\theta_n}$, $n = 0, 1, \ldots, N-1$, estimate the coefficients of the polynomials $A(z)$ and $B(z)$ in the ARMA(p,q) model

$$H(z) = \frac{B(z)}{A(z)} = \sum_{i=0}^{q} b_i z^{-i} \Big/ \sum_{i=0}^{p} a_i z^{-i}; \qquad a_0 = 1. \qquad\qquad 11.88$$

This problem was first addressed by Cauchy [1821] and is called the *Cauchy rational interpolation problem*. The problem does not always have a solution. Pole/zero cancellations may occur that forbid the equations to be satisfied. The corresponding point z_i is then considered unreachable. The special case $A(z) = 1$ is known as the *Lagrange interpolation problem*.

This problem may be solved by decoupling its solution into two parts involving $A(z)$ and $B(z)$, respectively. Jacobi [1845] was the first to use such a technique to solve this problem. In this section we discuss a technique discovered by Kumaresan to solve the problem. The interested reader is referred to [Kumaresan, 1990] for a more detailed discussion, alternate techniques, and additional references.

Divided Differences

The divided difference technique has been used extensively for manipulating and solving interpolation problems [Hildebrand, 1956; and Milne-Thompson, 1933]. The *divided differences* of order $0, 1, 2, \ldots$ of a function $A(z)$ are defined recursively by

$$A[z_0] = A(z_0)$$

$$A[z_0, z_1] = \frac{A[z_1] - A[z_0]}{z_1 - z_0} \qquad 11.89$$

$$A[z_0, z_1, \cdots, z_n] = \frac{A[z_1, z_2, \ldots, z_n] - A[z_0, z_1, \ldots, z_{n-1}]}{z_n - z_0},$$

where the complex numbers z_i are assumed distinct. It can be shown [Hildebrand, 1956] that

$$A[z_0, z_1, \ldots, z_n] = \sum_{i=0}^{n} \frac{A(z_i)}{L(z_i)}, \qquad 11.90$$

where $L(z_i)$ is a product of frequency differences:

$$L(z_i) = \prod_{\substack{j=1 \\ j \neq i}}^{n} (z_i - z_j). \qquad 11.91$$

The nth order divided difference of a polynomial of degree less than n is identically zero. For example, when $A(z) = a + bz + cz^2$, the divided difference table is that shown in Table 11.1.

Solving for A(z)

The divided difference method is applied to polynomials with positive powers of z. Let us therefore rewrite our problem with $H(z)$ written as

$$H(z) = z^{p-q} \left(\sum_{j=0}^{q} b_{q-j} z^j \middle/ \sum_{j=0}^{p} a_{p-j} z^j \right) = z^{p-q} \frac{\hat{B}(z)}{\hat{A}(z)} = z^{p-q} \hat{H}(z) \qquad 11.92$$

**Table 11.1 THE DIVIDED DIFFERENCES TABLE
OF A DEGREE TWO POLYNOMIAL**

z	$A[z_i]$	$A[z_i, z_j]$	$A[z_i, z_j, z_k]$	$A[z_i, z_j, z_k, z_l]$
z_0	$a + bz_0 + cz_0^2$			
		$b + c(z_1 + z_0)$		
z_1	$a + bz_1 + cz_1^2$		c	
		$b + c(z_1 + z_2)$		0
z_2	$a + bz_2 + cz_2^2$		c	
		$b + c(z_2 + z_3)$		
z_3	$a + bz_3 + cz_3^2$			

and rewrite our problem as $\hat{H}(z) = \hat{B}(z)/\hat{A}(z)$ or as

$$\hat{H}(z)\hat{A}(z) = \hat{B}(z). \qquad 11.93$$

The order of $\hat{B}(z)$ is q, so $z^i \hat{B}(z)$ is of order $q + i < p + q$ for $i < p$. This means that the nth divided difference of $z^i \hat{B}(z)$ is zero for $n \geq p + q$. But the divided differences of $\hat{B}(z)$ are the divided differences of $\hat{H}(z)\hat{A}(z)$. Therefore,

$$\sum_{j=0}^{n} \frac{\hat{H}(z_j)\hat{A}(z_j)z_j^i}{L(z_j)} = 0; \qquad i = 0, 1, \ldots, p-1. \qquad 11.94$$

Using $\hat{A}(z) = \sum_{k=0}^{p} \hat{a}_k z^k$, we have the homogeneous equation

$$\sum_{j=0}^{n} \sum_{k=0}^{p} \frac{\hat{H}(z_j)z_j^{i+k}}{L(z_j)} \hat{a}_k = 0. \qquad 11.95$$

This equation produces the linear system $\mathbf{Ha} = \mathbf{0}$, where $\mathbf{a} = [\hat{a}_0, \hat{a}_1, \ldots, \hat{a}_p]^T$ and $\mathbf{H} = \{H_{ij}\}$:

$$H_{ij} = \sum_{k=0}^{n} \frac{\hat{H}(z_k)z_k^{i+j}}{L(z_k)}; \qquad i = 0, 1, \ldots, p-1, \quad j = 0, 1, \ldots, p. \qquad 11.96$$

The matrix $\mathbf{H}$ is Hankel. If $n > p + q$, a least squares solution may be used to solve this standard linear equation.

Solving for B(z)

Once $\hat{A}(z)$ is known, the numerator $\hat{B}(z)$ is computed using

$$\hat{A}(z_i)\hat{H}(z_i) = \hat{B}(z_i)$$
$$= \hat{B}_0 + \hat{B}_1 z_i + \cdots + \hat{B}_q z_i^q \qquad 11.97$$
$$= b_q + b_{q-1} z_i + \cdots + b_1 z_i^q.$$

When written out for $i = 0, 1, \ldots, q$, this identity produces the Vandermonde system

$$
\begin{bmatrix}
1 & z_0 & z_0^2 & \cdots & z_0^q \\
1 & z_1 & z_1^2 & \cdots & z_2^q \\
\vdots & \vdots & \vdots & & \vdots \\
1 & z_q & z_q^2 & \cdots & z_q^q
\end{bmatrix}
\begin{bmatrix}
\hat{B}_0 \\
\hat{B}_1 \\
\vdots \\
\hat{B}_q
\end{bmatrix}
=
\begin{bmatrix}
\hat{A}(z_0)\hat{H}(z_0) \\
\hat{A}(z_1)\hat{H}(z_1) \\
\vdots \\
\hat{A}(z_q)\hat{H}(z_q)
\end{bmatrix} .
\qquad 11.98
$$

This row Vandermonde system may be solved using the algorithm of Björk and Pereyra [1970]. In the least squares version of this method, the QR factorization of the Hankel system may be obtained in the same way as the QR factorization of a Toeplitz system. The QR factorization of the row Vandermonde system may only be solved using a fast algorithm if the complex numbers y_i all lie on the unit circle (which is the case here). This comes from the fact that in this case the Grammian is an Hermitian Toeplitz matrix. A fast algorithm to compute this QR factorization is described in [Demeure 1989]. When p is unknown, the rank of an overfitted version of the matrix $\mathbf{H}$ allows for the determination of p [Kumaresan, 1988]. This frequency-domain method is strikingly similar to Prony's time-domain method.

Example 11.5

Suppose we want the coefficients of the ARMA(1,1) model

$$
H(z) = \frac{b_0 + b_1 z^{-1}}{1 + a_1 z^{-1}} = \frac{b_1 + b_0 z}{a_1 + z} = \hat{H}(z).
\qquad 11.99
$$

The equation to be solved is $\hat{A}(z)\hat{H}(z) = \hat{B}(z)$. As $\hat{B}(z)$ is of order one, we take the second-order divided differences on $H(z)\hat{A}(z)$ to get the system $\mathbf{Ha} = \mathbf{0}$:

$$
\begin{bmatrix}
\displaystyle\sum_{j=0}^{2} \frac{H(z_j)}{L(z_j)} & \displaystyle\sum_{j=0}^{2} \frac{H(z_j)z_j}{L(z_j)}
\end{bmatrix}
\begin{bmatrix}
a_1 \\
1
\end{bmatrix}
= 0.
\qquad 11.100
$$

The equation for $\hat{B}(z)$ is

$$
\begin{bmatrix}
1 & z_0 \\
1 & z_1
\end{bmatrix}
\begin{bmatrix}
b_1 \\
b_0
\end{bmatrix}
=
\begin{bmatrix}
(a_0 + z_0)H(z_0) \\
(a_0 + z_1)H(z_1)
\end{bmatrix} .
\qquad 11.101
$$

∎

REFERENCES

Aigrain, P. R., and E. M. Williams [1949]. "Synthesis of n-Reactive Networks for Desired Transient Response," *J Appl Phys* **20**, p. 297 (1949).

Akaike, H. [1974]. "A New Look at the Statistical Model Identification," *IEEE Trans Automatic Control* **19**, pp. 716–723 (1974).

Akaike, H. [1973]. "Information Theory and an Extension of the Maximum Likelihood Principle," *Proc 2nd Intl Symp IT,* suppl., Problems of Control and Inform. Theory, pp. 267–281 (1973).

Baker, Jr., G. A., and P. Graves-Morris [1981]. "Padé Approximants," vol. 13–14 in *Encyclopedia of Mathematics* (Reading, MA: Addison-Wesley, 1981).

Barrodale, I., and C. Phillips [1975]. "Solution of an Overdetermined System of Linear Equations in the Chebyshev Norm," *ACM Trans on Mathematical Software* **1**:3, pp. 264–270 (September 1975).

Bartlett, M. S. [1954]. "A Note on the Multiplying Factors for Various χ^2 Approximations," *J Roy Math Soc,* ser. B, **16**, pp. 296–298 (1954).

Björk, A., and V. Pereyra [1970]. "Solution of Vandermonde Systems of Equations," *Math of Comp* **24**:112, pp. 893–903 (October 1970).

Bresler, Y., and A. Macovski [1986]. "Exact Maximum Likelihood Parameter Estimation of Superimposed Exponential Signals in Noise," *IEEE Trans ASSP* **34**:5, pp. 1081–1089 (October 1986).

Cauchy, A. L. [1821]. "Sur la Formule de Lagrange Relative à l'Interpolation," *Analyse Algébrique* (Paris: 1821).

Chan, Y. T., J. M. M. Lavoie, and J. B. Plant [1981]. "A Parameter Estimation Approach to Estimation of Frequencies of Sinusoids," *IEEE Trans ASSP* **29**:2, pp. 214–219 (April 1981).

Conte, S. D., and C. de Boor [1972]. *Elementary Numerical Analysis*, 2nd ed. (New York: McGraw Hill, 1972), pp. 74–79.

Demeure, C. J. [1989]. "Fast QR Factorization of Vandermonde Matrices," *Journ Lin Algebra and its Appl,* special issue on Linear Systems and Control, **124**, pp. 165–194 (1989).

Demeure, C. J. [1988]. "Fast Algorithms for Linear Prediction and Modal Analysis," Ph.D. dissertation, University of Colorado (December 1988).

Dennis, Jr., I. E., and R. B. Schnabel [1983]. *Numerical Methods for Unconstrained Optimization and Nonlinear Equations* (Englewood Cliffs, NJ: Prentice-Hall, 1983).

Eckart, C., and G. Young [1936]. "The Approximation of One Matrix by Another of Lower Rank," *Psychometrika* **1**, pp. 211–218 (1936).

Evans, A. G., and R. Fischl [1973]. "Optimal Least Squares Time-Domain Synthesis of Recursive Digital Filters," *IEEE Trans Audio Electroacoustics* **21**, pp. 61–65 (February 1973).

Garbow, B. S., J. M. Boyle, J. J. Dongarra, and C. B. Moler [1972]. *Matrix Eigensystem Routines: EISPACK Guide Extension* (New York: Springer-Verlag, 1972).

Golub, G. H., and C. Reinsch [1970]. "Singular Value Decomposition and Least Squares Solutions," *Numerische Mathematik* **14**, pp. 403–420 (1970).

Golub, G. H., and C. F. Van Loan [1983]. *Matrix Computations,* Chap. 6, pp. 136–154 (Baltimore: Johns Hopkins University Press, 1983).

Golub, G. H., and C. F. Van Loan [1980]. "An Analysis of the Total Least Squares Problem," *SIAM J Num Anal* **17**:6, pp. 883–893 (1980).

Grenander, O., and S. Szegö [1958]. *Toeplitz Forms and their Applications* (Berkeley, CA: University of California Press, 1958).

Hildebrand, F. B. [1956]. *Introduction to Numerical Analysis,* p. 379 (New York: McGraw-Hill, 1956).

Hocking, R. R., and L. L. Leslie [1967]. "Selection of the Best Subset in Regression Analysis," *Technometrics* **9**, pp. 537–540 (1967).

Hua, Y., and T. K. Sarkar [1988]. "Matrix Pencil Method for Estimating Parameters of Exponentially Damped/Undamped Sinusoids in Noise," submitted to *IEEE Trans ASSP*, 1988.

Van Huffel, S., and J. Vandewalle [1988]. "The Partial Total Least Squares Algorithm," *J Comp and Appl Math* **21**, pp. 333–341 (1988).

Van Huffel, S., and J. Vandewalle [1985]. "The Use of Total Linear Least Squares Techniques for Identification and Parameter Estimation," pp. 1167–1172 *Proc. VIIth IFAC/IFORS Symp on Identification and System Parameter Estimation,* York, U.K., July 1985.

Jacobi, C. [1845]. "Uber die Darstellung eines Reibe Gegebener Werte Hurch eine Gebrochene Rationale Function," *Gesmmelte Werke* **3**, pp. 481–511 (1845).

Jenkins, M. A. [1985]. "Zeros of a Real Polynomial," *ACM Trans Mathematical Software* **1**:2, pp. 178–189 (June 1985).

Jenkins, M. A., and J. F. Traub [1972]. "Zeros of a Complex Polynomial," *Communications of the ACM* **15**, pp. 97–99 (February 1972).

Kay, S. M. [1988]. *Modern Spectral Estimation: Theory and Application* (Englewood Cliffs, NJ: Prentice-Hall, 1988).

Kay, S. M. [1984]. "Accurate Frequency Estimation at Low Signal-to-Noise Ratios," *IEEE Trans ASSP* **ASSP-32**, pp. 540–547 (June 1984).

Kay, S. M., and C. J. Demeure [1984]. "The High Resolution Spectrum Estimator: A Subjective Entity," *Proc IEEE* **72**, pp. 1815–1816 (December 1984).

Kopec, G. E., A. V. Oppenheim, and J. M. Tribolet [1977]. "Speech Analysis by Homomorphic Prediction," *IEEE Trans ASSP* **ASSP-27**:1, pp. 40–49 (February 1977).

Kumaresan, R. [1990]. "On a Frequency Domain Analog of Prony's Method," *IEEE Trans ASSP* **38**:1, pp. 168–170 (January 1990).

Kumaresan, R. [1982]. "Estimating the Parameters of Exponentially Damped Undamped Sinusoidal Signals in Noise," Ph.D. dissertation, University of Rhode Island, Kingston, RI (August 1982).

Kumaresan, R., and A. K. Shaw [1985]. "High Resolution Bearing Estimation without Eigendecomposition," pp. 576–579, *Proc IEEE Intl Conf on ASSP*, Tampa, Florida, 1985.

Kumaresan, R., and D. W. Tufts [1982]. "Estimating the Parameters of Exponentially Damped Sinusoids and Pole-Zero Modeling in Noise," *IEEE Trans ASSP* **ASSP-30**:6, pp. 833–840 (December 1982).

Kumaresan, R., L. L. Scharf, and A. K. Shaw [1986]. "An Algorithm for Pole-Zero Modeling and Spectral Analysis," *IEEE Trans ASSP* **34**:3, pp. 637–640 (June 1986).

Kumaresan, R., D. W. Tufts, and L. L. Scharf [1984]. "A Prony Method for Noisy Data: Choosing the Signal Components and Selecting the Order in Exponential Signal Models," *Proc IEEE* **72**:2, pp. 230–233 (February 1984).

Lawley, D. N. [1956]. "Tests of Significance of the Latent Roots of the Covariance and Correlation Matrices," *Biometrica* **43**, pp. 128–136 (1956).

Liu, C. N., and D. T. Tang [1973]. "Enumerating Combinations of M Out of N Objects," *Communications of the ACM* **16**:6, p. 485 (August 1973).

McClellan, J. H., and D. Lee [1990]. "Exact Equivalence of Steiglitz-McBride Iteration and IQML," *IEEE Trans ASSP* (to appear 1990).

McEliece, R. J., and J. B. Shearer [1978]. "A Property of Euclid's Algorithm and an Application to Padé Approximation," *SIAM J Appl Math* **34**, pp. 611–615 (June 1978).

Milne-Thomson, L. M. [1933]. *The Calculus of Finite Differences* (London: MacMillan, 1933).

Moler, C. B., and G. W. Stewart [1973]. "An Algorithm for Generalized Matrix Eigenvalue Problems," *SIAM J Num Anal* **10**, pp. 241–256 (1973).

Padé, H. [1892]. "Sur la Représentation Approchée d'Une Function par des Fractions Rationnelles," *Annales de l'Ecolé Normale Supérieure* **3**, pp. 3–93 (1892).

Paulraj, A., R. Roy, and T. Kailath [1985]. "Estimation of Signal Parameters via Rotational Invariance Techniques—ESPRIT," pp. 83–89, *Proc XIXth Asilomar Conf on Circuits, Systems, and Computers*, Pacific Grove, CA, November 1985.

Pisarenko, V. F. [1973]. "The Retrieval of Harmonics from a Covariance Function," *Geophysics Journal Royal Astronomical Soc* **33**, pp. 347–366 (1973).

de Prony (Gaspard Riche), Baron R. [1795]. "Essai Expérimental et Analytique: sur les Lois de la Dilatabilité de Fluides Élastiques et Sur Celles de la Force Expansive de la Vapeur de l'Eau et de la Vapeur de l'Alcool, à Différentes Températures," *Journal de l'École Polytechnique, Paris* **1**:2, pp. 24–76 (1795).

Rahman, M. A., and K. Yu [1987]. "Total Least Squares Approach for Frequency Estimation Using Linear Prediction," *IEEE Trans ASSP* **ASSP-35**:10, pp. 1440–1454 (October 1987).

Rissanen, J. [1978]. "Modeling by Shortest Data Description," *Automatica* **14**, pp. 465–471 (1978).

Roy, R., and T. Kailath [1987]. "Total Least Squares ESPRIT," pp. 297–301, *Proc XXIst Asilomar Conf on Circuits, Systems, and Computers*, Pacific Grove, CA, November 1987.

Roy, R., A. Paulraj, and T. Kailath [1986]. "ESPRIT—A Subspace Rotation Approach to Estimation of Parameters of Cisoids in Noise," *IEEE Trans ASSP* **ASSP-34**:5, pp. 1340–1342 (October 1986).

Sakai, H. [1984]. "Statistical Analysis of Pisarenko's Method for Sinusoidal Frequency Estimation," *IEEE Trans ASSP* **ASSP-32**:1, pp. 95–101 (February 1984).

Schmidt, B. T., J. M. Boyle, J. J. Dongarra, B. S. Garbow, Y. Ikebe, V. C. Klema, and C. B. Moler [1976]. *Matrix Eigensystem Routines: EISPACK Guide,* 2nd ed. (New York: Springer-Verlag, 1976).

Schmidt, R. O. [1981]. "A Signal Subspace Approach to Multiple Emitter Location and Spectral Estimation," Ph.D. dissertation, Stanford University (1981).

Schwartz, G. [1978]. "Estimating the Dimension of a Model," *Ann Stat* **6**, pp. 461–464 (1978).

Shank, J. L. [1967]. "Recursion Filters for Digital Processing," *Geophysics* **32**, pp. 32–51 (1967).

Steiglitz, K., and L. E. McBride [1965]. "A Technique for the Identification of Linear Systems," *IEEE Trans Autom Control* **AC-xx**, pp. 461–464 (1965).

Trench, W. [1964]. "An Algorithm for the Inversion of Finite Toeplitz Matrices," *SIAM J Appl Math* **12**:3, pp. 515–522 (September 1964).

Tufts, D., and R. Kumaresan [1983]. "Frequency Estimation of Multiple Sinusoids: Making Linear Prediction like Maximum Likelihood," *Proc IEEE* **70**, pp. 975–990 (March 1983).

Tufts, D., A. C. Kot, and R. J. Vaccaro [1987]. "The Analysis of Threshold Behavior of SVD-Based Algorithms," pp. 550–554, *Proc XXIInd Asilomar Conf on Circuits, Signals, and Computers*, Pacific Grove, CA, November 1987.

Vaccaro, R. J., D. W. Tufts, and G. F. Boudreaux-Bartels [1988]. "Advances in Principal Component Signal Processing," *SVD and Signal Processing*, E. F. Deprettere, ed., pp. 115–146. (Amsterdam: North-Holland, 1988).

Wax, M., and T. Kailath [1985]. "Detection of Signals by Information Theoretic Criteria," *IEEE Trans ASSP* **ASSP-33**:2, pp. 387–392 (April 1985).

Zohar, S. [1974]. "The Solution of a Toeplitz Set of Linear Equations" *J Ass Comput Mach* **21**, pp. 272–276 (April 1974).

Zohar, S. [1969]. "Toeplitz Matrix Inversion: The Algorithm of W. F. Trench," *J Ass Comput Mach* **16**, pp. 592–601 (October 1969).

Zoltowsky, M. D. [1987]. "Signal Processing Applications of the Method of Total Least Squares," pp. 290–296 *Proc XXIst Asilomar Conf on Signals, Systems, and Computers*, Pacific Grove, CA, (November 1987).

Index